# Springer Textbooks in Earth Sciences, Geography and Environment

The Springer Textbooks series publishes a broad portfolio of textbooks on Earth Sciences, Geography and Environmental Science. Springer textbooks provide comprehensive introductions as well as in-depth knowledge for advanced studies. A clear, reader-friendly layout and features such as end-of-chapter summaries, work examples, exercises, and glossaries help the reader to access the subject. Springer textbooks are essential for students, researchers and applied scientists.

More information about this series at http://www.springer.com/series/15201

Jochen Hoefs

# Stable Isotope Geochemistry

Ninth Edition

Springer

Jochen Hoefs
Abteilung Isotopengeologie,
Geowissenschaftliches Zentrum
Universität Göttingen
Göttingen, Germany

ISSN 2510-1307 ISSN 2510-1315 (electronic)
Springer Textbooks in Earth Sciences, Geography and Environment
ISBN 978-3-030-77694-7 ISBN 978-3-030-77692-3 (eBook)
https://doi.org/10.1007/978-3-030-77692-3



This Springer imprint is published by the registered company Springer Nature Switzerland AG
The registered company address is: Gewerbestrasse 11, 6330 Cham, Switzerland

# Preface

The nine editions of *Stable Isotope Geochemistry* have appeared over a time span of nearly 50 years. The first edition (1973) appeared as a slim book of 135 pages. Due to the rapid development of the field, extended editions became necessary from time to time. The first 4 editions centered on the classical light elements, so-called CHONS. From the 5th edition 2004—with the introduction of the Multicollector-ICP-Mass Spectrometry—the book had to be enlarged, because many more elements—47 elements in the 9th edition—can be measured with the necessary high precision. The increase of the number of elements, together with advances in the calculation of equilibrium isotope fractionation using ab initio methods, has led to an unbelievable rise of publications, making again a rewriting with substantial major revisions and extensions necessary. To follow the recommendations of the International Union of Pure and Applied Chemistry (IUPAC), the term $\delta D$ has been replaced by $\delta^2H$ in the new edition.

The general structure of the book has been kept. Chapter 1 gives a general introduction to the theoretical and experimental principles.

Chapter 2 is divided in two parts: Part I discusses the "traditional" elements hydrogen, carbon, oxygen, nitrogen, and sulfur measured by gas-source mass-spectrometry. Part II presents the "non-traditional isotopes", measured predominantly by multi-collector inductively coupled mass-spectrometry. At places elements with close geochemical relationships are discussed successively. Chapter 2 gives an overview of natural isotope variations of 47 elements; Rb, Zr, and Nd, Eu and the HREE have been added in the new edition. Special emphasis has been given to studies that have been published over the last three years, while still summarizing the important discoveries made before that time.

As in earlier editions, the third part discusses natural variations of isotope compositions in the context of the classic geochemical "spheres". New subsections have been added such as: Meteorite-Earth relationships, volatile elements on the Moon, magnesium and iron relationships in magmatic rocks, subduction-zone metamorphism, metal isotope variations in ore deposits and in the ocean, and carbon isotope stratigraphy. Chapter 3 ends with two new subtitles: (i) forensic isotope geochemistry and (ii) medical investigations. A very long list of references with many new citations enables a quick access to the exponentially growing recent literature, although I have neglected a number of recent references, because the citation list encompasses already about 30% of the book.

Ilya Bindeman, Daniel Herwartz and Franck Poitrasson have reviewed the manuscript; to all three of them I owe my deepest thanks. Andreas Pack and Jens Fiebig helped me with some figures. Special thanks to Annett Büttner for support during final preparation of the manuscript.

In summary, I hope the new enlarged edition gives a timely overview of the recent progress in stable isotope geochemistry; for any shortcoming that remains I take responsibility.

Göttingen, Germany

Jochen Hoefs

# Contents


**1 Theoretical and Experimental Principles** . . . . . . 1
1.1 General Characteristics of Isotopes . . . . . . 1
1.2 Isotope Effects . . . . . . 4
1.3 Isotope Fractionation Processes . . . . . . 5
1.3.1 Isotope Exchange . . . . . . 6
1.3.1.1 Fractionation Factor (α) . . . . . . 7
1.3.1.2 The Delta Value (δ) . . . . . . 8
1.3.1.3 Evaporation–Condensation Processes . . . . . . 9
1.3.2 Kinetic Effects . . . . . . 10
1.3.3 Mass-Dependent and Mass-Independent Isotope Effects . . . . . . 12
1.3.3.1 Mass Dependent Effects . . . . . . 12
1.3.3.2 Mass Independent Effects . . . . . . 13
1.3.4 Nuclear Volume and Magnetic Isotope Effects . . . . . . 14
1.3.4.1 Nuclear Volume Effects . . . . . . 14
1.3.4.2 Magnetic Isotope Effects . . . . . . 15
1.3.5 Multiply Substituted Isotopologues . . . . . . 15
1.3.5.1 Position or Site-Specific Isotope Fractionations . . . . . . 18
1.3.6 Diffusion . . . . . . 19
1.3.7 Other Factors Influencing Isotopic Fractionations . . . . . . 21
1.3.8 Isotope Geothermometers . . . . . . 23
1.4 Basic Principles of Mass Spectrometry . . . . . . 28
1.4.1 Continuous Flow–Isotope Ratio Monitoring Mass Spectrometers . . . . . . 30
1.4.2 General Remarks on Sample Preparation Methods for Gases . . . . . . 32
1.4.3 Laser Extraction Techniques . . . . . . 33
1.4.4 High-Mass-Resolution Multiple-Collector IR Mass Spectrometer . . . . . . 34
1.4.5 Infrared Spectroscopy . . . . . . 34
1.4.6 Nuclear Magnetic Resonance (NMR) Spectroscopy . . . . . . 34

1.5 Standards . . . 35
1.6 Microanalytical Techniques . . . 38
1.6.1 Multicollector-ICP-Mass Spectrometry . . . 38
1.6.2 Secondary Ion Mass Spectrometry (SIMS) . . . 39
References . . . 40

**2 Isotope Fractionation Processes of Selected Elements** . . . 49
2.1 Hydrogen . . . 50
2.1.1 Methods . . . 51
2.1.2 Standards . . . 52
2.1.3 Fractionation Processes . . . 53
2.1.3.1 Water Fractionations . . . 53
2.1.3.2 Equilibrium Reactions . . . 55
2.1.3.3 Fractionations During Biosynthesis . . . 56
2.1.3.4 Other Fractionations . . . 57
2.2 Carbon . . . 58
2.2.1 Analytical Methods . . . 58
2.2.1.1 Standards . . . 58
2.2.2 Fractionation Processes . . . 59
2.2.2.1 Carbonate System . . . 59
2.2.2.2 Other Equilibrium Isotope Fractionations . . . 60
2.2.2.3 Organic Carbon System . . . 61
2.2.2.4 Interactions Between Carbonate-Carbon and Organic Carbon . . . 62
2.3 Nitrogen . . . 64
2.3.1 Analytical Methods . . . 64
2.3.2 Biological Nitrogen Isotope Fractionations . . . 65
2.3.3 Trophic Level Indicator . . . 67
2.3.4 Nitrogen Isotopes in the Mantle and Crust . . . 67
2.3.5 Nitrogen in the Ocean and in Sediments . . . 68
2.4 Oxygen . . . 70
2.4.1 Analytical Methods . . . 70
2.4.1.1 Water . . . 70
2.4.1.2 Carbonates . . . 71
2.4.1.3 Silicates . . . 72
2.4.1.4 Phosphates . . . 73
2.4.1.5 Sulfates . . . 73
2.4.1.6 Nitrates . . . 73
2.4.2 Standards . . . 74
2.4.3 Fractionation Processes . . . 74

2.4.3.1 Fractionation of Water . . . . . . . . . . . . . . . . 74
2.4.3.2 $CO_2$–$H_2O$ System . . . . . . . . . . . . . . . . . . . . 76
2.4.3.3 Mineral Fractionations . . . . . . . . . . . . . . . . 77
2.4.4 Triple Oxygen Isotope Compositions . . . . . . . . . . . . . . 79
2.4.5 Fluid-Rock Interactions . . . . . . . . . . . . . . . . . . . . . 80
2.5 Sulfur . . . . . . . . . . . . . . . . . . . . . . . . . . . . . . . . 83
2.5.1 Methods . . . . . . . . . . . . . . . . . . . . . . . . . . . . 84
2.5.2 Fractionation Mechanisms . . . . . . . . . . . . . . . . . . . 85
2.5.2.1 Equilibrium Reactions . . . . . . . . . . . . . . . . 85
2.5.2.2 Dissimilatory Sulfate Reduction . . . . . . . . . 86
2.5.2.3 Thermochemical Reduction of Sulfate . . . . 89
2.5.3 Quadruple Sulfur Isotopes . . . . . . . . . . . . . . . . . . . 90
2.6 Lithium . . . . . . . . . . . . . . . . . . . . . . . . . . . . . . . 94
2.6.1 Methods . . . . . . . . . . . . . . . . . . . . . . . . . . . . 95
2.6.2 Diffusion . . . . . . . . . . . . . . . . . . . . . . . . . . . . 95
2.6.3 Magmatic Rocks . . . . . . . . . . . . . . . . . . . . . . . . 96
2.6.4 Weathering . . . . . . . . . . . . . . . . . . . . . . . . . . . 97
2.6.5 Ocean Water . . . . . . . . . . . . . . . . . . . . . . . . . . 98
2.6.6 Meteoric Water . . . . . . . . . . . . . . . . . . . . . . . . . 99
2.7 Boron . . . . . . . . . . . . . . . . . . . . . . . . . . . . . . . . 99
2.7.1 Methods . . . . . . . . . . . . . . . . . . . . . . . . . . . . 100
2.7.2 Isotope Fractionation Mechanism . . . . . . . . . . . . . . . 101
2.7.3 Fractionations at High Temperatures . . . . . . . . . . . . . 103
2.7.4 Weathering Environment . . . . . . . . . . . . . . . . . . . . 104
2.7.5 Tourmaline . . . . . . . . . . . . . . . . . . . . . . . . . . . 104
2.8 Magnesium . . . . . . . . . . . . . . . . . . . . . . . . . . . . . 105
2.8.1 Calculated Isotope Fractionations . . . . . . . . . . . . . . 106
2.8.1.1 Mantle Rocks . . . . . . . . . . . . . . . . . . . . . . 107
2.8.1.2 Continental Crust . . . . . . . . . . . . . . . . . . . 108
2.8.2 Fractionations During Weathering . . . . . . . . . . . . . . . 108
2.8.3 Ocean Water . . . . . . . . . . . . . . . . . . . . . . . . . . 109
2.8.4 Carbonates . . . . . . . . . . . . . . . . . . . . . . . . . . . 109
2.8.5 Plants and Animals . . . . . . . . . . . . . . . . . . . . . . . 110
2.9 Calcium . . . . . . . . . . . . . . . . . . . . . . . . . . . . . . . 111
2.9.1 Analytical Techniques . . . . . . . . . . . . . . . . . . . . . 111
2.9.2 High Temperature Fractionations . . . . . . . . . . . . . . . 112
2.9.3 Weathering . . . . . . . . . . . . . . . . . . . . . . . . . . . 113
2.9.4 Fractionations During Carbonate Precipitation . . . . . . . 114
2.9.5 Variations of Ocean Water with Geologic Time . . . . . . 115
2.9.6 Plants, Animals and Humans . . . . . . . . . . . . . . . . . . 116

2.10 Strontium . . . 117
2.10.1 Silicates . . . 117
2.10.2 Carbonates and Sulfates . . . 118
2.10.3 Fluids and Plants . . . 119
2.11 Barium . . . 119
2.11.1 Magmatic Systems . . . 120
2.11.2 Ocean . . . 121
2.12 Silicon . . . 122
2.12.1 Equilibrium Isotope Fractionations . . . 123
2.12.2 High-Temperature Fractionations . . . 123
2.12.3 Chemical Weathering and Mineral Precipitation . . . 124
2.12.4 Fractionations in Ocean Water . . . 125
2.12.5 Cherts . . . 126
2.12.6 Plants . . . 126
2.13 Chlorine . . . 127
2.13.1 Methods . . . 127
2.13.2 Hydrosphere . . . 128
2.13.3 Mantle-Derived Rocks . . . 129
2.13.4 Applications in the Environment . . . 130
2.14 Bromine . . . 131
2.15 Potassium . . . 132
2.15.1 Mineral Isotope Fractionations . . . 132
2.15.2 Magmatic Environment . . . 133
2.15.3 Weathering Environment . . . 133
2.16 Rubidium . . . 134
2.17 Titanium . . . 135
2.17.1 Magmatic Fractionations . . . 135
2.18 Vanadium . . . 136
2.18.1 High-Temperature Fractionations . . . 136
2.18.2 Low-Temperature Fractionations . . . 137
2.19 Chromium . . . 138
2.19.1 Mantle Rocks . . . 139
2.19.2 River and Ocean Water . . . 140
2.19.3 Carbonates . . . 140
2.19.4 Paleo-Redox Proxy . . . 141
2.19.5 Anthropogenic Cr in the Environment . . . 141
2.20 Iron . . . 142
2.20.1 Analytical Methods . . . 142
2.20.2 Isotope Equilibrium Studies . . . 143
2.20.3 Meteorites . . . 144
2.20.4 Igneous Rocks . . . 145

2.20.5 Sediments . . . . . 146
2.20.6 Ocean and River Water . . . . . 147
2.20.7 Plants . . . . . 148
2.21 Nickel . . . . . 149
2.21.1 Mantle Rocks and Meteorites . . . . . 150
2.21.2 Water . . . . . 151
2.21.3 Plants . . . . . 151
2.22 Copper . . . . . 152
2.22.1 Magmatic Rocks . . . . . 152
2.22.2 Ore Deposits . . . . . 153
2.22.3 Low-Temperature Fractionations . . . . . 154
2.22.4 River and Ocean Water . . . . . 155
2.22.5 Plants . . . . . 155
2.23 Zinc . . . . . 156
2.23.1 Fractionations During Evaporation . . . . . 157
2.23.2 Mantle Derived Rocks . . . . . 157
2.23.3 Ore Deposits . . . . . 158
2.23.4 Ocean . . . . . 158
2.23.5 Plants and Animals . . . . . 159
2.23.6 Anthropogenic Contamination . . . . . 160
2.24 Gallium . . . . . 160
2.25 Germanium . . . . . 161
2.25.1 Ore Deposits . . . . . 162
2.25.2 Hydrosphere . . . . . 163
2.26 Selenium . . . . . 163
2.26.1 Fractionation Processes . . . . . 165
2.26.2 Natural Variations at High Temperatures . . . . . 165
2.26.3 Ocean . . . . . 166
2.27 Tellurium . . . . . 166
2.28 Zirconium . . . . . 167
2.29 Molybdenum . . . . . 168
2.29.1 Magmatic Rocks . . . . . 169
2.29.2 Molybdenites . . . . . 170
2.29.3 Sediments . . . . . 170
2.29.4 Palaeoredox Proxy . . . . . 172
2.30 Silver . . . . . 173
2.31 Cadmium . . . . . 174
2.31.1 Extraterrestrial Materials . . . . . 175
2.31.2 Marine Environment . . . . . 175
2.31.3 Ore Deposits and Anthropogenic Pollution . . . . . 176

2.32 Tin . . . 177
2.32.1 Magmatic Rocks . . . 177
2.32.2 Ore Deposits . . . 178
2.32.3 Tin in the Environment . . . 178
2.33 Antimony . . . 179
2.34 Cerium . . . 180
2.35 Neodymium . . . 180
2.36 Europium . . . 181
2.37 Heavy Rare Earth Elements (HREE) . . . 181
2.38 Rhenium . . . 182
2.39 Tungsten . . . 183
2.40 Palladium . . . 185
2.41 Platinum . . . 185
2.42 Ruthenium . . . 186
2.43 Iridium . . . 187
2.44 Osmium . . . 187
2.45 Mercury . . . 188
2.45.1 MDF and MIF Fractionation Processes . . . 189
2.45.2 Igneous Rocks and Ore Deposits . . . 190
2.45.3 Sediments . . . 191
2.45.4 Environmental Pollutant . . . 191
2.46 Thallium . . . 193
2.46.1 Igneous Rocks . . . 194
2.46.2 Fractionations in the Ocean . . . 195
2.47 Uranium . . . 195
2.47.1 Fractionation Processes . . . 196
2.47.2 Mantle-Derived Rocks . . . 197
2.47.3 Ore Deposits . . . 197
2.47.4 Rivers and the Ocean . . . 197
2.47.5 Paleo-Redox Proxy . . . 198
References . . . 199

**3 Variations of Stable Isotope Ratios in Nature** . . . 267
3.1 Extraterrestrial Materials . . . 267
3.1.1 Chondrites . . . 268
3.1.1.1 Oxygen . . . 268
3.1.1.2 Hydrogen . . . 270
3.1.1.3 Carbon . . . 271
3.1.1.4 Nitrogen . . . 272
3.1.1.5 Sulfur . . . 273
3.1.1.6 Metals . . . 273
3.1.1.7 Meteorite-Earth Relationship . . . 275

3.1.2 The Moon ... 275
3.1.2.1 Oxygen ... 275
3.1.2.2 Hydrogen ... 276
3.1.2.3 Other Volatile Elements ... 276
3.1.3 Mars ... 277
3.1.3.1 Oxygen ... 277
3.1.3.2 Hydrogen ... 278
3.1.3.3 Carbon ... 278
3.1.3.4 Sulfur ... 279
3.1.4 Venus ... 280
3.2 Mantle ... 280
3.2.1 Oxygen ... 281
3.2.2 Hydrogen ... 282
3.2.3 Carbon ... 284
3.2.4 Nitrogen ... 286
3.2.5 Sulfur ... 287
3.2.6 Stable Isotope Composition of the Core ... 288
3.3 Magmatic Rocks ... 289
3.3.1 Fractional Crystallization ... 290
3.3.2 Differences Between Volcanic and Plutonic Rocks ... 290
3.3.3 Low Temperature Alteration Processes ... 291
3.3.4 Assimilation of Crustal Rocks ... 291
3.3.5 Glasses from Different Tectonic Settings ... 292
3.3.5.1 Oxygen ... 292
3.3.5.2 Hydrogen ... 293
3.3.5.3 Carbon ... 294
3.3.5.4 Nitrogen ... 294
3.3.5.5 Sulfur ... 295
3.3.6 Magnesium and Iron ... 296
3.3.7 Lithium and Boron ... 296
3.3.8 Ocean Crust ... 298
3.3.9 Granitic Rocks ... 299
3.3.9.1 Whole-Rock Oxygen ... 299
3.3.9.2 Non-traditional Isotopes ... 299
3.3.9.3 Zircon ... 300
3.3.10 Volatiles in Magmatic Systems ... 301
3.3.10.1 Water ... 302
3.3.10.2 Carbon ... 303
3.3.10.3 Nitrogen ... 305
3.3.10.4 Sulfur ... 305
3.3.11 Isotope Thermometers in Geothermal Systems ... 306

3.4 Metamorphic Rocks . . . . . 307
3.4.1 Contact Metamorphism . . . . . 309
3.4.2 Regional Metamorphism . . . . . 310
3.4.3 Subduction Zone Metamorphism . . . . . 312
3.4.4 Lower Crustal Rocks . . . . . 313
3.4.5 Thermometry . . . . . 313
3.5 Ore Deposits and Hydrothermal Systems . . . . . 316
3.5.1 Origin of Ore Fluids . . . . . 318
3.5.1.1 Magmatic Water . . . . . 319
3.5.1.2 Metamorphic Water . . . . . 320
3.5.1.3 Formation Waters . . . . . 320
3.5.2 Wall-Rock Alteration . . . . . 320
3.5.3 Fossil Hydrothermal Systems . . . . . 321
3.5.4 Hydrothermal Carbonates . . . . . 322
3.5.5 Sulfur Isotope Composition of Ore Deposits . . . . . 323
3.5.5.1 The Importance of $fO_2$ and pH . . . . . 324
3.5.5.2 Magmatic Ore Deposits . . . . . 325
3.5.5.3 Porphyry Copper Deposits . . . . . 325
3.5.5.4 Recent and Fossil Sulfide Deposits at Mid-Ocean Ridges . . . . . 325
3.5.5.5 Biogenic Deposits . . . . . 326
3.5.5.6 Metamorphosed Deposits . . . . . 327
3.5.6 Metal Isotopes . . . . . 327
3.5.6.1 Copper . . . . . 327
3.5.6.2 Iron . . . . . 328
3.5.6.3 Zinc . . . . . 329
3.6 Hydrosphere . . . . . 329
3.6.1 Meteoric Water–General Considerations . . . . . 329
3.6.1.1 $\delta^2H$–$\delta^{18}O$ Relationship, Deuterium (D)—Excess . . . . . 331
3.6.1.2 $\delta^{17}O$–$\delta^{18}O$ Relationships, $^{17}O$ Excess . . . . . 333
3.6.1.3 Meteoric Waters in the Past . . . . . 334
3.6.2 Ice Cores . . . . . 334
3.6.3 Groundwater . . . . . 335
3.6.4 Rivers . . . . . 337
3.6.5 Isotope Fractionations During Evaporation . . . . . 337
3.6.6 Ocean Water . . . . . 338
3.6.6.1 Oxygen and Hydrogen Isotopes . . . . . 338
3.6.6.2 Metal Isotopes . . . . . 340
3.6.7 Pore Waters . . . . . 342
3.6.8 Formation Water . . . . . 343
3.6.9 Water in Hydrated Salt Minerals . . . . . 345

3.7 The Isotopic Composition of Dissolved and Particulate Compounds in Ocean and Fresh Waters . . . 346
3.7.1 Carbon Species in Water . . . 346
3.7.1.1 Bicarbonate in Ocean Water . . . 346
3.7.1.2 Particulate Organic Matter (POM) . . . 348
3.7.1.3 Carbon Isotope Composition of Pore Waters . . . 348
3.7.1.4 Carbon in Fresh Waters . . . 349
3.7.2 Silicon . . . 350
3.7.3 Nitrogen . . . 351
3.7.4 Oxygen . . . 351
3.7.5 Sulfate . . . 352
3.7.6 Phosphate . . . 354
3.8 Isotopic Composition of the Ocean During Geologic History . . . 354
3.8.1 Oxygen . . . 355
3.8.2 Carbon . . . 357
3.8.3 Sulfur . . . 359
3.8.4 Lithium . . . 361
3.8.5 Boron . . . 361
3.8.6 Calcium . . . 362
3.9 Atmosphere . . . 362
3.9.1 Atmospheric Water Vapour . . . 363
3.9.2 Nitrogen . . . 364
3.9.2.1 Nitrous Oxide . . . 365
3.9.3 Oxygen . . . 366
3.9.3.1 Evolution of Atmospheric Oxygen . . . 368
3.9.4 Carbon Dioxide . . . 369
3.9.4.1 Carbon . . . 369
3.9.4.2 Oxygen . . . 370
3.9.4.3 Long Term Variations in the $CO_2$ Concentration and Isotope Composition . . . 372
3.9.5 Carbon Monoxide . . . 375
3.9.6 Methane . . . 375
3.9.7 Hydrogen . . . 376
3.9.8 Sulfur . . . 377
3.9.9 Perchlorate . . . 379
3.9.10 Metal Isotopes . . . 380
3.10 Biosphere . . . 380
3.10.1 Living Organic Matter . . . 380
3.10.1.1 Bulk Carbon . . . 380
3.10.1.2 Position Specific Isotope Composition . . . 383
3.10.1.3 Hydrogen . . . 383

3.10.1.4 Oxygen . . . 385
3.10.1.5 Nitrogen . . . 386
3.10.1.6 Sulfur . . . 386
3.10.1.7 Metals in Plants . . . 387
3.10.2 Indicators of Diet and Metabolism . . . 387
3.10.3 Tracing Anthropogenic Organic Contaminant Sources . . . 388
3.10.4 Marine Versus Terrestrial Organic Matter . . . 389
3.10.5 Fossil Organic Matter . . . 390
3.10.6 Oil . . . 391
3.10.7 Coal . . . 393
3.10.7.1 Black Carbon . . . 394
3.10.8 Natural Gas . . . 394
3.10.8.1 Biogenic Gas . . . 396
3.10.8.2 Thermogenic Gas . . . 396
3.10.8.3 Abiogenic Methane . . . 397
3.10.8.4 Isotope Clumping in Methane . . . 398
3.10.8.5 Nitrogen in Natural Gas . . . 399
3.10.8.6 Isotope Signatures of Early Life on Earth . . . 399
3.11 Sedimentary Rocks . . . 400
3.11.1 Fractionations During Weathering . . . 400
3.11.2 Clastic Sediments . . . 402
3.11.3 Clay Minerals . . . 403
3.11.4 Biogenic Silica and Cherts . . . 405
3.11.4.1 Biogenic Silica . . . 405
3.11.4.2 Cherts . . . 406
3.11.5 Marine Carbonates . . . 406
3.11.5.1 Oxygen . . . 406
3.11.5.2 Carbon . . . 410
3.11.6 Diagenesis . . . 412
3.11.6.1 Burial Pathway . . . 412
3.11.6.2 Meteoric Pathway . . . 413
3.11.7 Limestones . . . 413
3.11.7.1 Carbon Isotope Stratigraphy . . . 413
3.11.8 Dolomites . . . 414
3.11.9 Freshwater Carbonates . . . 416
3.11.10 Phosphates . . . 417
3.11.11 Iron Oxides . . . 418
3.11.11.1 Oxygen . . . 418
3.11.11.2 Iron . . . 419
3.11.11.3 Fe–Mn Crusts . . . 420

3.11.12 Sedimentary Sulfur and Pyrite . . . . . 420
3.11.12.1 Sulfur . . . . . 420
3.11.12.2 Iron . . . . . 422
3.12 Palaeoclimatology . . . . . 423
3.12.1 Continental Records . . . . . 423
3.12.1.1 Tree Rings . . . . . 423
3.12.1.2 Organic Matter . . . . . 424
3.12.1.3 Hydroxyl-Bearing Minerals . . . . . 424
3.12.1.4 Lake Sediments . . . . . 425
3.12.1.5 Speleothems . . . . . 425
3.12.1.6 Phosphates . . . . . 426
3.12.2 Ice Cores . . . . . 426
3.12.2.1 Correlations of Ice-Core Records . . . . . 428
3.12.2.2 Gas-Inclusions in Ice Cores . . . . . 429
3.12.3 Marine Records . . . . . 430
3.12.3.1 Corals . . . . . 431
3.12.3.2 Conodonts . . . . . 432
3.12.3.3 Characteristic Climatic Events . . . . . 432
3.12.3.4 Clumped Isotope Thermometry . . . . . 434
3.13 Additional Applications . . . . . 435
3.13.1 Forensic Isotope Geochemistry . . . . . 435
3.13.2 Medical Studies . . . . . 436
References . . . . . 439

**Index** . . . . . 499

# List of Figures


Fig. 1.1 Plot of number of protons (Z) and number of neutrons (N) in stable (*filled circles*) and unstable (*open circles*) nuclides. . . . 2

Fig. 1.2 Number of stable isotopes of elements with even and odd number of protons (radioactive isotopes with half-lives greater than $10^9$ years are included) . . . . . . . . . . . . . . . . . . . . . . . 3

Fig. 1.3 Schematic potential energy *curve* for the interaction of two atoms in a stable molecule or between two molecules in a liquid or solid (after Bigeleisen 1965) . . . . . . . . . . . . . . . . . 5

Fig. 1.4 $\delta^{18}O$ in a cloud vapour and condensate plotted as a function of a fraction of remaining vapour in a cloud for a Rayleigh process. The temperature of the cloud is shown on the lower axis. The increase in fractionation with decreasing temperature is taken into account (after Dansgaard 1964) . . . . . . . . . . . . . . . 11

Fig. 1.5 Calibrations of the clumped isotope thermometer (after Petersen et al. 2019, courtesy Jens Fiebig) . . . . . . . . . . . . 17

Fig. 1.6 Arrhenius plot of diffusion coefficients versus reciprocal temperatures for various minerals. Data from phases reacted under *wet conditions* are given as *solid lines*, whereas *dry conditions* are represented by *dashed lines*. Note that the rates for dry systems are generally lower and have higher activation energies (steeper slopes) (after Cole and Chakraborty 2011). . . . 21

Fig. 1.7 $CO_2$-graphite partial exchange experiments in a Northrop and Clayton plot at 700, 800, 1000 and 1200 °C. The connecting line in experiment at 1200 °C has a plain slope and defines the intercept more precisely than the experiment at 700 °C (after Scheele and Hoefs 1992). . . . . . . . . . . . . . . . . . . . . . . . . . 27

Fig. 1.8 Schematic representation of a gas-source mass spectrometer for stable isotope measurements during the 1960 and 70s. P denotes pumping system, V denotes a variable volume. . . . . . 29

Fig. 1.9 Schematic diagram of an elemental analyser-isotope ratio-mass spectrometer for the determination of carbon and nitrogen isotopes. . . . . . . . . . . . . . . . . . . . . . . . . . . . . . . . . . . . . 32

Fig. 1.10 Relationship between $^{18}O$ ($^{16}O$) content in percent and $\delta^{18}O$ in per mill. . . . . . . . . . . . . . . . . . . . . . . . . . . . . . . . . . 36

Fig. 1.11 Precision of various oxygen isotope methods as a function of sample weight or size (from Bindeman 2008) . . . . . . . . . . . . 39
Fig. 2.1 $\delta^2$H variation ranges of geologically important reservoirs. . . . . . 51
Fig. 2.2 $\delta^2$H-values versus time for two beakers that have equal surface areas and equal volumes undergoing isotopic exchange in sealed systems. In both experiments at 21 and 52 °C isotope ratios progress toward an average value of −56‰ via exchange with ambient vapour: *solid curves* are calculated, *points* are experimental data (after Criss 1999) . . . . . . . . . . . . . . . . . . . . 53
Fig. 2.3 Experimentally determined fractionation factors between liquid water and water vapour from 1 to 350 °C (after Horita and Wesolowski 1994) . . . . . . . . . . . . . . . . . . . . . . . . . . . . 54
Fig. 2.4 Global relationship between monthly means of $\delta^2$H and $\delta^{18}$O in precipitation, derived for all stations of the IAEA global network. *Line* indicates the global Meteoric Water Line (GMWL) (after Rozanski et al. 1993). . . . . . . . . . . . . . . . . . . . 55
Fig. 2.5 D/H fractionations between $H_2O$–$H_2$, $H_2O$–$H_2S$ and $H_2O$–$CH_4$ (from calculated data of Richet et al. 1977) . . . . . . . . . . . . . . . 55
Fig. 2.6 Carbon isotope fractionation between various geologic compounds and $CO_2$ (after Chacko et al. 2001) . . . . . . . . . . . . . 60
Fig. 2.7 Histogram of $\delta^{13}$C values of $C_3$ and $C_4$ plants (after Cerling and Harris 1999). . . . . . . . . . . . . . . . . . . . . . . . . . . . . . . . . 62
Fig. 2.8 $\delta^{13}$C-values of important geological reservoirs . . . . . . . . . . . . . 63
Fig. 2.9 $\delta^{15}$N-values of important geological reservoirs . . . . . . . . . . . . . 68
Fig. 2.10 Oxygen isotope fractionation factors between liquid water and water vapour in the temperature range 0–350 °C (after Horita and Wesolowski 1994) . . . . . . . . . . . . . . . . . . . . . 75
Fig. 2.11 Oxygen isotope fractionation between pure water and solutions of various ions (after O'Neil and Truesdell 1991). . . . . . . . . . . . 76
Fig. 2.12 Oxygen isotope fractionations between dissolved inorganic carbon (DIC) and water as function of pH and temperatures (after Beck et al. 2005). . . . . . . . . . . . . . . . . . . . . . . . . . . . . . 77
Fig. 2.13 Oxygen isotope fractionations between various minerals and calcite (after Chacko et al. 2001) . . . . . . . . . . . . . . . . . . . . . 79
Fig. 2.14 $\Delta^{17}$O- and $\delta^{18}$O-values of major Earth's reservoirs (courtesy Andreas Pack) . . . . . . . . . . . . . . . . . . . . . . . . . . . . . 80
Fig. 2.15 $\delta^{18}O_{(feldspar)}$ versus $\delta^{18}O_{(quartz)}$ and versus $\delta^{18}O_{(pyroxene)}$ plots of disequilibrium mineral pair arrays in granitic and gabbroic rocks. The arrays indicate open-system conditions from circulation of hydrothermal meteoric fluids (after Gregory et al. 1989) . . . . . . . . . . . . . . . . . . . . . . . . . . . . . . . . . . . . 82
Fig. 2.16 $\delta^{18}$O values of important geological reservoirs . . . . . . . . . . . . . 83
Fig. 2.17 $\delta^{34}$S-values of important geological reservoirs . . . . . . . . . . . . . 84

Fig. 2.18 Equilibrium fractionations among sulfur compounds relative to $H_2S$ (*solid lines* experimentally determined, *dashed lines* extrapolated or theoretically calculated (after Ohmoto and Rye 1979) . . . 86
Fig. 2.19 Rayleigh plot for sulfur isotope fractionations during reduction of sulfate in a closed system. Assumed fractionation factor 1.025, assumed composition of initial sulfate: +10‰) . . . 88
Fig. 2.20 Compilation of $\Delta^{33}S$ versus age for rock samples. Note large $\Delta^{33}S$ before 2.45 Ga, indicated by *vertical line*, small but measurable $\Delta^{33}S$ after 2.45 Ga (Farquahar et al. 2007). . . . 91
Fig. 2.21 Similarity of S-MIF signatures in modern sulfates and Archean barites (**a**) and sulfides from Archean records (**b**) (from Lin et al. 2018) . . . 92
Fig. 2.22 Lithium isotope variations in major geological reservoirs . . . 95
Fig. 2.23 Boron isotope variations in geologically important reservoirs . . . 100
Fig. 2.24 **a** Distribution of aqueous boron species versus pH; **b** $\delta^{11}B$ of the two dominant species $B(OH)_3$ and $B(OH)^-_4$ versus pH (after Hemming and Hanson 1992) . . . 102
Fig. 2.25 $\delta^{26}Mg$ values of important geological reservoirs . . . 106
Fig. 2.26 $\delta^{44/40}Ca$-values of important geological reservoirs . . . 112
Fig. 2.27 $\delta^{88/86}Sr$-values of important geological reservoirs . . . 117
Fig. 2.28 $^{138/134}Ba$ isotope variations in some geological reservoirs . . . 120
Fig. 2.29 $\delta^{30}Si$-values of important geological reservoirs . . . 122
Fig. 2.30 $\delta^{37}Cl$ values of important geological reservoirs . . . 128
Fig. 2.31 Potassium isotope variations in geological reservoirs . . . 132
Fig. 2.32 Vanadium isotope variations in some geological reservoirs . . . 136
Fig. 2.33 $\delta^{53}Cr$-values of important geological reservoirs (ocean water: Scheiderich et al. (2015) observed large variations in oceanic surface waters) . . . 138
Fig. 2.34 $\delta^{56}Fe$-values of important geological reservoirs . . . 143
Fig. 2.35 $\delta^{60/58}Ni$ isotope variations in important geological reservoir . . . 150
Fig. 2.36 $\delta^{65}Cu$-values of important geological reservoirs . . . 153
Fig. 2.37 $\delta^{66}Zn$-values of important geological reservoirs . . . 156
Fig. 2.38 $\delta^{74/70}Ge$ isotope variations of geological reservoirs . . . 162
Fig. 2.39 $\delta^{82/76}Se$-values of important geological reservoirs . . . 164
Fig. 2.40 $\delta^{98/95}Mo$-values of important geological reservoirs . . . 171
Fig. 2.41 $\delta^{114/110}Cd$-values of important geological reservoirs . . . 175
Fig. 2.42 $^{186}W/^{184}W$ compositions of some important reservoirs . . . 183
Fig. 2.43 $\delta^{202/198}Hg$ and $\Delta^{201}Hg$ values of important geological reservoirs . . . 190
Fig. 2.44 $\delta^{205}Tl$-values of important geological reservoirs . . . 193

Fig. 2.45 $\delta^{238}$U-values of important geological reservoirs . . . . . . . . . . . . . 196
Fig. 3.1 $^{17}$O versus $^{18}$O isotopic composition of Ca–Al rich inclusions (CAI) from chondrites (Clayton 1993) . . . . . . . . . . . . . . . . . . . 269
Fig. 3.2 Carbon compounds in primitive meteorites. Species classified as interstellar on the basis of C-isotopes are *coloured*. Only a minor fraction of organic carbon is interstellar (after Ming et al. 1989) . . . . . . . . . . . . . . . . . . . . . . . . . . . . . . . . . 271
Fig. 3.3 Three oxygen isotope plot of lunar, Martian rocks and HED meteorites supposed to be fragments of asteroid Vesta (after Wiechert et al. 2001) . . . . . . . . . . . . . . . . . . . . . . . . . 278
Fig. 3.4 Hydrogen isotope variations in mantle-derived minerals and rocks (modified after Bell and Ihinger 2000) . . . . . . . . . . . . 283
Fig. 3.5 Carbon isotope variations of diamonds (*arrows* indicate highest and lowest $\delta^{13}$C-values (modified after Cartigny 2005) . . . . . . . . . . . . . . . . . . . . . . . . . . . . . . . . . . . . 285
Fig. 3.6 Nitrogen isotope variations in mantle derived materials (modified after Marty and Zimmermann 1999) . . . . . . . . . . . . . . 287
Fig. 3.7 Sulfur isotope compositions of high- and low-S peridotites . . . . 288
Fig. 3.8 Plot of $\delta^{18}$O-values versus Mg numbers for oceanic basalts (*filled circles*) and continental basalts (*open circles*). The *shaded field* denotes the ±2 σ range of a MORB mean value of +5.7‰. The *clear vertical field* denotes the range for primary basaltic partial melts in equilibrium with a peridotitic source (Harmon and Hoefs 1995) . . . . . . . . . . . . . . . . . . . . . . . . 292
Fig. 3.9 **a** Forsterite variations in an olivine phenocryst from the Kilauea lava lake. **b** Predicted Mg and Fe isotope fractionation induced by interdiffusion. Solid lines represent initial conditions, dotted curves are calculated diffusion profiles (after Teng et al. 2011) . . . . . . . . . . . . . . . . . . . . . . . . . . . . . . 297
Fig. 3.10 Histogram of $\delta^{18}$O-values for igneous zircons (**a** *Archean*, **b** *Proterozoic*, **c** *Phanerozoic*) (after Valley et al. 2005) . . . . . . . 301
Fig. 3.11 Isotopic composition of thermal waters and associated local ground waters. Lines connect corresponding thermal waters to local groundwaters (Giggenbach 1992) . . . . . . . . . . . . . . . . . . 303
Fig. 3.12 S-isotope degassing scenarios at high and low pressures and at high and low oxygen fugacities (De Moor et al. 2013) . . . . . . . . 305
Fig. 3.13 Coupled C–O trends showing decreasing values of $\delta^{13}$C and $\delta^{18}$O with increasing metamorphic grade from contact metamorphic localities (Baumgartner and Valley 2001) . . . . . . . 308
Fig. 3.14 Plot of $\delta^{18}$O of quartz versus $\delta^{18}$O magnetite (*solid squares*) and of biotite versus muscovite (*open squares*) from rocks whose peak metamorphic conditions range from greenschist through granulite facies (after Kohn 1999) . . . . . . . . . . . . . . . . 314

Fig. 3.15 Frequency distribution of calcite-graphite fractionations (Δ) with increasing metamorphic grade (after Des Marais 2001). . . . 316
Fig. 3.16 Plot of $\delta^2$H versus $\delta^{18}$O of waters of different origin . . . . . . . . . 318
Fig. 3.17 C- and O-isotope compositions of calcites and siderites from the Bad Grund and Lautenthal deposits, Harz (after Zheng and Hoefs 1993). . . . . . . . . . . . . . . . . . . . . . . . . . 323
Fig. 3.18 Influence of $fO_2$ and pH on the sulfur isotope composition of sphalerite and barite at 250 °C and $\delta^{34}S_{\Sigma S}$ = 0‰ (modified after Ohmoto 1972). . . . . . . . . . . . . . . . . . . . . . . . . . . . . . . 324
Fig. 3.19 Schematic O-isotope fractionation of water in the atmosphere (after Siegenthaler 1979). . . . . . . . . . . . . . . . . . . . . . . . . . 330
Fig. 3.20 Average $\delta^2$H-values of the annual precipitation from oceanic islands as a function of the amount of annual rainfall. The island stations are distant from continents, within 30° of the equator and at elevations <120 m (after Lawrence and White 1991) . . . . . . . . . . . . . . . . . . . . . . . . . . . . . . . . . . . . . . . . . . 331
Fig. 3.21 Correlations of $\delta^2$H and $\delta^{18}$O values of Greenland (GISP-2) and Antarctic (Vostok) ice cores covering the last glacial-interglacial cycles (http://www.gisp2.sr.unh.edu/GISP2/DATA/Bender.html) . . . . . . . . . . . . . . . . . . . . . . . . . . . . . . . . 336
Fig. 3.22 $\delta^2$H versus $\delta^{18}$O values of the Dead Sea and its water sources as an example of an evaporative environment (after GAT 1984) . . . . . . . . . . . . . . . . . . . . . . . . . . . . . . . . . . . . . . . . . . 338
Fig. 3.23 Salinity versus $\delta^{18}$O relationships in modern ocean surface and deep waters (after Railsback et al. 1989) . . . . . . . . . . . . . . . 339
Fig. 3.24 Comparison of measured and modeled $\delta^{18}$O values of surface ocean waters. Characteristic features are: tropical maxima, equatorial low- and high-latitude minima, enrichment of the Atlantic relative to the Pacific (after Delaygue et al. 2000). . . . . 340
Fig. 3.25 $\delta^2$H versus $\delta^{18}$O values for formation waters from the midcontinental region of the United States (after Taylor 1974) . . . . . . . . . . . . . . . . . . . . . . . . . . . . . . . 344
Fig. 3.26 Vertical profiles of dissolved $CO_2$, $\delta^{13}$C, dissolved $O_2$ and $\delta^{18}$O in the North Atlantic (Kroopnick et al. 1972) . . . . . . . 347
Fig. 3.27 $\delta^{13}$C records of total dissolved carbon from pore waters of anoxic sediments recovered in various DSDP sites (after Anderson and Arthus 1983). . . . . . . . . . . . . . . . . . . . . . . . 349
Fig. 3.28 Carbon isotopic composition of total dissolved carbon in large river systems. *Data source* Amazon: Longinelli and Edmond (1983), Rhine: Buhl et al. (1991), St. Lawrence: Yang et al. (1996). . . . . . . . . . . . . . . . . . . . . . . . . . . . . . . . . . . . . . . . . . . . . 350
Fig. 3.29 Frequency distribution of $\delta^{34}$S-values in river sulfate . . . . . . . . . 353
Fig. 3.30 $\delta^{18}$O-values for cherts, carbonates and shales with geologic age (courtesy I. Bindeman). . . . . . . . . . . . . . . . . . . . . . . . . . . . 356

Fig. 3.31 $\delta^{13}C$-values for marine carbonates over time. Note persistent values of 0–3‰ for the last 600 Ma, anomaleous variability at 0.6–0.8 Ga and 2.0–2.3 Ga correlative with snowball earth episodes (Shields and Veizer 2002) . . . . . . . . . . . . . . . . . . . . . 358
Fig. 3.32 Marine sulfate $\delta^{34}S$ curve of marine barite for 130 Ma to present (Paytan et al. 2004) . . . . . . . . . . . . . . . . . . . . . . . . . . 360
Fig. 3.33 $\delta^{17}O$ versus $\delta^{18}O$ plot of atmospheric oxygen species (Thiemens 2006). . . . . . . . . . . . . . . . . . . . . . . . . . . . . . . . . . 364
Fig. 3.34 $\delta^{56}Fe$ values of pyrite and iron oxides versus time showing three evolutionary stages of the ocean (Anbar and Rouxel 2007) . . . . . . . . . . . . . . . . . . . . . . . . . . . . . . . . . . . . . . . 369
Fig. 3.35 Relationship between atmospheric $CO_2$ concentration and $\delta^{13}C_{(CO2)}$ (after Keeling 1958). . . . . . . . . . . . . . . . . . . . . . . 370
Fig. 3.36 Seasonal $\delta^{13}C$ variations of atmospheric $CO_2$ from five stations in the Northern Hemisphere. *Dots* denote monthly averages, *oscillating curves* are fits of daily averages (after Keeling et al. 1989) . . . . . . . . . . . . . . . . . . . . . . . . . . . . . . . . . . . . . . . . 371
Fig. 3.37 $\delta^{18}O$ seasonal record of atmospheric $CO_2$ from three stations: Point Barrow 71.3°N, Mauna Loa 19.5°S, South Pole 90.0°S (after Ciais et al. 1998). . . . . . . . . . . . . . . . . . . . . . . . . . . . . 372
Fig. 3.38 Law Dome ice core $CO_2$ and $\delta^{13}C$ record for the last 1000 years (after Trudinger et al. 1999). . . . . . . . . . . . . . . . . . . . . . . . 374
Fig. 3.39 S-isotope composition of **a** natural and **b** anthropogenic sulfur sources in the atmosphere. *DMS* Dimethylsulfide . . . . . . . . . . . . 378
Fig. 3.40 Generalized scheme of hydrogen isotope changes in plants (Sachse et al. 2012) . . . . . . . . . . . . . . . . . . . . . . . . . . . . . . . . . 384
Fig. 3.41 Petroleum-type curves of different oil components from the North Sea showing a positive oil-oil correlation and a negative source rock—oil correlation (*SAT* saturated hydrocarbons, *AROM* aromatic hydrocarbons, *NOS'S* heterocompounds, *ASPH* asphaltenes (Stahl 1977)) . . . . . . . . . . . . . . . . . . . . . . . . 392
Fig. 3.42 $\delta^{13}C$ and $\delta^{2}H$ variations of natural gases of different origins (after Whiticar 1999). . . . . . . . . . . . . . . . . . . . . . . . . . . . . . . . 397
Fig. 3.43 Predicted (*bars*) and measured (*crosses*) oxygen isotope composition of separated minerals from Haitian weathering profiles. The range of predicted $\delta^{18}O$-values are calculated assuming a temperature of 25 °C and a meteoric water $\delta^{18}O$-value of −3.1‰ (after Bird et al. 1992) . . . . . . . . . . 401
Fig. 3.44 Histogram of $\delta^{18}O$-values of quartz in sandstone from 6 to 10 μm spots by ion microprobe. Mixed analyses are on the boundary of detrital quartz and quartz overgrowth (Kelly et al. 2007). . . . . . . . . . . . . . . . . . . . . . . . . . . . . . . . . 403

Fig. 3.45 $\delta^2H$ and $\delta^{18}O$ values of kaolinites and related minerals from weathering and hydrothermal environments. The Meteoric Water Line, kaolinite weathering and supergene/hypogene (S/H) lines are given for reference (after Sheppard and Gilg 1996) . . . . 404
Fig. 3.46 $\Delta^{18}O$ and $\Delta^{13}C$ differences from equilibrium isotope composition of extant calcareous species (after Wefer and Berger 1991) . . . . 408
Fig. 3.47 $\delta^{13}C$-values of benthic foraminifera species. The $\delta^{13}C$-value for the dissolved bicarbonate in deep equatorial water is shown by the vertical line (after Wefer and Berger 1991) . . . . 411
Fig. 3.48 Carbon and oxygen isotope composition of some recent and Pleistocene dolomite occurences (after Tucker and Wright 1990) . . . . 416
Fig. 3.49 Carbon and oxygen isotope compositions of freshwater carbonates from recently closed lakes (after Talbot 1990). . . . . 417
Fig. 3.50 Dansgaard-Oeschger events in the time period from 45,000 to 30,000 years before present from GRIP and NGRIP ice core data (http://en.wikipedia.org/wiki/Image:Grip-ngrip-do18-closeup.png) . . . . 428
Fig. 3.51 Composite $\delta^{18}O$ fluctuation in the foraminifera species G saculifer from Caribbean cores (Emiliani 1978) . . . . 430
Fig. 3.52 Global deep-sea isotope record from numerous DSDP and ODP cores (Zachos et al. 2001). . . . . 433

# Theoretical and Experimental Principles

# 1

## 1.1 General Characteristics of Isotopes

Isotopes are atoms whose nuclei contain the same number of protons but a different number of neutrons. The term “isotopes” is derived from Greek (meaning equal places) and indicates that isotopes occupy the same position in the periodic table.

It is convenient to denote isotopes in the form $^{m}_{n}E$ where the super-script “m” denotes the mass number (i.e., sum of the number of protons and neutrons in the nucleus) and the subscript “n” denotes the atomic number of an element E. For example, $^{12}_{6}C$ is the isotope of carbon which has six protons and six neutrons in its nucleus. The atomic weight of each naturally occurring element is the average of the weights contributed by its various isotopes.

Isotopes can be divided into two fundamental kinds, stable and unstable (radioactive) species. The number of stable isotopes is about 300; whilst over 1200 unstable ones have been discovered so far. The term “stable” is relative, depending on the detection limits of radioactive decay times. In the range of atomic numbers from 1 (H) to 83 (Bi), stable nuclides of all masses except 5 and 8 are known. Only 21 elements are pure elements, in the sense that they have only one stable isotope. All other elements are mixtures of at least two isotopes. The relative abundance of different isotopes of an element may vary substantially. In copper, for example, $^{63}Cu$ accounts for 69% and $^{65}Cu$ for 31% of all copper nuclei. For the light elements, however, one isotope is predominant, the others being present only in trace amounts.

The stability of nuclides is characterized by several important rules, two of which are briefly discussed here. The first is the so-called symmetry rule, which states that in a stable nuclide with low atomic number, the number of protons is approximately equal to the number of neutrons, or the neutron-to-proton ratio, N/Z, is approximately equal to unity. In stable nuclei with more than 20 protons or neutrons, the N/Z ratio is always greater than unity, with a maximum value of about 1.5 for the heaviest stable nuclei. The electrostatic Coulomb repulsion of the



J. Hoefs, *Stable Isotope Geochemistry*, Springer Textbooks in Earth Sciences, Geography and Environment, https://doi.org/10.1007/978-3-030-77692-3_1

positively charged protons grows rapidly with increasing Z. To maintain the stability in the nuclei more neutrons (which are electrically neutral) than protons are incorporated into the nucleus (see Fig. 1.1).

The second rule is the so-called "Oddo-Harkins" rule, which states that nuclides of even atomic numbers are more abundant than those with odd numbers. As shown in Table 1.1, the most common of the four possible combinations is even-even, the least common odd-odd.

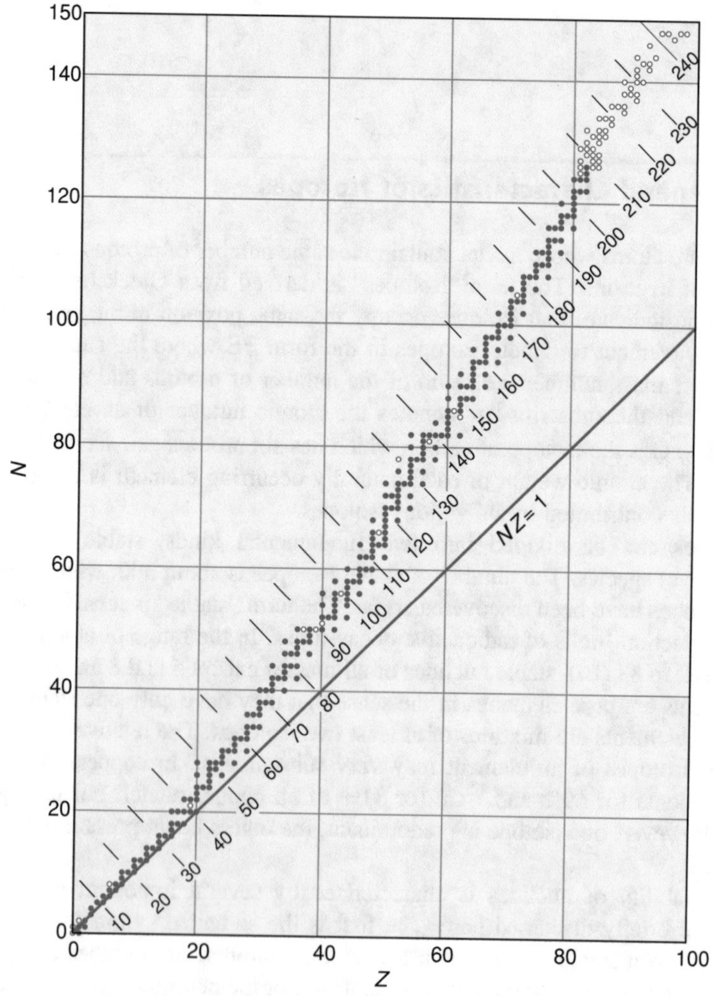


**Fig. 1.1** Plot of number of protons (Z) and number of neutrons (N) in stable (*filled circles*) and unstable (*open circles*) nuclides

Table 1.1 Types of atomic nuclei and their frequency of occurrence

| Z–N combination | Number of stable nuclides |
|---|---|
| Even–even | 160 |
| Even–odd | 50 |
| Odd–even 50 | 56 |
| Odd–odd | 5 |

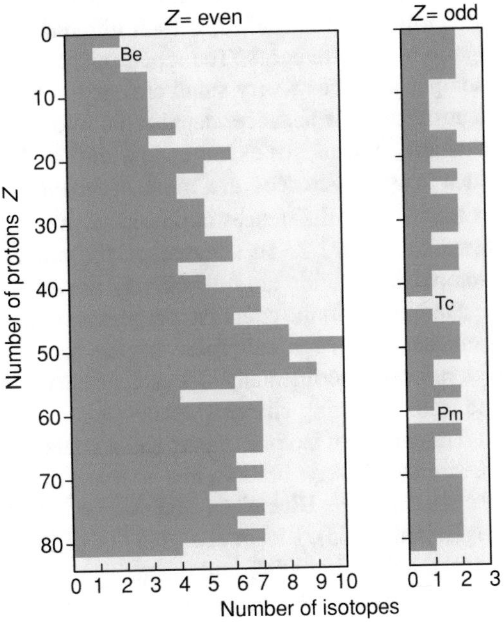


Fig. 1.2 Number of stable isotopes of elements with even and odd number of protons (radioactive isotopes with half-lives greater than $10^9$ years are included)

The same relationship is demonstrated in Fig. 1.2, which shows that there are more stable isotopes with even than with odd proton numbers.

Radioactive isotopes can be classified as being either artificial or natural. Only the latter are of interest in geology, because they are the basis for radiometric dating methods. Radioactive decay processes are spontaneous nuclear reactions and may be characterized by the radiation emitted, i.e. $\alpha$, $\beta$ and/or $\gamma$-emission. Decay processes may also involve electron capture.

Radioactive decay is one process that produces variations in isotope abundance. A second cause of differences in isotope abundance is isotope fractionation caused by small chemical and physical differences between the isotopes of an element. It is exclusively this important process that will be discussed in the following chapters.

## 1.2 Isotope Effects

Differences in chemical and physical properties arising from variations in atomic mass of an element are called "isotope effects". It is well known that the electronic structure of an element essentially determines its chemical behaviour, whereas the nucleus is more or less responsible for its physical properties. Because all isotopes of a given element contain the same number and arrangement of electrons, a far-reaching similarity in chemical behaviour is the logical consequence. But this similarity is not unlimited; certain differences exist in physicochemical properties due to mass differences. The replacement of any atom in a molecule by one of its isotopes produces a very small change in chemical behaviour. The addition of one neutron can, for instance, depress the rate of chemical reaction considerably. Furthermore, it leads, for example, to a shift of the lines in the Raman and IR spectra. Such mass differences are most pronounced among the lightest elements. For example, some differences in physicochemical properties of $H_2^{16}O$, $D_2^{16}O$, $H_2^{18}O$ are listed in Table 1.2. To summarize, the properties of molecules differing only in isotopic substitution are qualitatively the same, but quantitatively different.

Differences in the chemical properties of the isotopes of H, C, N, O, S, and other elements have been calculated by the methods of statistical mechanics and also determined experimentally. These differences can lead to considerable separation of the isotopes during chemical reactions.

The theory of isotope effects and a related isotope fractionation mechanism will be discussed very briefly. For a more detailed introduction to the theoretical background, see Bigeleisen and Mayer (1947), Urey (1947), Melander (1960), Bigeleisen (1965), Richet et al. (1977), O'Neil (1986), Criss (1999), Chacko et al. (2001), Schauble (2004) and others.

Differences in the physicochemical properties of isotopes arise as a result of quantum mechanical effects. Figure 1.3 shows schematically the energy of a diatomic molecule as a function of the distance between the two atoms. According to the quantum theory, the energy of a molecule is restricted to certain discrete energy levels. The lowest level is not at the minimum of the energy curve, but above it by an amount 1/2 hν, where h is Planck's constant and ν is the frequency with which the atoms in the molecule vibrate with respect to one another. Thus, even in the

**Table 1.2** Characteristic physical properties of $H_2^{16}O$, $D_2^{16}O$, and $H_2^{18}O$

| Property | $H_2^{16}O$ | $D_2^{16}O$ | $H_2^{18}O$ |
|---|---|---|---|
| Density (20 °C, in g cm$^{-3}$) | 0.997 | 1.1051 | 1.1106 |
| Temperature of greatest density (°C) | 3.98 | 11.24 | 4.30 |
| Melting point (760 Torr, in °C) | 0.00 | 3.81 | 0.28 |
| Boiling point (760 Torr, in °C) | 100.00 | 101.42 | 100.14 |
| Vapor pressure (at 100 °C, in Torr) | 760.00 | 721.60 | |
| Viscosity (at 20 °C, in centipoise) | 1.002 | 1.247 | 1.056 |

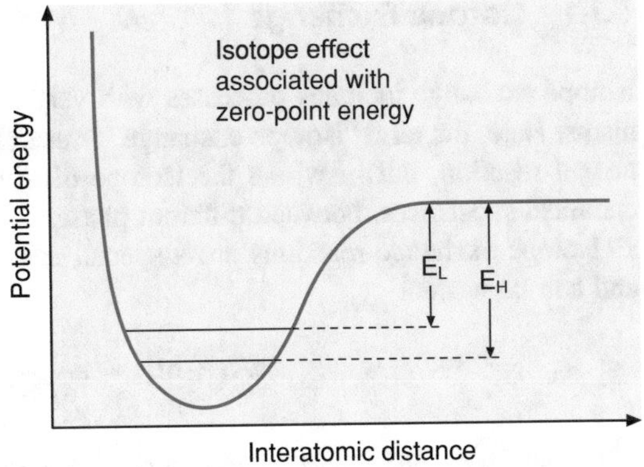


**Fig. 1.3** Schematic potential energy *curve* for the interaction of two atoms in a stable molecule or between two molecules in a liquid or solid (after Bigeleisen 1965)

ground state at a temperature of absolute zero, the vibrating molecule would possess a certain zero point energy above the minimum of the potential energy curve of the molecule. It vibrates with its fundamental frequency, which depends on the mass of the isotopes. In this context, it is important to note that vibrational motions dominate chemical isotope effects. Therefore, molecules of the same chemical formula that have different isotopic species will have different zero-point energies: the molecule of the heavy isotope will have a lower zero-point energy than the molecule of the light isotope, because it has a lower vibrational frequency. This is shown schematically in Fig. 1.3, where the upper horizontal line ($E_L$) represents the dissociation energy of the light molecule and the lower line ($E_H$), that of the heavy one. $E_L$ is actually not a line, but an energy interval between the zero-point energy level and the "continuous" level. This means that the bonds formed by the light isotope are weaker than bonds involving the heavy isotope. Thus, during a chemical reaction, molecules bearing the light isotope will, in general, react slightly more readily than those with the heavy isotope.

## 1.3 Isotope Fractionation Processes

The partitioning of isotopes between two substances or two phases of the same substance with different isotope ratios is called "isotope fractionation". The main phenomena producing isotope fractionations are

1. isotope exchange reactions (equilibrium isotope distribution),
2. kinetic processes, which depend primarily on differences in reaction rates of isotopic molecules.

### 1.3.1 Isotope Exchange

Isotope exchange includes processes with very different physico-chemical mechanisms. Here, the term "isotope exchange" is used for all situations in which there is no net reaction, but in which the isotope distribution changes between different chemical substances, between different phases, or between individual molecules.

Isotope exchange reactions are a special case of general chemical equilibrium and can be written

$$aA_1 + bB_2 = aA_2 + bB_1 \tag{1.1}$$

where the subscripts indicate that species A and B contain either the light or heavy isotope 1 or 2, respectively. For this reaction the equilibrium constant is expressed by

$$K = \frac{\left(\frac{A_2}{A_1}\right)^a}{\left(\frac{B_2}{B_1}\right)^b} \tag{1.2}$$

where the terms in parentheses may be, for example, the molar ratios of any species. Using the methods of statistical mechanics, the isotopic equilibrium constant may be expressed in terms of the partition functions Q of the various species

$$k = \left(\frac{Q_{A2}}{Q_{A1}}\right) / \left(\frac{Q_{B2}}{Q_{B1}}\right) \tag{1.3}$$

Thus, the equilibrium constant then is simply the quotient of two partition function ratios, one for the two isotopic species of A, the other for B.

The partition function is defined by

$$Q = \sum_i (g_i \exp(-E_i/kT)) \tag{1.4}$$

where the summation is over all the allowed energy levels, $E_i$, of the molecules and $g_i$ is the degeneracy or statistical weight of the ith level [of $E_i$], k is the Boltzmann constant and T is the temperature. Urey (1947) has shown that for the purpose of calculating partition function ratios of isotopic molecules, it is convenient to introduce the so-called reduced partition function, which is the ratio of its partition function to that of the corresponding isolated atom. This reduced partition function ratio can be manipulated in exactly the same way as the normal partition function ratio. The partition function of a molecule can be separated into factors corresponding to each type of energy: translation, rotation, and vibration

$$Q_2/Q_1 = (Q_2/Q_1)_{trans} \times (Q_2/Q_1)_{rot} \times (Q_2/Q_1)_{vib} \tag{1.5}$$

The difference of the translation and rotation energy is more or less the same among the compounds appearing at the left- and right-hand side of the exchange reaction Eq. (1.1), except for hydrogen, where rotation must be taken into account. This leaves differences in vibrational energy as the predominant source of isotope effects. The vibrational energy term can be separated into two components. The first is related to the zero-point energy difference and accounts for most of the variation with temperature. The second term represents the contributions of all the other bound states and is not very different from unity. The complications which may occur relative to this simple model are mainly that the oscillator is not perfectly harmonic, so an "anharmonic" correction has to be added.

For geologic purposes the dependence of the equilibrium constant on temperature is the most important property (Eq. 1.4). In principle, isotope fractionation factors for isotope exchange reactions are also slightly pressure-dependent because isotopic substitution makes a minute change in the molar volume of solids and liquids. Experimental studies up to 20 kbar by Clayton et al. (1975) have shown that the pressure dependence for oxygen is, however, less than the limit of analytical detection. Thus, as far as it is known today, the pressure dependence seems with the exception of hydrogen to be of no importance for crustal and upper mantle environments (but see Polyakov and Kharlashina 1994).

Isotope fractionations tend to become zero at very high temperatures. However, isotope fractionations do not decrease to zero monotonically with increasing temperatures. At higher temperatures, fractionations may change sign (called crossover) and may increase in magnitude, but they must approach zero at very high temperatures. Such crossover phenomena are due to the complex manner by which thermal excitation of the vibration of atoms contributes to an isotope effect (Stern et al. 1968).

For ideal gas reactions, there are two temperature regions where the behavior of the equilibrium constant is simple: at low temperatures (generally much below room temperature) the natural logarithm of K (ln K) follows $\sim 1/T$ where T is the absolute temperature and at high temperatures the approximation becomes $\ln K \sim 1/T^2$.

The temperature ranges at which these simple behaviors are approximated depend on the vibrational frequencies of the molecules involved in the reaction. For the calculation of a partition function ratio for a pair of isotopic molecules, the vibrational frequencies of each molecule must be known. When solid materials are considered, the evaluation of partition function ratios becomes even more complicated, because it is necessary to consider not only the independent internal vibrations of each molecule, but also the lattice vibrations.

#### 1.3.1.1 Fractionation Factor ($\alpha$)

For isotope exchange reactions in geochemistry, the equilibrium constant K is often replaced by the fractionation factor $\alpha$. The fractionation factor is defined as the ratio of the numbers of any two isotopes in one chemical compound A divided by the corresponding ratio for another chemical compound B:

$$\alpha_{A-B} = \frac{R_A}{R_B} \tag{1.6}$$

If the isotopes are randomly distributed over all possible positions in the compounds A and B, then $\alpha$ is related to the equilibrium constant K by

$$\alpha = K^{1/n} \tag{1.7}$$

where "n" is the number of atoms exchanged. For simplicity, isotope exchange reactions are written such that only one atom is exchanged. In these cases, the equilibrium constant is identical to the fractionation factor. For example, the fractionation factor for the exchange of $^{18}O$ and $^{16}O$ between water and $CaCO_3$ is expressed as follows:

$$H_2^{18}O + \frac{1}{3}CaC^{16}O_3 \Leftrightarrow H_2^{16}O + \frac{1}{3}CaC^{18}O_3 \tag{1.8}$$

with the fractionation factor $\alpha_{CaCO_3-H_2O}$ defined as:

$$\alpha_{CaCo_3-H_2O} = \frac{\left(\frac{^{18}O}{^{16}O}\right)_{CaCo_3}}{\left(\frac{^{18}O}{^{16}O}\right)_{H_2o}} = 1.031 \text{ at } 25^\circ C \tag{1.9a}$$

It has become common practice in recent years to replace the fractionation factor $\alpha$ by the $\varepsilon$-value (or isotope enrichment factor) which is defined as

$$\varepsilon = \alpha - 1 \tag{1.9b}$$

because $\varepsilon \times 1000$ approximates the fractionation in parts per thousand, similar to the $\delta$-value (see below).

#### 1.3.1.2 The Delta Value ($\delta$)

In isotope geochemistry, it is common practice to express isotopic composition in terms of "delta"–($\delta$) values. For two compounds "A" and "B" whose isotopic compositions have been measured in the laboratory by conventional mass spectrometry:

$$\delta_A = \left(\frac{R_A}{R_{St}} - 1\right) \times 10^3 (‰) \tag{1.10}$$

and

$$\delta_B = \left(\frac{R_B}{R_{St}} - 1\right) \times 10^3 (‰) \qquad (1.11)$$

where $R_A$ and $R_B$ are the respective isotope ratio measurements for the two compounds and $R_{St}$ is the defined isotope ratio of a standard sample.

For the two compounds A and B, the δ-values and fractionation factor α are related by:

$$\delta_A - \delta_B = \Delta_{A-B} \approx 10^3 \ln \alpha_{A-B} \qquad (1.12)$$

Table 1.3 illustrates the closeness of the approximation. Considering experimental uncertainties in isotope ratio determinations (typically $\geq 0.1‰$), these approximations are excellent for differences in δ-values of less than about 10 and for δ-values that are relatively small in magnitude.

#### 1.3.1.3 Evaporation–Condensation Processes

Of special interest in stable isotope geochemistry are evaporation–condensation processes, because differences in the vapor pressures of isotopic compounds lead to significant isotope fractionations. For example, from the vapor pressure data for water given in Table 1.2, it is evident that the lighter molecular species are preferentially enriched in the vapor phase, the extent depending upon the temperature. Such an isotopic separation process can be treated theoretically in terms of fractional distillation or condensation under equilibrium conditions as is expressed by the Rayleigh (1896) equation. For a condensation process this equation is

$$\frac{R_V}{R_{V_0}} = f^{\alpha-1} \quad (f = \text{fraction of residual liquid}) \qquad (1.13)$$

where "$R_{vo}$" is the isotope ratio of the initial bulk composition and "$R_v$" is the instantaneous ratio of the remaining vapour (v); "f" is the fraction of the residual

**Table 1.3** Comparison between δ, α, and $10^3 \ln \alpha_{A-B}$

| $^{\delta}A$ | $^{\delta}B$ | $^{\Delta}A\text{–}B$ | $^{\alpha}A\text{–}B$ | $103 \ln\alpha_{A-B}$ |
|---|---|---|---|---|
| 1.00 | 0 | 1.00 | 1.001 | 1.00 |
| 5.00 | 0 | 5.00 | 1.005 | 4.99 |
| 10.00 | 0 | 10.00 | 1.01 | 9.95 |
| 15.00 | 0 | 15.00 | 1.015 | 14.98 |
| 20.00 | 0 | 20.00 | 1.02 | 19.80 |
| 10.00 | 5.00 | 5.00 | 1.00498 | 4.96 |
| 20.00 | 15.00 | 5.00 | 1.00493 | 4.91 |
| 30.00 | 15.00 | 15.00 | 1.01478 | 14.67 |
| 30.00 | 20.00 | 10.00 | 1.00980 | 9.76 |
| 30.00 | 10.00 | 20.00 | 1.01980 | 19.61 |

vapour, and the fractionation factor α is given by $R_l/R_V$ (l = liquid). Similarly, the instantaneous isotope ratio of the condensate ($R_l$) leaving the vapour is given by

$$\frac{R_l}{R_{V_0}} = \alpha f^{\alpha-1} \tag{1.14}$$

and the average isotope ratio of the separated and accumulated condensate ($R_l$) at any time of condensation is expressed by

$$\frac{\overline{R}_l}{R_{V_0}} = \frac{1-f^{\alpha}}{1-f} \tag{1.15}$$

For a distillation process the instantaneous isotope ratios of the remaining liquid and the vapor leaving the liquid are given by

$$\frac{R_l}{R_{l_0}} = f^{(\frac{1}{\alpha}-1)} \tag{1.16}$$

and

$$\frac{\overline{R}_v}{R_{l_0}} = \frac{1}{\alpha} f^{(\frac{1}{\alpha}-1)} \tag{1.17}$$

The average isotope ratio of the separated and accumulated vapor is expressed by

$$\frac{\overline{R}_v}{R_{l_0}} = \frac{1-f^{\frac{1}{\alpha}}}{1-f} \text{(f = fraction of residual liquid)} \tag{1.18}$$

Any isotope fractionation occuring in such a way that the products are isolated from the reactants immediately after formation will show a characteristic trend in isotopic composition. As condensation or distillation proceeds the residual vapor or liquid will become progressively depleted or enriched with respect to the heavy isotope. A natural example is the fractionation between oxygen isotopes in the water vapor of a cloud and the raindrops released from the cloud. The resulting decrease of the $^{18}O/^{16}O$ ratio in the residual vapor and the instantaneous isotopic composition of the raindrops released from the cloud are shown in Fig. 1.4 as a function of the fraction of vapor remaining in the cloud.

### 1.3.2 Kinetic Effects

Besides equilibrium effects, the second main phenomena producing fractionations are kinetic isotope effects, which are associated with incomplete and unidirectional

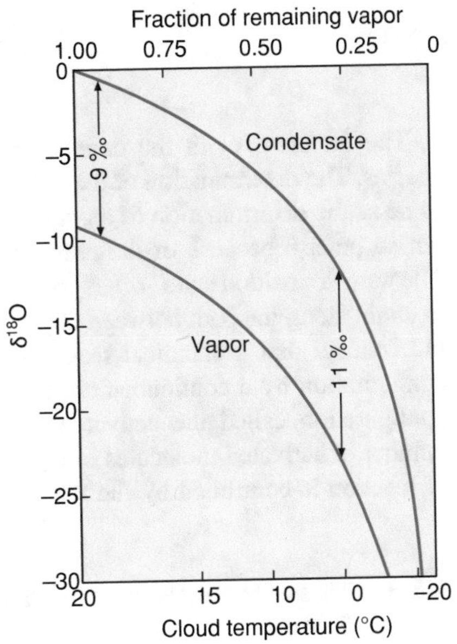


**Fig. 1.4** $\delta^{18}O$ in a cloud vapour and condensate plotted as a function of a fraction of remaining vapour in a cloud for a Rayleigh process. The temperature of the cloud is shown on the lower axis. The increase in fractionation with decreasing temperature is taken into account (after Dansgaard 1964)

processes like evaporation, dissociation reactions, biologically mediated reactions, and diffusion. The latter process is of special significance for geological purposes, which warrants separate treatment (Sect. 1.3.3). A kinetic isotope effect also occurs when the rate of a chemical reaction is sensitive to atomic mass at a particular position in one of the reacting species.

The theory of kinetic isotope fractionations has been discussed by Bigeleisen and Wolfsberg (1958), Melander (1960), and Melander and Saunders (1980). Knowledge of kinetic isotope effects is very important, because it can provide unique informations about details of reaction pathways.

Quantitatively, many observed deviations from simple equilibrium processes can be interpreted as consequences of the various isotopic components having different rates of reaction. Isotope measurements taken during unidirectional chemical reactions generally show a preferential enrichment of the lighter isotope in the reaction products. The isotope fractionation introduced during the course of a unidirectional reaction may be considered in terms of the ratio of rate constants for the isotopic substances. Thus, for two competing isotopic reactions

$$\begin{gathered} k1 \rightarrow k2, \\ A1 \rightarrow B1, \text{ and } A2 \rightarrow B2 \end{gathered} \tag{1.19}$$

the ratio of rate constants for the reaction of light and heavy isotope species $k_1/k_2$, as in the case of equilibrium constants, is expressed in terms of two partition function ratios, one for the two reactant isotopic species, and one for the two isotopic species of the activated complex or transition state $A^X$:

$$\frac{k_1}{k_2} = \left[\frac{Q^*_{(A_2)}}{Q^*_{(A_1)}} \Big/ \frac{Q^*_{(A_2^X)}}{Q^*_{(A_1^X)}}\right] \frac{v_1}{v_2} \tag{1.20}$$

The factor $v_1/v_2$ in the expression is a mass term ratio for the two isotopic species. The determination of the ratio of rate constants is, therefore, principally the same as the determination of an equilibrium constant, although the calculations are not so precise because of the need for detailed knowledge of the transition state. The term "transition state" refers to the molecular configuration that is most difficult to attain along the path between the reactants and the products. This theory follows the concept that a chemical reaction proceeds from some initial state to a final configuration by a continuous change, and that there is some critical intermediate configuration called the activated species or transition state. There are a small number of activated molecules in equilibrium with the reacting species and the rate of reaction is controlled by the rate of decomposition of these activated species.

### 1.3.3 Mass-Dependent and Mass-Independent Isotope Effects

#### 1.3.3.1 Mass Dependent Effects

At thermodynamic equilibrium isotope distributions are strictly governed by relative mass differences among different isotopes of an element. Mass dependent relationships hold for many kinetic processes as well. Thus, it has been a common belief that for most natural reactions, isotope effects arise solely because of isotopic mass differences. This means that for an element with more than two isotopes, such as oxygen or sulfur, the enrichment of $^{18}O$ relative to $^{16}O$ or $^{34}S$ relative to $^{32}S$ is expected to be approximately twice as large as the enrichment of $^{17}O$ relative to $^{16}O$ or as the enrichment of $^{33}S$ relative to $^{32}S$. Therefore, for many years interest in measuring more than one isotope ratio of a specific element was limited.

It is common practice to describe mass dependent isotope fractionation processes by a single linear curve on a three-isotope-plot (Matsuhisa et al. 1978). The resulting straight lines are referred to as terrestrial mass fractionation lines and deviations from it are used as indicating non-mass dependent isotope effects. Recent analytical improvements of multiple isotope elements have demonstrated, however, that different mass-dependent processes (e.g. diffusion, metabolism, high-temperature equilibrium processes) can deviate by a few percent and follow slightly different mass-dependent fractionation laws (Young et al. 2002; Miller 2002; Farquhar et al. 2003). These very small differences are measurable and have been documented for oxygen (Luz et al. 1999), for magnesium (Young et al. 2002), for sulfur (Farquhar et al. 2003) and for mercury (Blum 2011). As already shown by Matsuhisa et al. (1978), mass dependent fractionation laws for three or more isotopes are different for equilibrium and kinetic processes, the latter having shallower slopes than those produced by equilibrium exchange.

The three-isotope plots used in the recent literature are based on the approximation of a power law function to linear format. To describe how far a sample plots off the mass dependent fractionation line, the traditional terms to identify mass-independent effects, $\Delta^{17}O$, $\Delta^{25}Mg$, $\Delta^{33}S$ etc. have been redefined in several ways, see discussion by Assonov and Brenninkmeijer (2005). The simpliest definition is given by:

$$\begin{aligned}\Delta^{17}O &= \delta^{17}O - \lambda\delta^{18}O \\ \Delta^{25}Mg &= \delta^{25}Mg - \lambda\delta^{26}Mg \text{ or} \\ \Delta^{33}S &= \delta^{33}S - \lambda\delta^{34}S,\end{aligned}$$

where λ is an arbitrary reference slope that is often chosen similar to an important mass-dependent process. Positive Δ-values imply that samples or reservoirs are enriched compared with expectations from mass fractionation, negative Δ-values imply depletions. The value of the coefficient λ depends on the molecular mass, which for oxygen may range from 0.53 for atomic oxygen to 0.500 for species with high molecular weight. Recent progress in high-precision measurement of isotope ratios allows to distinguish λ-values in the third decimal, which has obscured the difference between mass-dependent and mass-independent fractionations at small Δ-values. The multiple applications of triple oxygen isotope geochemistry have been presented in Reviews of Mineralogy and Geochemistry 86, edited by Bindeman and Pack (2021).

#### 1.3.3.2 Mass Independent Effects

A few processes in nature do not follow the above mass-dependent fractionations. Deviations from mass-dependent fractionations were first observed for oxygen in meteorites (Clayton et al. 1973) and in ozone (Thiemens and Heidenreich 1983) and for sulfur in sulfides older than 2.45 Ga (Farquhar et al. 2000). These Mass Independent Fractionations (MIF) describe relationships that violate the mass-dependent rules $\delta^{17}O \approx 0.5\ \delta^{18}O$ or $\delta^{33}S \approx 0.5\ \delta^{34}S$ and produce isotopic compositions with much larger $\Delta^{17}O$ and $\Delta^{33}S$ anomalies compared to purely mass dependent processes.

A number of experimental and theoretical studies have focused on the causes of mass-independent fractionation effects, but as summarized by Thiemens (1999) the mechanism for mass independent fractionations remains uncertain. The best studied reaction is the formation of ozone in the stratosphere. Mauersberger et al. (1999) demonstrated experimentally that it is not the symmetry of a molecule that determines the magnitude of $^{17}O$ enrichment, but it is the difference in the geometry of the molecule. Gao and Marcus (2001) presented an advanced model, which has led to a better understanding of non-mass dependent isotope effects.

Mass-independent isotopic fractionations are widespread in the earth's atmosphere and have been observed in $O_3$, $CO_2$, $N_2O$ and CO, which are all linked to reactions involving stratospheric ozone (Thiemens 1999; Thiemens and Lin 2021).

For oxygen, it is a characteristic marker in the atmosphere (see Sect. 3.9). The discovery of chemically produced mass-independent oxygen isotope composition in ozone opened the view to investigate multi-isotope fractionation in other natural systems as summarized by Thiemens et al. (2012). MIF probably also play a role in the atmosphere of Mars and in the presolar nebula (Thiemens 1999).

Oxygen isotope measurements in meteorites by Clayton et al. (1973) (Sect. 3.1) first demonstrated that the effect is of significance in the formation of the solar system. There are numerous terrestrial solid reservoirs where mass independent isotope variations have been observed. For instance, Farquhar et al. (2000) and Bao et al. (2000) reported mass-independent oxygen isotope fractionations in terrestrial sulfates. A positive $^{17}O$-excess in sulfate has been found to be almost ubiquitous in desert environments (Bao et al. 2001).

Significant mass independent sulfur isotope fractionations have been reported first by Farquhar et al. (2000) in sulfides older than 2.4 Ga, whereas these fractionations do not occur in measurable amounts in sulfides younger than 2.4 Ga (see Fig. 2.24). A characteristic feature is that anomalous enrichments in $^{33}S$ are coupled with anomalous depletions in $^{36}S$ and vice versa (see Fig. 2.21). Smaller, but clearly resolvable MIFs have been measured in volcanic aerosol sulfates in polar ice (Baroni et al. 2007). Photolysis of $SO_2$ to sulfuric acid is thought to be the source reaction for these sulfur MIFs. These findings indicate that non-mass dependent isotope fractionations are more abundant than originally thought and represents a characteristic form of isotopic fingerprint.

The third element with significant mass-independent fractionations is mercury. As summarized by Blum and Johnson (2017) different types of mass-independent fractionation processes exist: Photochemical reduction of Hg(II) and photochemical decomposition of methylmercury may induce MIF fractionations of odd Hg isotopes. Even isotopes MIF fractionations seem to occur during photochemical oxidation of Hg(0) in the upper atmosphere (see Sect. 2.40).

## 1.3.4 Nuclear Volume and Magnetic Isotope Effects

### 1.3.4.1 Nuclear Volume Effects

For heavy elements, mass-independent isotope fractionations are considered to be due to nuclear volume fractionations (Fujii et al. 2009). Bigeleisen (1996), Schauble (2007, 2013), Estrade et al. (2009), and others demonstrated that isotope variations of very heavy elements are driven by differences in nuclear volumes and shapes which affect the electronic structure of atoms and molecules. Nuclear volume fractionations may be estimated using first principles quantum mechanical calculations (Schauble 2007). The magnitude of nuclear volume fractionations is very small for the light elements, but increases with nuclear weight.

The binding energy between electrons and nuclei depend on the distribution of protons inside the nucleus. Nuclear volume increases with the number of neutrons, but the increase caused by an odd isotope is slightly smaller than for an isotope with an even number (Bigeleisen 1996). Thus, nuclear volume effects are expected to generate odd–even isotope fractionation patterns (Schauble 2007; Fujii et al. 2009).

#### 1.3.4.2 Magnetic Isotope Effects

In contrast to nuclear volume effects that select isotopes due to their different masses, magnetic isotope effects separate isotopes by spin and related magnetic moment (Bucharenko 1995, 2001, 2013; Epov et al. 2011). Magnetic isotope effects seem to occur solely in kinetic reactions, not under equilibrium conditions. The magnetic isotope effect separates isotopes with and without unpaired nuclear spin due to coupling between nuclear spin and electronic spin. Thus, magnetic isotope effects distinguish between isotopes with odd and even numbers. The large mass-independent isotope fractionations of biological mercury compounds seem to be especially susceptible to magnetic isotope fractionations (Bucharenko 2013; Dauphas and Schauble 2016).

### 1.3.5 Multiply Substituted Isotopologues

In stable isotope geochemistry, generally bulk isotopic compositions of natural samples are given (e.g. $\delta^{13}C$, $\delta^{18}O$…). In measured gases, bulk compositions mainly depend on abundances of molecules containing one rare isotope (e.g. $^{13}C^{16}O^{16}O$ or $^{12}C^{18}O^{16}O$). These socalled isotopologues are molecules that differ from one another only in isotope composition. There also exist in very low concentration multiply substituted isotopologues such as $^{13}C^{18}O^{16}O$ or $^{12}C^{18}O^{17}O$. Table 1.4 gives the stochastic abundances of isotopologues of $CO_2$.

Already Urey (1947) and Bigeleisen and Mayer (1947) recognized that multiply substituted isotopologues have unique thermodynamic properties different from singly substituted isotopologues of the same molecule. Natural distributions of

**Table 1.4** Stochastic abundances of $CO_2$ isotopologues (Eiler 2007)

| Mass | Isotopologue | Relative abundance |
|---|---|---|
| 44 | $^{12}C^{16}O_2$ | 98.40% |
| 45 | $^{13}C^{16}O_2$ | 1.11% |
| | $^{12}C^{17}O^{16}O$ | 748 ppm |
| 46 | $^{12}C^{18}O^{16}O$ | 0.040% |
| | $^{13}C^{17}O^{16}O$ | 8.4 ppm |
| | $^{12}C^{17}O_2$ | 0.142 ppm |
| 47 | $^{13}C^{18}O^{16}O$ | 44.4 ppm |
| | $^{12}C^{17}O^{18}O$ | 1.50 ppm |
| | $^{13}C^{17}O_2$ | 1.60 ppb |
| 48 | $^{12}C^{18}O_2$ | 3.96 ppm |
| | $^{13}C^{17}O^{18}O$ | 16.8 ppb |
| 49 | $^{13}C^{18}O_2$ | 44.5 ppb |

multiply substituted isotopologues can thus provide unique constraints on geological, geochemical and cosmochemical processes (Wang et al. 2004). In particular, triple oxygen isotopes can be applied as a temperature proxy (Passey and Levin 2021 and others).

"Normal" gas-source mass spectrometers do not allow meaningful abundance measurements of these very rare species. However, if some demands on high abundance sensitivity, high precision and high mass resolving power are met, John Eiler and his group (e.g. Eiler and Schauble 2004; Affek and Eiler 2006; Eiler 2007) have reported precise (<0.1‰) measurements of $CO_2$ with mass 47 ($\Delta_{47}$-values) with an especially modified, but normal gas-source mass-spectrometer. The $\Delta_{47}$-values are defined as ‰ differences between the measured abundance of all molecules with mass 47 relative to the abundance of 47 expected for the stochastic distribution. Huntington et al. (2009) and Daeron et al. (2016) described the technical details of the method and discussed potential errors and accuracies. The main limitation of the analytical method for its wide application is the need for a relatively large amount (5–10 mg) of pure sample necessary for a precise measurement. With further improved analytical setup, Fiebig et al. (2019) succeeded in the combined analysis of the abundances of mass 47 and mass 48 isotopologues of $CO_2$, allowing the determination of precise paleotemperatures as well as for the identification of rate-limiting kinetic processes.

This new technique is termed "clumped isotope geochemistry" (Eiler 2007) because the respective species are produced by clumping two rare isotopes together. "Clumping" results in a statistical overabundance of multiply substituted isotopologues relative to a purely random distribution of all isotopes. Deviations from stochastic distributions are generally within 1% and may result from all processes of isotope fractionation observed in nature. Thus, processes that lead to isotope fractionations of bulk compositions also lead to fractionations of multiply substituted isotopologues (Eiler 2007).

Clumped isotope studies have been undertaken on $O_2$, $CO_2$ and $CH_4$ gases. So far the most used application is the measurement of $CO_2$ liberated by acid-digestion from carbonates. Schauble et al. (2006) calculated an $\sim$0.4‰ excess of $^{13}C^{18}O^{16}O$ groups in carbonate groups at room temperature relative to what would be expected in a stochastic mixture of carbonate isotopologues with the same bulk $^{13}C/^{12}C$, $^{18}O/^{16}O$ and $^{17}O/^{16}O$ ratios. The excess amount of $^{13}C^{18}O^{16}O$ decreases with increasing temperature (Ghosh et al. 2006). Various attempts to calibrate the carbonate clumped isotope thermometer, which is a true single mineral thermometer, has led to controversial results. Calibrations on synthetic carbonates (Ghosh et al. 2006; Tang et al. 2014 and others) cannot be applied a priori to natural biogenic carbonates (Tripati et al. 2010; Henkes et al. 2013; Came et al. 2014, and others). The reasons for the discrepancies in temperature calibrations are not well understood, but obviously reflect a combination of a variety of factors (Tang et al. 2014). In an international effort, Petersen (2019) reprocessed 14 published clumped isotope calibration relationships. Figure 1.5 shows the calibration of the clumped isotope thermometer as agreed by the clumped isotope community (Petersen et al. 2019).

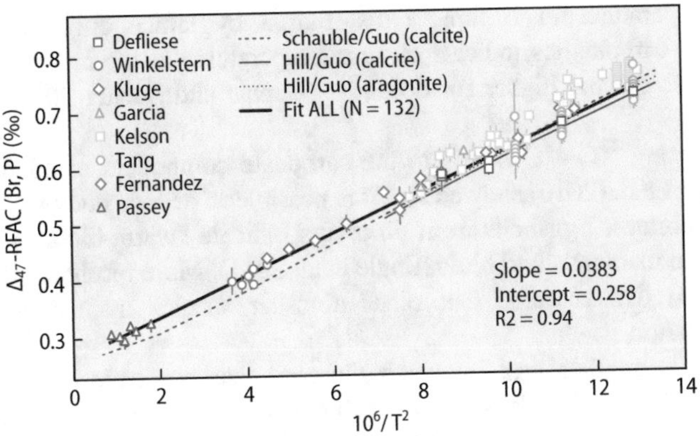


Fig. 1.5 Calibrations of the clumped isotope thermometer (after Petersen et al. 2019, courtesy Jens Fiebig)

The advantage of this thermometer is the potential to determine formation temperatures of carbonates without knowing the isotope composition of the fluid. But the clumped isotope proxy has not only been used as paleothermometer, but also as recorder of past diagenetic conditions (Stolper et al. 2018). Clumped isotopes have been contributing to palaeoclimatology, (i.e. temperatures of foraminifera and other marine organisms, Tripati et al. 2010 and others), to estimate temperatures of paleosol carbonates (Quade et al. 2011), of speleothems (Affek et al. 2008), and to constrain the diagenetic history of carbonates (Huntington et al. 2010; Stolper et al. 2018). Dissolution— recrystallisation of carbonates can alter the clumped isotope composition such that the carbonate reflects deeper burial temperatures.

Clumped isotope paleotemperatures derived from primary carbonates are susceptible to alteration via C–O bond reordering, involving the breakage of existing C–O bonds and their reformation by solid state diffusion of C and O atoms through the solid mineral lattice. This process can reset primary clumped isotope compositions without changing texture, isotope ratios and trace element compositions. Dennis and Schrag (2010) have used carbonatites to test the isotope integrity of the clumped isotope composition over long timescales and concluded reordering of C-and O-atoms is sufficiently slow to enable the use of clumped paleothermometry on timescales of $10^8$ years. Henkes et al. (2014) demonstrated that calcites being exposed to temperatures of about 100 °C for $10^6$–$10^8$ years will not be affected by solid state C–O reordering. The solid-state reordering can be used as a geospeedometer which centers on how clumped isotope compositions evolve during cooling and how the "frozen-in" clumped isotope composition relate to the rate of cooling (Passey and Henkes 2012).

The clumped isotope geothermometer may be also relevant to processes such as dolomitization (Ferry et al. 2011) and burial diagenesis (Huntington et al. 2011). The key question in using clumped temperatures of deeply buried carbonates is the closure temperature for isotopic redistribution by lattice diffusion. Studies of marbles and carbonatites indicate closure temperatures in the order of 200 °C for calcite and somewhat higher for dolomite (Dennis and Schrag 2010; Ferry et al. 2011).

By analyzing $^{13}C–^{18}O$ bonds in the carbonate component of apatite of vertebrates, Eagle et al. (2010) showed that it is possible to deduce body temperatures of extinct vertebrates. Eggshells from dinosaurs indicate "warm-blooded" body temperatures like mammals and birds (Eagle et al. 2011). More recent studies by Eagle et al. (2015) demonstrated that other dinosaur species maintained a variable thermoregulation.

Methane is another gas, in which clumped isotopes have been investigated (Stolper et al. 2014). Besides the 3 most abundant isotopologues, $^{12}CH_4$, $^{13}CH_4$ and $^{12}CH_3D$, there are 7 more isotopologues with the heaviest mass 21 for $^{13}CD_4$. In a first attempt, Stolper et al. (2014) presented data for the isotologues $^{13}CH_3D$ and $^{12}CH_2D_2$ which for equilibrated systems can be used as a geothermometer. Clumped isotopes yield consistent temperatures of formation for low-temperature biogenic and high-temperature thermogenic methane.

For the simple molecule $O_2$, clumped isotopes ($^{17}O^{18}O$, $^{18}O^{18}O$) may be used as an indicator of biological origin, not formation temperatures (Yeung et al. 2015).

#### 1.3.5.1 Position or Site-Specific Isotope Fractionations

Site-specific isotope fractionations describe differences between the isotope composition of a site in a molecule and the isotope composition it would have if the molecule had randomly distributed isotopes (Galimov 2006; Eiler 2013). Examples are the distribution of $^{15}N$ in the central position and the terminal position in $N_2O$. Nitrifying bacteria enrich $^{15}N$ in the central position, whereas $N_2O$ from denitrifying bacteria and other natural sources of $N_2O$ do not show site-specific fractionations.

Other characteristic site-specific fractionations are $^{13}C$ and D fractionations occurring during the synthesis of organic molecules. Abelson and Hoering (1961) were the first to analyse the $\delta^{13}C$-value of isolated amino acids separately and showed that the terminal carboxyl groups in most amino acids where significantly enriched in $^{13}C$ relative to other C positions. Blair et al. (1985) demonstrated that in acetate ($CH_3COOH$), the methyl group ($CH_3$) and the carboxyl group (COOH) differ by up to 20‰ in $^{13}C$. Gilbert et al. (2016) developed a method for analyzing intramolecular carbon isotope distributions of propane. They defined a site preference index as the isotope difference between terminal ($CH_3$-group) and central ($CH_2$-group) carbon position of propane. Piasecki et al. (2018) demonstrated that propane varies in its site-specific structure, which is correlated with increasing maturity. With further advances in analytical techniques many more applications of site-specific isotope fractionations will arise.

### 1.3.6 Diffusion

Ordinary diffusion can cause significant isotope fractionations. In general, light isotopes are more mobile and hence diffusion can lead to a separation of light from heavy isotopes. For gases, the ratio of diffusion coefficients is equivalent to the inverse square root of their masses (i.e. Graham's law). Consider the isotopic molecules of carbon in $CO_2$ with masses $^{12}C^{16}O^{16}O$ and $^{13}C^{16}O^{16}O$ having molecular weights of 44 and 45. Solving the expression equating the kinetic energies (1/2 m $v^2$) of both species, the ratio of velocities is equivalent to the square root of 45/44 or 1.01, which means that the average velocity of $^{12}C^{16}O^{16}O$ molecules is about 1% higher than the average velocity of $^{13}C^{16}O^{16}O$ molecules in the same system. This isotope effect, however, is more or less limited to ideal gases, where collisions between molecules are infrequent and intermolecular forces negligible. The carbon isotope fractionation of soil-$CO_2$ due to diffusional movement, for instance, has been estimated to be around 4‰ (Cerling 1984; Hesterberg and Siegenthaler 1991).

Distinctly different from ordinary diffusion is the process of thermal diffusion where a temperature gradient results in a mass transport. The greater the mass difference the more pronounced is the tendency of the two species to separate by thermal diffusion. A natural example of thermal diffusion has been presented by Severinghaus et al. (1996) who observed a small isotope depletion of $^{15}N$ and $^{18}O$ in air from a sand dune relative to the free atmosphere. This observation is contrary to the expectation that heavier isotopes in unsaturated zones of soils would be enriched by gravitational settling. Such thermally driven diffusional isotope effects have also been described in air bubbles from ice cores (Severinghaus et al. 1998; Severinghaus and Brook 1999; Grachiev and Severinghaus 2003).

In solutions and solids the relationships are much more complicated than in gases. The term "solid state diffusion" generally includes volume diffusion and diffusion mechanisms where the atoms move along paths of easy diffusion such as grain boundaries and surfaces. Diffusive-penetration experiments indicate a marked enhancement of diffusion rates along grain boundaries which are orders of magnitude faster than for volume diffusion. Thus, grain boundaries can act as pathways of rapid exchange. Volume diffusion is driven by the random temperature dependent motion of an element or isotope within a crystal lattice, and it depends on the presence of point defects, such as vacancies or interstitial atoms within the lattice.

The flux F of elements or isotopes diffusing through a medium is proportional to the concentration gradient (dc/dx) such that:

$$F = -D(dc/dx) \quad \text{(Fick, s first law)} \tag{1.21}$$

where "D" represents the diffusion coefficient, and the minus sign denotes that the concentration gradient has a negative slope, i.e. elements or isotopes move from points of high concentration towards points of low concentration. The diffusion coefficient "D" varies with temperature according to the Arrhenius relation

$$D = D_o e^{(-E_a/RT)} \tag{1.22}$$

where "$D_o$" is a temperature-independent factor, "$E_a$" is the activation energy, "R" is the gas constant and T is in Kelvins.

In recent years there have been several attempts to determine diffusion coefficients, mostly utilizing Secondary Ion Mass Spectrometry (SIMS), where isotope compositions have been measured as a function of depth below a crystal surface after exposing the crystal to solutions or gases greatly enriched in the heavy isotopic species.

A plot of the logarithm of the diffusion coefficient versus reciprocal temperature yields a linear relationship over a significant range of temperature for most minerals. Such an Arrhenius plot for various minerals is shown in Fig. 1.6, which illustrates the variability in diffusion coefficients for different minerals. The practical application of this fact is that the different minerals in a rock will exchange oxygen at different rates and become closed systems to isotopic exchange at different temperatures. As a rock cools from the peak of a thermal event, the magnitude of isotope fractionations between exchanging minerals will increase. The rate at which the coexisting minerals can approach equilibrium at the lower temperature is limited by the volume diffusion rates of the respective minerals.

Several models for diffusive transport in and among minerals have been discussed in the literature, one is the "Fast Grain Boundary/FGB) model" of Eiler et al. (1992, 1993). The FGB model considers the effects of diffusion between non-adjacent grains and shows that, when mass balance terms are included, closure temperatures become a strong function of both the modal abundances of constituent minerals and the differences in diffusion coefficients among all coexisting minerals.

Surprisingly large fractionations by chemical and thermal diffusion at very high temperatures have been reported by Richter et al. (1999, 2003, 2007, 2009) and others observing large isotope variations of Mg, Ca, Fe, Si and O in silicate melts subjected to thermal gradients Diffusion experiments between molten basalt and rhyolite also demonstrated considerable isotope fractionations of Li, Ca and Ge (the latter used as a Si analogue). Especially for Li, diffusion processes occuring at high temperatures seem to be of first order importance (see p. xx). Thus the notion that isotope fractionations above 1000 °C appear to be negligible has to be reconsidered, although the physical mechanisms are not clear. Under magmatic and metamorphic conditions, diffusion thus may cause significant isotope fractionations. As shown by Dauphas et al. (2010), Teng et al. (2011), Sio et al. (2013) and Oeser et al. (2015) negatively correlated Fe and Mg isotope compositions in zoned olivines can be explained by diffusive separation of Fe and Mg isotopes as Mg diffused out and Fe diffused in during magmatic differentiation.

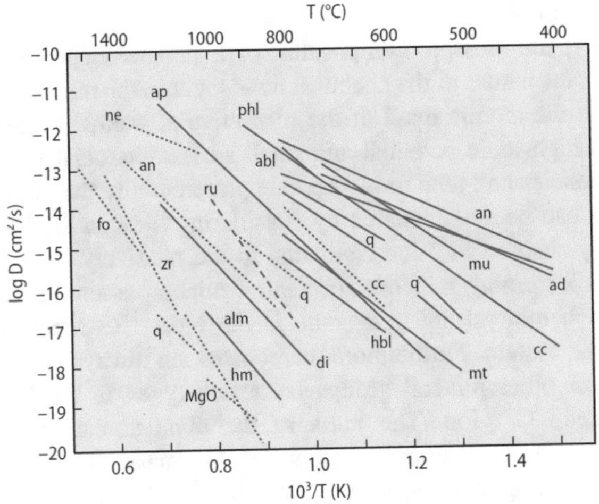


**Fig. 1.6** Arrhenius plot of diffusion coefficients versus reciprocal temperatures for various minerals. Data from phases reacted under *wet conditions* are given as *solid lines*, whereas *dry conditions* are represented by *dashed lines*. Note that the rates for dry systems are generally lower and have higher activation energies (steeper slopes) (after Cole and Chakraborty 2011)

### 1.3.7 Other Factors Influencing Isotopic Fractionations

a. Pressure

It is commonly assumed that temperature is the main variable determining the isotopic fractionation and that the effect of pressure is negligible, because molar volumes do not change with isotopic substitution. This assumption is generally fulfilled, except for hydrogen. Driesner (1997), Horita et al. (1999, 2002) and Polyakov et al. (2006) have shown, however, that for isotope exchange reactions involving water, changes of pressure can influence isotope fractionations. Driesner (1997) calculated hydrogen isotope fractionations between epidote and water and observed at 400 °C a change from −90‰ at 1 bar to −30‰ at 4000 bars. Horita et al. (1999, 2002) presented experimental evidence for a pressure effect in the system brucite $Mg(OH)_2$—water. Theoretical calculations indicate that pressure effects largely result on water rather than effects on brucite. Thus, it is likely that D/H fractionations of any hydrous mineral is subject to similar pressure effects (Horita et al. 2002). These pressure effects have to be taken into account when calculating the hydrogen isotope composition of the fluid from the mineral composition.For very high pressures in the deep earth, Shahar et al. (2016) presented experimental and theoretical evidence for pressure effects on iron isotope compositions of iron alloys.

b. Chemical composition

Qualitatively, the isotopic composition of a mineral depends to a very high degree upon the nature of the chemical bonds within the mineral and to a smaller degree upon the atomic mass of the respective elements. In general, bonds to ions with a high ionic potential and small size are associated with high vibrational frequencies and tend to incorporate preferentially the heavy isotope. This relationship can be demonstrated by considering the bonding of oxygen to the small highly charged $Si^{4+}$ ion compared to the relatively large $Fe^{2+}$ ion of the common rock-forming minerals. In natural mineral assemblages: quartz is the most $^{18}O$-rich mineral and magnetite is the most $^{18}O$-deficient given equilibration in the system. Furthermore, carbonates are always enriched in $^{18}O$ relative to most other mineral groups because oxygen is bonded to the small, highly charged $C^{4+}$ ion. The mass of the divalent cation is of secondary importance to the C–O bonding. However, the mass effect is apparent in $^{34}S$ distributions among sulfides, where, for example, ZnS always concentrates $^{34}S$ relative to coexisting PbS.

Compositional effects in silicates are complex and difficult to deduce, because of the very diverse substitution mechanisms in silicate minerals (Kohn and Valley 1988). The largest fractionation effect is clearly related to the NaSi = CaAl substitution in plagioclases which is due to the higher Si to Al ratio of albite and the greater bond strength of the Si–O bond relative to the Al–O bond. In pyroxenes, the jadeite ($NaAlSi_2O_6$)—diopside ($CaMgSi_2O_6$) substitution also involves Al, but Al in this case replaces an octahedral rather than tetrahedral site. Chacko et al. (2001) estimate that at high temperatures the Al-substitution in pyroxenes is about 0.4‰ per mole Al substitution in the tetrahedral site. The other very common substitutions, the Fe–Mg and the Ca–Mg substitutions, do not generate any significant difference in oxygen isotope fractionation (Chacko et al. 2001).

c. Crystal structure

Structural effects are secondary in importance to those arising from the primary chemical bonding: the heavy isotope being concentrated in the more closely-packed or well-ordered structures. The $^{18}O$ and D fractionations between ice and liquid water arise mainly from differences in the degree of hydrogen bonding (order). A relatively large isotope effect associated with structure is observed between graphite and diamond (Bottinga 1969). With a modified increment method, Zheng (1993a) has calculated this structural effect for the $SiO_2$ and $Al_2SiO_5$ polymorphs and demonstrated that $^{18}O$ will be enriched in the high-pressure forms. In this connection it should be mentioned, however, that Sharp (1995) by analyzing natural $Al_2SiO_5$ minerals observed no differences for kyanite versus sillimanite. Other examples of structural effects on isotope fractionations are carbonate minerals (Zheng and Böttcher 2016).

d. Sorption

The term "sorption" is used to indicate the uptake of dissolved species by solids irrespective of the mechanism. Isotope fractionations during sorption depend on mineral surface chemistry and on compositions of the solution. During physical sorption, when the element in question is not structurally incorporated, isotope fractionations should be small, whereas during chemical sorption, when the element is incorporated by stronger bonds, isotope fractionations are larger. Considering the large range of possible sorbents (oxide/hydroxides, phyllosilicates, biologic surfaces etc.), knowledge of isotope fractionations on solid/water interfaces are of crucial importance to understand the isotope geochemistry of metals. Experimental determinations on the fractionation of metal isotopes during absorption onto metal oxide phases have been presented by a number of studies (i.e. Teutsch et al. 2005; Gelabert et al. 2006, and others). Most studies show small (<1‰) isotope fractionations as metal ions are removed from solution onto oxide surfaces, except Mo where absorption on oxides causes about 2‰ fractionation in $^{98}Mo/^{95}Mo$ ratios. Generally, elements that are present as cations in a solution (Fe, Cu, Zn) exhibit enrichment of the heavier isotope on the surface of solids, which is consistent with shorter metal-oxide bonds and lower coordination numbers for the metal at the surface relative to the aqueous ion. Thus, the heavier isotope should concentrate in the species in which it is most strongly bound, for example, enrichment of absorbed tetrahedral metal with shorter metal–oxygen bonds compared to octahedral metal in solution.

Metal cations in solution that form soluble oxyanions such as Ge, Se, Mo, and U enrich the lighter isotope on Fe/Mn oxide surfaces. The molecular mechanism responsible for the sign and size of metal isotope fractionation between solids and aqueous phases remain poorly understood. Wasylenki et al. (2011) postulated that largest isotope effects occur when a trace solution species with different coordination than the major solution species absorb. Kashiwabara et al. (2011) argued similarly by stating that small isotope fractionations are associated with little changes in local structures during absorption.

### 1.3.8 Isotope Geothermometers

Isotope thermometry has become well established since the classic paper of Urey (1947) on the thermodynamic properties of isotopic substances. The partitioning of two stable isotopes of an element between two mineral phases can be viewed as a special case of element partitioning between two minerals. The most important difference between the two exchange reactions is the pressure-insensitivity of isotope partitioning due to the negligible $\Delta V$ of reaction for isotope exchange. This represents a considerable advantage relative to the numerous types of other geothermometers, all of which exhibit a pressure dependence.

The necessary condition to apply an isotope geothermometer is isotope equilibrium, which is most readily achieved at high temperatures, where isotope geothermometers are, however, less sensitive that at low temperatures. Isotope exchange equilibrium should be established during reactions whose products are in **chemical** and **mineralogical** equilibrium. Demonstration that the minerals in a rock are in oxygen isotope equilibrium is strong evidence that the rock is in chemical equilibrium. To break Al–O and Si–O bonds and allow re-arrangement towards oxygen isotope equilibrium needs sufficient energy to effect chemical equilibrium as well.

Theoretical studies show that the fractionation factor α for isotope exchange between minerals is a linear function of $1/T^2$, where "T" is temperature in degrees Kelvin. Bottinga and Javoy (1973) demonstrated that O-isotopic fractionation between anhydrous mineral pairs can be expressed in terms of a relationship of the form:

$$1000\ln \alpha = A/T^2 + B/T^2 + B/T + C,$$

One drawback to isotope thermometry in slowly cooled metamorphic and magmatic rocks is that, temperature estimates are often significantly lower than those from other geothermometers. This results from isotopic resetting associated with retrograde isotope exchange between coexisting phases or with transient fluids. During cooling in closed systems, volume diffusion may be the principal mechanism by which isotope exchange occurs between coexisting minerals.

Giletti (1986) proposed a model in which experimentally-derived diffusion data can be used in conjunction with measured isotope ratios to explain disequilibrium isotope fractionations in slowly-cooled, closed-system mineral assemblages. This approach describes diffusional exchange between a mineral and an infinite reservoir whose bulk isotopic composition is constant during exchange. However, mass balance requires that loss or gain of an isotope from one mineral must be balanced by a change in the other minerals still subject to isotopic exchange. Numerical modeling by Eiler et al. (1992) has shown that closed-system exchange depends not only on modal proportions of all of the minerals in a rock, but also on oxygen diffusivity in minerals, grain size, grain shape and cooling rate. As shown by Kohn and Valley (1998) there is an important water fugacity dependence as well. In the presence of fluids further complications may arise because isotope exchange may also occur by solution-reprecipitation or chemical reaction rather than solely by diffusion.

Three different methods have been used to determine the equilibrium fractionations for isotope exchange reactions:

(a) theoretical calculations
(b) experimental determinations in the laboratory, and
(c) empirical or semi-empirical calibrations.

Method (c) is based on the idea that the calculated "formation temperature" of a rock (calculated from other geothermometers) serves as a calibration to the measured isotopic fractionations, assuming that all minerals were at equilibrium.

However, because there is evidence that equilibrium is not always attained or retained in nature, such empirical calibrations should be regarded with caution.

Nevertheless, rigorous applications of equilibrium criteria to rock-type and the minerals investigated can provide important information on mineral fractionations (Kohn and Valley 1998; Sharp 1995; Kitchen and Valley 1995).

(A) Theoretical calculations

Calculations of equilibrium isotope fractionation factors have been particularly successful for gases. Richet et al. (1977) calculated the partition function ratios for a large number of gaseous molecules. They demonstrated that the main source of error in the calculation is the uncertainty in the vibrational molecular constants.

The theory developed for perfect gases could be extended to solids if the partition functions of crystals could be expressed in terms of a set of vibrational frequencies that correspond to its various fundamental modes of vibration (O'Neil 1986). By estimating thermodynamic properties from elastic, structural and spectroscopic data, Kieffer (1982) and subsequently Clayton and Kieffer (1991) calculated oxygen isotope partition function ratios and from these calculations derived a set of fractionation factors for silicate minerals. The calculations have no inherent temperature limitations and can be applied to any phase for which adequate spectroscopic and mechanical data are available. They are, however, limited in accuracy as a consequence of the approximations needed to carry out the calculations and the limited accuracy of the spectroscopic data.

Isotope fractionations in solids depend on the nature of the bonds between atoms of an element and the nearest atoms in the crystal structure (O'Neil 1986). The correlation between bond strength and oxygen isotope fractionation was investigated by Schütze (1980), who developed an "increment" method for predicting oxygen isotope fractionations in silicate minerals. Richter and Hoernes (1988) applied this method to the calculation of oxygen isotope fractionations between silicate minerals and water. Zheng (1991, 1993b, c) extended the increment method by using parameters of crystal chemistry with no empirical factor. The fractionation factors calculated using these methods over the temperature range 0–1200 °C are in relatively good agreement with experimental calibrations.

Ongoing advances in computer capacity and new development of software, have opened the possibility to calculate equilibrium isotope fractionations using first principles (or ab initio) methods on the base of density functional theory (Méheut et al. 2007; Schauble et al. 2009; Schauble 2011; Kowalski and Jahn 2011; Kowalski et al. 2013 and others). Calculations of equilibrium isotope fractionations can be carried out for fluids and solids at high P and T conditions with a reasonable precision (i.e. Kowalski and Jahn 2011; Kowalski et al. 2013). Although different approaches have been used in these calculations, all methods require knowledge of the vibrational spectrum of a system.

B. Experimental calibrations

In general, experimental calibrations of isotope geothermometers have been performed between 250 and 800 °C. The upper temperature limit is usually determined by the stability of the mineral being studied or by limitations of the experimental apparatus, whereas the lower temperature limit is determined by the decreasing rate of exchange.

Various experimental approaches have been used to determine fractionation factors. The three most common techniques are described below.

(a) Two-direction approach

This method is analogous to reversing reactions in experimental petrology and is the only method by which the attainment of equilibrium can be convincingly demonstrated. Equilibrium fractionations are achieved by starting on opposite sides of the equilibrium distribution.

(b) Partial-exchange technique

The partial exchange technique is used when rates of isotopic exchange are relatively low and is based on the assumption that the rates of isotope exchange for companion exchange experiments are identical. Experimental runs have to be the same in every respect except in the isotopic compositions of the starting materials. Rates of isotope exchange reactions in heterogeneous systems are relatively high at first (surface control) and then become progressively lower with time (diffusion control). Four sets of experiments are shown in Fig. 1.7 for the $CO_2$—graphite system (after Scheele and Hoefs 1992). Northrop and Clayton (1966) presented a set of equations to describe the kinetics of isotope exchange reactions and developed a general equation for the partial exchange technique. At low degrees of exchange the fractionations determined by the partial exchange technique are often larger than the equilibrium fractionations (O'Neil 1986).

(c) Three-isotope method

This method, introduced by Matsuhisa et al. (1978) and later modified by Matthews et al. (1983), uses the measurement of both $^{17}O/^{16}O$ and $^{18}O/^{16}O$ fractionations in a single experiment that has gone to equilibrium. The initial $^{18}O/^{16}O$ fractionation for the mineral—fluid system is selected to be close to the assumed equilibrium, while the initial $^{17}O/^{16}O$ fractionation is chosen to be very different from the equilibrium value. In this way the change in the $^{17}O/^{16}O$ fractionations monitor the extent of isotopic exchange and the $^{18}O/^{16}O$ fractionations reflect the equilibrium value. Some limitations of the method have been discussed by Cao and Bao (2017) by arguing that the method should be preferably used to study kinetic isotope effects.

Most of the published data on mineral fractionations have been determined by exchange of single minerals with water. This approach is limited by two factors: (i) many minerals are unstable, melt, or dissolve in the presence of

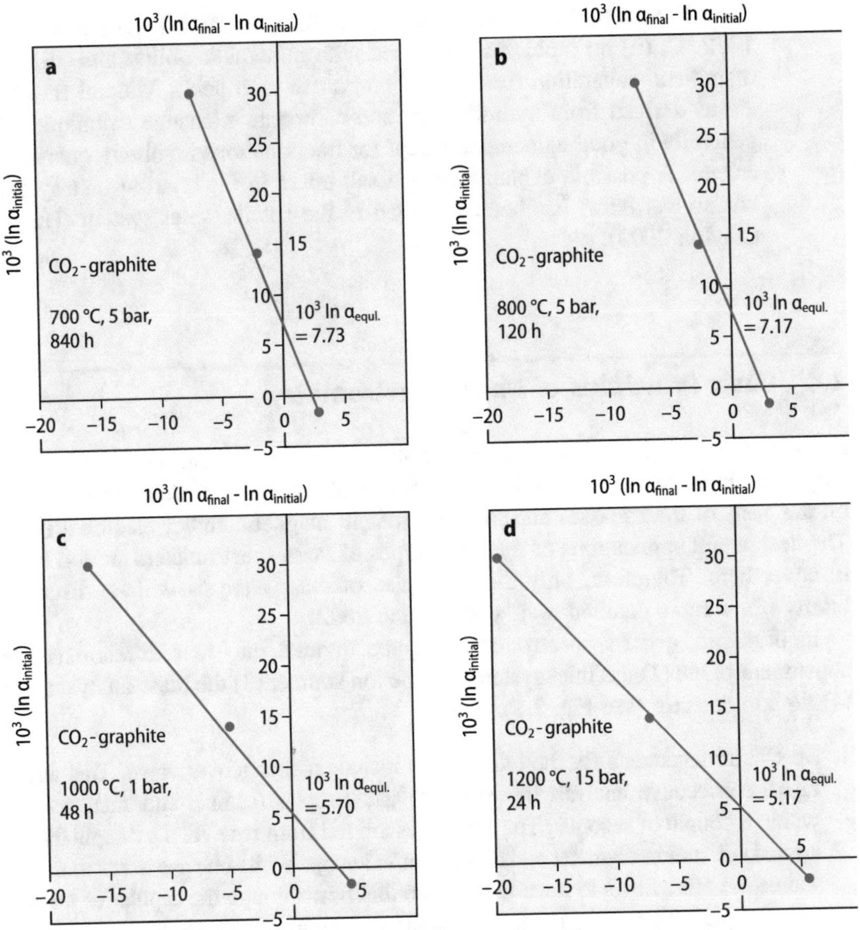


**Fig. 1.7** $CO_2$-graphite partial exchange experiments in a Northrop and Clayton plot at 700, 800, 1000 and 1200 °C. The connecting line in experiment at 1200 °C has a plain slope and defines the intercept more precisely than the experiment at 700 °C (after Scheele and Hoefs 1992)

water and (ii) the temperature dependence of the fractionation factor for aqueous systems is complicated as a consequence of the high vibrational frequencies of the water molecule. An alternative approach to the experimental determination of isotope fractionation between minerals was first employed by Clayton et al. (1989) and Chiba et al. (1989) who demonstrated that both limitations can be avoided by using $CaCO_3$, instead of $H_2O$, as the common exchange medium. These studies showed that most common silicates undergo rapid oxygen isotope exchange with $CaCO_3$ at temperatures above 600 °C and pressures of 15 kbars.

Advantages of the carbonate-exchange technique are: (i) experiments up to 1400 °C, (ii) no problems associated with mineral solubility and (iii) ease of mineral separation (reaction of carbonate with acid). Mineral fractionations derived from hydrothermal and carbonate exchange techniques are generally in good agreement except for fractionations involving quartz and calcite. A possible explanation is a salt effect in the quartz-water system, but no salt effect has been observed in the calcite-water system (Hu and Clayton 2003).

## 1.4 Basic Principles of Mass Spectrometry

Mass spectrometric methods are by far the most effective means of measuring isotope abundances. A mass spectrometer separates charged atoms and molecules on the basis of their masses and their motions in magnetic and/or electrical fields. The design and applications of the many types of mass spectrometers are too broad to cover here. Therefore, only the principles of mass analysis will be discussed briefly (for a more detailed review see Brand 2002).

In principle, a mass spectrometer may be divided into four different central constituent parts: (1) the inlet system, (2) the ion source, (3) the mass analyzer, and (4) the ion detector (see Fig. 1.8).

1. Special arrangements for the *inlet system* include a changeover valve. This allows rapid, consecutive analysis between two gas samples (sample and standard gas) within a couple of seconds. The two gases are fed from reservoirs by capillaries of around 0.1 mm in diameter and about 1 m in length. While one gas flows to the ion source the other flows to a waste pump so that flow through the capillaries remains uninterrupted. To avoid a mass discrimination, isotope abundance measurements of gaseous substances are carried out utilizing viscous gas flow. During viscous gas flow, the free path length of molecules is small, molecule collisions are frequent (causing the gas to be well mixed), and no mass separation takes place. At the end of the viscous-flow inlet system, there is a "leak", a constriction in the flow line.

   The smallest amount of sample that can be analyzed with high precision using the dual inlet system is limited by the maintenance of viscous flow conditions. This is generally in the order of 15–20 mbar (Brand 2002). When trying to reduce sample size, it is necessary to concentrate the gas into a small volume in front of the capillary.
2. The *ion source* is that part of the mass spectrometer where ions are formed, accelerated, and focused into a narrow beam. In the ion source, the gas flow is always molecular. Ions of gaseous samples are most reliably produced by electron bombardment. A beam of electrons is emitted by a heated filament, usually tungsten or rhenium, and is accelerated by electrostatic potentials to an

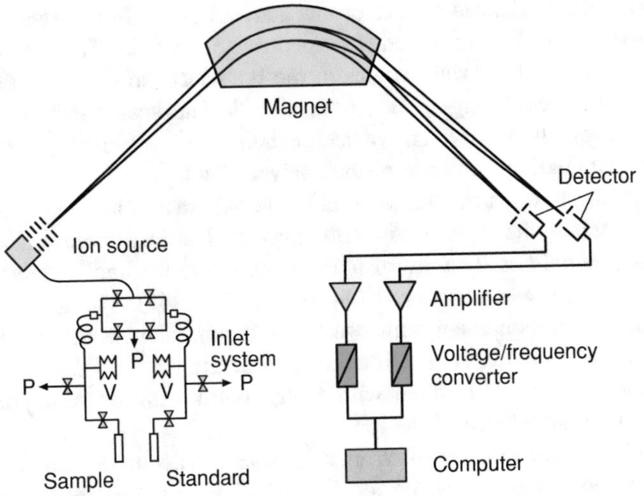


Fig. 1.8 Schematic representation of a gas-source mass spectrometer for stable isotope measurements during the 1960 and 70s. P denotes pumping system, V denotes a variable volume

energy between 50 and 150 eV before entering the ionization chamber, which maximizes the efficiency of single ionization. Following ionization any charged molecule can be further fragmented into several pieces depending on the energy the ion has acquired, producing a mass spectrum of a specific compound.
To increase the ionization probability, a homogeneous weak magnetic field is used to keep the electrons on a spiral path. At the end of the ionization chamber, electrons are collected in a positively charged trap, where the electron current is measured and kept constant by the emission regulator circuitry.
The ionized molecules are drawn out of the electron beam by action of an electric field, subsequently accelerated by up to several kV and their path shaped into a beam which passes through an exit slit into the analyzer. Thus, the positive ions entering the magnetic field are essentially monoenergetic, i.e., they will possess the same kinetic energy, given by the equation:

$$1/2\mathrm{Mv}^2 = \mathrm{eV}. \tag{1.23}$$

The efficiency of the ionization process determines the sensitivity of the mass spectrometer which generally is on the order of 1000–2000 molecules per ion (Brand 2002).

3. The *mass analyzer* separates the ion beams emerging from the ion source according to their m/e (mass/charge) ratios. As the ion beam passes through the magnetic field, the ions are deflected into circular paths, the radii of which are proportional to the square root of m/e. Thus, the ions are separated into beams, each characterized by a particular value of m/e.

In 1940, Nier introduced the sector magnetic analyzer. In this type of analyzer, deflection takes place in a wedge-shaped magnetic field. The ion beam enters and leaves the field at right angles to the boundary, so the deflection angle is equal to the wedge angle, for instance, 60°. The sector instrument has the advantage of its source and detector being comparatively free from the mass-discriminating influence of the analyzer field.

4. After passing through the magnetic field, the separated ions are collected in *ion detectors*, where the input is converted into an electrical impulse, which is then fed into an amplifier. The use of multiple detectors to simultaneously integrate the ion currents was introduced by Nier et al. (1947). The advantage of the simultaneous measurement with two separate amplifiers is that relative fluctuations of the ion currents as a function of time are the same for all m/e beams. Each detector channel is fitted with a high ohmic resistor appropriate for the mean natural abundance of the ion current of interest.
   Modern isotope ratio mass spectrometers have at least three Faraday collectors, which are positioned along the focal plane of the mass spectrometer. Because the spacing between adjacent peaks changes with mass and because the scale is not linear, each set of isotopes often requires its own set of Faraday cups.

### 1.4.1 Continuous Flow—Isotope Ratio Monitoring Mass Spectrometers

Between the early 1950s, when the dual viscous flow mass spectrometer was introduced by A. Nier, until the mid 80s only minor modifications have been made on the hardware of commercial mass spectrometers. Special efforts had been undertaken to reduce the sample size for isotope measurements. This has led to a modification of the classic dual inlet technique to the continuous-flow isotope ratio monitoring mass spectrometer in which the gas to be analyzed is a trace gas in a stream of carrier gas which achieves viscous flow conditions. Today the majority of gas mass spectrometers is sold with the continuous flow system instead of the dual inlet system.

The classical off-line procedures for sample preparations are time consuming and analytical precision depends on the skill of the investigator. With on-line techniques using a combination of an elemental analyzer directly coupled to the mass spectrometer many problems of the off-line preparation can be overcome and minimized. Differences in both techniques are summarized in Table 1.5.

This new generation of mass-spectrometers is often combined with chromatographic techniques. The sample size required for an isotope measurement has been drastically reduced to the nano–or even picomolar range (Merritt and Hayes 1994). Important features of the GC-IRMS technique are (Brand 2002):

i. ion currents are measured in the order in which molecules emerge from a GC column without significant capability of modifying their intensity relative to the reference gas. Chromotagraphy separates not only different chemical species, but also the different isotope species, which means that the isotope composition of a compound varies across the peak of the chemical species after

**Table 1.5** Differences between the offline and online techniques

| Offline method (dual inlet) | Online method (continuous flow) |
|---|---|
| Offline sample preparation | Online sample preparation |
| Offline purification of gases | Purification of gases by GC column |
| Large sample size (mg) | Small sample size (micrograms) |
| Direct inlet of sample gas | Sample gas inlet via carrier gas |
| Pressure adjust of both gases | No pressure adjust, linearity and stability of the system are necessary conditions |
| Sample/standard changes (>6 times) | One peak per sample |
| δ-value calculated from statistical mean | δ-value calculated by peak integration and reference gas |
| System calibration on a monthly basis | System calibration on a daily basis and during the run |
| Little problems with homogeneity of sample | Problems with homogeneity of sample |

elution. Therefore, each peak must be integrated over its entire width to obtain the true isotope ratio

ii. the time for measurement of the isotope signals is restricted by the width of the chromatographic peak. For sharply defined peaks this can mean <5s.

iii. absolute sensitivity is much more important than with the dual inlet system. Because sample sizes required for chromatography are significantly smaller, it is often important to use a significantly large set of samples in order to obtain a statistically sound data base.

Standardization has to be accomplished through the use of an added internal standard whose isotopic composition has been determined using conventional techniques.

The development of this technique has proceeded along several independent paths with two principal lines being elemental analyzer-IRMS and capillary gas chromatography-IRMS. In elemental analyzers, samples are combusted to $CO_2$, $N_2$, $SO_2$ and $H_2O$, which are either chemically trapped or separated on GC columns. There are two types of elemental analyzer: for carbon, nitrogen and sulfur, the sample is combusted in an oxygen containing atmosphere, for hydrogen and oxygen, the sample undergoes high temperature thermal conversion. The advantages of these techniques are an automated preparation with low costs per sample and a large sample through-put at comparable or even better precisions. Figure 1.9 shows a schematic diagram of an elemental analyser-IRMS.

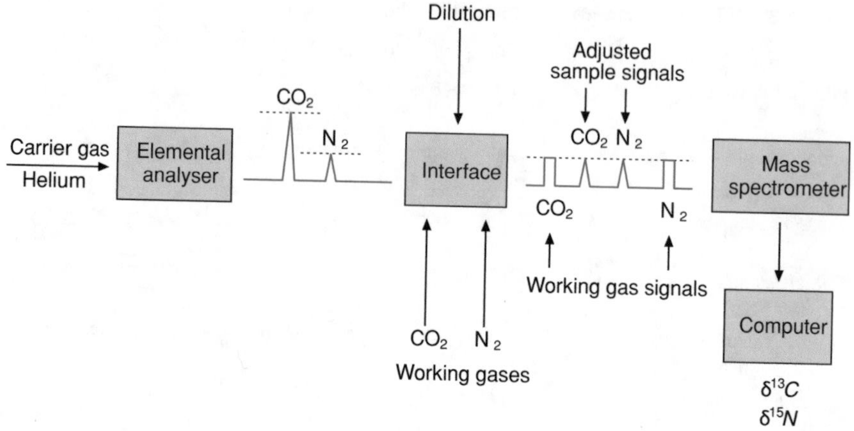


Fig. 1.9 Schematic diagram of an elemental analyser-isotope ratio-mass spectrometer for the determination of carbon and nitrogen isotopes

### 1.4.2 General Remarks on Sample Preparation Methods for Gases

Isotopic differences between samples to be measured are often extremely small. Therefore, great care has to be taken to avoid any isotope fractionation during chemical or physical treatment of the sample. The quality of a stable isotope analysis is determined by the purity of the gas prepared from the sample, quantitative yield, blank and memory effects.

To convert geologic samples to a suitable form for analysis, many different chemical preparation techniques must be used. These techniques all have one general feature in common: any preparation procedure providing a yield of less than 100% may produce a reaction product that is isotopically different from the original specimen because the different isotopic species have different reaction rates.

A quantitative yield of a pure gas is usually necessary for the mass spectrometric measurement in order to prevent not only isotope fractionation during sample preparation, but also interference in the mass spectrometer. Contamination with gases having the same molecular masses and having similar physical properties may be a serious problem. This is especially critical with $CO_2$ and $N_2O$, (Craig and Keeling 1963), and $N_2$ and CO. When $CO_2$ is used, interference by hydrocarbons and a CS + ion may also pose a problem.

Contamination may result from incomplete evacuation of the vacuum system and/or from degassing of the sample. The system blank should be normally <1% of the amount of gas prepared from a sample for analysis. For very small sample sizes, the blank may ultimately limit the analysis. Memory effects result from samples that have previously been analyzed. They will become noticeable when samples having widely different isotopic compositions are analyzed consecutively.

How gases are transferred, distilled, or otherwise processed in vacuum lines is briefly discussed under the different elements. A more detailed description can be found in the recently published "Handbook of stable isotope analytical techniques" edited by De Groot (2004).

All errors due to chemical preparation limit the overall precision of an isotope ratio measurement to usually 0.1–0.2‰, while modern mass spectrometer instrumentation enables a precision better than 0.02‰ for light elements other than hydrogen. Larger uncertainties are expected when elements present in a sample at very low concentration are extracted by chemical methods (e.g., carbon and sulfur from igneous rocks).

Commercial combustion elemental analyzers perform a "flash combustion" converting samples to $CO_2$, $H_2O$, $N_2$ and $SO_2$ simultaneously. These different gases are then chemically trapped, converted or separated on GC columns and measured in a continuous flow mass-spectrometer. This technique allows the determination of several isotope ratios from the same component, increasing the possibilities of isotope fingerprinting of organic and inorganic compounds containing isotopes of more than one element of interest. Because of very high combustion temperatures, the quantitative conversion of the sample material is guaranteed.

By coupling chromatographic techniques with isotope ratio mass spectrometers mixtures of organic compounds can be analysed separately (compound-specific stable isotope analysis). This method has been first introduced for carbon by Matthews and Hayes (1978) and later modified for the separate analysis of hydrogen, nitrogen, chlorine and oxygen compounds. A recent review of the technique has been published by Elsner et al. (2012).

### 1.4.3 Laser Extraction Techniques

Further advances in techniques have been achieved in the introduction of laser assisted extraction, which is based on the fact that the energy of the laser beam is absorbed efficiently by a number of natural substances. The absorption characteristics depends on the structure, composition, and crystallinity of the sample. High-energy, finely-focussed laser beams have been used for some years for Ar isotope analysis, the first well-documented preparation techniques with $CO_2$ and Nd:YAG laser systems for stable isotope determinations have been described by Crowe et al. (1990), Kelley and Fallick (1990) and Sharp (1990). Their results show that sub-milligram quantities of mineral can be analyzed for oxygen, sulfur and carbon. In order to achieve precise and accurate measurements, the samples have to be evaporated completely because steep thermal gradients during laser heating induce isotopic fractionations (Elsenheimer and Valley 1992). The thermal effects of $CO_2$ and Nd-YAG laser assisted preparation techniques require that sample sections be cut into small pieces before total evaporation. Thermal effects can be overcome by vaporizing samples with ultraviolet (UV) KrF and ArF lasers (Wiechert and Hoefs 1995; Fiebig et al. 1999; Wiechert et al. 2002).

Nearly all laboratories today use, however, $CO_2$-lasers for sample extractions.

### 1.4.4 High-Mass-Resolution Multiple-Collector IR Mass Spectrometer

Gas source isotope ratio mass spectrometers in use till now are single focusing magnetic sector spectrometers. The first double-focusing large-radius gas-source mass spectrometer with maximal mass resolution power and sensitivity measuring rare multiply substituted isotopologues was described by Young et al. (2016). In the case of methane, they showed that $^{13}CH_3D/^{12}CH_4$ and $^{12}CH_2D_2/^{12}CH_4$ can be measured with precisions of 0.1 and 0.5‰.

### 1.4.5 Infrared Spectroscopy

An alternative method to isotope ratio mass spectrometry (IRMS) is infrared spectroscopy, which offers certain advantages such as low costs, easy installation and maintenance and potential for field-based applications (continuous measurements of water vapor and atmospheric gases). Cavity ring-down spectroscopy (CRDS) uses the infrared absorption spectrum of gaseous molecules to measure the concentration and isotope composition of water vapor, $CO_2$, $CH_4$ and $N_2O$. Laser light is injected into a precisely aligned optical cavity consisting of very high reflectance mirrors. When steady state conditions are reached, the laser is switched off and the light intensity in the cavity is measured as it "rings-down". Water isotope compositions ($\delta D$, $\delta^{18}O$ and $\Delta^{17}O$) are measured simultaneously and continuously.

For water, CRDS has become the method of choice, because it is fast, easy to handle, no sample preparation is needed and operable under field conditions. As has been demonstrated by Brand et al. (2009), Gupta et al. (2009) and Maithani and Pradham (2020), the CRDS method reaches precisions and accuracies comparable to the IRMS method. Its ease of use and lower cost provides advantages over IRMS.

### 1.4.6 Nuclear Magnetic Resonance (NMR) Spectroscopy

Quantitative $^2H$- and $^{13}C$-NMR spectroscopy have been developed to measure the $^2H/^1H$ and $^{13}C/^{12}C$ ratios of specific sites in organic compounds (i.e. Robins et al. 2002, Gilbert et al. 2009). The measurement of the $^{13}C/^{12}$ ratio of carbonates has been described by Pironti et al. (2017). Limitations of the method are large (10–50 mg) sample sizes and long integration times. Precisions that can be reached are in the range of 1‰.

## 1.5 Standards

The accuracy with which *absolute* isotope abundances can be measured is substantially poorer than the precision with which *relative* differences in isotope abundances between two samples can be determined. Nevertheless, the determination of absolute isotope ratios is very important, because these numbers form the basis for the calculation of the relative differences, the δ-values. Table 1.6 summarizes absolute isotope ratios of primary standards used by the international stable isotope community.

To compare isotope data from different laboratories an internationally accepted set of standards is necessary. Irregularities and problems concerning standards have been evaluated by Friedman and O'Neil (1977), Gonfiantini (1978, 1984), Coplen et al. (1983), Coplen (1996) and Coplen et al. (2006). The accepted unit of isotope ratio measurements is the delta value (δ) given in per mil (‰). The δ-value is defined as

$$\delta \text{in} ‰ = \frac{R_{(\text{Sample})} - R_{(\text{Standard})}}{R_{(\text{Standard})}} \times 1000 \qquad (1.24)$$

where "R" represents the measured isotope ratio. If $\delta_A > \delta_B$, it is convenient to speak of A being enriched in the rare or "heavy" isotope compared to B. Unfortunately, not all of the δ-values cited in the literature are given relative to a single universal standard, so that often several standards of one element are in use. To convert δ-values from one standard to another, the following equation may be used

$$\delta_{X-A} = \left[\left(\frac{\delta_{B-A}}{10^3} + 1\right)\left(\frac{\delta_{X-B}}{10^3} + 1\right) - 1\right] \times 10^3 \qquad (1.25)$$

**Table 1.6** Absolute isotope ratios of international standards (after Hayes 1983)

| Standard | Ratio source | Accepted value ($\times 10^6$) (with 95% confidence interval) | |
|---|---|---|---|
| SMOW | D/H | 155.76 ± 0.10 | Hagemann et al. (1970) |
| | $^{18}O/^{16}O$ | 2,005.20 ± 0.43 | Baertschi (1976) |
| | $^{17}O/^{16}O$ | 373 ± 15 | Nier (1950) by Hayes (1983) |
| PDB | $^{13}C/^{12}C$ | 11,237.2 ± 2.9 | Craig (1957) |
| | $^{18}O/^{16}O$ | 2 067.1 ± 2.1 | |
| | $^{17}O/^{16}O$ | 379 ± 15 | |
| Air nitrogen | $^{15}N/^{14}N$ | 3,676.5 ± 8.1 | Junk and Svec (1958) |
| Canyon Diablo Troilite (CDT) | $^{34}S/^{32}S$ | 45,004.5 ± 9.3 | Jensen and Nakai (1962) |

where X represents the sample, A and B different standards.

For different elements a convenient "working standard" is used in each laboratory. However, all values measured relative to the respective "working standard" are reported in the literature relative to a universal standard.

As an example for the relationship between the content of an isotope in % and the δ-value in ‰, Fig. 1.10 demonstrates that large changes in the δ-value only involve very small changes in the heavy isotope content (in this case the $^{18}O$ content). An ideal standard used worldwide as the "zero-point" on a δ-scale should satisfy the following requirements:

1. be homogeneous in composition,
2. be available in relatively large amounts,
3. be easy to handle for chemical preparation and isotopic measurement, and
4. have an isotope ratio near the middle of the natural range of variation.

Among the reference samples now used, relatively few meet all of these requirements. For instance, the situation for the SMOW standard is rather confusing. The SMOW standard was originally a hypothetical water sample with an isotopic composition very similar to average untreated ocean water (Craig 1961), but being defined in terms of a water sample distributed by the National Bureau of Standards (NBS-1). Later, the IAEA distributed a distilled water sample named V-SMOW (Vienna-SMOW) which is very close to but not identical in isotope composition to the original SMOW standard. The worldwide standards now in general use are given in Table 1.7.

The problems related to standards are discussed by an IAEA advisory group, which meet from time to time. As a result of these meetings the quality and availibility of the existing standards and the need of new standards have been discussed and agreed on.

A further advancement comes from interlaboratory comparison of two standards having different isotopic composition that can be used for a normalization procedure correcting for all proportional errors due to mass spectrometry and to sample

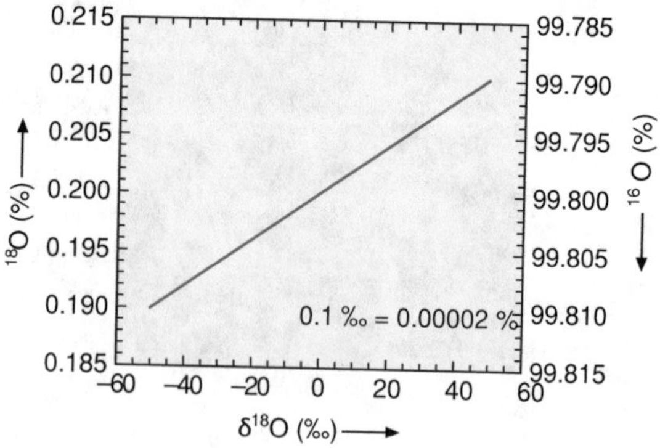


Fig. 1.10 Relationship between $^{18}O$ ($^{16}O$) content in percent and $\delta^{18}O$ in per mill

**Table 1.7** Worldwide standards in use for the isotopic composition of hydrogen, boron, carbon, nitrogen, oxygen, silicium, sulfur, chlorine and of selected metals (Möller et al. 2012)

| Element | Standard | Standard |
|---|---|---|
| H | Standard Mean Ocean Water | V-SMOW |
| B | Boric acid (NBS) | SRM 951 |
| C | Belemnitella americana from the Cretaceous, Peedee formation, South Carolina | V-PDB |
| N | Air nitrogen | $N_2$ (atm.) |
| O | Standard Mean Ocean Water | V-SMOW |
| Si | Quartz sand | NBS-28 |
| S | Troilite (FeS) from the Canyon Diablo iron meteorite | V-CDT |
| Cl | Seawater chloride | SMOC |
| Mg | | DSM-3<br>NIST SRM 980 |
| Ca | | NIST SRM 915a |
| Cr | | NIST SRM 979 |
| Fe | | IRMM-014 |
| Cu | | NIST SRM 976 |
| Zn | | JMC3-0749 |
| Mo | | NIST 3134 |
| Tl | | NIST SRM 997 |
| U | | NIST SRM 950a |
| Ge | | NIST SRM 3120a |

preparation. Ideally, the two standard samples should have isotope ratios as different as possible, but still within the range of natural variations. There are, however, some problems connected with data normalization, which are still under debate. For example, the $CO_2$ equilibration of waters and the acid extraction of $CO_2$ from carbonates are indirect analytical procedures, involving temperature-dependent fractionation factors (whose values are not beyond experimental uncertainties) with respect to the original samples and which might be reevaluated on the normalized scale.

For metal isotopes, standards generally come from two institutions: the Institute for Reference Materials and Mesurements (IRMM) in Belgium and from the National Institute for Standards and Technology (NIST) in the USA. IRMM and NIST mostly supply standard materials in the form of a purified metal or a salt that are easy to dissolve. Some laboratories use natural samples as standards, which has the advantage that samples and standards have to follow the same chemical purification steps. So far, for some elements, there is no consensus on one wordwide used standard, which complicates direct comparison of datasets. Vogl and Pritzkow (2010) have listed currently available reference materials that will be presented under the specific elements.

**Table 1.8** Gases most commonly used in isotope ratio in mass spectrometry

| Element | Gas |
|---|---|
| H | $H_2$ |
| C | $CO_2$, CO |
| N | $N_2$, $N_2O$ |
| O | $CO_2$, CO, $O_2$ |
| S | $SO_2$, $SF_6$ |
| Si | $SiF_4$ |

Table 1.8 summarizes which gases are used for mass-spectrometric analysis of the various elements.

## 1.6 Microanalytical Techniques

In recent years microanalytical techniques, which permit relatively precise isotopic determinations on a variety of samples that are orders of magnitude smaller than those used in conventional techniques, have become increasingly important. Different approaches have been used in this connection, which generally reveal greater isotope heterogeneity than conventional analytical approaches. As a rule of thumb: the smaller the scale of measurement the larger the sample heterogeneity.

For oxygen, Fig. 1.11 demonstrates the improvement in analytical techniques involving lasers and ion microprobes and the enormous reduction in sample sizes. There is of course a reduction in precision with decreasing sample sizes, which is fortunately surprisingly small.

### 1.6.1 Multicollector-ICP-Mass Spectrometry

Advances in multiple collector-ICP-MS (MC-ICP-MS) techniques have enabled the research on natural variations of a wide range of transition and heavy metal systems which so far could not have been measured with the necessary precision (Albarede et al. 2004). The MC-ICP-MS technique combines the strength of the ICP technique (high ionization efficiency for nearly all elements) with the high precision of thermal ion source mass-spectrometry equipped with an array of Faraday collectors (Becker 2005; Vanhaecke et al. 2009). The ICP source allows the analysis of samples introduced either as a solution or as an aerosol produced by laser ablation.

Accuracy and precision of MC-ICP-MS mainly depends on 2 factors: (i) quantitative removal of molecular interferences. All MC-ICP-MS instruments need Ar as the plasma support gas in a similar manner to that commonly used in conventional ICP-MS. Argon interferences are thus an inherent feature of this technique, which have to be circumvented by using desolvating nebulisers and other techniques, (ii) corrections for isobaric elemental interferences and for instrumental mass bias

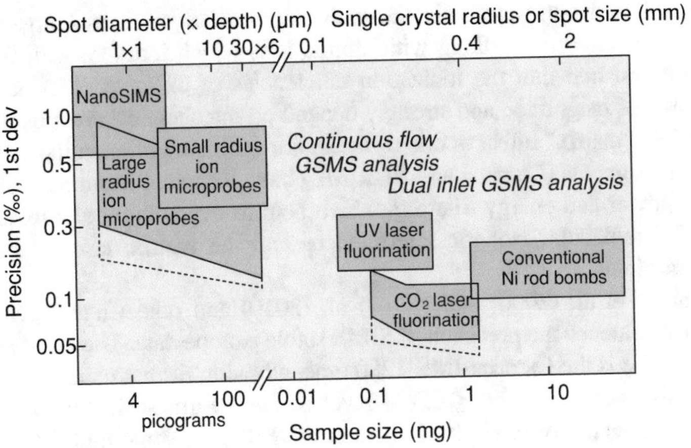


**Fig. 1.11** Precision of various oxygen isotope methods as a function of sample weight or size (from Bindeman 2008)

depending on the purity and the matrix of the sample. The uptake of elements from solution and ionisation in a plasma allows correction for instrument-dependent mass fractionations by addition of external spikes or the comparison of standards with samples under identical operating conditions. 3 methods for correction exist: (i) standard-sample bracketing, (ii) external normalization in which a standard solution of a similarly behaving element is added to both standard and sample and (iii) a double spike technique, in which an enriched spike containing a known mixture of two stable isotopes of the analyzed element is added. In order to achieve the highest precision and accuracy, samples need to be chemically purified.

Mass bias corrections may not be necessary during ablation by femto second lasers as reported by Horn and von Blanckenburg (2007) and Poitrasson and d'Abzac (2017). These authors demonstrated that deep UV femtosecond laser MC-ICP-MS measurements are free of isotopic fractionations.

### 1.6.2 Secondary Ion Mass Spectrometry (SIMS)

Two different types of SIMS are generally used: the Cameca f-series and the SHRIMP (Sensitive High mass Resolution Ion MicroProbe) series (Valley and Graham 1993; Valley et al. 1998; McKibben and Riciputi 1998). Analysis by the ion microprobe is accomplished by sputtering a sample surface using a finely focused primary ion beam producing secondary ions which are extracted and analyzed in the secondary mass spectrometer. The main advantages of this technique are its high sensitivity, high spatial resolution and its small sample size. Sputter pits for a typical 30 min SIMS analyses have a diameter of 10–30 μm and a depth of 1–6 μm, a spatial resolution that is an order of magnitude better than laser techniques. Disadvantages are that the sputtering process produces a large variety of molecular secondary ions along with atomic ions which interfere with the atomic ions of interest and that the ionization efficiencies of different elements vary by many orders of magnitude and strongly depend on the chemical composition of the sample. This "matrix" effect is one of the major problems of quantitative analysis. The two instruments (Cameca and SHRIMP) have technical features, such as high resolving power and energy filtering, which help to overcome the problems of the presence of molecular isobaric interferences and the matrix dependence of secondary ion yields.

Fitzsimons et al. (2000) and Kita et al. (2010) and others have reviewed the factors that influence the precision of SIMS stable isotope data. The latest version of ion microprobe is the Cameca-IMS-1280 type, allowing further reduction in sample and spot size achieving the $\leq 0.3‰$ level at the 10 μm scale for O, S and Fe isotope ratios (Huberty et al. 2010; Kita et al. 2010). In some minerals like magnetite, hematite, sphalerite and galena the authors recognized analytical artefacts related to crystal orientation effects.

## References


Abelson PH, Hoering TC (1961) Carbon isotope fractionation in formation of amino acids by photosynthetic organisms. Proc Natl Acad Sci USA 47:623

Affek HP, Eiler JM (2006) Abundance of mass 47 $CO_2$ in urban air, car exhaust and human breath. Geochim Cosmochim Acta 70:1–12

Affek HP, Bar-Matthews M, Ayalon A, Matthews A, Eiler JM (2008) Glacial/interglacial temperature variations in Soreq cave speleothems as recorded by "clumped isotope" thermometry. Geochim Cosmochim Acta 72:5351–5360

Albarede F, Telouk P, Blichert-Toft J, Boyet M, Agrinier A, Nelson B (2004) Precise and accurate isotope measurements using multi-collector ICPMS. Geochim Cosmochim Acta 68:2725–2744

Assonov SS, Brenninkmeijer CA (2005) Reporting small $\Delta^{17}O$ values: existing definitions and concepts. Rapid Commun Mass Spectrometry 19:627–636

Baertschi P (1976) Absolute $^{18}O$ content of standard mean ocean water. Earth Planet Sci Lett 31:341–344

Bao H, Thiemens MH, Farquahar J, Campbell DA, Lee CC, Heine K, Loope DB (2000) Anomalous $^{17}O$ compositions in massive sulphate deposits on the Earth. Nature 406:176–178

Bao H, Thiemens MH, Heine K (2001) Oxygen-17 excesses of the Central Namib gypcretes: spatial distribution. Earth Planet Sci Letters 192:125–135

Baroni M, Thiemens MH, Delmas RJ, Savarino J (2007) Mass-independent sulfur isotopic composition in stratospheric volcanic eruptions. Science 315:84–87

Becker JS (2005) Recent developments in isotopic analysis by advanced mass spectrometric techniques. J Anal at Spectrom 20:1173–1184

Bigeleisen J (1965) Chemistry of isotopes. Science 147:463–471

Bigeleisen J (1996) Nuclear size and shape effects in chemical reactions. Isotope chemistry of heavy elements. J Am Chem Soc 118:3676–3680

Bigeleisen J, Mayer MG (1947) Calculation of equilibrium constants for isotopic exchange reactions. J Chem Phys 15:261–267

Bigeleisen J, Wolfsberg M (1958) Theoretical and experimental aspects of isotope effects in chemical kinetics. Adv Chem Phys 1:15–76

Bindeman I (2008) Oxygen isotopes in mantle and crustal magmas as revealed by single crystal analysis. Rev Miner Geochem 69:445–478

Bindeman I, Pack A (eds) (2021) Triple oxygen isotope geochemistry. Rev Mineral Geochem 86:1–488

Blair N, Leu A, Munoz E, Olsen J, Kwong E, Desmarais D (1985) Carbon isotopic fractionation in heterotrophic microbial metabolism. Appl Environ Microbiol 50:996–1001

Blum JD (2011) Applications of stable mercury isotopes to biogeochemistry. In: Baskaran M (ed) Handbook of environmental isotope geochemistry. Springer, pp 229–246

Blum JD, Johnson MW (2017) Recent developments in mercury stable isotope analysis. Rev Mineral Geochem 82:733–757

Bottinga Y (1969) Carbon isotope fractionation between graphite, diamond and carbon dioxide. Earth Planet Sci Lett 5:301–307

Bottinga Y, Javoy M (1973) Comments on oxygen isotope geothermometry. Earth Planet Sci Lett 20:250–265

Brand W (2002) Mass spectrometer hardware for analyzing stable isotope ratios. In: de Groot P (ed) Handbook of stable isotope analytical techniques. Elsevier, New York

Brand W, Geilmann H, Crosson ER, Rella CW (2009) Cavity ring-down spectroscopy versus high-temperature conversion isotope ratio mass spectrometry; a case study on $\delta^2H$ and $\delta^{18}O$ of pure water samples and alcohol/water mixtures. Rapid Comm Mass Spectrometry 23:1879–1884

Bucharenko AI (1995) MIE versus CIE: comparative analysis of magnetic and classical isotope effects. Chem Rev 95:2507–2528

Bucharenko AI (2001) Magnetic isotope effect: nuclear spin control of chemical reactions. J Phys Chem A 105:9995–10011

Bucharenko AI (2013) Mass-independent isotope effects. J Phys Chem B 117:2231–2238

Came RE, Brand U, Affek HP (2014) Clumped isotope signatures in modern brachiopod carbonate. Chem Geol 377:20–30

Cao X, Bao H (2017) Redefining the utility of the three-isotope method. Geochim Cosmochim Acta 212:16–32

Cerling TE (1984) The stable isotopic composition of modern soil carbonate and its relationship to climate. Earth Planet Sci Lett 71:229–240

Chacko T, Cole DR, Horita J (2001) Equilibrium oxygen, hydrogen and carbon fractionation factors applicable to geologic systems. Rev Miner Geochem 43:1–81

Chiba H, Chacko T, Clayton RN, Goldsmith JR (1989) Oxygen isotope fractionations involving diopside, forsterite, magnetite and calcite: application to geothermometry. Geochim Cosmochim Acta 53:2985–2995

Clayton RN, Kieffer SW (1991) Oxygen isotope thermometer calibrations. In: Taylor HP, O'Neil JR, Kaplan IR (eds) Stable isotope geochemistry: a tribute to Sam Epstein. Geochem Soc Spec Publ 3:3–10

Clayton RN, Grossman L, Mayeda TK (1973) A component of primitive nuclear composition in carbonaceous meteorites. Science 182:485–488

Clayton RN, Goldsmith JR, Karel KJ, Mayeda TK, Newton RP (1975) Limits on the effect of pressure in isotopic fractionation. Geochim Cosmochim Acta 39:1197–1201

Clayton RN, Goldsmith JR, Mayeda TK (1989) Oxygen isotope fractionation in quartz, albite, anorthite and calcite. Geochim Cosmochim Acta 53:725–733

Cole DR, Chakraborty S (2011) Rates and mechanisms of isotopic exchange. In: Stable isotope geochemistry. Rev Mineral Geochem 43:83–223

Coplen TB (1996) New guidelines for the reporting of stable hydrogen, carbon and oxygen isotope ratio data. Geochim Cosmochim Acta 60:3359–3360

Coplen TB, Kendall C, Hopple J (1983) Comparison of stable isotope reference samples. Nature 302:236–238

Coplen TB, Brand W, Gehre M, Gröning M, Meijer HA, Toman B, Verkouteren RM (2006) New guidelines for $\delta^{13}C$ measurements. Anal Chem 78:2439–2441

Craig H (1957) Isotopic standards for carbon and oxygen and correction factors for mass-spectrometric analysis of carbon dioxide. Geochim Cosmochim Acta 12:133–149

Craig H (1961) Standard for reporting concentrations of deuterium and oxygen-18 in natural waters. Science 133:1833–1834

Craig H, Keeling CD (1963) The effects of atmospheric $N_2O$ on the measured isotopic composition of atmospheric $CO_2$. Geochim Cosmochim Acta 27:549–551

Criss RE (1999) Principles of stable isotope distribution. Oxford University Press

Crowe DE, Valley JW, Baker KL (1990) Micro-analysis of sulfur isotope ratios and zonation by laser microprobe. Geochim Cosmochim Acta 54:2075–2092

Daeron M, Blamart D, Peral M, Affek HP (2016) Absolute isotope abundance ratios and the accuracy of $\Delta_{47}$ measurements. Chem Geol 442:83–96

Dansgaard W (1964) Stable isotope in precipitation. Tellus 16:436–468

Dauphas N, Schauble EA (2016) Mass fractionation laws, mass-independent effects and isotope anomalies. Ann Rev Earth Planet Sci 44:709–783

Dauphas N, Teng FZ, Arndt NT (2010) Magnesium and iron isotopes in 2.7 Ga Alexo komatiites: mantle signatures, no evidence for Soret diffusion and identification of diffusive transport in zoned olivine. Geochim Cosmochim Acta 74:3274–3291

De Groot PA (2004) Handbook of stable isotope analytical techniques. Elsevier Amsterdam, Europe

Dennis KJ, Schrag DP (2010) Clumped isotope thermometry of carbonatites as an indicator of diagenetic alteration. Geochim Cosmochim Acta 74:4110–4122

Driesner T (1997) The effect of pressure on deuterium-hydrogen fractionation in high-temperature water. Science 277:791–794

Eagle RA, Schauble EA, Tripati AK, Tütken T, Hulbert RC, Eiler JM (2010) Body temperatures of modern and extinct vertebrates from $^{13}C$–$^{18}O$ bond abundances in bioapatite. PNAS 107:10377–10382

Eagle RA, Tütken T, Martin TS, Tripati AK, Fricke HC, Connely M, Cifelli RL, Eiler JM (2011) Dinosaur body temperatures determined from the ($^{13}C$–$^{18}O$) ordering in fossil biominerals. Science 333:443–445

Eagle RA et al (2015) Isotopic ordering in eggshells reflects body temperatures and suggests differing thermophysiology in two Cretaceous dinosaurs. Nat Commun 6:8296

Eiler JM (2007) The study of naturally-occuring multiply-substituted isotopologues. Earth Planet Sci Lett 262:309–327

Eiler JM (2013) The isotopic anatomies of molecules and minerals. Ann Rev Earth Planet Sci 41:411–441

Eiler JM, Schauble E (2004) $^{18}O^{13}C^{16}O$ in earth's atmosphere. Geochim Cosmochim Acta 68:4767–4777

Eiler JM, Baumgartner LP, Valley JW (1992) Intercrystalline stable isotope diffusion: a fast grain boundary model. Contr Mineral Petrol 112:543–557

Eiler JM, Valley JW, Baumgartner LP (1993) A new look at stable isotope thermometry. Geochim Cosmochim Acta 57:2571–2583

Elsenheimer D, Valley JW (1992) In situ oxygen isotope analysis of feldspar and quartz by Nd-YAG laser microprobe. Chem Geol 101:21–42

Elsner M, Jochmann MA, Hofstetter TB, Hunkeler D, Bernstein A, Schmidt T, Schimmelmann A (2012) Current challenges in compound-specific stable isotope analysis of environmental organic contaminants. Anal Bioanal Chem 403:2471–2491

Epov VN, Malinovskiy D, Vanhaecke F, Begue D, Donard OF (2011) Modern mass spectrometry for studying mass-independent fractionation of heavy stable isotopes in environmental and biological sciences. J Anal Spectrom 26:1142–1156

Estrade N, Carignan J, Sonke JE, Donard O (2009) Mercury isotope fractionation during liquid-vapor evaporation experiments. Geochim Cosmochim Acta 73:2693–2711

Farquhar J, Bao H, Thiemens M (2000) Atmospheric influence of Earth's earliest sulfur cycle. Science 289:756–759

Farquhar J, Johnston DT, Wing BA, Habicht KS, Canfield DE, Airieau S, Thiemens MH (2003) Multiple sulphur isotope interpretations for biosynthetic pathways: implications for biological signatures in the sulphur isotope record. Geobiology 1:27–36

Ferry JM, Passey BH, Vasconcelos C, Eiler JM (2011) Formation of dolomite at 40–80°C in the Latemar carbonate buildup, Dolomites, Italy from clumped isotope thermometry. Geology 39:571–574

Fiebig J, Wiechert U, Rumble D, Hoefs J (1999) High-precision in-situ oxygen isotope analysis of quartz using an ArF laser. Geochim Cosmochim Acta 63:687–702

Fiebig J, Bajnai D, Löffler N, Methner K, Krsnik E, Mulch A, Hofmann S (2019) Combined high-precision $\Delta_{48}$ and $\Delta_{47}$ analysis of carbonates. Chem Geol 522:186–191

Fitzsimons ICW, Harte B, Clark RM (2000) SIMS stable isotope measurement: counting statistics and analytical precision. Min Mag 64:59–83

Friedman I, O'Neil JR (1977) Compilation of stable isotope fractionation factors of geochemical interest. In: Data of geochemistry, 6th edn. Geological States Geological Survey Professional Paper 440-KK

Fujii T, Moynier F, Albarede F (2009) The nuclear field shift effect in chemical exchange reactions. Chem Geol 267:139–156

Galimov EM (2006) Isotope organic geochemistry. Org Geochem 37:1200–1262

Gao YQ, Marcus RA (2001) Strange and unconventional isotope effects in ozone formation. Science 293:259–263

Gelabert A, Pokrovsky OS, Viers J, Schott J, Boudou A, Feurtet-Mazel A (2006) Interaction between zinc and marine diatom species: surface complexation and Zn isotope fractionation. Geochim Cosmochim Acta 70:839–857

Ghosh P et al (2006) $^{13}C$–$^{18}O$ bonds in carbonate minerals: a new kind of paleothermometer. Geochim Cosmochim Acta 70:1439–1456

Gilbert A, Silvestre V, Robins R, Remaud G (2009) NMR spectroscopy for the determination of the intramolecular distribution of $^{13}C$ in glucose at natural abundance. Anal Chem 81:8975–8985

Gilbert A, Yamada K, Suda K, Ueno Y, Yoshida N (2016) Measurement of position-specific $^{13}C$ isotopic composition of propane at the nanomole level. Geochim Cosmochim Acta 177:205–216

Giletti BJ (1986) Diffusion effect on oxygen isotope temperatures of slowly cooled igneous and metamorphic rocks. Earth Planet Sci Lett 77:218–228

Gonfiantini R (1978) Standards for stable isotope measurements in natural compounds. Nature 271:534–536

Gonfiantini R (1984) Advisory group meeting on stable isotope reference samples for geochemical and hydrological investigations. Report Director General IAEA Vienna

Grachev AM, Severinghaus JP (2003) Laboratory determination of thermal diffusion constants for $^{29}N/^{28}N_2$ in air at temperatures from –60 to 0 °C for reconstruction of magnitudes of abrupt climate changes using the ice core fossil-air paleothermometer. Geochim Cosmochim Acta 67:345–360

Gupta P, Noone D, Galewsky J, Sweeney C, Vaughn BH (2009) Demonstration of high-precision continuous measurements of water vapor isotopologues in laboratory and remote field deployments using wavelength-scanned cavity ring-down spectroscopy (WS-CRDS) technology. Rapid Comm Mass Spectrometry 23:2534–2542

Hagemann R, Nief G, Roth E (1970) Absolute isotopic scale for deuterium analysis of natural waters. Absolute d/h Ratio for SMOW. Tellus 22:712–715

Hayes JM (1983) Practice and principles of isotopic measurements in organic geochemistry. In: Organic geochemistry of contemporaneous and ancient sediments. Great Lakes Section, SEPM, Bloomington, Ind, pp 5-1–5-31

Henkes GA, Passey BH, Wanamaker AD, Grossman EI, Ambrose WG, Carroll ML (2013) Carbonate clumped isotope composition of modern marine mollusk and brachiopod shells. Geochim Cosmochim Acta 106:307–325

Henkes GA, Passey BH, Grossman EL, Shenton BJ, Perez-Huerta A, Yancey TE (2014) Temperature limits of preservation of primary calcite clumped isotope paleotemperatures. Geochim Cosmochim Acta 139:362–382

Hesterberg R, Siegenthaler U (1991) Production and stable isotopic composition of $CO_2$ in a soil near Bern, Switzerland. Tellus 43B:197–205

Horita J, Driesner T, Cole DR (1999) Pressure effect on hydrogen isotope fractionation between brucite and water at elevated temperatures. Science 286:1545–1547

Horita J, Cole DR, Polyakov VB, Driesner T (2002) Experimental and theoretical study of pressure effects on hydrous isotope fractionation in the system brucite-water at elevated temperatures. Geochim Cosmochim Acta 66:3769–3788

Horn I, von Blanckenburg F (2007) Investigation on elemental and isotopic fractionation during 196 nm femtosecond laser ablation multiple collector inductively coupled plasma mass spectrometry. Spectrochimica Acta Part B 62:410–422

Hu G, Clayton RN (2003) Oxygen isotope salt effects at high pressure and high temperature and the calibration of oxygen isotope thermometers. Geochim Cosmochim Acta 67:3227–3246

Huberty JM, Kita NT, Kozdon R et al (2010) Crystal orientation effects in $\delta^{18}O$ for magnetite and hematite by SIMS. Chem Geol 276:269–283

Huntington KW, Eiler JM et al (2009) Methods and limitations of “clumped” $CO_2$ isotope ($\Delta_{47}$) analysis by gas-source isotope ratio mass spectrometry. J Mass Spectrom 44:1318–1329

Huntington KW, Budd DA, Wernicke BP, Eiler JM (2011) Use of clumped-isotope thermometry to constrain the crystallization temperature of diagenetic calcite. J Sediment Res 81:656–669

Huntington KW, Wernicke BP, Eiler JM (2010) Influence of climate change and uplift on Colorado Plateau paleotemperatures from carbonate clumped isotope thermometry. Tectonics 29 TC3005.https://doi.org/10.1029/2009TC002449

Jensen ML, Nakai N (1962) Sulfur isotope meteorite standards, results and recommendations. In: Jensen ML (ed) Biogeochemistry of sulfur isotopes. NSF Symp Vol, p 31

Junk G, Svec H (1958) The absolute abundance of the nitrogen isotopes in the atmosphere and compressed gas from various sources. Geochim Cosmochim Acta 14:234–243

Kashiwabara T, Takahashi Y, Tanimizu M, Usui A (2011) Molecular-scale mechanisms of distribution and isotopic fractionation of molybdenum between seawater and ferromanganese oxides. Geochim Cosmochim Acta 75:5762–5784

Kelley SP, Fallick AE (1990) High precision spatially resolved analysis of $\delta^{34}S$ in sulphides using a laser extraction technique. Geochim Cosmochim Acta 54:883–888

Kieffer SW (1982) Thermodynamic and lattice vibrations of minerals: 5. Application to phase equilibria, isotopic fractionation and high-pressure thermodynamic properties. Rev Geophys Space Phys 20:827–849

Kita NT, Hyberty JM, Kozdon R, Beard BL, Valley JW (2010) High-precision SIMS oxygen, sulfur and iron stable isotope analyses of geological materials: accuracy, surface topography and crystal orientation. Surf Interface Anal 43:427–431

Kitchen NE, Valley JW (1995) Carbon isotope thermometry in marbles of the Adirondack Mountains, New York. J Metamorphic Geol 13:577–594

Kohn MJ, Valley JW (1998) Obtaining equilibrium oxygen isotope fractionations from rocks: theory and examples. Contr Mineral Petrol 132:209–224

Kowalski PM, Jahn S (2011) Prediction of equilibrium Li isotope fractionation between minerals and aqueous solutions at high P and T: an efficient ab initio approach. Geochim Cosmochim Acta 75:6112–6123

Kowalski PM, Wunder B, Jahn S (2013) Ab initio prediction of equilibrium boron isotope fractionation between minerals and aqueous fluids at high P and T. Geochim Cosmochim Acta 101:285–301

Luz B, Barkan E, Bender ML, Thiemens MH, Boering KA (1999) Triple-isotope composition of atmospheric oxygen as a tracer of biosphere productivity. Nature 400:547–550

Maithani S, Pradham M (2020) Cavity ring-down spectroscopy and its applications to environmental, chemical and biomedical systems. J Chem Soc 132:114

Matsuhisa Y, Goldsmith JR, Clayton RN (1978) Mechanisms of hydrothermal crystallization of quartz at 250 °C and 15 kbar. Geochim Cosmochim Acta 42:173–182

Matthews DE, Hayes JM (1978) Isotope-ratio-monitoring gas chromatography-mass spectrometry. Anal Chem 50:1465–1473

Matthews A, Goldsmith JR, Clayton RN (1983) Oxygen isotope fractionation involving pyroxenes: the calibration of mineral-pair geothermometers. Geochim Cosmochim Acta 47:631–644

Mauersberger K, Erbacher B, Krankowsky D, Günther J, Nickel R (1999) Ozone isotope enrichment: isotopomer-specific rate coefficients. Science 283:370–372

McKibben MA, Riciputi LR (1998) Sulfur isotopes by ion microprobe. In: applications of microanalytical techniques to understanding mineralizing processes. Rev Econ Geol 7:121–140

Méheut M, Lazzari M, Balan E, Mauri F (2007) Equilibrium isotopic fractionation in the kaolinite, quartz, water system: prediction from first principles calculations density-functional theory. Geochim Cosmochim Acta 71:3170–3181

Melander L (1960) Isotope effects on reaction rates. Ronald, New York

Melander L, Saunders WH (1980) Reaction rates of isotopic molecules. Wiley, New York

Merritt DA, Hayes JM (1994) Nitrogen isotopic analyses of individual amino acids by isotope-ratio-monitoring gas chromatography/mass spectrometry. J Am Soc Mass Spectrom 5:387–397

Miller MF (2002) Isotopic fractionation and the quantification of $^{17}O$ anomalies in the oxygen three-isotope system: an appraisal and geochemical significance. Geochim Cosmochim Acta 66:1881–1889

Möller K, Schoenberg R, Pedersen RB, Weiss D, Dong S (2012) Calibration of new certified reference materials ERM-AE633 and ERM-AE647 for copper and IRMM-3702 for zinc isotope amount ratio determinations. Geostand Geoanal Res 36:177–199

Nier AO (1950) A redetermination of the relative abundances of the isotopes of carbon, nitrogen, oxygen, argon and potassium. Phys Rev 77:789

Nier AO, Ney EP, Inghram MG (1947) A null method for the comparison of two ion currents in a mass spectrometer. Rev Sci Instrum 18:294

Northrop DA, Clayton RN (1966) Oxygen isotope fractionations in systems containing dolomite. J Geol 74:174–196

O'Neil JR (1986) Theoretical and experimental aspects of isotopic fractionation. In: Stable isotopes in high temperature geological processes. Rev Mineral 16:1–40

Oeser M, Dohmen R, Horn I, Schuth S, Weyer S (2015) Processes and time scales of magmatic evolution as revealed by Fe–Mg chemical and isotopic zoning in natural olivines. Geochim Cosmochim Acta 154:130–150

Passey BJ, Henkes GA (2012) Carbonate clumped isotope bond reordering and geospeeedometry. Earth Planet Sci Lett 351–352:223–236

Passey BJ, Levin NE (2021) Triple oxygen isotopes in carbonates, biological apatites and continental paleoclimate reconstruction. Rev Mineral Geochem 86:429–462

Petersen SV, 29 others (2019) Effects of improved $^{17}O$ correction on interlaboratory agreement in clumped isotope calibrations, estimates of mineral-specific offsets, and temperature dependence of acid digestion fractionation. Geochem Geophys Geosys 20:3495–3519

Piasecki A, Sessions A, Lawson M, Ferreira AA, Santos Neto EV, Ellis GS, Lewan MD, Eiler JM (2018) Position-specific $^{13}C$ distribitions within propane from experiments and natural gas samples. Geochim Cosmochim Acta 220:110–124

Pironti C, Cucciniello R, Lamin F, Tonon A, Motta O, Proto A (2017) Determination of the $^{13}C/^{12}C$ carbon isotope ratio in carbonates and bicarbonates by $^{13}$CNMR spectroscopy. Anal Chem 89:11413–11418

Poitrasson F, d'Abzac FX (2017) Femto second laser ablation inductively coupled plasma source mass spectrometry for elemental and isotopic analysis: are ultrafast lasers worthwhile? JAAS 32:1075–1091

Polyakov VB, Kharlashina NN (1994) Effect of pressure on equilibrium isotope fractionation. Geochim Cosmochim Acta 58:4739–4750
Polyakov VB, Horita J, Cole DR (2006) Pressure effects on the reduced partition function ratio for hydrogen isotopes in water. Geochim Cosmochim Acta 70:1904–1913
Quade J, Breecker DO, Daeron M, Eiler J (2011) The paleoaltimetry of Tibet: an isotopic perspective. Am J Sci 311:77–115
Rayleigh JWS (1896) Theoretical considerations respecting the separation of gases by diffusion and similar processes. Philos Mag 42:493
Richet P, Bottinga Y, Javoy M (1977) A review of H, C, N, O, S, and Cl stable isotope fractionation among gaseous molecules. Ann Rev Earth Planet Sci 5:65–110
Richter FM (2007) Isotopic fingerprints of mass transport processes. Geochim Cosmochim Acta 71:A839
Richter R, Hoernes S (1988) The application of the increment method in comparison with experimentally derived and calculated O-isotope fractionations. Chem Erde 48:1–18
Richter FM, Liang Y, Davis AM (1999) Isotope fractionation by diffusion in molten oxides. Geochim Cosmochim Acta 63:2853–2861
Richter FM, Davis AM, DePaolo D, Watson BE (2003) Isotope fractionation by chemical diffusion between molten basalt and rhyolite. Geochim Cosmochim Acta 67:3905–3923
Richter FM, Dauphas N, Teng FZ (2009) Non-traditional fractionation of non-traditional isotopes: evaporation, chemical diffusion and Soret diffusion. Chem Geol 258:92–103
Robins R, Billault I, Duan J, Guiet S, Pionnier S, Zhang BL (2002) Measurement of $^{2}H$ distribution in natural products by quantitative $^{2}H$-NMR: an approach to understanding metabolism and enzyme mechanism? Phytochem Rev 2:87–102
Schauble EA (2004) Applying stable isotope fractionation theory to new systems. Rev Mineral Geochem 55:65–111
Schauble EA (2007) Role of nuclear volume in driving equilibrium stable isotope fractionation of mercury, thallium and other very heavy elements. Geochim Cosmochim Acta 71:2170–2189
Schauble EA (2011) First principles estimates of equilibrium magnesium isotope fractionation in silicate, oxide, carbonate and hexaaquamagnesium(2+) crystals. Geochim Cosmochim Acta 75:844–869
Schauble EA (2013) Modeling nuclear volume isotope effects in crystals. PNAS 110:17714–17719
Schauble EA, Ghosh P, Eiler JM (2006) Preferential formation of $^{13}C$–$^{18}O$ bonds in carbonate minerals, estimated using first-principles lattice dynamics. Geochim Cosmochim Acta 70:2510–2519
Schauble E, Méheut M, Hill PS (2009) Combining metal stable isotope fractionation theory with experiments. Elements 5:369–374
Scheele N, Hoefs J (1992) Carbon isotope fractionation between calcite, graphite and $CO_2$. Contr Mineral Petrol 112:35–45
Schütze H (1980) Der Isotopenindex—eine Inkrementmethode zur näherungsweisen Berechnung von Isotopenaustauschgleichgewichten zwischen kristallinen Substanzen. Chemie Erde 39:321–334
Severinghaus JP, Brook EJ (1999) Abrupt climate change at the end of the last glacial period inferred from trapped air in polar ice. Science 286:930–934
Severinghaus JP, Bender ML, Keeling RF, Broecker WS (1996) Fractionation of soil gases by diffusion of water vapor, gravitational settling and thermal diffusion. Geochim Cosmochim Acta 60:1005–1018
Severinghaus JP, Sowers T, Brook EJ, Alley RB, Bender ML (1998) Timing of abrupt climate change at the end of the Younger Dryas interval from thermally fractionated gases in polar ice. Nature 391:141–146
Shahar A, Schauble EA, Caracas R, Gleason AE, Reagan MM, Xiao Y, Shu J, Mao W (2016) Pressure-dependent isotopic composition of iron alloys. Science 352:580–582

Sharp ZD (1990) A laser-based microanalytical method for the in-situ determination of oxygen isotope ratios of silicates and oxides. Geochim Cosmochim Acta 54:1353–1357

Sharp ZD (1995) Oxygen isotope geochemistry of the $Al_2SiO_5$ polymorphs. Am J Sci 295:1058–1076

Sio CK, Dauphas N, Teng FZ, Chaussidon M, Helz RT, Roskosz M (2013) Discerning crystal growth from diffusion profiles in zoned olivine by in-situ Mg–Fe isotope analysis. Geochim Cosmochim Acta 123:302–321

Stern MJ, Spindel W, Monse EU (1968) Temperature dependence of isotope effects. J Chem Phys 48:2908

Stolper DA, Sessions AL, Ferreira AA, Santos Neto EV, Schimmelmann A, Shusta SS, Valentine DL, Eiler JM (2014) Combined $^{13}C$–D and D–D clumping in methane: methods and preliminary results. Geochim Cosmochim Acta 126:169–191

Tang J, Dietzel M, Fernandez A, Tripati AK, Rosenheim BE (2014) Evaluation of kinetic effects on clumped isotope fractionation ($\Delta_{47}$) during inorganic calcite precipitation. Geochim Cosmochim Acta 134:120–136

Teng FZ, Dauphas N, Helz RT, Gao S, Huang S (2011) Diffusion-driven magnesium and iron isotope fractionation in Hawaiian olivine. Earth Planet Sci Lett 308:317–324

Teutsch N, von Gunten U, Hofstetter TB, Halliday AN (2005) Adsorption as a cause for isotope fractionation in reduced groundwater. Geochim Cosmochim Acta 69:4175–4185

Thiemens MH (1999) Mass-independent isotope effects in planetary atmospheres and the early solar system. Science 283:341–345

Thiemens MH, Heidenreich JE (1983) The mass independent fractionation of oxygen—a novel isotope effect and its cosmochemical implications. Science 219:1073–1075

Thiemens MH, Lin M (2021) Discoveries of mass independent isotope effects in the solar system: past, present and future. Rev Mineral Geochem 86:35–95

Thiemens MH, Chakraborty S, Dominguez G (2012) The physical chemistry of mass-independent isotope effects and their observation in nature. Ann Rev Phys Chem 63:155–177

Tripati AK, Eagle RA, Thiagarajan N, Gagnon AC, Bauch H, Halloran PR, Eiler JM (2010) $^{13}C$–$^{18}O$ isotope signaturesand "clumped isotope" thermometry in foraminifera and coccoliths. Geochim Cosmochim Acta 74:5697–5717

Urey HC (1947) The thermodynamic properties of isotopic substances. J Chem Soc 1947:562

Valley JW, Graham C (1993) Cryptic grain-scale heterogeneity of oxygen isotope ratios in metamorphic magnetite. Science 259:1729–1733

Valley J, Graham CM, Harte B, Eiler JM, Kinney PD (1998) Ion microprobe analysis of oxygen, carbon and hydrogen isotope ratios. In: applications of microanalytical techniques to understanding mineralizing processes. Rev Econ Geol 7:73–98

Vanhaecke F, Balcaen L, Malinovsky D (2009) Use of single-collector and multi-collector ICP-mass spectrometry for isotope analysis. J Anal Spectrom 24:863–886

Vogl J, Pritzkow W (2010) Isotope reference materials for present and future isotope research. J Anal Spectrom 25:923–932

Wang Z, Schauble EA, Eiler JM (2004) Equilibrium thermodynamics of multiply substituted isotopologues of molecular gas. Geochim Cosmochim Acta 68:4779–4797

Wiechert U, Hoefs J (1995) An excimer laser-based microanalytical preparation technique for in-situ oxygen isotope analysis of silicate and oxide minerals. Geochim Cosmochim Acta 59:4093–4101

Wiechert U, Fiebig J, Przybilla R, Xiao Y, Hoefs J (2002) Excimer laser isotope-ratio-monitoring mass spectrometry for in situ oxygen isotope analysis. Chem Geol 182:179–194

Yeung LY, Ash JL, Young ED (2015) Biological signatures in clumped isotopes of $O_2$. Science 348:431–434

Young ED, Galy A, Nagahara H (2002) Kinetic and equilibrium mass-dependent isotope fractionation laws in nature and their geochemical and cosmochemical significance. Geochim Cosmochim Acta 66:1095–1104

Young ED, Rumble D, Freedman P, Mills M (2016) A large-radius high-mass-resolution multiple-collector isotope ratio mass spectrometer for analysis of rare isotopologues of $O_2$, $N_2$, $CH_4$ and other gases. Inter J Mass Spectr 401:1–10

Zheng YF (1991) Calculation of oxygen isotope fractionation in metal oxides. Geochim Cosmochim Acta 55:2299–2307

Zheng YF (1993a) Oxygen isotope fractionation in $SiO_2$ and $Al_2SiO_5$ polymorphs: effect of crystal structure. Eur J Mineral 5:651–658

Zheng YF (1993b) Calculation of oxygen isotope fractionation in anhydrous silicate minerals. Geochim Cosmochim Acta 57:1079–1091

Zheng YF (1993c) Calculation of oxygen isotope fractionation in hydroxyl-bearing minerals. Earth Planet Sci Lett 120:247–263

Zheng YF, Böttcher ME (2016) Oxygen isotope fractionation in double carbonates. Isot Environ Health Stud 52:29–46

# Isotope Fractionation Processes of Selected Elements

# 2

## Part I: "Traditional" Isotopes

The foundations of stable isotope geochemistry were laid in 1947 by Urey's classic paper on the thermodynamic properties of isotopic substances and by Nier's development of the isotope ratio mass spectrometer (IRMS). Before discussing details of the naturally occurring variations in stable isotope ratios, it is useful to describe some generalities that are pertinent to the field of non-radiogenic isotope geochemistry as a whole.

1. Isotope fractionation is pronounced when the mass differences between the isotopes of a specific element are large relative to the mass of the element. Therefore, isotope fractionations are especially large for the light elements. Recent developments in analytical techniques have opened the possibility to detect small variations for elements with much higher mass numbers. The heaviest element for which natural isotope fractionations have been reported is uranium, the last relatively stable element in the periodic table.
2. All elements that form solid, liquid, and gaseous compounds stable over a wide temperature range are likely to have variations in isotopic composition. Generally, the heavy isotope is concentrated in the phase in which it is more tightly bound. Heavier isotopes tend to concentrate in compounds in which they are present in the highest oxidation state and lowest coordination number.
3. Mass balance effects can cause isotope fractionations because modal proportions of substances can change during a chemical reaction. They are especially important for elements in situations where these coexist in molecules of reduced and oxidized compounds. Under closed conditions, conservation of mass in an n component system can be described by

$$\delta_{(\mathrm{system})} = \sum x_i \delta_i \tag{2.1}$$




J. Hoefs, *Stable Isotope Geochemistry*, Springer Textbooks in Earth Sciences, Geography and Environment, https://doi.org/10.1007/978-3-030-77692-3_2

where "$x_i$" is the mole fraction of the element in question for each of n phases within the system.

4. Isotopic variations can also be caused by kinetic effects, especially in gases and in biological systems. During biological reactions (e.g. photosynthesis, bacterial processes) the lighter isotope is often enriched in the reaction product relative to the substrate. Most of the fractionations observed in biological reactions take place during the so-called rate determining step, which is the slowest step. It commonly involves a large reservoir, where the material actually used is small compared to the size of the reservoir.

All are mass-dependent as the magnitude of fractionation is dependent on the mass difference between coexisting compounds. In this connection, the reader is also referred to mass-independent isotope fractionations that are chiefly induced by photochemical effects such as ozone formation and discussed on p. 13.

## 2.1 Hydrogen

Until 1931 it was assumed that hydrogen consists of only one isotope. Urey et al. (1932) detected the presence of a second stable isotope, which was called deuterium (In addition to these two stable isotopes there is a third naturally occurring but radioactive isotope, $^3H$, tritium, with a half-life of approximately 12.5 years). Rosman and Taylor (1998) gave the following average abundances of the stable hydrogen isotopes

$$^1H \quad 99.9885\%$$
$$^2H \quad 0.0115\%$$

In earlier editions of this book, hydrogen isotope ratios were given as δD-values. To be consistent with the definition for all the other isotope systems and following the recommendations of the International Union of Pure and Applied Chemistry (IUPAC), hydrogen isotope ratios are written as $\delta^2H$-values.

The isotope geochemistry of hydrogen is particular interesting, for two reasons:

(1) Hydrogen is omnipresent in terrestrial environments occurring in different oxidation states in the forms of $H_2O$, $H_3O^+$, $OH^-$, $H_2$ and $CH_4$, even at great depths within the Earth. Therefore, hydrogen is envisaged to play a major role, directly or indirectly, in a wide variety of naturally occurring geological processes.

(2) Hydrogen has by far the largest relative mass difference between its stable isotopes, 100%. Consequently, hydrogen exhibits the largest variations in stable isotope ratios of all elements.

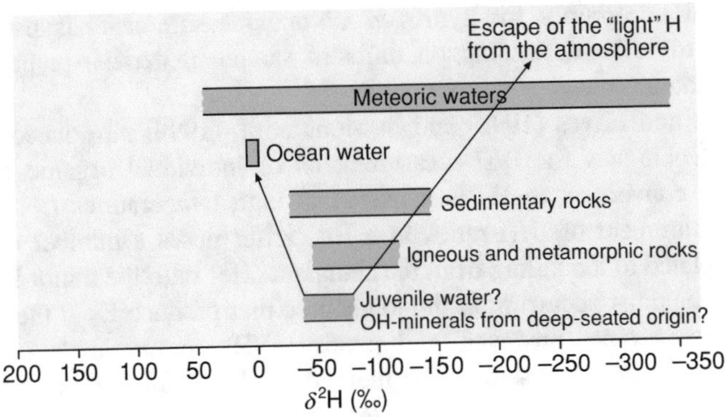


**Fig. 2.1** $\delta^2H$ variation ranges of geologically important reservoirs

The ranges of hydrogen isotope compositions of some geologically important reservoirs are given in Fig. 2.1. It is noteworthy that all rocks on Earth have somewhat similar hydrogen isotope compositions, which is a characteristic feature of hydrogen, but not of the other elements. The reason for this overlap in isotope composition for rocks is likely due to the enormous amounts of water that have been cycled through the outer shell of the Earth.

### 2.1.1 Methods

The D/H ratio of water is measured on $H_2$-gas. There are two different preparation techniques: (i) equilibration of milliliter-sized samples with hydrogen gas, followed by mass-spectrometric measurement and back calculation of the D/H of the equilibrated $H_2$ (Horita 1988). Due to the very large fractionation factor (0.2625 at 25 °C) the measured $H_2$ is very much depleted in D, which complicates the mass-spectrometric measurement. (ii) water is converted to hydrogen by passage over hot metals (uranium: Bigeleisen et al. 1952; Friedman 1953; Godfrey 1962; zinc: Coleman et al. 1982; Vennemann and O'Neil 1993; chromium: Gehre et al. 1996) or glassy carbon (Sharp et al. 2001). This is still the classic method and commonly used.

A difficulty in measuring D/H isotope ratios is that, along with the $H_{2+}$ and $HD^+$ formation in the ion source, $H_{3+}$ is produced as a by-product of ion–molecule collisions. Therefore, a $H_{3+}$ correction has to be made. The amount of formed is directly proportional to the number of $H_2$ molecules and $H^+$ ions. Generally, the current measured for hydrogen from ocean water is on the order of 16% of the total mass 3. The relevant procedures for correction have been evaluated by Brand (2002). The $H_3$ factor is typically measured before each analytical session which is applied for ther following session.

Analytical uncertainty for hydrogen isotope measurements is usually in the range ±0.5 to ±3‰ depending on different sample materials, preparation techniques and laboratories.

Burgoyne and Hayes (1998) and Sessions et al. (1999) introduced the continuous flow techniques for D/H measurements of individual organic compounds. Quantitative conversion to $H_2$ is achieved at high temperatures (>400 °C). The precise measurement of D/H ratios in a He carrier poses a number of analytical problems, related to the tailing from the abundant $^4He^+$ onto the minor $HD^+$ peak as well as on reactions occurring in the ion source that produce $H_{3+}$; these problems have been overcome, however, and precise D/H measurements of individual organic compounds are possible. The method of choice for hydrous minerals and glasses are continuous flow techniques described by Sharp et al. (2001) and Martin et al. (2017).

An alternative to mass-spectrometry represents the direct measurement of D/H, $^{17}O/^{16}O$ and $^{18}O/^{16}O$ isotope compositions of water vapour by laser absorption spectroscopy, also called Cavity Ring-Down Spectroscopy (CRDS) (Kerstel et al. 2002; Brand et al. 2009a, b; Schmidt et al. 2010 and others). The CRDS technique is fast, easy in operation, and allows the direct analysis of hydrogen and oxygen isotopes in water vapour with high precisions comparable to the classic continuous flow techniques (Brand et al. 2009a, b).

SIMS techniques for in-situ measurements of solids have ben described by Shimizu et al. (2019).

## 2.1.2 Standards

There is a range of standards for hydrogen isotopes. The primary reference standard, the zero point of the δ-scale, is V-SMOW, which is virtually identical in isotopic composition with the earlier defined SMOW, being a hypothetical water sample orginally defined by Craig (1961b).

V-SMOW has a D/H ratio that is higher than most natural samples on Earth, thus $\delta^2H$-values in the literature are generally negative. The other standards, listed in Table 2.1, are generally used to verify the accuracy of sample preparation and mass spectrometry.

**Table 2.1** Hydrogen isotope standards

| Standards | Description | $\delta^2H$-value |
|---|---|---|
| V-SMOW | Vienna standard mean | 0 |
| | Ocean water | |
| GISP | Greenland ice sheet | |
| | Precipitation | −189.9 |
| V-SLAP | Vienna standard light | |
| | Antarctic precipitation | −428 |
| NBS-30 | Biotite | 65 |

## 2.1.3 Fractionation Processes

### 2.1.3.1 Water Fractionations

The most effective processes in the generation of hydrogen isotope variations in the terrestrial environment are phase transitions of water between vapor, liquid, and ice through evaporation/precipitation and/or boiling/condensation in the atmosphere, at the Earth's surface, and in the upper part of the crust. Differences in the hydrogen isotope composition arise due to vapor pressure differences of water and, to a smaller degree, to differences in freezing points. Because the vapor pressure of HDO is slightly lower than that of $H_2O$, the concentration of D is lower in the vapor than in the liquid phase. In a simple, but elegant experiment Ingraham and Criss (1998) have monitored the effect of vapor pressure on the rate of isotope exchange between water and vapor, which is shown in Fig. 2.2. Two beakers with isotopically differing waters were juxtaposed in a sealed box to monitor the exchange process at different temperatures (in this case 21 and 52 °C). As shown in Fig. 2.2 in the 52 °C experiment the isotopic composition of the water changes rapidly and nearly reaches equilibrium in only 27 days.

Horita and Wesolowski (1994) have summarized experimental results for the hydrogen isotope fractionation between liquid water and water vapor in the temperature range 0 to 350 °C (see Fig. 2.3). Hydrogen isotope fractionations decrease rapidly with increasing temperatures and become zero at 220 to 230 °C. Above the crossover temperature, water vapor is more enriched in deuterium than liquid water. Fractionations again approach zero at the critical temperature of water (Fig. 2.3).

From experiments, Lehmann and Siegenthaler (1991) determined the equilibrium hydrogen isotope fractionation between ice and water to be +21.2‰. However, under natural conditions ice will not necessarily be formed in isotopic equilibrium with the bulk water, depending mainly on the freezing rate.

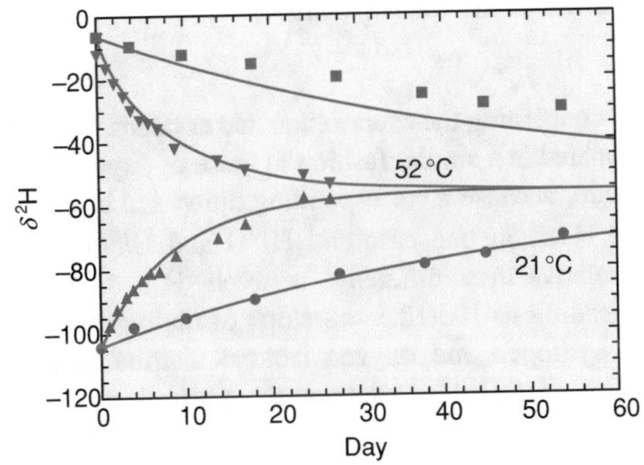


Fig. 2.2 $\delta^2H$-values versus time for two beakers that have equal surface areas and equal volumes undergoing isotopic exchange in sealed systems. In both experiments at 21 and 52 °C isotope ratios progress toward an average value of −56‰ via exchange with ambient vapour: *solid curves* are calculated, *points* are experimental data (after Criss 1999)

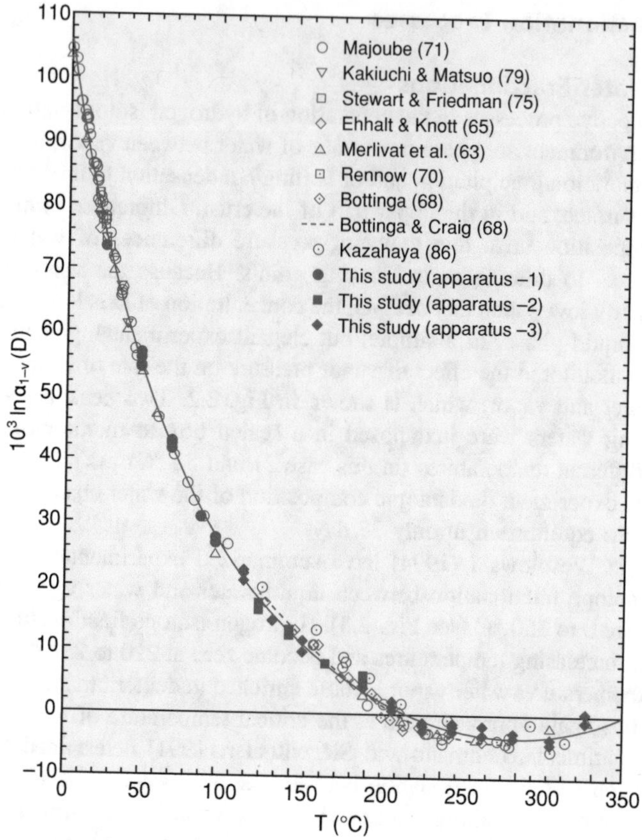


**Fig. 2.3** Experimentally determined fractionation factors between liquid water and water vapour from 1 to 350 °C (after Horita and Wesolowski 1994)

In all processes concerning the evaporation and condensation of water, hydrogen isotopes are fractionated in a similar fashion to those of oxygen isotopes, albeit with a different magnitude, because a corresponding difference in vapor pressures exists between $H_2O$ and HDO in one case and $H_2^{16}O$ and $H_2^{18}O$ in the other. This is explained by the relative mass difference between $^2H$ and $^1H$ (100%) versus $^{18}O$ and $^{16}O$ (12.5%) leading to 100/12.5 = 8 slope of mass-dependent fractionation.

Therefore, the hydrogen and oxygen isotope distributions are correlated for meteoric waters. Craig (1961a) first defined the generalized relationship:

$$\delta^2H = 8\delta^{18}O + 10$$

which describes the interdependence of H$^-$ and O-isotope ratios in meteoric waters on a global scale.

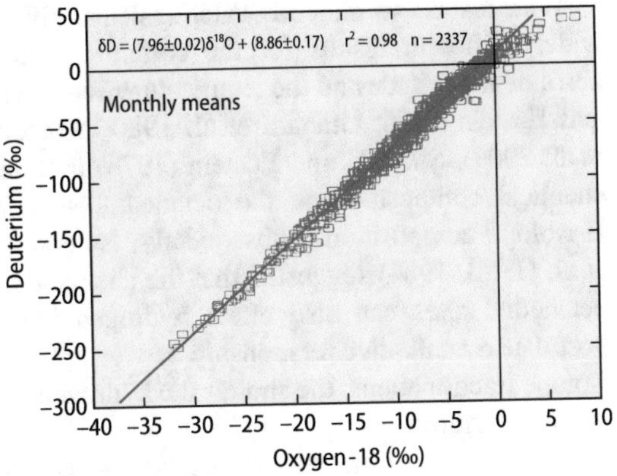


**Fig. 2.4** Global relationship between monthly means of $\delta^2H$ and $\delta^{18}O$ in precipitation, derived for all stations of the IAEA global network. *Line* indicates the global Meteoric Water Line (GMWL) (after Rozanski et al. 1993)

This relationship, shown in Fig. 2.4, is called in the literature the "Global Meteoric Water Line (GMWL)".

Neither the coefficient 8 nor the so-called deuterium excess d of 10 are actually constant in nature. Both may vary due to a superposition of equilibrium and kinetic isotope effects depending on the conditions of evaporation, vapor transport and precipitation and, as a result, offer insight into climatic processes. The deuterium excess d is a valuable tool to derive information on relative humidities (see discussion on p. 242).

#### 2.1.3.2 Equilibrium Reactions

D/H fractionations among gases are extraordinarily large, as calculated by Bottinga (1969) and Richet et al. (1977) and plotted in Fig. 2.5. Even in magmatic systems, fractionation factors are sufficiently large to affect the $\delta^2H$-value of dissolved water in melts during degassing of $H_2$, $H_2S$ or $CH_4$. The oxidation of $H_2$ or $CH_4$ to $H_2O$ and $CO_2$ may also have an effect on the isotopic composition of water dissolved in melts due to the large fractionation factors.

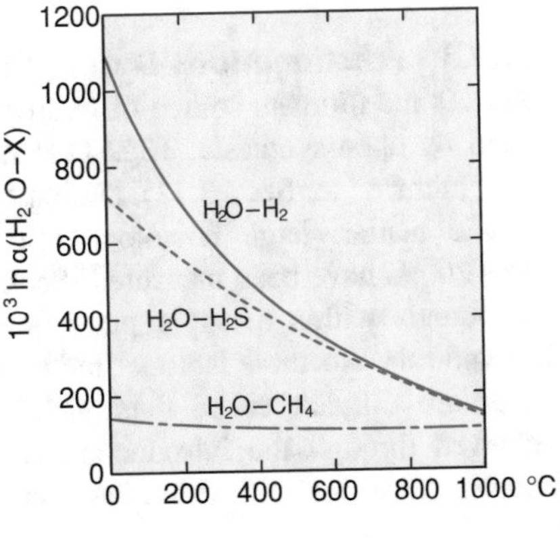


**Fig. 2.5** D/H fractionations between $H_2O$–$H_2$, $H_2O$–$H_2S$ and $H_2O$–$CH_4$ (from calculated data of Richet et al. 1977)

With respect to mineral–water systems, different experimental studies obtained widely different results for the common hydrous minerals with respect to the absolute magnitude and the temperature dependence of D/H fractionations (Suzuoki and Epstein 1976; Graham et al. 1980; Vennemann and O'Neil 1996; Saccocia et al. 2009). Suzuoki and Epstein (1976) first demonstrated the importance of the chemical composition of the octahedral sites in crystal lattices to the mineral H-isotope composition. Subsequently, isotope exchange experiments by Graham et al. (1980, 1984) suggested that the chemical composition of sites other than the octahedral sites can also affect hydrogen isotope compositions. These authors postulate a qualitative relationship between hydrogen-bond distances and hydrogen isotope fractionations: the shorter the hydrogen bond, the more depleted the mineral is in deuterium.

On the basis of theoretical calculations, Driesner (1997) proposed that many of the discrepancies between the experimental studies were due to pressure differences at which the experiments were carried out. Thus for hydrogen, pressure is a variable that must be taken into account in fluid-bearing systems. Later, Horita et al. (1999) also presented experimental evidence for a pressure effect between brucite and water.

Chacko et al. (1999) developed an alternative method for the experimental determination of hydrogen isotope fractionation factors. Instead of using powdered minerals as starting materials, these authors carried out exchange experiments with large single crystals and then analyzed the exchanged rims with the ion probe. Although the precision of the analytical data is less than that for conventional bulk techniques, this technique allows the determination of fractionation factors in experiments in which isotopic exchange occurs by a diffusional process.

In summary, as discussed by Vennemann and O'Neil (1996), discrepancies between published experimental calibrations in individual mineral–water systems are difficult to resolve, which limits the application of D/H fractionations in mineral–water systems to estimate $\delta^2$H-values of coexisting fluids. As shown by Méheut et al. (2010) first-principles calculations of D/H fractionations may reproduce experimental calculations within a range of about 15‰. These authors also demonstrated that internal fractionations between inner-surface and inner hydroxyl groups may be large and even opposite in sign.

#### 2.1.3.3 Fractionations During Biosynthesis

Water is the ultimate source of hydrogen in all naturally organic compounds produced by photosynthesis. Thus D/H ratios in organic matter contain information about climate (see Sect. 3.11). During biosynthetic hydrogen conversion of water to organic matter, large H-isotope fractionations with $\delta^2$H-values between −400 and +200‰ have been measured (Sachse et al. 2012). $\delta^2$H-variations in individual compounds within a single plant or organism can be related to differences in biosynthesis. Accurate isotope fractionation factors among organic molecules and water are difficult to be determined, although tremendous progress has been achieved through the introduction of the compound specific hydrogen isotope analysis (Sessions et al. 1999; Sauer et al. 2001; Schimmelmann et al. 2006), which allows the $\delta^2H$ analysis of individual biochemical compound. Further details are discussed in Sect. 3.10.1.2.

Using a combination of experimental calibration and theoretical calculation Wang et al. (2009a, b) estimated equilibrium factors for various H positions in molecules such as alkanes, ketones, carboxyl acids and alcoholes. By summing over individual H positions, equilibrium fractionations relative to water are −90 to −70‰ for n-alkanes and about −100‰ for pristane and phytane. Wang et al. (2013a, b) extended his approach to cyclic compounds and observed total equilibrium fractionations of −100 to −65‰ for typical cyclic paraffins being similar to linear hydrocarbons. These numbers, however, are very different to typical biosynthetic fractionations that are between −300 and −150‰ due to kinetic isotope fractionations. These data reflect the weighted average of D contents of compounds across all non-equivalent molecular positions.

Xie et al. (2018) developed a method to analyze intramolecular hydrogen isotope variations in hydrocarbons. They used propane as the smallest alkane with chemically non-equivalent hydrogen positions: the central methylene ($-CH_2-$) and the terminal methyl group ($-CH_3-$). As concluded by Xie et al. (2018), hydrogen isotope fractionations between these two positions can be used potentially as a geothermometer.

#### 2.1.3.4 Other Fractionations

In salt solutions, isotopic fractionations can occur between water in the “hydration sphere” and free water (Truesdell 1974). The effects of dissolved salts on hydrogen isotope activity ratios in salt solutions can be qualitatively interpreted in terms of interactions between ions and water molecules, which appear to be primarily related to their charge and radius. Hydrogen isotope activity ratios of all salt solutions studied so far are appreciably higher than H-isotope composition ratios. As shown by Horita et al. (1993), the D/H ratio of water vapor in isotope equilibrium with a solution increases as salt is added to the solution. Magnitudes of the hydrogen isotope effects are in the order $CaCl_2 > MgCl_2 > MgSO_4 > KCl \sim NaCl > NaSO_4$ at the same molality.

Isotope effects of this kind are relevant for the proper interpretation of isotope fractionations in aquous salt solutions and for an understanding of the isotope composition of clay minerals and absorption of water on mineral surfaces. The tendency for clays and shales to act as semipermeable membranes is well known. This effect is also known as “ultrafiltration”. Coplen and Hanshaw (1973) postulated that hydrogen isotope fractionations may occur during ultrafiltration in such a way that the residual water is enriched in deuterium due to its preferential adsorption on the clay minerals and its lower diffusivity.

## 2.2 Carbon

Carbon occurs in a wide variety of compounds on Earth, from reduced organic compounds in the biosphere to oxidized inorganic compounds like $CO_2$ and carbonates. The broad spectrum of carbon-bearing compounds involved in low- and high-temperature geological settings can be assessed on the basis of carbon isotope fractionations.

Carbon has two stable isotopes (Rosman and Taylor 1998)

$^{12}C$ 98.93% (reference mass for atomic weight scale)

$^{13}C$ 1.07%

and radiation produced $^{14}C$ with a half-life of 5700 years.

The naturally occurring variations in carbon isotope composition are greater than 120‰, neglecting extraterrestrial materials. Heavy carbonates with $\delta^{13}C$-values > +20‰ and light methane of < −100‰ have been reported in the literature.

### 2.2.1 Analytical Methods

The gases used in $^{13}C/^{12}C$ measurements are $CO_2$ or CO obtained during pyrolysis. For $CO_2$ the following preparation methods exist:

(a) Carbonates are reacted with 100% phosphoric acid at temperatures between 20 and 90 °C (depending on the type of carbonate) to liberate $CO_2$ (see also "oxygen").

(b) Organic compounds are generally oxidized at high temperatures (850–1050 °C) in a stream of oxygen or by an oxidizing agent like CuO. For the analysis of individual compounds in complex organic mixtures, a gas chromatography—combustion—isotope ratio mass-spectrometry (GC-C-IRMS) system is used, first described by Matthews and Hayes (1978). This device can measure individual carbon compounds in mixtures of sub-nanogram samples with a precision of better than ±0.5‰.

#### 2.2.1.1 Standards

As the commonly used international reference standard PDB (Pee Dee Belemnite) has been exhausted for several decades, there is a need for introducing new standards. Several other standards are in use today, nevertheless published $\delta^{13}C$-values are given relative to the V-PDB-standard (Table 2.2).

Table 2.2 $\delta^{13}C$-values of NBS-reference samples relative to V-PDB

| | | |
|---|---|---|
| NBS-18 | Carbonatite | −5.00 |
| NBS-19 | Marble | +1.95 |
| NBS-20 | Limestone | −1.06 |
| NBS-21 | Graphite | −28.10 |
| NBS-22 | Oil | −30.03 |

### 2.2.2 Fractionation Processes

The two main terrestrial carbon reservoirs, organic matter and sedimentary carbonates, have distinctly different isotopic characteristics because of the operation of two different reaction mechanisms:

(1) Isotope equilibrium exchange reactions within the inorganic carbon system "atmospheric $CO_2$—dissolved bicarbonate or carbonate ions—solid carbonate" lead to an enrichment of $^{13}C$ in carbonates.
(2) Kinetic isotope effects during photosynthesis concentrate the light isotope $^{12}C$ in the synthesized organic material.

#### 2.2.2.1 Carbonate System

The inorganic carbonate system is comprised of multiple chemical species linked by a series of equilibria:

$$CO_{2(aq)} + H_2O \leftrightarrow H_2CO_3 \tag{2.2}$$

$$H_2CO_3 \leftrightarrow H^+ + HCO_3^- \tag{2.3}$$

$$HCO_3^- \leftrightarrow H^+ + CO_3^{2-} \tag{2.4}$$

The carbonate ($CO_3^{2-}$) ion can combine with divalent cations to form solid minerals, calcite and aragonite being the most common

$$Ca^{2+} + CO_3^{2-} = CaCO_3 \tag{2.5}$$

An isotope fractionation is associated with each of these equilibria, the $^{13}C$-differences between the species depend only on temperature, although the relative abundances of the species are strongly dependent on pH. Several authors have reported isotope fractionation factors for the system dissolved inorganic carbon (DIC)—gaseous $CO_2$ (Vogel et al. 1970; Mook et al. 1974; Zhang et al. 1995). The major problem in the experimental determination of the fractionation factor is the separation of the dissolved carbon phases ($CO_{2aq}$, $HCO_3^-$, $CO_3^-$). The generally accepted carbon isotope equilibrium values between calcium carbonate and dissolved bicarbonate are derived from inorganic precipitate data of Rubinson and

Clayton (1969), Emrich et al. (1970), and Turner (1982). What is often not adequately recognized is the fact that systematic C-isotope differences exist between calcite and aragonite. Rubinson and Clayton (1969) and Romanek et al. (1992) found calcite and aragonite to be 0.9 and 2.7‰ enriched in $^{13}C$ relative to bicarbonate at 25 °C. Another complicating factor is that shell carbonate—precipitated by marine organisms—is frequently not in isotopic equilibrium with the ambient dissolved bicarbonate. Such so-called "vital" effects can be as large as a few permil (see discussion on p. 13).

#### 2.2.2.2 Other Equilibrium Isotope Fractionations

Carbon isotope fractionations under equilibrium conditions are important not only at low-temperature, but also at high temperatures within the system carbonate, $CO_2$, graphite, and $CH_4$. Of these, the calcite-graphite fractionation has become a useful geothermometer (e.g., Valley and O'Neil 1981; Scheele and Hoefs 1992; Kitchen and Valley 1995) (see discussion on p. 13). More recently carbon isotope fractionations factors have been determined experimentally in the temperature range from 300 to 1200 °C in the system $CH_4$–$CO_2$–CO (Kueter et al. 2019). A one-phase geothermometer has been described by measuring the clumped isotope composition of methane, in specific that of the two mass-18 components $^{13}CH_3D$ and $^{12}CH_2D_2$ (Douglas et al. 2017; Young et al. 2017). The clumped isotope temperature dependence for methane formed in isotope equilibrium has been empirically calibrated by methane samples with known formation temperatures (see Fig. 2 in Douglas et al. 2017).

Figure 2.6 summarizes carbon isotope fractionations between various geologic materials and gaseous $CO_2$ (after Chacko et al. 2001).

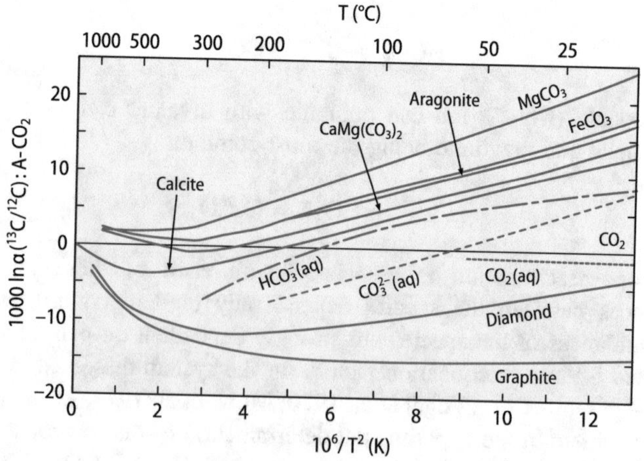


Fig. 2.6 Carbon isotope fractionation between various geologic compounds and $CO_2$ (after Chacko et al. 2001)

#### 2.2.2.3 Organic Carbon System

Early reviews by O'Leary (1981) and Farquhar et al. (1989) have provided the biochemical background of carbon isotope fractionations during photosynthesis, with more recent accounts by Hayes (2001), Freeman (2001) and Galimov (2006). The main isotope-discriminating steps during biological carbon fixation are (i) the uptake and intracellular diffusion of $CO_2$ and (ii) the biosynthesis of cellular components. Such a two-step model was first proposed by Park and Epstein (1960):

$$CO_{2(\text{external})} \leftrightarrow CO_{2(\text{internal})} \rightarrow \text{organic molecule}$$

From this simplified scheme, it follows that the diffusional process is reversible, whereas the enzymatic carbon fixation is irreversible. The two-step model of carbon fixation clearly suggests that isotope fractionation is dependent on the partial pressure of $CO_2$ of the system. With an unlimited amount of $CO_2$ available to a plant, the enzymatic fractionation will determine the isotopic difference between the inorganic carbon source and the final bioproduct. Under these conditions, $^{13}C$ fractionations may vary from −17 to −40‰ (O'Leary 1981). When the concentration of $CO_2$ is the limiting factor, the diffusion of $CO_2$ into the plant is the slow step in the reaction and carbon isotope fractionation of the plant decreases.

Atmospheric $CO_2$ first moves through the stomata, dissolves into leaf water and enters the outer layer of photosynthetic cells, the mesophyll cell. Mesophyll $CO_2$ is directly converted by the enzyme ribulose biphosphate carboxylase/oxygenase ("Rubisco") to a 6 carbon molecule, that is then cleaved into 2 molecules of phosphoglycerate (PGA), each with 3 carbon atoms (plants using this photosynthetic pathway are therefore called $C_3$ plants). Most PGA is recycled to make ribulose biphosphate, but some is used to make carbohydrates. Free exchange between external and mesophyll $CO_2$ makes the carbon fixation process less efficient, which causes the observed large $^{13}C$-depletions of $C_3$ plants.

$C_4$ plants incorporate $CO_2$ by the carboxylation of phosphoenolpyruvate (PEP) via the enzyme PEP carboxylase to make the molecule oxaloacetate which has 4 carbon atoms (hence $C_4$). The carboxylation product is transported from the outer layer of mesophyll cells to the inner layer of bundle sheath cells, which are able to concentrate $CO_2$, so that most of the $CO_2$ is fixed with relatively little carbon fractionation (see Fig. 2.7).

The final carbon isotope composition of naturally synthesized organic matter depends on a complex set of parameters. (i) the $^{13}C$-content of the carbon source, (ii) isotope effects associated with the assimilation of carbon, (iii) isotope effects associated with metabolism and biosynthesis and (iv) cellular carbon budgets (Hayes 1993, 2001; Galimov 2006).

Even more complex is C-isotope fractionation in aquatic plants. Factors that control the $\delta^{13}C$ of phytoplankton include temperature, availability of $CO_{2(aq)}$, light intensity, nutrient availability, pH and physiological factors such as cell size and growth rate (Laws et al. 1995, 1997; Bidigare et al. 1997; Popp et al. 1998 and others). In particular the relationship between the C-isotope composition of

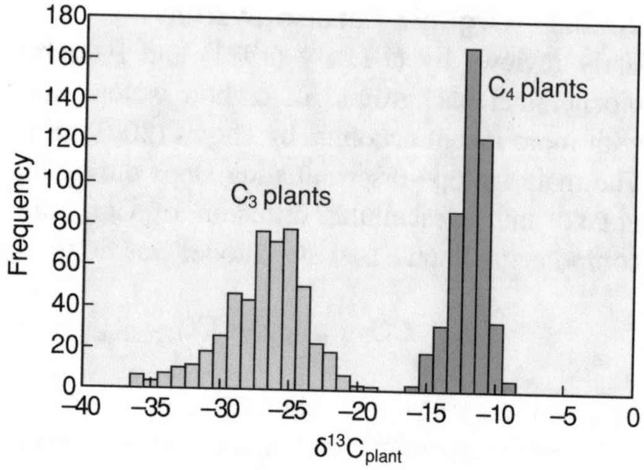


**Fig. 2.7** Histogram of $\delta^{13}C$ values of $C_3$ and $C_4$ plants (after Cerling and Harris 1999)

phytoplankton and the concentration of oceanic dissolved $CO_2$ has been subject of considerable debate because of its assumed potential as a palaeo-$CO_2$ barometer (see discussion p. 400).

Since the pioneering work of Park and Epstein (1960) and Abelson and Hoering (1961), it is well known that $^{13}C$ is not uniformly distributed among the total organic matter of plant material, but varies between carbohydrates, proteins and lipids. The latter class of compounds is considerably depleted in $^{13}C$ relative to the other products of biosynthesis. Although the causes of these $^{13}C$-differences are not entirely clear, kinetic isotope effects seem to be more plausible (De Niro and Epstein 1977; Monson and Hayes 1982) than thermodynamic equilibrium effects (Galimov 1985, 2006). The latter author argued that $^{13}C$-concentrations at individual carbon positions within organic molecules are principally controlled by structural factors. Approximate calculations suggested that reduced C–H bonded positions are systematically depleted in $^{13}C$, while oxidized C–O bonded positions are enriched in $^{13}C$. Many of the observed relationships are qualitatively consistent with that concept. However, it is difficult to identify any general mechanism by which thermodynamic factors should be able to control chemical equilibrium within a complex organic structure. Experimental evidence presented by Monson and Hayes (1982) suggests that kinetic effects will be dominant in most biological systems.

High resolution gas source mass spectrometry opens the possibility to determine the isotope composition of a particular atom in a molecule (Eiler et al. 2014; Piasecki et al. 2016). Propane as the simplest organic molecule that could record site-specific carbon isotope variations shows isotope variations that can be related with the maturity of the precursor material (Piasecki et al. 2018).

#### 2.2.2.4 Interactions Between Carbonate-Carbon and Organic Carbon

Variations in $^{13}C$ content of some important carbon compounds are schematically demonstrated in Fig. 2.8: The two most important carbon reservoirs in the crust, marine carbonates and the biogenic organic matter, are characterized by very

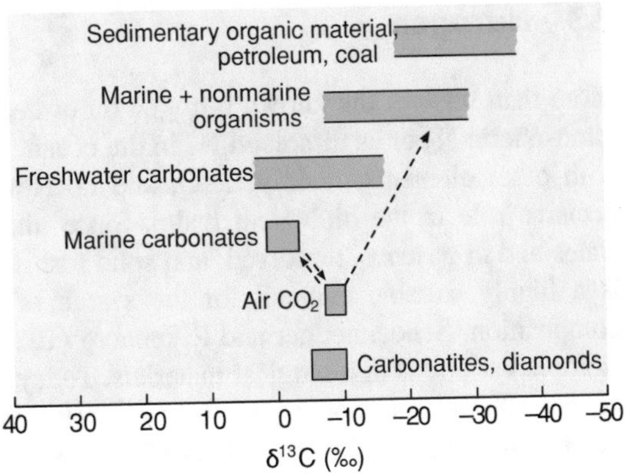


**Fig. 2.8** $\delta^{13}$C-values of important geological reservoirs

different isotopic compositions: the carbonates being isotopically heavy with a mean $\delta^{13}$C-value around 0‰ and organic matter being isotopically light with a mean $\delta^{13}$C-value around −25‰. For these two sedimentary carbon reservoirs a binary isotope mass balance must exist such that:

$$\delta^{13}C_{input} = f_{org}\delta^{13}C_{org} + (1 - f_{org})\delta^{13}C_{carb} \tag{2.6}$$

If $\delta_{input}$, $\delta_{org}$, $\delta_{carb}$ can be determined for a specific geologic time, $f_{org}$ can be calculated, where $f_{org}$ is the fraction of organic carbon entering the sediments. It should be noted that $f_{org}$ is defined in terms of the global mass balance and is independent of biological productivity referring to the burial rather than the synthesis of organic matter. That means that large $f_{org}$ values might be a result of particular high productivity of organic matter or of particular high levels of organic matter preservation.

The $\delta^{13}$C-value for the input carbon cannot be measured precisely but can be estimated with a high degree of certainty. As will be shown later, mantle carbon has an isotopic composition around −5‰ and estimates of the global average isotope composition for crustal carbon also fall in that range. Assigning −5‰ to $\delta^{13}$C-input, a modern value for $f_{org}$ is calculated as 0.2 or expressed as the ratio of $C_{org}/C_{carb}$ = 20/80. As will be shown later (Sect. 3.7.2) $f_{org}$ has obviously changed during specific periods of the Earth's history affecting $\delta^{13}$C of carbonates and organic matter according to Eq. (2.6) (e.g., Hayes et al. 1999). With each molecule of organic carbon being buried, a mole of oxygen is released to the atmosphere. Hence, knowledge of $f_{org}$ is of great value in reconstructing the crustal redox budget.

## 2.3 Nitrogen

More than 99% of the known nitrogen on or near the Earth's surface is present as atmospheric $N_2$ or as dissolved $N_2$ in the ocean. Only a minor amount is combined with other elements, mainly C, O, and H. Nevertheless, this small part plays a decisive role in the biological realm. Since nitrogen occurs in various oxidation states and in gaseous, dissolved, and solid forms ($N_2$, $NO_3^-$, $NO_2^-$, $NH_3$, $NH_4^+$), it is a highly suitable element for the search of natural variations in its isotopic composition. Schoenheimer and Rittenberg (1939) were the first to report nitrogen isotopic variations in biological materials. Today, the range of reported $\delta^{15}N$-values covers 100‰, from about −50 to +50‰. However, most δ-values fall within the much narrower spread from −10 to +20‰, as described in reviews of the exogenic nitrogen cycle by Heaton (1986), Owens (1987), Peterson and Fry (1987) and Kendall (1998).

Nitrogen consists of two stable isotopes, $^{14}N$ and $^{15}N$. Atmospheric nitrogen, given by Rosman and Taylor (1998) has the following composition:

$$^{14}N \quad 99.63\%$$
$$^{15}N \quad 0.37\%.$$

### 2.3.1 Analytical Methods

$N_2$ is used for $^{15}N/^{14}N$ isotope ratio measurements, the standard is atmospheric $N_2$. Various preparation procedures have been described for the different nitrogen compounds (Bremner and Keeney 1966; Owens 1987; Velensky et al. 1989; Kendall and Grim 1990, and others). In the early days of nitrogen isotope studies, the extraction and combustion techniques potentially involved chemical treatments that could have introduced isotopic fractionations. More recently, simplified techniques for combustion have come into routine use, so that a precision of 0.1–0.2‰ for $\delta^{15}N$ determinations can be achieved. Organic nitrogen-compounds are combusted to $CO_2$, $H_2O$ and $N_2$ in an elemental analyzer. The cryogenically purified $N_2$ is trapped on molecular sieves for analysis.

More recently methods have been described that are based on the isotope analysis of $N_2O$. Measurements of bulk $\delta^{15}N$-values yield qualitative rather quantitative information on the nitrogen cycle, special techniques are necessary for a separate analysis of nitrate and nitrite in samples containing both species. Sigman et al. (2001) measured $N_2O$ generated by denitrifying bacteria lacking $N_2O$ reductase. McIlvin and Altabet (2005) introduced an alternative approach of the bacteria method. Nitrate is first reduced with a Cd catalyst to nitrite followed by sodium azide treatment in order to reduce nitrite to $N_2O$. This method allows sequential analysis of nitrate and nitrite, but azide is toxic and has to be handled with great care.

Compound-specific analysis of amino acids has been described by McClelland and Montoya (2002) studying 16 amino acids in planktonic consumers and their food sources. Some amino acids, like glutamate and aspartate, show $^{15}$N-enrichments with increased trophic level, while others like phenylalamine, serine and threonine record the N-isotope composition of the system in which organism exist. $^{15}$N differences between the two groups can be attributed to differences in metabolic pathways.

Different preparation techniques have been used for nitrogen in mantle derived samples with N-concentrations being too low to be analysed by conventional techniques. For these samples, static mass spectrometry, in which the gas is left under static conditions in the ion source, a method developed for noble gas analysis and adopted for nitrogen, has been used. As an alternative, Bebout et al. (2007) described a continuous flow technique for nanomole quantities of nitrogen.

### 2.3.2 Biological Nitrogen Isotope Fractionations

To understand the processes leading to the nitrogen isotope distribution in the geological environment, a short discussion of the biological nitrogen cycle is required. Atmospheric nitrogen, the most abundant form of nitrogen, is the least reactive species of nitrogen. It can, however, be converted to "fixed" nitrogen by bacteria and algae, which, in turn, can be used by biota for degradation to simple nitrogen compounds such as ammonium and nitrate. Thus, microorganisms are responsible for all major conversions in the biological nitrogen cycle, which generally is divided into fixation, nitrification, and denitrification. Other bacteria return nitrogen to the atmosphere as $N_2$.

The term *fixation* is used for processes that convert unreactive atmospheric $N_2$ into reactive nitrogen such as ammonium, usually involving bacteria. Fixation commonly produces organic materials with $\delta^{15}$N-values slightly less than 0‰ ranging from –3 to +1 (Fogel and Cifuentes 1993) and occurs in the roots of plants by symbiontic bacteria.

*Nitrification* is a multi-step oxidation process mediated by several different autotrophic organisms. Nitrate is not the only product of nitrification, different reactions produce various nitrogen oxides as intermediate species. Nitrification can be described as two partial oxidation reactions, each of which proceeds separately: oxidation by Nitrosomas ($NH_4 \rightarrow NO_2^-$) followed by oxidation by Nitrobacter ($NO_2 \rightarrow NO_3$). Because the oxidation of nitrite to nitrate is generally rapid, most of the N-isotope fractionations is caused by the slow oxidation of ammonium by Nitrosomas. However, as shown by Casciotti (2009) the second oxidation step from nitrite to nitrate is accompanied by an inverse kinetic isotope fractionation, such that nitrite becomes progressively depleted in $^{15}$N as the oxidation reaction proceeds.

*Denitrification* (reduction of more oxidized forms to more reduced forms of nitrogen) is a multi-step process with various nitrogen oxides as intermediate compounds resulting from biologically mediated reduction of nitrate.

Table 2.3 Nitrogen isotope fractionations for microbial cultures (after Casciotti 2009)

| | | |
|---|---|---|
| $N_2$ fixation | $N_2 \rightarrow N_{org}$ | −2 to +2‰ |
| $NH_4^+$ assimilation | $NH_4^+ \rightarrow N_{org}$ | +14 to +27‰ |
| $NH4^+$ oxidation (nitrification) | $NH_4^+ \rightarrow NO_2^-$ | +14 to 38‰ |
| Nitrite oxidation (nitrification) | $NO_2^- \rightarrow NO_3^-$ | −12.8‰ |
| Nitrate reduction (denitrification) | $NO_3^- \rightarrow NO_2^-$ | +13 to +30‰ |
| Nitrite reduction (denitrification) | $NO_2^- \rightarrow NO$ | +5 to +10‰ |
| Nitrous oxide reduction (denitrification) | $N_2O \rightarrow N_2$ | +4 to +13‰ |
| Nitrate reduction (nitrate assimilation) | $NO_3^- \rightarrow NO_2^-$ | +5 to +10 |

Denitrification takes place in poorly aerated soil and in suboxic water bodies, especially in oxygen minimum zones of the ocean. There is debate about the relative contributions of denitrification in sediments versus in the ocean. Denitrification supposedly balances the natural fixation of nitrogen, if it did not occur, then atmospheric nitrogen would be exhausted in less than 100 million years. Denitrification causes the $\delta^{15}N$-values of the residual nitrate to increase exponentially as nitrate concentrations decrease. Experimental investigations have demonstrated that fractionation factors may change from 10 to 30‰, with the largest values obtained under lowest reduction rates. Nitrogen isotope fractionations during denitrification in the ocean involves a greater fractionation than in sediments. Table 2.3 gives a summary of measured N-isotope fractionations.

Noteworthy is the inverse kinetic fractionation during nitrite oxidation, which is different from all other microbial processes in which N-isotope fractionation is involved. Casciotti (2009) argued that the inverse fractionation effect is due to the reverse reaction at the enzyme level.

One very important finding in the nitrogen cycle is the discovery of anaerobic ammonium oxidation, briefly called anammox, a dissimilatory process involving the reaction of ammonia with nitrite

$$NH_4^+ + NO_2^- \rightarrow N_2 + 2\ H_2O$$

which has been demonstrated using sediment incubations (Thamdrup and Dalsgaard 2002) and later shown to be the major N-loss process in oxygen minimum zone waters (i.e. Lam et al. 2009). Nitrogen isotope fractionations during anammox appear to be very similar to denitrifying bacteria. Another less well explored nitrogen pathway is the dissimilatory nitrate reduction to ammonium, which may provide the ammonium utilized by anammox bacteria (Lam et al. 2009). Nitrogen isotope fractionations are not well known, but may be comparable to denitrifying bacteria.

So far, only kinetic isotope effects have been considered, but isotopic fractionations associated with equilibrium exchange reactions have been demonstrated for the common inorganic nitrogen compounds (Letolle 1980). Of special importance in this respect is the ammonia volatilization reaction:

$$NH_{3gas} \leftrightarrow NH_4^+ aq$$

for which isotope fractionation factors of 1.025–1.035 have been determined (Kirshenbaum et al. 1947; Mariotti et al. 1981). Experimental data by Nitzsche and Stiehl (1984) indicate fractionation factors of 1.0143 at 250 °C and of 1.0126 at 350 °C. During the solution of atmospheric $N_2$ in ocean water, a very small $^{15}N$-enrichment of about 0.1‰ occurs (Benson and Parker 1961).

### 2.3.3 Trophic Level Indicator

The determination of the food source and trophic position of an organism is fundamental in food web science. Nitrogen isotopes have been very useful to quantify trophic structure. Typically, a $^{15}N$ enrichment of 2.5 to 3.4‰ is observed from diet to consumer (Peterson and Fry 1987; Hedges and Reynard 2007; Post 2002 and others). The enrichment results from amination- and de-animation reactions in the consumer. The N isotopic composition of a consumer alone is generally not sufficient to infer the trophic position without an isotopic baseline. If the $\delta^{15}N$-values of the primary producer at the base of the food web and of the consumer are known, then the trophic level can be calculated based on the number of diet-consumer fractionations.

To investigate past food and trophic positions, bulk collagen $\delta^{15}N$-values are generally analyzed which represent average values of the constituent amino acids. Compound-specific analysis of individual amino acids help to clarify the complex metabolic processes (McMahon and McCarthy 2016; O'Connell 2017; Whiteman et al. 2019). This analytical approach relies on the observation that N isotopes in source amino acids remain more or less unaltered during assimilation by consumers, whereas trophic amino acids are biochemically transformed becoming enriched in $^{15}N$. The most used amino acid pairing is *trophic* glutamic acid vs *source* phenylalanine. Since trophic amino acids experience nitrogen isotope fractionation at each trophic level, their $\delta^{15}N$-values increase with trophic steps and consequently larger $\Delta^{15}N_{(glu\text{-}phe)}$ offsets indicate higher trophic positions.

### 2.3.4 Nitrogen Isotopes in the Mantle and Crust

Nitrogen is generally regarded as a volatile element with chemical similarities to noble gases. For a long time, the dominant nitrogen reservoir was regarded to be the atmosphere, which is true, if only the Earth's surface is considered. Budget estimates of N for the earth as a whole indicate, however, that the dominant reservoir is in the mantle. The average content and speciation of nitrogen in the mantle is poorly constrained. Estimates for average concentrations vary between 0.3 and 36 ppm (Busigny and Bebout 2013).

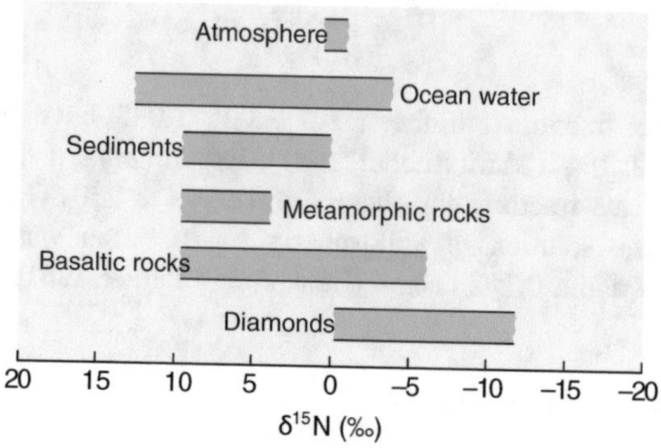


**Fig. 2.9** $\delta^{15}$N-values of important geological reservoirs

Mantle nitrogen extracted from MORB glasses (Marty and Humbert 1997; Marty and Zimmermann 1999) and from diamonds (Javoy et al. 1986; Cartigny et al. 1997, Cartigny 2005; Cartigny and Marty 2013) has an average $\delta^{15}$N-value of around −5‰ with considerable scatter. Nitrogen isotope values extracted from peridotite xenoliths and mineral separates show large variations with phlogopites being depleted and clinopyroxene and olivine being enriched in $^{15}$N (Yokochi et al. 2009). Positive $\delta^{15}$N values measured in some MORB samples may reflect the occurrence of subducted nitrogen.

In the crust, during metamorphism of sediments, there is a significant loss of ammonium during devolatilisation, which is associated with a nitrogen fractionation, leaving behind $^{15}$N-enriched residues (Haendel et al. 1986; Bebout and Fogel 1992; Jia 2006; Plessen et al. 2010). Thus, high-grade metamorphic rocks and granites are relatively enriched in $^{15}$N and typically have $\delta^{15}$N-values between 8 and 10‰. Sadofsky and Bebout (2000) have examined the nitrogen isotope fractionation among coexisting micas, but could not find any characteristic difference between biotite and white mica.

In summary, nitrogen in sediments and crustal rocks has positive $\delta^{15}$N-values around 6‰, whereas in mantle-derived rocks $\delta^{15}$N-values are around −5‰.

Figure 2.9 gives an overview about the nitrogen isotope variations in some important reservoirs.

### 2.3.5 Nitrogen in the Ocean and in Sediments

Nitrogen isotope studies may evaluate the source and fate of nitrogen in the ocean (Casciotti 2016). Nitrogen in the ocean is present in different redox states (nitrate, nitrite, ammonium). Biological processes in the water column may transform one nitrogen compound to the other, which is associated with N-isotope fractionations. Nitrogen fixation is regarded as the dominant process for primary production that causes little N isotope fractionation. Thus, nitrogen produced by this process should have a $\delta^{15}N$-value close to zero. However, average oceanic $\delta^{15}N$ is close to 5‰ as measured in nitrate, the N-isotope enrichment resulting from denitrification. Denitrification occurring in oxygen depleted zones preferentially reduces $^{14}N$, the remaining nitrate thus becomes progressively enriched in $^{15}N$. Upwelling of such $^{15}N$ enriched water masses causes the production of relatively $^{15}N$-rich phytoplankton particles that sink to the seafloor. The nitrogen isotope composition of sedimentary organic material, thus, can serve as an indicator of water column nitrogen reactions and of nutrient dynamics (e.g. Farrell et al. 1995; Adler et al. 2016).

Nitrogen isotopes in particulate organic nitrogen depends on (i) the isotopic composition of dissolved nitrate, on (ii) isotope fractionation that occurs during nitrogen uptake by phytoplankton and (iii) on the relative contribution of nitrate versus $N_2$-assimilating primary producers. In the photic zone phytoplankton preferentially incorporates $^{14}N$, which results in a corresponding $^{15}N$-enrichment in the residual nitrate. The N-isotope composition of settling organic detritus, thus varies depending on the extent of nitrogen utilization: low $^{15}N$ contents indicate low relative utilization, high $^{15}N$ contents indicate a high utilization.

Denitrification is believed to be enhanced in interglacial times compared to glacial times. Ganeshram et al. (2000) showed that $\delta^{15}N$ values during interglacials are about 2 to 3‰ heavier than $\delta^{15}N$ values during glacial times. This relationship has been used as a recorder of paleoproductivity.

Nitrogen isotopes in marine sediments, thus, may reflect nutrient cycles of ancient oceans. However, diagenetic reactions on the seafloor and deeper in the sediments may alter the primary nitrogen isotope composition. Nevertheless, Tesdall et al. (2013) argue that although diagenetic effects have to be considered, diagenesis is a secondary effect and therefore bulk sedimentary nitrogen isotope records from the seafloor and subseafloor sediments monitor past changes of the marine nitrogen cycle. They presented a global database of more than 2300 bulk sediment $\delta^{15}N$ measurements and demonstrated that $\delta^{15}N$-values range from 2.5 to 16.6‰ with a mean value of 6.7‰, which is higher than the average 5‰ of nitrate in the ocean (http://www.ncdc.noaa.gov/paleo/pubs/nicopp/nicopp.html).

Kast et al. (2019) presented an age record of foraminifera shell-bound nitrogen from 70 to 30 Ma (see Fig. 1 by Kast et al. 2019). In the Paleocene, $\delta^{15}N$ values were high, suggesting expanded suboxia and denitrification. Between 57 and 50 Ma ago, $\delta^{15}N$ values declined drastically and from 50 to 35 Ma $\delta^{15}N$ was lower than modern values.

In recent years there has been increasing interest to reconstruct the biogeochemical nitrogen cycle over geologic time (Ader et al. 2016; Stüeken et al. 2016). On the basis of the limited number of nitrogen isotope data, it has been concluded that abiotic nitrogen sources were insufficient to sustain a large biosphere, favoring an early origin of biological nitrogen fixation and supporting current views that nitrogen isotopes may also indicate a stepwise oxidation of the hydro- and atmosphere.

For a long time, denitrification was thought to be the only mechanism that reduces nitrate to $N_2$. The anaerobic ammonia oxidation (anammox) is another mechanism in which bacteria use ammonium to convert nitrite to $N_2$. Brunner et al.

(2013) demonstrated that the magnitude of N isotope fractionation associated with the anammox reaction fall in the same range as denitrification. They further showed that anammox may be responsible for the large fractionations between nitrate and nitrite in oxygen minimum zones.

In sediments, with increasing thermal degradation of organic matter, ammonium ($NH_4$) is liberated which can replace potassium in clay minerals. The nitrogen in the crystal lattice of clay minerals and micas is, thus, derived from decomposing organic matter reflecting the N-isotope composition of its source (Williams et al. 1995).

## 2.4 Oxygen

Oxygen is the most abundant element on Earth. It occurs in gaseous, liquid and solid compounds, most of which are thermally stable over large temperature ranges. These facts make oxygen one of the most interesting elements in isotope geochemistry.

Oxygen has three stable isotopes with the following abundances (Rosman and Taylor 1998)

$^{16}O$ 99.757%
$^{17}O$ 0.038%
$^{18}O$ 0.205%

Because of the higher abundance and the greater mass difference, the $^{18}O/^{16}O$ ratio is normally determined, which may vary in natural samples by about 10% or in absolute numbers from about 1: 475–1: 525. More recently, with improved analytical techniques, the precise measurement of the $^{17}O/^{16}O$ ratio also became of interest (see p. 13).

### 2.4.1 Analytical Methods

$CO_2$ is the gas generally used for mass-spectrometric analysis. CO and $O_2$ have also been used in high temperature conversion of organic material and in laser-based preparation techniques. For the determination of the $^{17}O$ content $O_2$ has to be used. A wide variety of methods have been described to liberate oxygen from the various oxygen-containing compounds.

#### 2.4.1.1 Water

The $^{18}O/^{16}O$ ratio of water is usually determined by equilibration of a small amount of $CO_2$ with a surplus of water at a constant temperature. For this technique, the exact value of the fractionation for the $CO_2/H_2O$ equilibrium at a given temperature is of crucial importance. A number of authors have experimentally determined this fractionation at 25 °C with variable results. A value of 1.0412 was proposed at the 1985 IAEA Consultants Group Meeting to be the best estimate.

It is also possible to quantitatively convert all water oxygen directly to $CO_2$ by reaction with guanidine hydrochloride (Dugan et al. 1985) which has the advantage that it is not necessary to assume a value for the $H_2O$–$CO_2$ isotope fractionation in order to obtain the $^{18}O/^{16}O$ ratio. Sharp et al. (2001) described a technique reducing $H_2O$ by reaction with glassy carbon at 1450 °C. O'Neil and Epstein (1966) first described the reduction of water with $Br_5F$. For the precise measurement of $^{17}O$ and $^{18}O$ the method later was modified using $CoF_3$ (Baker et al. 2002; Barkan and Luz 2005). As mentioned under Sect. 2.1.1, an alternative method to mass spectrometry is the direct determination of oxygen isotope ratios by CRDS (Brand et al. 2009a, b and others), which has the advantage of being applicable directly in the field and has high temporal resolution, that can be useful for isotope hydrologic studies. Steig et al. (2014) developed a CRDS instrument that allows the precise measure ment of the $^{17}O/^{16}O$ ratio.

#### 2.4.1.2 Carbonates

The standard procedure for the isotope analysis of carbonates is the reaction with 100% phosphoric acid at 25 °C first described by McCrea (1950). The following reaction equation:

$$MeCO_3 + H_3PO_4 \rightarrow MeHPO_4 + CO_2 + H_2O$$

where Me is a divalent cation, shows that only two-thirds of the carbonate oxygen present is in the product $CO_2$, which carries a significant isotope effect on the order of 10‰, but varies by up to a few‰ depending on the cation, the reaction temperature and the preparation procedure. The so-called acid fractionation factor must be precisely known to back-calculate the oxygen isotope ratio of the carbonate. This can be done by measuring the $\delta^{18}O$-value of the carbonate by fluorination with $BrF_5$, first described by Sharma and Clayton (1965).

Experimental details of the phosphoric acid method vary significantly among different laboratories. The two most common varieties are the "sealed vessel" and the "acid bath" methods. In the latter method the $CO_2$ generated is continuously removed, while in the former it is not. Swart et al. (1991) demonstrated that the two methods have a systematic $^{18}O$ difference between 0.2 and 0.4‰ over the temperature range 25 to 90 °C. Of these the "acid-bath" method probably provides the more accurate results. A further modification of this technique is referred to as the "individual acid bath", in which contaminations from the acid delivery system are minimized. Wachter and Hayes (1985) demonstrated that careful attention must be given to the phosphoric acid. In their experiments best results were obtained by using a 105% phosphoric acid and a reaction temperature of 75 °C. This high reaction temperature should not be used when attempting to discriminate between mineralogically distinct carbonates by means of differential carbonate reaction rates.

Modern carbonate methods include an on-line gas preparation and introduction system (gas chromatography-based Gas-Bench and Kiel-Device) that involve automatic acidification of 0.1–0.4 mg of carbonates under thermally controlled constant conditions.

Because some carbonates like magnesite or siderite react very sluggishly at 25 °C, higher reaction temperatures are necessary to extract $CO_2$ from these minerals. Reaction temperatures have varied up to 90 or even 150 °C (Rosenbaum and Sheppard 1986; Böttcher 1996), but there still exist considerable differences in the fractionation factors determined by various workers. Crowley (2010) showed that for minerals of the $CaCO_3$–$MgCO_3$ group the oxygen isotope composition of $CO_2$ is a linear function of the reciprocal of reaction temperature. Deviations from this relationship may be attributed to structural state and differences in chemical composition.

Another uncertainty exists for fractionations between aragonite and calcite. Different workers have reported fractionations from negative to positive. Nevertheless, there seems to be a general agreement that the fractionation factor for aragonite is about 0.6‰ higher than for calcite (Tarutani et al. 1969; Kim and O'Neil 1997), although Grossman and Ku (1986) have reported a value of up to 1.2‰. The dolomite-calcite fractionation may vary depending on specific composition (Land 1980). Table 2.4 reports acid fractionation factors for various carbonates.

The measurement of the $^{17}O$ composition of $CO_2$ requires additional analytical steps due to the isobaric interference of $^{13}C^{16}O^{16}O$ and $^{12}C^{17}O^{16}O$. Different fluorination techniques have been described for the liberation of oxygen from $CO_2$ (Brenninkmeijer and Rockmann 1998, Hofmann and Pack 2010, and others). Mahata et al. (2013) determined the $^{17}O$ content of $CO_2$ by analyzing $O_2$ that has been exchanged with $CO_2$ over hot platinum. An alternative technique has been reported by Adnew et al. (2019), analyzing fragment ions of $CO_2$ with an ultra-high resolution mass spectrometer.

#### 2.4.1.3 Silicates

Oxygen in silicates and oxides is usually liberated through fluorination with $F_2$, $BrF_5$ or $ClF_3$ in nickel-tubes at 500–650 °C (Taylor and Epstein 1962; Clayton and Mayeda 1963; and Borthwick and Harmon 1982) or by heating with a laser (Sharp 1990). Decomposition by carbon reduction at 1000–2000 °C may be suitable for

**Table 2.4** Acid fractionation factors for various carbonates determined at 25 °C (modified after Kim et al. 2007)

| Mineral | α | References |
|---|---|---|
| Calcite | 10.30 | Kim et al. (2007) |
| Aragonite | 10.63 | Kim et al. (2007) |
| | 11.14 | Gilg et al. (2007) |
| Dolomite | 11.75 | Rosenbaum and Sheppard (1986) |
| Magnesite | 10.79 (50 °C) | Das Sharma et al. (2002) |
| Siderite | 11.63 | Carothers et al. (1988) |
| Witherite | 10.57 | Kim and O'Neil (1997) |

quartz and iron oxides but not for all silicates (Clayton and Epstein 1958). The oxygen is converted to $CO_2$ over heated graphite or diamond. For an analysis of the three isotope ($^{16}O$, $^{17}O$, $^{18}O$) $O_2$ has to be the analyte gas. Care must be taken to ensure quantitative oxygen yields, which can be a problem in the case of highly refractive minerals like olivine and garnet. Low yields may result in anomalous $^{18}O/^{16}O$ ratios, high yields are often due to excess moisture in the vacuum extraction line.

Today, infrared-laser fluorination, first described by Sharp (1990), most commonly is used for mineral analysis. Alternatively, UV lasers have been used by Wiechert and Hoefs (1995) and Wiechert et al. (2002). A precise SIMS method with a reproducibility of better than 0.3‰ from 15 μm mineral spots has been described by Kita et al. (2009).

#### 2.4.1.4 Phosphates

Phosphates are first dissolved, then precipitated as silver phosphate (Crowson et al. 1991). $Ag_3PO_4$ is preferred because it is non-hygroscopic and can be precipitated rapidly without numerous chemical purification steps (O'Neil et al. 1994). This $Ag_3PO_4$ is then fluorinated (Crowson et al. 1991), reduced with C either in a furnace (O'Neil et al. 1994) or with a laser (Wenzel et al. 2000) or pyrolyzed (Vennemann et al. 2002). Because $PO_4$ does not exchange oxygen with water at room temperature (Kolodny et al. 1983), the isotopic composition of the $Ag_3PO_4$ is that of the $PO_4$ component of the natural phosphate. As summarized by Vennemann et al. (2002) conventional fluorination remains the most precise and accurate analytical technique for $Ag_3PO_4$. Laser techniques on bulk materials have also been attempted (Cerling and Sharp 1996; Kohn et al. 1996; Wenzel et al. 2000); but because fossil phosphates invariably contain oxygen in other sites (carbonate ion or hydroxyl ion) that may also be more prone to diagenetic changes, chemical processing and analysis of a specific component ($CO_3^{2-}$ or $PO_4^{2-}$) has to be orderly performed.

#### 2.4.1.5 Sulfates

Sulfates are precipitated as $BaSO_4$, and then reduced with carbon at 1000 °C to produce $CO_2$ and CO. The CO is either measured directly or converted to $CO_2$ by electrical discharge between platinum electrodes (Longinelli and Craig 1967). Total pyrolysis by continuous flow methods has made the analysis of sulfate oxygen more precise and less time-consuming than the off-line methods.

#### 2.4.1.6 Nitrates

Oxygen isotopes in nitrate may be measured by high-temperature combustion with graphite (Revesz et al. 1997). Since this method is labour-intensive, Sigman et al. (2001) used cultured denitryfing bacteria for the reduction of nitrate. In the analyzed $N_2O$ only one of six oxygen atoms present in the initial nitrate will be measured, therefore potential oxygen isotope fractionations must be adequately taken into account (Casciotti et al. 2002).

**Table 2.5** $\delta^{18}O$-values of commonly used O-isotope standards (data for sulfate and nitrate are from Brand et al. 2009a, b)

| Standard | Material | VPDB scale | VSMOW scale |
|---|---|---|---|
| NBS-19 | Marble | −2.20 | |
| NBS-20 | Limestone | −4.14 | |
| NBS-18 | Carbonatite | −23.00 | |
| NBS-28 | Quartz | | 9.60 |
| NBS-30 | Biotite | | 5.10 |
| GISP | Water | | −24.75 |
| SLAP | Water | | −55.50 |
| NBS-127 | Ba sulfate | | 8.59 |
| USGS 35 | Na nitrate | | 56.81 |

### 2.4.2 Standards

Two different δ-scales are in use: $\delta^{18}O_{(VSMOW)}$ and $\delta^{18}O_{(VPDB)}$, because of two different categories of users, who have traditionally been engaged in O-isotope studies. The VPDB scale is used in low-temperature studies of carbonate. The original PDB standard was prepared from a Cretaceous belemnite from the Pee Dee Formation and was the laboratory working standard used at the University of Chicago in the early 1950s when the paleotemperature scale was developed. The original supply of this standard has long been exhausted, therefore secondary standards have been introduced (see Table 2.5), whose isotopic compositions have been calibrated relative to VPDB. All other oxygen isotope analyses (waters, silicates, phosphates, sulfates, high-temperature carbonates) are given relative to VSMOW.

The conversion equations of $\delta^{18}O_{(VPDB)}$ versus $\delta^{18}O_{(VSMOW)}$ and vice versa (Coplen et al. 1983) are:

$$\delta^{18}O_{(VSMOW)} = 1.03091\delta^{18}O_{(VPDB)} + 30.91$$
$$\delta^{18}O_{(VPDB)} = 0.97002\delta^{18}O_{(VSMOW)} - 30.91$$

Table 2.5 gives the $\delta^{18}O$-values of commonly used oxygen isotope standards on both scales.

### 2.4.3 Fractionation Processes

Out of the numerous possibilities to fractionate oxygen isotopes in nature, the following are of special significance.

#### 2.4.3.1 Fractionation of Water

Knowledge of the oxygen isotope fractionation between liquid water and water vapor is essential for the interpretation of the isotope composition of different water

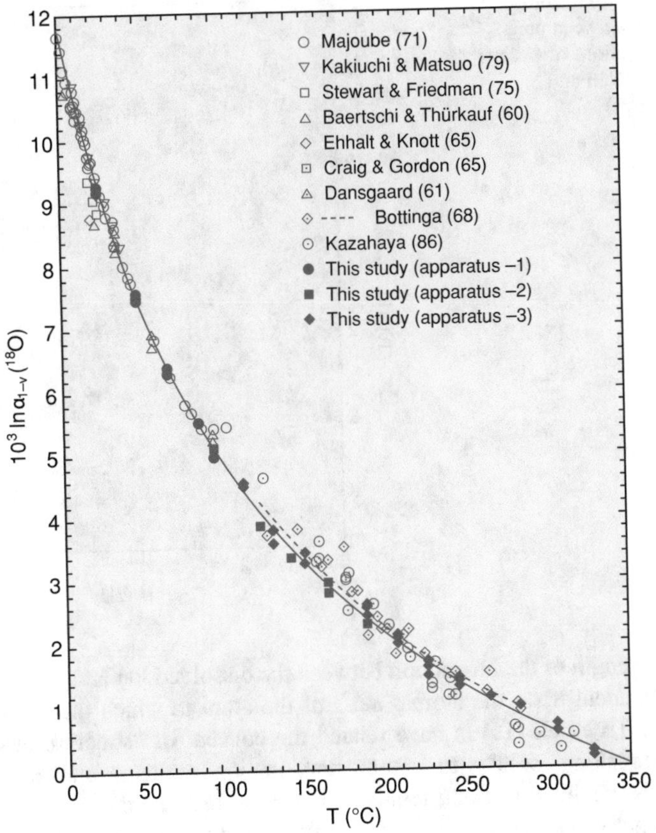


**Fig. 2.10** Oxygen isotope fractionation factors between liquid water and water vapour in the temperature range 0–350 °C (after Horita and Wesolowski 1994)

types. Fractionation factors experimentally determined in the temperature range from 0 to 350 °C have been summarized by Horita and Wesolowski (1994) and are shown in Fig. 2.10.

Addition of salts to water also affects isotope fractionations. The presence of ionic salts in solution changes the local structure of water around dissolved ions. Taube (1954) first demonstrated that the $^{18}O/^{16}O$ ratio of $CO_2$ equilibrated with pure $H_2O$ decreased upon the addition of $MgCl_2$, $AlCl_3$ and HCl, remained more or less unchanged for NaCl, and increased upon the addition of $CaCl_2$. The changes vary roughly linearly with the molality of the solute (see Fig. 2.11).

To explain this different fractionation behavior, Taube (1954) postulated different isotope effects between the isotopic properties of water in the hydration sphere of the cation and the remaining bulk water. The hydration sphere is highly ordered, whereas the outer layer is poorly ordered. The relative sizes of the two layers are dependent upon the magnitude of the electric field around the dissolved

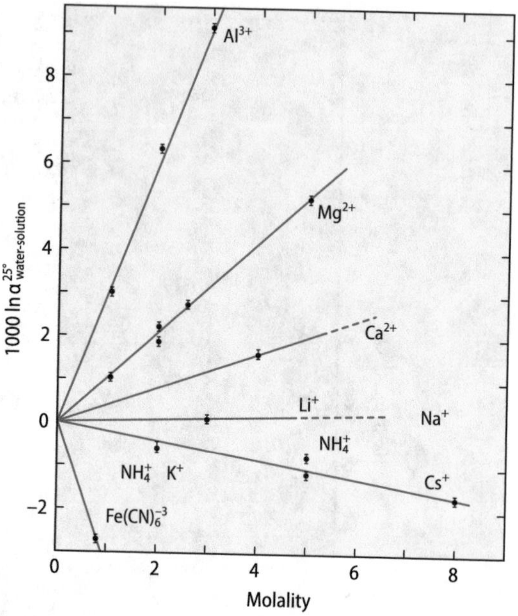


**Fig. 2.11** Oxygen isotope fractionation between pure water and solutions of various ions (after O'Neil and Truesdell 1991)

ions. The strength of the interaction between the dissolved ion and water molecules is also dependent upon the atomic mass of the atom to which the ion is bonded. O'Neil and Truesdell (1991) have related the concept of "structure-making" and "structure-breaking" solutes to isotope fractionation: structure makers yield more positive isotope fractionations relative to pure water whereas structure breakers produce negative isotope fractionations. Any solute that results in a positive isotope fractionation is one that causes the solution to be more structured as is the case for the structure of ice, when compared to solutes that lead to less structured forms, in which cation–$H_2O$ bonds are weaker than $H_2O$–$H_2O$ bonds.

As already treated in Sect. 2.1, isotope fractionations, the hydration of ions may play a significant role in hydrothermal solutions and volcanic vapors (Driesner and Seward 2000). Such isotope salt effects may change the oxygen isotope fractionation between water and other phases by several permil.

#### 2.4.3.2 $CO_2$–$H_2O$ System

Of equal importance is the oxygen isotope fractionation in the $CO_2$–$H_2O$ system. Early work concentrated on the oxygen isotope partitioning between gaseous $CO_2$ and water (Brenninkmeijer et al. 1983). In more recent work by Usdowski et al. (1991), Beck et al. (2005) and Zeebe (2007), it has been demonstrated that the oxygen isotope composition of the individual carbonate species are isotopically different, which is consistent with experimental work of McCrea (1950) and Usdowski and Hoefs (1993). Table 2.6 summarizes the equations for the temperature dependence between 5 and 40 °C (Beck et al. 2005).

**Table 2.6** Experimentally determined oxygen isotope fractionation factors relative to water for the aqueous system $CO_2$–$H_2O$ between 5 and 40 °C according to $10^3 \ln \alpha = A(10^6/T^{-2}) + B$ (Beck et al. 2005)

| | A | B |
|---|---|---|
| $HCO_3^-$ | 2.59 | 1.89 |
| $CO_3^{2-}$ | 2.39 | −2.70 |
| $CO_{2(aq)}$ | 2.52 | 12.12 |

The oxygen isotope fractionation (1000 lnα) between aqueous $CO_2$ and water at 25 °C is 41.6, decreasing to 24.7 at high pH values when $CO_3^{2-}$ is the dominant species (see Fig. 2.12). The pH dependence of the oxygen isotope composition in the carbonate-water system has important implications in the derivation of oxygen isotope temperatures.

#### 2.4.3.3 Mineral Fractionations

The oxygen isotope composition of a rock depends on the $^{18}O$ contents of the constituent minerals and the mineral proportions. Garlick (1966) and Taylor (1968) arranged coexisting minerals according to their relative tendencies to concentrate $^{18}O$. The list given in Table 2.7 has been augmented by data from Kohn et al. (1998a, b, c).

This order of decreasing $^{18}O$-contents has been explained in terms of the bond-type and strength in the crystal structure. Semi-empirical bond-type calculations have been developed by Garlick (1966) and Savin and Lee (1988) by assuming that oxygen in a chemical bond has similar isotopic behavior regardless of the mineral in which the bond is located. This approach is useful for estimating fractionation

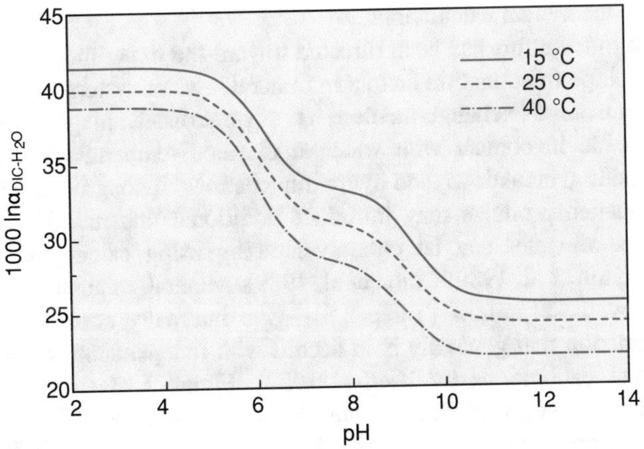


**Fig. 2.12** Oxygen isotope fractionations between dissolved inorganic carbon (DIC) and water as function of pH and temperatures (after Beck et al. 2005)

**Table 2.7** Sequence of minerals in the order (bottom to top) of their increasing tendency to concentrate $^{18}O$

| |
|---|
| Quartz |
| Dolomite |
| K-feldspar, albite |
| Calcite |
| Na-rich plagioclase |
| Ca-rich plagioclase |
| Muscovite, paragonite, kyanite, glaucophane |
| Orthopyroxene, biotite |
| Clinopyroxene, hornblende, garnet, zircon |
| Olivine |
| Ilmenite |
| Magnetite, hematite |

factors. The accuracy of this approach is limited due to the assumption that the isotope fractionation depends only upon the atoms to which oxygen is bonded and not upon the structure of the mineral, which is not strictly true. Kohn and Valley (1998a, b) determined empirically the effects of cation substitutions in complex minerals such as amphiboles and garnets spanning a large range in chemical compositions. Although isotope effects of cation exchange are generally <1‰ at T > 500 °C, they increase considerably at lower temperatures. Thus; use of amphiboles and garnets for thermometry requires exact knowledge of chemical compositions.

On the basis of these systematic tendencies of $^{18}O$ enrichment found in nature, significant temperature information can be obtained up to temperatures of 1000 °C, and even higher, if calibration curves can be worked out for the various mineral pairs. The published literature contains many calibrations of oxygen isotope geothermometers, most are determined by laboratory experiments, although some are based on theoretical calculations.

Although much effort has been directed toward the experimental determination of oxygen isotope fractionation factors in mineral—water systems, the use of water as an oxygen isotope exchange medium has several disadvantages. Some minerals become unstable in contact with water at elevated temperatures and pressures, leading to melting, breakdown and hydration reactions. Incongruent dissolution and ill-defined quench products may introduce additional uncertainties. Most of the disadvantages of water can be circumvented by using calcite as an exchange medium (Clayton et al. 1989; Chiba et al. 1989). Mineral–mineral fractionations—determined by these authors (Table 2.8)—give internally consistent geothermometric information that generally is in accord with independent estimates, such as the theoretical calibrations by Kieffer (1982). Figure 2.13 summarizes mineral fractionations relative to calcite presented by Chacko et al. (2001).

Rather than considering individual mineral pairs, Vho et al. (2019) presented more recently an internally consistent database in which published semi-empirical, experimental and natural studies have been included.

**Table 2.8** Coefficients A for mineral—pair fractionations $(1000 \ln \alpha_{X-Y} = A/T^2)10^6$ (after Chiba et al. 1989)

| | Cc | Ab | An | Di | Fo | Mt |
|---|---|---|---|---|---|---|
| Qtz | 0.38 | 0.94 | 1.99 | 2.75 | 3.67 | 6.29 |
| Cc | | 0.56 | 1.61 | 2.37 | 3.29 | 5.91 |
| Ab | | | 1.05 | 1.81 | 2.73 | 5.35 |
| An | | | | 0.76 | 1.68 | 4.30 |
| Di | | | | | 0.92 | 3.54 |
| Fo | | | | | | 2.62 |

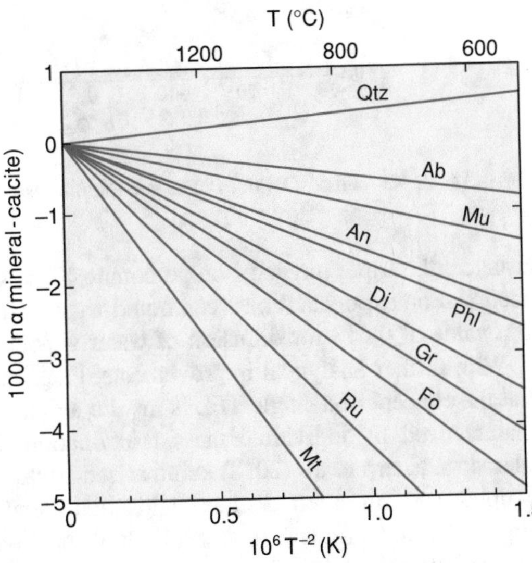


**Fig. 2.13** Oxygen isotope fractionations between various minerals and calcite (after Chacko et al. 2001)

### 2.4.4 Triple Oxygen Isotope Compositions

Since the natural oxygen isotope ratio of $^{17}O/^{16}O$ is close to one half of the $^{18}O/^{16}O$ ratio, in the past it was generally assumed that there was no need to measure the rare $^{17}O$. However, with improvements in analytical techniques, it became clear that the precise measurement of $^{17}O$ contents may give additional informations on fractionation processes in the earth's reservoirs. In a diagram $\delta^{17}O$ versus $\delta^{18}O$ values, all terrestrial rocks and minerals plot on an array with a coefficient $\lambda$ 0.52 $\times$ which was called the Terrestrial Fractionation Line (TFL). Miller et al. (1999) and Rumble et al. (2007) showed that rocks and minerals fall on slopes $\lambda$ between 0.524 and 0.526, whereas the Meteoric Water Line (MWL) has a slope of 0.528 (Luz and Barkan 2010). Slopes and intercepts of TFL and MWL vary, because the triple oxygen isotope intercept $\theta$ depends on the fractionation process (kinetic or equilibrium), on the substances involved and on temperature. The $\theta$-value connects the two fractionation factors for $^{17}O$ and $^{18}O$ $((^{17}O/^{16}O\alpha_{A-B} = {^{18}O/^{16}O\alpha_{A-B}})^{\theta})$ and varies between 0.509 which is the lower limit for kinetic fractionations and 0.535 which is the equilibrium high temperature limit (Young et al. 2002). For water, for

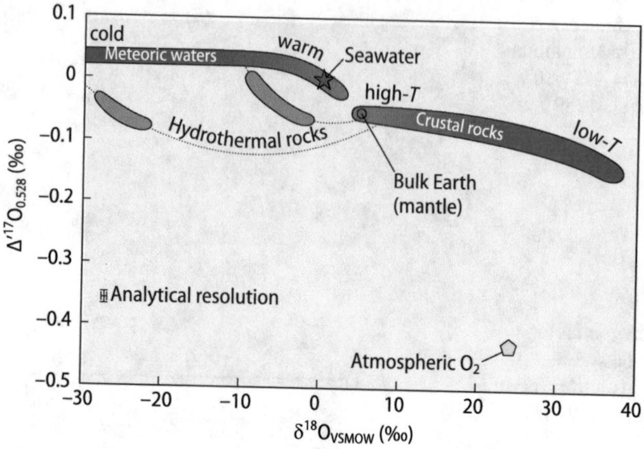


**Fig. 2.14** $\Delta^{17}O$- and $\delta^{18}O$-values of major Earth's reservoirs (courtesy Andreas Pack)

instance, the triple oxygen isotope composition is characterized by an equilibrium fractionation exponent $\lambda$ between liquid water and water vapour of 0.529 compared to a value of 0.518 for diffusion of water vapour.

With further analytical improvements, Pack and Herwartz (2014) demonstrated that the concept of a single TFL is invalid and that different reservoirs on Earth are characterized by individual mass fractionation lines with individual slopes and intercepts. Sharp et al. (2018) summarized triple oxygen isotope data of measured samples and observed distinct fields and trends in $\Delta^{17}O$–$\delta^{18}O$ diagrams (see Fig. 2.14). Meteoric waters generally have positive $\Delta^{17}O$-values with more or less no overlap with rocks.

Pristine mantle has a rather constant $\Delta^{17}O$-value of $-0.04 \pm 0.01$‰. Atmospheric $O_2$ has a unique $\Delta^{17}O$-composition, which may interact with rocks leaving a unique fingerprint (Pack 2021). Rocks that interact with meteoric waters may shift to low $\Delta^{17}O$-values at low fluid/rock ratios and to high values at higher fluid/rock ratios.

Besides the small mass-dependent fractionations, mass-independent fractionations can be much larger and are related to the transfer of stratospheric ozone to the tropospheric atmosphere, see recent reviews by Thiemens and Lin (2021), Pack (2021) and Brinjikji and Lyons (2021).

### 2.4.5 Fluid-Rock Interactions

Oxygen isotope ratio analysis provides a powerful tool for the study of water/rock interaction. The geochemical effect of such an interaction between water and rock or mineral is a shift of the oxygen isotope ratios of the rock and/or the water away from their initial values. Detailed studies of the kinetics and mechanisms of oxygen isotope exchange between minerals and fluids show that there are three possible exchange mechanisms (Matthews et al. 1983a, b; Giletti 1985).

(1) Solution-precipitation. During a solution-precipitation process, larger grains grow at the expense of smaller grains. Smaller grains dissolve and recrystallize on the surface of larger grains which decreases the overall surface area and lowers the total free energy of the system. Isotopic exchange with the fluid occurs while material is in solution.
(2) Chemical reaction. The chemical activity of one component of both fluid and solid is so different in the two phases that a chemical reaction occurs. The breakdown of a finite portion of the original crystal and the formation of new crystals is implied. The new crystals would form at or near isotopic equilibrium with the fluid.
(3) Diffusion. During a diffusion process isotopic exchange takes place at the interface between the crystal and the fluid with little or no change in morphology of the reactant grains. The driving force is the random thermal motion of the atoms within a concentration or activity gradient.

In the presence of a fluid phase coupled dissolution—reprecipitation is known to be a much more effective process than diffusion. This has been first demonstrated experimentally by O'Neil and Taylor (1967) and later re-emphasized by Cole (2000) and Fiebig and Hoefs (2002).

The first attempts to quantify isotope exchange processes between water and rocks have been undertaken by Sheppard et al. (1971) and Taylor (1974). By using a simple closed-system material balance equation these authors were able to calculate cumulative fluid/rock ratios.

$$W/R = \frac{\delta_{rock_f} - \delta_{rock_i}}{\delta_{H_2O_i} - (\delta_{rock_f} - \Delta)}, \tag{2.7}$$

where $\Delta = \delta_{rockf} - \delta H_2O_f$.

The equation requires adequate knowledge of both the initial (i) and final (f) isotopic states of the system and describes the interaction of one finite volume of rock with a fluid. The utility of such "zero-dimensional" equations has been questioned by Baumgartner and Rumble (1988), Blattner and Lassey (1989), Nabelek (1991), Bowman et al. (1994) and others. Only under special conditions do one-box models yield information on the amount of fluid that actually flowed through the rocks. If the rock and the infiltrating fluid were not far out of isotopic equilibrium, then the calculated fluid/rock ratios rapidly approach infinity. Therefore, the equations are sensitive only to small fluid/rock ratios. Nevertheless, the equations can constrain fluid sources. More sophisticated one-dimensional models like the chromatographic or continuum mechanics models (i.e. Baumgartner and Rumble 1988) are physically more plausible and can describe how the isotopic composition of the rock and of the fluid change with time and space. The diffusion–advection equations are based on partial differential equations that must be solved numerically. Examples of fluid-rock interactions in contact metamorphic environments have been presented by Nabelek and Labotka (1993), Bowman et al. (1994)

and application to contrasting lithologies by Bickle and Baker (1990) and Cartwright and Valley (1991).

Criss et al. (1987) and Gregory et al. (1989) developed a theoretical framework that describes the kinetics of oxygen isotope exchange between minerals and coexisting fluids. Figure 2.14 shows characteristic patterns in δ–δ plots for some hydrothermally altered granitic and gabbroic rocks. The $^{18}O/^{16}O$ arrays displayed on Fig. 2.15 cut across the 45° equilibrium lines at a steep angle as a result of the much faster oxygen isotope exchange of feldspar compared to that of quartz and pyroxene. If a low-$^{18}O$ fluid such as meteoric or ocean water is involved in the exchange process, the slopes of the disequilibrium arrays can be regarded as "isochrons" where, with continued exchange through time the slopes become less steep and approach the 45° equilibrium line. These "times" represent the duration of a particular hydrothermal event.

Figure 2.16 summarizes the naturally observed oxygen isotope variations in important geological reservoirs.

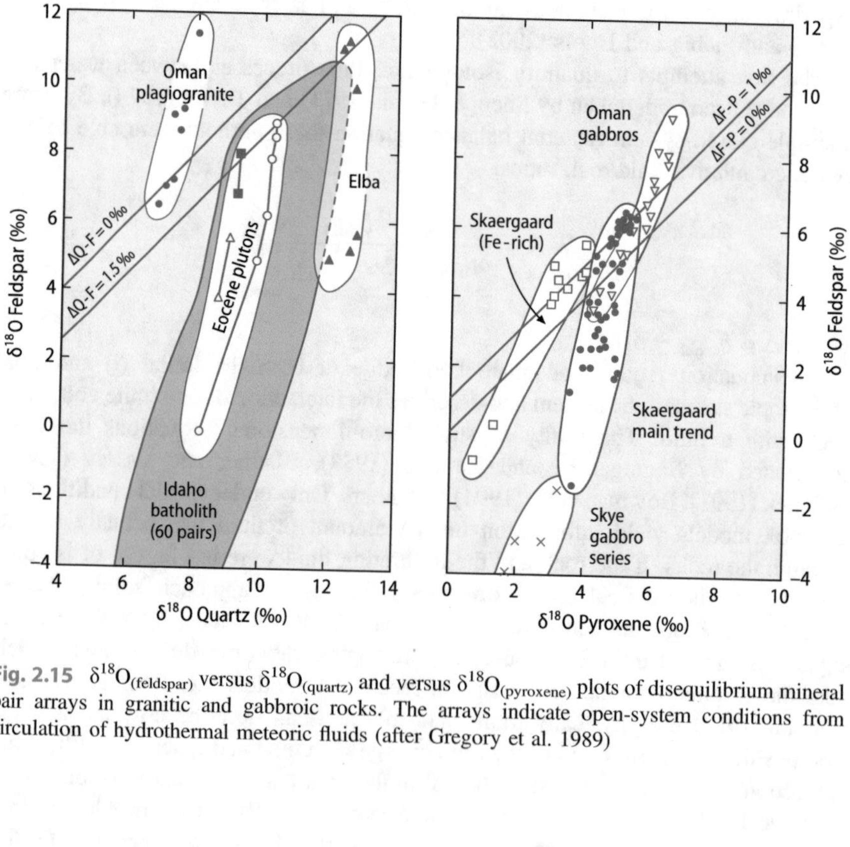


**Fig. 2.15** $\delta^{18}O_{(feldspar)}$ versus $\delta^{18}O_{(quartz)}$ and versus $\delta^{18}O_{(pyroxene)}$ plots of disequilibrium mineral pair arrays in granitic and gabbroic rocks. The arrays indicate open-system conditions from circulation of hydrothermal meteoric fluids (after Gregory et al. 1989)

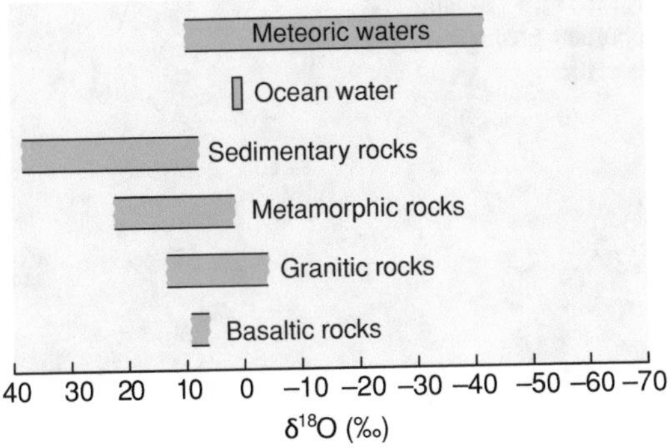


**Fig. 2.16** $\delta^{18}O$ values of important geological reservoirs

## 2.5 Sulfur

Sulfur has four stable isotopes with the following abundances (De Laeter et al. 2003).

$^{32}S$ 95.04%

$^{32}S$ 0.75%

$^{32}S$ 4.20%

$^{32}S$ 0.01%

Sulfur is present in nearly all environments. It may be a major component in ore deposits, where sulfur is the dominant nonmetal, and as sulfates in evaporites. It occurs as a minor component in igneous and metamorphic rocks, throughout the biosphere in organic substances, in marine waters and sediments as both sulfide and sulfate. These occurrences cover the whole temperature range of geological interest. Thus, it is quite clear that sulfur is of special interest in stable isotope geochemistry.

Thode et al. (1949) and Trofimov (1949) were the first to observe wide variations in the abundances of sulfur isotopes. Variations on the order of 180‰ have been documented with the "heaviest" sulfates having $\delta^{34}S$-values of > +120‰ (Hoefs, unpublished results), and the "lightest" sulfides having $\delta^{34}S$-values of around −65‰. Some of the naturally occurring S-isotope variations are summarized in Fig. 2.17. Reviews of the isotope geochemistry of sulfur have been published by Rye and Ohmoto (1974), Ohmoto and Rye (1979), Ohmoto (1986), Ohmoto and Goldhaber (1997), Seal et al. (2000), Canfield (2001a) and Seal (2006).

For many years the reference standard commonly referred to was sulfur from troilite of the Canyon Diablo iron meteorite (CDT). As Beaudoin et al. (1994) have pointed out, the original CDT is not homogeneous and may display variations in $^{34}S$ up to 0.4‰. Therefore, a new reference scale, Vienna-CDT (V-CDT) has been introduced by an advisory committee of IAEA in 1993, recommending an artificially prepared $Ag_2S$ (IAEA-S-1) with a $\delta^{34}S_{VCDT}$ of −0.3‰ as the new international standard reference material.

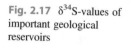

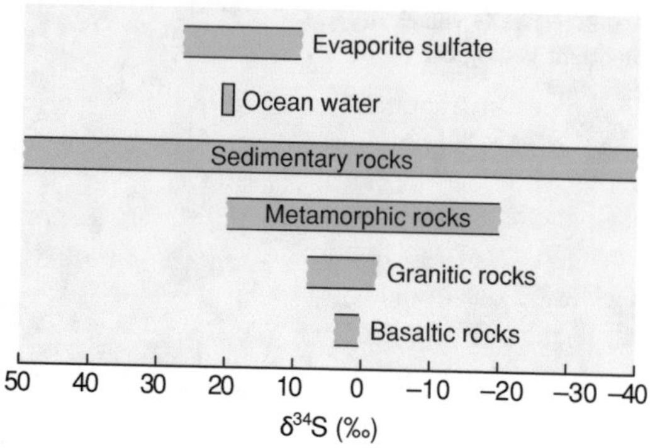


**Fig. 2.17** $\delta^{34}S$-values of important geological reservoirs

### 2.5.1 Methods

The gas conventionally used for gas-source mass-spectrometric measurement is $SO_2$. The introduction of on-line combustion methods (Giesemann et al. 1994) has reduced multistep off-line preparations to one single preparation step, namely the combustion in an elemental analyzer. Sample preparations have become less dependent on possibly fractionating wet-chemical extraction steps and less time-consuming, thereby reducing minimum sample gas to <1 mg.

Puchelt et al. (1971) and Rees (1978) first described a method using $SF_6$ instead of $SO_2$ which has some distinct advantages: it has no mass spectrometer memory effect and because fluorine is monoisotopic, no corrections of the raw data of measured isotope ratios are necessary. Comparison of $\delta^{34}S$-values obtained using the conventional $SO_2$ and the laser $SF_6$ technique has raised serious questions about the reliability of the $SO_2$ correction for oxygen isobaric interferences (Beaudoin and Taylor 1994). Therefore, the $SF_6$ technique has been revitalized (Hu et al. 2003), demonstrating that $SF_6$ is an ideal gas for measuring $^{33}S/^{32}S$, $^{34}S/^{32}S$ and $^{36}S/^{32}S$ ratios, especially for mass-independent isotope variations, despite the high toxicity of the gases.

Microanalytical techniques such as laser microprobe (Kelley and Fallick 1990; Crowe et al. 1990; Hu et al. 2003) and secondary ion microprobe (Chaussidon et al. 1987, 1989; Eldridge et al. 1988, 1993; Kozdon et al. 2010; Ushikubo et al. 2014) have become promising tools for determining sulfur isotope ratios. Hauri et al. (2016) reported multiple sulfur isotope data using NanoSIMS.

More recently the use of MC-ICP-MS techniques has been described by Bendall et al. (2006), Craddock et al. (2008) and Paris et al. (2013). Amrani et al. (2009) developed a GC–MC-ICP-MS method for the analysis of individual sulfur organic compounds. Due to low detection limits, sample sizes are orders of magnitude smaller than for $SO_2$ and $SF_6$ (Grotheer et al. 2017). GC–MC-ICP-MS requires no chemical pretreatment and allows for simultaneous collection of the individual 4 sulfur isotopes, but is associated with lower precision than the other methods described above.

## 2.5.2 Fractionation Mechanisms

Two types of fractionation mechanisms are responsible for the naturally occurring sulfur isotope variations:

(a) Kinetic isotope effects during microbial processes. Micro-organisms have long been known to fractionate isotopes during their sulfur metabolism, particularly during dissimilatory sulfate reduction, which produces the largest fractionations in the sulfur cycle,
(b) Various chemical exchange reactions under kinetic and equilibrium conditions between both sulfate and sulfides and the different sulfides themselves.

### 2.5.2.1 Equilibrium Reactions

There have been a number of theoretical and experimental determinations of sulfur isotope fractionations between coexisting sulfide phases as a function of temperature. Theoretical studies of fractionations among sulfides have been undertaken by Sakai (1968) and Bachinski (1969), who reported reduced partition function ratios and bond strengths of sulfide minerals and described the relationship of these parameters to isotope fractionation. In a manner similar to that for oxygen in silicates, there is a relative ordering of $^{34}S$-enrichment among coexisting sulfide minerals (Table 2.9). Considering the three most common sulfides (pyrite, sphalerite and galena) under conditions of isotope equilibrium, pyrite is always the most $^{34}S$ enriched mineral and galena the most $^{34}S$ depleted, sphalerite has an intermediate enrichment in $^{34}S$.

The experimental determinations of sulfur isotope fractionations between various sulfides do not exhibit good agreement. The most suitable mineral pair for temperature determination is the sphalerite—galena pair. Rye (1974) has argued that the Czamanske and Rye (1974) fractionation curve gives the best agreement with filling temperatures of fluid inclusions over the temperature range from 370 to 125 °C. By contrast, pyrite—galena pairs do not appear to be suitable for a temperature determination, because pyrite tends to precipitate over larger time periods

**Table 2.9** Equilibrium isotope fractionation factors of sulfides with respect to $H_2S$

| Mineral | Chemical composition | A |
|---|---|---|
| Pyrite | $FeS_2$ | 0.40 |
| Sphalerite | $ZnS$ | 0.10 |
| Pyrrhotite | $FeS$ | 0.10 |
| Chalcopyrite | $CuFeS_2$ | −0.05 |
| Covellite | $CuS$ | −0.40 |
| Galena | $PbS$ | −0.63 |
| Chalcosite | $Cu_2S$ | −0.75 |
| Argentite | $Ag_2S$ | −0.80 |

The temperature dependence is given by $A/T^2$ (after Ohmoto and Rye 1979)

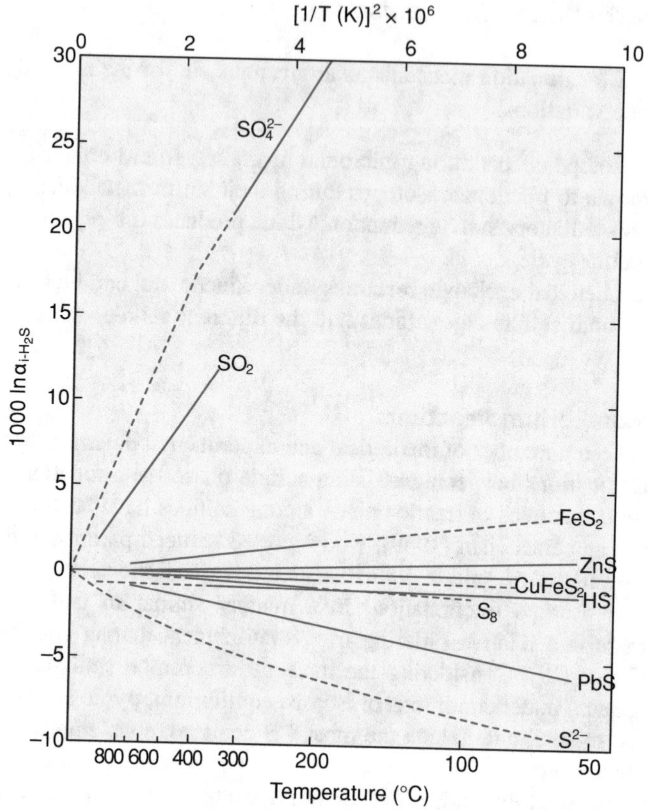


Fig. 2.18 Equilibrium fractionations among sulfur compounds relative to $H_2S$ (*solid lines* experimentally determined, *dashed lines* extrapolated or theoretically calculated (after Ohmoto and Rye 1979)

of ore deposition than galena, implying that these two minerals may frequently not be contemporaneous. The equilibrium isotope fractionations for other sulfide pairs are generally so small that they are not useful as geothermometers. Ohmoto and Rye (1979) critically examined the available experimental data and presented a summary of what they believe to be the best S-isotope fractionation data. These S-isotope fractionations relative to $H_2S$ are shown in Fig. 2.18.

Sulfur isotope temperatures from ore deposits often have been controversial; one of the reasons are strong $^{34}S$ zonations in sulfide minerals that have been observed by laser probe and ion probe measurements (McKibben and Riciputi 1998).

#### 2.5.2.2 Dissimilatory Sulfate Reduction

Dissimilatory sulfate reduction is conducted by a large group of microorganisms (over 100 species are known so far, Canfield 2001a), that gain energy for their growth by reducing sulfate while oxidizing reduced carbon (or $H_2$). Sulfate reducers are widely distributed in anoxic environments. They can tolerate temperatures from −1.5 to over 100 °C and salinities from fresh water to brines. Since the pioneering isotope fractionation studies with living cultures (Harrison and Thode 1957a, b; Kaplan and Rittenberg 1964) it is well known that sulfate reducing bacteria produce $^{32}$S-depleted sulfide. Despite decades of intense research, the factors that determine the magnitude of sulfur isotope fractionation during bacterial sulfate reduction are still under debate. The magnitude of isotope fractionation depends on the cellular rate of sulfate reduction with the highest fractionation at low rates and the lowest fractionation at high rates. Kaplan and Rittenberg (1964) and Habicht and Canfield (1997) suggested that fractionations depend on the specific rate (cell$^{-1}$ time$^{-1}$) and not so much on absolute rates (volume$^{-1}$ time$^{-1}$). What is clear, however, is that the rates of sulfate reduction are controlled by the availability of both of dissolved organic compounds and of sulfate. Canfield (2001b) observed no influence of sulfate concentration on sulfur isotope fractionations for natural populations. Another parameter that has been assumed to be important is temperature insofar as it regulates in natural populations the sulfate-reducing community (Kaplan and Rittenberg 1964; Brüchert et al. 2001). Furthermore, differences in fractionation with temperature relate to differences in the specific temperature response to internal enzyme kinetics as well as cellular properties and corresponding exchange rates of sulfate in and out of the cell of mesophilic sulfate reducing bacteria. For different types (including thermophilic) of sulfate-reducers, Canfield et al. (2006) found high fractionations in the low and high temperature range, and lowest fractionations in the intermediate temperature range.

The reaction chain during anaerobic sulfate reduction has been described in detail by Goldhaber and Kaplan (1974). In general, the rate-limiting step is the breaking of the first S–O bond, namely the reduction of sulfate to sulfite. Early laboratory studies with pure cultures of mesophilic sulfate reducing bacteria produced sulfide depleted in $^{34}$S by 4 up to 47‰ (Harrison and Thode 1957a, b; Kaplan and Rittenberg 1964; Kemp and Thode 1968; McCready et al. 1974; McCready 1975; Bolliger et al. 2001) and for decades this maximum value was considered to be a possible limit for the microbial dissimilatory process (e.g. Canfield and Teske 1996). More recently, sulfur isotope fractionations have been determined from incubations with sediments containing natural populations covering a wide spectrum of environments (from rapidly metabolizing microbial mats to slowly metabolizing coastal sediments; Habicht and Canfield 1997, 2001; Canfield 2001a). Sim et al. (2011) found that the type of organic electron donor is essential in controlling the magnitude of sulfur isotope fractionations of pure culture sulfate reducing bacteria, with complex substrates leading to sulfur isotope discrimination exceeding 47‰.

Naturally occurring sulfides in sediments and euxinic waters are commonly depleted in $^{34}$S by up to 65‰ (Jørgensen et al. 2004), covering the range of experiments with sulfate reducing bacteria (Sim et al. 2011). Recent studies have demonstrated that natural populations are able to fractionate S-isotopes by up to 70‰ under in situ conditions (Wortmann et al. 2001; Rudnicki et al. 2001; Canfield et al. 2010).

Another factor of great importance for the preserved sulfur isotope compositions of natural sulfides is whether sulfate reduction took place in a system open or closed with respect to dissolved sulfate. An "open" system has an infinite reservoir of sulfate in which continuous removal from the source produces no detectable loss of material. Typical examples are the Black Sea and local oceanic deeps. In such cases, $H_2S$ is extremely depleted in $^{34}S$ while consumption and change in $^{34}S$ remain negligible for the sulfate (Neretin et al. 2003). In a "closed" system, the preferential loss of the lighter isotope from the reservoir has a feedback on the isotopic composition of the unreacted source material. The changes in the $^{34}S$-content of residual sulfate and of the $H_2S$ are modeled in Fig. 2.19, which shows that $\delta^{34}S$-values of the residual sulfate steadily increase with sulfate consumption (a linear relationship on the log-normal plot). The curve for the derivative $H_2S$ is parallel to the sulfate curve at a distance which depends on the magnitude of the fractionation factor. As shown in Fig. 2.18, $H_2S$ may become isotopically heavier than the original sulfate when about 2/3 of the reservoir has been consumed. The $\delta^{34}S$-curve for "total" sulfide asymptotically approaches the initial value of the original sulfate. It should be noted, however, that apparent "closed-system" behavior of covarying sulfate and sulfide $\delta^{34}S$-values might be also explained by "open-system" differential diffusion of the different sulfur isotope species (Jørgensen et al. 2004).

In marine sediments typically 90% of the sulfide produced during sulfate reduction is reoxidized (Canfield and Teske 1996). The pathways of sulfide oxidation are still poorly known, and include both biological and abiological oxidation to sulfate, elemental sulfur and other intermediate compounds (Fry et al. 1988). Reoxidation of sulfide often occurs via compounds in which sulfur has intermediate oxidation states (sulfite, thiosulfate, elemental sulfur, polythionates) that do not accumulate, but are readily transformed and can be anaerobically disproportionated by bacteria. Therefore, Canfield and Thamdrup (1994) suggested that through a repeated cycle of sulfide oxidation to sulfur intermediates like elemental sulfur and

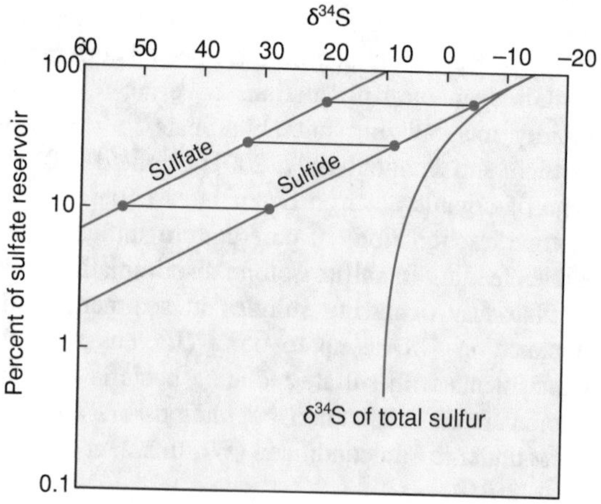


**Fig. 2.19** Rayleigh plot for sulfur isotope fractionations during reduction of sulfate in a closed system. Assumed fractionation factor 1.025, assumed composition of initial sulfate: +10‰)

subsequent disproportionation, bacteria can additionally generate $^{34}S$ depletions that may add on the isotopic composition of marine sulfides. Distinct sulfur isotope fractionation between generated sulfide and sulfate was also shown to occur upon bacterial disproportionation of other intermediates like thiosulfate and sulfite (Habicht et al. 1998).

For a long time it was ssumed that sulfur isotope fractionations during micobiological oxidation of sulfide are small. As shown by Pellerin et al. (2019), however, there are sulfide oxidation microorganism such as *Desulphovibrio alkaliphilus* which generate $^{34}S$ enrichments by 12.5‰ and more in the produced sulfate. As summarized by Pellerin et al. (2019) sulfur isotope fractionations in sediments are the combined result of 3 microbial metabolism: microbial sulfate reduction, disproportionation of external sulfur intermediates and microbial sulfide oxidation.

Finally, it should be mentioned that sulfate is labeled with two biogeochemical isotope systems, sulfur and oxygen. Coupled isotope fractionations of both sulfur and oxygen isotopes have been investigated in experiments (Mizutani and Rafter 1973; Böttcher et al. 2001) and in naturally occurring sediments and aquifers (Fritz et al. 1989; Ku et al. 1999; Aharon and Fu 2000; Wortmann et al. 2001). Böttcher et al. (1998) and Brunner et al. (2005) argued that a single characteristic $\delta^{34}S$–$\delta^{18}O$ fractionation slope for bacterial reduction processes does not exist, but that the isotope covariations depend on cell-specific sulfate reduction rates and associated oxygen isotope exchange rates with cellular water. Despite the extremely slow abiotic oxygen isotope exchange of sulfate with ambient water, $\delta^{18}O$ in sulfate obviously depend on the $\delta^{18}O$ of water via an exchange of sulfite with water. Böttcher et al. (1998) and Antler et al. (2013, 2017) demonstrated how the fractionation slopes depend on the net sulfate reduction rate: higher rates result in a lower slope meaning that sulfur isotopes increase faster relative to oxygen isotopes. The critical parameter for the evolution of oxygen and sulfur isotopes in sulfate is the relative difference in rates of sulfate reduction and of intracellular sulfite oxidation. Furthermore, Böttcher et al. (2001) argued that the disproportionation of sulfur intermediates in highly biologically active sediments may superimpose on the dominant sulfate reduction trend.

As an additional parameter, the triple oxygen isotope composition of sedimentary sulfates can be analyzed and as summarized by Bao (2015) sulfate may carry direct signals of ancient atmospheric $O_2$ and $O_3$. Hemingway et al. (2020) argued, however, that direct incorporation of atmospheric oxygen into sedimentary sulfate might be excluded.

#### 2.5.2.3 Thermochemical Reduction of Sulfate

In contrast to bacterial reduction, thermochemical sulfate reduction is an abiotic process with sulfate being reduced to sulfide under the influence of heat rather than bacteria (Trudinger et al. 1985; Krouse et al. 1988). The crucial question, which has been the subject of a controversial debate, is whether thermochemical sulfate reduction can proceed at temperatures as low as about 100 °C, just above the limit of microbiological reduction (Trudinger et al. 1985). There is increasing evidence from natural occurrences that the reduction of aqueous sulfates by organic compounds can occur at temperatures as low as 100 °C, given enough time for the reduction to proceed (Krouse et al. 1988; Machel et al. 1995). S isotope fractionations during thermochemical reduction generally should be smaller than during bacterial sulfate reduction, although experiments by Kiyosu and Krouse (1990) have indicated S-isotope fractionations of 10–20‰ in the temperature range of 200 to 100 °C. Experiments by Watanabe et al. (2009) and Oduro et al. (2011) at high temperatures using amino acids as organic substrates have yielded mass-independent S-isotope fractionations due to a magnetic isotope effect.

To summarize, bacterial sulfate reduction is characterized by large and heterogeneous $^{34}S$-depletions over very small spatial scales, whereas thermogenic sulfate reduction leads to smaller and "more homogeneous" $^{34}S$-depletions.

### 2.5.3 Quadruple Sulfur Isotopes

With respect to quadruple S isotope investigations, a distinction has to be made between large mass-independent S isotope fractionations observed in Archean sulfides and sulfates (Farquhar et al. 2000 and following papers) and much smaller mass-dependent S fractionations being characteristic for biosynthetic pathways. (Farquhar et al. 2003; Johnston 2011; Johnston et al. 2005; Ono et al. 2006, 2007). For many years, it was thought that $\delta^{33}S$ and $\delta^{36}S$ values carry no additional information, because sulfur isotope fractionations follow strictly mass-dependent fractionation laws. By studying all sulfur isotopes with very high precision, it was demonstrated that bacterial sulfate reduction follows a mass-dependent relationship that is slightly different from that expected by equilibrium fractionations. On plots $\Delta^{33}S$ versus $\delta^{34}S$, mixing of two sulfur reservoirs is non-linear in these coordinates; as a result, samples with the same $\delta^{34}S$-value can have different $\Delta^{33}S$ and $\Delta^{36}S$ values. This opens the possibility to distinguish between different fractionation mechanisms and biosynthetic pathways, even when S isotope fractionations are identical (Ono et al. 2006, 2007).

Bacterial sulfate reduction shows slightly different fractionation relationships compared to sulfur disproportionation reactions and sulfide oxidizing bacteria. On plots $^{33}S/^{32}S$ versus $^{34}S/^{32}S$, microbial communities range from 0.508 to 0.514 for sulfate reduction with higher values for oxidative bacteria (Tostevin et al. 2014). For instance, multiple S-isotope measurements of 1.8 Ga sulfates indicate the earliest initiation of microbial S disproportionation (Johnston et al. 2005). In another example, Canfield et al. (2010) demonstrated that S-isotope systematics in an euxinic lake in Switzerland clearly favour microbial reduction as the only reduction pathway. Thus, multiple sulfur isotope analyses have great potential in identifying the presence or absence of specific metabolisms in modern environment or may represent a proxy when a particular sulfur metabolism develops in the geologic record.

Large mass-independent S isotope fractionations measured in Archean sulfides and sulfates are a distinctive feature of sedimentary rocks older than 2.4 Ga. It is

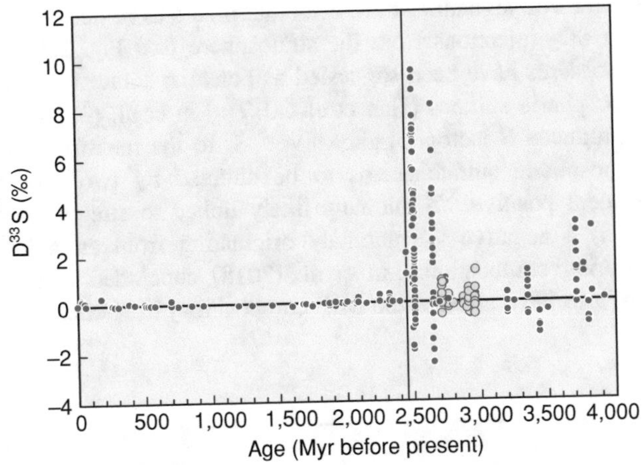


**Fig. 2.20** Compilation of $\Delta^{33}S$ versus age for rock samples. Note large $\Delta^{33}S$ before 2.45 Ga, indicated by *vertical line*, small but measurable $\Delta^{33}S$ after 2.45 Ga (Farquahar et al. 2007)

generally agreed that they indicate the near absence of $O_2$ and the presence of a reducing gas (likely $CH_4$ and/or $H_2$) in the Archean atmosphere. The geologic record of $\Delta^{33}S$ is shown in Fig. 2.20, which is characterized by time dependent magnitudes and signs of MIF-S indicating a temporal structure: $\leq 4‰$ $\Delta^{33}S$ anomalies in early Archean sulfides, even smaller variations in the middle Archean and very large ($\approx 12‰$) variations in the late Archean (see Fig. 2.24). The record of large magnitude $\Delta^{33}S$ values for sulfides terminates abruptly at approximately 2.4 Ga. Besides $\Delta^{33}S$, $\Delta^{36}S$ records also have received a great deal of attention, demonstrating that $\Delta^{36}S$ is preferentially negative down to values lower than $-8‰$.

Experiments that have verified the large $\Delta^{33}S$ and $\Delta^{36}S$ values in the Archean geologic record involve gaseous $SO_2$ (Farquhar et al. 2000; Claire et al. 2014). The specific chemical reaction that produced the effect noted in Archean samples is unknown, but gas phase reactions involving $SO_2$ are likely candidates. Farquhar and Wing (2003) and others demonstrated that photolysis of atmospheric $SO_2$ produces mass-independent S isotope fractionations, if atmospheric $O_2$ concentrations are very low. Farquhar et al. (2007) and Halevy et al. (2010) attributed these variations to changes in the composition and oxidation state of volcanic sulfur gases. When $\Delta^{36}S$ values are combined with $\Delta^{33}S$ values, many Archean sulfides plot close to an array with a $\Delta^{36}S/\Delta^{33}S$ slope of around $-1$ (Ono et al. 2006, and others) which implies similarities in the MIF producing atmospheric reactions. Much steeper slopes—observed in the Archean Dresser Formation—may suggest a different atmospheric reservoir (Wacey et al. 2015). In summary characteristic slopes in Archean samples and in products of laboratory photochemical experiments may be used as fingerprints (Farquhar et al. 2013).

Besides MIF isotope signatures in sedimentary rocks older than 2.5 Ga, modern stratospheric sulfates may also acquire S-MIF patterns. Sulfates from ice cores and volcanic ash with MIF signatures have been interpreted to be derived from volcanic eruptions with $SO_2$ injections from the stratosphere (see Fig. 2.21). In addition, combustion processes have been suggested as potential sources of S-MIF compositions in tropospheric sulfates (Han et al. 2017). Lin et al. (2018) added a fifth, cosmogenic produced S isotope, radioactive $^{35}$S, to the measurements of sulfate areosols. Tropospheric sulfate seems to be affected by two processes: (i) an altitude-dependent positive $^{33}$S anomaly likely linked to stratospheric $SO_2$ photolysis, and (ii) a negative $^{36}$S anomaly originating from an unknown S-MIF mechanism during combustion. Lin et al. (2018) concluded that the unknown process might also have affected the MIF compositionj in Archean rocks.

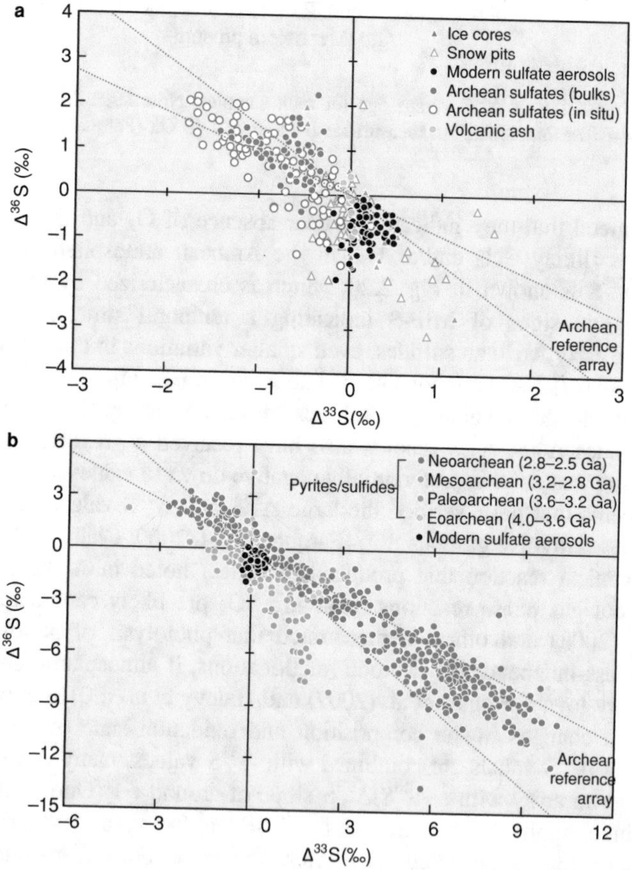


**Fig. 2.21** Similarity of S-MIF signatures in modern sulfates and Archean barites (**a**) and sulfides from Archean records (**b**) (from Lin et al. 2018)

# Part II: "Non-traditional" Isotopes

## Introductory Remarks

Since the introduction of precise metal isotope analytical techniques by Maréchal et al. (1999) and Zhu et al. (2000), numerous publications have followed as summarized in the recent book on "non-traditional isotopes" edited by Teng et al. (2017). Measured isotope variations at low temperatures are on the order of several ‰, much more than originally expected on the basis of the relatively small mass differences among isotopes of heavier elements. The magnitude of fractionations depends on several factors such as the participation of redox reactions and biologically mediated reactions.

Redox reactions represent presumably the most important source of metal isotope variations. As is the case for the light elements C and S, the reduced species of the metal is generally isotopically lighter than the oxidized species, except for uranium where the dominant nuclear volume effect causes an enrichment of the reduced species. Thus, the isotope composition of redox sensitive metals may help to understand redox systematics in the environment. Attempts to reconstruct redox conditions in the geologic past is one of the most common subjects in the recent literature on metal isotopes.

Although equilibrium fractionations have been documented for some transition metals (i.e. Fe), they should be small and may be overwhelmed by kinetic fractionations in low-temperature and biological systems (Schauble 2004). Kinetic fractionations have also been noted at high temperatures during diffusive mass transfer for elements like Li, Mg and Fe, which in the case of Li can be large occurring on spatial scales from µm to m.

Equilibrium fractionations of metal isotopes depend primarily on vibration frequencies. As a complete set of frequencies is generally not availiable, advances in computer capacity have allowed to calculate vibrational frequencies of simple molecules and crystalline compounds using first-principles electronic structure theory, which have the advantage that isotope effects can be calculated self-consistent and errors cancel out when calculating isotope frequency shifts (Schauble et al. 2009). Another advantage of first principles vibrational models is that output data can be compared with measured data and thus can test the accuracy of the model (e.g. Polyakov et al. 2007). Examples of calculations applying equilibrium fractionations derived by ab initio methods have been published by Blanchard et al. (2009) for Fe and by Rustad et al. (2010) for Mg and Ca.

Coordination numbers of metals in liquids and solids are another important parameter governing isotope fractionations of cations (Schauble (2004). The lighter isotope preferentially occupies the higher coordinated site. As an example, hematite is isotopically lighter than magnetite for Fe isotopes, because hematite is in octahedral coordination, whereas it is in octahedral and tetrahedral coordination in magnetite.

Precipitation and dissolution of minerals are generally associated with metal isotope fractionations, the sign and magnitude of isotope fractionations depend on experimental conditions, in specific whether kinetic or equilibrium conditions prevail. Minerals (i.e. carbonates) precipitating from aqueous solutions under kinetically controlled conditions are generally depleted in heavy metal isotopes in contrast to what is expected for equilibrium fractionations which depend on bond energies (Hofmann et al. 2013).

Sorption of metals on particle surfaces are an additional parameter causing metal isotope fractionations. The direction and magnitude of isotope fractionation is, however, highly metal specific. For example, sorption of isotopically light Mo on Fe–Mn oxides can be regarded as the most important fractionation mechanism of Mo (Wasylenki et al. 2008). In contrast Fe (II) sorbed on goethite particles is isotopically heavy compared to dissolved Fe (II) (Beard et al. 2010). For Cr, a negligible isotope fractionation is reported during sorption (Ellis et al. 2004).

Another characteristic feature of metal isotopes is their fractionation in plants and animals that can be used to understand transport mechanism. Generally, heavy metal isotopes are depleted in plants relatively to the soil in which they grew. Weiss et al. (2005) first showed that plants are generally depleted in heavy metal isotopes compared to their growth solution and that shoots of plants are isotopically lighter than roots. The uptake and transformation of metals within organisms can lead to further characteristic isotope fractionations that depend on the specific metal, its chemical speciation and on the type of organisms.

## 2.6 Lithium

Lithium has two stable isotopes with the following abundances (Rosman and Taylor 1998):

$$^{6}\text{Li} \quad 7.59\%$$
$$^{7}\text{Li} \quad 92.41\%$$

Lithium is one of the rare elements where the lighter isotope is less abundant than the heavier one. In order to be consistent with the other isotope systems, lithium isotope ratios are reported as $\delta^7$Li-values.

The large relative mass difference between $^6$Li and $^7$Li of about 16% is a favorable condition for their fractionation in nature. Taylor and Urey (1938) found a change of 25% in the Li-isotope ratio when Li-solutions percolate through a zeolite column. Thus, fractionation of Li-isotopes might be expected in geochemical settings in which cation exchange processes are involved. Li is only present in the 1 + valence state, so redox reactions do not influence its isotope composition. Reviews about natural Li isotope variations have been given by Burton and Vigier (2011), Tomascak et al. (2016), Penniston-Dorland et al. (2017).

Lithium isotope geochemistry is characterized by a difference close to 30‰ between ocean water ($\delta^7$Li of +31‰) and bulk silicate earth with a $\delta^7$Li-value of 3.2‰ (Seitz et al. 2007). In this respect lithium isotope geochemistry is very similar to that of boron (see p. 128). The isotopic difference between the mantle and the

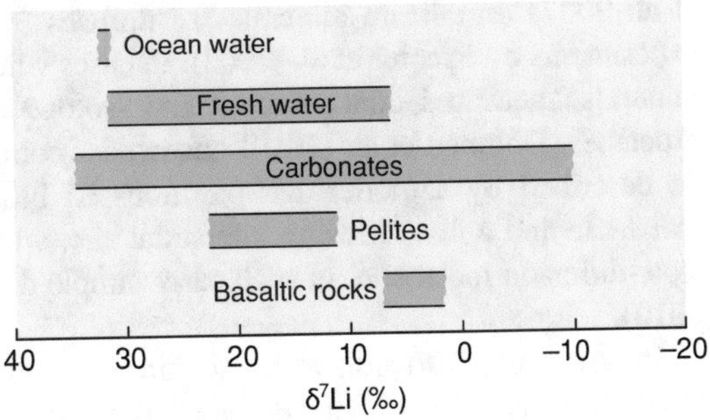


**Fig. 2.22** Lithium isotope variations in major geological reservoirs

ocean can be used as a powerful tracer to constrain water/rock interactions (Tomascak 2004). Figure 2.22 gives an overview of Li-isotope variations in major geological reservoirs.

### 2.6.1 Methods

Li isotopes have been analysed with TIMS (James and Palmer 2000), ion microprobe (Kasemann et al. 2005a, b) and multicollector ICP-MS techniques first described by Tomascak et al. (1999) and modified by Millot et al. (2004) and Jeffcoate et al. (2004). In order to avoid interferences and matrix effects, Li has to be separated from the rest of the sample. During elution, a complete 100% yield is necessary, since even a small loss of Li may shift the $\delta^7$Li value by several ‰.

In most studies Li isotope values are given relative to L-SVEC which, however, has been exhausted and replaced by IRMM-016 which is considered to be identical in isotope composition with L-SVEC. James and Palmer (2000) have determined nine international rock standards ranging from basalt to shale relative to NIST L-SVEC standard. In addition, Jeffcoate et al. (2004) and Gao and Casey (2011) presented $\delta^7$Li values for other reference materials.

### 2.6.2 Diffusion

Li isotope variations have been interpreted—like other isotope systems—in terms of isotope equilibrium between minerals and fluids; however, the analysis of high-temperature samples and experimental studies have shown that Li isotope variations may be very often kinetically controlled due to the large differences in $^6$Li and $^7$Li diffusivities that may far exceed Li isotope variations produced by equilibrium processes. The large Li mobility makes the preservation of primary compositions difficult; in contrast this complication supports the potential for Li to be used as a geospeedometer for short-term processes. Diffusive Li isotope fractionation has been reported to occur on a meter to micrometer scale during cooling processes (Lundstrom et al. 2005; Teng et al. 2006; Jeffcoate et al. 2007; Parkinson et al. 2007). In silicate minerals $^{6}Li$ diffuses 3% faster than $^{7}Li$, consistent with experiments by Richter et al. (2003). In cases where Li can occupy multiple sites in minerals, measured concentration and isotope profiles reflect multiple diffusion processes. Dohmen et al. (2010) observed a complex diffusion behaviour, that can be described by a model that partitions Li between two sites: a slow-diffusing metal-site and a fast-diffusing interstitial site, with Li strongly partitioning into the slow-diffusion metal site. In such cases simple diffusion models fail (Dohmen et al. 2010).

In summary, diffusion at magmatic and metamorphic temperatures is a very effective mechanism for generating large variations in $^{7}Li/^{6}Li$ ratios (Lundstrom et al. 2005; Teng et al. 2006; Rudnick and Ionov 2007 and others). As reviewed by Marschall and Tang (2020) diffusive fractionation of Li isotopes can be used to determine timescales of short-term geologic processes using arrested diffusion profiles.

## 2.6.3 Magmatic Rocks

High temperature equilibrium Li isotope fractionations have been investigated experimentally (Wunder et al. 2006, 2007) and theoretically (Kowalski and Jahn 2011). Calculated fractionation factors between staurolite, spodumene, mica and aqueous fluids are in good agreement with experimentally derived fractionation factors.

Li isotope compositions in mantle minerals show controversial results. Peridotite whole rocks and minerals have a large range in $\delta^{7}Li$ values between −20 to +10‰ that have been attributed to kinetic isotope fractionations due to Li diffusion (Brooker et al. 2004; Lundstrom et al. 2005; Jeffcoate et al. 2007; Pogge von Strandmann et al. 2011; Lai et al. 2015 and others) and/or to addition of distinct Li from metasomatizing melts and fluids (Nishio et al. 2004, Ionov et al. 2007). As Li diffusion in clinopyroxene (cpx) is faster than in olivine (Dohmen et al. 2010) and cpx is more frequently affected by metasomatic overprint; a pristine mantle Li isotope signature is more likely to be found in olivine than in cpx. Thus, when discussing Li isotope variations in mantle derived rocks, secondary processes such as overprinting by fluids and melts and isotope fractionations due to diffusional exchange have to be considered and camouflage the pristine mantle composition.

Although olivine and clinopyroxene phenocrysts are frequently affected by diffusion, mantle-derived basalts (MORB and OIB) have a relatively uniform composition with $\delta^{7}Li$ values of 4 ± 2‰ (Tomaszak 2004; Elliott et al. 2004). Because Li isotopes may be used as a tracer to identify the existence of recycled material in the mantle, systematic studies of arc lavas have been undertaken (Moriguti and Nakamura 1998; Tomascak et al. 2000; Leeman et al. 2004 and others). However, most arc lavas have $\delta^{7}Li$ values that are indistinguishable from those of MORB. On the other hand, Tang et al. (2014) have demonstrated that lower $\delta^{7}Li$ values can be attributed to the incorporation of subducted terrigenous sediment in the source of arc lavas from the Lesser Antilles.

Li isotope distribution through the oceanic crust reflects the varying conditions of seawater alteration with depth (Chan et al. 2002; Gao et al. 2012). At low temperatures, altered volcanic rocks have heavier Li isotope compositions than MORB whereas at higher temperatures in deeper parts of the oceanic crust $\delta^7$Li-values become similar to MORB. Gao et al. (2012) concluded that the Li isotope pattern in drilled oceanic sections reflects variations in water/rock ratios in combination with increasing downhole temperatures.

During fluid-rock interaction, Li as a fluid-mobile element will enrich in aqueous fluids. It might therefore be expected that $\delta^7$Li enriched seawater, incorporated into altered oceanic crust, should be removed during subduction zone metamorphism. Continuous dehydration of pelagic sediments and altered oceanic crust results in $^7$Li-depleted rocks and in $^7$Li enriched fluids. A subducting slab therefore should introduce large amounts of $^7$Li into the mantle wedge. To quantitatively understand this process Li isotope fractionation factors between minerals and coexisting fluids must be known (Wunder et al. 2006, 2007).

Fluids from marine hydrothermal systems typically show $\delta^7$Li values between +3 and +11‰ (Chan et al. 1993; Foustoukos et al. 2004) implying that Li in these fluids cannot have been derived entirely from seawater or from unaltered MORB (Tomaszak et al. 2006).

Granites of various origin have an average $\delta^7$Li-value slightly lighter than the mantle (Teng et al. 2004, 2009). I- and S-type granites do not vary in a systematic way, but largely inherit the Li isotope composition of their source rocks.

Considering the small Li isotope fractionation at high temperature during igneous differentiation processes (Tomascak 2004), continental crust, in theory, should not be too different in Li isotope composition from the mantle. Because this is not the case, the isotopically light crust must have been modified by secondary processes, such as weathering, hydrothermal alteration and prograde metamorphism (Teng et al. 2007a, b). By analyzing metamorphic rocks that range from very low grade to eclogite facies, Romer and Meixner (2014) observed with progressive metamorphism an overall loss in Li concentration and a decrease in $\delta^7$Li-values. Li isotope fractionations depend on protolith chemistry and relative contributions of exchangeable Li that might become lost during low-grade metamorphism or might become incorporated in newly formed minerals.

### 2.6.4 Weathering

Because Li is mainly hosted in silicate minerals—Li is orders of magnitude more concentrated in silicates than in carbonates—and not involved in biological reactions, Li can be regarded as one of the best proxies for silicate weathering. The best evidence for Li isotope fractionation during weathering is the systematic $^7$Li enrichment of natural waters relative to their source rocks (Burton and Vigier 2011; Dellinger et al. 2016 and others). During weathering $^7$Li is preferentially mobilized, whereas $^6$Li becomes enriched in the weathering residue. The major control of Li isotopic composition is the balance between primary mineral dissolution and secondary mineral formation, where $^6$Li is preferentially taken up by the solid, driving the fluid to heavy values (Wimpenny et al. 2010). The magnitude of fractionation seems to depend on the extent of weathering: large Li isotope fractionations seem to occur during superficial weathering while little fractionation is noted during prolonged weathering in stable environments (Millot et al. 2010a, b). Li measuremenets in ancient zircons (Ushikubo et al. 2008) were connected to the onset of early weathering in the Hadean.

What complicates a quantitative understanding of Li isotope behavior during weathering is the fact that Li in clays and other secondary minerals exists in two forms; (i) structurally bound and (ii) exchangeable (Hindshaw et al. 2019; Pogge von Strandheim et al. 2020). While the structurally bound Li has $\delta^7$Li values lower than −21‰, the exchangeable Li has $\delta^7$Li values between 0 and −12‰. Since secondary minerals contain exchangeable and structural Li, their $\delta^7$Li-values fall between the two endmembers. Preferential dissolution of primary minerals does not generate significant Li isotope fractionations. Wimpenny et al. (2010) demonstrated that dissolution of basaltic glass and olivine does not result in measurable Li isotope fractionation. Secondary mineral formation and adsorption on mineral surfaces are regarded to be the major process responsible for the high $\delta^7$Li values in waters. Considerable Li isotope fractionations, for instance, have been observed during chemical sorption of Li on the surface of gibbsite (Pistiner and Henderson 2003) or on clay minerals (Zhang et al. 1998; Millot et al. 2010a, b).

As shown by a large number of studies of global rivers (i.e. Amazon: Huh et al. 1998; Dellinger et al. 2016; Orinoco: Huh et al. 2001; Ganges–Brahmaputra: Manaka et al. 2017; Pogge von Strandmann et al. 2017; Yangtzee: Wang et al. 2015a, b, c; Congo: Henchiri et al. 2016), the range of $\delta^7$Li values in river waters can be quite large (from +6 to +33‰, Huh et al. 1998). These studies have shown that low $\delta^7$Li values occur at very high weathering intensities (e.g. lowland tropical rivers) and at very low weathering intensities (e.g. rapidly eroding mountain streams), because secondary clays are either dissolving themselves or never form (Frings 2019).

### 2.6.5 Ocean Water

Lithium is a conservative element in the ocean with a residence time of about one million year. Its isotope composition ($\delta^7$Li: 31‰) is maintained by inputs of dissolved Li from rivers (average $\delta^7$Li +23‰, Huh et al. 1998) and high-temperature hydrothermal fluids at ocean ridges on one side and low temperature removal of Li into oceanic basalts and marine sediments at the other. Precipitation of carbonates does not play a major role due to the low Li-concentrations of carbonates. This fractionation pattern explains, why the Li isotope composition of seawater is heavier than its primary sources (continental weathering: 23‰; Huh et al. 1998) and high-temperature hydrothermal fluids (6–10‰, Chan et al. 1993).

Any variance in Li sources and sinks during geologic history should cause secular variations in the isotope composition of oceanic Li. And, indeed, Misra and

Froelich (2012) reconstructed the Li isotope composition of ocean water for the last 68 Ma and observed an 9‰ increase from the Paleocene to the present requiring changes in continental weathering and/or low temperature ocean crust alteration (see p. 388). By extending this approach, Wanner et al. (2014) presented a model that revealed a close relationship between $\delta^7$Li and $CO_2$ consumption by silicate weathering. In pore waters, Li isotope compositions are determined by two divergent processes (Chan and Kastner 2000). Low temperature alteration of volcanic material produces $^7$Li-depleted clay minerals leading to $^7$Li enriched pore waters. Exchange reactions of clay minerals preferentially liberate $^6$Li driving pore waters to isotopically lighter compositions.

### 2.6.6 Meteoric Water

Rainwaters collected monthly over a one year period have extremely variable Li isotope compositions (Millot et al. 2010a, b). A surprising result is that Li derived from seawater is not the most important source: variable contributions from seawater spray depend—as expected—on the distance from the coast, but non-marine sources (crustal, anthropogenic, biogenic) may become dominant. High $\delta^7$Li-values, for instance, have been explained by anthropogenic contamination from fertilizers in agriculture (Millot et al. 2010a, b). The anthropogenic use of Li has dramatically increased in recent years, such as in mobile electronic devices, in electric powered vehicles and in therapeutic drugs. For the area of Seoul in Korea, Choi et al. (2019) have analyzed river, waste and tap water. In the downstream part of the Han River, they observed an increase in Li-concentrations by a factor of 6, while Li isotope values decrease from 30.1 to 19.2‰, which indicates increasing amounts of waste waters characterized by high Li concentrations and low $\delta^7$Li-values. In less contaminated areas, groundwaters may indicate mixing between rainwater of variable $\delta^7$Li and waters that are affected by water/rock interaction being isotopically lighter than rainwaters.

## 2.7 Boron

Boron has two stable isotopes with the following abundances (Rosman and Taylor 1998).

$$^{10}\mathrm{B} \quad 19.9\%$$
$$^{11}\mathrm{B} \quad 80.1\%$$

The large mass difference between $^{10}$B and $^{11}$B and large chemical isotope effects between different species (Bigeleisen 1965) make boron a very promising element to study for isotope variations. The utility of boron isotopes as a geochemical tracer stems from the high mobility of boron during high- and low-temperature fluid-related processes, showing a strong affinity for silicate melts and aqueous fluids.

Boron isotope geochemistry is characterized by distinct isotope signatures:

(i) A strong enrichment of $^{11}B$ in seawater (+39.6‰, Foster et al. 2010).
(ii) A depletion of $^{11}B$ of the average continental crust (−10‰) relative to pristine mantle

The lowest $\delta^{11}B$-values of around − 70‰ have been measured for certain coals (Williams and Hervig 2004), whereas the most enriched $^{11}B$-reservoir has been found in brines from Australia and Israel (Dead Sea) which have $\delta^{11}B$-values of up to 60‰ (Vengosh et al. 1991a, b). A very characteristic feature of boron geochemistry is the isotopic composition of ocean water with a constant $\delta^{11}B$-value of 39.6‰ (Foster et al. 2010), which is about 50‰ heavier than the average continental crust of $-10 \pm 2$‰ (Chaussidon and Albarede 1992). Isotope variations of boron in some geological reservoirs are shown in Fig. 2.23.

## 2.7.1 Methods

Three different methods have been used in recent years for boron isotope analysis: (i) thermal ionisation mass-spectrometry (TIMS), either with positively charged (P-TIMS) or negatively charged (N-TIMS) ions, (ii) multi-collector-ICP mass spectrometry and (iii) secondary ion mass spectrometry (SIMS).

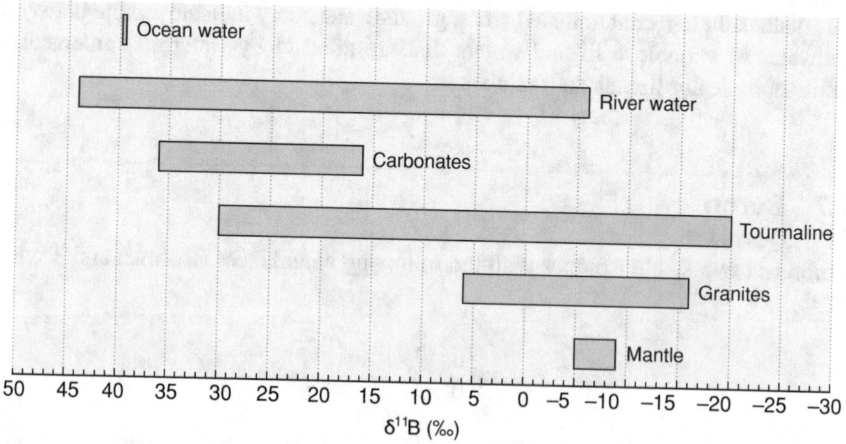


Fig. 2.23 Boron isotope variations in geologically important reservoirs

(i) Two different methods have been developed for TIMS. The positive thermal ionization technique uses $Na_2BO_{2^+}$ ions (McMullen et al. 1961). Subsequently, Spivack and Edmond (1986) modified this technique by using $CS_2BO_{2^+}$ ions (measurement of the masses 308 and 309). The substitution of $^{133}Cs$ for $^{23}Na$ increases the molecular mass and reduces the relative mass difference of its isotopic species, which limits the thermally induced mass dependent isotopic fractionation. This latter method has a precision of about ±0.25‰, which is better by a factor of 10 than the $Na_2BO_{2^+}$ method. In negative ion mode (N-TIMS), boron isotopes are analysed as $BO_2^-$ (masses 42 and 43). N-TIMS has the advantage that no chemical separation of boron from the sample matrix is required.

(ii) Lécuyer et al. (2002) first described the use of MC-ICP-MS for B isotopic measurements of waters, carbonates, phosphates and silicates with an external reproducibilty of ±0.3‰; improvement in reproducibility has been achieved by Guerrot et al. (2011) and Louvat et al. (2011). Le Roux et al. (2004) introduced an in situ laser ablation ICP-MS method at the nanogram level. The amount of boron measured are two orders of magnitude lower than P-TIMS and acid solution ICP-MS methods.

(iii) Chaussidon and Albarede (1992), performed boron isotope determinations with an ion-microprobe having an analytical uncertainty of about ±2‰. Significant improvements with SIMS analysis have been described by Rollion and Erez (2010).

As analytical techniques have been consistently improved in recent years, the number of boron isotope studies has increased rapidly. A recent review has been given by Marschall and Foster (2018). $\delta^{11}B$-values are generally given relative NBS boric acid SRM 951, which is prepared from a Searles Lake borax. This standard has a $^{11}B/^{10}B$ ratio of 4.04558 (Palmer and Slack 1989).

## 2.7.2 Isotope Fractionation Mechanism

(a) pH dependence of isotope fractionations

Boron is generally bound to oxygen or hydroxyl groups in either triangular (e.g., $BO_3$) or tetrahedral (e.g., $B(OH)_4^-$) coordination. The dominant isotope fractionation process occurs in aqueous systems via an equilibrium exchange process between boric acid ($B(OH)_3$) and coexisting borate anion ($B(OH)_4^-$). At low pH-values trigonal $B(OH)_3$ predominates, at high pH-values tetrahedral $B(OH)_4^-$ is the primary anion. The pH-dependence of the two boron species and their related isotope fractionation is shown in Fig. 2.24 (after Hemming and Hanson 1992). The pH dependence has been used reconstructing past ocean pH-values by measuring

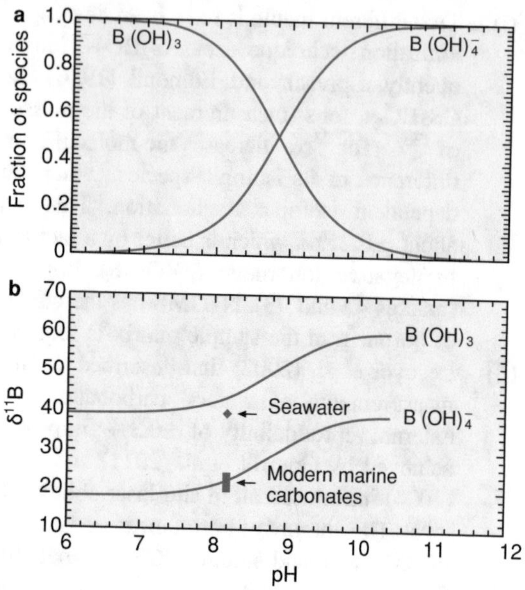


**Fig. 2.24** **a** Distribution of aqueous boron species versus pH; **b** $\delta^{11}B$ of the two dominant species $B(OH)_3$ and $B(OH)^-{}_4$ versus pH (after Hemming and Hanson 1992)

the boron isotope composition of carbonates e.g. foraminifera. This relies on the fact that mainly the charged species $B(OH)_4{}^-$ is incorporated into carbonate minerals with small to insignificant fractionations (Hemming and Hanson 1992; Sanyal et al. 2000). In corals, Rollion-Bard et al. (2011), however, observed both coordination species in the coral microstructure.

There has been debate about the mechanisms of boron incorporation in carbonates. Early investigations suggested the sole incorporation of $B(OH_4)^-$ into the anion site of aragonite and calcite, which seems to be the case for aragonite. For calcite, there is increasing evidence that both, $B(OH)_3$ and $B(OH_4)^-$, are incorporated into the mineral (Branson 2017 and others). Controvery has also existed about the size of B fractionation between ocean water and carbonates. Today, the generally accepted fractionation factor for boron in seawater is 27.2‰ at 25 °C (Klochko et al. 2006). Offsets from predicted boron isotope fractionations may be caused by a variety of factors such as kinetic fractionations during incorporation of boron, the incorporation of both boron species into the lattice or biological factors during carbonate precipitation.

The boron isotope approach has been not only used to indirectly estimate the seawater pH from $\delta^{11}B$ of foraminifera, but to estimate from the pH proxy the past atmospheric $CO_2$ concentrations (i.e. Pearson and Palmer 1999, 2000; Pagani et al. 2005; Henehan et al. 2013; Foster and Rae 2016; Rasbury and Heming 2017). An increase in atmospheric $CO_2$ results in increased dissolved $CO_2$ in ocean water, which in turn causes a reduction in oceanic pH, well known as ocean acidification. Comparison with the $CO_2$ ice core record demonstrates that boron isotopes, indeed, can be used to reconstruct past atmospheric $CO_2$ concentrations (Rae 2017). Shao et al. (2019) built a composite pH/$pCO_2$ curve for surface oceans, showing that various parts of the ocean were releasing $CO_2$ to the atmosphere over the last 25,000 years. On longer time scales, $\delta^{11}B$ records demonstrate coupling between $CO_2$ and climate over the last 60 Myr (Pearson et al. 2000; Anagnoustou et al. 2016 and others).

(b) Adsorption

Significant isotope fractionations may occur when aqueous boric acid absorbs on solid surfaces, as shown by Lemarchand et al. (2005) and others. Boron isotopic compositions are controlled by ion exchange rates at the mineral/water interface. The extent of B isotope fractionation depends on the aqueous speciation of boron and on the structure of surface complexes. High values of B isotope fractionation are observed at low pH, lower values are observed at high pH, which is due to the change in B-coordination from trigonal to tetrahedral.

### 2.7.3 Fractionations at High Temperatures

Experimental studies of boron isotope fractionation between hydrous fluids, melts and minerals have shown that $^{11}B$ preferentially partitions into the fluid relative to minerals or melts (Palmer et al. 1987; Williams et al. 2001; Wunder et al. 2005; Liebscher et al. 2005), ranging from about 33‰ for fluid-clay (Palmer et al. 1987), to about 6‰ for fluid-muscovite at 700 °C (Wunder et al. 2005) and to a few‰ for fluid-melt above 1000 °C (Hervig et al. 2002). The main fractionation effect seems to be due to the change from trigonal boron in neutral pH hydrous fluid to tetrahedrally coordinated boron in most rock forming minerals. Ab initio calculations of Kowalski et al. (2013a, b) have confirmed that B isotope fractionations are driven by B coordination. In addition, Kowalski et al. (2013a, b) showed that the magnitude of B isotope fractionation correlates with B–O bond lengths.

Boron like lithium is a useful tracer for mass transfer in subduction zones. Both elements are mobilized by fluids and melts and display considerable isotope fractionation during dehydration reactions. Concentrations of B are low in mantle derived materials, whereas they are high in sediments and altered oceanic crust. Any input of fluid and melt from the subducting slab into the overlying mantle has a strong impact on the isotope composition of the mantle wedge and on magmas generated there. Thus, arc volcanic rocks have distinct $\delta^{11}B$-values relative to the mantle. As summarized by De Hoog and Savov (2017) arc lavas show a 25‰ variation from –9 to +16‰ with an average $\delta^{11}B$-value of +3.2‰. As suggested by Palmer (2017) the likely source of the heavy boron isotope values is from serpentinite dehydration.

The boron isotopic of unaltered MORB glasses have been investigated in several studies. Summarizing available data, Marschall et al. (2017) concluded that glasses have a homogeneous mean B isotope composition of –7.1‰. Since B isotope fractionation during mantle melting and crystal fractionation appears to be small, the average MORB glass value may reflect the B isotope composition of the depleted mantle and the bulk silicate Earth (Marschall et al. 2017).

I-type granites derived from mantle-derived or magmatic rocks are on average isotopically heavier than S-type granites derived from melting of sedimentary rocks (Trumbull and Slack 2017). This distinction also holds for hydrothermal ore deposits that are associated with the respective granite magmatism.

### 2.7.4 Weathering Environment

According to Gaillardet and Lemarchand (2018) boron isotope variations in the weathering environment may reach up to 70‰. Rivers and vegetation are enriched in $^{11}B$, while clay minerals and boron absorbed on organic and inorganic surfaces are depleted in $^{11}B$. Dissolved boron in rivers shows a wide range in $\delta^{11}B$ values with a mean value of about 10‰ (Rose et al. 2000; Lemarchand et al. 2002; Chetelat et al. 2009a). The latter authors estimated that 70% of the boron is transported in suspended form having a $\delta^{11}B$ value close to the average crust (–7‰), whereas the $\delta^{11}B$ value of dissolved boron is controlled by incorporation into secondary phases (Gaillardet and Lemarchand 2018). Rivers draining active volcanic areas show low $\delta^{11}B$ values due to hydrothermal input.

Boron concentrations in rain water are very low, but improved analytical techniques have allowed the very precise determination of B isotope values (Chetelat et al. 2009b; Millot et al. 2010a, b). $\delta^{11}B$ values in rain show a large variation depending on the sampling site (coastal vs. inland). Near coastal stations reflect the marine origin of boron, variably influenced by evaporation–condensation fractionation processes, inland stations may be affected by crustal, anthropogenic and biogenic boron sources.

Boron is widely used in industry; most commonly in the form of sodium perborate as an oxidative bleaching agent in cleaning products. The abundant use results in boron accumulation in waste effluents. Borate minerals and synthetic borate products are characterized by a narrow range in $\delta^{11}B$-values that are distinctly different from boron isotope values in unpolluted groundwater (Vengosh et al. 1994; Barth 1998). Thus, boron isotopes may identify or even quantify a potential contamination of surface waters.

### 2.7.5 Tourmaline

Tourmaline is the most abundant reservoir of boron in metamorphic and magmatic rocks. Tourmaline is stable over a very large p–T range and forms where crustal rocks interact with fluids or melts. Thus, its isotope composition provides a record of fluids and melts from which it crystallized. Swihart and Moore (1989), Palmer and Slack (1989), Slack et al. (1993), Smith and Yardley (1996) and Jiang and Palmer (1998) analyzed tourmaline from various geological settings and observed a large range in $\delta^{11}B$-values which reflects the different origins of boron and its high mobility during fluid related processes.

Boron isotope compositions of hydrothermal tourmalines from ore deposits vary from about +35‰ to −27‰ (Marschall and Jiang 2011; Trumbull et al. 2020). Noteworthy is the fact that large ranges of $\delta^{11}B$-values exist within each investigated ore deposit type. High $\delta^{11}B$-values can be related to seawater, whereas low $\delta^{11}B$-values are either derived from nonmarine evaporites or produced by interaction between rocks and fluids during metamorphic dehydration. Tourmalines in most granites and pegmatites show $\delta^{11}B$-values around −10‰ close to the average composition of the continental crust (Marschall and Jiang 2011).

Since volume diffusion of B isotopes is insignificant in tourmalines (Nakano and Nakamura 2001), isotopic heterogeneities of zoned tourmalines should be preserved up to at least 600 °C. By using the SIMS method, Marschall et al. (2008) demonstrated that boron isotopes in zoned tourmalines, indeed, may reflect different stages of tourmaline growth. Finally, it is noteworthy that the large chemical variability of tourmaline can be used as a fingerprint for a large number of other isotope systems including O, H, Si, Mg and Li (Marschall and Jiang 2011).

### 2.8–2.11 Alkaline Earth Elements

The alkaline earth elements **magnesium (2.8), calcium (2.9), strontium (2.10), barium (2.11)** belong to group 2 of the periodic table. They have similar properties forming divalent cations in minerals and solution. No redox reactions affect isotope fractionation, this leaves coordination as the main factor of equilibrium isotope fractionation.

## 2.8 Magnesium

Since the oxidation state of magnesium in natural compounds always is two, it might be expected that the natural range of Mg isotope composition is comparably small. On the other hand, Mg is incorporated during growth of biogenic $CaCO_3$ and plays an essential role during photosynthesis indicating that biological fractionations may play an important role for Mg isotopes.

Magnesium is composed of three isotopes (Rosman and Taylor 1998).

| | |
|---|---|
| $^{24}Mg$ | 78.99% |
| $^{25}Mg$ | 10.00% |
| $^{26}Mg$ | 11.01% |

Early investigations on Mg isotope variations have been limited by an uncertainty of 1–2‰. Catanzaro and Murphy (1966) for instance concluded that terrestrial Mg isotope variations are restricted to a few ‰. The introduction of multicollector-inductively coupled-plasma mass spectrometry (MC-ICP-MS) increased the precision by one order of magnitude and has initiated a new search of natural isotope variations (Galy et al. 2001, 2002). Factors affecting the accuracy of Mg isotopes measured by MC-ICP-MS have been summarized by Teng and Yang (2013). $\delta^{25}$Mg and $\delta^{26}$Mg values are reported relative to the DSM-3 standard (Galy et al. 2003; Oeser et al. 2014; Teng et al. 2015a). Teng et al. (2015a, b) reported Mg isotope compositions for a large number of reference materials, the long-term reproducibility for $\delta^{25}$Mg was 0.05‰ and for $\delta^{26}$Mg 0.07‰. Precision and accuracy are matrix dependent, matching sample and standard compositions is required for high-precision and high-accuracy Mg isotope analysis.

One of the advantages of the MC-ICPMS technique is the ability to measure $^{25}$Mg/$^{24}$Mg and $^{26}$Mg/$^{24}$Mg ratios independently many times smaller than the magnitude of the natural variations. The relationship between $^{25}$Mg/$^{24}$Mg and $^{26}$Mg/$^{24}$Mg ratios is diagnostic of kinetic versus equilibrium fractionations: for equilibrium processes the slope on a three-isotope diagram should be close to 0.521, for kinetic processes the slope should be 0.511 (Young and Galy 2004).

Figure 2.25 summarizes the natural $\delta^{26}$Mg isotope variations relative to DSM-3. Recently, Teng (2017) has reviewed Mg isotope geochemistry.

## 2.8.1 Calculated Isotope Fractionations

Calculations by Schauble (2011) yield systematic $^{26}$Mg isotope fractionations among silicates, carbonates and oxides with enrichments ín the order magnesite, dolomite, forsterite, orthoenstatite, diopside, periclase and spinel. Fractionations are controlled by the coordination number of Mg in minerals: tetrahedral sites tend to have higher $^{26}$Mg/$^{24}$Mg ratios than octahedral sites. The coordination number of Mg in olivine, orthopyroxene, clinopyroxene, hornblende and biotite is 6, thus limited isotope fractionations among these minerals are observed, however, in

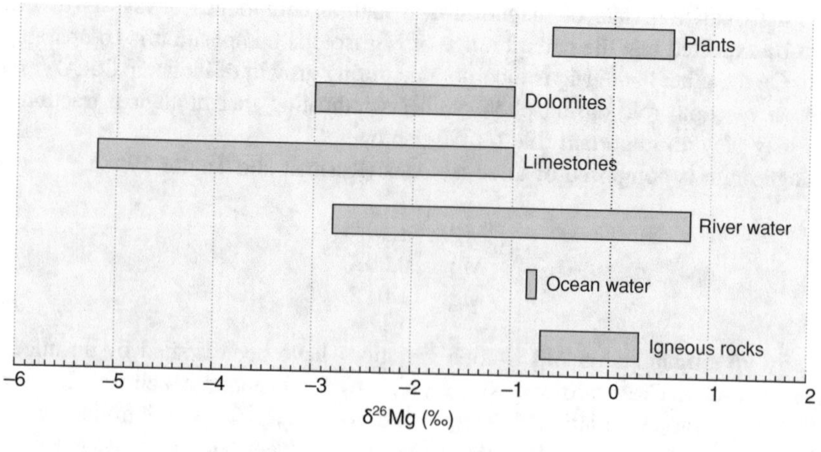


Fig. 2.25 $\delta^{26}$Mg values of important geological reservoirs

spinel the coordination is 4 and in garnet it is 8. Thus, pyrope is depleted in heavy Mg isotopes relative to pyroxenes and olivine, whereas spinel is enriched in heavy Mg isotopes relative to the sixfold coordination. Experimentally determined equilibrium isotope fractionations between spinel, forsterite and magnesite by Macris et al. (2013) are consistent with the postulated dependence on coordination numbers. Chemical diffusion experiments in silicate melts showed fractionations of $^{26}Mg/^{2}$ Mg by as much as 7‰ (Richter et al. 2008).

#### 2.8.1.1 Mantle Rocks

Peridotites have rather constant Mg isotope compositions suggesting a homogeneous mantle with a $\delta^{26}$Mg-value of −0.25‰ (Teng et al. 2010; Hu et al. 2016; Wang et al. 2016a, b) that has been used as the average Mg isotope composition of the mantle and bulk Earth. Many peridotites, however, are overprinted by metasomatic melt/rock interactions which may disturb the primary Mg isotope composition (Stracke et al. 2018). Extensively altered peridotites may show very large ranges in Mg isotope composition, as shown by De Obeso et al. (2020) for serpentinization and carbonation of the Oman ophiolite.

Studies by Teng et al. (2007a, b), Wiechert and Halliday (2007), Young et al. (2009), Handler et al. (2009) and Bourdon et al. (2010) have demonstrated slight or no differences between basalts and peridotite. Teng et al. (2007a, b) studied the behavior of Mg isotopes during basalt differentiation on samples from the Kilauea Iki lava lake. They found that highly differentiated basalts and olivine-rich cumulates have nearly identical Mg isotope compositions to the primitive magma indicating a lack of Mg isotope fractionation during crystal-melt fractionation. On the other hand, inter-mineral Mg isotope differences imply that fractional crystallization and preferential melting should cause Mg isotope fractionations between melt and residue. Very low $\delta^{26}$Mg-values have been interpreted to indicate metasomatism by carbonate-rich fluids. Basaltic rocks from eastern China with low $\delta^{26}$Mg-values have been interpreted to indicate crustal carbonates recycled into the mantle (Yang et al. 2012a, b; Huang et al. 2015; Tian et al. 2016; He et al. 2019).

Diffusion processes play an important role in causing Mg isotope variations in mantle rocks. Inter-mineral disequilibrium fractionations reflect mantle metasomatic processes or subsolidus Mg–Fe exchange (Xiao et al. 2013; Hu et al. 2016). Coupled Mg and Fe isotope fractionations induced by diffusion have been found in zoned olivines (Teng et al. 2011; Sio et al. 2013; Oeser et al. 2014), resulting from interdiffusion exchange of Mg and Fe. Lighter Fe isotopes diffuse in and lighter Mg isotopes diffuse out of the olivine, thereby causing negatively coupled isotope variations from rim to core (see Fig. 3.9).

The Mg isotope composition of the Moon and chondrites are indistinguishable from Earth, suggesting a homogeneous Mg isotope distribution in the solar system and no Mg isotope fractionation during the Moon-forming seem event (Sedaghatpour and Jacobsen 2019). In an earlier paper, Sedaghatpour et al. (2013) showed that low-Ti basalts are similar to terrestrial basalts whereas high-Ti basalts tend to have lighter Mg isotope compositions reflecting source heterogeneities produced during differentiation of the lunar magma ocean.

#### 2.8.1.2 Continental Crust

The upper continental crust is heterogeneous in Mg isotope composition and on average slightly heavier than the mantle (Shen et al. 2009; Li et al. 2010; Teng et al. 2013). First studies on the Mg isotope composition of the continental crust analyzed the composition of granites (Shen et al. 2009). Separated hornblendes and biotites and whole rocks suggest limited Mg isotope fractionation during granite differentiation. A detailed study of the upper crust by Li et al. (2010) analyzing different types of granites, loess and shale found large Mg isotope variations with a mantle-like average. The largest variations (more than 2.5‰) occur in sediments (Huang et al. 2013; Li et al. 2010; Telus et al. 2012; Teng 2017), which are due to mixing of isotopically light carbonates with heavy silicates. Yang et al. (2016) report Mg isotope data for the deep continental crust which indicate larger variations than for granites; the bulk continental crust has, however, a mantle like Mg isotope composition.

Clastic sediments are, overall, enriched in heavy Mg isotopes with $\delta^{26}$Mg values up to 0.92‰ (Li et al. 2010). During subduction, clastic sediments generally retain their Mg isotope composition (Li et al. 2014), thus recycling of clastic sediments will introduce Mg enriched in heavy isotopes into the mantle. Carbonates on the other hand are significantly depleted in heavy Mg isotopes. Light isotope values in basalts from the North China Craton have been interpreted to indicate recycling of carbonates derived from oceanic crust (Yang et al. 2012a, b).

### 2.8.2 Fractionations During Weathering

The behaviour of Mg isotopes during weathering is rather complex (Wimpenny et al. 2010; Ryu et al. 2011; Huang et al. 2012). Mg is soluble and mobile during weathering, potentially inducing small fractionations during dissolution and precipitation of minerals. Wimpenny et al. (2010) and Huang et al. (2012) observed that light Mg isotopes are preferentially released during dissolution of basalt leading to enriched residues.

Compared to dissolution, the behaviour of Mg isotopes during secondary formation of Mg minerals is even more complex (Huang et al. 2012). Soil and clays are generally heavier than their parent rocks (Tipper et al. 2006a, b, 2010; Opfergelt et al. 2012; Pogge von Strandmann et al. 2014) suggesting that heavy Mg isotopes are preferentially incorporated into the structure of clay minerals or absorbed in soils. Ryu et al. (2021) investigated the behavior of Mg during basalt weathering and soil development by sampling soils with increasing age. The youngest soils display Mg isotope ratios similar to basalt; with increasing age, soils become enriched in heavy Mg isotopes due to plant-related Mg recycling and progressive mineral transformations to goethite, gibbsite and kaolin minerals.

The complex behaviour of Mg during weathering results in large Mg isotope variations of river waters. As summarized by Li et al. (2012) $\delta^{26}$Mg values range from −3.80 to +0.75‰ reflecting differences of catchment lithologies particularly in the proportions of carbonate to silicate rocks. Tipper et al. (2006a) on the other hand observed a total variation in $^{26}$Mg of 2.5‰ and concluded that the lithology in the drainage area is of limited significance; instead, the major part of the variability may be attributed to fractionations in the weathering environment.

### 2.8.3 Ocean Water

The dominant Mg source to the ocean is riverine input with an average $\delta^{26}$Mg-value of −1.09‰ (Tipper et al. 2006b); major sinks are removal by hydrothermal fluids, dolomite formation and low-temperature clay formation during alteration of the oceanic crust. Because of its relatively long mean residence time, ocean water has a constant isotope composition of −0.80‰ that is slightly heavier than average river water resulting from Mg uptake into silicate minerals during weathering. Mg removal from seawater by hydrothermal interaction with the oceanic crust forming smectites and, at higher temperatures, chlorite does not cause a measurable Mg isotope fractionation. Dolomitisation, however, affects the ocean water, driving seawater to heavier values.

By analyzing pore waters from a large range of oceanographic settings, Higgins and Schrag (2010) demonstrated, that although Mg concentrations in pore waters do not vary in many deep-sea sediments, profiles of $\delta^{26}$Mg values are highly variable, which is best explained by precipitation of Mg-minerals in sediments or underlying crust.

### 2.8.4 Carbonates

Mg is present in $CaCO_3$ in the form of high Mg calcite (4 to ≈30 mol%), as low Mg calcite (≤4 mol%) and to a minor extent as aragonite (≤0.6 mol%). First-principles molecular dynamics calculations by Wang et al. (2019a, b, c) indicated that compared to aqueous $Mg^{2+}$, carbonates are depleted in heavy Mg isotopes with aragonite being the lightest carbonate mineral.

Marine organisms produce a wide range of $\delta^{26}$Mg values from −5 to −1‰ that are species dependent (Hippler et al. 2009; Li et al. 2012). Since the extent of Mg substitution in $CaCO_3$ is temperature dependant, Mg/Ca ratios are used as a thermometer for oceanic temperatures. The Mg/Ca temperature dependence, however, does not play a major role in determining Mg isotope ratios; the observed variability can instead be attributed to mineralogy (Hippler et al. 2009). Mg isotope fractionations between carbonates and water follows the sequence aragonite < dolomite < magnesite < calcite (Saenger and Wang 2014).

Vital effects in low-Mg calcite organisms exhibit no clear temperature dependence affecting the Mg isotope composition (Wang et al. 2013a, b). Most recent benthic and planktonic foraminifera show nearly identical $\delta^{26}$Mg ratios (Pogge von Strandmann 2008), making them suitable for investigating past isotopic variations of ocean water. Pogge von Strandmann et al. (2014) measured Mg isotopes of single-species planktonic foraminifera over the past 40 Ma and concluded that seawater Mg has changed from $\delta^{26}$Mg of −0.83‰ at present to 0‰ at 15 Ma.

In cave carbonates, equilibrium fractionations have been observed for low-Mg calcite speleothems (Galy et al. 2002). Mg isotope fractionations between speleothems and associated drip waters result in characteristic differences between both phases indicating near equilibrium conditions. Immenhauser et al. (2010) presented a complete data set of Mg isotopes on solid and liquid phases from a cave. They demonstrated that Mg isotope fractionations depend on a complex interplay of solution residence times, precipitation rates and adsorption effects.

Dolomite is the major rock-forming Mg-bearing mineral that forms under specific environmental conditions. During dolomite formation, Mg isotope ratios are affected by a variety of factors, making the application of Mg isotopes as a proxy for their depositional and diagenetic environment challenging. Geske et al. (2015) reported Mg isotope compositions of dolomite from various environments with a total range from −2.49 to −0.45‰. As observed by Azmy et al. (2013), early diagenetic dolomite inherits its isotope signature from precursor carbonates and diagenetic fluids. Later formed diagenetic dolomite phases may be slightly enriched in $^{26}$Mg suggesting that temperature is not the decisive factor, but instead the Mg-isotope composition of the interacting fluid upon diagenesis.

### 2.8.5 Plants and Animals

Magnesium is an essential plant nutrient that is central to photosynthesis. Black et al. (2008) investigated the Mg isotope distribution in wheat and observed a slight enrichment of the whole plant in $^{25}$Mg and $^{26}$Mg relative to the nutrient solution. These results have been confirmed by Boulou-Bi et al. (2010). Most of the plant Mg is bound in leaves, but the decisive process for the enrichment of $^{26}$Mg occurs at the root level. From roots to leaves or shoots a slight $^{26}$Mg depletion is observed (Boulou-Bi et al. 2010).

Mg plays a fundamental role in the formation of chlorophyll, in which it is embedded in a porphyrin or chlorin ring. Chlorophylls may have different forms (a–f) with chlorophyll-a being the main pigment used by most oxygen producing organisms. Using ab initio calculations Moynier and Fujii (2017) showed that the different forms of chlorpophyll differ in their Mg isotope composition. The biological process linked to the incorporation of Mg into the chlorophyll molecule induces Mg isotope fractionation; the sign and size of isotope fractionations depend on species and environmental conditions (Black et al. 2006, 2008; Ra and Kitagawa 2007; Ra 2010). Ra (2010) observed a 2.4‰ variation in $\delta^{26}$Mg of phytoplankton from different regions in the northwestern Pacific and related them to different growth rates and phytoplankton heterogeneities.

Mg isotopes may help reconstructing food webs of extinct animals. By analyzing bone and teeth apatite in mammals, Martin et al. (2014, 2015) demonstrated that soft tissues become enriched in heavy Mg isotopes whereas light Mg isotopes are preferentially excreted in feces. $\delta^{26}$Mg increases from herbivores to higher level consumers which is probably due to a $^{26}$Mg enrichment in muscle relative to bone.

## 2.9 Calcium

Calcium has six stable isotopes in the mass range of 40–48 being the largest relative mass difference except hydrogen and helium. Taylor and Rosman (1998) gave the following abundances.

| | |
|---|---|
| $^{40}$Ca | 96.94% |
| $^{42}$Ca | 0.647% |
| $^{43}$Ca | 0.135% |
| $^{44}$Ca | 2.08% |
| $^{46}$Ca | 0.004% |
| $^{48}$Ca | 0.187% |

Calcium plays an essential role in biological processes such as the calcification of organisms, and the formation of bones. Its wide natural distribution and the large relative mass difference suggest a large isotope fractionation, which may be caused by mass-dependent fractionations and by radiogenic growth (radioactive decay of $^{40}$K to $^{40}$Ca, half life of about 1.3 Ga). Due to the low K/Ca ratio of the mantle, the radiogenic $^{40}$Ca content of the mantle can be regarded as being virtually constant. Since the crust is enriched in K/Ca relative to the mantle, the Ca isotope composition of the crust is more affected by $^{40}$K decay, and, indeed, as demonstrated by Caro et al. (2010), Archean K-rich, Ca-poor rocks show enlarged $^{44}$Ca/$^{40}$Ca variations.

### 2.9.1 Analytical Techniques

Early studies on natural Ca isotope variations found no differences or ambigous results. By using a double-spike TIMS technique and by using a mass-dependent law for correction of instrumental mass fractionation, Russell et al. (1978) were the first to demonstrate that differences in the $^{44}$Ca/$^{40}$Ca ratio are clearly resolvable to a level of 0.5‰. More recent investigations by Skulan et al. (1997) and by Zhu and MacDougall (1998), also using the TIMS technique, have improved the precision to about 0.10–0.15‰.

MC-ICP-MS techniques have been described by Halicz et al. (1999) using "hot plasma" and by Fietzke et al. (2004) using "cool plasma". SIMS techniques with high spatial resolution and uncertainties of about 0.3‰ have been developed by Rollion-Bard et al. (2007) and Kasemann et al. (2008). A combined MC-ICPMS and TIMS approach has been presented by Schiller et al. (2012).

Comparing data obtained with different methods and from different laboratories, complications may arise from the use of different δ-values, either $\delta^{44/40}$Ca or $\delta^{44/42}$Ca, and from the use of different standards. Ca isotope data obtained by TIMS are generally reported as $\delta^{44/40}$Ca, whereas Ca isotope compositions obtained by MC-ICPMS are reported as $\delta^{44/42}$Ca-values. Data expressed as $\delta^{44/42}$ can be converted to $\delta^{44/40}$Ca by multiplying a factor of 2.048.

By initiating a laboratory exchange of internal standards, Eisenhauer et al. (2004) have suggested to use NIST SRM 915a as international standard. As the original SRM 915a is not any more available, SRM 915a has been replaced by SRM 915b which is 0.72‰ heavier than SRM 915a (Heuser and Eisenhauer 2008). In the following all data are given as $\delta^{44/40}$Ca-values. As shown in reviews by DePaolo (2004), Nielsen et al. (2011a, b, c), Fantle and Tipper (2014) and Gussone et al. (2016), the natural variation range in $\delta^{44/40}$Ca-values is about 5‰. Figure 2.26 shows natural Ca-isotope variations of important geological reservoirs.

## 2.9.2 High Temperature Fractionations

Calcium as a lithophile element does not partition into planetary cores, therefore Ca isotopes may reveal genetic links between Earth and meteorites. According to Simon and de Paolo (2010) and Valdes et al. (2014), Earth, Moon, Mars and differentiated asteroids are indistinguishable from ordinary chondrites, whereas enstatite chondrites are slightly enriched in heavier Ca isotopes and carbonaceous chondrites are variably depleted in heavier Ca isotopes. Ca isotopes, thus, suggest that ordinary chondrites are representative for the material that formed the terrestrial planets.

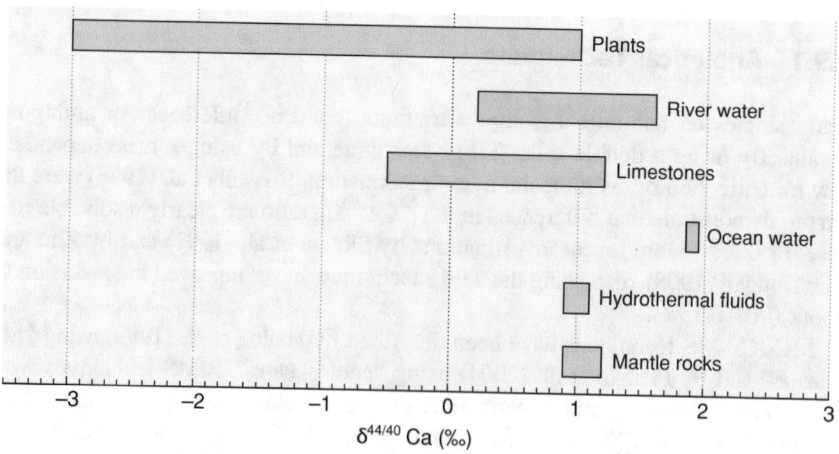


**Fig. 2.26** $\delta^{44/40}$Ca-values of important geological reservoirs

Huang et al. (2010), Kang et al. (2016, 2017) and Zhao et al. (2017a, b) analysed a suite of terrestrial mantle xenoliths, peridotites, ocean island basalts, komatiites and carbonatites. More recent studies by Amsellen et al. (2019), Chen et al. (2019a, b) and Ionov et al. (2019) demonstrated that mantle rocks generally display a homogeneous Ca isotope composition with $\delta^{44/40}$Ca-values of 0.94 ± 0.10‰. Deviations from this value observed in some mantle rocks may be attributed to kinetic isotope fractionations and/or metasomatism by melts. Since metasomatism tends to decrease $\delta^{44/40}$Ca ratios in the mantle, there has been debate whether Ca isotopes can distinguish between carbonate and silicate type of mantle metasomatism (e.g. Ionov et al. 2019; Kang et al. 2019). Sun et al. (2021) reported uniform Ca isotope compositions of carbonatites and associated silicate rocks comparable to basalts with Ca isotope fractionations during late stages of carbonatite evolution.

Pronounced kinetic isotope effects in high-temperature rocks have been reported by Antonelli et al. (2019). At temperatures around 900 °C they observed in whole rock samples a 4‰ variation in $\delta^{44/40}$Ca-ratios, in individual minerals 8‰ and between mineral pairs, for example garnet-plagioclase, variations from −1.5 to +1.5‰ inferring grain boundary diffusion during slow cooling. Ca isotope fractionations during partial melting and crystal fractionation are controversially discussed. Zhang et al. (2018a, b, c) reported no Ca isotope fractionation during crystal fractionation. Valdes et al. (2019), on the other hand, demonstrated that cogenetic samples from an ultramafic–mafic-anorthosite complex show considerable Ca isotope fractionations during fractional crystallization. They further demonstrated that plagioclase is considerable depleted in $\delta^{44/40}$Ca-ratios relative to olivine and pyroxene.

Huang et al. (2010) and Kang et al. (2016) measured the Ca isotope composition of coexisting clino- and orthopyroxene in mantle peridotites. $\delta^{44}$Ca-values of orthopyroxene are heavier than clinopyroxene with large variations that depend on the Ca/Mg ratio in orthopyroxene. First principles calculations by Feng et al. (2014) reached very similar conclusions. Combined with data from low-temperature Ca-minerals, Huang et al. (2010) inferred that inter-mineral fractionations are controlled by Ca–O bond strengths. Thus, the Ca-mineral with a shorter Ca–O bond yields a heavier $\delta^{44}$Ca-value.

### 2.9.3 Weathering

Chemical weathering of silicates controls long-term atmospheric $CO_2$ concentrations coupling the cycles of carbon and calcium. Dissolution of silicates and carbonates does not strongly fractionate Ca isotopes (Fantle and Tipper 2014). Ca ions released during dissolution may be taken up by vegetation, may precipitate as secondary minerals or can be absorbed by clays, oxyhydroxides and humic acids. As shown by Ockert et al. (2013), the absorption of $Ca^{2+}$ on clay minerals favors light Ca isotopes over heavy ones. The largest Ca isotope fractionation in the weathering environment, however, is the uptake by plants.

Ca isotope analysis of rivers represents another approach to identify weathering processes (Tipper et al. 2008, 2010; Fantle and Tipper 2014). From an extensive data compilation, Fantle and Tipper (2014) concluded that the average Ca isotope value of carbonates is 0.60‰, and of silicates is 0.94‰, whereas average river water has a value of 0.88‰. Since most of the Ca in river water originates from the dissolution of carbonates and not from silicates, the Ca isotope difference between carbonates and rivers remains unexplained.

### 2.9.4 Fractionations During Carbonate Precipitation

A large number of studies—summarized by Gussone and Dietzel (2016)—have investigated isotope fractionations during earth alkaline (Mg, Ca, Sr, Ba) precipitation experiments in which usually depletions of heavy isotopes in solid phases are observed. Isotope fractionations depend on many factors, such as temperature, precipitation rate, coordination environment. Marriott et al. (2004) presented a model of carbonate precipitation which is based on isotope fractionation at equilibrium, whereas Gussone et al. (2003) favored non-equilibrium isotope fractionations. DePaolo (2011) presented a surface-kinetic model, in which calcite precipitating from aqueos solutions does not form at isotope equilibrium.

Experiments on inorganic precipitation of calcite and aragonite (Marriott et al. 2004; Gussone et al. 2003) have demonstrated that Ca isotope fractionation correlates with temperature with an offset of aragonite of about −0.5‰ relative to calcite. During biogenic precipitation, the Ca isotope composition of shells depend on the chemistry of the solution, in which the organisms live and on the process by which Ca is precipitated (Griffith et al. 2008a, b). Calcification processes differ among different types of organisms: foraminifera precipitate carbonate in vacuoles from pH-modified seawater, corals pump seawater through various tissues to the site of precipitation. Each step in these processes may cause differences in Ca isotope fractionation.

The magnitude of Ca isotope fractionation during biogenic carbonate precipitation as well as the mechanism—either isotope equilibrium or kinetic effects—remain a matter of debate. Studies by Nägler et al. (2000), Gussone et al. (2005) and Hippler et al. (2006) reported temperature dependent Ca isotope fractionations precipitated in natural environments or under cultured laboratory conditions with a slope of about 0.02‰/°C. Temperature dependent fractionations, however, have not been found in all shell secreting organisms (Lemarchand et al. 2004; Sime et al. 2005). Sime et al. (2005) analyzed 12 species of foraminifera and found negligible temperature dependence for all 12 species. These contradictory results indicate a complex physiological control on Ca uptake by calcifying organisms (Eisenhauer et al. 2009). Thus, the Ca isotope composition of biogenic carbonates seems not to be a straightforward paleothermometer (Griffith and Fantle 2020).

Distinct Ca isotope fractionations have been observed in dolomites (Holmden 2009). For primary dolomite a $^{44}$Ca depletion relative to coexisting pore fluids has been observed in siliciclastic sediments (Wang et al. 2013a, b), whereas for dolomite formed during recrystallization more or less no Ca isotope fractionation has been reported (Holmden 2009).

Ca sulfates in evaporites show a relatively large range in $^{44/40}$Ca isotope ratios (Blättler and Higgins 2014), which is explained by Rayleigh fractionation during sulfate precipitation as the Ca isotope composition in the brine increases due to the preferential incorporation of $^{40}$Ca in the solid.

### 2.9.5 Variations of Ocean Water with Geologic Time

Zhu and MacDougall (1998) have made the first attempt to investigate the global Ca cycle. They found a homogeneous isotope composition of the ocean, but distinct isotope differences of the sources and sinks. The marine Ca-cycle is characterized by inputs from hydrothermal fluids at oceanic ridge systems and from dissolved Ca delivered by continental weathering and by output through $CaCO_3$ precipitation, the latter causing the main Ca isotope fractionation. Dissolution of silicate and carbonate rocks during weathering does not strongly fractionate Ca isotopes (Hindshaw et al. 2011). Hydrothermal solutions to the ocean at ocean ridges are about 1‰ depleted in $\delta^{44/40}$Ca values relative to seawater (Amini et al. 2008).

Since the first study of Zhu and MacDougall (1998), several studies have investigated secular changes in the Ca isotope composition of the ocean: De La Rocha and de Paolo (2000), Fantle and de Paolo (2005) and Fantle (2010) for the Neogene, Steuber and Buhl (2006) for the Cretaceous; Farkas et al. (2007) for the late Mesozoic; and Kasemann et al. (2005a, b) for the Neoproterozoic. Model simulations of the Ca cycle by Farkas et al. (2007) indicated that the observed Ca isotope variations can be produced by variable Ca input fluxes to the oceans. Maximum measured temporal variations in selected age periods are around 1‰ in $^{44/40}$Ca isotope ratios (see also p. 13 about ocean water history).

High resolution records with 0.3‰ excursions for the Permian–Triassic boundary from southern China have been reported by Payne et al. (2010) and by Hinojosa et al. (2012). Shifts in isotope composition could be due to changes in mineralogy (i.e. calcite/aragonite) or to a change in ocean pH-values. By comparing $\delta^{44}$Ca-values of conodont apatite with coexisting carbonates, Hinojosa et al. (2012) found a comparable shift in apatite, which argues against a shift in mineralogy, but favors changes of ocean acidification.

In this context, it is interesting to note, that Griffith et al. (2008c, 2011) proposed that pelagic barite, containing about 400 ppm Ca, might be an additional recorder of Ca seawater isotope composition through time showing an offset of about 2‰ from seawater.

In recent years Ca isotopes have also been used to constrain the degree of diagenesis (Lau et al. 2017; Ahm et al. 2018; Griffith and Fantle 2020). Since seawater has the highest $\delta^{44/40}$Ca value of any reservoir and biogenic primary carbonates are offset from seawater by about −1.3‰ and Ca isotope fractionations during carbonate recrystallization is close to zero‰, diagenetic overprint shifts the $\delta^{44/40}$Ca of marine carbonate closer to that of seawater (Fantle et al. 2020).

### 2.9.6 Plants, Animals and Humans

Vegetation shows the widest range in Ca isotope values, larger than variations caused by carbonate precipitation. Studies on higher plants by Page et al. (2008), Wiegand et al. (2005) and Holmden and Belanger (2010) demonstrated systematic Ca isotope fractionations between roots, stemwood and leaves: fine roots yield the lowest $\delta^{44}$Ca-values, stemwood are intermediate and leaves have the highest δ-values. Experiments under controlled plant growth conditions allow the identification of 3 different Ca isotope fractionation steps (Cobert et al. 2011; Schmitt et al. 2013): (i) preferential $^{40}$Ca uptake in the roots, (ii) preferential adsorption of $^{40}$Ca on the cell walls during transfer from the roots to the leaves, (iii) additional $^{40}$Ca fractionation in the storage organs, which seems to be controlled by the physiology of the plant.

Overall variation in $^{44}$Ca values from bottom to top in trees is about 0.8‰ (Cenki-Tok et al. 2009; Holmden and Belanger 2010). The magnitude of Ca isotope fractionation depends on species and on season (Hindshaw et al. 2013). The preferential uptake of light Ca-isotopes into plants results in an enrichment of Ca in soil solutions. Thus, vegetation controls the Ca isotope composition of soil pools (Cenki-Tok et al. 2009).

Ca isotope measurements of diet, soft tissues and bone show that bone is considerably lighter than soft tissue and diet. As much as 4‰ variation in $^{44}$Ca/$^{40}$Ca ratios is observed in single organisms (Skulan and DePaolo 1999). Ca isotopes of bone apatite in animals suggest that Ca isotope composition gets increasingly light as trophic levels increases. Thus, Ca isotopes have great potential to study vertebrate metabolism (Skulan and De Paolo 1999; Reynard et al. 2010; Heuser et al. 2011).

Bones in humans are continuously replaced. In healthy adults the rates of bone formation and the rates of bone loss are equal, and the net bone mineral balance is zero. Bone formation favors the lighter Ca isotopes depleting soft tissues in lighter isotopes, bone resorption releases the lighter isotopes back into soft tissue with no Ca isotope fractionation (Heuser et al. 2011; Morgan et al. 2012). In urine Ca isotopes shift to heavier δ-values during bone formation and to lighter δ-values during bone resorption. The latter process has been confirmed in bed rest studies (Heuser et al. 2011; Morgan et al. 2012). Ca isotopes, thus, are very sensitive tracers to assess net bone losses which characterize osteoporosis and some cancers (see also p. 13).

## 2.10 Strontium

Sr has 4 stable isotopes.

$^{84}$Sr 0.56%
$^{86}$Sr 9.86%
$^{87}$Sr 7.00%
$^{88}$Sr 82.58%

Isotopes of Sr mainly have been used as a geochronometer. Due to radioactive decay of $^{87}$Rb to $^{87}$Sr, the $^{87}Sr/^{86}Sr$ ratio of a sample together with the Rb/Sr concentration ratio carries geochronologic information. Conventional $^{87}Sr/^{86}Sr$ measurements by thermal ionisation mass-spectrometry (TIMS) use the $^{88}Sr/^{86}Sr$ ratio for internal instrumental mass fractionation correction. Normalization to a fixed $^{88}Sr/^{86}Sr$ ratio assumes that this ratio is constant for natural samples. However, as shown by Fietzke and Eisenhauer (2006), this is not the case. MC-ICP-MS and double spike TIMS methods document $^{88}Sr/^{86}Sr$ variations in terrestrial and meteoritic samples (Fietzke and Eisenhauer 2006; Krabbenhöft et al. 2009; Neymark et al. 2014). Figure 2.27 demonstrates the range of natural variations of $\delta^{88/86}$Sr-values relative to the $SrCO_3$ standard SRM987.

### 2.10.1 Silicates

Earth, Mars and Moon have indistinguishable bulk Sr isotope compositions, exceptions are some carbonaceous chondrites being depleted in heavy Sr isotopes (Moynier et al. 2010). The bulk Earth has a $\delta^{88/86}$Sr-value of 0.27‰, which more recently has been changed to 0.30‰ (Amsellem et al. 2018).

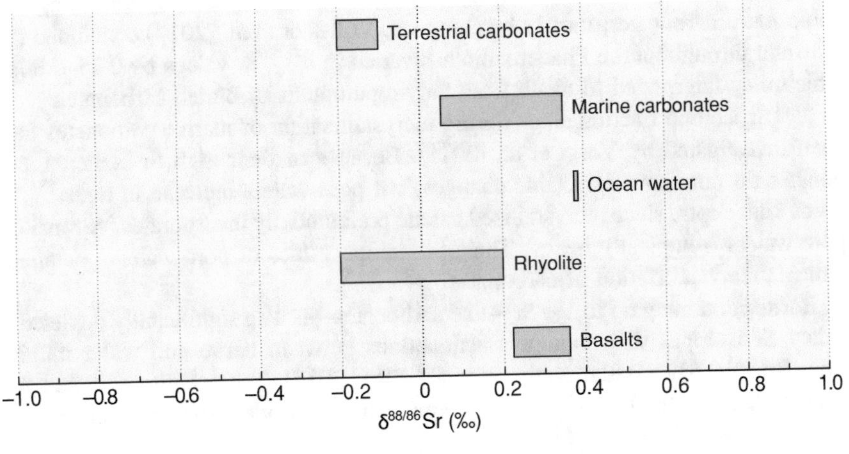


**Fig. 2.27** $\delta^{88/86}$Sr-values of important geological reservoirs

Early measurements by Halicz et al. (2008) and Charlier et al. (2012) indicated that basaltic rocks have a rather uniform value of +0.3‰ whereas more evolved rocks—andesites to rhyolites—have lighter values from −0.2 to +0.2‰. Charlier et al. (2012) interpreted the observed Sr isotope variations in differentiated rocks as resulting from isotope fractionations during fractional crystallization in which $^{88}Sr$ becomes enriched in plagioclase and K-feldspar. Amsellem et al. (2018) demonstrated on a suite of MORBs, OIBs and komatiites nearly constant isotope values around 0.3‰, except a few Barberton komatiites with higher Sr isotope values that possibly reflect Archean seawater alteration.

### 2.10.2 Carbonates and Sulfates

One of the main Sr isotope fractionation processes is the preferential uptake of lighter Sr isotopes during carbonate precipitation which is comparable with fractionations occurring in the Ca and Mg isotope systems. Sr isotope fractionations during inorganic precipitation of calcite depend primarily on precipitation rates resulting in larger fractionations at higher rates (Böhm et al. 2012). Carbonate precipitating organisms generally fractionate $^{88}Sr/^{86}Sr$ ratios by –0.1 to –0.2‰ relative to ocean water; the magnitude of Sr isotope fractionation is species dependent. Larger depletions in heavy isotopes have been observed in planktonic foraminifera (Böhm et al. 2012; Stevenson et al. 2014). Reports of temperature dependicies in corals (Fietzke and Eisenhauer 2006), Rüggeburg et al. 2008) have not been confirmed by more recent studies (Raddatz et al. 2013; Fruchter et al. 2016). Sr isotope ratios in corals have been suggested to reflect the composition of sea water with an offset of −0.2‰.

Knowledge of the magnitude of Sr fractionations during carbonate precipitation opens the possibility to quantify the output carbonate flux from the ocean (Krabbenhöft et al. 2010), which is not possible on the basis of $^{87}Sr/^{86}Sr$ ratios because ocean water and carbonates are very similar in $^{87}S/^{86}Sr$ ratios. By analysing biogenic fossil carbonates, mostly brachiopods, Vollstädt et al. (2014) concluded that seawater throughout the Phanerozoic has varied in $\delta^{88/86}Sr$ values by 0.25–0.60‰, which they interpreted to result from varying amounts of buried carbonates.

$^{88/86}Sr$ isotope fractionations during recrystallization of marine carbonates have been investigated by Voigt et al. (2015). Because of their high Sr contents, carbonates do not show detectable changes, but pore waters increase in their $^{88/86}Sr$ ratios with depth, since recrystallized calcite preferentially incorporates $^{86}Sr$ making pore waters isotopically heavy. Thus, Sr isotope values of pore waters potentially indicate recrystallization of carbonates.

Barite containing up to 1% Sr—like carbonates—is also significantly depleted in heavy Sr isotopes with isotope fractionations between barite and water ranging from 0 to −0.6‰ (Widanagamage et al. 2014, 2015). These authors argued that kinetic effects rather than equilibrium temperature dependent isotope fractionations control Sr isotope fractionations.

### 2.10.3 Fluids and Plants

By analysing Sr dissolved in rivers, the behaviour of $^{88/86}Sr$ during weathering has been investigated (Krabbenhöft et al. 2010; de Souza et al. 2010; Pearce et al. 2015; Chao et al. 2015). Krabbenhöft et al. (2010) demonstrated that large rivers are quite variable in $\delta^{88}Sr$. Pearce et al. (2015) estimated that the flux weighted $\delta^{88/86}Sr$-value of riverine input to the ocean is 0.32‰. They suggested that variations in the proportion of carbonate to silicate weathering plays a significant role in determining riverine values. However, as shown by Pearce et al. (2015), rivers draining silicate rocks have distinctly heavier Sr isotope values than their bedrocks pointing to fractionations during weathering. Shalev et al. (2017) suggested that the $^{88}Sr$ enrichment of rivers relative to carbonate and silicate source rocks is due to carbonate precipitation on the continents.

Hydrothermal fluids along the Mid Atlantic Ridge have an endmember composition of 0.24‰, which is similar to the average composition of the oceanic crust (Pearce et al. 2015).

Plants are isotopically lighter by 0.2–0.5‰ than corresponding soils (De Souza et al. 2010; Oeser and von Blanckenburg 2020). $\delta^{88}Sr$ values become enriched during translocation in plants (Oeser and von Blanckenburg 2020). Foliar tissues (leaves, flowers) are isotopically depleted relative to roots and stem.

## 2.11 Barium

Barium consists of 7 naturally occurring isotopes:

| | |
|---|---|
| $^{130}Ba$ | 0.11 |
| $^{132}Ba$ | 0.10 |
| $^{134}Ba$ | 2.42 |
| $^{135}Ba$ | 6.59 |
| $^{136}Ba$ | 7.85 |
| $^{137}Ba$ | 11.23 |
| $^{138}Ba$ | 71.70 |

Barium in nature occurs as discrete minerals such as barite and witherite ($BaCO_3$), but also may substitute potassium in common minerals, especially feldspars and calcium in carbonates. Distinct differences in Ba-isotope compositions have been first reported for barites and Ba carbonates of different origins (Von Allmen et al. 2010). By measuring $^{137}Ba/^{134}Ba$ ratios with a MC-ICP-MS technique, Von Allmen et al. (2010), Böttcher et al. (2012) and Pretet et al. (2015) reported that Ba minerals vary by up to 0.5‰. In more recent studies $^{138}Ba/^{134}Ba$ ratios (i.e. Horner et al. 2015, 2017) have been determined, which can be transferred into $^{137}Ba/^{134}Ba$ ratios by using of factor of 1.33 (Pretet et al. 2015). Ba isotope cosmochemistry and geochemistry has been reviewed by Charbonnier et al.

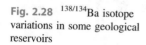

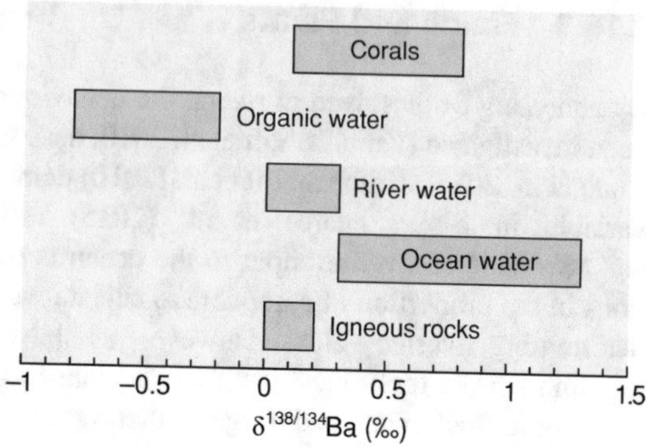


**Fig. 2.28** $^{138/134}$Ba isotope variations in some geological reservoirs

(2018). Figure 2.28 gives an overview of the natural observed $^{138/134}$Ba isotope variations.

Using modified MC-ICP-MS techniques Miyazaki et al. (2014) and Nan et al. (2015) further improved Ba isotope measurements. Van Zuilen et al. (2016a) presented Ba isotope ratios for 12 geological reference material by interlaboratory comparison. Van Zuilen et al. (2016b) showed experimentially that diffusion-driven Ba isotope fractionations can be expected in marine sediments. Synthesis experiments to precipitate $BaCO_3$ and $BaSO_4$ resulted in solid phases depleted in heavy isotopes compared to aqueous solutions (von Allmen et al. 2010; Mavromatis et al. 2016; Böttcher et al. 2018a, b). The magnitude of isotope fractionation is temperature insensitive, but at least for carbonates depends on growth rates. Fast growth rates yield small fractionations, slow growth rates yield larger fractionations.

## 2.11.1 Magmatic Systems

As a large lithophile element, Ba is highly incompatible during mantle melting and thus will become strongly enriched in the crust relative to the mantle. The Ba isotope composition of the upper continental crust has been investigated by Nan et al. (2018). They showed that granites, loess, glacial diamictites and river sediments varied in $^{137}$Ba/$^{134}$Ba ratios from −0.47 to 0.35‰. I-, S- and A-type granites differ in Ba isotope composition which may result from variable contributions of crustal materials and from isotope fractionations of Ba-rich minerals during late stages of differentiation.

As shown by Deng et al. (2021) Ba isotopes can be fractionated during granite differentiation. Higher differentiated albite granites are enriched in heavy Ba isotopes relative to K-feldspar granites. Biotites are enriched in heavy Ba isotopes relative to K-feldspar; coexisting biotite and muscovite indicate non-equilibrium conditions.

Basalts analyzed by Nan et al. (2015) showed small resolvable variations in Ba isotope composition. Nielsen et al. (2018) observed 0.1‰ variations in different types of MORB having variable radiogenic and trace element ratios, which they explained by sedimentary contamination.

Ba as a fluid mobile element should be also well suited to trace fluid/melt interaction in subduction zone volcanic rocks. Wu et al. (2020a, b) postulated that Ba isotopes may distinguish between fluid and melt derived from different subducted components. Nielsen et al. (2020) demonstrated that a small negative Ba isotope fractionation occurs during Ba mobilization in subduction zones. Guo et al. (2020a, b) showed experimentally that light Ba isotopes favor aqueous fluids over silicate melts.

### 2.11.2 Ocean

Although not being a primary nutrient, dissolved Ba in ocean water shows a nutrient-type behaviour with low concentrations in surface and elevated concentrations in deep water. Since barium takes part in biological and chemical processes, barium concentrations and isotopes in ocean water may be used as a proxy of primary productivity, nutrient cycling, and water mass mixing. The main Ba input to the ocean is from rivers, but vent fluids also play a role (Hsieh and Henderson 2017). The main output is by precipitation of barite preferentially removing light Ba isotopes to the solid.

Seawater profiles (Horner et al. 2015; Cao et al. 2016, 2020) indicate distinct patterns of water-mass mixing. These authors concluded that organic carbon, $BaSO_4$ cycle and Ba isotope fractionation in seawater are closely coupled. As shown by Bates et al. (2017) Ba isotopes in conjunction with Ba concentration allow the distinction between small scale barite precipitation inducing Ba isotope fractionation and large-scale water mass mixing inducing no fractionation. Bates et al. (2017) demonstrated on 4 vertical water profiles that a combination of both processes can explain the observed barium isotope distribution.

Hsieh et al. (2021) observed in vent fluids a wide range of Ba isotope compositions with the lightest values being close to the source rocks. During subsequent mixing with seawater, barite precipitation removes isotopically light Ba from vent fluids. Hsieh et al. (2021) estimated the hydrothermal input to Atlantic deep water as to be in the range from 3 to 9%.

Marine carbonates are assumed to preserve the Ba isotope composition of seawater, but Ba incorporated in or absorbed on detrital minerals may complicate its application for environmental reconstructions. Corals from different oceanic localities show considerable Ba isotope variations which also suggest a heterogeneous Ba isotope composition of seawater (Pretet et al. 2015). Corals grown in aquarium experiments indicate vital isotope fractionation effects. Hemsing et al. (2018) observed a constant fractionation of −0.21‰ between coral aragonite and seawater allowing the reconstruction of the Ba seawater composition.

## 2.12 Silicon

Silicon has three stable isotopes with the following abundances (Rosman and Taylor 1998):

$^{28}Si$ 92.23%
$^{29}Si$ 4.68%
$^{30}Si$ 3.09%

Because of its high abundance on Earth, silicon is a very interesting element to be investigated for isotope variations. The biogeochemical cycle of silicon is of special importance, because it is coupled with the marine carbon cycle being of crucial importance for the global climate. Under natural conditions, silicon is in general bound to oxygen. Thus, redox reactions at the earth,s surface do not occur, which should exclude large Si isotope fractionations. Early investigations by Douthitt (1982) and Ding (1996) observed a total range of $\delta^{30}Si$ values in the order of 6‰. This range has extended to about 12‰ with the lowest $\delta^{30}Si$ value of −5.7‰ in siliceous cements (silcretes) (Basile-Doelsch et al. 2005) and the highest of +6.1‰ for rice grains (Ding et al. 2005). A recent review has been published by Poitrasson (2017). Aspects of low temperature Si geochemistry have been discussed by Frings et al. (2016) and Sutton et al. (2018).

In early studies, silicon isotope ratios have been generally measured by fluorination methods (Douthitt 1982; Ding 1996). This method, however, is time consuming and potentially hazardous, therefore, more recently MC-ICP-MS techniques have been introduced (Engstrom 2006). Chmeleff et al. (2008) have shown that a UV-femtosecond laser ablation system coupled with MC-ICP-MS gives $\delta^{29}Si$- and $\delta^{30}Si$-values with very high precision. Determinations with SIMS have been carried out by Robert and Chaussidon (2006), Heck et al. (2011) and others. Independent of the method used, the commonly used standard is NBS-28 quartz. Figure 2.29 summarizes the naturally occurring silicon isotope variations.

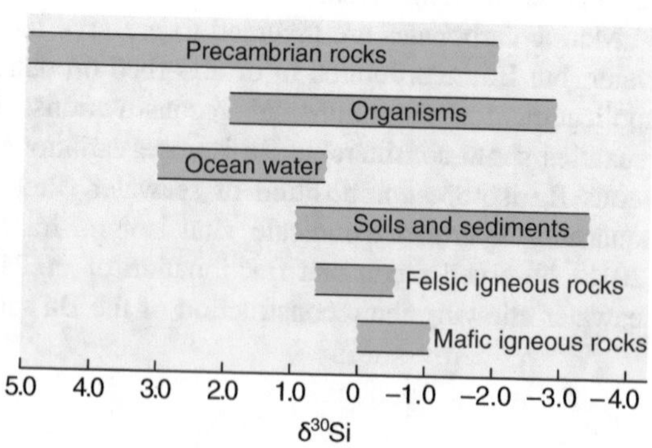


**Fig. 2.29** $\delta^{30}Si$-values of important geological reservoirs

### 2.12.1 Equilibrium Isotope Fractionations

First-principle calculations based on density functional theory have been carried out by Méheut et al. (2007, 2009) to determine the silicon (and oxygen) isotope fractionations between quartz, enstatite, forsterite, lizardite and kaolinite. Based on an earlier study by Grant (1954), it was originally suggested that Si isotope fractionation should highly depend on the degree of polymerization of the $SiO_4^{4-}$ tetrahedra. Later, Méheut and Schauble (2014) showed that the cationic content, in specific Al, of a mineral plays an important role on its silicon isotope fractionation. They could explain the fractionation behavior of the investigated minerals, except forsterite suggesting that other parameter also are important. Huang et al. (2014) and Qin et al. (2016) calculated Si isotope fractionations among the major minerals in mantle rocks and in granites, showing isotope fractionations are strongly correlated with Si–O bond lengths. The order of $^{30}$Si enrichment is quartz > albite > anorthite > olivine, zircon > enstatite > diopside. Because of the small Si isotope fractionations among minerals, experimentally determined Si isotope fractionation are scarce. A notable exception is a study by Trail et al. (2019), investigating the quartz-zircon system.

### 2.12.2 High-Temperature Fractionations

A number of studies have estimated the $\delta^{30}$Si-value of the bulk silicate Earth to be −0.29‰ (Fitoussi et al. 2009; Savage et al. 2011, 2014; Armytage et al. 2011; Zambardi et al. 2013). This value is identical with that of the Moon, but isotopically heavier than all types of meteorites. Ordinary chondrites have $\delta^{30}$Si values being on average 0.15‰ lighter than the bulk silicate earth (Armytage et al. 2011; Zambardi et al. 2013). The difference is best explained by Si isotope fractionation during Earth's core formation (Georg et al. 2007). High pressure, high temperature experiments by Shahar et al. (2009) indicated a 2‰ fractionation between metal and silicate melts (see discussion on p. 315). Similar findings have been reported by Ziegler et al. (2010) by measuring silicon isotope fractionations between Si in metal and silicates in enstatite achondrites. But, as demonstrated by Huang et al. (2014), Si isotope fractionations decrease with increasing pressure, thus silicon isotope fractionations obtained experimentally at relatively low pressures may not be applicable to the high-pressure conditions of core formation.

Using a continuous accretion model, the Si isotope fractionation can be used to constrain the amount of Si that entered the Earth's core (Chakrabarti and Jacobsen 2010; Zambardi et al. 2013 and others). Estimated percentages vary somewhat depending on model assumptions, but generally are between 6 and 12%.

Equilibrium isotope fractionations among mantle minerals are very small, but may become significant between minerals with different Si coordination numbers, such as Mg-perovskite in 6-coordination and olivine in 4-coordination. No differences in Si isotope composition are observed between ultramafic rocks and basalts indicating no isotope fractionation during partial melting (Savage et al. 2011,

2014). As shown on rocks from the Hekla volcano, Iceland, magmatic differentiation may cause Si isotope fractionation (Savage et al. 2011).

Felsic igneous rocks are slightly heavier than mafic igneous rocks and exhibit small, but systematic $^{30}$Si variations increasing with the silicon contents of igneous rocks and minerals. The order of $^{30}$Si enrichment in common minerals correlates with the order of $^{18}$O enrichment (Qin et al. 2016). However, silicon isotope variations are relatively small and generally not very sensitive to sedimentary input. Nevertheless, as argued by Pringle et al. (2016), Si isotopes can be used as tracers for the presence of crustal material in oceanic island basalts. By combining Si- and O-isotope data from zircons, Trail et al. (2018) presented evidence that coupled Si- and O-data can constrain sedimentary input into felsic melts.

High-temperature hydrothermal fluids acquire isotope compositions close to those of igneous rocks. Subsequent precipitation of amorphous silica at lower temperatures lead to the loss of isotopically depleted silica from the solution (Geilert et al. 2015). At the Geysir geothermal field in Iceland, Geilert et al. (2015) showed that silicon isotope fractionation during deposition of amourphous silica correlates inversely with temperature, but is essentially a function of precipitation rate.

As shown by Savage et al. (2013a, b) the average silicon isotope composition of the upper and lower continental crust is close to the composition of the bulk silicate earth. Silicon isotope ratios of quartzites and sandstones are in the range of felsic magmatic and metamorphic rocks reflecting their detrital derivation (Andre et al. 2006). In contrast, microcrystalline quartz from silcretes and clay minerals formed by weathering processes incorporate preferentially light Si isotopes relative to igneous minerals. Shales display $\delta^{30}$Si-values between −0.82 and 0.00‰ reflecting various degrees of physical and chemical weathering, diagenetic processes and source lithologies (Savage et al. 2013b).

### 2.12.3 Chemical Weathering and Mineral Precipitation

Considerable Si isotope fractionation takes place during chemical weathering (Ziegler et al. 2005a, b; Basile-Doelsch et al. 2005; Georg et al. 2006; Cardinal et al. 2010; Opfergelt et al. 2012; Pogge von Strandmann et al. 2014). During dissolution of primary silicate minerals, silicon partitions in about equal proportions into the dissolved phase that is isotopically enriched and into solid secondary phases that are isotopically depleted (Ziegler et al. 2005a, b; Georg et al. 2006). Soils show relatively large silicon isotope variations reflecting the simultaneous action of weathering of primary phases, precipitation of clay minerals and, as shown by Oelze et al. (2014), absorption of silica on Fe phases and amorphous Al-hydroxides. Oelze et al. (2014) demonstrated that preferential adsorption of $^{28}$Si on Al-hydroxides may be the main cause for the light isotope signature of clay minerals. Thus, weathering processes can be regarded as one of the main fractionation mechanism separating silicon isotopes into an isotopically heavy dissolved phase and an isotopically light residue.

Rivers integrate the complex reactions occurring during weathering and biological processes. Both the formation of secondary weathering products and the uptake of silicic acid by plants and diatoms enriches the water in the heavy isotope. For the Yangtze river, Ding et al. (2004) measured for dissolved silicon a $\delta^{30}$Si range from 0.7 to 3.4‰, whereas the suspended matter has a more constant composition from 0 to −0.7‰. For the Congo, Cardinal et al. (2010) measured low $\delta^{30}$Si values close to zero‰ for small tributaries rich in organic carbon ("black water") and high $\delta^{30}$Si values close to 1‰ in large tributaries. As summarized by Frings et al. (2016), the range of $\delta^{30}$Si values in river water worldwide is almost 5‰ which is due to fractionation processes during weathering but not to isotope variations of source rocks. The mean value for rivers is 1.35‰ (Frings et al. 2016). In recent years a number of studies have investigated Si isotopes in glacial systems (Hatton et al. 2019, 2020; Hirst et al. 2020), which lack the interference from primary productivity. Hatton et al. (2019) demonstrated that glacial meltwaters are less depleted in $\delta^{30}$Si than non-glacial rivers (0.16 vs. 1.38‰).

Georg et al. (2009) presented $\delta^{30}$Si values of dissolved Si in groundwaters. Of special interest is the observation that $\delta^{30}$Si decreases by about 2‰ along the groundwater flow path of 100 km deciphering complex Si-cycling, weathering and diagenetic reactions.

### 2.12.4 Fractionations in Ocean Water

Silicic acid is an important nutrient in the ocean that is required for the growth of mainly diatoms and radiolaria. Silicon incorporation into siliceous organisms is associated with Si isotope fractionation, because $^{28}$Si is preferentially removed as the organisms form biogenic silica (De la Rocha et al. 1998, 2003, 2006; Hendry et al. 2010; Egan et al. 2012; Hendry and Brzezinski 2014). Both field studies and laboratory culture experiments have revealed that diatoms incorporate lighter Si isotopes with a more or less constant silicon isotope fractionation of −1.1‰ (De la Rocha et al. 1997, 1998; Varela et al. 2004, and others). Culture experiments by Demarest et al. (2009) demonstrate that Si isotopes also fractionate during biogenic silica dissolution whereby the lighter isotopes are preferentially released into seawater with a fractionation of about 0.55‰. Thus, dissolution acts in the opposite sense to production and reduces the net silicon fractionation considerably.

Dissolved silicon in the oceans exhibit a relatively large range in $\delta^{30}$Si-values from 0.5 to 4.4‰ (Poitrasson 2017). Surface water in tropical oceans are particularly enriched in heavy Si isotopes as a result of diatoms blooms that consume dissolved silicon. Thus, an increase in opal formation by diatoms results in more positive $\delta^{30}$Si-values, whereas a decrease results in more negative δ-values. In this manner variations in $^{30}$Si contents of diatoms may provide information on changes of oceanic silicon cycling (De la Rocha et al. 1998). This makes $\delta^{30}$Si-values in sediments a promising proxy for past silica concentrations. $\delta^{30}$Si sedimentary records over the last glacial cycles indicate that glacial values are generally 0.5–1‰

lower than interglacial values which is interpreted in terms of paleoproductivity (Frings et al. 2016).

Diatoms as surface dwellers give a surface water signal only. Sponges, however, can be found throughout the water column. The $\delta^{30}Si$ of sponges is thus a potential proxy to quantify changes in oceanic Si concentrations (Hendry et al. 2010; Wille et al. 2010). As shown by these authors $^{30}Si$ fractionations during biosilification of sponges depends on silica concentrations in sea water with larger $^{30}Si$ depletions as silica concentrations increase. Thus, $\delta^{30}Si$ values of fossil silicified sponges may be used as a proxy for the reconstruction of palaeo Si-concentrations during the past (Hendry et al. 2010; Wille et al. 2010).

### 2.12.5 Cherts

Modern cherts are formed via biological precipitation of siliceous organisms while Precambrian cherts are formed by inorganic precipitation. Silicon isotopes in cherts have been used to infer the Si isotope composition of sea water, but it is still unclear whether Si isotope compositions of sedimentary chert deposit reflect primary sedimentary conditions. For example, Marin-Carbonne et al. (2014) and Zakharov et al. (2021) identified a complex diagenetic history of transformation of biogenic and abiogenic signatures into microcrystalline quartz requiring careful in situ SIMS work. This is because chert forms by diagenetic dissolution-reprecipation processes, meaning that crystalline phases increase relative to amorphous phases inducing silicon isotope fractionations (Tatzel et al. 2015).

A wide range of $\delta^{30}Si$ values from −0.8 to +5.0‰ have been reported for Precambrian cherts (Robert and Chaussidon 2006; Marin-Carbonne et al. 2014; Stefurak et al. 2015 and others), much larger than for Phanerozoic cherts. These authors observed a positive correlation of $\delta^{18}O$ with $\delta^{30}Si$ values, which they interpreted as reflecting temperature changes in the ocean from about 70 °C at 3.5 Ga to about 20 °C at 0.8 Ga years ago. In contrast, cherts within banded iron formations exhibit largely negative $\delta^{30}Si$-values from −2.5 to −0.5‰ (Andre et al. 2006; Van den Boorn et al. 2010) reflecting different sources of silica. These authors argued that variations in $\delta^{30}Si$ are best explained by mixing between hydrothermal fluids and seawater. Lamina-scale Si isotope heterogeneity within individual chert layers up to 2.2‰ may reflect the dynamics of hydrothermal systems (see "chert" section in Chap. 3.

### 2.12.6 Plants

Silicon is an important element for vascular plants favouring growth. Silicon is taken up by terrestrial plants from soil solution, transported into the xylem and deposited as hydrated amorphous silica to form phytoliths that are restored to the soil by decomposition of plant material. Already Douthitt (1982) noted that Si uptake by plants leads to Si isotope fractionation. Plants preferentially incorporate the light Si isotopes; Si concentrations and $\delta^{30}$Si-values increase from soil and roots through the stem and leaves. $\delta^{30}$Si values range from −2.7 (rice stem) to 6.1‰ (rice grains) (Ding et al. 2005, 2008; Sun et al. 2016a, b) with large interplant fractionations between low values in roots and high values in leaves and corn. Phytoliths persisting after the decay of plants in soils may be used for palaeoenvironmental reconstructions (Prentice and Webb 2016), however, the extent of secondary isotope alteration has to be considered before their use as an environmental indicator.

### 2.13–2.14 The Halogens, Chlorine and Bromine

The five elements in group 17 of the periodic table: fluorine, chlorine, bromine, iodine and astatine have some chemical properties in common. The halogens form geochemically a rather coherent group showing anionic character. With respect to stable isotope variations, only chlorine and bromine are suitable, the other three elements are mono-isotopic. Chlorine and bromine are commonly found in ocean water. The most abundant solid compounds are salts built during evaporation of water bodies. Another important group are organohalogens, dominated by chlorinated hydrocarbons that are widely distributed in the environment.

## 2.13 Chlorine

Chlorine has two stable isotopes with the following abundances (Coplen et al. 2002):

$$^{35}\mathrm{Cl} \quad 75.78\%$$
$$^{37}\mathrm{Cl} \quad 24.22\%$$

Natural isotope variations in chlorine isotope ratios might be expected due to the mass difference between $^{35}$Cl and $^{37}$Cl as well as to variations in coordination of chlorine in the vapor, aqueous and solid phases. Schauble et al. (2003) calculated equilibrium fractionation factors for some geochemically important species. They showed that the magnitude of fractionations systematically varies with the oxidation state of Cl, but also depends on the oxidation state of elements to which Cl is bound with larger fractionations for 2+ cations than for 1+ cations.

### 2.13.1 Methods

Measurements of Cl-isotope abundances have been made by different techniques. The first measurements by Hoering and Parker (1961) used gaseous chlorine in the form of HCl. The 81 samples measured exhibited no significant variations relative to the standard ocean chloride. In the early eighties a new technique has been developed by Kaufmann et al. (1984), that uses methylchloride ($CH_3Cl$). The chloride-containing sample is precipitated as AgCl, reacted with excess methyliodide, and separated by gas chromatography. The total analytical precision reported is near ±0.1‰ (Long et al. 1993; Eggenkamp 1994; Sharp et al. 2007). The technique requires relatively large quantities of chlorine (>1 mg), which precludes the analysis of materials with low chlorine concentrations. Magenheim et al. (1994) described a method involving the thermal ionization of $Cs_2Cl^+$, which, as argued by Sharp et al. (2007), is very sensitive to analytical artefacts and therefore might lead to erroneous results. In any case both methods are laborintensive and rely on offline chemical conversion reactions. Recent attempts use continuous flow mass-spectrometry (Shouakar-Stash et al. 2005) or use MC-ICPMS techniques (Van Acker et al. 2006). SIMS techniques have been described for glasses by Layne et al. (2004), Godon et al. (2004) and more recently by Manzini et al. (2017).

$\delta^{37}$Cl-values are generally given relative to seawater chloride termed SMOC (Standard Mean Ocean Chloride). The current knowledge about chlorine isotope geochemistry has been summarized in a book by Eggenkamp (2014) and in a review article by Barnes and Sharp (2017). A summary of the observed natural chlorine isotope variations is presented in Fig. 2.30. Ransom et al. (1995) gave a natural variation range in chlorine isotope composition of about 15‰ with subduction zone pore waters having $\delta^{37}$Cl values as low as −8‰ whereas minerals in which Cl substitutes OH have $\delta^{37}$Cl values as high as 7‰.

## 2.13.2 Hydrosphere

Chloride ($Cl^-$) is the major anion in surface- and mantle-derived fluids. It is the most abundant anion in ocean water and in hydrothermal solutions and is the dominant metal complexing agent in ore forming environments (Banks et al. 2000). Despite its variable occurrence, chlorine isotope variations in natural waters commonly are small and close to the chlorine isotope composition of the ocean. This is also true for chlorine from fluid inclusions in hydrothermal minerals which indicate no significant differences between different types of ore deposits such as

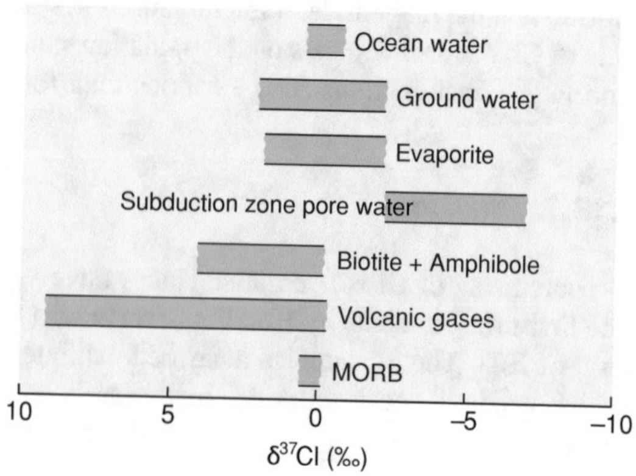


**Fig. 2.30** $\delta^{37}$Cl values of important geological reservoirs

Mississippi-Valley and Porphyry Copper type deposits (Eastoe et al. 1989; Eastoe and Gilbert 1992). $^{37}Cl$ depletions detected in some pore waters have been attributed to processes such as ion filtration, alteration and dehydration reactions and clay mineral formation (Long et al. 1993; Eggenkamp 1994; Eastoe et al. 2001; Hesse et al. 2006). A pronounced downward depletion of −4‰ in pore waters has been presented by Hesse et al. (2006). Even lower $\delta^{37}Cl$-values have been reported in pore waters from subduction-zone environments (Ransom et al. 1995; Spivack et al. 2002). The downward depletion trend might be explained by mixing of shallow ocean water with a deep low $^{37}Cl$ fluid of unknown origin. Barnes and Sharp (2017) proposed that kinetic fractionations associated with the formation of organochlorine compounds may explain the very light $\delta^{37}Cl$ values.

Relatively large isotopic differences have been found in slow flowing groundwater, where Cl-isotope fractionation is attributed to a diffusion process (Kaufmann et al. 1984, 1986; Desaulniers et al. 1986). Desaulniers et al. (1986) for instance investigated a ground water system, in which chloride diffused upward from saline into fresh water deposits by demonstrating that $^{35}Cl$ moved about 1.2‰ faster than $^{37}Cl$. Some deep aquifers in sedimentary basins may remain isolated for millions of years (Holland et al. 2013). As shown by Giunta et al. (2017) in such aquifers correlated Cl- and Br-isotope fractionations may occur due to the Earth's gravity field.

Cl isotope fractionations between salt minerals and brine have been determined by Eggenkamp et al. (1995), Eggenkamp et al. (2016), Eastoe et al. (1999, 2007). Eggenkamp et al. (2016) showed that chlorine and bromine isotope fractionation during precipitation from a saturated salt solution depends on the respective cation of the precipitated salt. NaCl is enriched in $^{37}Cl$ by 0.3‰ relative to the brine, whereas potassium and magnesium chloride show little fractionation relative to the brine. 83 halite samples from different geological periods varied from −0.24 to +0.51‰ in $\delta^{37}Cl$ values (Eggenkamp et al. 2019). They concluded that the Cl isotope composition of the ocean has been constant over the last 2 billion years.

### 2.13.3 Mantle-Derived Rocks

Controversial results have been reported for chlorine isotopes in mantle-derived rocks. According to Magenheim et al. (1995) $\delta^{37}Cl$-values for MORB glasses show a surprisingly large range. By questioning the findings of Magenheim et al. (1995), Sharp et al. (2007) argued that the mantle and the crust have very similar isotopic composition. A possible explanation for this apparent discrepancy might be related to analytical artifacts of the TIMS technique (Sharp et al. 2007). Bonifacie et al. (2008) also observed small Cl-isotope variations only in mantle derived rocks. They demonstrated that $\delta^{37}Cl$ values correlate with chlorine concentrations: Cl-poor basalts have low $\delta^{37}Cl$ values representing the composition of uncontaminated mantle derived magmas, whereas Cl-rich basalts are enriched in $^{37}Cl$ being contaminated by ocean water. In contrast to MORB, John et al. (2010) reported larger $\delta^{37}Cl$ variations in OIB glasses which they interpreted as being due to subducting sediments that have developed high $\delta^{37}Cl$-values by expelling $^{37}Cl$ depleted pore fluids.

Barnes et al. (2009) have investigated the serpentinization process in the oceanic lithosphere and interpreted chlorine isotope data to reflect a record of multiple fluid events. Slightly positive $\delta^{37}$Cl-values represent typical seawater-hydration conditions under low temperature conditions; negative $\delta^{37}$Cl-values result from interaction with porefluids from overlying sediments.

Volcanic gases and associated hydrothermal waters have a large range in $\delta^{37}$Cl-values from –2 to +12‰ (Barnes et al. 2006). To evaluate chlorine isotope fractionations in volcanic systems, HCl liquid–vapor experiments performed by Sharp (2006) yield large isotope fractionations of dilute HCl at 100 °C. $^{37}$Cl enrichments in fumaroles seem to be due to isotope fractionations between between $Cl^-$ in aquatic solution and HCl gas.

Unusually high chlorine isotope values ($\delta^{37}$Cl from 0 to +81‰) have been reported for lunar basalts, glasses and apatite (Sharp et al. 2010; Boyce et al. 2015; Stephant et al. 2019; Barnes et al. 2019 and others). Depletions in chlorine content and isotope enrichment are generally interpreted to result from degassing of the lunar magma ocean early in the Moon's history.

### 2.13.4 Applications in the Environment

Chlorine isotope studies have been performed to understand the environmental chemistry of anthropogenic organic compounds, such as chlorinated organic solvents or biphenyls. The primary goal of such studies is to identify sources and biodegradation processes in the environment. To do this successfully, chorine isotope values should differ among compounds and manufacturers and indeed the range of reported $\delta^{37}$Cl-values is from about –5 to +6‰ with distinct signatures from different suppliers (Van Warmerdam et al. 1995; Jendrzewski et al. 2001) (see also Sect. 3.10.3).

Perchlorate is another anthropogenic compound, which may contaminate surface and ground waters. The widespread occurrence of perchlorate in the environment makes it desirable to distinguish between a synthetic and a natural origin (Böhlke et al. 2005). The occurrence of natural perchlorate is limited to extremely dry environments, such as the Atacama desert. Synthetic perchlorate is produced by electrolyte oxidation reactions, whereas natural perchlorate is formed by photochemical reactions involving atmospheric ozone. Böhlke et al. (2005) showed that natural perchlorate has the lowest $\delta^{37}$Cl-values on Earth (as low as −15‰), whereas synthetic perchlorate has more "normal" $\delta^{37}$Cl-values. During microbial reduction of perchlorate, large kinetic isotope effects have been observed by Sturchio et al. (2003) and Ader et al. (2008), which may document in situ bioremediation (see also Sect. 3.9.9).

## 2.14 Bromine

Bromine has two stable isotopes with nearly equal abundances (Berglund and Wieser 2011).

$$^{79}Br \quad 50.69\%$$
$$^{81}Br \quad 49.31\%$$

Eggenkamp and Coleman (2000) measured Br isotope values in the form of gaseous $CH_3Br$. Xiao et al. (1993) used positive thermal ionization mass spectrometry for the measurement of $Cs_2Br^+$. Bromine in organic compounds have been analysed with MC-ICP-MS techniques (Hitzfeld et al. 2011; Holmstrand et al. 2010). The standard in use is SMOB (Standard Mean Ocean Bromine).

In general, bromide concentrations in geological settings are too low for a precise isotope measurement, a notable exception are sedimentary formation waters. The most common natural form of bromine is the bromide anion ($Br^-$). Bromine isotope fractionations between salts and brine are small (Eggenkamp et al. 2016). By using first-principles calculations, Gao and Liu (2021) demonstrated that Br isotope fractionations between salts and brines are small, much smaller than Br isotope variations observed in natural samples. Eggenkamp et al. (2019) determined the Br isotope composition in halite of varying geological ages. Br isotope values range from −0.24 to +1.08‰. Systematic variations depending on geological age have not been observed.

During evaporative salt precipitation, the heavy bromine isotope becomes depleted in salts relative to the fluid, whereas the heavy chlorine isotope becomes enriched in the salt relative to the fluid (Hanlon et al. 2017). Although higher oxidation states of bromine exist in nature, little is known about the Br isotope composition of bromine oxyanions. From calculated Br isotope fractionations among substances with different Br oxidation states, Gao and Liu (2021) concluded that the relatively large Br isotope variations have to be explained by changes of redox conditions.

Of special interest are high bromine concentrations in very saline deep groundwaters from old crystalline shields. Shouakar-Stash et al. (2007) and Stotler et al. (2010), observed very large Br-isotope variations from −0.80 to +3.35‰ that do not indicate a simple marine origin, but more likely complex water/rock interactions.

Another interesting aspect of bromine isotope geochemistry is that of all brominated organic compounds in the stratosphere, methyl bromide is the most important contributor to stratospheric ozone depletion. $CH_3Br$ may originate from natural and anthropogenic sources. Horst et al. (2013) determined the Br isotope composition of methyl bromide at two locations in Sweden. Subarctic samples in northern Sweden were more negative than samples in the Stockholm area. The $CH_3Br$ concentration in northern Sweden was 2–3 times lower than in the Stockholm area, possibly indicating industrial contamination of the latter area. $CH_3Br$ emissions from plants are about 2‰ depleted in $^{81}Br$ relative to bromine in the plant (Horst et al. 2014).

# 2.15 and 2.16 Alkali Elements (Potassium and Rubidium)

## 2.15 Potassium

Potassium has Two Stable Isotopes.

$$^{39}K \quad 93.26\%$$
$$^{41}K \quad 6.73\%$$

and a naturally occurring radioactive isotope $^{40}K$ (0.12%), which decays dually to stable $^{40}Ca$ and $^{40}Ar$. Potassium has only one oxidation state (+1). Because of its wide occurrence in the earth's crust and its key role in the biosphere, there is wide geochemical interest in determining K stable isotope variations. Taylor and Urey (1938) observed a 10% variation in $^{41}K/^{39}K$ ratios when K solutions percolate through a zeolite ion exchange column with $^{39}K$ preferentially eluted from the column.

Wang and Jacobsen (2016a) and Li et al. (2016a, b, c) described precise MC-ICP-MS techniques which eliminated argon-hydride interferences as well as the large $^{40}Ar$ peak from the Ar plasma. Further improvement in analytical precision have been reported by Morgan et al. (2018), Chen et al. (2019a, b) and Xu et al. (2019). Figure 2.31 shows potassium isotope variations of some important geological reservoirs.

### 2.15.1 Mineral Isotope Fractionations

Using density functional theory, Li et al. (2019a, b) calculated reduced partition function ratios of 17 K-bearing minerals. At low temperatures, the most enriched K mineral is alunite ($KAl_3(SO_4)_2(OH)_6$) and the most depleted K-mineral is

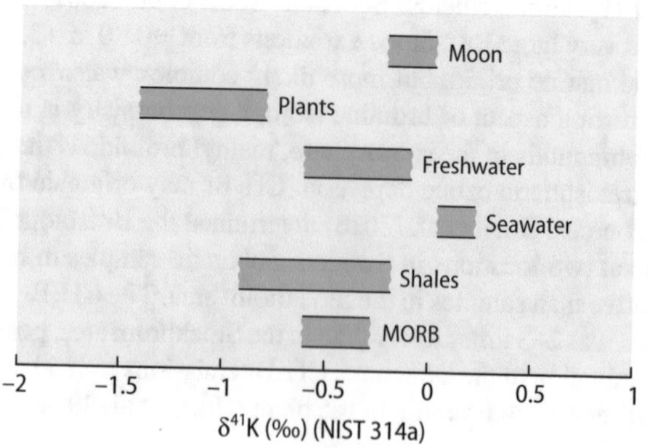


Fig. 2.31 Potassium isotope variations in geological reservoirs

djerfisherite ($K_6CuFe_{24}S_{26}Cl$). Fluids are enriched in $^{41}K$ indicating higher K isotope ratios in seawater relative to silicate minerals. In high-temperature environments, phengite is slightly enriched in $^{41}K$ relative to omphacite and amphibole, due to differences in coordination numbers (phengite (6), omphacite (7–8), amphibole (8)) (Liu et al. 2020).

Li et al. (2017) investigated the controlling factors of K isotope fractionations between aqueous solutions and K salts at 25 °C. They showed that K isotope fractionation decreases with increasing distances between the K atom and the negatively charged neighboring atoms. They reported total variations in $^{41}K/^{39}K$ ratios up to 1.4‰ with seawater and sylvite being enriched and plants being depleted in $^{41}K$. A characteristic feature of K isotopes is a 0.6‰ difference between seawater and basalts (Wang and Jacobsen 2016a, b). Potassium isotopes may, thus, be used as a tracer of oceanic crust recycled to the mantle (Parendo et al. 2017; Hu et al. 2020).

### 2.15.2 Magmatic Environment

During magmatic differentiation, no K isotope fractionation seems to occur and no differences in basalts from different geotectonic settings have been observed (Tuller-Ross et al. 2019a, b). Pristine basalts have an average $^{41}K$-value of −0.43‰ relative to the SRM 3141a standard, which agrees with the reported value for the bulk silicate earth (Wang and Jacobsen 2016a, b). Any differences in K isotopes reported for basaltic rocks are due to low-temperature alteration processes or due to the presence of crustal materials in magmas as for instance observed in continental basalts from NE China by Sun et al. (2019). Hu et al. (2021) presented K isotope evidence for sedimentary input to arc lavas from Martinique, Lesser Antilles.

Large K isotope fractionations have been observed during metamorphic dehydration in subduction zones (Liu et al. 2020). Oceanic crust may show quite variable K-isotope values, high values may indicate uptake of isotopically heavy seawater, low values may indicate signatures from weathering and/or clay formation. As demonstrated by Huang et al. (2019, 2020), the upper continental crust has a variation range from −0.68 to −0.12‰, with an average value which is similar to the mantle.

The Moon being volatile-element depleted is enriched in heavy $^{41}K$ by about 0.4‰ compared to the Earth (Wang and Jacobsen 2016b) which supports the Moon-forming giant impact event.

### 2.15.3 Weathering Environment

Potassium isotopes are good tracers to evaluate fractionation processes during weathering, because about 90% of K in riverine dissolved loads is derived from silicate weathering. Li et al. (2019a, b) evaluated K isotope fractionation by analyzing dissolved loads and sediments from major rivers in China. Dissolved potassium varies considerably in $^{41}K/^{39}K$ ratios which correlates negatively with the chemical weathering intensity. Li et al. (2019a, b) derived an average K fractionation of −0.55‰ between clay minerals and fluid. They estimated the global riverine runoff to have a $\delta^{41}$K-value of −0.22‰ and concluded that the δ-value of seawater is sensitive to continental weathering intensity.

Chen et al. (2020) and Teng et al. (2020) investigated fractionation processes during weathering and showed that potassium behaves similar to lithium in sofar that weathered products are depleted and the fluids are enriched in heavy K isotopes.

Wang et al. (2020) analyzed the K isotope composition of major rivers worldwide. They observed large $^{41}$K-variations from −0.59 to −0.08 with an average of −0.38‰. The composition of seawater with a $\delta^{41}$K-value of +0.12‰, thus, cannot be explained solely by the riverine input. Hu et al. (2020) argued that uptake of light potassium during submarine diagenetic alteration of subducting plates is essential for generating the relatively heavy seawater composition.

Santiago-Ramos et al. (2018) investigated pore-fluid profiles from ODP sites. $\delta^{41}$K-values vary systematically from site to site depending on sedimentation rates. Ramos and Higgins (2015) suggested that diffusive K fractionation in pore fluids may be, in part, responsible for the $^{41}$K enrichment of seawater.

## 2.16 Rubidium

Like potassium, rubidium belongs to the alkali metal group and is a volatile, lithophile, incompatible element. Rubidium has only one oxidation state (+1), thus isotope fractionations due to redox reactions will not play a role.

Rubidium consists of two isotopes.

| | |
|---|---|
| $^{85}$Rb | 72.165 |
| $^{87}$Rb | 27.835 |

$^{87}$Rb is radioactive and decays to $^{87}$Sr with a half-life of about $5 \times 10^{10}$ y. Due to its volatile nature, mass-dependent isotope fractionations may occur during evaporation. Nebel et al. (2011) first investigated Rb isotope variations between the Earth and chondrites and found within their reached analytical uncertainty no differences. With an improved technique for the chemical purification of Rb, Pringle and Moynier (2017) analyzed different classes of chondrites and achondrites, terrestrial and lunar igneous rocks. Terrestrial rocks define a narrow range in Rb isotope composition. Volatile depleted planetary bodies (howardite-eucrite-diogenite parent body, thermally metamorphosed meteorites) are enriched in $^{87}$Rb compared to chondrites, suggesting volatile loss by evaporation. In addition, lunar samples are slightly enriched in $^{87}$Rb compared to terrestrial samples. Zhang et al. (2021) observed significant Rb isotope fractionations during granite weathering related to adsorption and desorption of clay minerals.

## 2.17 Titanium

Although titanium is an abundant element on Earth, it has received little attention in isotope geochemistry, mainly because of two reasons. (i) a precise and accurate analytical method was missing and (ii) titanium behaves conservative during weathering and resides in minerals that formed at high temperatures where limited isotope fractionations are to be expected. Recently, Millet and Dauphas (2014) described a precise determination of $^{49}Ti/^{47}Ti$ ratios using a double spike technique.

Titanium has 5 stable isotopes.

| | |
|---|---|
| $^{46}Ti$ | 8.01% |
| $^{47}Ti$ | 7.33% |
| $^{48}Ti$ | 73.81% |
| $^{49}Ti$ | 5.50% |
| $^{51}Ti$ | 5.35% |

### 2.17.1 Magmatic Fractionations

Millet et al. (2016) and Greber et al. (2017a) investigated the Ti isotope composition of chondrites, terrestrial and lunar igneous rocks. Chondrites have uniform Ti isotope compositions being identical with estimates for the bulk silicate earth. $\delta^{49/47}Ti$ values in terrestrial samples vary by about 2.0‰ (Anguelova et al. 2019; Deng et al. 2019); lunar basalts show very little variations. Primitive MORB, island arc, and intraplate basalts are identical within analytical uncertainty, whereas significant isotope variations have been observed in peridotites by Anguelova et al. (2019), which they interpreted to reflect metasomatic interactions.

Differentiated magmas vary and show a positive correlation with $SiO_2$ contents resulting presumably from preferential depletion of heavy Ti isotopes in Fe-Ti oxides during fractional crystallization (Millet et al. 2016). Johnson et al. (2019) estimated an isotope fractionation of 0.39‰ between silicates and Ti-oxide at 1000 °C. Hoare et al. (2020) explored Ti isotope fractionation by analyzing alkaline, calc-alkaline and tholeiitic magma series. Alkaline differentiation suites show the largest range in Ti isotope composition followed by tholeiitic and calc-alkaline magmas, because alkaline magmas have high initial melt $TiO_2$ contents enabling early saturation of ilmenite plus titanomagnetite.

Greber et al. (2017b) demonstrated that Ti isotope ratios of shales are more or less constant since 3.5 billion years with $\delta^{49/47}Ti$ values being about 0.6‰ heavier than those of basalts, which they interpreted to reflect a felsic crust and indicating plate tectonics since 3.5 Ga. In contrast, Deng et al. (2019) argued that Ti isotopes cannot serve as a direct indicator for plate tectonics from 3.5 Ga ago, but have to be combined with other informations on silica contents of crustal rocks.

## 2.18 Vanadium

Vanadium has two stable isotopes

$$^{50}V \quad 0.24\%$$
$$^{51}V \quad 99.76\%$$

Since vanadium exists in four valence states ($2^+$, $3^+$, $4^+$, $5^+$), it is highly sensitive to reduction–oxidation reactions inducing isotope fractionations. Using first principles calculations, Wu et al. (2015) calculated equilibrium fractionation factors of V isotopes among V(III), V(IV) and V(V) species in waters. The heavy isotope $^{51}V$ will become enriched up to 6.4‰ at 25 °C in the order V(III) < V(IV) < V(V). Fractionations of 1.5‰ have been calculated among $5^+$ valence species which is caused by different bond length and coordination numbers. Furthermore, calculations indicate that the light V isotope is preferentially absorbed on goethite.

Nielsen et al. (2011a, b, c) and Prytulak et al. (2011) described a precise MC-ICP-MS technique and reported a $\delta^{51}V$ isotope variation of 1.2‰ for various reference samples. Recently, Schuth et al. (2017) performed in-situ V isotope analysis using a femtosecond laser coupled to MC-ICP-MS and observed 1.8‰ variations among different $V^{(IV)}$ and $V^{(V)}$ minerals. Figure 2.32 presents $\delta^{51}V$ isotope variation in some geological reservoirs.

### 2.18.1 High-Temperature Fractionations

Nielsen et al. (2014) postulated that V in the silicate earth is 0.8‰ enriched relative to carbonaceous and ordinary chondrites implying that the bulk Earth cannot be entirely reconstructed by mixing chondritic meteorites in various proportions.

Prytulak et al. (2013a, b) observed a 1‰ variation in mafic and ultramafic rocks. Secondary alteration reactions do not appear to induce V isotope fractionations. Qi et al. (2019) analyzed peridotites and komatiites reporting a mean $\delta^{51}V$ value of $-0.91 \pm 0.09$‰ which indicates a homogeneous mantle. For MORB, Wu et al. (2018) report a mean value of $-0.84 \pm 0.02$‰ that is slightly heavier than for peridotites. Due to its multiple oxidation states in melts and minerals, V isotopes

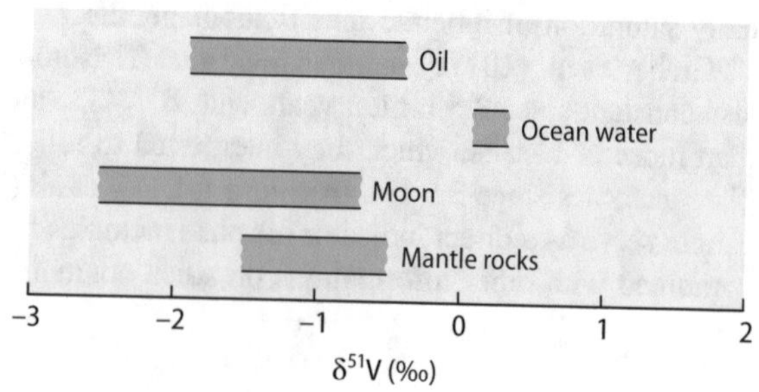


Fig. 2.32 Vanadium isotope variations in some geological reservoirs

vary during magmatic differentiation. Prytulak et al. (2017a, b) reported 2‰ variations during magmatic differentiation. They argued that differences in coordination among minerals and melt cause the observed fractionations. Evolved MORB lavas show shifts to heavier values with increasing differentiation. Removal of Fe-Ti oxide during crystallization seems to be responsible for the shift.

Sossi et al. (2018a, b, c) experimentally determined V-isotope fractionations between magnetite and melt at 800 °C under varying $fO_2$. Magnetite is depleted in $^{51}V$ relative to the melt presumebly to the VI-fold coordination in magnetite relative to the IV- and V-fold coordination in the melt. Sossi et al. (2018a) proposed that V isotopes may be used as a fingerprint for the redox state of magmas.

### 2.18.2 Low-Temperature Fractionations

Vanadium isotopes in fresh and seawater vary depending on speciation, $E_H$ and $p_H$. $V^V$ as soluble $H_2VO_4^-$ and $HVO_4^{2-}$ appears to be the dominant species. Wu et al. (2019) and Neddermeyer (2019) first described a method to analyze the V isotope composition of ocean water. They concluded that the deep ocean is isotopically homogeneous. Dissolved and particulate fractions in river water are quite variable, possibly due to adsorption processes (Schuth et al. 2019).

V isotope compositions of marine sediments are controlled by isotope fractionations between V species bound to particulates and dissolved in seawater. Sediments deposited under a range of redox conditions show variable V isotope compositions. Wu et al. (2020a, b) observed a direct link between the authigenic sedimentary components and the overlying redox conditions of the water column. Authigenic V in oxic sediments is controlled by isotope fractionations during absorption of $V^V$-vanadate on Fe–Mn hydoxides. V isotopes in sediments deposited under oxygen-deficient seawater may depend on $V^{IV}$-vanadyl absorbed on organic particulates (Wu et al. 2020a, b).

V is enriched in organic matter, especially in crude oils. Ventura et al. (2015) demonstrated that oils show a large variation in V isotopes reflecting the isotope differences of primary source rocks. Comparisons between vanadium concentrations and V isotope ratios yield distinct clusters of oils derived from terrestrial/lacustrine or marine/carbonate source rocks. As shown by Gao et al. (2018) V isotopes in oils are significantly modified during maturation and biodegradation.

## 2.19 Chromium

Chromium has 4 naturally occurring isotopes with the following abundances (Rosman and Taylor 1998)

$^{50}Cr$ 4.35%

$^{52}Cr$ 83.79%

$^{53}Cr$ 9.50%

$^{54}Cr$ 2.36%

Early interest in Cr isotopes was based on the application of the $^{53}Mn$–$^{53}Cr$ chronometer to date early solar system events (i.e. Lugmair and Shukolyukov 1998). $^{53}Cr$ is a radiogenic product of extinct $^{53}Mn$ which has a half-life of 3.7 Myr.

In contrast to meteorites, Cr isotope variations on Earth are due to mass-dependent isotope fractionations. Chromium exists inf two oxidation states, Cr(III) as a cation $Cr^{3+}$, and Cr(VI), as an oxyanion ($CrO_4^{2-}$ or $HCrO_4^-$) having different chemical behaviors: $Cr^{3+}$ is the dominant form in most minerals and in water under reducing conditions, whereas Cr(VI) is stable under oxidizing conditions. Cr(VI) in chromate is highly soluble, mobile and toxic, whereas trivalent chromium, existing as a cation, is largely insoluble and immobile. These properties make Cr isotope investigations very suitable to detect and quantify redox changes in different geochemical reservoirs. A recent review has been presented by Qin and Wang (2017).

Cr isotope variations have been measured by TIMS (Ellis et al. 2002) and by MC-ICP-MS techniques (Halicz et al. 2008; Schoenberg et al. 2008), $\delta^{53/52}Cr$-values are given relative to the NIST SRM 979 standard. Figure 2.33 summarizes average Cr-isotope compositions in important reservoirs.

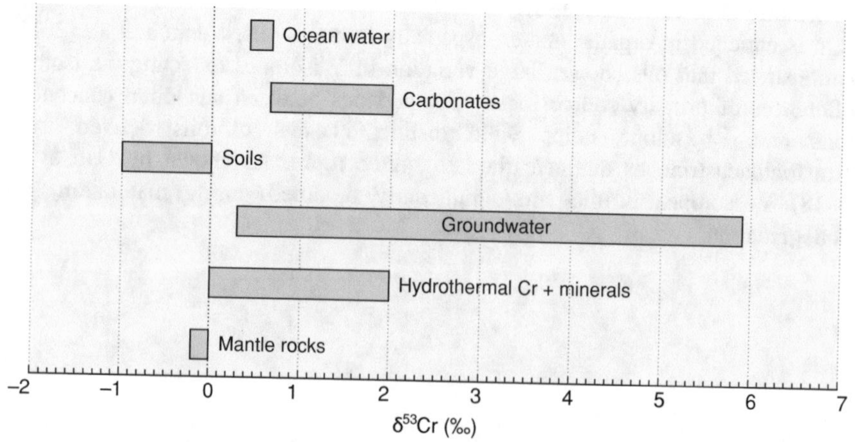


**Fig. 2.33** $\delta^{53}Cr$-values of important geological reservoirs (ocean water: Scheiderich et al. (2015) observed large variations in oceanic surface waters)

Schauble et al. (2004) predicted Cr isotope fractionations >1‰ between Cr species with different oxidation states. Cr isotope fractionations between $CrO_4^{2-}$ and $Cr(H_2O)_6^{3+}$ complexes have been calculated to be 7‰ at 0 °C, with chromate being enriched in $^{53}Cr$. However, since isotope equilibration between Cr(VI) and Cr (III) species at low temperatures is slow (Zink et al. 2010), it appears that isotope disequilibrium between Cr-species is common and, therefore, natural Cr isotope fractionations probably are kinetically controlled.

### 2.19.1 Mantle Rocks

Chromium is a compatible and moderately siderophile element that tends to be concentrated in the mantle. Mantle xenoliths, ultramafic cumulates and basalts have, as first shown by Schoenberg et al. (2008), a mean $\delta^{53}$Cr-value of − 0.12‰ with relatively little variations. Shen et al. (2018) investigated Cr isotope fractionations among major mantle minerals by coupling the measurement of natural mineral samples with theoretical ionic modeling and Cr valence determinations by XANES spectroscopy. A general $^{53}$Cr order of spinel > pyroxene >= olivine was obtained.

Farkas et al. (2013) observed in chromites from peridotites a mean $\delta^{53}$Cr-value of 0.08‰, slightly heavier than in mantle xenoliths. As shown by Wang et al. (2016b) basalts and their metamorphosed equivalents have very similar Cr isotope compositions supporting the suggestion of Schoenberg et al. (2008) that limited Cr isotope fractionation occurs under high temperature conditions. Xia et al. (2017) observed, however, a large range of $^{53}$Cr-values (from −1.36 to 0.75‰) in mantle xenoliths from Mongolia, which they interpreted as indicating kinetic isotope fractionations during melt percolation. After corrections for fractionation effects during partial melting and metasomatism, Xia et al. (2017) estimated a mean $\delta^{53}$Cr-value of −0.14‰ for the pristine, fertile upper mantle, which agrees with the estimated value for the bulk silicate earth by Schoenberg et al. (2008). Nevertheless, as demonstrated by Xia et al. (2017), large scale heterogeneities in Cr isotope composition may exist in the mantle. During serpentinization of ultramafic rocks, $^{53}$Cr will become enriched (Farkas et al. 2013; Wang et al. 2016b). Thus, oxidative secondary aqueous alteration of ultramafic rocks shifts the primary mantle composition towards heavier $^{53}$Cr-values.

Moynier et al. (2011a, b) published $\delta^{53}$Cr-values of various chondritic meteorites that are up to 0.3‰ lighter than the bulk silicate earth. Moynier et al. (2012) suggested that the difference results from Cr-isotope fractionation during core formation. Qin et al. (2015), Bonnand et al. (2016) and Schoenberg et al. (2016), however, found no difference in Cr isotope composition between chondrites and bulk silicate earth providing evidence that Cr isotopes are not fractionated during core formation.

Sossi et al. (2018b) measured a Cr-isotope difference between the Earth's mantle and the Moon, the latter being depleted in $^{53}$Cr relative to the former, being consistent with the loss of Cr-bearing oxidized vapor phase from the Moon. As argued by Sossi et al. (2018b) the required temperatures are much lower than predicted by models implying that Cr was removed from the Moon following cooling rather than during the impact event.

### 2.19.2 River and Ocean Water

During weathering, oxidation of Cr(III) leads to a $^{53}$Cr enrichment in the resulting Cr(VI), leaving soils depleted in $^{53}$Cr. Thus, river water is enriched in heavy Cr-isotopes relative to mantle and crustal rocks (Bonnand et al. 2013; Frei et al. 2014). Rivers yield a large spread in $\delta^{53}$Cr-values with rivers draining serpentinized ultramafic rocks showing the heaviest Cr-isotope values (Farkas et al. 2013; D' Arcy et al. 2016). River water along the Connecticut River varies from −0.17 to 0.92‰ with seasonal variations in some, but not all tributaries (Wu et al. 2017a, b).

Seawater also exhibits a large range in Cr isotope composition from 0.4 to 1.6‰ (Bonnand et al. 2013; Scheiderich et al. 2015; Pereira et al. 2016; Rickli et al. 2019). Surface waters have higher Cr isotope values than deep waters, the latter having a homogeneous composition of about 0.5–0.6‰. As demonstrated by Scheiderich et al. (2015) and Rickli et al. (2019), Cr isotopes in ocean water strongly correlate with Cr concentrations: high concentrations correlate with low Cr-isotope values. Oxic sediments have Cr-isotope compositions close to the average upper continental crust, whereas anoxic sediments may reflect the Cr-isotope composition of deep ocean water (Gueguen et al. 2016).

Cr(VI) accounts for most of dissolved Cr in seawater, reduction to Cr(III) by biological activity may occur in oxygen-deficient zones (Janssen et al. 2020). By analyzing hexa- and tri-valent Cr in North-Pacific oxygen-deficient zones, Huang et al. (2021) demonstrated that trivalent Cr is isotopically lighter than total dissolved chromium and residual hexavalent Cr.

As shown by Reinhard et al. (2014) Cr isotope values in anoxic marine sediments from the Cariaco Basin did not change over the last 14.5 kyr suggesting that the Cr cycle has been broadly in steady state during that time period.

### 2.19.3 Carbonates

Carbonates encompass the range of Cr-isotopes in seawater (Bonnand et al. 2013). Cr isotopes in marine carbonates, thus, may be a sensitive tracer of weathering of the continental crust as well as of variations of hydrothermal input (Frei et al. 2011). During inorganic precipitation of carbonates, Cr(VI) will become enriched by about 0.3‰ (Rodler et al. 2015). Biogenic carbonates in the ocean fractionate, however, during biological processes; Cr isotopes fractionate by −0.5 to +0.3‰ depending on species (Pereira et al. 2016; Frei et al. 2018; Bruggmann et al. 2019). Wang et al. (2017a, b) observed large variations in $\delta^{53}$Cr-values in foraminifera, even within single species questioning the use of Cr isotope as a seawater archive.

The use of Cr isotopes in carbonates for paleoenvironmental reconstructions is based on the assumption that Cr in aquatic systems is dominated by Cr(VI). Fang et al. (2020) presented evidence that Cr(III) is the dominant species in their investigated carbonates, and concluded that Cr(VI) has been reduced before or after carbonate precipitation. Wang et al. (2021) observed significant Cr isotope fractionations during diagenesis and dolomitization limiting the use of chromium isotopes in carbonates as a paleoredox indicator.

### 2.19.4 Paleo-Redox Proxy

Before the emergence of oxygenic photosynthesis, Cr in rocks mainly exists as Cr (III). With the appearance of free oxygen, Cr(III) will be oxidized during weathering to soluble Cr(VI), which will be transported as dissolved oxyanion to the ocean. This relationship has been used as a Cr isotope redox proxy. Thus, Frei et al. (2009), Frei and Polat (2013), Crowe et al. (2013), Planavsky et al. (2014), Cole et al. (2016), Gilleaudeau et al. (2016), Canfield et al. (2018), and Wei et al. (2020) applied Cr-isotopes to deduce the oxygenation history of the Earth's hydro- and atmosphere by investigating iron-rich sediments, carbonates and shales. Shales are a special case insofar because major Cr-containing components consist of detrital phases probably reflecting igneous $\delta^{53}$Cr-values, which have to be corrected (Frank et al. 2019). Combined together, these studies have suggested that signs of oxidative weathering environments might be as old 3.0 Ga, and that the Great Oxidation Event did not lead to a unidirectional increase of oxygen, but instead is better characterized by punctuated fine-scale fluctuations in atmospheric oxygen concentrations.

### 2.19.5 Anthropogenic Cr in the Environment

Extensive industrial use of hexavalent chromate has led to a widespread Cr contamination of soils and groundwater. Reduction of Cr(VI) to Cr(III) may proceed by a variety of abiogenic and microbial processes. All reduction mechanisms induce Cr isotope fractionations with the lighter isotope enriched in the product (Dossing et al. 2011; Sikora et al. 2008). The magnitude of Cr isotope fractionation varies from −1 to −5‰ for naturally occurring reductants (Basu and Johnson 2012; Kitchen et al. 2012; Zhang et al. 2018a, b, c; Joe-Wong et al. 2021).

Since isotope fractionation during Cr(VI) reduction is little affected by adsorption (Ellis et al. 2004), $^{53}$Cr/$^{52}$Cr ratios in soils and groundwaters can be used as an indicator of Cr(VI) reduction and pollution. Groundwaters have $\delta^{53}$Cr-values ranging from 0.3 to 5.9‰ (Ellis et al. 2002, 2004; Berna et al. 2010; Zink et al. 2010; Izbicki et al. 2012). These authors observed an increase up to 6‰ in $^{53}$Cr/$^{52}$Cr ratios during the reduction of chromate. In experiments with *Shewanella*, Sikora et al. (2008) observed a Cr isotope fractionation of about 4‰ during dissimilatory Cr(VI) reduction. There are other genera of anaerobic and aerobic bacteria that produce comparable isotope fractionations during Cr(VI) reduction (Han et al. 2012). These findings can be applied to quantify Cr(VI) reduction at sites undergoing active remediation.

## 2.20 Iron

Iron has 4 stable isotopes with the following abundances

$$\begin{array}{ll} ^{54}\mathrm{Fe} & 5.84\% \\ ^{56}\mathrm{Fe} & 91.76\% \\ ^{57}\mathrm{Fe} & 2.12\% \\ ^{58}\mathrm{Fe} & 0.28\% \end{array}$$

Iron is the third most abundant element on Earth that participates in a wide range of biotically- and abiotically-controlled redox processes in low- and high-temperature environments. Iron has a variety of important bonding partners and ligands, forming sulfide, oxide and silicate minerals as well as complexes with water. As is well known, bacteria can use Fe during both dissimilatory and assimilatory redox processes.

Because of its high abundance and its important role in high and low temperature processes, isotope studies of iron have received the most attention of the transition elements. Since the first investigations on Fe isotope variations by Beard and Johnson (1999), the number of studies on Fe isotope variations has increased exponentially. Reviews on Fe-isotope geochemistry have been given by Anbar (2004a, b), Beard and Johnson (2004), Dauphas and Rouxel (2006), Anbar and Rouxel (2007), Dauphas et al. (2017) and by Johnson et al. (2020). Figure 2.29 summarizes Fe-isotope variations in important geological reservoirs.

### 2.20.1 Analytical Methods

By using the double-spike TIMS technique, Johnson and Beard (1999) described an analytical procedure with a precision of 0.2‰. With the introduction of MC-ICP-MS techniques and their ability to measure Fe isotope ratios with little drift, most researchers have concentrated on MC-ICP-MS (Belshaw et al. 2000; Weyer and Schwieters 2003; Arnold et al. 2004a, b; Schoenberg and von Blanckenburg 2005; Dauphas et al. 2009a, b; Craddock and Dauphas 2010; Millet et al. 2012). In-situ SIMS techniques have been described by Sio et al. (2013). Horn et al. (2006) and Oeser et al. (2014) introduced the use of in-situ femtosecond laser MC-ICP-MS.

Iron isotope analysis is highly challenging, because of interferences from $^{40}Ar^{14}N^+$, $^{40}Ar^{16}O^+$ and $^{40}Ar\ ^{16}OH^+$ at masses 54, 56 and 57 respectively. Nevertheless δ-values can be measured routinely with a precision of ±0.1‰ or better (Craddock and Dauphas 2010).

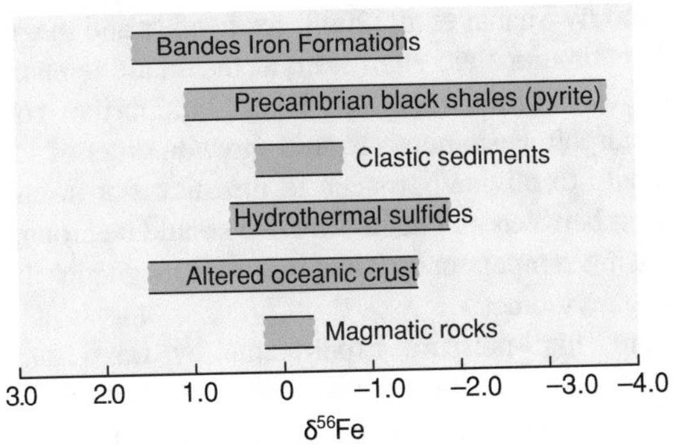


Fig. 2.34 $\delta^{56}$Fe-values of important geological reservoirs

Literature data have been presented either in the form of $^{57}Fe/^{54}Fe$ or as $^{56}Fe/^{54}Fe$ ratios. In the following all data are given as $\delta^{56}$Fe values. $\delta^{57}$Fe values would be 1.5 times greater than $\delta^{56}$Fe values, because only mass-dependent fractionations are expected. Fe isotope ratios are generally reported relative to the IRMM-14 standard, an ultra-pure synthetic Fe metal, or are given to the average composition of various rock types (Beard et al. 2003; Craddock and Dauphas 2010; He et al. 2015). Relative to IRMM-14, igneous rocks have an average composition of $\delta^{56}$Fe of 0.09‰. The maximum range in $\delta^{56}$Fe-values is more than 5‰, with low values for sedimentary pyrite and high values in iron oxides from banded iron formations. A summary of some important geological reservoirs is presented in Fig. 2.34.

## 2.20.2 Isotope Equilibrium Studies

Equilibrium Fe isotope fractionations for mineral–mineral and mineral-fluid systems have been studied by 2 different approaches: (i) calculations of β-factors based on vibrational spectroscopy and density functional theory. As reviewed by Blanchard et al. (2017), theoretical iron isotope studies have used a variety of theoretical approaches i.e. Schauble et al. (2001), Anbar et al. (2005), Polyakov et al. (2007), Blanchard et al. (2009), Rustad and Dixon (2009), Rustad et al. (2010) and (ii) isotope exchange experiments (Skulan et al. 2002; Welch et al. 2003; Shahar et al. 2008; Beard et al. 2010; Saunier et al. 2011; Wu et al. 2011; Frierdich et al. 2014, 2019; Sossi and O'Neill 2017).

Experimental studies at low temperatures focused on the Fe isotope fractionation in aqueous fluids (Johnson et al. 2002; Welch et al. 2003; Hill et al. 2009) and between an aqueous fluid and a mineral (Johnson et al. 2002; Skulan et al. 2002). A major problem for low-temperature experiments is the achievement of isotope equilibrium. Experimental studies at high temperatures, conducted by Schüßler et al. (2007) for equilibrium isotope fractionations between iron sulfide (pyrrhotite) and silicate melt and by Shahar et al. (2008) for fayalite and magnetite demonstrate that Fe isotope fractionations are significant at magmatic temperatures and potentially can be used as a geothermometer. Under equilibrium conditions common igneous and metamorphic Fe-minerals should show an order of $^{56}$Fe depletion from hematite to magnetite to olivine/pyroxene to ilmenite. For instance, at 800 °C Fe isotope fractionation between magnetite-ilmenite should be around 0.5‰ becoming larger with decreasing temperatures. Thus, the pair magnetite-ilmenite potentially may serve as a geothermometer.

High-temperature, high-pressure experiments by Sossi and O'Neill (2017) demonstrated that the coordination environment and the oxidation state of iron control Fe isotope fractionations. For $Fe^{2+}$-bearing minerals, $\delta^{56}$Fe fractionations decrease from IV-fold to VIII-fold coordination and increase with increasing $Fe^{3+}/Fe_{total}$ ratios.

### 2.20.3 Meteorites

Fe isotopes in meteorites have been used to investigate processes associated with core formation. Iron meteorites are considered to represent remnants of metallic cores of differentiated planetary bodies. Magmatic Fe meteorites are on average 0.10‰ heavier in $\delta^{56}$Fe isotope composition than chondritic meteorites (Poitrasson et al. 2005; Ni et al. 2020). Whether core formation leads to Fe isotope fractionation or not is still a matter of debate. Poitrasson et al. (2009) and Hin et al. (2012) experimentally determined no Fe isotope fractionation between Fe–Ni alloy and silicate liquid at temperatures up to 2000 °C. Ni et al. (2020) showed experimentally that the iron isotope composition of iron meteorites can be explained by core crystallization; they observed that solid metal becomes about 0.1‰ enriched in heavy Fe isotopes relative to liquid metal.

Carbonaceous and ordinary chondrites have a uniform bulk Fe isotope composition close to zero‰ (Craddock and Dauphas 2010; Wang et al. 2013b), whereas the individual Fe components in meteorites are isotopically variable. Chondrules display the largest variation, metals and sulfides show smaller variations (Needham et al. 2009). As shown by Williams et al. (2006) Fe isotope differences between metal and troilite are in the range of 0.5‰—the metal phase being heavier than the sulfide phase troilite, potentially reflecting equilibrium fractionations.

Terrestrial basalts are on average 0.1‰ enriched in $\delta^{56}$Fe values relative to chondrites, whereas basalts from Mars and Vesta have the same isotope composition as chondrites. The non-chondritic composition of terrestrial basalts have been interpreted as resulting from vaporization into space during the Moon forming impact (Poitrasson et al. 2004), or from Fe isotope fractionations during core formation (Schoenberg and von Blanckenburg 2006; Elardo and Shahar 2017) or from iron isotope fractionation during partial melting of the mantle to produce the Earth's crust (Weyer and Ionov 2007; Dauphas et al. 2014 and others).

The bulk iron isotope composition of the Moon is not well constrained. As shown by Weyer et al. (2005) and Liu et al. (2010a, b), low Ti-basalts have $\delta^{56}$Fe values that are 0.1‰ lower than high Ti basalts, possibly reflecting different mantle sources. The lack of plate tectonics on the Moon obviously has allowed the survival of heterogeneities in the lunar mantle. Poitrasson et al. (2019) concluded that the Fe isotope composition of the bulk Moon is indistinguishable from the Earth, but heavier than other planetary bodies.

### 2.20.4 Igneous Rocks

Early studies indicated that all terrestrial igneous rocks have homogeneous Fe isotope compositions (Beard and Johnson 1999, 2004). Later studies suggested that igneous processes such as partial melting and crystal fractionation may lead to measurable Fe isotope variations. Weyer et al. (2005) and Weyer and Ionov (2007) reported that the Fe isotope composition in mantle peridotites is about 0.1‰ lighter than in basalts. Mantle xenoliths exhibit Fe isotope variations between minerals in a single sample and between samples suggesting a heterogenous lithospheric mantle (Macris et al. 2015). Spinel, olivine and possibly orthopyroxene seem to indicate Fe isotope equilibrium, whereas clinopyroxene appears to have been affected by late stage metasomatism (Williams et al. 2005; Weyer and Ionov 2007; Dziony et al. 2014; Macris et al. 2015).

Because $Fe^{3+}$ is more incompatible than $Fe^{2+}$ during partial melting and given the fact that $Fe^{3+}$ has higher $\delta^{56}Fe$ values than $Fe^{2+}$, liquids should become enriched relative to the solid residue. Dauphas et al. (2009a, b) presented a quantitative model that relates the iron isotope composition of basalts to the degree of partial melting. Williams and Bizimis (2014) reported that MORB and OIB are enriched by 0.1–0.2‰ relative to unmetasomatized mantle peridotites. They explained the fractionation effect by the more incompatible nature of $Fe^{3+}$ during partial melting and by preferential melting of isotopically heavier pyroxene than isotopically lighter olivine. As demonstrated by Williams and Bizimis (2014), modeling of partial melting—assuming even large degrees of mantle melting—cannot explain the large spread of $\delta^{56}Fe$-values of OIB. They argued that heterogeneous mantle sources containing both light and heavy $^{57}Fe$ components best explain the large variation in MORB and OIB. In some areas like Samoa, basalts show enriched $\delta^{56}Fe$ values up to 0.3‰ (Konter et al. 2017). Heavy iron isotope values have been reported from other OIB areas, but, as argued by Soderman et al. (2021), cannot be ascribed to a unique source.

Small Fe isotope variations between MORB and OIB have been reported by Teng et al. (2013) that can be explained by fractional crystallization of OIBs. Teng et al. (2008) demonstrated that Fe isotopes fractionate during magmatic differentiation on whole-rock and on crystal scales. They observed that iron in basalts becomes isotopically heavier as more olivine crystallizes, implying that differences in the redox state of Fe play a decisive role. Zoned olivine crystals yield $^{56}Fe$ isotope fractionations of up to 1.6‰, which they interpreted as being due to diffusion between olivines and evolving melt (Teng et al. 2011; Sio et al. 2013; Oeser et al. 2015).

In granitic rocks $\delta^{56}$Fe values are generally positively correlated with $SiO_2$ contents (Poitrasson and Freydier 2005; Heimann et al. 2008). These authors suggested that exsolution of fluids has removed light Fe isotopes causing the enrichment of $SiO_2$-rich granitoids. Telus et al. (2012) argued that exsolution alone cannot explain the high iron isotope values in all granitoids, instead fractional crystallization seems to be the major cause of enrichment.

In separated minerals from I-type granites, Wu et al. (2017a, b) observed the following order of $^{56}$Fe enrichment: feldspar (containing more than 1000 ppm Fe) > pyrite > magnetite > biotite ≈ hornblende. Variations of intermineral Fe fractionations depend on mineral compositions. Plagioclase-magnetite and alkali feldspar-magnetite fractionations depend on albite and orthoclase content respectively, which might be explained by differences in Fe–O bond strength due to different $Fe^{3+}$ contents.

### 2.20.5 Sediments

During weathering, Fe is dissolved by ligands and/or bacteria. Iron isotope fractionation may occur during Fe mobilization by Fe reduction or ligand-promoted dissolution or during immobilization of Fe oxy/hydroxides (Fantle and de Paolo 2005; Yesavage et al. 2012 and others). $\delta^{56}$Fe values of bulk and HCl-extractable Fe become isotopically lighter as the extent of weathering proceeds; exchangeable Fe is more depleted in $^{56}$Fe than Fe in ironhydroxides.

Theoretical calculations and experimental determinations show that Fe(III) bearing phases tend to be enriched in heavy Fe isotopes compared to Fe(II) bearing phases. The largest Fe isotope fractionations have been attributed to redox effects (Johnson et al. 2008). For example, Fe isotope fractionations between Fe(II) and Fe (III) species at 25 °C yield a 2.5–3‰ $^{54}$Fe depletion in the Fe(III) species. As discussed by Crosby et al. (2005), Fe isotope fractionation results from isotope exchange between Fe(II) and Fe(III) at oxide surfaces explaining why Fe isotope fractionations are very similar for microbial dissimilatory Fe(III) reduction, microbial Fe(II) oxidation and equilibrium between dissolved Fe(II) and Fe(III) species in abiotic systems. This hampers the assertion of Fe isotopes as biosignatures.

Marine sediments reflect the average Fe isotope composition of the continental crust, deviations from the mean value are due to biogeochemical processes in the sediments. Observed natural Fe isotope variations of around 5‰ have been attributed to a large number of processes, which can be divided into inorganic reactions and into processes initiated by micro-organisms. Mechanism governing inorganic Fe isotope fractionation include precipitation of Fe bearing minerals (Skulan et al. 2002; Butler et al. 2005), isotope exchange between different ligand species (Hill and Schauble 2008; Dideriksen et al. 2008; Wiederhold et al. 2006) and adsorption of dissolved Fe(II) to Fe(III) surfaces (Icopini et al. 2004; Crosby et al. 2007; Jang et al. 2008). Changes in bond partners and/or coordination number also have an effect on isotope fractionation (Hill et al. 2009, 2010), implying that Fe isotope compositions reflect both the redox state and the solution chemistry.

Up to 1‰ fractionation can result from inorganic precipitation of Fe-containing minerals (oxides, carbonates, sulfides) (Anbar and Rouxel 2007). Larger Fe isotope fractionations occur during biogeochemical redox processes, which include dissimilatory Fe(III) reduction (Beard et al. 1999; Icopini et al. 2004; Crosby et al. 2007), anaerobic photosynthetic Fe(II) oxidation (Croal et al. 2004), abiotic Fe (II) oxidation (Bullen et al. 2001) and sorption of aqueous Fe(II) on Fe (III) hydroxides (Balci et al. 2006). Controversy exists about the potential role of abiotic versus microbiological fractionations. This complicates the ability to use iron isotopes to identify microbiological processing in the rock record (Balci et al. 2006). As argued by Johnson et al. (2008) microbiological reduction of $Fe^{3+}$ produces much larger quantities of iron with distinct $\delta^{56}Fe$ values than abiological processes. Thus, a number of studies have interpreted negative $\delta^{56}Fe$ values in sediments to reflect dissimilatory iron reduction (DIR) (e.g. Bergquist and Boyle 2006; Severmann et al. 2006, 2008, 2010; Teutsch et al. 2009). Coupled Fe and S isotope intergrain variations in pyrite have been used as a proxy for microbial dissimilatory Fe(III) and sulfate reduction (Archer and Vance 2006).

In summary, negative $\delta^{56}Fe$-values in sedimentary rocks may reflect ancient DIR (Yamaguchi et al. 2005; Johnson et al. 2008). Especially large iron isotope fractionations have been found in Proterozoic and Archean banded iron formations (BIFs) and shales (Rouxel et al. 2005; Yamaguchi et al. 2005). In particular BIFs have been used to reconstruct Fe cycling through Archean oceans and the rise of $O_{2\,(atm)}$ during the Proterozoic (see discussion under Sect. 3.8.4 and Fig. 3.30). The pattern shown in Fig. 3.30 distinguishes three stages of Fe isotope evolution, which might reflect redox changes in the Fe cycle (Rouxel et al. 2005). Interplays of the Fe-cycle with the C- and S-record might reflect changing microbial metabolisms during the Earth's history (Johnson et al. 2008).

### 2.20.6 Ocean and River Water

Dissolved and particulate iron in water occur not only in two oxidation states but in a wide range of chemical species that interact by adorption/desorption, and/or precipitation/dissolution processes. All these processes potentially may fractionate Fe isotopes that may modify the iron isotope composition of waters.

Iron in the ocean is an important micronutrient; the growth of phytoplankton is often limited by low Fe concentrations. Potential iron sources to the ocean are (i) atmospheric dust, (ii) hydrothermal vents, (iii) reducing sediments along continental margins and (iv) oxic ocean floor sediments. In the North Atlantic for example, Fe isotope data point to Saharan dust aerosol as the dominant Fe source (Mead et al. 2013). In hydrothermal vents Fe is depleted in heavy Fe isotopes relative to basaltic rocks (Sharma et al. 2001; Severmann et al. 2004; Bennett et al. 2009) which has been explained by precipitation of isotopically heavy Fe minerals (i.e. amphiboles, pyrite) (Rouxel et al. 2004a, b, 2016).

Because of its very low concentration, the Fe isotope composition of ocean water is not easily determined. Radic et al. (2011) and John and Adkins (2012) were among the first presenting dissolved and particulate Fe isotope data in depth profiles from the Pacific and Atlantic. Water profiles characterized by positive $\delta^{56}$Fe values mainly reflect the continental input with slight transformations in the water column. John and Adkins (2012) demonstrated that dissolved iron in the upper 1500 m of the North Atlantic is homogeneous with $\delta^{56}$Fe values between 0.30 and 0.45‰, whereas in the deeper ocean $\delta^{56}$Fe-values increase to 0.70‰. In the Southern Ocean, Abadie et al. (2017) demonstrated for intermediate depths $\delta^{56}$Fe depleted values, contrasting with heavier values for deep waters. The iron cycle in ocean water thus depends in a sensitive way to continental erosion, dissolved/particulate interactions and deep-water upwelling.

Iron isotope data of bulk rivers as well as particulate, dissolved and colloidal Fe have been reported by Bergquist and Boyle (2006), Ingri et al. (2006), Poitrasson et al. (2014), Dos Santos Pinheiro et al. (2014) and others, indicating a large variability. Rivers rich in clastic suspended detrital material, like the white waters of the Amazon have a Fe isotope composition close to the continental crust (Poitrasson et al. 2014). Rivers rich in organic material contain a large portion in dissolved Fe form and are depleted in heavy Fe isotopes with significant annual variations (Dos Santos Pinheiro et al. 2014).

Fluids in diagenetic systems are variable in Fe isotope compsition with a preferential depletion in $^{56}$Fe (Severmann et al. 2006) reflecting the interaction of $Fe^{3+}$ with $Fe^{2+}$ during bacterial iron and sulfate reduction. Processes dominated by sulfate reduction produce high $\delta^{56}$Fe values in porewaters, whereas the opposite occurs when dissimilatory iron reduction is the major pathway (Severmann et al. 2006). Fe isotope compositions of pore fluids may reflect the extent of Fe recycling during early diagenesis (Homoky et al. 2011). Fe(II) in pore waters, formed by bacterial Fe(III) reduction, may be reoxidized during sediment suspension events. The resulting fine grained isotopically light FeOOH may be transported back to the deep ocean, a process that has been termed “benthic iron shuttle” (Severmann et al. 2008).

### 2.20.7 Plants

Although sufficient supply of Fe is essential for all living organisms, iron is one of the most limiting nutrients, because iron in soils exists predominantly in the nearly insoluble Fe(III) form. Therefore, higher plants developed different strategies to make iron available. Guelke and von Blanckenburg (2007) presented evidence that Fe isotope signatures in plants reflect two different strategies that plants have developed to incorporate Fe from the soil. Group I plants induce chemical reactions in the rhizosphere and reduce iron before uptake by incorporating light isotopes in the roots with further depletion during transport to leaves and seeds. Group II plants transport Fe(III) complexes into plant roots via a specific membrane transport system that do not fractionate Fe relative to Fe in soils (Guelke et al. 2010;

Guelke-Stelling and von Blanckenburg 2012). As shown by Kiczka et al. (2010) Fe isotopes may fractionate during remobilization of Fe from old into new plant tissues which may change the Fe isotope composition of leaves and flowers over the season.

## 2.21 Nickel

Nickel can occur in oxidation states from 4+ to 0, but the 2+ state is essentially the only natural oxidation state. Thus, redox controlled reactions do not play an important role; instead chemical precipitation, adsorption in aqueous systems and crystallization of Ni-sulfides in magmatic systems might induce isotope fractionations. Fujii et al. (2011a, b) investigated theoretically and experimentally Ni isotopes fractionations between inorganic Ni-species and organic ligands and observed Ni isotope fractionations up to 2.5‰ controlled by organic ligands. First principles calculations have been used by Liu et al. (2018) to calculate Ni isotope fractionations of Ni sulfides. Ni isotope fractionations depend on average Ni–S bond lengths and show a negative correlation with average Ni–Ni bond lengths. Since nickel is a bioessential trace element, playing vital roles in enzymes, biological processes might cause isotope fractionations.

Ni has five stable isotopes

| | |
|---|---|
| $^{58}$Ni | 68.08 |
| $^{60}$Ni | 26.22 |
| $^{61}$Ni | 1.14 |
| $^{62}$Ni | 3.63 |
| $^{64}$Ni | 0.93 |

The initial interest in Ni isotope geochemistry was focussed on the identification of extinct $^{60}$Fe in the early solar system that decays with a short half life (2.62 Ma) to $^{60}$Ni. Thus, from $^{60}$Ni abundances in meteorites, the prior presence of $^{60}$Fe may be addressed (Elliott and Steele 2017).

Ni isotopes generally are reported as $\delta^{60/58}$Ni values relative to the NIST SRM 986 standard. Gueguen et al. (2013) described an analytical procedure for Ni isotope determinations and determined Ni isotope ratios for various geological reference materials. Ni isotope variations, observed so far, range within 3.5‰. The lightest Ni isotope ratios have been observed for magmatic sulfides, the heaviest values have been presented for C-rich sediments and Fe/Mn crusts (Estrade et al. 2015) (see Fig. 2.35).

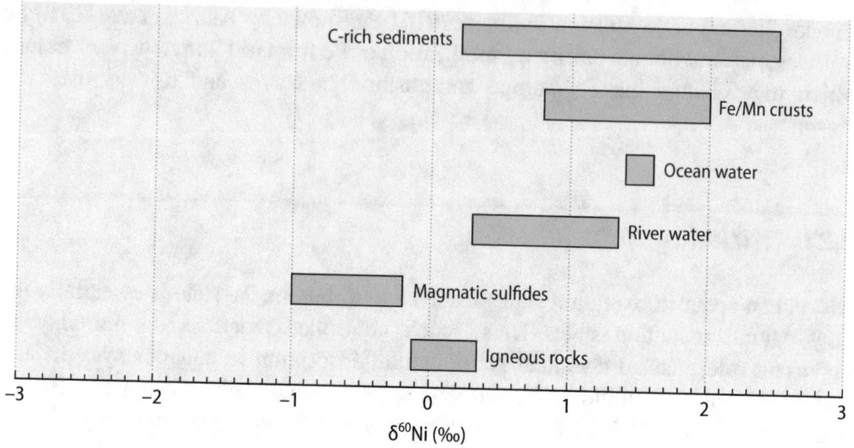


**Fig. 2.35** $\delta^{60/58}$Ni isotope variations in important geological reservoir

### 2.21.1 Mantle Rocks and Meteorites

First measurements by Cameron et al. (2009) indicated that Ni isotope variations in the mantle and the continental crust are negligible. More recently, Gueguen et al. (2013) and Hofmann et al. (2014) reported Ni isotope fractionations up to 1‰ among komatiites and associated Ni–sulfide mineralisations, the latter being depleted in heavy Ni isotopes. Gall et al. (2017) reported isotope variations between different mantle minerals indicating that olivine and orthopyroxene are slightly depleted in heavier isotopes relative to clinopyroxene and resulting in Ni isotope variations that are controlled by differences in modal abundances. Saunders et al. (2020) report nickel isotope variations for peridotites ranging from 0.02 to 0.26‰. Based on mantle xenoliths and komatiites Gall et al. (2017) concluded that the bulk silicate earth has a $\delta^{60}$Ni value of 0.23‰ and is chondritic in composition. Given the observed heterogeneities, Saunders et al. (2020) argued that the value of 0.23‰ is not representative for the bulk silicate Earth.

To investigate potential Ni isotope fractionation between core and mantle, Lazar et al. (2012) determined Ni isotope fractionations between Ni metal and Ni talc silicate. Since the metal phase has been found to be enriched in light Ni isotopes, Lazar et al. (2012) suggested that Ni isotope fractionations might have occurred during Earth's core segregation. On the other hand, Gall et al. (2017) concluded that core formation did not generate measurable Ni isotope fractionations.

In iron meteorites, Ni isotopes—in conjunction with Fe isotopes—have been used to interprete the cooling and crystallization histories of metal rich meteorites (Cook et al. 2007; Weyrauch et al. 2017; Chernonozhkin et al. 2017). High resolution profiles reveal lighter Fe isotopes in kamacite (Fe-rich phase) than in taenite (Ni rich phase), while Ni isotopes behave vice versa: heavier Ni isotopes are enriched in kamacite compared to taenite. Ni and Fe isotope compositions along the interphase of the metal phases indicate diffusion driven fractionations during the formation of kamacite by the replacement of taenite.

### 2.21.2 Water

Dissolved Ni compounds in rivers vary by about 1‰ (Cameron and Vance 2014), and are heavier than average continental rocks. To compensate the enrichment, Ni isotope fractionation during weathering should lead to a light Ni isotope reservoir on the continents and indeed as shown by Ratié et al. (2015) clays and iron oxides are isotopically depleted in heavy Ni isotopes. Wasylenki et al. (2015) and Gueguen et al. (2018) demonstrated experimentally that lighter Ni isotopes preferentially enrich on ferrihydrite by about 0.3‰. Calcite also preferentially adsorbs light Ni isotopes (Castillo-Alvarez et al. 2020); the measured 0.5‰ fractionation between absorbed and aqueous nickel probably results from the change in coordination environment. From a latheritic weathering profile, Spivak-Birndorf et al. (2018) concluded that sorption of Ni to Fe oxyhydroxides and to a lesser extent copre-cipitation with smectitic clays may explain the enrichment of heavy Ni isotopes in rivers.

Ni dissolved in the ocean has a mean $\delta^{60}$Ni-value of 1.44‰ (Cameron and Vance 2014) being heavier than riverine Ni input. Beneath 500 m the nickel isotope compositions are constant with a $\delta^{60}$Ni-value of 1.31‰; surface ocean waters are enriched with $\delta^{60}$Ni-values up to 1.74‰ (Archer et al. 2020). A major imbalance exists between the Ni inputs to the ocean that are about half of the sedimentary output fluxes: input values are isotopically light at about 0.8‰ while sedimentary outputs are isotopically similar to or heavier than seawater. Little et al. (2020) suggested that diagenetic remobilization of isotopically heavy Ni balances the oceanic Ni budget.

For the water column and sediments of the Black Sea, Vance et al. (2016) demonstrated that light Ni isotopes are extracted by sequestration in sulfide species, to particulates and the sediment. Porter et al. (2014) reported Ni isotope variations between 0.15 and 2.5‰ in sediments rich in organic carbon; Ciscato et al. (2018), on the other hand, found that C-org rich sediments have Ni isotope values close to seawater. The latter authors argued that variable Ni isotope values are controlled by differences of oceanic sources.

### 2.21.3 Plants

Nickel is an essential micronutrient for most higher plants; a minor group of plants can be classified as hyper-accumulating Ni-plants (Deng et al. 2014). The uptake of Ni in the roots of plants mainly controls Ni isotope fractionation enriching plants in heavy isotopes relative to soils (Estrade et al. 2015). Deng et al. (2014) related Ni-isotope fractionations to uptake and translocation mechanisms within plants.

Ni plays an essential role in the metabolism of methanogenic archaea. Biological uptake during methanogenic growth produces substantial Ni isotope fractionations resulting in isotopically light cells and heavy residual media (Cameron et al. 2009). As postulated by these authors biological fractionations of Ni may provide a tracer for elucidating the nature of early life.

## 2.22 Copper

Copper occurs in nature in two oxidation states, Cu(I)$^+$ and Cu(II)$^{++}$ and rarely in the form of elemental Cu. The major Cu-containing minerals are sulfides (chalcopyrite, bornite, chalcosite and others), and, under oxidizing conditions, secondary copper minerals in the form of oxides and carbonates. Cu(I) is the common form in sulfide minerals, whereas Cu(II) is dominant in aqueous solution. Copper is a nutrient element, although toxic for all aquatic photosynthetic microorganisms. Copper may form a great variety of complexes with very different coordinations such as square, trigonal and tetragonal complexes. These properties are ideal prerequisites for relatively large isotope fractionations.

Copper has two stable isotopes

$$
\begin{array}{ll}
{}^{63}\mathrm{Cu} & 69.1\% \\
{}^{65}\mathrm{Cu} & 30.9\%.
\end{array}
$$

Early work of Shields et al. (1965) using the TIMS technique has indicated a total variation of ~12‰ with the largest variations in low temperature secondary minerals. Later studies using MC-ICP-MS techniques, by Maréchal et al. (1999), Maréchal and Albarede (2002), Zhu et al. (2002), Ruiz et al. (2002), measured a variation range of nearly 10‰, which is larger than for Fe. Most samples so far analysed, however, vary between $\delta^{65}$Cu values from +2 to −2‰ (see Fig. 2.36). Since the commonly used Cu standard NIST SRM 976 is no longer available, new certified reference materials are ERM-AE633 and ERM-AE647 (Möller et al. 2012). A review of Moynier et al. (2017) has summarized Cu isotope geochemistry.

### 2.22.1 Magmatic Rocks

Early measurements on Cu isotopic fractionations at magmatic temperatures indicated negligible Cu isotope fractionations of magmatic rocks. By analysing native copper grains and whole rock copper in peridotite, Ikehata and Hirata (2012) reported Cu isotope values close to 0‰ with no differences between Cu metal grains and whole rock copper; thus, the Cu isotope composition of mantle and crust appear to have $\delta^{65}$Cu values close to zero‰ (Li et al. 2009a, b).

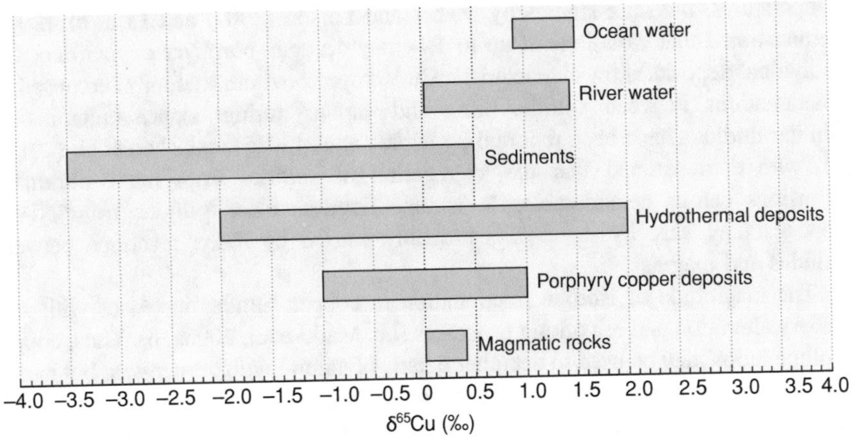


**Fig. 2.36** $\delta^{65}$Cu-values of important geological reservoirs

Since copper is sensitive to redox reactions, Cu isotope variations in magmatic rocks potentially could be ascribed to recycled crustal material and oxidative metsomatism. In a large set of peridotites, basalts and subduction related andesites and dacites, Liu et al. (2015) demonstrated that MORBs, OIBs and non-metasomatized peridotites show a limited range in Cu isotope composition indicating negligible Cu isotope fractionation during partial melting. Oxidized arc systems show a comparable small range (Wang et al. 2019a, b, c). As shown by Huang et al. (2016a) fractional crystallization of silicates also does not cause measurable Cu isotope fractionation. On the other hand, as demonstrated by Liu et al. (2015) metasomatized peridotites and as demonstrated by Huang et al. (2016b) hydrothermal alterations under high temperatures crust may result in significant Cu isotope fractionations.

Savage et al. (2015) postulated that the Earth's mantle is fractionated relative to bulk Earth which they explained by a segregation of a sulfide melt from the mantle during differentiation. Variable extents of magmatic sulfide segregation can change the Cu isotope composition of magmas and can lead to Cu isotope heterogeneity in the mantle (Zou et al. 2019).

### 2.22.2 Ore Deposits

Different types of Cu-ore deposits have been investigated by Larson et al. (2003), Rouxel et al. (2004a, b), Mathur et al. (2005, 2010a, b), Markl et al. (2006a, b) and Li et al. (2010). Early studies showed very limited Cu-isotope variations at high temperatures, but later studies by Maher and Larson (2007) and Li et al. (2010) demonstrated that variations of up to 4‰ may occur in porphyry copper deposits. Individual deposits show characteristic Cu isotope zonations that may be caused by fractionations between sulfide, brine and vapour during copper precipitation. Similar findings have been reported by Ripley et al. (2015) and Zhao et al. (2017a, b), who demonstrated that the segregation of sulfides from mafic–ultramafic intrusions causes considerable Cu isotope fractionations. Sulfides from Ni–Cu deposits may vary by up to 2‰ probably caused by redox reactions between sulfides and magmas.

The magnitude of isotope fractionation in copper sulfides increases with secondary alteration and reworking processes (i.e. Markl et al. 2006a, b). Thus, copper isotope ratios may be used to decipher details of natural redox processes, but hardly can be used as reliable fingerprints for the source of copper because the variation caused by redox processes within a single deposit is usually much larger than the inter-deposit variation. Experiments by Maher et al. (2011) indicated that the magnitude of Cu-isotope fractionation in an ore deposit depend on the pH of the mineralizing fluid and the partitioning of Cu between vapour and liquid. Extremely large differences in Cu isotope compositions between hypogene and supergene Cu minerals in porphyry copper deposits can be used as a tool for exploration (Mathur et al. 2009).

### 2.22.3 Low-Temperature Fractionations

Low-temperature processes are the major source of Cu isotope variations; in specific: (i) variation of redox conditions, (ii) adsorption on mineral surfaces and organic matter (Pokrovsky et al. 2008; Balistrieri et al. 2008), (iii) inorganic and organic complexation to ligands (Pokrovsky et al. 2008), (iv) biological fractionation by plants and micro-organisms (Weinstein et al. 2011; Coutaud et al. 2017).

Cu isotope fractionations for inorganic Cu species in water have been calculated by ab initio methods (Fujii et al. 2013, 2014; Sherman 2013). Experimental investigations have demonstrated that redox reactions between Cu (I) and Cu(II) species are the principal process that fractionates Cu isotopes (Ehrlich et al. 2004; Zhu et al. 2002). During precipitation of copper compounds without redox change, the heavier Cu isotope is preferentially incorporated; however, during Cu(II) reduction precipitated Cu(I) species are 3–5‰ lighter than dissolved Cu(II) species. Pokrovsky et al. (2008) observed a change in sign of Cu isotope fractionations during adsorption from aqueous solutions depending on the kind of surface, either organic or inorganic: on biological cell surfaces a depletion of $^{65}$Cu, whereas on hydroxide surfaces an enrichment of $^{65}$Cu is observed. In contrast to abiotic reactions, bacteria preferentially incorporate the lighter Cu isotope into their cells, regardless of experimentally conditions (Navarette et al. 2011). This was questioned by Coutaud et al. (2017) by demonstrating that Cu isotope fractionations between biofilms and aqueous solution change from short term enrichment to long term depletions.

### 2.22.4 River and Ocean Water

Oxidative weathering of sulfides is the major process that releases Cu into the terrestrial and marine environment. During weathering $^{65}Cu$ becomes enriched in the dissolved phase and thus depleted in the residual phase.

Rivers exhibit a large range of Cu isotope values (Vance et al. 2008; Little et al. 2014a, b). Particle-bound Cu is isotopically lighter than dissolved Cu species. In sites contaminated by acid mine drainage, Borrok et al. (2008) and Kimball et al. (2009) demonstrated systematic copper isotope fractionations between ore minerals and stream water that may be used for ore-prospecting.

In the ocean Cu plays an important role as a micronutrient, although dissolved $Cu^{2+}$ is toxic. By far the largest part of dissolved copper is complexed to natural organic ligands resulting in very low concentrations of the free $Cu^{2+}$ ion. Dissolved Cu in ocean water is heavier than the dissolved riverine input implying an isotopically light output process which may be caused by scavenging of light Cu to particulate material, preferentially to Fe–Mn oxides (Vance et al. 2008; Little et al. 2014a, b; Takano et al. (2014). As shown by Little et al. (2018) the vertical distribution of Cu in the ocean is somewhat unusual, being neither a nutrient-type nor scavenged-type distribution. Deep ocean water has a homogeneous Cu isotope composition of +0.65‰. The particulate phase is characterized by two Cu-pools, (i) a lithogenic pool of about 0‰ and a labile pool liberated by weak-acid leaching at about 0.4‰ (Little et al. 2018).

### 2.22.5 Plants

Copper is an essential micronutrient for plant growth. Cu isotopes may be used to elucidate Cu uptake and to investigate Cu isotope fractionations between root and stem and different plant tissues (Li et al. 2016a, b, c). Studies by Weinstein et al. (2011), Jouvin et al. (2012) and Ryan et al. (2013) demonstrated that different uptake strategies lead to different Cu isotope fractionations in plants. Tomatoe and oat grown under controlled solution cultures yield Cu isotope fractionations which support previous findings for Fe uptake in strategy 1 and 2 plants (Ryan et al. 2013). Tomatoes preferentially fractionate light $^{63}Cu$ by about 1‰, which is attributed to Cu reduction whereas oat shows minimal Cu fractionation suggesting that Cu uptake and transport is not redox selective.

## 2.23 Zinc

Zinc has 5 stable isotopes of mass 64, 66, 67, 68 and 70 with the following abundances:

$^{64}$Zn 48.63%
$^{66}$Zn 27.90%
$^{67}$Zn 4.10%
$^{68}$Zn 18.75%
$^{70}$Zn 0.62%

Most Zn isotope data have been acquired by MC-ICP-MS, first described by Maréchal et al. (1999) and recently summarized by Moynier et al. (2017). The commonly used Zn isotope standard, JMC-Lyon standard, in the past, is not longer available. Möller et al. (2012) calibrated IRMM-3702 as the new certified Zn standard, which has a $\delta^{66}$Zn-value of 0.29‰ relative to the JMC-Lyon standard. In Fig. 2.37 natural Zn isotope variations given as $^{66}$Zn/$^{64}$Zn ratios are summarized.

Processes fractionating zinc isotopes are (i) evaporation-condensation processes in which the vapor phase is depleted in the heavier isotopes relative to the solid phase and (ii) sorption processes (Cloquet et al. 2008). Zn sorption on Fe hydroxides causes an enrichment of heavy Zn isotopes in the solid (Juillot et al. 2008), Zn-sorption on organic matter causes variable Zn isotope fractionations depending on the type of organic matter and pH (Jouvin et al. 2009). Little et al. (2014a, b) and Bryan et al. (2015) quantified Zn isotope fractionations during adsorption on Mn oxyhydroxide and demonstrated preferential enrichment of heavy Zn isotopes on ferromanganese crusts. The magnitude of isotope fractionation depends on the structure of Zn-complexes on the surface of the solid.

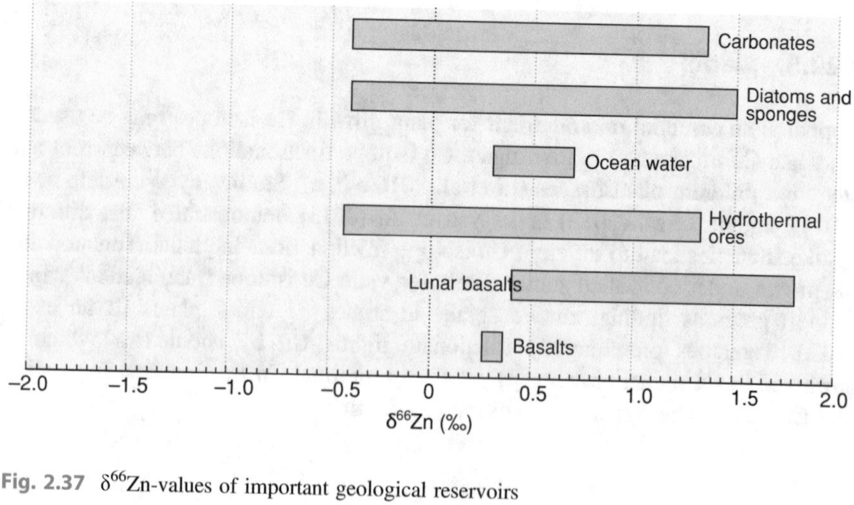


**Fig. 2.37** $\delta^{66}$Zn-values of important geological reservoirs

In water, Zn isotope fractionation depends on the ligands present, especially on dissolved phosphate and carbonate. Ab initio calculations of Zn isotope fractionations between aqueous sulfide, chloride and carbonate species by Black et al. (2011) and Fujii et al. (2011a, b) indicate that Zn sulfide complexes are isotopically depleted in heavy Zn isotopes relative to $Zn^{2+}$ and Zn chlorides, whereas carbonates are more enriched than chlorides.

### 2.23.1 Fractionations During Evaporation

Evaporation–condensation processes may cause large fractionations in meteorites (Luck et al. 2005; Wombacher et al. 2008; Moynier et al. 2011a, b). As an example, tektites being extremely depleted in volatile elements show an enrichment in heavy Zn isotopes which is attributed to kinetic fractionations during evaporation (Moynier et al. 2009). In lunar rocks large $^{66}Zn$ enrichments have been interpreted as indicating whole-scale evaporation of the Moon (Paniello et al. 2012) or evaporation during magma ocean differentiation (Day and Moynier 2014; Kato et al. 2015a, b).

Significant amounts of Zn may be emitted by degassing from volcanoes. Fumarolic gases and condensates from the Merapi volcano have a relatively large range in Zn isotope compositions. Gaseous Zn samples are enriched in lighter Zn isotopes whereas condensates are enriched in the heavier isotopes (Toutain et al. 2008).

### 2.23.2 Mantle Derived Rocks

The mantle represented by non-metasomatized peridotites has, depending on estimates by Sossi et al. (2018c), Wang et al. (2017a, b) and Huang et al. (2019) a $\delta^{66}Zn$-value between 0.16 and 0.20‰. Metasomatized peridotites and pyroxenites display relatively large variations in Zn isotope compositions with depletions and enrichments relative to the primordial Zn isotope value (Doucet et al. 2016). Enrichments were ascribed to the incorporation of isotopically heavy carbonates (Liu et al. 2016; Wang et al. 2017a, b); depletions were attributed to kinetic isotope fractionations due to preferential diffusion of $^{64}Zn$ during melt-rock interactions (Wang et al. 2017a, b; Huang et al. 2019).

MORB and ocean island basalts have slightly heavier Zn isotope values than the mantle (Huang et al. 2018). A literature survey by Wang et al. (2017a, b) indicates that basalts are about 0.1‰ enriched in heavy Zn isotopes relative to non-metasomatized peridotites. Most arc lavas have slightly enriched $\delta^{66}Zn$-values probably due to isotope fractionation during partial melting (Huang et al. 2018). By studying two chemically diverse suites of volcanic rocks from Hawaii and Iceland, Chen et al. (2013) concluded that Kilauea basalts show small, but systematic Zn isotope enrichment with increasing degree of differentiation.

Zn isotope fractionations among separate minerals are close to zero, except for spinels that are enriched relative to olivine. The same coordination number of Zn in olivine and pyroxenes, replacing Mg and Fe, precludes large intermineral Zn isotope fractionations. In spinels, however, Zn substitutes Mg and Fe on tetrahedral instead of octahedral sites. Relatively large Zn isotope fractionations between olivine and pyroxene in metasomatized lherzolites reported by Wang et al. (2017a, b) may reflect kinetic isotope fractionations during melt-rock interactions.

Zinc isotopes have been applied to trace recycled crustal components in mantle derived rocks based on the Zn isotope difference of crustal rocks, in particular carbonates, and the mantle (Liu et al. 2016; Wang et al. 2017a, b; Yang and Liu 2019; Beunon et al. 2020).

### 2.23.3 Ore Deposits

By analyzing sphalerites from ore deposits, Mason et al. (2005), Wilkinson et al. (2005), Kelley et al. (2009), Gagnevin et al. (2012) and Zhou et al. (2014) observed Zn isotope variations of about 1.5‰. These studies indicate that early precipitated sphalerites have higher Zn-isotope values than late precipitates. The variations have been related to kinetic fractionations during rapid sphalerite precipitation. Gagnevin et al. (2012) explained relatively large Zn-isotope variations at the millimetre scale by mixing of hot hydrothermal fluids with cool brines containing biogenic sulfide. By measuring shear zone-hosted ore bodies at the Balmat mine, New York, Matt et al. (2020) interpreted Zn isotope differences as a result of large-scale remobilization during upper amphibolite metamorphism. Mondillo et al. (2018) investigated Zn isotope fractionations of secondary zinc minerals (willemite-$ZnSiO_4$; smithsonite-$ZnCO_3$ and others) formed from the precursor mineral sphalerite; compared to primary sphalerites, negative and positive Zn isotope fractionations have been observed with willemite showing the largest compositional range.

### 2.23.4 Ocean

Zinc is an essential micronutrient for phytoplankton, its concentration is controlled by phytoplankton uptake and remineralization. Light Zn isotopes are preferentially incorporated into phytoplankton organic matter, leaving residual Zn in surface water enriched in Zn isotopes (John et al. 2007a, b; Andersen et al. 2011; Hendry and Andersen 2013). Zn removal from surface waters occurs via 2 processes (i) incorporation of Zn into organic matter involving no isotope fractionation and (ii) metabolic uptake of light Zn isotopes into the cells of phytoplankton enriching surface waters in heavy Zn isotopes (Zhao et al. 2014).

Isotopic fractionations during uptake and adsorption processes lead to considerable zinc isotope variations in surface ocean waters from −1.1 to +0.9‰ (Bermin et al. 2006; Conway and John 2014; Zhao et al. 2014; John et al. 2018). The bulk isotope composition of dissolved Zn in the ocean below 1000 m is homogeneous with $\delta^{66}$Zn-values of 0.53‰, which is heavier than the input from river water (Little et al. 2014a, b; Balistrieri et al. 2008; Chen et al. 2008; Borrok et al. 2009). To balance the isotopically heavy composition of dissolved zinc in the ocean with respect to Zn inputs, an isotopically light sink is required. Organic-rich continental margin sediments may provide the necessary light Zn sink (Little et al. 2016). As demonstrated by Vance et al. (2016), euxinic sediments in the Black Sea are enriched in light Zn isotopes.

Relatively large Zn isotope fractionations have been reported in hydrothermal vent fluids (John et al. 2008). Low-temperature fluids have heavier $\delta^{66}$Zn-values than high temperature fluids. Cooling of vent fluids leads to—due to kinetic isotope fractionations—precipitation of isotopically light sphalerite causing an enrichment of heavy isotope in the fluid.

Variations of Zn isotopes in marine carbonates have been interpreted to reflect changes of ancient Zn seawater isotope compositions (Pichat et al. 2003; Kunzmann et al. 2013). The reliability of modern carbonates as a proxy of seawater Zn isotope composition has been investigated by Zhao et al. (2021): Skeletal calcite is considerable enriched in heavy Zn isotopes relative to seawater, which agrees with precipitation experiments by Mavromatis et al. (2019), whereas skeletal aragonite closely reflects modern ocean water composition, indicating that aragonite carbonate phases may potentially be used as a seawater Zn isotope proxy.

### 2.23.5 Plants and Animals

Zinc is a vital element for most organisms; it plays an essential role in various biochemical processes. The largest variation of Zn isotopes has been found in land plants (Viers et al. 2007; Weiss et al. 2005). As shown by Moynier et al. (2008) and Viers et al. (2007), Zn isotopes fractionate during incorporation into roots and during transport within plants. The magnitude of fractionation is species dependent (Viers et al. 2007) and may depend on the size of the plant. Caldelas and Weiss (2017) reviewed zinc isotopic fractionations during uptake and transport to roots, root-to-shoot transport and remobilization. The mechanisms of Zn isotope fractionations are not well understood, but may depend on surface absorption, solution speciation and membrane-controlled uptake.

Jaouen et al. (2013) reported systematic $^{66}$Zn enrichments of plants relative to bones in herbivores. For a diet with a $\delta^{66}$Zn-value of 0.2‰, the spread of $^{66}$Zn-values in sheep has been from –0.5 in liver to + 0.5‰ in bone, muscle and serum (Balter et al. 2010). Jaouen et al. (2016) demonstrated that Zn isotopes can be used as dietary indicator. They showed that Zn isotopes in bone and enamel allow a clear distinction between herbivores and carnivores. Bourgon et al. (2020) analyzed enamel of Late Pleistocene teeth and observed a trophic carnivore-herbivore difference of 0.6‰ with omnivores having intermediate values. Zinc isotopes are thus suitable to assess the food of mammals as well as trophic relationships in past food webs.

### 2.23.6 Anthropogenic Contamination

Due to anthropogenic activities, many environmental systems are polluted with zinc. The potential of using zinc isotopes to trace Zn contaminations was demonstrated by Cloquet et al. (2008), Sonke et al. (2008), Chen et al. (2008) and Weiss et al. (2007). Less polluted waters have higher $\delta^{66}$Zn-values than polluted ones. Chen et al. (2008) measured Zn isotope variations along a transsect of the Seine. Variations along the river transect showed an increase in Zn concentrations with highest values in the region of Paris.

John et al. (2007a, b) measured the Zn isotope composition of various man-made Zn products. They showed that the range of $\delta^{66}$Zn values of industrial products is smaller than of Zn ores indicating Zn isotope homogenezation during processing and ore purification. During smelting of Zn ores, studies by Sonke et al. (2008) and Yin et al. (2016a, b) have shown that slag residues are enriched in heavy Zn isotopes while fine dust emitted tends to be isotopically light.

By analyzing peat profiles, Weiss et al. (2007) concluded that Zn isotopes may identify atmospheric sources such as zinc derived from mining and smelting. Bigalke et al. (2010) measured isotopically light Zn in highly polluted soils around a smelter. They pointed out that light Zn isotope signatures may be also generated by uptake of light Zn from plants.

## 2.24 Gallium

Gallium primarily exists in the trivalent state ($Ga^{3+}$) and is generally regarded as having chemical properties similar to aluminum. Unlike the major element Al, the trace element Ga may behave differently and the Ga/Al ratio has been used to investigate its transport during weathering processes and behavior in the ocean. Ga plays a role in biological reactions, but its exact nature is unknown. Due to its increased usage in high-tech applications and in medical fields, Ga potentially may cause environmental problems.

Gallium has two stable isotopes.

$$^{69}\text{Ga} \quad 60.10\%$$
$$^{71}\text{Ga} \quad 39.90\%$$

Yuan et al. (2016), Zhang et al. (2016) and Kato et al. (2017) described precise MC-ICP-MS methods to measure Ga isotope variations. A remarkable difference in Ga isotope composition has been reported between chondrites and the Earth. Isotope fractionation during mantle-core formation may be a likely process to explain the systematic difference (Kato et al. 2015a, b). During magmatic differentiation, no Ga isotope fractionation seems to occur (Kato et al. 2015a, b).

Absorption on oxides, clays and carbonates may control the behavior of Ga in surficial environments. Ga isotope fractionation during Ga absorption on calcite and goethite has been investigated by Yuan et al. (2018). They showed that the light Ga isotope preferentially becomes enriched on the surface of the solid. Yuan et al. (2018) concluded that Ga isotope fractionations are governed by changes of Ga coordination and Ga-O bond length during absorption of $Ga(OH)_4^-$ on solid surfaces.

As gallium is a moderately volatile element, Ga isotopes may undergo kinetic isotope fractionation during evaporation, which has been verified by Wimpenny et al. (2020) in samples that have undergone evaporation at very high temperatures.

## 2.25 Germanium

Because of nearly identical ionic radii, Ge may replace Si in silicates and thus is expected to show an isotope fractionation behaviour similar to silicon. But germanium is also moderately siderophile and chalcophile which leads to considerable enrichments in sulfides and metal.

Ge has 5 stable isotopes with the following abundances (Rosman and Taylor 1998):

| | |
|---|---|
| $^{70}$Ge | 20.84% |
| $^{72}$Ge | 27.54% |
| $^{73}$Ge | 7.73% |
| $^{74}$Ge | 36.28% |
| $^{76}$Ge | 7.61% |

Early investigations using the TIMS method had an uncertainty of several‰. Over the past few years advances have been made in the MC-ICP-MS technique with a long-term external reproducibility of 0.2–0.4‰ (Rouxel et al. 2006; Siebert et al. 2006a). Even better reproducibility has been reported by Luais (2012) and Escoube et al. (2012). $\delta^{74}$Ge-values are commonly given as $^{74}$Ge/$^{70}$Ge ratios relative to NIST SRM 3120a standard. A total variation from −4.0 to +4.7‰ for earth materials and meteorites have ben reported (El Korh et al. 2017; Rouxel and Luais 2017) (see Fig. 2.38).

Li et al. (2009a, b) and Li and Liu (2010) calculated isotope fractionation factors among Ge-bearing phases and predicted that relative to Ge-oxides, sulfides will be depleted in heavy Ge isotopes. During adsorption on goethite, Pokrovsky et al. (2014) demonstrated experimentally that goethite will become depleted in heavy Ge isotopes by 1.7‰ relative to aqueous solutions. Thus, adsorption phenomena will affect the Ge isotope composition of ocean water.

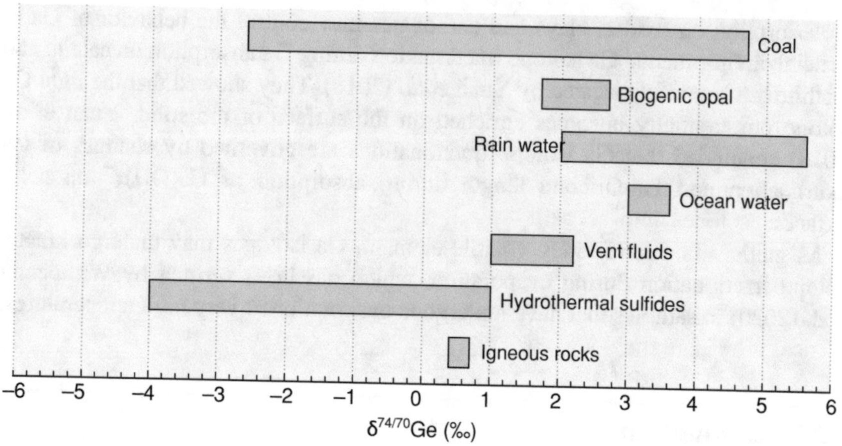


**Fig. 2.38** $\delta^{74/70}$Ge isotope variations of geological reservoirs

Mantle derived rocks show a narrow range in Ge isotope compositions with an average of 0.53‰. $\delta^{74}$Ge values from 0.37 to 0.62‰ reported for basalts may reveal isotope variations in the mantle source (Escoube et al. 2012; Luais 2012). During late stages of retrogression, water/rock interactions may lead to an enrichment of heavy Ge isotopes (El Korh et al. 2017).

## 2.25.1 Ore Deposits

Although Ge concentrations in ore deposits are generally very low, they can reach several housand ppm in sphalerite. Sulfides from various hydrothermal ore deposits range from −5 to +2‰ (Belissont et al. 2014; Meng et al.2015). Calculated fractionation factors between sphalerite-like sulfides and fluid at 25 °C yield fractionations of 12.2 to 11.5‰ resulting in very light sphalerite compositions compared to fluids (Li et al. 2009a, b).

Differences in temperatures of precipitation and reservoir effects seem to be the major factors explaining the large range in $\delta^{74}$Ge values.

Relatively high Ge concentrations have been reported in coal seams. Qi et al. (2011) observed $\delta^{74}$Ge variations of more than 7‰ in coals and their combustion products. They showed that coal combustion fractionates Ge isotopes, with soot being more depleted in $^{74}$Ge than slags.

### 2.25.2 Hydrosphere

Ge isotopes of suspended material in rivers are indistinguishable from silicate source rocks; $\delta^{74}$Ge-values of dissolved phases show a large range from 0.9 to 5.6‰, being considerably heavier than crust and mantle. Baronas et al. (2017, 2018) interpreted the enrichment as being due to the preferential incorporation of light Ge isotopes into secondary weathering phases and to a lesser extent into organisms.

Guillermic et al. (2017) presented first germanium isotope data for seawater. Deep waters are relatively homogeneous with a $\delta^{74/70}$Ge value of 3.14 ± 0.38‰. Vertical profiles in the Southern Ocean indicate enrichment of heavy Ge isotopes in surface waters with a minimum in $\delta^{74/70}$Ge values at the depth of maximum remineralization. Guillermic et al. (2017) suggested a combination of two processes to explain Ge isotope distributions: (i) Ge isotope fractionations during siliceous phytoplankton formation and (ii) mixing of different water masses. It is noteworthy in this connection that deep sea sponges show a $^{74}$Ge depletion of about 0.9‰ relative to coexisting sea water.

By analyzing low- and high-temperature vent fluids, Escoube et al. (2015) showed that vent fluids are about 1.5‰ lighter than the Ge isotope composition of the ocean. Vent fluids become depleted in $^{74}$Ge relative to seawater mainly due to Ge adsorption on precipitated Fe-oxyhydroxides.

In summary, Ge isotopes are characterized by a depletion of $^{74}$Ge in hydrothermal sulfides and an enrichment of $^{74}$Ge in ocean water and marine sediments.

**2.26–2.27 Selenium and tellurium.**

Selenium and tellurium belong to the same group as sulfur showing a close relationship in their geochemical properties. They are both chalcophile trace elements with multiple oxidation states.

## 2.26 Selenium

Selenium is an essential trace element for animals and humans having a narrow concentration range between sufficiency and toxicity (Schilling et al. 2011). It occurs in four oxidation states that differ in their nutritional and toxic behaviour. Selenium to some extent is chemically similar to sulfur, therefore, one might expect relatively large fractionations of selenium isotopes in nature. Six stable selenium isotopes are known with the following abundances (Coplen et al. 2002)

$^{74}Se$ 0.89%
$^{76}Se$ 9.37%
$^{77}Se$ 7.63%
$^{78}Se$ 23.77%
$^{80}Se$ 49.61%
$^{82}Se$ 8.73%

Because of the 7% mass difference between $^{76}Se$ and $^{82}Se$ and numerous microbial and inorganic Se redox transformations, interest in selenium isotope studies has grown in recent years. An early study by Krouse and Thode (1962), using $SeF_6$ gas, required relatively large quantities of Se, limiting the applications of selenium isotopes. Johnson et al. (1999) developed a double-spike solid-source technique that corrects for fractionations during sample preparation and mass spectrometry, yielding an overall reproducibility of ±0.2‰. This technique brought sample requirements down to submicrogram levels. Even lower Se amounts (10 ng) are required for measurements with the MC-ICP-MS technique (Rouxel et al. 2002) using a commercial Se solution as standard. Early TIMS studies published Se isotope as $^{80}Se/^{76}Se$ ratios (Johnson et al. 1999; Herbel et al. 2000); more recent work has given Se isotope values as $^{82}Se/^{78}Se$ (Stüeken et al. 2013) or as $^{82}Se/^{76}Se$ ratios (Mitchell et al. 2012; Layton-Matthews et al. 2013). $^{82}Se/^{78}Se$ ratios can be transformed to $^{82}Se/^{76}Se$ ratios by multiplication with a factor of 1.5. $\delta^{82}Se$-values are generally given relative to the NIST SRM 3149 standard distributed as a solution. Figure 2.39 summarizes Se isotope variations as $^{82}Se/^{76}Se$ ratios in specific reservoirs. Stüeken (2017) has reviewed Se isotope geochemistry.

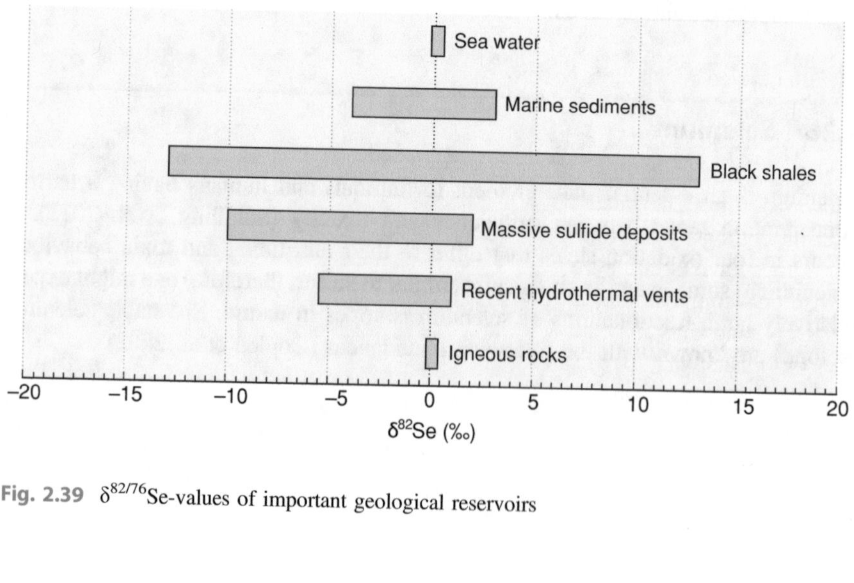


**Fig. 2.39** $\delta^{82/76}Se$-values of important geological reservoirs

## 2.26.1 Fractionation Processes

As shown by calculations of Li and Liu (2011) equilibrium isotope exchange reactions can lead to large Se isotope fractionations. Because kinetic fractionations during biological reduction seem to be dominant in the Se-cycle, equilibrium isotope fractionations seem to be of less importance under natural conditions.

Selenium oxyanions can be reduced by certain microbes. Reduction proceeds in 3 steps with Se(IV) and Se(0) species as stable intermediates (Johnson 2004). Se isotope fractionation experiments by Herbel et al. (2000) indicate about 5‰ fractionations during reduction of selenate to selenite. Schilling et al. (2020) demonstrated that microbial reduction of Se(VI) results in larger isotope fractionation than reduction of Se(IV) confirming different metabolic pathways of Se(VI) and Se(IV). Johnson and Bullen (2003) investigated Se isotope fractionations induced by inorganic reduction of selenate by Fe(II)–Fe(III) hydroxide sulfate ("green rust"). The overall fractionation is 7.4‰, which is larger than during bacterial selenate reduction. This indicates that the magnitude of Se isotope fractionations depends on the specific reaction mechanism.

Se isotope fractionation during adsorption is another important mechanism causing isotope fractionations. Mitchell et al. (2013) determined Se isotope fractionations during sorption to iron oxides and to iron sulfides showing that fractionations caused by iron oxides are generally small, whereas fractionations associated with sulfides are much larger. Xu et al. (2020) investigated Se(IV) and Se(VI) adsorption on metal oxides. Se(IV) absorption causes Se isotope fractionation, whereas Se(VI) does not. According to Xu et al. (2020) the contrasting Se isotope fractionation is caused by a structural difference between dissolved and absorbed Se(IV)/Se(VI) on metal oxide surfaces.

## 2.26.2 Natural Variations at High Temperatures

Mantle-derived rocks have $\delta^{82}$Se compositions close to zero. Rouxel et al. (2002) measured several igneous rocks and a few iron meteorites, which all lie within 0.6‰ of the NIST-SRM 3149 standard. Kurzawa et al. (2017) and Yierpan et al. (2018) described analytical methods to determine precise Se isotope data of basalts and peridotites. The latter have been analyzed by Varas-Reus et al. (2019) with varying compositions from refractive harzburgites to highly fertile lherzolite. All peridotites have a more or less constant $\delta^{82/76}$Se values of −0.03‰, indicating that Se isotope compositions are unaffected by melt depletion and melt enrichment processes and thus may be regarded as representative for the bulk silicate Earth (Varas-Reus et al. 2019). In contrast, basalts from world wide locations have slightly heavier Se isotope values (Yierpan et al. 2018), whereas MORB from the Pacific-Antarctic ridge are lighter than peridotites (Yierpan et al. 2019). For Mariana arc lavas Kurzawa et al. (2019) found a large range in Se isotope ratios, which they interpreted as being due to fluid- and melt-like subduction components.

Selenium may become enriched in recent hydrothermal vent sulfides, in which Se may be derived from leaching of igneous rocks or from Se-rich organic sediments. Layton-Matthews et al. (2013) reported a wide range of $\delta^{82}$Se values in ancient seafloor hydrothermal deposits. Very negative values are probably due to Se loss from carbonaceous shales during hydrothermal activity.

### 2.26.3 Ocean

Although Se and S share similar geochemical behavior, in the oceanic environment, Se behaves different to S, where it exists as Se(VI) and Se(IV) oxyanions and—most important—as dissolved organic Se. Mitchell et al. (2012) observed in marine shales with low organic carbon contents a small range in $\delta^{82}$Se values, whereas in $C_{org}$ rich black shales with high Se concentrations larger Se isotope variations do occur (Wen and Carigman 2011). In a profile of very Se-rich carbonaceous shales, Zhu et al. (2014) observed a very large range in $^{82/76}$Se-values from −14.2 to +11.4‰, suggesting multiple cycles of oxidation and reduction.

Selenium isotopes may be used to trace redox changes in the atmosphere and ocean over geological time (Wen et al. 2014; Pogge von Strandmann et al. 2015, Stüeken et al. 2015a, b, c). As suggested by Stüeken et al. (2015a), Se isotopes in marine sediments may reflect ancient productivity and redox conditions in the water column. Positive Se isotope values should indicate high biological productivity and/or anoxic conditions, whereas negative values should indicate oxic conditions and low productivity. In an attempt to reconstruct the biogeochemical Se-cycle, Stüeken et al. (2015b) interpreted a shift from positive Se isotope values in the Proterozoic to negative ones in the Paleozoic as indicating the oxidation of Se(IV) to Se(VI) during the late Proterozoic. Se isotope data from the 2.5 Ga Mount Mc Rae shale (Australia) may indicate oxygenic photosynthesis before the Great Oxidation Event as suggested by Stüeken et al. (2015c). In more recent studies, Kipp et al. (2017) and Stüeken and Kipp (2020) showed that significant Se isotope fractionations are absent till 2.5 Ga, but appear around 2.3 Ga.

## 2.27 Tellurium

Tellurium is a very rare element on Earth, but may become enriched in Au, Ag and Cu-Ni-PGE mineralizations. It is economically valuable, since it is used in photovoltaic technology and nanotechnology. Tellurium occurs in nature in four oxidation states: as two oxyanions, tellurate and tellurite, and in two reduced forms, as native tellurium and as metal telluride. The two oxyanions are soluble and mobile, whereas in the reduced states tellurium has low solubility. Tellurium has 8 stable isotopes with the following abundances

| | |
|---|---|
| $^{120}Te$ | 0.10% |
| $^{122}Te$ | 2.60 |
| $^{123}Te$ | 0.91 |
| $^{124}Te$ | 4.82 |
| $^{125}Te$ | 7.14 |
| $^{126}Te$ | 19.0 |
| $^{128}Te$ | 31.6 |
| $^{130}Te$ | 33.7 |

By measuring $^{130/122}Te$-ratios in gaseous $TeF_6$, Smithers and Krouse (1968) first demonstrated that inorganic and microbiological reductions of tellurite to elemental tellurium causes isotope fractionations with depletions of the heavy isotope in the reaction product. Due to considerable memory effects and other problems during chemical preparation, the method has been abandoned. Fehr et al. (2004) introduced a MC-ICP-MS method for tellurium that has been extended by Fornadel et al. (2014) and Fukami et al. (2018). By measuring $^{130}Te/^{125}Te$ ratios, Fornadel et al. (2014) demonstrated that tellurides and native tellurium in ore deposits reveal isotope differences up to 1.64‰ with significant variations within individual deposits. By measuring $^{130}Te/^{126}Te$ ratios, Wassermann and Johnson (2020) observed similar ranges of isotope variations in standard reference materials, sediments and soils.

As for sulfur and selenium, oxidized tellurium compounds are isotopically enriched relative to reduced species. First-principles thermodynamic calculations by Fornadel et al. (2017) suggest Te isotope fractionations at 100 °C as large as 4‰ between Te(IV) and Te(II) or Te(0) compounds and smaller fractionations between Te(I) or Te(II) containing minerals.

Because abiotic and microbial redox reactions fractionate Te isotopes, Te isotopes potentially may become an indicator of paleoredox conditions.

## 2.28 Zirconium

Zirconium has five stable isotopes

| | |
|---|---|
| $^{90}Zr$ | 51.45% |
| $^{91}Zr$ | 11.22% |
| $^{92}Zr$ | 17.15% |
| $^{94}Zr$ | 17.38% |
| $^{96}Zr$ | 2.80% |

Zirconium is resistant to secondary alteration effects and as high-field strength element behaves incompatible during mantle melting. In the accessory phases zircon and baddeleyite ($ZrO_2$), zirconium plays a key role for the understanding of the Earth's evolution.

Inglis et al. (2018), Ibanez-Mejia and Tissot (2019), Zhang et al. (2019) and Tompkins et al. (2020) described a double-spike MC-ICPMS technique measuring $^{94/}$Zr/$^{90}$Zr ratios relative Zr solutions. By measuring 22 geological reference samples, Tian et al. (2020a) report a total variation range of 0.6‰ in $^{94/90}$Zr ratios. Different types of basalt show very limited Zr isotope variations, but may vary with magmatic differentiation (Inglis et al. 2019). Ocean island basalts show a limited range in zirconium isotope ratios, as do E-MORB samples. By analyzing komatiites around the world, Tian et al. (2020b) concluded that the mantle has a constant zirconium isotope composition since at least 3.55 Ga.

As discussed by Inglis et al. (2019) highly evolved rocks from Hekla volcano in Iceland are isotopically enriched relative to less evolved rocks. The most evolved lavas from Hekla have a $^{94}$Zr/$^{90}$Zr ratio 0.5‰ higher than less evolved lavas. Inglis et al. (2019) argued that Zr isotope fractionation is determined by incorporation of light Zr isotopes within the eightfold coordinated sites of zircon, driving residual melts with a lower coordination chemistry towards heavier values. In single zircon grains from continental arc plutonic rocks, Guo et al. (2020a, b) observed large internal Zr isotope zoning with lighter Zr isotopes in the core and heavier Zr isotopes towards the rim, suggesting preferential incorporation of lighter Zr isotopes from the melt, driving the residual melt to heavier values. Contrasting results and interpretations have been presented by Ibanez-Mejia and Tissot (2019). By analyzing single zircon and baddeleyite crystals from a gabbroic igneous cumulate, they observed that zircon and baddeleyite are isotopically heavy relative to the melt from which they crystallize, driving differentiated melts towards isotopically light compositions.

To better understand the origin of the Zr fractionations, Méheut et al. (2020) investigated the main driving forces causing the relatively large fractionations of more than 5‰ in zircons. Méheut et al. (2021) concluded that equilibrium effects cannot be responsible; instead diffusion processes—in particular the development of Zr diffusive boundary layers in magmas during fractional crystallization—may be the most likely mechanism.

## 2.29 Molybdenum

Molybdenum is relatively unreactive under oxygenated conditions making it the most abundant transition metal in the ocean, despite very low concentrations in the crust. Under anoxic-sulfidic conditions, molybdenum is readily removed from ocean water, leading to characteristic Mo-enrichment in sediments. Mo is an essential cofactor for enzymes in nearly all organisms.

Mo consists of 7 stable isotopes that have the following abundances:

$^{92}Mo$ 15.86%,
$^{94}Mo$ 9.12%,
$^{95}Mo$ 15.70%,
$^{96}Mo$ 16.50%,
$^{97}Mo$ 9.45%,
$^{98}Mo$ 23.75%,
$^{100}Mo$ 9.62%.

Either $^{97}Mo/^{95}Mo$ or $^{98}Mo/^{95}Mo$ ratios have been reported in the literature. Thus, care has to be taken when comparing Mo isotope values. In low temperature geochemistry of Mo, isotope data, given in the following as $\delta^{98}Mo$ values, are generally reported relative to internal laboratory standards calibrated against ocean water (Mean Ocean Molybdenum (MOMo), Barling et al. 2001; Siebert et al. 2003). In high-temperature Mo geochemistry, $\delta^{98}Mo$ values are commonly reported relative to the NIST SRM 3134 standard (Burkhardt et al. 2011; Willbold and Elliott 2017). Nägler et al. (2014) proposed that NIST SRM 3134 should be accepted as international standard with a $\delta^{98}Mo$ value of +0.25‰ relative to MOMo.

Mo isotope geochemistry has been summarized by Kendall et al. (2017). What makes Mo particular interesting, is its use as a potential proxy for the redox history of the oceans and the atmosphere (Barling et al. 2001; Siebert et al. 2003; Wille et al. 2007; Dahl et al. 2010a, b; Herrmann et al. 2012; Scott and Lyons 2012 besides others). Figure 2.35 summarizes natural Mo isotope variations.

### 2.29.1 Magmatic Rocks

Mo isotope studies in meteorites have been used to investigate genetic relationships between meteorites and the terrestrial planets (Burkhardt et al. 2011, 2014). As demonstrated by Burkhardt et al. (2014), silicates in differentiated planetary bodies are isotopically heavy compared to the Mo isotopic composition of bulk chondrites and iron meteorites. Experiments by Hin et al. (2013) have shown that Mo isotopes fractionate between metal and silicate liquids. More recent experiments by Hin et al. (2019) have demonstrated that Mo isotope fractionation depends on the $Mo^{6+}/Mo^{4+}$ ratio in silicate liquids. Hin et al. (2019) concluded that core formation does not account for the Mo isotope composition of the upper mantle.

Mo isotope compositions of igneous rocks vary by more than 1‰ (Burkhardt et al. 2014; Freymuth et al. 2015; Greber et al. 2014, 2015; Voegelin et al. 2014; Yang et al. 2015a, b, Bezard et al. 2016; Willbold and Elliott 2017). Bezard et al. (2016) reported 0.4‰ variations in MORB which they interpreted to reflect heterogeneities in the mantle. Elevated $\delta^{98}Mo$ values correspond to enriched mantle sources suggesting that recycled crustal material is isotopically heavy compared to depleted mantle. On the other hand, Liang et al. (2017) reported constant MORB isotope compositions, whereas ocean island basalts display large isotope variabilities, even within a single locality.

Mo isotopes, thus, may provide a tracer for the recycling of crustal material into the mantle. In arc lavas a large range of Mo isotope values have been observed by Freymuth et al. (2015), König et al. (2016), Willbold and Elliott (2017) and Gaschnig et al. (2019). Heavy values have been reported for fluid dominated samples, light values are characterized by sediment subduction and sediment melt components.

Voegelin et al. (2014) and Yang et al. (2015a, b) investigated Mo isotope behavior during magmatic differentiation. While Voegelin et al. (2014) observed a 0.3‰ Mo isotope fractionation from basalt to dacite, Yang et al. (2015a, b) observed no Mo isotope fractionation during magmatic differentiation of Hekla volcano, Iceland. Voegelin et al. (2014) showed that amphibole and biotite crystallizing from a silicate melt are depleted in heavy Mo isotopes by about 0.5‰.

I-, S- and A-type granites record a range of about 1‰ in $\delta^{98}$Mo values with significant overlaps among the different granite types (Yang et al. 2017). Thus, Mo isotopes may not be effective to discriminate granite sources; isotope variations seem to be due to source heterogeneities and hydrothermal alterations.

### 2.29.2 Molybdenites

Large Mo isotope variations have been found in molybdenites ($MoS_2$), an accessory mineral in many magmatic rocks (Hannah et al. 2007; Mathur et al. 2010a, b). The total range is about 4‰ (Breillat et al. 2016). According to Mathur et al. (2010a, b) Mo isotope variations depend on the type of ore deposit; molybdenites from porphyry coppers have lighter Mo isotope composition relative to other ore deposits. Greber et al. (2011) observed isotope variations of 1.35‰ in a single molybdenite deposit which is larger than the overall Mo isotope variation in igneous rocks. By analysing molybdenites from the well-known porphyry copper deposit of Questa, New Mexico, Greber et al. (2014) subdivided three stages during which Mo isotope fractionations may occur, all lead to molybdenites being heavier than the magmatic source. This implies that Mo isotope compositions of molydenites are not necessarily representative of the average isotope composition of igneous rocks.

### 2.29.3 Sediments

Marine sediments show a large range in Mo isotope composition (Siebert et al. 2006a, b; Poulson et al. 2006 and others). The magnitude of Mo isotope fractionation between seawater and sediments correlates with the redox state of the depositional environment. Oxic settings are characterized by the largest Mo isotope fractionation of around 3‰, the most reducing anoxic setting reflects a Mo isotope composition close to seawater (see Fig. 2.40). Environments of intermediate redox state have variable Mo isotope values.

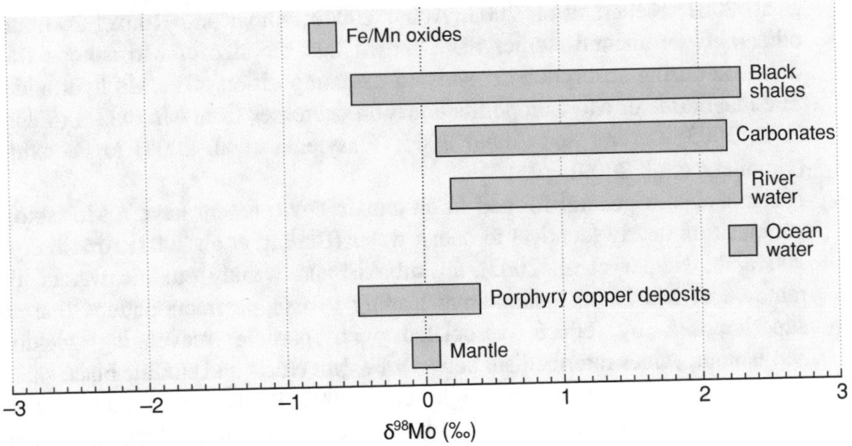


Fig. 2.40 $\delta^{98/95}$Mo-values of important geological reservoirs

As summarized by Poulson Brucker et al. (2009) Mo in sediments originates from 3 different sources:

(1) The isotope composition of Mo input from rivers has been investigated by Archer and Vance (2008), Neubert et al. (2011) and Wang et al. (2015a, b, c). They found a large range of $\delta^{98}$Mo values from 0.2 to 2.3‰ with an average of 0.8‰ that are heavier than the average continental crust. Horan et al. (2020) investigated the importance of the lithology and the formation of secondary minerals on the Mo isotope composition of rivers. They concluded that varying source lithologies account for 1‰ variability and the partitioning between dissolved load and weathering products accounts for another 1‰ variability. Revels et al. (2021) found little evidence for lithological or mineralogical control on the Mo isotope composition of Amazonian rivers, but suggested that Mo isotope variations are best explained by variations of the weathering intensity. Along streams no significant modification of Mo isotope signatures is observed (Neubert et al. 2011).

(2) Mo associated with biological material that is delivered to the seafloor. The relationship between organic matter and Mo is complex, because Mo is not only incorporated into cells, but is also absorbed to organic material in the water column (Poulson Brucker et al. 2009; Kowalski et al. 2013a, b). As demonstrated by Kowalski et al. (2013a, b), Mo isotope fractionations in tidal systems of the North Sea are caused by biological activity. Zerkle et al. (2011) reported cyanobacterial assimilation of Mo that produce considerable isotope fractionations comparable to those in sedimentary organic matter.

(3) Mo absorbed to Fe/Mn oxides under oxic conditions and Mo bounded through complexation with sulfides under anoxic conditions. Absorbed Mo has a light composition ($\delta^{98}$Mo −0.7‰) being 3‰ depleted relative to seawater (Barling

et al. 2001; Siebert et al. 2003, Anbar 2004b; Anbar and Rouxel 2007 and others). Experimental studies have shown that the size of Mo isotope fractionation during absorption depends on the composition of Fe/Mn hydroxides. The magnitude of Mo isotope fractionation decreases from Mn oxides (Siebert et al. 2003; Barling and Anbar 2004; Wasylenki et al. 2008) to Fe oxides (Goldberg et al. 2009).

Black shales in general formed in an anoxic environment have a Mo isotope composition nearly identical to ocean water (Barling et al. 2001; Arnold et al. 2004a, b; Nägler et al. 2005). In suboxic and weakly euxinic waters, the removal of Mo is not quantitative leading to isotope fractionations that are superimposed by effects associated with particle scavenging yielding Mo-isotope values intermediate between Fe–Mn crusts and euxinic black shales (McManus et al. 2002, 2006; Nägler et al. 2005; Poulson et al. 2006; Siebert et al. 2003, 2006b). Thus, the Mo isotope composition of black shales only reflects the seawater composition when a critical sulfidity is reached (Neubert et. al. 2008).

### 2.29.4 Palaeoredox Proxy

Because of its long residence time, Mo in ocean water has a uniform isotope composition with a $\delta^{98}$Mo value of 2.3‰ (Anbar 2004b; Anbar and Rouxel 2007). The Mo isotope composition of ancient oceans has been inferred from black shales assuming that the C-org rich sediments accumulated in euxinic settings (Gordon et al. 2009). However, not all black shales represent euxinic conditions. In recent Black Sea sediments, incomplete removal of Mo from seawater may lead to a Mo isotope depletion of $^{98}$Mo in anoxic sediments (Neubert et al. 2008). Therefore, when reconstructing paleoenvironments it is important to distinguish between euxinic and non-euxinic black shales. During any particular age period, the most enriched $\delta^{98}$Mo-value in black shales should provide the best estimate of seawater $\delta^{98}$Mo. In a compilation of Mo-isotope values from black shales, Dahl et al. (2010a, 2010b) postulated two episodes of global ocean oxygenation: the emergence of the Ediacaran fauna at around 550 Ma, and the diversification of vascular plants at around 400 Ma. However, as argued by Gordon et al. (2009) the reconstruction of the Mo isotope composition of ancient oceans from organic rich-shales requires independent evidence of local euxinia with sufficient sulfidity (Nägler et al. 2011).

Mo isotopes have been used to search for evidence of free $O_2$ in the Archean atmosphere. As shown by Anbar and Rouxel (2007), Lyons et al. (2014) and others Mo isotopes in Archaen rocks indicate episodic increases in atmospheric $O_2$ levels ("whiffs of oxygen") (Kurzweil et al. 2015). In younger rocks Mo isotopes may indicate variations in the extent of global oceanic anoxic events (Dickson et al. 2016; Goldberg et al. 2016). As an example, for the Toarcian oceanic Anoxic Event in the Early Jurassic, Dickson et al. (2017) observed a larger extent of global seafloor euxinia shifting the recent seawater $\delta^{98}$Mo-value from 2.3‰ to about 1.4‰.

As shales are not as ubiquitous in the geologic record as carbonates, carbonates may be regarded as alternative lithology for the reconstruction of past ocean chemistry. Voegelin et al. (2009, 2010) observed a large spread in $\delta^{98}$Mo-values of biogenic carbonates, which they attributed to vital effects, but inorganic carbonates closely approach modern ocean Mo-values. As shown by Romaniello et al. (2016), Mo isotope values in shallow-water carbonates are typically lighter than seawater, but approach seawater Mo isotope compositions under very reducing conditions. Thus, knowledge about early diagenetic conditions of pore waters are required for the reconstruction of seawater Mo-isotope compositions, restricting carbonates to C-and S-rich sediments.

As reported by Thoby et al. (2019), Precambrian carbonates lie generally above 1‰ in $\delta^{98}$Mo-values, whereas shales from the same age period show lower crustal values. Thoby et al. (2019) attributed this discrepancy to low Mo concentrations in Precambrian seawater and to a general absence of euxinic conditions.

## 2.30 Silver

Silver has two stable isotopes

$$^{107}\mathrm{Ag} \quad 48.6\%$$
$$^{109}\mathrm{Ag} \quad 51.4\%$$

Silver isotope investigations have been of interest, because of the extinct isotope $^{107}$Pd decaying to $^{107}$Ag with a half life of 6.5 Mys. During early Earth history, the decay causes large isotope Ag isotope variations in certain meteorites, as reported by Chen and Wasserburg (1983) and others.

More recently, improvement in MC-ICP-MS techniques (Woodland et al. 2005; Schönbächler et al. 2007; Luo et al. 2010) has led to the detection of mass-dependent natural Ag isotope variations of up to 1‰ in silver-containing ores and whole rocks (Tessalina 2015). Fujii and Albarede (2018) calculated reduced partition functions for $^{107}$Ag and $^{109}$Ag in different species to discriminate between different silver sources and demonstrated that native silver should be fractionated relative to hydrated $Ag^+$.

A range of about 1.5‰ has been found in orogenic gold systems by Voisey et al. (2019), who suggested that silver isotope variations are primarily related to redox reactions such as oxidation of Ag(0) in native gold to $Ag^+$ in solution. Even larger variations up to 3‰ have been reported for native silver from world-wide deposits by Mathur et al. (2018), relating Ag fractionations to the reduction of $Ag^+$ to Ag(0) during precipitation on mineral surfaces.

Variable $^{109}$Ag/$^{107}$Ag isotope ratios among silver ores from different types of ore deposits have been used as a provenance tool to detect time-dependent changes in Ag isotope composition of Roman (Albarede et al. 2016) and medieval silver coins (Desaulty et al. 2011; Desaulty and Albarede 2013).

Another important application of silver isotopes are the investigation of silver nanoparticles in the environment, which due to their antibacterial properties are widely used. Although silver nanoparticles are generally thought to originate from human activities, they can also form naturally via the reduction of $Ag^+$ in natural waters mediated by dissolved organic matter and sunlight. Lu et al. (2016) reported that the formation and dissolution of silver nanoparticles may cause significant silver isotope fractionations. They further demonstrated that anthropogenic nanoparticles during dissolution show isotope fractionation behavior that is different from naturally formed nanoparticles.

## 2.31 Cadmium

Cadmium has 8 stable isotopes:

| | |
|---|---|
| $^{106}Cd$ | 1.25% |
| $^{108}Cd$ | 0.89 |
| $^{110}Cd$ | 12.49 |
| $^{111}Cd$ | 12.80 |
| $^{112}Cd$ | 24.13 |
| $^{113}Cd$ | 12.22 |
| $^{114}Cd$ | 28.73 |
| $^{116}Cd$ | 7.49 |

Either $^{114}Cd/^{110}Cd$ or $^{112}Cd/^{110}Cd$ ratios have been reported in the literature; analytical techniques are MC-ICP-MS (Wombacher et al. 2003; Li et al. 2018) or double-spike TIMS (Schmitt et al. 2009). Comparing datasets from different laboratories is difficult, because no generally agreed standard exists. Different laboratories have used different commercially available Cd-solutions. Rehkämper et al. (2011), Abouchami et al. (2013) and Li et al. (2018) suggested NIST SRM 3108 as certified reference material. δ-values reported here are $^{114/110}Cd$ ratios given relative to SRM 3108 (see Fig. 2.41). Cd isotope variations in important geological reservoirs have been reviewed by Rehkämper et al. (2011).

Cd isotope variations are generated mainly by two fractionation processes: (i) partial evaporation/condensation processes in planetary objects and during refining of ore minerals, and (ii) biological utilization of Cd in the oceanic water column. Small Cd isotope fractionations have been calculated for different Cd complexes in hydrothermal fluids as shown by Yang et al. (2015a, b). Quantum chemical calculations have indicated Cd isotope enrichments in the order hydroxide $\geq$ nitrate $\geq$ hydrate $\geq$ chloride $\geq$ hydrogensulfide.

Rocks and minerals show rather constant Cd isotope compositions (Wombacher et al. 2003, 2008). Schmitt et al. (2009) observed in basalts and loess very small differences, suggesting small Cd isotope differences between mantle and crustal rocks.

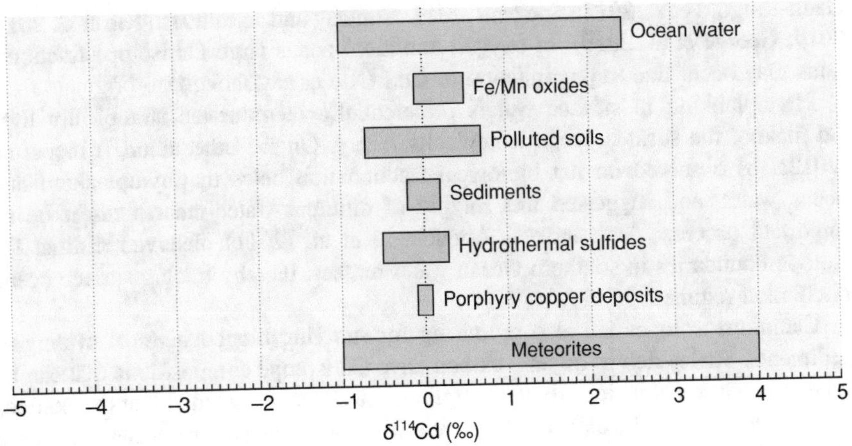


**Fig. 2.41** $\delta^{114/110}$Cd-values of important geological reservoirs

## 2.31.1 Extraterrestrial Materials

Cd isotope variations in extraterrestrial material may be caused by kinetic fractionations during evaporation/condensation processes. Carbonaceous chondrites have relative constant Cd isotope compositions (Rehkämper et al. 2011). In contrast, ordinary chondrites and many enstatite chondrites show very large Cd isotope variations with a range in $\delta^{114}$Cd values from −8 to +16‰ (Wombacher et al. 2008). The large range of Cd isotopes in ordinary chondrites obviously results from evaporation/condensation processes, which has been supported by experiments evaporating Cd in vacuo (Wombacher et al. 2004). The Moon seems to have the same Cd isotope composition as the Earth. Lunar soils are enriched in heavy Cd isotopes, indicating kinetically controlled cadmium loss from the soils.

## 2.31.2 Marine Environment

Rivers are thought to be the most important source of marine Cd, whereas atmospheric aerosols are of lesser importance. Rivers in Siberia, analyzed by Lambelet et al. (2013), show a Cd isotope composition close to the continental crust implying that weathering does not produce a measurable Cd isotope fractionation.

Cd in the ocean is a micronutrient, its distribution resembles that of phosphate. Cd isotope distribution is characterized by uptake into biological material in surface waters and its regeneration in deepwater. Large Cd isotope variations are observed in oceanic surface waters, the most $^{114}$Cd enriched values are observed in waters most depleted in Cd concentration. Rather uniform $\delta^{114}$Cd values of 0.3‰ were determined for waters below 1000 m water depth (Lacan et al. 2006; Rippberger et al. 2007; Horner et al. 2010; Abouchami et al. 2011; Yang et al. 2012a, b;

Gault-Ringold et al. 2012; Xue et al. 2013; Conway and John 2015; Xie et al. 2017, 2019; George et al. 2019). In oxygen minimum zones large Cd isotope fractionations may occur due to precipitation of CdS (Xie et al. 2019).

Phytoplankton in surface waters preferentially incorporates isotopically light Cd making the surface ocean isotopically heavy. On the other hand, Yang et al. (2012a, b) observed no net biological fractionation between phytoplankton and ocean water, and suggested that mixing of different water masses might be an important process. And, indeed, Abouchami et al. (2011) observed distinct Cd isotope boundaries in southern Ocean water masses, thereby tracing surface ocean circulation regimes.

Cadmium isotopes are also promising for studying biogeochemical cycling in sediments. Shales rich in organic carbon have Cd isotope compositions that can be related to marine organic carbon. Georgiev et al. (2015) concluded that Cd isotopes in sediments record the extent of nutrient utilization of the surface ocean at the time of deposition.

Carbonates precipitated from ocean water show very little Cd isotope fractionation and therefore might be used as a tracer for the Cd isotope composition of oceans in the past (Horner et al. 2011). Schmitt et al. (2009) and Horner et al. (2010) reported Cd isotopes for Fe–Mn crusts and demonstrated that nearly all samples were indistinguishable from oceanic deep waters. Thus, Fe–Mn crusts might potentially be used as a proxy of ancient deep-water Cd isotope composition (Wasylenki et al. 2014).

### 2.31.3 Ore Deposits and Anthropogenic Pollution

Cadmium in ore deposits occurs mainly as a minor element in Zn–Pb deposits, where it is primarily hosted in sphalerite. Different types of ore deposits show relatively large variations in Cd isotope ratios: high-temperature deposits show little variations, low-temperature deposits show much larger variations (Wen et al. 2016; Zhu et al. 2017a, b).

The weathering of Zn–Pb deposits releases large amounts of Cd to the environment. Cd may enter cultured soils via mineral phosphate fertilizer. Although Cd is generally toxic for plants and humans, plants may take up small amounts of Cd entering the food chain and potentially causing health problems. As shown by Wiggenhauser et al. (2016) and Inseng et al. (2018), Cd is significantly fractionated in soil–plant systems. Wei et al. (2016) compared Cd isotope fractionations of Cd-tolerant and Cd-sensitive plants. Cd isotope fractionation patterns differ, which may reflect different mechanisms of Cd accumulation.

Soils sampled near ore refineries may be enriched in Cd concentration exhibiting characteristic δ-values (Cloquet et al. 2006). Since Cd isotopes fractionate during evaporation, measurable Cd isotope fractionations should occur during coal burning and sulfide smelting and refining, and, indeed, Shiel et al. (2010) observed a 1‰ fractionation in $\delta^{114}$Cd values during smelting of Zn and Pb ores. Wen et al. (2015) reported Cd isotope ratios in contaminated soils from the Jinding mining area,

China, demonstrating that Cd pollution extended up to 5 km from the mine. Thus, Cd isotope ratios are useful indicators to identify their anthropogenic origin.

Mineral phosphate fertilizer deliver Cd into agricultural soils in which Cd may enter the food chain. Although Cd is not a plant nutrient, some plants may take up small amount of Cd. As shown by Wei et al. (2016) Cd isotopes become depleted in heavier isotopes during transport from nutrient solution to plants. Within plants, Cd isotopes may fractionate between leaf and stem. In wheat, grains are enriched in heavy Cd isotopes relative to straw and root (Wiggenhauser et al. 2016).

## 2.32 Tin

Tin has 10 stable isotopes, more than any other element, covering the mass range from 112 to 124.

| | |
|---|---|
| $^{112}Sn$ | 0.97 |
| $^{114}Sn$ | 0.66 |
| $^{115}Sn$ | 0.34 |
| $^{116}Sn$ | 14.54 |
| $^{117}Sn$ | 7.68 |
| $^{118}Sn$ | 24.22 |
| $^{119}Sn$ | 8.59 |
| $^{120}Sn$ | 32.58 |
| $^{122}Sn$ | 4.63 |
| $^{124}Sn$ | 5.79 |

Tin is a chalcophile and highly volatile element. It exists mainly in two oxidation states, Sn(II) and Sn(IV). Polyakov et al. (2005) concluded from synchrotron radiation experiments that large tin isotope fractionations should be found between tin compounds of different oxidation states. Roskosz et al. (2020) confirmed the expected large range in tin isotope fractionations between $Sn^{2+}$ and $Sn^{4+}$ compounds, even at magmatic temperatures.

Early studies using TIMS could not detect measurable Sn isotope fractionations due to the high ionization potential of Sn. However, with the introduction of MC-ICP-MS, precise Sn isotope measurements become possible measuring either $^{124}Sn/^{116}Sn$, $^{124}Sn/^{120}Sn$ or $^{122}Sn/^{118}Sn$ ratios (Clayton et al. 2002; Haustein et al. 2010; Yamazaki et al. 2013; Creech et al. 2017a, b, c; Brügmann et al. 2017; Schulze et al. 2017; Wang et al. 2018; Yao et al. 2018).

### 2.32.1 Magmatic Rocks

During partial melting $Sn^{4+}$, enriched in heavy Sn isotopes, is more incompatible than $Sn^{2+}$ resulting in heavy $Sn^{4+}$ rich melt and light $Sn^{2+}$ rich residue. Thus, basalts are enriched in heavy tin isotopes relative to peridotites (Wang et al. 2018; Roskosz et al. 2020). Basalts show very small isotope variations, whereas peridotites are more variable and relative to basalts more depleted in Sn isotope ratios. The degree of depletion increases with the degree of melt depletion. Badullovich et al. (2017) investigated Sn isotope compositions in basalts from the Kilauea lava lake. Sn isotopes do not fractionate during crystallization of silicates, but decrease to lighter Sn isotope values upon ilmenite precipitation, which may be caused by the coordination change in the melt and ilmenite. Komatiites have the same Sn isotope composition as Hawaiian basalts (Badullovich et al. 2017).

### 2.32.2 Ore Deposits

Tin ores from major tin provinces (South Dakota, Cornwall, Erzgebirge, Bolivia) show a large range in tin isotopic composition (Yao et al. 2018; Berger et al. 2019). Cassiterite ($SnO_2$) is the major tin mineral, but tin also occurs in complex sulfide minerals. Yao et al. (2018) compared the tin isotope composition of early-precipitating cassiterite ($SnO_2$) with late-precipitating stannite ($Cu_2FeSnS_4$), the two most common tin minerals in hydrothermal ore deposits. Cassiterite is consistently more enriched in heavy Sn isotopes than stannite. Precipitation of isotopically enriched cassiterite can be related to oxidation of dissolved Sn complexes causing isotope depletion in the remaining solution and thus in stannite. Wang et al. (2019a, b, c) showed experimentally that tin isotopes in ore deposits fractionate to a greater extent during redox reactions than during vapor formation.

Fingerprinting the origin of tin ores used for the production of bronze is complex; difficulties may arise due to potential isotope fractionation processes during smelting processes (Haustein et al. 2010 and others). By analyzing Bronze Age artefacts from Serbia and Romania, Mason et al. (2016) concluded that Sn isotope signatures of bronzes can be traced to different tin ore provinces. Three isotopically distinct ore sources may have been used to produce Late Bronze age artefacts in the central Balkan region (Mason et al. 2020).

### 2.32.3 Tin in the Environment

Organotin compounds are used in industry, most prominently in the production of polyvinyl chloride as heat and light stabilizer. Due to their widespread use, large amounts of organotin compounds have entered the environment. Investigating Sn isotope fractionations during methylation reactions, Malinovskiy et al. (2009) demonstrated that under irradiation of UV light, synthesis and decomposition of methyltin is accompanied by mass-dependent and mass-independent tin isotope fractionations.

## 2.33 Antimony

Antimony has two stable isotopes

$^{121}Sb$ 57.21%
$^{123}Sb$ 42.79%

In nature, antimony occurs mainly as sulfide, particularly as stibnite, $Sb_2S_3$; oxides are far less common. Antimony is moderately volatile and occurs mainly in two oxidation states, Sb(V) and Sb(III). Antimony is widely used in industry, it enters into the environment through combustion of fossil fuel and mining activities.

The most extensive study about Sb isotope variations has been presented by Rouxel et al. (2003) using a MC-ICP-MS technique. More recently, modified MC-ICP techniques have been published by Tanimizu et al. (2011) and Lobo et al. (2013). In a recent review, Wen et al. (2018) have summarized analytical methods and environmental applications.

By analysing water samples and a suite of sedimentary and magmatic rocks including hydrothermal sulfides from deep-sea vents, Rouxel et al. (2003) observed a total range in $^{123}Sb/^{121}Sb$ ratios of 1.6‰ with the largest variations occurring in hydrothermal sulfides. Redox changes from Sb being reduced in vent fluids to oxidized Sb in seawater may cause the Sb fractionations, which have been confirmed experimentally during the reduction of Sb(V) to Sb(III).

Resongles et al. (2015) have found that two rivers in France have distinct Sb isotope compositions depending on country rocks and mine wastes polluting the rivers.

An interesting aspect of Sb isotope geochemistry is its potential use of provenancing ancient pre-Roman and Roman glass. Sb had been added to obtain colour and opacity in glass. Lobo et al. (2013) demonstrated that different Sb sources had been used for glass production in the Roman era.

### 2.34–2.36 Rare Earth Elements (REE)

Investigations of the stable isotope fractionation of REE are difficult to perform because the similar chemical properties of the REEs hampers complete separation of individual REEs by standard ion exchange chromatography. Hu et al. (2018) discussed the stable isotope composition of Ce, Nd, Sm, Eu, Gd, Dy and Yb. Stable isotope fractionations show correlations with the general slope of REE patterns for igneous rocks. Nakada et al. (2013a, b) showed that isotopes of REEs fractionate during absorption on Fe–Mn oxides. For cerium, lighter isotopes preferentially enrich in the solid, whereas for Nd and Sm heavier isotopes enrich in the solid.

## 2.34 Cerium

The Rare Earth Element (REE) cerium has the unique property of forming tetravalent cations under oxic conditions, in contrast to most other REEs that occur in the trivalent state. This redox sensitive behavior can be used to estimate the redox state of the palaeo-environment. Cerium has four stable isotopes.

| | |
|---|---|
| $^{136}$Ce | 0.19 |
| $^{138}$Ce | 0.25 |
| $^{140}$Ce | 88.48 |
| $^{142}$Ce | 11.08 |

Precise MC-ICP-MS techniques to measure $^{142}$Ce/$^{140}$Ce ratios have been described by Ohno and Hirata (2013), Laycock et al. (2016) and Nakada et al. (2019). A TIMS technique has been published by Bonnand et al. (2019).

Nakada et al. (2019) have measured the Ce isotope composition of geochemical standards. Igneous rocks showed no isotope fractionation, sediments on the other hand showed relatively large Ce isotope variations. Adsorption and precipitation experiments by Nakada et al. (2013a, 2016, 2017) have demonstrated measurable depletions of heavy Ce isotopes in solids (Fe and Mn oxides) relative to coexisting liquids. As shown by Nakada et al. (2013b), the direction of Ce isotope fractionation during absorption is opposite to Nd and Sm isotopes, in which heavier isotopes are enriched in the solid.

## 2.35 Neodymium

Neodymium has 7 naturally occurring isotopes, two ($^{142}$Nd and $^{143}$Nd) are radiogenic, five are stable. $^{142}$Nd is derived from the decay of relatively short lived $^{146}$Sm, $^{143}$Nd from the decay of long-lived $^{147}$Sm

| | |
|---|---|
| $^{142}$Nd | 27.13 |
| $^{143}$Nd | 12.18 |
| $^{144}$Nd | 23.80 |
| $^{145}$Nd | 8.30 |
| $^{146}$Nd | 17.19 |
| $^{148}$Nd | 5.76 |
| $^{150}$Nd | 5.64 |

Wakaki and Tanaka (2012), using the TIMS technique, first demonstrated that natural Nd isotope fractionations do occur, which have been confirmed by Ma et al. (2013a, b), Saji et al. (2016) and McCoy-West et al. (2020a) using MC-ICPMS techniques.

McCoy-West et al. (2017) presented Nd isotope data for chondritic meteorites and a range of terrestrial rocks. They concluded that the Nd isotope composition of chondrites and the bulk silicate earth (BSE) are indistinguishable and that MORBs are slightly enriched, which, as they argued, may be due to Nd partitioning into sulfide or S-rich metal in the core.

Recent studies by McCoy-West et al. (2018, 2020b) and Grattage et al. (2018) found a total variation range in $^{146/144}$Nd ratios of about 0.4‰. MORB $\delta^{146/144}$Nd values are constant with −0.025‰, despite variable $^{143}$Nd/$^{144}$Nd ratios, which may suggest that source heterogeneities in $\delta^{146/144}$Nd have been overprinted by processes in the magma chamber. Gabbros from the lower oceanic crust are slightly lighter than MORBs and OIBs (McCoy-West et al. 2020b). During basalt weathering Nd isotopes fractionate resulting in fluids being enriched relative to the bedrock.

## 2.36 Europium

Europium consists of two stable isotopes

| | |
|---|---|
| $^{151}$Eu | 48.03 |
| $^{153}$Eu | 51.97 |

REE generally exist in the trivalent state, Eu, however, can also exist in the divalent state, where, for example, it may replace Ca in plagioclase. Crystallisation of plagioclase thus depletes magmas of europium relative to the other REEs (europium anomaly).

Moynier et al. (2006) observed no difference in Eu isotope composition between meteorites and terrestrial rocks. Carvalho et al. (2017) described a MC-ICP-MS method for the precise analysis of europium isotopes in water. Various freshwater samples varied within analytical precision.

## 2.37 Heavy Rare Earth Elements (HREE)

Out of the group of heavy rare earth elements, isotope variations of dysprosium (Dy: 7 isotopes in the mass range from 156 to 164), erbium (Er: 6 isotopes in the mass range from 162 to 170) and ytterbium (Yb: 7 isotopes in the mass range 168 to 176) have been reported in terrestrial and meteoritic samples by Albalat et al. (2012) and Shollenberger and Brennecka (2020). Measured by TIMS and MC-ICP-MS techniques, differences in isotope compositions have been explained by differences in volatility and in nucleosynthetic pathways. Although Dy, Er and Yb have overall similar crystallochemical properties, they differ in their temperatures of condensation (Dy and Er 1650 °K, Yb 1490 °K). By measuring terrestrial basalts, garnets and different classes of meteorites, Albalat et al. (2012) showed that the range of isotope fractionation for Yb is twice as much as for Er. For terrestrial rocks, differences in isotopic fractionation between Yb and Er may be due to the presence of $Yb^{2+}$ besides the common $Yb^{3+}$. In leachates of the Murchison meteorite, Shollenberger and Brennecka (2020) explained smaller Yb isotope anomalies compared to Dy and Er by differences in volatility.

## 2.38 Rhenium

Rhenium is a siderophile element, which preferentially partitions into metal phases. It is a very rare element, but may become enriched in molybdenite and to a lesser extent in sulfidic copper minerals. Rhenium occurs in oxidation states from −1 to +7, the most common are +4 and +7. It is soluble under oxic conditions as the oxyanion $Re^{VII}O_4^-$. Under anoxic conditions $Re^{VII}$ is reduced to $Re^{IV}$, thereby becoming insoluble. Rhenium as a redox sensitive element, thus, behaves similarly to molybdenum and uranium, and becomes enriched in black shales.

Rhenium is composed of two isotopes

$$^{185}Re \quad 37.4\%$$
$$^{187}Re \quad 62.6\%$$

$^{187}Re$ is decaying to $^{187}Os$ with a very long half-life of about $4 \times 10^{10}$ years.

A precise MC-ICP-MS technique to measure Re isotope variations has been presented by Miller et al. (2009) and Dellinger et al. (2020). Miller et al. (2015) have carried out first principle calculations of equilibrium mass-dependent and nuclear volume isotope effects among Re(VII) and Re(IV) species. They predicted measurable Re isotope variations at low to moderate temperatures. And indeed, Re isotopes vary by about 0.3‰ in a weathering profile of the New Albany shale. Dickson et al. (2020) determined rhenium concentrations and isotope compositions of ocean water, showing uniform isotope ratios, irrespective of water depth and water mass.

Rhenium as a siderophile element preferentially partitions into metal phases during planetary differentiation. In the metal fraction of different classes of meteorites, Liu et al. (2017) observed Re isotope differences of 0.14‰.

Despite Re concentrations being very low, future Re isotope investigations might contribute to the understanding of the Earth,s paleoredox history.

## 2.39 Tungsten

Tungsten has 5 naturally occurring isotopes with the following abundances.

| | |
|---|---|
| $^{180}W$ | 0.12 |
| $^{182}W$ | 26.50 |
| $^{183}W$ | 14.31 |
| $^{184}W$ | 30.64 |
| $^{186}W$ | 28.43 |

Tungsten isotope studies have focused on the application of the short-lived decay of $^{182}Hf$ to $^{182}W$ with a half-life of about 6 million years. W isotopes, thus, have been used generally as a chronometer to date meteorites and the differentiation of asteroids and terrestrial planets (Lee and Halliday 1996; Kleine et al. 2009 and others).

Since W is highly refractory and moderately siderophile, tungsten preferentially partitions into a metallic core. During partial melting W behaves incompatibly and fluid mobile causing enrichment in the crust relative to the mantle. The most common oxidation state of tungsten is +6, although it might occur in all oxidation states from −2 ($WC_2$) to +6 ($WO_3$). Measurable mass-dependent stable isotope fractionations, thus, might be expected and indeed, by using double-spike MC-ICP-MS techniques, Breton and Quitté (2014), Abraham et al. (2015) and Krabbe et al. (2017) demonstrated that tungsten isotopes ratios reveal small isotope variations. Kurzweil et al. (2018) described an improved MC-ICP technique, with which $^{186}W/^{184}W$ measurements of common rock standards yield a total spread of 0.155‰ (see Fig. 2.42). Mazza et al. (2020) report tungsten isotope data as $^{184}W/^{183}W$ ratios; Wasylenki et al. (2020) on the other hand report $^{183}W/^{182}W$ ratios.

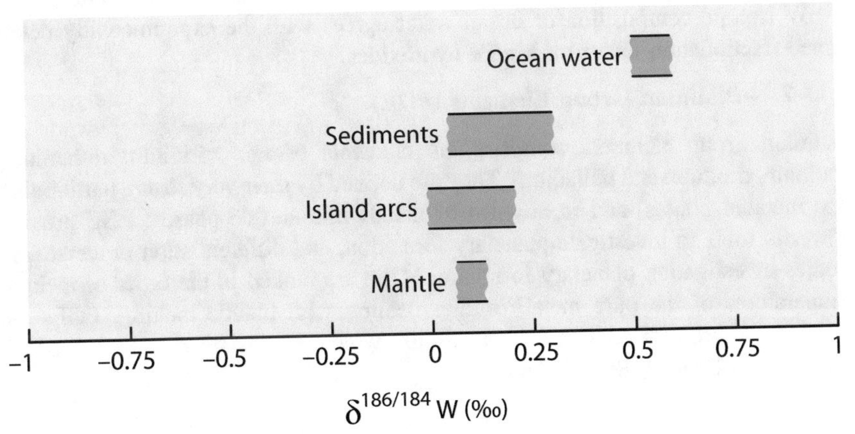


**Fig. 2.42** $^{186}W/^{184}W$ compositions of some important reservoirs

Chondrites, iron meteorites and terrestrial samples show a very narrow range in W isotope ratios, which can be interpreted to indicate that core formation on Earth does not cause measurable isotope fractionation. Kurzweil et al. (2019) analyzed volcanic rocks from different igneous settings. MORB and OIB basalts show a narrow range and are isotopically indistinguishable favoring a homogeneous mantle. Subduction related rocks are more variable in $^{186}W/^{184}W$ ratios and generally isotopically heavy due to a subducted sedimerntary component. Variable tungsten isotope ratios in arc lavas indicate pronounced W isotope fractionations in low temperature sedimentary environments. Mazza et al. (2020) showed that W isotope ratios may be used as a tracer for slab dehydration depending on the amount of aqueous fluid and on slab temperatures. Low-temperature, shallow water-rich subduction zones have heavy W isotope signatures; high-temperature, deep dry subduction zones have light tungsten isotope ratios. Compared to modern igneous rocks, very old rocks from Isua, Greenland show larger W isotope variations implying multistage mobilization processes by metasomatic fluids (Kurzweil et al. 2020).

Sorption to mineral particles appears to control tungsten isotope fractionations in soils and the ocean. Kashiwabara et al. (2017) and Wasylenki et al. (2020) investigated tungsten isotope fractionations during absorption on Fe and Mn (oxy) hydroxides. Lighter W isotopes fractionate preferentially on ferrihydrite and Mn oxide/hydroxide. Therefore, the W isotopic composition of seawater should be enriched in heavy isotopes relative to the input. As reported by Kurzweil et al. (2018), manganese crusts from the Pacific and the Atlantic show differences in W isotope ratios, implying an isotopically heterogeneous W distribution in the modern ocean. For the North Pacific Ocean, Fujiwara et al. (2020) described in a vertical profile a constant $\delta^{186/184}W$ isotope value of 0.55‰ relative to NIST SRM 3163. As tungsten concentrations in seawater are extremely low, a 3000-fold preconcentration of seawater was necessary for an analysis. Fujiwara et al. (2020) concluded that the W isotope composition of ocean water agrees with the experimentally determined fractionation factor on Mn/Fe hydroxides.

## 2.40–2.44 Platinum Group Elements (PGE)

Platinum group elements comprise the elements osmium, iridium, ruthenium, platinum, rhodium and palladium. They are defined by their very strong partitioning into metallic phases, and in absence of metals into sulfide phases. PGE provide powerful tools to investigate planetary formation, and differentiation processes. In studies investigating planetary formation PGEs are ranked in the order of melting temperatures of the pure metal. Studies dealing with mantle melting, PGEs are listed in the order of relative incompatibility which is considered to be Pd < Pt < Rh (monoisotopic) < Ru < Ir and Os.

## 2.40 Palladium

Palladium has six stable isotopes

| | |
|---|---|
| $^{102}Pd$ | 102 |
| $^{104}Pd$ | 11.14 |
| $^{105}Pd$ | 22.33 |
| $^{106}Pd$ | 27.33 |
| $^{108}Pd$ | 26.46 |
| $^{110}Pd$ | 11.72 |

Palladium has the lowest density and melting point of all PGE elements leading to small differences in the geochemical behavior relative to its closest associate platinum. Its preferred partitioning into the Earth,s core leads to a strong depletion in the silicate Earth. Palladium can occur in a number of oxidation states, but exists primarily in the Pd(0) and Pd(2+) form.

Differences in oxidation state and siderophile character relative to other PGEs may lead to isotope fractionatios during planetary differentiation processes. Creech et al. (2017a, b, c) described a MC-ICP-MS technique by measuring $^{106}Pd/^{105}Pd$ ratios in a variety of terrestrial and extraterrestrial materials. Measured $\delta^{106/105}Pd$-values of the investigated samples were within analytical uncertainty, nevertheless palladium seems to be an interesting element to search for small isotope variations.

## 2.41 Platinum

As a highly siderophile element, platinum should be concentrated in the Earth's core. Platinum has six stable isotopes with the following abundance (Creech et al. 2013).

| | |
|---|---|
| $^{190}Pt$ | 0.01 |
| $^{192}Pt$ | 0.79 |
| $^{194}Pt$ | 32.81 |
| $^{195}Pt$ | 33.79 |
| $^{196}Pt$ | 25.29 |
| $^{198}Pt$ | 7.31 |

($^{190}Pt$ is radioactive having a very long half-life ($\sim 10^{11}$ years))

Platinum exists in different oxidation states, the most common are $Pt^0$, $Pt^{2+}$ and $Pt^{4+}$. Thus, redox reactions potentially fractionate Pt isotopes. Creech et al. (2013, 2014) described a precise MC-ICPMS technique measuring $^{198}Pt/^{194}Pt$ ratios relative to the IRMM 010 Standard. 11 samples including mantle, igneous rocks and ores yielded a total variation range of 0.4‰. As shown more recently by Creech et al. (2017a, b, c), $\delta^{198}$Pt values among chondrites, achondrites, iron meteorites and mantle rocks vary by more than 0.5‰. Chondrites are isotopically similar to mantle rocks, whereas primitive achondrites are isotopically enriched indicating metal-silicate fractionation. Thus, during core formation heavy Pt isotopes should retain in the mantle and light isotopes should preferentially enrich in the core. Post-Archean mantle rocks are isotopically depleted relative to Archean rocks which has been interpreted by Creech et al. (2017a, b, c) to indicate the addition of late-veneer material.

The behavior of platinum in the marine environment is not well known; the largest Pt sink in the ocean are Fe–Mn oxyhydroxides, which during absorption should fractionate Pt isotopes. Potentially, Pt isotopes may be used as a redox tracer in the marine environment.

## 2.42 Ruthenium

Ruthenium has seven stable isotopes with the following abundances

| | |
|---|---|
| $^{96}$Ru | 5.54 |
| $^{98}$Ru | 1.87 |
| $^{99}$Ru | 12.76 |
| $^{100}$Ru | 12.60 |
| $^{101}$Ru | 17.06 |
| $^{102}$Ru | 31.55 |
| $^{104}$Ru | 18.62 |

Ruthenium occurs as a metal phase or as an alloy with other platinum group elements, but is also found in sulfides and chromites from large igneous complexes such as Bushveld. Ruthenium can exhibit a large range of oxidation states (from −2 to +8), but essentially occurs as Ru(0) in metals and $Ru^{4+}$ in sulfides. Ruthenium isotopes, thus, may become interesting to investigate the separation and crystallization of metallic cores of terrestrial planets as well as the formation of ore deposits.

Hopp et al. (2016) developed a MC-ICP-MS method to measure precise $^{102}$Ru/$^{99}$Ru ratios. Besides commercially available Ru standard solutions, they investigated chromitites from 3 different localities. Chromites vary by about 1‰ in $\delta^{102/99}$Ru values.

Hopp et al. (2018) reported Ru isotopic variations in magmatic iron meteorites which may be related to Ru isotope fractionations during progressing crystallization of planetary cores. Bermingham and Walker (2017) showed that recent basalts and an Archean sample have the same Ru isotope composition indicating that Ru is well mixed.

As shown in different meteorite types, Ru isotopes are characterized by considerable mass independent isotope variations (Fischer-Gödde et al. 2015; Bermingham et al. 2018). Nucleosynthetic Ru isotope fractionations have been also reported for terrestrial rocks (Fischer-Gödde et al. 2020), which has been interpreted that the composition of Ru isotopes may indicate the existence of an unknown building block of Earth, that does not match any known meteorite composition. As postulated by Fischer-Gödde et al. (2020), Ru and possibly other HFS elements may record the composition of the last material added to the Earth after core formation.

## 2.43 Iridium

Iridium has two stable isotopes

| | |
|---|---|
| $^{191}Ir$ | 37.22 |
| $^{193}Ir$ | 62.78 |

Iridium is best known for "its anomaly": unusually high concentrations in a thin layer of clay interpreted to indicate an impact event 65 Ma ago. Natural isotope variations of iridium have not been reported so far. In a recent paper Zhu et al. (2017a, b) determined the absolute isotope ratio of Ir by MC-ICP-MS.

## 2.44 Osmium

Besides the two radiogenic isotopes $^{186}Os$ ($^{190}Pt$–$^{186}Os$) and $^{187}Os$ ($^{187}Re$–$^{187}Os$), osmium has five stable isotopes.

| | |
|---|---|
| $^{184}Os$ | 0.02% |
| $^{188}Os$ | 13.21% |
| $^{189}Os$ | 16.11% |
| $^{190}Os$ | 26.21% |
| $^{192}Os$ | 40.74% |

Osmium is a refractive, highly siderophile and chalcophile element. Osmium isotopes are suitable to study planetary differentiation and core formation. Since Os partitions into sulfides, osmium behaves compatible during mantle melting, where sulfides remain as a residual phase in the source. Nanne et al. (2017) described high-precision MC-ICP-MS and N-TIMS measurements using a $^{188}Os$–$^{190}Os$ double spike technique. Measured $^{190}Os/^{188}Os$ isotope ratios of a small number of terrestrial and extraterrestrial samples are all within analytical uncertainty. Small isotope variations in chromitites have been suggested to be caused by Os remobilization during hydrothermal alteration (Nanne et al. 2017).

## 2.45 Mercury

Mercury has seven stable isotopes with the following abundances (Rosman and Taylor 1998)

| | |
|---|---|
| $^{196}Hg$ | 0.15 |
| $^{198}Hg$ | 9.97 |
| $^{199}Hg$ | 16.87 |
| $^{200}Hg$ | 23.10 |
| $^{201}Hg$ | 13.18 |
| $^{202}Hg$ | 29.86 |
| $^{204}Hg$ | 6.87 |

Due to the relative uniform isotope abundances in the mass range $^{198}Hg$ to $^{204}Hg$, several possibilities exist for the measurement of Hg isotope ratios; in most studies δ-values are given as $^{202}Hg/^{198}Hg$ ratios. Since the first description of a precise MC-ICP-MS technique (Lauretta et al. 2001), the number of Hg-isotope studies has grown exponentially. Reviews have been presented by Bergquist and Blum (2009), Yin et al. (2010), Blum (2011), Blum et al. (2014) and Blum and Johnson (2017). The large interest in Hg isotopes relies on two factors: (i) due to its ability to be transported over long distances in the atmosphere, mercury is a global pollutant and (ii) large mass independent isotope fractionations have been observed besides mass-dependent fractionations (Bergquist and Blum 2007; Sonke 2011 and others).

The biogeochemical cycle of Hg is complex including different redox states and various chemical speciations affecting its mobility and toxicity. Hg has two common oxidation states: Hg(0) exists primarily in gaseous form and Hg(II) exists as highly particle-reactive gaseous, aqueous and solid species. Dissolved Hg(II) has affinities for sulfides and organic matter.

Mercury can exist as stable HgS (cinnabar) and in the form of Hg–S complexes, in methylated form (methylmercury) and in gaseous and aerosol phases in the atmosphere. Emissions are dominated by anthropogenic activity (coal combustion), but inputs from volcanic and hydrothermal emissions are also significant. Atmospheric Hg can be converted into methylmercury by bacteria that may accumulate in aquatic food webs potentially causing severe health problems.

Large $\delta^{202/198}$Hg-isotope fractionations have been observed in natural samples (Bergquist and Blum 2009 and others), far larger than anticipated. The natural Hg isotope variation encompasses 7‰, from $\delta^{202}$Hg −5 to +2.5‰ relative to NIST 3133 (Zambardi et al. 2009; Moynier et al. 2020).

Bucharenko (2001) and Schauble (2007) demonstrated that isotope variations are controlled by nuclear volume and magnetic shift isotope effects.

## 2.45.1 MDF and MIF Fractionation Processes

Generally, chemical and biological processes cause mass-dependent fractionations of mercury compounds. There are a few processes such as photochemical reduction of Hg(II) and photochemical decomposition of methylmercury that induce mass independent fractionations of odd Hg isotopes. As summarized by Blum and Johnson (2017), there are 4 different types of mercury isotope fractionation processes.

(i) Mass-dependent fractionations, reported as $\delta^{202}$Hg-values, occur in all biological and abiological natural reactions. Most equilibrium and kinetic processes for Hg are mass dependent fractionations (MDF), occurring for instance during biogeochemical reactions and during microbial transformations (Kritee et al. 2007, 2009); as for other elements, MDF depend on the type of organism, temperature, growth rate etc.

(ii) Odd-mass independenpent fractionations, reported as $\Delta^{199}$Hg/$\Delta^{201}$Hg ratios between 1.0 and 1.3, seem to be caused by magnetic isotope effects during kinetic photochemical reduction of Hg(II) and methylmercury bond to organic ligands in water (Bergquist and Blum 2009; Blum and Johnson, 2017).

(iii) Odd-mass independent fractionations with $\Delta^{199}$Hg/$\Delta^{201}$Hg ratios = 1.6 appear to be caused by the nuclear volume effect during evaporation of Hg(0) and during dark reduction of organic matter. As predicted by Bucharenko et al. (2004) and Schauble (2007) and confirmed in experiments by Zheng and Hintelmann (2010), nuclear volume effects have been reported for the Hg liquid–vapor transition (Estrade et al. 2009; Ghosh et al. 2013). The ratio $\Delta^{199}$Hg/$\Delta^{201}$Hg thus seems to be diagnostic of the process causing the MIF (Bergquist and Blum 2009).

(iv) Even mass independent fractionations reported as $\Delta^{200}$Hg/$\Delta^{204}$Hg ratios of −0.5 to −0.6 appear to be related to photochemical oxidation of Hg(0) in the upper atmosphere (Chen et al. 2012; Rolison et al. 2013). The effect likely affects both even and odd isotopes, but is termed even MIF as it is clearly observed in the even isotopes (Blum and Johnson 2017). The mechanism for the even isotope mass independent fractionations remain, however, unclear.

For the calculation of odd and even numbered MIF values, Blum and Bergquist (2007) gave the following definitions.

$$\Delta^{199}\mathrm{Hg} = \delta^{199}\mathrm{Hg} - \left(\delta^{202}\mathrm{Hg} \times 0.2520\right)$$
$$\Delta^{200}\mathrm{Hg} = \delta^{200}\mathrm{Hg} - \left(\delta^{202}\mathrm{Hg} \times 0.5024\right)$$
$$\Delta^{201}\mathrm{Hg} = \delta^{201}\mathrm{Hg} - \left(\delta^{202}\mathrm{Hg} \times 0.7520\right)$$
$$\Delta^{204}\mathrm{Hg} = \delta^{204}\mathrm{Hg} - \left(\delta^{202}\mathrm{Hg} \times 1.4930\right)$$

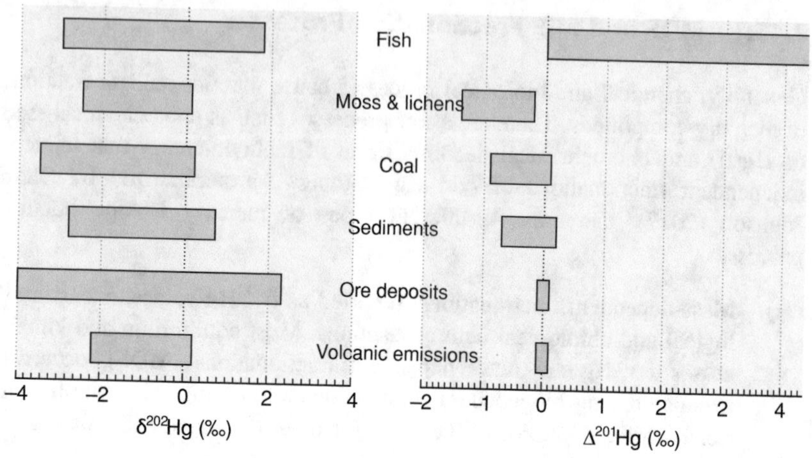


**Fig. 2.43** $\delta^{202/198}$Hg and $\Delta^{201}$Hg values of important geological reservoirs

Figure 2.43 summarizes MDF and MIF Hg isotope variations in important reservoirs (modified from Bergquist and Blum 2009).

## 2.45.2 Igneous Rocks and Ore Deposits

Isotope variations of mercury in common magmatic rocks are much larger than originally thought and on average lighter than in sedimentary rocks. Due to its chalcophile nature, Moynier et al. (2020) postulated that most of the original terrestrial mercury may have been partitioned into the core or due to its high volatility may have been lost during accretion. Thus, the present-day mantle composition of Hg has been affected by core formation or mantle degassing. Moynier et al. (2020) reported for a small set of magmatic rocks a range of $\delta^{202}$Hg-values from −4.95 to −2.35‰ and suggested that the variation range is due to magmatic degassing. As concluded by Moynier et al. (2020) the mantle, thus, must have a Hg isotope composition lower than −2.35‰.

Chondritic meteorites fall within the general range of terrestrial igneous rocks (Meier et al. 2016; Moynier et al. 2020) and exhibit large MDF- and odd MIF Hg fractionations for carbonaceous chondrites and no MIF fractionations for ordinary chondrites.

A large range also occurs in Hg ore deposits and in hydrothermal springs (Smith et al. 2008). Smith et al. (2008) postulated that boiling of hydrothermal fluids and separation of a Hg-bearing vapour phase are responsible for the observed isotope variations. Sherman et al. (2009) investigated the Guaymas and Yellowstone hydrothermal systems. They reported considerable isotope fractionations, in the

Guyamas system solely being mass-dependent, whereas at Yellowstone small mass-independent fractionations occur which may be due to the presence of light facilitating photochemical reactions.

Hg is a common minor element in hydrothermal ore deposits, specifically enriched in sphalerites. Sphalerites collected from more than 100 Zn deposits in China show a large range in $\delta^{202}$Hg-values from −1.9 to +0.7‰ and a $\Delta^{199}$Hg range from −0.24 to 0.18‰, indicating that Hg-MIFs can be transported deep into the Earth's crust during recycling of crustal material (Yin et al. 2016a, b). Even larger ranges in $\delta^{202}$Hg-values have been observed by Xu et al. (2018) and Liu et al. (2021) in sediment hosted Pb–Zn deposits.

### 2.45.3 Sediments

Hg isotopes may be used to investigate Hg sources in marine sediments. Mercury enters the ocean via atmospheric deposition and terrestrial runoff. A fraction of the entered mercury is reduced to gaseous Hg(0) which escapes back to the atmosphere. The remaining mercury will be scavenged by adsorption to sinking organic particles. Sediments, in which Hg mainly originates from atmospheric deposition, have positive $\Delta^{199}$Hg and slightly negative $\delta^{202}$Hg values, whereas sediments that receive Hg through terrestrial runoff tend to have negative $\Delta^{199}$Hg and more negative $\delta^{202}$Hg values (Thibodeau and Bergquist 2017). As shown by Zheng et al. (2018), sediments deposited under oxic conditions can be distinguished from those deposited under $H_2S$-rich conditions. Sapropels—sediments deposited during periods of high primary productivity—may record the Hg isotopic composition of the ocean by quantitative sequestration of Hg to organic matter (Gehrke et al. 2009).

Anomaleous enrichments of Hg in sedimentary sequences have been interpreted to be associated with volcanic activity, in specific with the occurrence of Large Igneous Provinces (LIPs). Hg enrichments may be related to direct volcanic gas emissions or to magmas intruding into organic-rich sediments. They are characterized by different MIF signatures.

The MIF Hg-signature in sedimentary sequences has been used as a proxy for a relationship of LIP magmatism coupled with mass extinctions (Grasby et al. 2019; Shen et al. 2019 and others). There is, however, debate whether the anomaleously high Hg concentrations in sedimentary sections that contain fossil extinction records are a proxy for past volcanism. As summarized by Grasby et al. (2019), not all Hg anomalies are directly related to volcanism, but may reflect continental terrestrial sources (Them et al. 2019).

### 2.45.4 Environmental Pollutant

The geochemical cycle of mercury is characterized by atmospheric transport over long distances. Mercury exists in 3 species in the atmosphere: (i) elemental Hg ($Hg^0$) having a residence time of about 1 year in the atmosphere, (ii) divalent reactive gaseous $Hg^{2+}$ and (iii) Hg bound to particles. These species are linked together by abundant oxidation and reduction processes. $Hg^0$ comprises more than 90% of total atmospheric Hg and is relatively stable allowing large scale mixing, whereas the other two species are much more reactive and deposit readily.

Besides natural inputs from volcanic and hydrothermal emissions, anthropogenic sources dominate Hg emissions with coal combustion being the largest contributor. Because elementary Hg is extremely volatile, mercury easily exchanges between water and air and between land and air, resulting in global dispersion. Hg isotope values of biota from deep-sea trenches indicate that surface derived mercury has infiltrated the deepest locations of the ocean (Blum et al. 2020).

Hg MDF and MIF isotope signatures in moss, peat, coal and soils demonstrate that a large part of the Hg surface reservoir has been affected by anthropogenic activities offering the possibility to quantify the relative contributions of Hg deposition from local, regional and global sources (Sonke 2011). As suggested by Kritee et al. (2007, 2009), Hg isotopes may distinguish between different sources of mercury emissions based on the magnitude of isotope fractionations.

Mosses and lichens are passive filters of atmospheric particulates, which may monitor atmospheric Hg emissions. Carignan et al. (2009) demonstrated that they are characterized by negative MIF. Snow samples also may be regarded as good collectors of atmospheric Hg particulates (Sherman et al. 2010).

Sonke et al. (2010) investigated mercury pollution from two European metal refiniries and showed that heavy Hg isotopes are preferentially retained in slag residues. Ma et al. (2013a, b) investigated Hg emissions from a heavy metal smelter in Manitoba. Hg isotope variations observed in sediment cores can be explained by mixing of a natural endmember ($\delta^{202}$Hg −2.4‰) and an anthropogenic endmember emitted from the smelter ($\delta^{202}$Hg −0.9). Sediment cores 5 and 73 km away from the smelter reveal decreasing Hg concentrations and characteristic shifts in Hg isotope values. Even at the distance of 73 km 70% of the Hg in the sediments originated from the smelter. By investigating Hg pollution and Hg isotope fractionation in the vicinity of Ag, Au and Hg-bearing mines, Stetson et al. (2009) and Yin et al. (2013) have reached similar conclusions.

At the global scale, anthropogenic emissions are dominated by coal fired power plants. Biswas et al. (2008) and Sun et al. (2016a, b) demonstrated that coal deposits in the United States, China and Kazakhstan have characteristic Hg isotope values that can be used to discriminate among Hg sources. $\delta^{202}$Hg values in coal vary by 4.7‰ and $\Delta^{199}$Hg by 1.0‰. Combining the two variables may result in a characteristic fingerprint for coal deposits.

Other applications that have investigated Hg isotopes in the environment track mercury in fish in the Minamata Bay, Japan (known as the Minamata disease) (Balogh et al. 2015) or determine mercury sources in the Pearl River estuary (Yin et al. 2015) and in the Great Lakes (Lepak et al. 2015).

## 2.46 Thallium

The geochemical behaviour of thallium is largely controlled by its large ionic radius, which makes it highly incompatible during magmatic processes. Tl exists in two valence states as $Tl^{+}$ and $Tl^{3+}$. Because of its high redox potential, the oxidized form is uncommon in natural environments, but seems to play an important role during adsorption processes. Furthermore, Tl is a highly volatile element favoring kinetic fractionations during degassing processes.

Thallium has two stable isotopes with masses 203 and 205.

$^{203}$Tl 29.52
$^{205}$Tl 70.48

The small relative mass difference between the two Tl isotopes predicts little Tl isotope fractionations. However, the so far observed Tl isotope variation is larger than 3‰ (Rehkämper et al. 2002; Nielsen et al. 2006, 2017). Particularly large isotope variations are observed in the marine environment that are characterized by light $\delta^{205}$Tl values in altered basalts and heavy values in ferromanganese crusts. Responsible for the large variation are Tl isotope fractionations between seawater and Fe–Mn oxyhydroxides and fractionations during low temperature alterations of the oceanic crust. As discussed by Schauble (2007), nuclear field shift effects are mainly responsible for the observed Tl fractionations.

The generally used standard is NIST 997 Tl metal. It is important to note that Tl isotope ratios are commonly given in the ε-notation (variations in parts per 10,000); to be consistent within the book, Tl isotope ratios are given here as δ-values. Reviews about the Tl isotope geochemistry have been published by Nielsen and Rehkämper (2011) and Nielsen et al. (2017). Figure 2.44 summarizes natural Tl isotope variations.

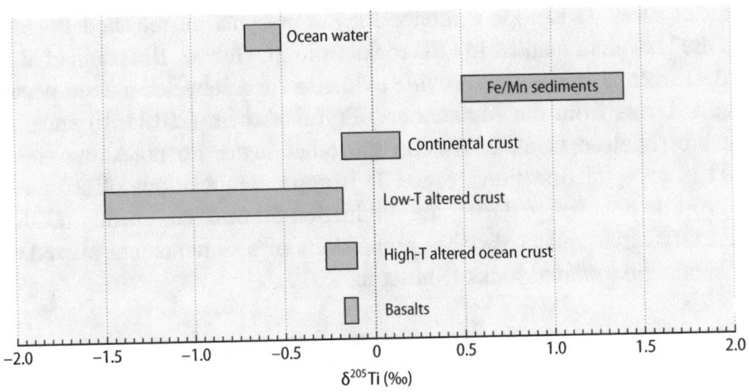


**Fig. 2.44** $\delta^{205}$Tl-values of important geological reservoirs

### 2.46.1 Igneous Rocks

During magmatic processes (crystal fractionation, partial melting etc.) little fractionations seem to occur (Prytulak et al. 2017a, b). By analyzing MORB glasses from different basins, Nielsen et al. (2006) concluded that the upper mantle has a homogeneous Tl isotope composition. Nielsen et al. (2005, 2006, 2007) demonstrated that the continental crust does not differ from the mantle. A large range in Tl isotope composition has been observed by Horton et al. (2021) in carbonatites. Horton et al. (2021) postulated that the large spread is due to simultaneous igneous differentiation and metasomatism.

By analysing igneous rocks in the vicinity of porphyry copper deposits, Baker et al. (2010) reported a variation range of about 0.6‰ due to hydrothermal alteration processes. At temperatures around 300 to 400 °C, hydrothermal alterations have a limited effect on Tl isotope values, whereas at lower temperatures Tl isotope fractionations increase considerably (Fitzpayne et al. 2018). As indicated in late magmatic/hydrothermal veins, fluids released during degassing are enriched in $^{205}$Tl (Hettmann et al. 2014).

Because Tl is a volatile trace element, it becomes enriched in volcanic condensates. As shown by Baker et al. (2009) gaseous volcanic emissions are more variable in Tl isotope composition than igneous rocks, but have a mean value being indistinguishable from the estimated mantle composition. The larger variability may result from partial evaporation during mantle degassing.

Rader et al. (2018) investigated Tl-isotope compositions of coexisting minerals from igneous, metamorphic and metasomatic systems. Reported $\delta^{205}$Tl values range from −1.2 to +1.8‰. Sulfides display the heaviest $\delta^{205}$Tl values, Fe-rich micas have the lowest values.

Since most geochemical reservoirs, except Fe–Mn marine sediments and low-temperature seawater altered basalts, are more or less invariant in Tl isotope composition, admixing of small amounts of Fe–Mn sediments or low-temperature altered oceanic crust into the mantle should induce small Tl isotope fractionations in mantle derived rocks. As discussed by Nielsen et al. (2006, 2007), OIB samples from Hawaii show Tl isotope evidence for the presence of recycled Fe–Mn sediments in the Hawaiian mantle. In OIB rocks from St. Helena, Blusztain et al. (2018) observed Tl isotope values that provide evidence for a recycled altered ocean crust component. Lavas from the Mariana arc, Prytulak et al. (2013a, b) and from the Aleutian arc (Nielsen et al. 2016), on the other hand, do not show subduction related Tl isotope fractionations. Since Tl isotope compositions of sediments and oceanic crust do not change during the subduction process, contrasting results in arc rocks and OIBs may reflect variable proportions of sediments and altered oceanic crust in subduction related rocks (Shu et al. 2019).

### 2.46.2 Fractionations in the Ocean

No significant Tl isotope fractionations occur during weathering. Dissolved and particulate components in river water do not differ from those of the continental crust (Nielsen et al. 2005). The oceans, however, are depleted in $^{205}$Tl compared to the continental crust. A systematic 2‰ difference between Fe–Mn crusts enriched in $^{205}$Tl and seawater has been observed by Rehkämper et al. (2002), which seems to be due to a fractionation effect during adsorption of Tl onto Fe–Mn particles (Rehkämper et al. 2004).

Variations of Tl concentrations and isotope compositions of seawater over time may depend on different rates of Tl removal via scavenging on Fe–Mn oxyhydoxides and via uptake during low temperature alteration of oceanic crust (Nielsen et al. 2009, 2011a, b, c; Owens et al. 2017). Nielsen et al. (2009) observed that growth layers of two Fe–Mn crusts from the Pacific Ocean show a systematic change of Tl isotope composition with age, which they explained by time-dependent changes in Tl-isotope composition of seawater. Low Tl isotope ratios during the age range between 55 and 45 Ma might be explained by a fourfold increase of Fe–Mn oxide precipitation compared to present day.

Mn oxide burial rates are related to the extent of bottom water anoxia (Them et al. 2018). The potential to use Tl isotopes as a paleoredox proxy has been shown by Nielsen et al. (2011a, b, c). Early diagenetic pyrite deposited in sediments beneath an oxic water column has Tl isotope ratios heavier than seawater, whereas pyrite deposited under euxinic conditions has a Tl isotope composition close to seawater, due to reduced precipitation of Fe/Mn oxides in a sulfidic water column. Thus, thallium isotopes may be used to track the global burial history of manganese oxides (Owens 2019).

## 2.47 Uranium

Natural uranium is mainly composed of two long-lived radioactive isotopes:

$$^{235}\text{U} \quad 0.72\%$$

$$^{238}\text{U} \quad 99.27\%$$

In the past uranium isotopes have been widely used as a chronological tool. Present day isotope fractionation between $^{235}$U and $^{238}$U has been considered to be insignificant. The ratio $^{238}$U/$^{235}$U has been assumed to be a constant with a value of 137.88. However, precise measurements by Hiess et al. (2012) on a suite of uranium-bearing minerals commonly used for U–Pb geochronology, e.g. zircons, exhibit isotope variations in $\delta^{238}$U values larger than 5‰.

Uranium exists in two oxidation states having different solubilities. Under oxidizing conditions, U is typically present as soluble hexavalent uranyl ion $UO_2^{2+}$, under reducing conditions U occurs in the tetravalent state, forming relatively

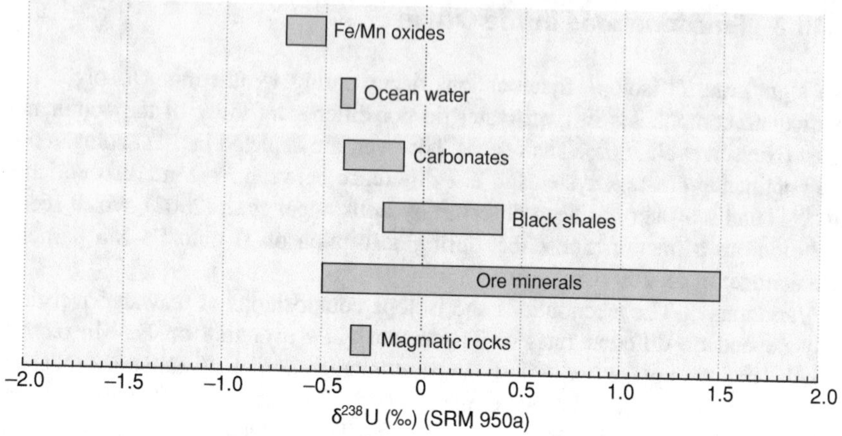


**Fig. 2.45** $\delta^{238}$U-values of important geological reservoirs

insoluble complexes. These properties favor natural isotope variations. Fractionations occur due to mass-independent nuclear volume fractionations, resulting from the differences in nuclear size and shape (Bigeleisen 1996; Schauble 2007; Abe et al. 2008).

Using MC-ICP-MS techniques, Stirling et al. (2007), Weyer et al. (2008), Bopp et al. (2009), Montoya-Pino et al. (2010) reported $\delta^{238}$U variations of more than 1‰ in various rock types (see Fig. 2.45). Several standards have been in use; in a recent review Andersen et al. (2017) used CRM-112a as the primary standard.

## 2.47.1 Fractionation Processes

Uranium isotope fractionations are mainly attributed to abiogenic or biogenic reduction of U(VI) to U(IV). Theoretically calculated and experimentally determined equilibrium isotope fractionations between dissolved U(IV) and U(VI) result in 1.6‰ at 25 °C with U(IV) being enriched in $^{238}$U relative to U(VI) (Fujii et al. 2006; Wang et al. 2015a, b, c). A similar range has been observed during oxidation of tetravalent uranium by dissolved oxygen at low pH (Wang et al. 2015a, b, c). Induced isotope fractionations during uranium reduction are opposite in direction observed during reduction of nitrate, sulfate and chromate: $^{238}$U preferentially partitions into U(IV) phases, whereas $^{235}$U is enriched in U(VI) phases. Thus, heavy $\delta^{238}$U values are observed for black shales, which contain the reduced form of U, and light isotope values are observed for Fe/Mn oxides.

Diverse microorganisms are capable of reducing U(VI) to U(IV). Experiments using sulfate or metal-reducing bacteria to reduce uranium from oxidized solution have shown $^{238}$U enrichments of about 1‰ in the reduced phase (Basu et al. 2014; Stirling et al. 2015; Stylo et al. 2015).

Not only micro-organisms fractionate uranium during reduction. As shown experimentally by Brown et al. (2018), U isotope fractionation has been observed during abiotic reduction of aqueous uranium onto the surfaces of synthetic iron monosulfide. Adsorption processes may cause significant U isotope fractionations between seawater and Fe–Mn oxides. As redox conditions do not change during uranium absorption, a difference in the coordination environment between dissolved and absorbed U seems to be responsible for the isotope fractionation (Brennacka et al. 2011).

## 2.47.2 Mantle-Derived Rocks

Assuming that the U content of the core is negligible, the U isotope composition of the bulk Earth can be estimated from the mean isotopic composition of the mantle and of the continental crust resulting in a $\delta^{238}$U value of −0.34‰ relative CRM-112a for the bulk Earth (Andersen et al. 2017). MORB, ocean island basalts (OIB) and island arc basalt (IAB) differ in their U isotope compositions with MORBs being slightly heavier than OIBs and IABs (Andersen et al. 2015). Since partial melting presumably will not cause U isotope fractionations, heterogeneaus U isotope values of the mantle should reflect variations in the mantle source. Low uranium isotope values of arc lavas, compared to MORB, may be explained by U isotope fractionation during mobilization from subducted slabs (Freymuth et al. 2019).

As reported by Weyer et al. (2008), Telus et al. (2012) and Tissot and Dauphas (2015), the uranium isotope composition of the bulk crust is indistinguishable from the mantle.

## 2.47.3 Ore Deposits

Large differences in uranium isotope composition have been observed among uranium ores of different origin (Bopp et al. 2009; Brennecka et al. 2010; Uvarova et al. 2014; Murphy et al. 2014; Bhattacharyya et al. 2017): magmatic ores vary from −0.7 to −0.3‰ whereas sandstone-type low temperature ores have $\delta^{238}$U-values around +0.4‰. Isotope variations seem to be controlled by the isotope composition of the U source and the efficiency of U reduction. Up to 5‰ fractionations have been observed in U mineralised sediment—groundwater systems (Murphy et al. 2014) with $^{238}$U preferentially enriching in the sediment, leading to depletions in the groundwater.

## 2.47.4 Rivers and the Ocean

The U isotope compositions of rivers have been investigated by Stirling et al. 2007, Tissot and Dauphas (2015), Noordmann et al. (2016) and Andersen et al. (2017).

Individual rivers show a large variation from −0.70 to 0.06‰, reflecting presumably differences in the composition of catchment rocks. Andersen et al. (2017) estimated a mean value of −0.34‰ for rivers worldwide, indistinguishable from the continental crust.

Due to its long residence time, modern seawater has a uniform isotope composition with a $\delta^{238}$U-value value of −0.39‰ (Andersen et al. 2014). Uranium in the ocean occurs mainly in the soluble U(VI) form. Under anoxic conditions, sediments enrich $^{238}$U shifting ocean water to lighter U isotope values.

The isotopic signature of U in seawater reflects changes in the global uranium flux between $^{238}$U-depleted oxic and $^{238}$U-enriched anoxic settings, indicating the relative proportions of oxic and anoxic U removal (Dahl et al. 2014). By studying phosphorites, Kolodny et al. (2017) investigated the uranium isotope composition of U(IV) and U(VI) species within the same sample. Tetravalent U comprises up to 80% of the sample with higher $\delta^{238}$U- than $\delta^{235}$U-values, confirming that the fractionation between $^{235}$ and $^{238}$U is redox related.

### 2.47.5 Paleo-Redox Proxy

Uranium has been increasingly used as a redox proxy for the reconstruction of past ocean anoxia. Sediments in anoxic environments are slightly enriched in $^{238}$U compared to oxic settings. By analyzing black shales and carbonates (Montoya-Pino et al. 2010; Brennecka et al. 2011; Kendall et al. 201; Andersen et al. 2014; Noordmann et al. 2015; Lau et al. 2016; Zhang et al. 2018a, b, c; Clarkson et al. 2018; Abshire et al. 2020 and others), uranium isotopes as a paleo-redox tracer have been investigated.

To deduce the U isotope composition of shales, uranium fractionations between shales and ocean water have to be known. Anoxic conditions in the ocean cause more U to be scavenged by anoxic sediments, preferentially sequestering $^{238}$U from seawater leading to a decrease of the $^{238}$U/$^{235}$U ratio of seawater. Montoya-Pino et al. (2010) first demonstrated that U isotope variations in black shales can be used to quantify the extent of marine anoxia. Black shales from the Cretaceous (Oceanic Anoxic Event 2) are systematically lighter in $^{238}$U than modern Black Sea shales which has been interpreted to indicate a threefold increase of oceanic anoxia relative to the present ocean.

As shown by Brennecka et al. (2011) and Lau et al. (2016), uranium concentrations and isotope ratios abruptly decrease across the end-Permian extinction followed by a gradual return to pre-existing values. These trends have been interpreted to imply a 100-fold increase in the extent of seafloor anoxia. Stylo et al. (2015) argued that the U isotope composition of the rock record can be used as a specific "paleo-bioredox" proxy rather than a general redox-proxy. Wang et al. (2016a) demonstrated a remarkable stable redox state of the ocean for the last 70 million years.

For carbonates, it is assumed that they directly record the U isotope composition of ancient seawater. Uranium speciation in seawater seems to control U isotope fractionation in inorganic carbonates (Chen et al. 2016). Stirling et al. (2007), Weyer et al. (2008) and others demonstrated that limited U isotope fractionation takes place during U isotope incorporation into calcium carbonate. Recent high precision measurements by Chen et al. (2018) reveal an enrichment up to 0.1‰ in biogenic carbonates relative to seawater. Livermore et al. (2020) showed that low-magnesium shells of brachiopods may be suitable for the reconstruction of the U isotope composition of ancient oceans.

On the other hand, as observed by Romaniello et al. (2013), $\delta^{238}U$ values of ancient carbonates are affected by diagenetic processes and may become enriched in $^{238}U$ due to U accumulation under anoxic pore water conditions. Similar observations have been made by Hood et al. (2016) describing large variations in U isotope compositions among different carbonate components within a single sample making a careful petrographic analysis necessary.

## References

Abadie C, Lacan F, Radic A, Pradoux C, Poitrasson F (2017) Iron isotopes reveal distinct dissolved iron sources and pathways in the intermediate versus deep Southern Ocean. PNAS 114:858–863

Abe M, Suzuki T, Fujii Y, Hada M, Hirao K (2008) An ab initio molecular orbital study of the nuclear volume effects in uranium isotope fractionations. J Chem Phys 129:164309

Abelson PH, Hoering TC (1961) Carbon isotope fractionation in formation of amino acids by photosynthetic organisms. PNAS 47:623

Abouchami W, Galer S et al (2013) A common reference material for cadmium isotope studies—NIST SRM 3108. Geostand Geoanal Res 37:5–17

Abouchami W, Galer S, de Baar H, Alderkamp A, Middag R, Laan P, Feldmann H, Andreae M (2011) Modulation of the southern-ocean cadmium isotope signature by ocean circulation and primary productivity. Earth Planet Sci Lett 305:83–91

Abraham K, Barling J, Siebert C, Belshaw N, Gall L, Halliday AN (2015) Determination of mass-dependent variations in tungsten stable isotope compositions of geological reference materials by double-spike and MC-ICPMS. J Anal at Spectrom 30:2334–2343

Abshire ML, Romaniello SJ, Kuzminov AM, Cofrancesco J, Severmann S, Riedinger N (2020) Uranium isotopes as a proxy for primary depositional redox conditions in organic-rich marine systems. Earth Planet Sci Lett 529:115878

Ader M, Chaudhuri S, Coates JD, Coleman M (2008) Microbial perchlorate reduction: a precise laboratory determination of the chlorine isotope fractionation and its possible biochemical basis. Earth Planet Sci Lett 269:604–612

Ader M, Thomazo C, Sansjofre P, Busigny V, Papineau D, Laffont R, Cartigny P, Halverson GP (2016) Interpretation of the nitrogen isotope composition of Precambrian sedimentary rocks: assumptions and perspectives. Chem Geol 429:93–110

Adler M, Thomazo C, Sansjovre P, Busigny V, Papineau D, Laffont R, Cartigny P, Halverson GP (2016) Interpretation of the nitrogen isotopic composition of Precambrian sedimentary rocks: assumptions and perspectives. Chem Geol 429:93–110

Adnew G, Hofmann ME, Paul D, Laskar A, Surma J, Albrecht N, Pack A, Schwieters J, Koren G, Peters W, Röckmann T (2019) Determination of the triple oxygen and carbon isotopic composition of $CO_2$ from atomic ion fragments formed in the ion source of the 253 Ultra high-resolution isotope ratio mass spectrometer. Rapid Commun Mass Spectrom 33:1363–1380

Aharon P, Fu B (2000) Microbial sulfate reduction rates and sulfur and oxygen isotope fractionation at oil and gas seeps in deepwater Gulf of Mexico. Geochim Cosmochim Acta 64:233–246

Ahm AS, Bjerum CJ, Blättler CL, Swart PK, Higgins JA (2018) Quantifying early marine diagenesis in shallow-water carbonate sediments. Chem Geol 236:140–159

Albarede F, Blichert-Toft J, Rivoal M, Telouk P (2016) A glimpse into the Roman finances of the Second Punic War through silver isotopes. Geochem Persp Lett 2:128–133

Altabet E, Telouk P, Albarede F (2012) Er and Yb isotope fractionation in planetary materials. Earth Planet Sci Lett 355–356:39–50

Amini M, Eisenhauer A, Böhm F, Fietzke J, Bach W, Garbe-Schoenberg D, Rosner M, Bock B, Lackschewitz K, Hauff F (2008) Calcium isotope ($\delta^{44/40}$Ca) fractionation along hydrothermal pathways, Logatchev field (Mid-Atlantic Ridge, 14°45′N). Geochim Cosmochim Acta 72:4107–4122

Amrani A, Sessions AL, Adkins JF (2009) Compound-specific $\delta^{34}$S analysis of volatile organics by coupled GC/Multicollector-ICPMS. Anal Chem 81:9027–9034

Amsellem E, Moynier F, Day JM, Moreira M, Puchtel IS, Teng FZ (2018) The stable strontium isotope composition of ocean island basalts, mid-ocean ridge basalts, and komatiites. Chem Geol 483:595–602

Amsellen E, Moynier F, Puchtel IS (2019) Evolution of the Ca isotope composition of the mantle. Geochim Cosmochim Acta 258:195–206

Anagnoustou E et al (2016) Changing atmospheric $CO_2$ concentration was the primary driver of early Cenzoic climate. Nature 533:380–384

Anbar AD (2004a) Iron stable isotopes: beyond biosignatures. Earth Planet Sci Lett 217:223–236

Anbar AD (2004b) Molybdenum stable isotopes: observations, interpretations and directions. Rev Mineral Geochem 55:429–454

Anbar AD, Jarzecki AA, Spiro TG (2005) Theoretical investigation of iron isotope fractionation between $Fe(H_2O)_6^{3+}$ and $Fe(H_2O)_6^{2+}$: implications for iron stable isotope geochemistry. Geochim Cosmochim Acta 69:825–837

Anbar AD, Rouxel O (2007) Metal stable isotopes in paleoceanography. Ann Rev Earth Planet Sci 35:717–746

Andersen MB, Vance D, Archer C, Anderson RF, Ellwood MJ, Allen CS (2011) The Zn abundance and isotopic composition of diatom frustules, a proxy for Zn availability in ocean surface seawater. Earth Planet Sci Lett 301:137–145

Andersen MB, Romaniello S, Vance D, Little SH, Herdman R, Lyons TW (2014) A modern framework for the interpretation of $^{238}U/^{235}U$ in studies of ancient ocean redox. Earth Planet Sci Lett 400:184–194

Andersen MB, Elliott T, Freymuth H, Sims KW, Niu Y, Kelley KA (2015) The terrestrial uranium cycle. Nature 517:356–359

Andersen MB, Stirling CH, Weyer S (2017) Uranium Isotope Fractionation. Rev Mineral Geochem 82:799–850

Andre L, Cardinal D, Alleman LY, Moorbath S (2006) Silicon isotopes in 3.8 Ga west Greenland rocks as clues to the Eoarchaean supracrustal Si cycle. Earth Planet Sci Lett 245:162–173

Anguelova M, Fehr H, Takazawa E, Schönbächler M (2019) Titanium isotope heterogeneity in the Earth's mantle. Goldschmidt2019 Abstr

Antler G, Turchyn AV, Rennie V, Herut B, Sivan O (2013) Coupled sulphur and oxygen isotope insight into bacterial sulphate reduction in the natural environment. Geochim Cosmochim Acta 118:98–117

Antler G, Turchyn AV, Ono S, Sivan O, Bosak T (2017) Combined $^{34}$S, $^{33}$S and $^{18}$O isotope fractionations record different intracellular steps of microbial sulfate reduction. Geochim Cosmochim Acta 203:364–380

Antonelli MA, Schiller M, Schauble EA, Mittal T, dePaolo DJ, Chacko T, Grew ES, Tripoli B (2019) Kinetic and equilibrium Ca isotope effects in high-T rocks and minerals. Earth Planet Sci Lett 517:71–82

Archer C, Vance D (2006) Coupled Fe and S isotope evidence for Archean microbial Fe(III) and sulphate reduction. Geology 34:153–156

Archer C, Vance D (2008) The isotopic signature of the global riverine molybdenum flux and anoxia in the ancient oceans. Nat Geosci 1:597–600

Archer C, Vance D, Milne A, Lohan MC (2020) The oceanic biogeochemistry of nickel and its isotopes: new data from the South Atlantic and the Southern Ocean biogeochemical divide. Earth Planet Sci Lett 535:116118

Armytage RM, Savage PS, WIlliams HM, Halliday AN (2011) Silicon isotopes in meteorites and planetary core formation. Geochim Cosmochim Acta 75:3662–3676

Arnold GL, Anbar AD, Barling J, Lyons TW (2004a) Molybdenum isotope evidence for widespread anoxia in Mid-Proterozoic oceans. Science 304:87–90

Arnold GL, Weyer S, Anbar AD (2004b) Fe isotope variations in natural materials measured using high mass resolution multiple collector ICPMS. Anal Chem 76:322–327

Azmy K, Lavoie D, Wang Z, Brand U, Al-Aasm I, Jackson S, Girard I (2013) Magnesium-isotope and REE compositions of Lower Ordovician carbonates from eastern Laurentia: implications for the origin of dolomites and limestones. Chem Geol 356:64–75

Bachinski DJ (1969) Bond strength and sulfur isotope fractionation in coexisting sulfides. Econ Geol 64:56–65

Badullovich N, Moynier F, Creech J, Teng FZ, Sossi PA (2017) Tin isotopic fractionation during igneous differentiation and Earth's mantle composition. Geochem Persp Lett 5:24–28

Baker L, Franchi IA, Maynard J, Wright IP, Pillinger CT (2002) A technique for the determination of $^{18}O/^{16}O$ and $^{17}O/^{16}O$ isotopic ratios in water from small liquid and solid samples. Anal Chem 74:1665–1673

Baker RG, Rehkämper M, Hinkley TK, Nielsen SG, Poutain JP (2009) Investigation of thallium fluxes from subaerial volcanism—implications for the present and past mass balance of thallium in the oceans. Geochim Cosmochim Acta 73:6340–6359

Baker RG, Rehkämper M, Ihlenfeld C, Oates CJ, Coggon R (2010) Thallium isotope variations in an ore-bearing continental igneous setting: Collahuasi Formation, northern Chile. Geochim Cosmochim Acta 74:4405–4416

Balci N, Bullen TD, Witte-Lien K, Shanks WC, Motelica M, Mandernack KW (2006) Iron isotope fractionation during microbially simulated Fe(II) oxidation and Fe(III) precipitation. Geochim Cosmochim Acta 70:622–639

Balistrieri L, Borrok DM, Wanty RB, Ridley WI (2008) Fractionation of Cu and Zn isotopes during adsorption onto amorphous Fe(III) oxyhydroxide: experimental mixing of acid rock drainage and ambient river water. Geochim Cosmochim Acta 72:311–328

Balogh SJ, Tsui MT, Blum JD, Matsuyama A, Woerndle GE, Yano S, Tada A (2015) Tracking the fate of mercury in the fish and bottom sediments of Minamata Bay, Japan, using stable mercury isotopes. Environ Sci Techn 49:5399–5406

Balter V, Zazzo A, Moloney A, Moynier F, Schmidt O, Monahan F, Albarede F (2010) Bodily variability of zinc natural isotope abundances in sheep. Rapid Commun Mass Spectr 24:605–612

Banks DA, Green R, Cliff RA, Yardley BWD (2000) Chlorine isotopes in fluid inclusions: determination of the origins of salinity in magmatic fluids. Geochim Cosmochim Acta 64:1785–1789

Bao H (2015) Sulfate: a time capsule for Earth's $O_2$, $O_3$ and $H_2O$. Chem Geol 395:108–118

Barkan E, Luz B (2005) High precision measurements of $^{17}O/^{16}O$ and $^{18}O/^{16}O$ ratios in $H_2O$. Rapid Commun Mass Spectr 19:3737–3742

Barling J, Arnold GL, Anbar AD (2001) Natural mass-dependent variations in the isotopic composition of molybdenum. Earth Planet Sci Lett 193:447–457

Barling J, Anbar AD (2004) Molybdenum isotope fractionation during adsorption by manganese oxides. Earth Planet Sci Lett 217:315–329

Barnes JD, Paulick H, Sharp ZD, Bach W, Beaudoin G (2009) Stable isotope ($\delta^{18}O$, $\delta D$, $\delta^{37}Cl$) evidence for multiple fluid histories in mid-Atlantic abyssal peridotites (ODP Leg 209). Lithos 110:83–94

Barnes JD, Sharp ZD, Fischer TP (2006) Chlorine stable isotope systematics and geochemistry along the Central American and Izu-Bonin-Mariana volcanic arc. Eos Trans AGU 87(52), Fall Meet Suppl V52B-08

Barnes JD, Sharp ZD (2017) Chlorine Isotope Geochemistry. Reviews Mineral Geochem 82:345–378

Barnes JD, Franchi IA, McCubbin FM, Anand M (2019) Multiple reservoirs of volatiles in the Moon revealed by the isotope composition of chlorine in lunar basalts. Geochim Cosmochim Acta 266:144–162

Baronas JJ, Hammond DE, McManus J, Wheat CG, Siebert C (2017) A global Ge isotope budget. Geochim Cosmochim Acta 203:265–283

Baronas JJ, Torres MA, West AJ, Rouxel O, Georg B, Bouchez J, Gaillardet J, Hammond DE (2018) Ge and Si isotope signatures in rivers: a quantitative multi-proxy approach. Earth Planet Sci Lett 503:194–215

Barth S (1998) Application of boron isotopes for tracing source of anthropogenic contamination in groundwater. Water Res 32:685–690

Basile-Doelsch I, Meunier JD, Parron C (2005) Another continental pool in the terrestrial silicon cycle. Nature 433:399–402

Basu A, Johnson TM (2012) Determination of hexavalent chromium reduction using Cr stable isotopes: isotopic fractionation factors for permeable reactive barrier materials. Environ Sci Tech 46:5353–5360

Basu A, Sanford RA, Johnson TM, Lundstrom CC, Löffler FE (2014) Uranium isotopic fractionation factors during U(VI) reduction by bacterial isolates. Geochim Cosmochim Acta 136:100–113

Bates SL, Hendry KR, Pryer HV, Kinsley CW, Pyle KM, Woodward MS, Horner TJ (2017) Barium isotopes reveal role of ocean circulation on barium cycling in the Atlantic. Geochim Cosmochim Acta 204:286–299

Baumgartner LP, Rumble D (1988) Transport of stable isotopes. I. Development of a kinetic continuum theory for stable isotope transport. Contr Mineral Petrol 98:417–430

Beard BL, Johnson CM (1999) High-precision iron isotope measurements of terrestrial and lunar materials. Geochim Cosmochim Acta 63:1653–1660

Beard BL, Johnson CM (2004) Fe isotope variations in the modern and ancient Earth and other planetary bodies. Rev Mineral Geoch 55:319–357

Beard BL, Johnson CM, Cox L, Sun H, Nealson KH, Aguilar C (1999) Iron isotope biosphere. Science 285:1889–1892

Beard BL, Johnson CM, Skulan JL, Nealson KH, Cox L, Sun H (2003) Application of Fe isotopes to tracing the geochemical and biological cycling of Fe. Chem Geol 195:87–117

Beard BL, Handler RM, Scherer MM, Wu L, Czaja AD, Heimann A, Johnson CM (2010) Iron isotope fractionation between aqueous ferrous iron and goethite. Earth Planet Sci Lett 295:241–250

Beaudoin G, Taylor BE (1994) High precision and spatial resolution sulfur-isotope analysis using MILES laser microprobe. Geochim Cosmochim Acta 58:5055–5063

Beaudoin G, Taylor BE, Rumble D, Thiemens M (1994) Variations in the sulfur isotope composition of troilite from the Canyon Diablo iron meteorite. Geochim Cosmochim Acta 58:4253–4255

Bebout GE, Fogel ML (1992) Nitrogen isotope compositions of metasedimentary rocks in the Catalina Schist, California: implications for metamorphic devolatilization history. Geochim Cosmochim Acta 56:2839–2849

Bebout GE, Idleman BD, Li L, Hilkert A (2007) Isotope-ratio-monitoring gas chromatography methods for high-precision isotopic analysis of nanomole quantities of silicate nitrogen. Chem Geol 240:1–10

Beck WC, Grossman EL, Morse JW (2005) Experimental studies of oxygen isotope fractionation in the carbonic acid system at 15°, 25°, and 40 °C. Geochim Cosmochim Acta 69:3493–3503

Belissont R, Boiron MC, Luais B, Cathelineau M (2014) LA-ICP-MS analyses of minor and trace elements and bulk Ge isotopes in zoned Ge-rich sphalerites from the Noailhac-Saint-Salvy deposit (France): insights into incorporation mechanisms and ore deposition processes. Geochim Cosmochim Acta 126:518–540

Belshaw N, Zhu X, Guo Y, O'Nions K (2000) High precision measurement of iron isotopes by plasma source mass spectrometry. Inter J Mass Spectrom 197:191-195

Bendall C, Lahaye Y, Fiebig J, Weyer S, Brey GP (2006) In-situ sulfur isotope analysis by laser-ablation MC-ICPMS. Appl Geochem 21:782–787

Bennett SA, Rouxel O, Schmidt K, Garbe-Schönberg D, Statham PJ, German CR (2009) Iron isotope fractionation in a buyant hydrothermal plume, 5°S Mid-Atlantic Ridge. Geochim Cosmochim Acta 73:5619–5634

Benson BB, Parker PDM (1961) Nitrogen/argon and nitrogen isotope ratios in aerobic sea water. Deep Sea Res 7:237–253

Berger D, Brügmann G, Pernicka E (2019) On smelting cassiterite in geological and archaeological samples: preparation and implications for provenance studies on metal artefacts with tin isotopes. Archaeol Anthrol Sci 11:293–319

Berglund M, Wieser ME (2011) Isotopic compositions of the elements 2009 (IUPAC Technical Report). Pure Appl Chem 83:397–410

Bergquist BA, Blum JD (2007) Mass-dependent and –independent fractionations of Hg isotopes by photoreduction in aquatic systems. Science 318:417–420

Bergquist BA, Blum JD (2009) The odds and evens of mercury isotopes: applications of mass-dependent and mass-independent isotope fractionation. Elements 5:353–357

Bergquist BA, Boyle EA (2006) Iron isotopes in the Amazon River system: weathering and transport signatures. Earth Planet Sci Lett 248:54–68

Bermingham KR, Walker RJ (2017) The ruthenium isotopic composition of the oceanic mantle. Earth Planet Sci Lett 474:466–473

Bermingham KR, Worsham EA, Walker RJ (2018) New insights into Mo and Ru isotope variations in the nebula and terrestrial planet accretionary genetics. Earth Plnet Sci Lett 487:221–229

Bermin J, Vance D, Archer C, Statham PJ (2006) The determination of the isotopic composition of Cu and Zn in seawater. Chem Geol 226:280–297

Berna EC, Johnson TM, Makdisi RS, Basu A (2010) Cr stable isotopes as indicators of Cr (VI) reduction in groundwater: a detailed time-series study of a point-source plume. Environ Sci Technol 44:1043–1048

Beunon H, Mattielli N, Doucet LS, Moine B, Debret B (2020) Mantle heterogeneity through Zn systematics in oceanic basalts: evidence for a deep carbon cycling. Earth Sci Rev 205:103174

Bezard R, Fischer-Gödde M, Hamelin C, Brennecka GA, Kleine T (2016) The effects of magmatic processes and crustal recycling on the molybdenum stable isotope composition of Mid-Ocean Ridge Basalts. Earth Planet Sci Lett 453:171–181

Bickle MJ, Baker J (1990) Migration of reaction and isotopic fronts in infiltration zones: assessments of fluid flux in metamorphic terrains. Earth Planet Sci Lett 98:1–13

Bidigare RR et al (1997) Consistent fractionation of $^{13}C$ in nature and in the laboratory: growth-rate effects in some haptophyte algae. Global Biogeochem Cycles 11:279–292

Bigalke M, Weyer S, Kobza J, Wilcke W (2010) Stable Cu and Zn isotope ratios as tracers of sources and transport of Cu and Zn in contaminated soil. Geochim Cosmochim Acta 74:6801–6813

Bigeleisen J (1965) Chemistry of isotopes. Science 147:463–471

Bigeleisen J, Perlman ML, Prosser HC (1952) Conversion of hydrogenic materials for isotopic analysis. Anal Chem 24:1356

Bigeleisen J (1996) Nuclear size and shape effects in chemical reactions. Isotope chemistry of heavy elements. J Am Chem Soc 118:3676–3680

Biswas A, Blum JD, Bergquist BA, Keeler GJ, Xie Z (2008) Natural mercury isotope variation in coal deposits and organic soils. Environ Sci Technol 42:8303–8309
Bhattacharyya A, Cambell KM, Kelly SD, Roebbert Y, Weyer S, Bernier-Latmani R, Borch T (2017) Biogenic non-crystalline U(IV) revealed as major component in uranium ore deposits. Nat Commun 8:15538
Black JR, Casey WH, Yin QZ (2006) An experimental study of magnesium isotope fractionation in chlorophyll-a photosynthesis. Geochim Cosmochim Acta 70:4072–4079
Black JR, Epstein E, Rains WD, Yin QZ, Casey WD (2008) Magnesium isotope fractionation during plant growth. Environ Sci Tech 42:7831–7836
Black JR, Kavner A, Schauble EA (2011) Calculation of equilibrium stable isotope partition function ratios for aqueous zinc complexes and metallic zinc. Geochim Copsmochim Acta 75:769–783
Blättler CL, Higgins JA (2014) Calcium isotopes in evaporates record variations in Phanerozoic seawater $SO_4$ and Ca. Geology 42:711–714
Blanchard M, Poitrasson F, Meheut M, Lazzari M, Mauri F, Balan E (2009) Iron isotope fractionation between pyrite ($FeS_2$), hematite ($Fe_2O_3$) and siderite ($FeCO_3$): a first-principles density functional theory study. Geochim Cosmochim Acta 73:6565–6578
Blanchard M, Balan E, Schauble E (2017) Equilibrium fractionation of non-traditional isotopes: a molecular modeling perspective. Rev Mineral Geochem 82:27–63
Blattner P, Lassey KR (1989) Stable isotope exchange fronts, Damköhler numbers and fluid to rock ratios. Chem Geol 78:381–392
Blum JD, Sherman LS, Johnson MW (2014) Mercury isotopes in earth and environmental sciences. Ann Rev Earth Planet Sci 42:249–269
Blum JD (2011) Applications of stable mercury isotopes to biogeochemistry. In: Baskaran M (ed) Handbook of environmental isotope geochemistry. Springer, New York, pp 229–246
Blum JD, Johnson MW (2017) Recent developments in mercury stable isotope analysis. Rev Mineral Geochem 82:733–757
Blum JD, Drazen JC, Johnson MW, Popp BN, Motta LC, Jamieson AJ (2020) Mercury isotopes identify near-surface marine mercury in deep-sea trench biota. PNAS 117:29292–29298
Blusztajn J, Nielsen SG, Marschall HR, Shu Y, Ostrander CM, Hanyu T (2018) Thallium isotope systematics in volcanic rocks from St. Helena—constraints on the origin of the HIMU reservoir. Chem Geol 476:292–301
Bolliger C, Schroth MH, Bernasconi SM, Kleikemper J, Zeyer J (2001) Sulfur isotope fractionation during microbial reduction by toluene-degrading bacteria. Geochim Cosmochim Acta 65:3289–3299
Bonifacie M, Jendrzejewski N, Agrinier P, Humler E, Coleman M, Javoy M (2008) The chlorine isotope composition of the Earth's mantle. Science 319:1518–1520
Bonnand P, James RH, Parkinson IJ, Connelly DP, Fairchild IJ (2013) The chromium isotopic composition of seawater and marine carbonates. Earth Planet Sci Lett 382:10–20
Bonnand P, Williams HM, Parkinson LJ, Wood BJ, Halliday AN (2016) Stable chromium isotopic composition of meteorites and metal-silicate experiments: implications for fractionation during core formation. Earth Planet Sci Lett 435:14–21
Bonnand P, Israel C, Boyet M, Doucelance R, Auclair D (2019) Radiogenic and stable Ce isotope measurements by thermal ionization mass spectrometry. JAAS 34:504–516
Bopp CJ, Lundstrom CC, Johnson TM, Glessner JJ (2009) Variations in $^{238}U/^{235}U$ in uranium ore deposits: isotopic signatures of the U reduction process? Geology 37:611–614
Borrok DM, Nimick DA, Wanty RB, Ridley WI (2008) Isotope variations of dissolved copper and zinc in stream water affected by historical mining. Geochim Cosmochim Acta 72:329–344
Borrok DM, Wanty RB, Ridley WI, Lamothe PJ, Kimball BA, Verplanck PL, Runkel RL (2009) Application of iron and zinc isotopes to track the sources and mechanisms of metal loading in a mountain watershed. Appl Geochem 24:1270–1277
Borthwick J, Harmon RS (1982) A note regarding $ClF_3$ as an alternative to $BrF_5$ for oxygen isotope analysis. Geochim Cosmochim Acta 46:1665–1668

Bottinga Y (1969) Calculated fractionation factors for carbon and hydrogen isotope exchange in the system calcite-carbon dioxide-graphite-methane-hydrogen-water-vapor. Geochim Cosmochim Acta 33:49–64

Boulou-Bi EB, Poszwa A, Leyval C, Vigier N (2010) Experimental determination of magnesium isotope fractionation during higher plant growth. Geochim Cosmochim Acta 74:2523–2537

Bourdon B, Tipper ET, Fitoussi C, Stracke A (2010) Chondritic Mg isotope composition of the Earth. Geochim Cosmochim Acta 74:5069–5083

Bourgon N, 26 others (2020) Zinc isotopes in Late Pleistocene fossil teeth from a Southeast Asian cave setting preserve paleodietary information. PNAS 117:4675–4681

Bowman JR, Willett SD, Cook SJ (1994) Oxygen isotope transport and exchange during fluid flow. Am J Sci 294:1–55

Boyce JW, Treiman AH, Guan Y, Ma C, Eiler JM, Gross J, Greenwood JP, Stolper EM (2015) The chlorine isotope fingerprint of the lunar magma ocean. Scie Adv 1(8):e1500380

Böhlke JK, Sturchio NC, Gu B, Horita J, Brown GM, Jackson WA, Batista J, Hatzinger PB (2005) Perchlorate isotope forensics. Anal Chem 77:7838–7842

Böhm F, Eisenhauer A, Tang J, Dietzel M, Krabbenhöft A, Kisakürek B, Horn C (2012) Strontium isotope fractionation of planktic foraminifera and inorganic calcite. Geochim Cosmochim Acta 93:300–314

Böttcher ME (1996) $^{18}O/^{16}O$ and $^{13}C/^{12}C$ fractionation during the reaction of carbonates with phosphoric acid: effects of cationic substitution and reaction temperature. Isotopes Environ Health Stud 32:299–305

Böttcher ME, Brumsack HJ, Lange GJ (1998) Sulfate reduction and related stable isotope ($^{34}S$, $^{18}O$) variations in interstitial waters from the eastern Mediterranean. Proc Ocean Drill Progr Sci Res 160:365–373

Böttcher ME, Thamdrup B, Vennemann TW (2001) Oxygen and sulfur isotope fractionation during anaerobic bacterial disproportionation of elemental sulfur. Geochim Cosmochim Acta 65:1601–1609

Böttcher ME, Geprägs P, Neubert N, von Allmen K, Pretet C, Samankassou E, Tf N (2012) Barium isotope fractionation during experimental formation of the double carbonate BaMn $(CO_3)^{2-}$ at ambient temperature. Isotopes Environmental Health Studies 48:457–463

Böttcher ME, Neubert N, Escher P, von Allmen K, Samankassou E, Nägler TF (2018a) Multi-isotope (Ba, C, O) partitioning during experimental carbonatization of a hyper-alkaline solution. Chemie Erde 78:241–247

Böttcher ME, Neubert N, von Allmen K, Samankassou E, Nägler TF (2018b) Barium isotope fractionation during the experimental transformation of aragonite to witherite and of gypsum to barite, and the effect of ion (de)solvation. Isotopes in environm health studies 54:1–12

Brand W, Coplen TB et al (2009a) Comprehensive inter-laboratory calibration of reference materials for $\delta^{18}O$ versus VSMOW using various on-line high-temperature conversion techniques. Rapid Comm Mass Spectrom 23:999–1019

Brand W, Geilmann H, Crosson ER, Rella CW (2009b) Cavity ring-down spectroscopy versus high-temperature conversion isotope ratio mass spectrometry: a case study on $\delta^2H$ and $\delta^{18}O$ of pure water samples and alcohol/water mixtures. Rapid Comm Mass Spectrom 23:1879–1884

Brand W (2002) Mass spectrometer hardware for analyzing stable isotope ratios. In de Groot P (ed) Handbook of stable isotope analytical techniques. Elsevier, New York

Branson O (2017) Boron incorporation into marine $CaCO_3$. In: Marschall H (eds) Advances of boron isotope geochemistry. Springer

Breillat N, Guerrot C, Marcoux E, Negrel P (2016) A new global database of $\delta^{98}Mo$ in molybdenites: a literature review and new data. J Geochem Exploration 161:1–15

Bremner JM, Keeney DR (1966) Determination and isotope ratio analysis of different forms of nitrogen in soils, III. Soil Sci Soc Am Proc 30:577–582

Brennecka GA, Borg LE, Hutcheon ID, Sharp MA, Anbar AD (2010) Natural variations in uranium isotope ratios of uranium ore concentrates: understanding the $^{238}U/^{235}U$ fractionation mechanism Earth Planet Sci Lett 291:228–233

Brennecka GA, Wasylenki LE, Bargar JR, Weyer S, Anbar AD (2011) Uranium isotope fractionation during adsorption to Mn-oxyhydroxides. Environ Sci Technol 45:1370–1375
Brenninkmeijer CAM, Kraft MP, Mook WG (1983) Oxygen isotope fractionation between $CO_2$ and $H_2O$. Isotope Geosci 1:181–190
Brenninkmeijer CAM, Röckmann TA (1998) A rapid method for the preparation of $O_2$ from $CO_2$ for mass spectrometric measurement of $^{17}O/^{16}O$ ratios. Rapid Commun Mass Spectrom 12:479–483
Breton T, Quitté G (2014) High-precision measurements of tungsten stable isotopes and application to earth sciences. J Anal at Spectrom 29:2284–2293
Brinjikji M, Lyons JR (2021) Mass independent fractionation of oxygen isotopes in the atmosphere. Rev Mineral Geochem 86:197–216
Brooker R, Blundy J, James R (2004) Trace element and Li isotope systematics in zabargad peridotites: evidence of ancient subduction processes in the Red Sea mantle. Chem Geol 212:179–204
Brown ST, Basu A, Ding X, Christensen JN, DePaolo DJ (2018) Uranium isotope fractionation by abiotic reductive precipitation. PNAS 115:8688–8693
Bruggmann S, Klaebe RM, Paulukat C, Frei R (2019) Heterogeneity and incorporation of chromium isotopes in recent marine molluscs (Mytilus). Geobiology 17:417–435
Brunner B, Bernasconi SM, Kleikemper J, Schroth MH (2005) A model of oxygen and sulfur isotope fractionation in sulfate during bacterial sulfate reduction. Geochim Cosmochim Acta 69:4773–4785
Brunner B, Contreras S, 9 others (2013) Nitrogen isotope effects induced by anammox bacteria. PNAS 110:18994–18999
Bryan AL, Dong S, Wilkes EB, Wasylenki LE (2015) Zinc isotope fractionation during adsorption onto Mn oxyhydroxide at low and high ionic strength. Geochim Cosmochim Acta 157:182–197
Brüchert V, Knoblauch C, Jörgensen BB (2001) Controls on stable sulfur isotope fractionation during bacterial sulfate reduction in Arctic sediments. Geochim Cosmochim Acta 65:763–776
Brügmann G, Berger D, Pernicka E (2017) Determination of the tin stable isotopic composition in tin-bearing metals and minerals by MC-ICP-MS. Geostan Geoanal Res 41:437–448
Bucharenko AI (2001) Magnetic isotope effect: nuclear spin control of chemical reactions. J Phys Chem A 105:9995–10011
Burgoyne TW, Hayes JM (1998) Quantitative production of $H_2$ by pyrolysis of gas chromatographic effluents. Anal Chem 70:5136–5141
Burton KW, Vigier N (2011) Lithium isotopes as tracers in marine and terrestrial environments. In: Baskaran M (ed) Handbook environment isotope geochemistry. Springer, New York, pp 41–59
Busigny V, Bebout GE (2013) Nitrogen in the silicate earth: speciation and isotopic behavior during mineral-fluid interactions. Elements 9:353–358
Butler IB, Archer C, Vance D, Oldroyd A, Rickard D (2005) Fe isotope fractionation on FeS formation in ambient aqueous solution. Earth Planet Sci Lett 236:430–442
Burkhardt C, Kleine T, Oberli F, Pack A, Bourdon B, Wieler R (2011) Molybdenum isotope anomalies in meteorites: constraints on solar nebula evolution and origin oft he Earth. Earth Planet Sci Lett 312:390–400
Burkhardt C, Hin RC, Kleine T, Bourdon B (2014) Evidence for Mo isotope fractionation in the solar nebula and during planetary differentiation. Earth Planet Sci Lett 391:201–211
Caldelas C, Weiss DJ (2017) Zinc homeostasis and isotopic fractionation in plants: a review. Plant Soil 411:17–46
Cameron V, Vance D, Archer C, House CH (2009) A biomarker based on the stable isotopes of nickel. PNAS 106:10944–10948
Cameron V, Vance D (2014) Heavy nickel isotope compositions in rivers and oceans. Geochim Cosmochim Acta 128:195–211
Canfield DE (2001a) Biogeochemistry of sulfur isotopes. Rev Mineral 43:607–636

Canfield DE (2001b) Isotope fractionation by natural populations of sulfate-reducing bacteria. Geochim Cosmochim Acta 65:1117–1124

Canfield DE, Thamdrup B (1994) The production of $^{34}S$ depleted sulfide during bacterial disproportion to elemental sulfur. Science 266:1973–1975

Canfield DE, Farquhar J, Zerkle AL (2010) High isotope fractionations during sulfate reduction in a low-sulfate ocean analog. Geology 38:415–418

Canfield DE, Olsen CA, Cox RP (2006) Temperature and its control of isotope fractionation by a sulfate reducing bacterium. Geochim Cosmochim Acta 70:548–561

Canfield DE, Zhang S, Frank AB, Wang X, Wang H, Su J, Ye Y, Frei R (2018) Highly fractionated chromium isotopes in Mesoproterozoic-aged shales and atmospheric oxygen. Nature Comm 9:2871

Cao Z, Siebert C, Hathorne EC, Dai M, Frank M (2016) Constraining the oceanic barium cycle with stable barium isotopes. Earth Planet Sci Lett 434:1–9

Cao Z, Siebert C, Hathorne EC, Dai M, Frank M (2020) Corrigendum to "Constraining the oceanic barium cycle with stable barium isotopes". Earth Planet Sci Lett 434:1–9. Earth Planet Sci Lett 530:116003

Cardinal D, Gaillardet J, Hughes HJ, Opfergelt S, Andre L (2010) Contrasting silicon isotope signatures in rivers from the Congo Basin and the specific behaviour of organic-rich waters. Geophys Res Lett 37:12403

Carignan J, Estrade N, Sonke J, Donard O (2009) Odd isotope deficit in atmospheric Hg measured in lichens. Environ Sci Technol 43:5660–5664

Caro G, Papanastassiou DA, Wasserburg GJ (2010) $^{40}K/^{40}Ca$ isotopic constraints on the oceanic calcium cycle. Earth Planet Sci Lett 296:124–132

Cartigny P (2005) Stable isotopes and the origin of diamond. Elements 1:79–84

Cartigny P, Boyd SR, Harris JW, Javoy M (1997) Nitrogen isotopes in peridotitic diamonds from Fuxian, China: the mantle signature. Terra Nova 9:175–179

Cartigny P, Marty B (2013) Nitrogen isotopes and mantle geodynamics: the emergence of life and the atmosphere-crust-mantle connection. Elements 9:359–366

Cartwright I, Valley JW (1991) Steep oxygen isotope gradients at marble-metagranite contacts in the NW Adirondacks Mountains, N.Y. Earth Planet Sci Lett 107:148–163

Carvalho GG, Oliveira PV, Yang L (2017) Determination of europium isotope ratios in natural water by MC-ICP-MS. JAAS 32:987–995

Casciotti KL (2009) Inverse kinetic isotope fractionation during bacterial nitrite oxidation. Geochim Cosmochim Acta 73:2061–2076

Casciotti KL (2016) Nitrogen and oxygen isotopic studies of the marine nitrogen cycle. Ann Rev Mar Sci 8:379–407

Casciotti KL, Sigman DM, Galanter Hastings M, Böhlke JK, Hilkert A (2002) Measurement of the oxygen isotopic composition of nitrate in seawater and freshwater using the denitrifier method. Anal Chem 74:4905–4912

Castillo-Alvarez C, Quitte G, Schott J, Oelkers EH (2020) Experimental determination of Ni isotope fractionation during Ni adsorption from an aqueous fluid onto calcite surfaces. Geochim Cosmochim Acta 273:26–36

Catanzaro EJ, Murphy TJ (1966) Magnesium isotope ratios in natural samples. J Geophys Res 71:1271

Cenki-Tok B, Chabaux F, Lemarchand D, Schmitt A, Pierret M, Viville D, Bagard M, Stille P (2009) The impact of water-rock interaction and vegetation on calcium isotope fractionation in soil- and stream waters of a small, forested catchment (the Strengbach case). Geochim Cosmochim Acta 73:2215–2228

Cerling TE, Sharp ZD (1996) Stable carbon and oxygen isotope analyses of fossil tooth enamel using laser ablation. Palaeo Palaeo Palaeoecol 126:173–186

Cerling TE, Harris JM (1999) Carbon isotope fractionation between diet and bioapatite in ungulate mammals and implications for ecological and paleocogical studies. Oecologia 120:347–363

Chacko T, Cole DR, Horita J (2001) Equilibrium oxygen, hydrogen and carbon fractionation factors applicable to geologic systems. Rev Mineral Geochem 43:1–81
Chacko T, Riciputi LR, Cole DR, Horita J (1999) A new technique for determining equilibrium hydrogen isotope fractionation factors using the ion microprobe: application to the epidote-water system. Geochim Cosmochim Acta 63:1–10
Chakrabarti B, Jacobsen S (2010) Silicon isotopes in the inner solar system: implications for core formation, solar nebula processes and partial melting. Geochim Cosmochim Acta 74:6921–6933
Chan LH, Kastner M (2000) Lithium isotopic compositions of pore fluids and sediments in the Costa Rica subduction zone: implications for fluid processes and sediment contribution to the arc volcanoes. Earth Planet Sci Lett 183:275–290
Chan LH, Alt JC, Teagle DAH (2002) Lithium and lithium isotope profiles through the upper oceanic crust: a study of seawater-basalt exchange at ODP Sites 504B and 896A. Earth Planet Sci Lett 201:187–201
Chan LH, Edmond JM, Thompson G (1993) A lithium isotope study of hot-springs and metabasalts from midocean ridge hydrothermal systems. J Geophys Res 98:9653–9659
Chao HC, You CF, Liu HC, Chung CH (2015) Evidence for stable Sr isotope fractionation by silicate weathering in a small sedimentary watershed in southwestern Taiwan. Geochim Cosmochim Acta 165:324–341
Charbonnier Q, Moyner F, Bouchez J (2018) Barium isotope cosmochemistry and geochemistry. Science Bull 63:385–394
Charlier BL, Nowell GM, Parkinson IJ, Kelley SP, Pearson DG, Burton KW (2012) High temperature strontium stable isotope behaviour in the early solar system and planetary bodies. Earth Planet Sci Lett 329–330:31–40
Chaussidon M, Albarede F (1992) Secular boron isotope variations in the continental crust: an ion microprobe study. Earth Planet Sci Lett 108:229–241
Chaussidon M, Albarede F, Sheppard SMF (1987) Sulphur isotope heterogeneity in the mantle from ion microprobe measurements of sulphide inclusions in diamonds. Nature 330:242–244
Chaussidon M, Albarede F, Sheppard SMF (1989) Sulphur isotope variations in the mantle from ion microprobe analysis of microsulphide inclusions. Earth Planet Sci Lett 92:144–156
Chen CF, Dai W, Wang ZC, Liu YS, Li M, Becker H, Foley S (2019a) Calcium isotope fractionation during magmatic processes in the upper mantle. Geochim Cosmochim Acta 249:121–137
Chen JH, Wasserburg GJ (1983) The isotopic composition of silver and lead in two iron meteorites: Cape York and Grant. Geochim Cosmochim Acta 47:1725–1737
Chen JB, Gaillardet J, Louvat P (2008) Zinc isotopes in the Seine river waters, France: a probe of anthropogenic contamination. Environ Sci Technol 42:6494–6501
Chen JB, Hintelmann H, Feng XB, Dimcock B (2012) Unusual fractionation of both odd and even mercury isotopes in precipitation from Peterborough, ON, Canada. Geochim Cosmochim Acta 90:33–46
Chen H, Savage PS, Teng FZ, Helz RT, Moynier F (2013) Zinc isotopic fractionation during magmatic differentiation and the isotopic composition of bulk Earth. Earth Planet Sci Lett 369–370:34–42
Chen H, Tian Z, Tuller-Ross B, Korotev RL, Wang K (2019b) High-precision potassium isotopic analysis by MC-ICP-MS: an inter-laboratory comparison and refined K atomic weight. JAAS 34:160–171
Chen H, Liu XM, Wang K (2020) Potassium isotope fractionation during continental weathering of basalts. Earth Planet Sci Lett 539:116192
Chen X, Romaniello J, Herrmann AD, Wasylenki LE, Anbar AD (2016) Uranium isotope fractionation during coprecipitation with aragonite and calcite. Geochim Cosmochim Acta 188:189–207

Chen X, Romaniello SJ, Herrmann A, Samankassou E, Anbar AD (2018) Biological effects on uranium isotope fractionation ($^{238}U/^{235}U$) in primary biogenic carbonates. Geochim Cosmochim Acta 240:1–10
Chernnozhkin SM, Weyrauch M, Goderis S, Oeser M, McKibbin SJ, Horn I, Hecht L, Weyer S, Claeys P, Vanhaecke F (2017) Thermal equilibration of iron meteorite and pallasite parent bodies recorded at the mineral scale by Fe and Ni isotope systematics. Geochim Cosmochim Acta 217:95–111
Chetelat B, Liu CQ, Gaillardet J, Wang QL, Zhao ZQ, Liang CS, Xiao YK (2009a) Boron isotopes geochemistry of the Changjiang basin rivers. Geochim Cosmochim Acta 73:6084–6097
Chetelat B, Gaillardet J, Freydier F (2009b) Use of B isotopes as a tracer of anthropogenic emissions in the atmosphere of Paris, France. Appl Geochem 24:810–820
Chiba H, Chacko T, Clayton RN, Goldsmith JR (1989) Oxygen isotope fractionations involving diopside, forsterite, magnetite and calcite: application to geothermometry. Geochim Cosmochim Acta 53:2985–2995
Chmeleff J, Horn I, Steinhöfel G, von Blanckenburg F (2008) In situ determination of precise stable Si isotope ratios by UV-femtosecond laser ablation high-resolution multi-collector ICP-MS. Chem Geol 249:155–160
Choi HB, Ryu JS, Shin WJ, Vigier N (2019) The impact of anthropogenic inputs on lithium content in river and tap water. Nat Commun 10:5371
Ciscato ER, Bontognali TR, Vance D (2018) Nickel and its isotopes in organic-rich sediments: implications for oceanic budgets and a potential record of ancient seawater. Earth Planet Sci Lett 494:239–250
Claire MW, Kasting JF, Domagal-Goldman SD, Stueken EE, Buick R, Meadows VS (2014) Modeling the signature of sulphur mass-independent fractionation produced in the Archean atmosphere. Geochim Cosmochim Acta 141:365–380
Clarkson MO, Stirling CH, Jenkyns HC, Dickson AJ, Porcelli D, Moy CM, Pogge von Strandmann PA, Cooke IR, Lenton TM (2018) Uranium isotope evidence for two episodes of deoxygenation during Oceanic Anoxic Event 2. PNAS 115:2918–2923
Clayton RN, Anderson P, Gale NH, Gills G, Whitehouse MJ (2002) Precise determination of the isotopic composition of Sn using MC-ICP-MS. JAAS 17:1248–1256
Clayton RN, Epstein S (1958) The relationship between $^{18}O/^{16}O$ ratios in coexisting quartz, carbonate and iron oxides from various geological deposits. J Geol 66:352–373
Clayton RN, Goldsmith JR, Mayeda TK (1989) Oxygen isotope fractionation in quartz, albite, anorthite and calcite. Geochim Cosmochim Acta 53:725–733
Clayton RN, Mayeda TK (1963) The use of bromine pentafluoride in the extraction of oxygen from oxides and silicates for isotopic analysis. Geochim Cosmochim Acta 27:43–52
Cloquet C, Carignan J, Lehmann MF, Vanhaecke F (2008) Variation in the isotopic composition of zinc in the natural environment and the use of zinc isotopes in biogeosciences: a review. Anal Bioanal Chem 390:451–463
Cloquet C, Carignan J, Libourel G, Sterckeman T, Perdrix E (2006) Tracing source pollution in soils using cadmium and lead isotopes. Environ Sci Technol 40:2525–2530
Cobert F, Schmitt AD, Bourgeade P, Labolle F, Badot PM, Chabaux F, Stille P (2011) Experimental identification of Ca isotopic fractionations in higher plants. Geochim Cosmochim Acta 75:5467–5482
Cole DR (2000) Isotopic exchange in mineral-fluid systems IV: the crystal chemical controls on oxygen isotope exchange rates in carbonate-$H^2O$ and layer silicate-$H_2O$ systems. Geochim Cosmochim Acta 64:921–933
Cole DB, Reinhard CT, Wang X, Gueguen B, Halverson GP, Gibson T, Hodgskiss MS, McKenzie R, Lyons TW, Planavsky NJ (2016) A shale-hosted Cr isotope record of low atmospheric oxygen during the Proterozoic. Geology 44:555–558
Coleman ML, Sheppard TJ, Durham JJ, Rouse JE, Moore GR (1982) Reduction of water with zinc for hydrogen isotope analysis. Anal Chem 54:993–995

Conway TM, John SG (2014) The biogeochemical cycling of zinc and zinc isotopes in the North Atlantic Ocean. Glob Biogeochem Cycl 28:1111–1128
Conway TM, John SG (2015) Biogeochemical cycling of cadmium isotopes along a high-resolution section through the North Atlantic Ocean. Geochim Cosmochim Acta 148:269–283
Cook DL, Wadhwa M, Clayton RN, Dauphas N, Janney PE, Davis AM (2007) Mass-dependent fractionation of nickel isotopes in meteoritic metal. Meteorit Planet Sci 42:2067–2077
Coplen TB et al (2002) Isotope abundance variations of selected elements. Pure Appl Chem 74:1987–2017
Coplen TB, Hanshaw BB (1973) Ultrafiltration by a compacted clay membrane. I. Oxygen and hydrogen isotopic fractionation. Geochim Cosmochim Acta 37:2295–2310
Coplen TB, Kendall C, Hopple J (1983) Comparison of stable isotope reference samples. Nature 302:236–238
Coutaud M, Meheut M, Glatzel P, Pokrovski GS, Viers J, Rols JL, Pokrovsky OS (2017) Small changes in Cu redox state and speciation generate large isotope fractionation during adsorption and incorporation of Cu by a phototrophic biofilm. Geochim Cosmochim Acta 220:1–18
Craddock PR, Dauphas N (2010) Iron isotopic compositions of geological reference materials and chondrites. Geostand Geoanal Res 35:101–123
Craddock PR, Rouxel OJ, Ball LA, Bach W (2008) Sulfur isotope measurement of sulfate and sulfide by high-resolution MC-ICP-MS. Chem Geol 253:102–113
Craig H (1961a) Isotopic variations in meteoric waters. Science 133:1702–1703
Craig H (1961b) Standard for reporting concentrations of deuterium and oxygen-18 in natural waters. Science 133:1833–1834
Creech J, Baker J, Handler M, Schiller M, Bizzarro M (2013) Platinum stable isotope ratio measurements by double-spike multiple collector ICPMS. JAAS 28:853–865
Creech J, Baker J, Handler M, Bizzarro M (2014) Platinum stable isotope analysis of geological standard reference materials by double-spike MC-ICPMS. Chem Geol 363:293–300
Creech J, Baker J, Handler M, Lorand J, Storey M, Wainwright A, Luguet A, Moynier F, Bizzarro M (2017a) Late accretion history of the terrestrial planets inferred from platinum stable isotopes. Geoch Perspect Lett 3:94–104
Creech JB, Moynier F, Badullovich N (2017b) Tin stable isotope analysis of geological materials by double-spike MC-ICPMS. Chem Geol 457:61–67
Creech JB, Moynier F, Bizarro M (2017c) Tracing metal silicate segradation and late veneer in the Earth and the ureilite parent body with palladium stable isotopes. Geochim Cosmochim Acta 216:28-41
Criss RE, Gregory RT, Taylor HP (1987) Kinetic theory of oxygen isotopic exchange between minerals and water. Geochim Cosmochim Acta 51:1099–1108
Criss RE (1999) Principles of stable isotope distribution. Oxford University Press, Oxford
Croal LR, Johnson CM, Beard BL, Newman DK (2004) Iron isotope fractionation by Fe(II)-oxidizing photoautotrophic bacteria. Geochim Cosmochim Acta 68:1227–1242
Crosby HA, Johnson CM, Roden EE, Beard BL (2005) Fe(II)-Fe(III) electron/atom exchange as a mechanism for Fe isotope fractionation during dissimilatory iron oxide reduction. Environ Sci Tech 39:6698–6704
Crosby HA, Roden EE, Johnson CE, Beard BL (2007) The mechanisms of iron isotope fractionation produced during dissimilatory Fe(III) reduction by *Shewanella putrefaciens* and *Geobacter sulfurreducens*. Geobiology 5:169–189
Crowe DE, Valley JW, Baker KL (1990) Micro-analysis of sulfur isotope ratios and zonation by laser microprobe. Geochim Cosmochim Acta 54:2075–2092
Crowe SA, Dossing LN, Beukes NJ, Bau M, Kruger SJ, Frei R, Canfield DE (2013) Atmospheric oxygenation three billion years ago. Nature 501:535–538
Crowley SF (2010) Effect of temperature on the oxygen isotope composition of carbon dioxide prepared from carbonate minerals by reaction with polyphosphoric acid: an example of the rhombohedral $CaCO_3$-$MgCO_3$ group minerals. Geochim Cosmochim Acta 74:6406–6421

Crowson RA, Showers WJ, Wright EK, Hoering TC (1991) Preparation of phosphate samples for oxygen isotope analysis. Anal Chem 63:2397–2400

Czamanske GK, Rye RO (1974) Experimentally determined sulfur isotope fractionations between sphalerite and galena in the temperature range 600 °C to 275 °C. Econ Geol 69:17–25

D'Arcy J, Babechuk MG, Dossing LN, Gaucher C, Frei R (2016) Processes controlling the chromium isotopic composition of river water: constrains from basaltic river catchments. Geochim Cosmochim Acta 186:296–315

Dahl TW, Anbar AD, Gordon GW, Rosing MT, Frei R, Canfield DE (2010a) The behavior of molybdenum and its isotopes across the chemocline and in the sediments of sulfidic Lake Cadagno, Switzerland. Geochim Cosmochim Acta 74:144–163

Dahl TW, Hammarlund EU et al (2010b) Devonian rise in atmospheric oxygen correlated to the radiations of terrestrial plants and large predatory fish. PNAS 107:17911–17915

Dahl TW, Boyle RA, Canfield DE, Connelly JN, Gill BC, Lenton TM, Bizzarro M (2014) Uranium isotopes distinguish two geochemically distinct stages during the later Cambrian SPICE event. Earth Planet Sci Lett 401:313–326

Dauphas N, Craddock PR, Asimov PD, Bennett VC, Nutman A, Ohnenstetter D (2009a) Iron isotopes may reveal the redox conditions of mantle melting from Archean to Present. Earth Planet Sci Lett 288:255–267

Dauphas N, Pourmand A, Teng FZ (2009b) Routine isotopic analysis of iron by HR-MC-ICPMS: how precise and how accurate? Chem Geol 267:175–184

Dauphas N, Rouxel O (2006) Mass spectrometry and natural variations in iron isotopes. Mass Spectrom Rev 25:515–550

Dauphas N, John SG, Rouxel O (2017) Iron Isotope Systematics. Rev Mineral Geochem 82:415–510

De Hoog JC, Savov IP (2017) Boron isotopes as a tracer of subduction zone processes. In: Boron, the fifth element. Springer (2017)

De Laeter JR, Böhlke JK, De Bièvre P, Hidaka H, Peiser HS, Rosman KJR, Taylor PD (2003) Atomic weights of the elements: review 2000 (IUPAC Technical Report). Pure Appl Chem 75:683–2000

De La Rocha C (2003) Silicon isotope fractionation by marine sponges and the reconstruction of the silicon isotope composition of ancient deep water. Geology 31:423–426

De La Rocha CL, Brzezinski MA, De Niro MJ (1997) Fractionation of silicon isotopes by marine diatoms during biogenic silica formation. Geochim Cosmochim Acta 61:5051–5056

De La Rocha CL, Brzezinski MA, De Niro MJ, Shemesh A (1998) Silicon-isotope composition of diatoms as an indicator of past oceanic change. Nature 395:680–683

De La Rocha CL, De Paolo DJ (2000) Isotopic evidence for variations in the marine calcium cycle over the Cenozoic. Science 289:1176–1178

De Souza GF, Reynolds B, Kiczka M, Bourdon B (2010) Evidence for mass-dependent isotopic fractionation of strontium in a glaciated granitic watershed. Geochim Cosmochim Acta 74:2596–2614

Dellinger M, Gaillardet J, Bouchez J, Calmels D, Louvat P, Dosseto A, Gorge C, Alanoca L, Maurice L (2016) Riverine Li isotope fractionation in the Amazon River basin controlled by the weathering reactions. Geochim Cosmochim Acta 164:71–93

Dellinger M, Hilton RG, Nowell GM (2020) Measurements of rhenium isotopic composition in low-abundance samples. JAAS 35:377

Demarest MS, Brzezinski MA, Beucher CP (2009) Fractionation of silicon isotopes during biogenic silica dissolution. Geochim Cosmochim Acta 73:5572–5583

Deng G, Kang J, Nan X, Li Y, Guo J, Ding X, Huang F (2021) Barium isotope evidence for crystal-melt separation in granitic magma reservoirs. Geochim Cosmochim Acta 242:115–129

Deng Z, Chaussidon M, Savage P, Savage P, Robert F, Pik R, Moynier F (2019) Titanium isotopes as a tracer for the plume or island arc affinity of felsic rocks. PNAS 116:1132–1135

Deng T-H, Cloquet C, Tang Y-T, Sterckeman T, Echevarria G, Estrade N, Morel J-L, Qiu R-L (2014) Nickel and zinc isotope fractionation in hyperaccumulating and nonaccumulating plants. Environ Sci Tech 48:11926–11933

DeNiro MJ, Epstein S (1977) Mechanism of carbon isotope fractionation associated with lipid synthesis. Science 197:261–263

De Obeso JC, Santiago Ramos DP, Higgins JA, Keleman PB (2020) A Mg isotopic perspective on the mobility of magnesium during serpentinization and carbonation of the Oman ophiolite. J Geophys Res Solid Earth 126:e2020JB020237

DePaolo D (2004) Calcium isotope variationsproduced by biological, kinetic, radiogenic and nucleosynthetic processes. Rev Mineral Geochem 55:255–288

DePaolo D (2011) Surface kinetic model for isotopic and trace element fractionation during precipitation of calcite from aqueous solution. Geochim Cosmochim Acta 75:1039–1056

Desaulniers DE, Kaufmann RS, Cherry JO, Bentley HW (1986) $^{37}Cl$-$^{35}Cl$ variations in a diffusion-controlled groundwater system. Geochim Cosmochim Acta 50:1757–1764

Desaulty AM, Albarede F (2013) Copper, lead and silver isotopes solve a major economic conumdrum of Tudor and early Stuart Europe. Geology 41:135–138

Desaulty AM, Telouk P, Albalat E, Albarede F (2011) Isotopic Ag–Cu–Pb record of silver circulation through 16th-18 th century Spain. PNAS 108:9002–9007

Dickson AJ, Jenkyns HC, Porcelli D, van den Boorn S, Idiz E (2016) Basin-scale controls on the molybdenum isotope composition of seawater during Oceanic anoxic event 2 (late Cretaceous). Geochim Cosmochim Acta 178:291–306

Dickson AJ, Gill BC, Ruhl M, Jenkyns HC, Porcelli D, Idiz E, Lyons TW, van den Boorn SH (2017) Molybdenum-isotope chemostratigraphy and paleoceanography of the Toarcian Oceanic Anoxic Event (Early Jurassic). Paleoceanography 32:813–829

Dickson AJ, Hsieh YT, Bryan A (2020) The rhenium isotope composition of Atlantic Ocean seawater. Geochim Cosmochim Acta 287:221–228

Dideriksen K, Baker JA, Stipp SLS (2008) Equilibrium Fe isotope fractionaqtion between inorganic aqueous Fe(III) and the siderophore complex, Fe(III)-desferrioxamine B. Earth Planet Sci Lett 269:280–290

Ding T, Ma GR, Shui MX, Wan DF, Li RH (2005) Silicon isotope study on rice plants from the Zhejiang province, China. Chem Geol 218:41–50

Ding T, Wan D, Wang C, Zhang F (2004) Silicon isotope compositions of dissolved silicon and suspended matter in the Yangtze River, China. Geochim Cosmochim Acta 68:205–216

Ding TP, Zhou JX, Wan DF, Chen ZY, Wang CY, Zhang F (2008) Silicon isotope fractionation in bamboo and its significance to the biogeochemical cycle of silicon. Geochim Cosmochim Acta 72:1381–1395

Ding T (1996) Silicon isotope geochemistry. Geological Publishing House, Beijing

Dohmen R, Kasemann SA, Coogan L, Chakraborty S (2010) Diffusion of Li in olivine. Part 1: experimental observations and a multi species diffusion model. Geochim Cosmochim Acta 74:274–292

Dos Santos Pinheiro GM, Poitraason F, Sondag F, Cochonneau G, Cruz Vieira L (2014) Contrasting iron isotopic compositions in river suspended particulate matter: the Negro and the Amazon annual river cycles. Earth Planet Sci Lett 394:168–178

Dossing LN, Dideriksen K, Stipp SL, Frei R (2011) Reduction of hexavalent chromium by ferrous iron: a process of chromium isotope fractionation and its relevance to natural environments. Chem Geol 285:157–166

Doucet LS, Mattielli N, Ionov DA, Debouge W (2016) Zn isotopic heterogeneity in the mantle: a melting control? Earth Planet Sci Lett 451:232–240

Douglas PMJ, 22 others (2017) Methane clumped isotopes: progress and potential for a new isotopic tracer. Org Geochem 113:262–282

Douthitt CB (1982) The geochemistry of the stable isotopes of silicon. Geochim Cosmochim Acta 46:1449–1458

Driesner T (1997) The effect of pressure on deuterium-hydrogen fractionation in high-temperature water. Science 277:791–794

Driesner T, Seward TM (2000) Experimental and simulation study of salt effects and pressure/density effects on oxygen and hydrogen stable isotope liquid-vapor fractionation for 4–5 molal aqueous NaCl and KCl solutions to 400 °C. Geochim Cosmochim Acta 64:1773–1784

Dugan JP, Borthwick J, Harmon RS, Gagnier MA, Glahn JE, Kinsel EP, McLeod S, Viglino JA (1985) Guadinine hydrochloride method for determination of water oxygen isotope ratios and the oxygen-18 fractionation between carbon dioxide and water at 25 °C. Anal Chem 57:1734–1736

Dziony W, Horn I, Lattard D, Koepke J, Steinhoefel G, Schuessler J, Holtz F (2014) In-situ Fe isotope ratio determination in Fe-Ti oxides and sulphides from drilled gabbros and basalt from the IODP Hole 1256D in the eastern equatorial Pacific. Chem Geol 363:101–113

Eastoe CJ, Gilbert JM, Kaufmann RS (1989) Preliminary evidence for fractionation of stable chlorine isotopes in ore-forming hydrothermal deposits. Geology 17:285–288

Eastoe CJ, Guilbert JM (1992) Stable chlorine isotopes in hydrothermal processes. Geochim Cosmochim Acta 56:4247–4255

Eastoe CJ, Long A, Knauth LP (1999) Stable chlorine isotopes in the Palo Duro basin, Texas: evidence for preservation of Permian evaporate brines. Geochim Cosmochim Acta 63:1375–1382

Eastoe CJ, Long A, Land LS, Kyle JR (2001) Stable chlorine isotopes in halite and brine from the Gulf Coast Basin: brine genesis and evolution. Chem Geol 176:343–360

Eastoe CJ, Peryt TM, Petrychenko OY, Geisler-Cussey D (2007) Stable chlorine isotopes in Phanerozoic evaporates. Appl Geochem 22:575–588

Egan KE, Rickaby RE, Leng H, Hendry KE, Hemoso M, Sloane HJ, Bostock H, Halliday RN (2012) Diatom silicon isotopes as a proxy for silicic acid utilisation: a southern-ocean core top calibration. Geochim Cosmochim Acta 96:174–192

Eggenkamp HGM, Coleman M (2000) Rediscovery of classical methods and their application to the measurement of stable bromine isotopes in natural samples. Chem Geol 167:393–402

Eggenkamp HGM, Kreulen R, Koster van Groos AF (1995) Chlorine stable isotope fractionation in evaporates. Geochim Cosmochim Acta 59:5169–5175

Eggenkamp HGM (1994) $\delta^{37}$Cl: the geochemistry of chlorine isotopes. University of Utrecht, Thesis

Eggenkamp HGM (2014) The geochemistry of stable chlorine and bromine isotopes. Springer, New York

Eggenkamp HGM, Bonifacie M, Ader M, Agrinier P (2016) Experimental determination of stable chlorine and bromine isotope fractionation during precipitation of salt from a saturated solution. Chem Geol 433:46–56

Eggenkamp HGM, Louvat P, Agrinier P, Bonifacie M, Bekker A, Krupenik V, Griffioen J, Horita J, Brocks JJ, Bagheri R (2019) The bromine and chlorine isotope composition of primary halite deposits and their significance for the secular isotope composition of seawater. Geochim Cosmochim Acta 264:13–29

Ehrlich S, Butler I, Halicz L, Rickard D, Oldroyd A, Matthews A (2004) Experimental study of the copper isotope fractionation between aqueous Cu(II) and covellite, CuS. Chem Geol 209:259–269

Eiler JM et al (2014) Frontiers of stable isotope geoscience. Chem Geol 372:119–143

Eisenhauer A et al (2004) Proposal for an international agreement on Ca notation as result of the discussion from the workshops on stable isotope measurements in Davos (Goldschmidt 2002) and Nice (EUG 2003). Geostand Geoanal Res 28:149–151

Eisenhauer A, Kisakürek B, Böhm F (2009) Marine calcification: an alkali earth metal isotope perspective. Elements 5:365–368

Elardo SM, Shahar A (2017) Non-chondritic iron isotope composition in planetary mantles as a result of core formation. Nat Geosci 10:317–321

Eldridge CS, Compston W, Williams IS, Both RA, Walshe JL, Ohmoto H (1988) Sulfur isotope variability in sediment hosted massive sulfide deposits as determined using the ion microprobe SHRIMP. I. An example from the Rammelsberg ore body. Econ Geol 83:443–449

Eldridge CS, Williams IS, Walshe JL (1993) Sulfur isotope variability in sediment hosted massive sulfide deposits as determined using the ion microprobe SHRIMP. II. A study of the H.Y.C. deposit at McArthur River, Northern Territory, Australia. Econ Geol 88:1–26

El Korh AE, Luais B, Boiron MC, Deloule E, Cividini D (2017) Investigation of Ge and Ga exchange behavior and Ge isotope fractionation during subduction zone behavior. Chem Geol 449:165–181

Elliott T, Jeffcoate AB, Bouman C (2004) The terrestrial Li isotope cycle: light-weight constraints on mantle convection. Earth Planet Sci Lett 220:231–245

Elliott T, Steele RC (2017) The isotope geochemistry of Ni. Rev Mineral Geochem 82:511–542

Ellis AS, Johnson TM, Bullen TD (2002) Chromium isotopes and the fate of hexavalent chromium in the environment. Science 295:2060–2062

Ellis AS, Johnson TM, Bullen TD (2004) Using chromium stable isotope ratios to quantify Cr(VI) reduction: lack of sorption effects. Environ Sci Technol 38:3604–3607

Emrich K, Ehhalt DH, Vogel JC (1970) Carbon isotope fractionation during the precipitation of calcium carbonate. Earth Planet Sci Lett 8:363–371

Engstrom E, Rodushkin I, Baxter DC, Ohlander B (2006) Chromatographic purification for the determination of dissolved silicon isotopic compositions in natural waters by high-resolution multicollector inductively coupled mass spectrometry. Anal Chem 78:250–257

Escoube R, Rouxel OJ, Luais B, Ponzevera E, Donard OF (2012) An intercomparison study of the germanium isotope composition of geological reference materials. Geostand Geoanal Res 36:149–159

Escoube R, Rouxel OJ, Edwards K, Glazer B, Donard OFX (2015) Coupled Ge/Si and Ge isotope ratios as geochemical tracers of seafloor hydrothermal systems: case studies at Loihi Seamount and East Pacific Rise 9°50′N. Geochim Cosmochim Acta 167:93–112

Estrade N, Carignan J, Sonke JE, Donard O (2009) Mercury isotope fractionation during liquid-vapor evaporation experiments. Geochim Cosmochim Acta 73:2693–2711

Estrade N, Cloquet C, Echevarria G, Sterckeman T, Deng T-H, Tang Y-T, Morel J-L (2015) Weathering and vegetation controls on nickel isotope fractionation in surface ultramafic environments (Albania). Earth Planet Science Lett 423:24–35

Fang Z, Qin L, Liu W, Yao T, Chen X, Wie S (2020) Absence of hexavalent chromium in marine carbonates: implications for chromium isotopes as paleoenvironment proxy. National Sci Rev (in press)

Fantle MS (2010) Evaluating the Ca isotope proxy. Am J Sci 310:194–210

Fantle MS, de Paolo DJ (2005) Variations in the marine Ca cycle over the past 20 million years. Earth Planet Sci Lett 237:102–117

Fantle MS, Tipper ET (2014) Calcium isotopes in the global biogeochemical Ca cycle: implications for development of a Ca isotope proxy. Earth Sci Rev 129:148–177

Fantle MS, Barnes BD, Lau KV (2020) The role of diagenesis in shaping the geochemistry of the marine carbonate record. Ann Rev Earth Planet Sci 48:549–583

Farkas J, Buhl D, Blenkinsop J, Veizer J (2007) Evolution of the oceanic calcium cycle during the late Mesozoic: evidence from $\delta^{44/40}$ Ca of marine skeletal carbonates. Earth Planet Sci Lett 253:96–111

Farkas J, Chrastny V, Novak M, Cadkova E, Pasava J, Chakrabarti R, Jacobsen S, Ackerman L, Bullen TD (2013) Chromium isotope variations ($\delta^{53}/^{52}$Cr) in mantle-derived sources and their weathering products: implications for environmental studies and the evolution of $\delta^{53}/^{52}$Cr in the Earth's mantle over geologic time. Geochim Cosmochim Acta 123:74–92

Farquhar J, Bao H, Thiemens M (2000) Atmospheric influence of Earth's earliest sulfur cycle. Science 289:756–759

Farquhar J, Day JM, Hauri EH (2013) Anomaleous sulphur isotopes in plume lavas reveal deep mantle storage of Archaean crust. Nature 496:490–493

Farquhar GD, Ehleringer JR, Hubick KT (1989) Carbon isotope discrimination and photosynthesis. Ann Rev Plant Physiol Plant Mol Biol 40:503–537

Farquhar J, Johnston DT, Wing BA, Habicht KS, Canfield DE, Airieau S, Thiemens MH (2003) Multiple sulphur isotope interpretations for biosynthetic pathways: implications for biological signatures in the sulphur isotope record. Geobiology 1:27–36

Farquhar J, Kim ST, Masterson A (2007) Implications from sulfur isotopes of the Nakhla meteorite for the origin of sulfate on Mars. Earth Planet Sci Lett 264:1–8

Farrell JW, Pedersen TF, Calvert SE, Nielsen B (1995) Glacial-interglacial changes in nutrient utilization in the equatorial Pacific Ocean. Nature 377:514–517

Fehr MA, Rehkämper M, Halliday AN (2004) Application of MC-ICP-MS to the precise determination of tellurium isotope compositions in chondrites, iron meteorites and sulphides. Inter J Mass Spectr 232:83–94

Feng C, Qin T, Huang S, Wu Z, Huang F (2014) First principles investigations of equilibrium calcium isotope fractionation between clinopyroxene and Ca-doped orthopyroxene. Geochim Cosmochim Acta 143:132–142

Fiebig J, Hoefs J (2002) Hydrothermal alteration of biotite and plagioclase as inferred from intragranular oxygen isotope- and cation-distribution patterns. Eur J Mineral 14:49–60

Fietzke J, Eisenhauer A et al (2004) Direct measurement of $^{44}Ca/^{40}Ca$ ratios by MC-ICP-MS using the cool plasma technique. Chem Geol 206:11–20

Fietzke J, Eisenhauer A (2006) Determination of temperature-dependent stable strontium isotope ($^{88}Sr/^{86}Sr$) fractionation via bracketing standard MC-ICP-MS. Geochem Geophys Geosys 7(8). https://doi.org/10.1029/2006GC001243

Fischer-Gödde M, Burkhardt C, Kruijer TS, Kleine T (2015) Ru isotope heterogeneity in the solar protoplanetary disk. Geochim Cosmochim Acta 168:151–171

Fischer-Gödde M, Elfers BM, Münker C, Szilas K, Maier WD, Messling N, Morishita TT, van Kranendonk M, Smithies H (2020) Ruthenium isotope vestige of Earth' s pre-late veneer mantle preserved in Arcaean rocks. Nature 579:240–244

Fitzpayne A, Prytulak J, Wilkinson JJ, Cooke DR, Baker MJ, Wilkinson CC (2018) Assessingthallium elemental systematics and isotope ratio variations in porphyry ore systems: a case study of the Bingham Canyon District. Minerals 8:548. https://doi.org/10.3390/min8120548

Fogel ML, Cifuentes LA (1993) Isotope fractionation during primary production. In: Engel MH, Macko SA (eds) Organic geochemistry. Plenum Press, New York, pp 73–98

Foustoukos DI, James RH, Berndt ME, Seyfried WE (2004) Lithium isotopic systematics of hydrothermal vent fluids at Main Endeavour Field, Northern Juan de Fuca Ridge. Chem Geol 212:17–26

Fornadel AP, Spry GP, Jackson SE, Mathur RD, Chapman JB, Girard I (2014) Methods for the determination of stable Te isotopes of minerals in the system Au–Ag–Te by MC-ICP-MS. J Anal at Spectrom 29:623–637

Fornadel AP, Spry PG, Haghnegahdar MA, Schauble EA, Jackson SE, Mills SJ (2017) Stable Te isotope fractionation in tellurium-bearing minerals from precious metal hydrothermal ore deposits. Geochim Cosmochim Acta 202:215–230

Foster GL, Pogge von Strandmann PA, Rae JW (2010) Boron and magnesium isotopic compositions of seawater. Geochem Geophys Geosys 11. https://doi.org/10.1029/2010GC003201

Foster GL, Rae JW (2016) Reconstructing ocean pH with boron isotopes in foraminifera. Ann Rev Earth Planet Sci 44:207–237

Frank AB, Klaebe RM, Frei R (2019) Fractionation behavior of chromium isotopes during the sorption of Cr(VI) on kaolin and its implications for using black shales as a paleoredox archive. Geochem Geophys Geosys 20. https://doi.org/10.1029/2019GC008284

Freeman KH (2001) Isotopic biogeochemistry of marine organic carbon. Rev Mineral Geochem 43:579–605

Frei R, Gaucher C, Dossing LN, Sial AN (2011) Chromium isotopes in carbonates—a tracer for climate change and for reconstructing the redox state of ancient seawater. Earth Planet Sci Lett 312:114–125

Frei R, Gaucher C, Poulton SW, Canfield DE (2009) Fluctuations in Precambrian atmospheric oxygenation recorded by chromium isotopes. Nature 461:250–253

Frei R, Polat A (2013) Chromium isotope fractionation during oxidative weathering—implications from the study of a paleoproterozoic (ca. 1.9 Ga) paleosol, Schreiber Beach, Ontario,Canada. Precam Res 224:434–453

Frei R, Poiret D, Frei KM (2014) Weathering on land and transport of chromium to the ocean in a subtropical region (Misiones, NW Argentina): a chromium stable isotope perspective. Chem Geol 381:110–124

Frei R, Paulukat C, Bruggmann S, Klaebe RM (2018) A systematic look at chromium isotopes in modern shells—implications for paleo-environmental reconstructions. Biogeosciences 15:4905–4922

Freymuth H, Vils F, Willbold M, Taylor RN, Elliott T (2015) Molybdenum mobility and isotopic fractionation during subduction at the Mariana arc. Earth Planet Sci Lett 432:176–186

Freymuth H, Andersen MB, Elliott T (2019) Uranium isotope fractionation during slab dehydration beneath the Izu arc. Earth Planet Sci Lett 522:244–254

Friedman I (1953) Deuterium content of natural waters and other substances. Geochim Cosmochim Acta 4:89–103

Frierdich AJ, Beard BL, Scherer MM, Johnson CM (2014) Determination of the Fe(II) aq-magnetite equilibrium iron isotope fractionation factor using the three-isotope method and a multi-direction approach to equilibrium. Earth Planet Sci Lett 391:77–86

Frierdich AJ, Nebel O, Beard BL, Johnson CM (2019) Iron isotope exchange and fractionation between hematite (alpha-$Fe_2O_3$) and aqueous Fe(II): a combined three-isotope and reversal approach to equilibrium study. Geochim Cosmochim Acta 245:207–221

Frings PJ (2019) Palaeoweathering: how do weathering rates vary with climate? Elements 15:259–265

Frings PJ, Clymans W, Fontorbe G, de la Rocha C, Conley DJ (2016) The continental Si cycle and its impact on the ocean Si isotope budget. Chem Geol 425:12–36

Fritz P, Basharmel GM, Drimmie RJ, Ibsen J, Qureshi RM (1989) Oxygen isotope exchange between sulphate and water during bacterial reduction of sulphate. Chem Geol 79:99–105

Fruchter N, Eisenhauer A, Dietzel M, Fietzke J, Böhm F, Montagna P, Stein M, Lazar B, Rodolfo-Metalpa R, Erez J (2016) $^{88}Sr/^{86}Sr$ fractionation in inorganic aragonite and corals. Geochim Cosmochim Acta 178:268–280

Fry B, Ruf W, Gest H, Hayes JM (1988) Sulphur isotope effects associated with oxidation of sulfide by $O_2$ in aqueous solution. Chem Geol 73:205–210

Fujii Y, Higuchi N, Haruno Y, Nomura M, Suzuki T (2006) Temperature dependence of isotope effects in uranium chemical exchange reactions. J Nucl Sci Techn 43:400–406

Fujii T, Moynier F, Pons ML, Albarede F (2011a) The origin of Zn isotope fractionation in sulfides. Geochim Cosmochim Acta 75:7632–7643

Fujii T, Moynier F, Dauphas N, Abe M (2011b) Theoretical and experimental investigation of nickel isotope fractionation in species relevant to modern and ancient oceans. Geochim Cosmochim Acta 75:469–482

Fujii T, Moynier F, Abe M, Nemoto K, Albarede F (2013) Copper isotope fractionation between aqueous compounds relevant to low temperature geochemistry and biology. Geochim Cosmochim Acta 110:29–44

Fujii T, Moynier F, Blichert-Toft J, Albarede F (2014) Density functional theory estimation of isotope fractionation of Fe, Ni, Cu and Zn among species relevant to geochemical and biological environments. Geochim Cosmochim Acta 140:553–576

Fujii T, Albarede F (2018) $^{109}Ag$–$^{107}Ag$ fractionation in fluids with applications to ore deposits, archeometry and cosmochemistry. Geochim Cosmochem Acta 234:37–49

Fujiwara Y, Tsujisaka M, Takano S, Sohrin Y (2020) Determination of the tungsten isotope composition in seawater: the first vertical profile from the western North Pacific Ocean. Chem Geol 555:119835

Fukami Y, Kimura JI, Suzuki K (2018) Precise isotope analysis of tellurium by inductively coupled plasma mass spectrometry using a double spike method. JAAS 33:1233–1242

Gagnevin D, Boyce AJ, Barrie CD, Menuge JF, Blakeman RJ (2012) Zn, Fe, and S isotope fractionation in a large hydrothermal system. Geochim Cosmochim Acta 88:183–198

Galimov EM (2006) Isotope organic geochemistry. Org Geochem 37:1200–1262

Galimov EM (1985) The biological fractionation of isotopes. Academic Press Inc., Orlando

Gall L, Williams HM, Halliday AN, Kerr AC (2017) Nickel isotope composition of the mantle. Geochim Cosmochim Acta 199:196–209

Gaillardet J, Lemarchand D (2018) Boron in the weathering environment. In: Marschall H, Foster G (eds) Boron isotopes. Springer, pp 163–188

Galy A et al (2003) Magnesium isotope heterogeneity of the isotopic standard SRM980 and new reference materials for magnesium-isotope-ratio measurements. J Anal at Spectr 18:1352–1356

Galy A, Bar-Matthews M, Halicz L, O'Nions RK (2002) Mg isotopic composition of carbonate: insight from speleothem formation. Earth Planet Sci Lett 201:105–115

Galy A, Belshaw NS, Halicz L, O'Nions RK (2001) High-precision measurement of magnesium isotopes by multiple-collector inductively coupled plasma mass spectrometry. Inter J Mass Spectr 208:89–98

Ganeshram RS, Pedersen TF, Calvert SE, McNeill GW, Fontugue MR (2000) Glacial-interglacial variability in denitrification in the world's oceans: causes and consequences. Paleoceanography 15:361–376

Gao C, LiuY (2021) First-principles calculations of equilibrium bromine isotope fractionations. Geochim Cosmochim Acta 297:65-81

Gao Y, Casey JF (2011) Lithium isotope composition of ultramafic geological reference materials JP-1 and DTS-2. Geostand Geoanal Res 36:75–81

Gao Y, Vils F et al (2012) Downhole variation of lithium and oxygen isotopic compositions of oceanic crust at East Pacific Rise, ODP Site 1256. Geochem Geophys Geosystems 13. https://doi.org/10.1029/2012GC004207

Gao Y, Casey JF, Bernardo LM, Yang W, Bissada KK (2018) Vanadium isotope composition of crude oil: effects of source, maturation and biodegradation. In: From source to seep: geochemical applications in hydrocarbon systems. Spec Publ Geological Soc London 468:83–103

Garlick GD (1966) Oxygen isotope fractionation in igneous rocks. Earth Planet Sci Lett 1:361–368

Gaschnig RM, Reinhard CT, Planavsky NJ, Wang X, Asael D, Chauvel C (2017) The molybdenum isotope system as a tracer of slab input in subduction zones: an example from Martinique, Lesser Antilles Arc. Geochem Geophys Geosys 18:4674–4689

Gault-Ringold M, Adu T, Stirling C, Frew RD, Hunter KA (2012) Anomalous biogeochemical behaviour of cadmium in subantarctic surface waters: mechanistic constraints from cadmium isotopes. Earth Planet Sci Lett 341–344:94–103

Gehre M, Hoefling R, Kowski P, Strauch G (1996) Sample preparation device for quantitative hydrogen isotope analysis using chromium metal. Anal Chem 68:4414–4417

Gehrke GE, Blum JD, Meyers PA (2009) The geochemical behaviour and isotope composition of Hg in a mid-Pleistocene western Mediterranean sapropel. Geochim Cosmochim Acta 73:1651–1665

Geilert S, Vroon PZ, Keller NS, Gudbrandsson S, Stefansson A, van Bergen MJ (2015) Silicon isotope fractionation during silica precipitation from hot-spring waters: evidence from the Geysir geothermal field, Iceland. Geochim Cosmochim Acta 164:403–427

Georg RB, Reynolds BC, Frank M, Halliday AN (2006) Mechanisms controlling the silicon isotopic compositions of river water. Earth Planet Sci Lett 249:290–306

Georg RB, Halliday AN, Schauble EA, Reynolds BC (2007) Silicon in the Earth's core. Nature 447:1102–1106

Georg RB, Zhu C, Reynolds BC, Halliday AN (2009) Stable silicon isotopes of groundwater, feldspars and clay coating in the Navajo sandstone aquifer, Black Mesa, Arizona, USA. Geochim Cosmochim Acta 73:2229–2241

George E, Stirling C, Gault-Ringold M, Ellwood MJ, Middag R (2019) Marine biogeochemical cycling of cadmium isotopes in the extreme nutrient-depleted subtropical gyre of the South West Pacific Ocean. Earth Planet Sci Lett 514:84–95

Georgiev SV, Horner TJ, Stein H, Hannah JL, Bingen B, Rehkämper M (2015) Cadmium-isotopic evidence for increasing primary productivity during the late Permian anoxic event. Earth Planet Sci Lett 410:84–96

Geske A, Goldstein RH, Mavromatis V, Richter DK, Buhl D, Kluge T, John CM, Immenhauser A (2015) The magnesium isotope ($\delta^{26}$Mg) signature of dolomites. Geochim Cosmochim Acta 149:131–151

Ghosh S, Schauble EA, Lacrampe Coulome G, Blum JD, Bergquist BA (2013) Estimation of nuclear volume dependent fractionation of mercury isotopes in equilibrium liquid-vapor evaporation experiment. Chem Geol 366:5–12

Giesemann A, Jäger HA, Norman AL, Krouse HR, Brand WA (1994) On-line sulphur isotope determination using an elemental analyzer coupled to a mass spectrometer. Anal Chem 66:2816–2819

Giletti BJ (1985) The nature of oxygen transport within minerals in the presence of hydrothermal water and the role of diffusion. Chem Geol 53:197–206

Gilleaudeau GJ, Frei R, Kaufman AJ, Kah LC, Azmy K, Bartley JK, Chernyavskiy P, Knoll AH (2016) Oxygenation of the mid-Proterozoic atmosphere: clues from chromium isotopes in carbonates. Geochem Perspect Lett 2:178–187

Giunta T, Devauchelle O, Ader M, Locke R, Louvat P, Bonifacie M, Metivier F, Agrinier P (2017) The *gravitas* of gravitational isotope fractionation revealed in isolated aquifer. Geochem Persp Lett 4:53–58

Godfrey JD (1962) The deuterium content of hydrous minerals from the East Central Sierra Nevada and Yosemite National Park. Geochim Cosmochim Acta 26:1215–1245

Godon A, Webster JD, Layne GD, Pineau F (2004) Secondary ion mass spectrometryfor the determinationof $\delta^{37}$Cl. Part II: intsercalibration of SIMS and IRMSfor alumino-silicate glasses. Chem Geol 207:291–303

Goldberg T, Archer C, Vance D, Poulton SW (2009) Mo isotope fractionation during adsorption to Fe(oxyhdr)oxides. Geochim Cosmochim Acta 73:6502–6516

Goldberg T, Poulton SW, Wagner T, Kolonic SF, Rehkämper M (2016) Molybdenum drawdown during Cretaceous Oceanic Anoxic Event 2. Earth Planet Sci Lett 440:81–91

Goldhaber MB, Kaplan IR (1974) The sedimentary sulfur cycle. In: Goldberg EB (ed) The sea, vol IV. Wiley and Sons, New York

Gordon GW, Lyons TW, Arnold GL, Roe J, Sageman BB, Anbar AD (2009) When do black shales tell molybdenum isotope tales? Geology 37:535–538

Graham CM, Harmon RS, Sheppard SMF (1984) Experimental hydrogen isotope studies: hydrogen isotope exchange between amphibole and water. Am Mineral 69:128–138

Graham CM, Sheppard SMF, Heaton THE (1980) Experimental hydrogen isotope studies. I. Systematics of hydrogen isotope fractionation in the systems epidote-$H_2O$, zoisite-$H_2O$ and AlO(OH)-$H_2O$. Geochim Cosmochim Acta 44:353–364

Grant F (1954) The geological significance of variations in the abundances of the isotopes of silicon in rocks. Geochim Cosmochim Acta 5:225–242

Grasby SE, Them TR, Chen ZH, Yin RS, Ardakani OH (2019) Mercury as a proxy for volcanic emissions in the geologic record. Earth Sci Rev 196:102880

Grattage J, McCoyWest AJ, Gislason SG, Nowell GM, Burton KW (2018) Neodymium stable isotope behaviour accompanying basalt weathering. Goldschmidt 2018, Abstract

Greber ND, Hofmann BD, Voegelin AR, Villa IM, Nägler TF (2011) Mo isotope compositions in Mo-rich high- und low-T hydrothermal systems from the Swiss Alps. Geochim Cosmochim Acta 75:6600–6609

Greber ND, Pettke T, Nägler TF (2014) Magmatic-hydrothermal molybdenum isotope fractionation and its relevance to the igneous crustal signature. Lithos 190–191:104–110

Greber ND, Puchtel IS, Nägler TF, Mezger K (2015) Komatiites constrain molybdenum isotope composition of the Earth's mantle. Earth Planet Sci Lett 421:129–138

Greber ND, Dauphas N, Puchtel IS, Hofmann BA, Arndt NT (2017a) Titanium stable isotopic variations in chondrites, achondrites and lunar rocks. Geochim Cosmochim Acta 213:534–552

Greber ND, Dauphas N, Bekker A, Ptacek MP, Bindeman IN, Hofmann A (2017b) Titanium isotopic evidence for felsic crust and plate tectonics 3.5 billion years ago. Science 357:1271–1274

Gregory RT, Criss RE, Taylor HP (1989) Oxygen isotope exchange kinetics of mineral pairs in closed and open systems: applications to problems of hydrothermal alteration of igneous rocks and Precambrian iron formations. Chem Geol 75:1–42

Griffith EM, Paytan A, Eisenhauer A, Bullen TD, Thomas E (2011) Seawater calcium isotope ratios across the Eocene-Oligocene transition. Geology 39:683–686

Griffith EM, Paytan A, Kozdon R, Eisenhauer A, Ravelo AC (2008a) Influences on the fractionation of calcium isotopes in planktonic foraminifera. Earth Planet Sci Lett 268:124–136

Griffith EM, Payton A, Caldeira K, Bullen TD, Thomas E (2008b) A dynamic marine calcium cycle during the past 28 million years. Science 322:1671–1674

Griffith EM, Schauble EA, Bullen TD, Paytan A (2008c) Characterization of calcium isotopes in natural and synthetic barite. Geochim Cosmochim Acta 72:5641–5658

Griffith EM, Fantle S (2020) Calcium isotopes. Cambrige University Press

Grossman EL, Ku T-L (1986) Oxygen and carbon isotope fractionation in biogenic aragonite: temperature effects. Chem Geol 59:59–74

Grotheer H, Greenwood PF, McCulloch MT, Böttcher ME, Grice K (2017) $\delta^{34}$S character of organosulfur compounds in kerogen and bitumen fractions of sedimentary rocks. Org Geochem 110:60–64

Gueguen B, Rouxel O, Ponzevera E, Bekker A, Fouquet Y (2013) Nickel isotope variations in terrestrial silicate rocks and geological reference materials measured by MC-ICP-MS. Geostand Geoanal Res 37:297–317

Gueguen B, Reinhard CT, Algeo TJ, Peterson LC, Nielsen SG, Wang X, Rowe H, Planavsky NJ (2016) The chromium isotope composition of reducing and oxic marine sediments. Geochim Cosmochim Acta 184:1–19

Gueguen B, Sorensen JV, Lalonde SV, Pena Brandy J, Toner M, Rouxel O (2018) Variable Ni isotope fractionation between Fe-oxyhydroxides and implications for the use of Ni isotopes as geochemical tracers. Chem Geol 481:38–52

Guelke M, von Blanckenburg F (2007) Fractionation of stable iron isotopes in higher plants. Environ Sci Technol 41:1896–1901

Guelke M, von Blanckenburg F, Schoenberg R, Staubwasser M, Stuetzel H (2010) Determining the stable Fe isotope signature of plant-available iron in soils. Chem Geol 277:269–280

Guelke-Stelling M, von Blanckenburg F (2012) Fe isotope fractionation caused by translocation of iron during growth of bean and oat as models of strategy I and II plants. Plant Soil 352:217–231

Guerrot C, Millot R, Robert M, Negrel P (2011) Accurate and high-precision determination of boron isotopic ratios at low concentration by MC-ICP-MS (Neptune). Geostand Geoanal Res 35:275–284

Guo H, Li WY, Nan X, Huang F (2020a) Experimental evidence for light Ba isotopes favouring aqueous fluids over silicate melts. Geochem Persp Lett 16:6–11

Guo JL, Wang Z, Zhang W, Moynier F, Cui D, Hu Z, Ducea MN (2020b) Significant Zr isotope variations in single zircon grains recording magma evolution history. PNAS 117:21135–21131

Gussone N, Eisenhauer A, Heuser A et al (2003) Model for kinetic effects on calcium isotope fractionations ($\delta^{44}$Ca) in inorganic aragonite and cultured planktonic foraminifera. Geochim Cosmochim Acta 67:1375–1382

Gussone N, Böhm F, Eisenhauer A et al (2005) Calcium isotope fractionation in calcite and aragonite. Geochim Cosmochim Acta 69:4485–4494
Gussone N, Dietzel M (2016) Calcium isotope fractionation during mineral precipitation from aqueous solution. In: Advances in Ca isotope geochemistry. Springer, pp 75–144
Gussone N, Schmitt AD, Heuser A, Wombacher F, Dietzel M, Tipper E, Schiller M (2016) Calcium stable isotope geochemistry. Advances in isotope geochemistry. Springer
Guillermic M, Lalonde SV, Hendry KV, Rouxel OJ (2017) The isotopic composition of inorganic germanium in seawater and deep-sea sponges. Geochim Cosmochim Acta 212:99–118
Habicht KS, Canfield DE (1997) Sulfur isotope fractionation during bacterial sulfate reduction in organic-rich sediments. Geochim Cosmochim Acta 61:5351–5361
Habicht KS, Canfield DE, Rethmeier JC (1998) Sulfur isotope fractionation during bacterial reduction and disproportionation of thiosulfate and sulfite. Geochim Cosmochim Acta 62:2585–2595
Habicht KS, Canfield DE (2001) Isotope fractionation by sulfate-reducing natural populations and the isotopic composition of sulfide in marine sediments. Geology 29:555–558
Haendel D, Mühle K, Nitzsche HIM, Stiehl G, Wand U (1986) Isotopic variations of the fixed nitrogen in metamorphic rocks. Geochim Cosmochim Acta 50:749–758
Halevy I, Johnston DT, Schrag DP (2010) Explaining the structure of the Archean mass-independent sulfur isotope record. Science 329:204–207
Halicz L, Galy A, Belshaw N et al (1999) High-precision measurement of calcium isotopes in carbonates and related materials by multiple collector inductively coupled plasma mass spectrometry (MC-ICP-MS). J Anal at Spectr 14:1835–1838
Halicz L, Yang L, Teplyakov N, Burg A, Sturgeon R, Kolodny Y (2008) High precision determination of chromium isotope ratios in geological samples by MC-ICP-MS. J Anal at Spectrom 23:1622–1627
Han R, Qin L, Brown ST, Christensen JN, Beller HR (2012) Differential isotopic fractionation during Cr(VI) reduction by an aquifer-derived bacterium under aerobic versus denitrifying conditions. Appl Environ Microbiol 78:2462–2464
Han X, Guo Q, Strauss H, Liu C, Hu J, Guo Z, Wei R, Peters M, Tian L, Kong J (2017) Multiple sulfur isotope constraints on sources and formation processes of sulfate in Beijing PM2.5 aerosol. Environ Sci Technol 51:7794–7803
Handler MR, Baker JA, Schiller M, Bennett VC, Yaxley GM (2009) Magnesium stable isotope composition of Earth's upper mantle. Earth Planet Sci Lett 282:306–313
Hanlon C, Stotler R, Frape S, Gurgnne R (2017) Comparison of $\delta^{81}$Br and $\delta^{37}$Cl composition of volatiles, salt precipitates and associated water in terrestrial evaporative saline lake systems. Isotopes Environ Health Studies 53:446–465
Hannah JL, Stein HJ, Wieser ME, de Laeter JR, Varner MD (2007) Molybdenum isotope variations in molybdenite: vapor transport and Rayleigh fractionation of Mo. Geology 35:703–706
Harouaka K, Eisenhauer A, Fantle MS (2014) Experimental investigation of Ca isotopic fractionation during abiotic gypsum precipitation. Geochim Cosmochim Acta 129:157–176
Harrison AG, Thode HG (1957a) Kinetic isotope effect in chemical reduction of sulphate. Faraday Soc Trans 53:1648–1651
Harrison AG, Thode HG (1957b) Mechanism of the bacterial reduction of sulphate from isotope fractionation studies. Faraday Soc Trans 54:84–92
Hatton JE, Hendry KR, Hawkings JR, Wadham JL, Opfergelt S, Kohler TJ, Yde JC, Stibal M, Zarsky JD (2019) Silicon isotopes in Arctic and sub-Arctic meltwaters: the role of subglacial weathering in the silicon cycle. Proc R Soc A 475:20190098
Hatton JE, 11 others (2020) Silicon isotopic composition of dry and wet-based glaciers in Antarctica. Front Earth Sci 14:00286
Hauri EH, Papineau D, Wang J, Hillion F (2016) High-precision analysis of multiple sulfur isotope using NanoSIMS. Chem Geol 420:148–161

Haustein M, Gillis C, Pernicka E (2010) Tin isotopy—a new method for solving old questions. Archaeometry 52:816–832

Hayes JM (1993) Factors controlling $^{13}C$ contents of sedimentary organic compounds: principle and evidence. Mar Geol 113:111–125

Hayes JM, Strauss H, Kaufman AJ (1999) The abundance of $^{13}C$ in marine organic matter and isotopic fractionation in the global biogeochemical cycle of carbon during the past 800 Ma. Chem Geol 161:103–125

Hayes JM (2001) Fractionation of carbon and hydrogen isotopes in biosynthetic processes. In: Valley JW, Cole DR (eds) Stable isotope geochemistry. Rev Mineral Geochem 43:225–277

He Y, Ke S, Teng FZ, Wang T, Wu H, Lu Y, Li S (2015) High precision iron isotope analysis of geological standards by high resolution MC-ICPMS. Geostand Geoanal Res 39:341–356

He Y, Chen LH, Shi JH, Zeng G, Wang XJ, Xue XQ, Zhong Y, Erdmann S, Xie LW (2019) Light Mg isotopic composition in the mantle beyond the Big Mantle Wedge beneath eatern Asia. J Geophys Res: Solid Earth 124:8043–8056

Heaton THE (1986) Isotopic studies of nitrogen pollution in the hydrosphere and atmosphere: a review. Chem Geol 59:87–102

Heck PR, Huberty JM, Kita NT, Ushikubo T, Kozdon R, Valley JW (2011) SIMS analysis of silicon and oxygen isotope ratios for quartz from Archean and Paleoproterozoic banded iron formations. Geochim Cosmochim Acta 75:5879–5891

Hedges RM, Reynard LM (2007) Nitrogen isotopes and the trophic level of humans in archaeology. J Archaeolog Sci 34:1240–1262

Heimann A, Beard BL, Johnson CM (2008) The role of volatile exsolution and sub-solidus fluid/rock interactions in producing high $^{56}Fe/^{54}Fe$ ratios in siliceous igneous rocks. Geochim Cosmochim Acta 72:4379–4396

Hemingway JD, Olson H, Turchyn AV, Tipper ET, Bickle MJ, Johnston DT (2020) Triple oxygen isotope insight into terrestrial pyrite oxidation. PNAS 117:7650–7657

Hemming NG, Hanson GN (1992) Boron isotopic composition in modern marine carbonates. Geochim Cosmochim Acta 56:537–543

Hemsing F, Hsieh YT, Bridgestock L, Spooner PT, Robinson LF, Frank N, Henderson GM (2018) Barium isotopes in cold-water corals. Earth Planet Sci Lett 491:183–192

Henchiri S, Gaillardet J, Dellinger M, Bouchez J, Spencer RG (2016) Temporal variations of riverine dissolved lithium isotope signatures unveil contrasting weathering regimes in low-relief Central Africa. Geophys Res Lett 43:4391–4399

Hendry KR, Andersen MB (2013) The zinc isotopic composition of siliceous marine sponges: investigating nature's sediment traps. Chem Geol 354:33–41

Hendry KR, Georg RB, Rickaby R, Robinson LR, Halliday AN (2010) Deep ocean nutrients during the last glacial maximum deduced from sponge silicon isotopic compositions. Earth Planet Sci Lett 292:290–300

Hendry KR, Brzezinski MA (2014) Using silicon isotopes to understand the role of the Southern Ocean in modern and ancient biogeochemistry and climate. Quaternary Sci Rev 89:13–26

Henehan MJ, 11 others (2013) Calibration of the boron isotope proxy in the planktonic foraminifera *Globigerinoides ruber* for use in palaeo-$CO_2$ reconstruction. Earth Planet Sci Lett 364:111–122

Herbel MJ, Johnson TM, Oremland RS, Bullen TD (2000) Fractionation of selenium isotopes during bacterial respiratory reduction of selenium oxyanions. Geochim Cosmochim Acta 64:3701–3710

Herrmann AD, Kendall B, Algeo TJ, Gordon GW, Wasylenki LE, Anbar AD (2012) Anomalous molybdenum isotope trends in Upper Pennsylvanian euxinic facies: significance for the use of $\delta^{98}Mo$ as a global marine redox proxys. Chem Geol 324–325:87–98

Hervig RL, Moore GM, Williams LB, Peacock SM, Holloway JR, Roggensack K (2002) Isotopic and elemental partitioning of boron between hydrous fluid and silicate melt. Am Mineral 87:769–774

Hesse R, Egeberg PK, Frape SK (2006) Chlorine stable isotope ratios as tracer for pore-water advection rates in a submarine gas-hydrate field: implication for hydrate concentration. Geofluids 6:1–7

Hettmann K, Marks MA, Kreissig K, Zack T, Wenzel T. Rehkämper M, Jacob DE, Markl G (2014) The geochemistry of Tl and its isotopes during magmatic and hydrothermal processes: the peralkaline Ilimaussaq complex, southwest Greenland. Chem Geol 366:1–13

Heuser A, Eisenhauer A (2008) The calcium isotope composition ($\delta^{44/40}$Ca) of NIST SRM 915a and NIST SRM 1486. Geostand Newslett J Geostand Geoanal 32:311–315

Heuser A, Tütken T, Gussone N, Galer SJG (2011) Calcium isotopes in fossil bones and teeth—diagenetic versus biogenic origin. Geochim Cosmochim Acta 75:3419–3433

Hiess J, Condon DJ, McLean N, Noble SR (2012) $^{238}U/^{235}U$ systematics in terrestrial uranium-bearing minerals. Science 335:1610–1614

Higgins JA, Schrag DP (2010) Constraining magnesium cycling in marine sediments using magnesium isotopes. Geochim Cosmochim Acta 74:5039–5053

Hill P, Schauble E (2008) Modeling the effects of bond environment on equilibrium iron isotope fractionation in ferric aquo-chloro complexes. Geochim Cosmochim Acta 72:1939–1958

Hill P, Schauble E, Shahar A, Tonui E, Young ED (2009) Experimental studies of equilibrium iron isotope fractionation in ferric aquo-chloro complexes. Geochim Cosmochim Acta 73:2366–2381

Hill P, Schauble E, Young ED (2010) Effects of changing solution chemistry on $Fe^{3+}/Fe^{2+}$ isotope fractionation in aqueous Fe–Cl solution. Geochim Cosmochim Acta 74:6669–6705

Hin RC, Schmidt MW, Bourdon B (2012) Experimental evidence for the absence of iron isotope fractionation between metal and silicate liquids at 1GPA and 1250–1300 °C and its cosmochemical consequences. Geochim Cosmochim Acta 93:164–181

Hin RC, Burkhardt C, Schmidt MW, Bourdon B, Kleine T (2013) Experimental evidence for Mo isotope fractionation between metal and silicate liquids. Earth Planet Sci Lett 379:38–48

Hin RC, Burnham AD, Gianolio D, Walter MJ, Elliott T (2019) Molybdenum isotope fractionation between $Mo^{4+}$ and $Mo^{6+}$ in silicate liquid and metallic Mo. Chem Geol 504:177–189

Hindshaw RS, Reynolds BC, Wiederhold JG, Kiczka M, Kretzschmar R, Bourdon B (2013) Calcium isotope fractionation in alpine plants. Biogeochemistry 112:373–388

Hindshaw RS, Reynolds BC, Wiederhold JG, Kretzschmar R, Bourdon B (2011) Calcium isotopes in a proglacial weathering environment: Damma glacier, Switzerland. Geochim Cosmochim Acta 75:106–118

Hindshaw RS, Tosca R, Gout TL, Farnan I, Tosca NJ, Tipper ET (2019) Experimental constraints on Li isotope fractionation during clay formation. Geochim Cosmochim Acta 250:219–237

Hinojosa JL, Brown ST, Chen J, DePaolo DJ, Paytan A, Shen SZ, Payne J (2012) Evidence for end-Permian ocean acidification from calcium isotopes in biogenic apatite. Geology 40:743–746

Hippler D, Buhl D, Witbaard R, Richter DK, Immenhauser A (2009) Towards a better understanding of magnesium-isotope ratios from marine skeletal carbonates. Geochim Cosmochim Acta 73:6134–6146

Hippler D, Eisenhauer A, Nägler TF (2006) Tropical Atlantic SST history inferred from Ca isotope thermometry over the last 140 ka. Geochim Cosmochim Acta 70:90–100

Hirst C, Opfergelt S, Gaspard F, Hendry KR, Hatton JE, Welch S, McKnight DM, Lyons WB (2020) Silicon isotopes reveal a non-glacial source of silicon to Crescent Stream, McMurdo Dry Valleys. Antarctica. Front Earth Sci 2020:00229

Hitzfeld KL, Gehre M, Richnow HH (2011) A novel online approach to the determination of isotope ratios for organically bound chlorine, bromine and sulphur. Rapid Commun Mass Spectr 25:3114–3122

Hoare L, Klaver M, Saji NS, Gillies J, Parkinson IJ, Lissenberg J, Millet MA (2020) Melt chemistry and redox conditions control titanium isotope fractionationm during magmatic differentiation. Geochim Cosmochim Acta 282:38–54

Hoering T, Parker PL (1961) The geochemistry of the stable isotopes of chlorine. Geochim Cosmochim Acta 23:186–199

Hofmann A, Bekker A, Dirks P, Gueguen B, Rumble D, Rouxel O (2014) Comparing orthomagmatic and hydrothermal mineralization models for komatiite-hosted nickel deposits in Zimbabwe using multiple-sulfur, iron and nickel isotope data. Miner Deposita 49:75–100

Hofmann MEG, Pack A (2010) Technique for high-precision analysis of triple oxygen isotope ratios in carbon dioxide. Anal Chem 82:4357–4361

Holland G, Sherwood-Lollar B, Li L, Lacrampe-Couloume G, Slater GF (2013) Deep fracture fluids isolated in the crust since the Precambrian era. Nature 497:357–360

Holmden C (2009) Ca isotope study of Ordovician dolomite, limestone, and anhydrite in the Williston basin: implications for subsurface dolomitization and local Ca cycling. Chem Geol 268:180–188

Holmden C, Belanger N (2010) Ca isotope cycling in a forested ecosystem. Geochim Cosmochim Acta 74:995–1015

Holmstrand H, Unger M, Carrizo D, Andersson P, Gustafsson Ö (2010) Compound specific bromine isotope analysis of brominated diphenyl ethers using GC-ICP-MC-MS. Rapid Commun Mass Spectr 24:2135–2142

Homoky WB, Severmann S, Mills RA, Statham PJ, Fones GR (2011) Pore-fluid Fe isotopes reflect the extent of benthic Fe redox recycling: evidence from continental shelf and deep-sea sediments. Geology 37:751–754

Hood AS, Planavsky NJ, Wallace MW, Wang X, Bellefroid EJ, Gueguen B, Cole DB (2016) Integrated geochemical-petrographic insights from component-selective $\delta^{238}U$ of Cryogenian marine carbonates. Geology 44:935–938

Hopp T, Fischer-Gödde M, Kleine T (2016) Ruthenium stable isotope measurements by double spike MC-ICP-MS. JAAS 31:1515–1526

Hopp T, Fischer-Gödde M, Kleine T (2018) Ruthenium isotope fractionation in protoplanetary cores. Geochim Cosmochim Acta 223:75–89

Horan K, Hilton RG, McCoy-West AJ, Selby D, Tipper ET, Hawley S, Burton KW (2020) Unravelling the controls on the molybdenum isotope ratios of river waters. Geochem Persp Lett 13:1–6

Horita J (1988) Hydrogen isotope analysis of natural waters using an $H^2$-water equilibration method: a special implication to brines. Chem Geol 72:89–94

Horita J, Driesner T, Cole DR (1999) Pressure effect on hydrogen isotope fractionation between brucite and water at elevated temperatures. Science 286:1545–1547

Horita J, Wesolowski DJ (1994) Liquid-vapor fractionation of oxygen and hydrogen isotopes of water from the freezing to the critical temperature. Geochim Cosmochim Acta 58:3425–3437

Horita J, Wesolowski DJ, Cole DR (1993) The activity-composition relationship of oxygen and hydrogen isotopes in aqueous salt solutions. I. Vapor-liquid water equilibration of single salt solutions from 50 to 100 °C. Geochim Cosmochim Acta 57:2797–2817

Horn I, von Blanckenburg F, Schoenberg R, Steinhoefel G, Markl G (2006) In situ iron isotope ratio determination using UV-femtosecond laser ablation with application to hydrothermal ore formation processes. Geochim Cosmochim Acta 70:3677–3688

Horner TJ, Schönbächler M, Rehkämper M et al (2010) Ferromanganese crusts as archives of deep-water Cd isotope composition. Geochem Geophys Geosyst 11:Q04001

Horner T, Rickaby R, Henderson G (2011) Isotopic fractionation of cadmium into calcite. Earth Planet Sci Lett 312:243–253

Horner TJ, Kinsley CW, Nielsen SC (2015) Barium isotopic fractionation in seawater mediated by barite cycling and oceanic circulation. Earth Planet Sci Lett 430:511–522

Horner TJ, Pryer HV, Nielsen SG, Crockford PW, Gauglitz JM, Wing BA, Ricketts RD (2017) Pelagic barite precipitation at micromolar ambient sulfate. Nature Comm 8:1342

Horst A, Andersson P, Thornton BJ, Holmstrand H, Wishkerman A, Keppler F, Gustafsson Ö (2014) Stable bromine isotope composition of methyl bromide released from plant matter. Geochim Cosmochim Acta 125:186–195

Horst A, Thornton BJ, Holmstrand H, Andersson P, Crill PM, Gustafsson Ö (2013) Stable bromine isotopic composition of atmospheric $CH_3Br$. Tellus Ser B Chem Phys Meteor 65:21040

Horton F, Nielsen S, Shu Y, Gagnon A, Blusztajn J (2021) Thallium isotopes reveal brine activity during carbonatite magmatism. Geochem Geophys Geosys (in press, e009472)

Hsieh YT, Henderson GM (2017) Barium stable isotopes in the global ocean: tracer of Ba inputs and utilization. Earth Planet Sci Lett 473:269–278

Hsieh YT, Bridgestock L, Scheuermann PP, Seyfried WE, Henderson GM (2021) Barium isotopes in mid-ocean ridge hydrothermal vent fluids: a source of isotopically heavy Ba to the ocean. Geochim Cosmochim Acta 292:348–363

Hu GX, Rumble D, Wang PL (2003) An ultraviolet laser microprobe for the in-situ analysis of multisulfur isotopes and its use in measuring Archean sulphur isotope mass-independent anomalies. Geochim Cosmochim Acta 67:3101–3118

Hu Y, Teng FZ, Zhang HF, Xiao Y, Su BX (2016) Metasomatism-induced mantle magnesium isotopic heterogeneity: evidence from pyroxenites. Geochim Cosmochim Acta 185:88–111

Hu Y, Teng FZ, Plank T, Chauvel C (2020) Potassium isotopic heterogeneity in subducting oceanic plates. Sci Adv 6:eabb2472

Hu Y, Teng FZ, Chauvel C (2021) Potassium isotope evidence for sedimentary input to the mantle source of Lesser Antilles lavas. Geochim Cosmochim Acta 295:98–111

Hu JY, Dauphas N, Tissot FL, Yokochi R, Ireland TJ (2018) Beyond REE abundance patterns: REEstable isotopic compositions. Goldschmidt 2018 Abstract

Huang S, Farkas J, Jacobsen S (2010) Calcium isotopic fractionation between clinopyroxene and orthopyroxene from mantle peridotites. Earth Planet Sci Lett 292:337–344

Huang KJ, Teng FZ, Wei GJ, Ma JL, Bao ZY (2012) Adsorption- and desorption-controlled magnesium isotope fractionation during extreme weathering of basalt in Hainan Island, China. Earth Planet Sci Lett 359–360:73–83

Huang KJ, Teng FZ, Elsenouy A, Li WY, Bao ZY (2013) Magnesium isotope variations in loess: origins and implications. Earth Planet Sci Lett 374:60–70

Huang F, Wu Z, Huang S, Wu F (2014) First-principles calculations of equilibrium silicon isotope fractionation among mantle minerals. Geochim Cosmochim Acta 140:509–520

Huang J, Li SG, Xiao Y, Ke S, Li WY, Tian Y (2015) Origin of low $\delta^{26}$Mg Cenozoic basalts from South China Block and their geodynamic implications. Geochim Cosmochim Acta 164:298–317

Huang J, Liu SA, Wörner G, Yu H, Xiao Y (2016a) Copper isotope behavior during extreme magma differentiation and degassing: a case study on Laacher See phonolite tephra (East Eifel, Germany). Contr Mineral Petrol 171:76

Huang J, Liu SA, Gao Y, Xiao Y, Chen S (2016b) Copper and zinc isotope systematics of altered oceanic crust at IODP Site 1256 in the eastern equatorial Pacific. J Geophy Res Solid Earth 121:7086–7100

Huang J, Zhang XC, Chen S, Tang L, Wörner G, Yu H, Huang F (2018) Zinc isotope systematics of Kamchatka-Aleutian arc magmas controlled by mantle melting. Geochim Cosmochim Acta 238:85–101

Huang J, Ackerman L, Zhang XC, Huang F (2019) Mantle Zn isotopic heterogeneity caused by melt-rock reaction: evidence from Fe-rich peridotites and pyroxenites from the Bohemian Massif, Central Europe. J Geophys Res Solid Earth 124:3588–3604

Huang T, Moos SB, Boyle EA (2021) Trivalent chromium isotopes in the eastern tropical North Pacific oxygen-deficient zone. PNAS 118:e1918605118

Huang TY, Teng FZ, Rudnick RL, Chen XY, Hu Y, Liu YS, Wu FY (2020) Heterogeneous potassium isotopic composition of the upper continental crust. Geochim Cosmochim Acta 278:122–136

Huh Y, Chan L-H, Zhang L, Edmond JM (1998) Lithium and its isotopes in major world rivers; implications for weathering and the oceanic budget. Geochim Cosmochim Acta 62:2039–2051

Huh Y, Chan LH, Edmond JM (2001) Lithium isotopes as a probe of weathering processes: Orinoco river. Earth Planet Sci Lett 194:189–199

Ibanez-Mejia M, Tissot FL (2019) Extreme Zr stable isotope fractionation during magmatic fractional crystallization. Sci Adv 5:eaax8648

Icopini GA, Anbar AD, Ruebush SS, Tien M, Brantley SL (2004) Iron isotope fractionation during microbial reduction of iron: the importance of adsorption. Geology 32:205–208

Ikehata K, Hirata T (2012) Copper isotope characteristics of copper-rich minerals from the Horoman peridotite complex, Hokkaido, Northern Japan. Econ Geol 107:1489–1497

Immenhauser A, Buhl D, Richter D, Niedermayer A, Riechelmann D, Dietzel M, Schulte U (2010) Magnesium isotope fractionation during low-Mg calcite precipitation in a limestone cave—field study and experiments. Geochim Cosmochim Acta 74:4346–4364

Inglis EC, Creech JB, Deng Z, Moynier (2018) High-precision zirconium stable isotope measurements of geological reference materials as measured by double-spike MC-ICPMS. Chem Geol 493:544–552

Inglis EC, Moynier F, Creech J, Deng Z, Day JM, Teng FZ, Bizzarro M, Jackson M, Savage P (2019) Isotopic fractionation of zirconium during magmatic differentiation and the stable isotope composition of the silicate earth. Geochim Cosmochim Acta 250:311–323

Ingraham NL, Criss RE (1998) The effect of vapor pressure on the rate of isotopic exchange between water and vapour. Chem Geol 150:287–292

Ingri J, Malinovsky D, Rodushkin I, Baxter DC, Widerlund A, Andersson P, Gustafsson O, Forsling W, Ohlander B (2006) Iron isotope fractionation in river colloidal matter. Earth Planet Sci Lett 245:792–798

Ionov DA, Qi YH, Kang JT, Golovin AV, Oleinikov OB, Zheng W, Anbar AD, Zhang ZF, Huang F (2019) Calcium isotopic signatures of carbonatite and silicate metasomatism, melt percolation and crystal recycling in the lithospheric mantle. Geochim Cosmochim Acta 248:1–13

Izbicki JA, Bullen TD, Martin P, Schroth B (2012) Delta chromium-53/52 isotopic composition of native and contaminated groundwater, Mojave Desert, USA. Appl Geochem 27:841–853

James RH, Palmer MR (2000) The lithium isotope composition of international rock standards. Chem Geol 166:319–326

Jang JH, Mathur R, LiermannLJ Ruebush S, Brantley SL (2008) An iron isotope signature related to electron transfer between aqueous ferrous iron and goethite. Chem Geol 250:40–48

Janssen DJ, Rickli J, Quay PD, White AE, Nasemann P, Jaccard SI (2020) Biological control of chromium redox and stable isotope composition in the surface ocean. Global Biogeochem Cycles 34:e2019GB006397

Jaouen K, Pons ML, Balter V (2013) Iron, copper and zinc isotopic fractionation up mammal trophic chains. Earth Planet Sci Lett 374:164–172

Jaouen K, Beasley M, Schoeninger M, Hublin JJ, Richards MP (2016) Zinc isotope ratios of bones and teeth as new dietary indicators: results from a modern food web (Koobi Fory, Kenya). Sci Reports 6:26281

Javoy M, Pineau F, Delorme H (1986) Carbon and nitrogen isotopes in the mantle. Chem Geol 57:41–62

Jeffcoate AB, Elliott T, Kasemann SA, Ionov D, Cooper K, Brooker R (2007) Li isotope fractionation in peridotites and mafic melts. Geochim Cosmochim Acta 71:202–218

Jeffcoate AB, Elliott T, Thomas A, Bouman C (2004) Precise, small sample size determination of lithium isotope isotopic compositions of geological reference materials and moders seawater by MC-ICP-MS. Geostand Geoanal Res 28:161–172

Jendrzejewski N, Eggenkamp HGM, Coleman ML (2001) Characterisation of chlorinated hydrocarbons from chlorine and carbon isotopic compositions: scope of application to environmental problems. Appl Geochem 16:1021–1031

Jia Y (2006) Nitrogen isotope fractionations during progressive metamorphism: a case study from the Paleozoic Cooma metasedimentary complex, southeastern Australia. Geochim Cosmochim Acta 70:5201–5214

Jiang SY, Palmer MR (1998) Boron isotope systematics of tourmaline from granites and tourmalines: a synthesis. Eur J Mineral 10:1253–1265

Joe-Wong C, Weaver KL, Brown ST, Maher K (2021) Chromium isotope fractionation during reduction of chromium(vi) by iron(II/III)-bearing clay minerals. Geochim Cosmochim Acta 292:235–253

John SG, Geis RW, Saito MA, Boyle EA (2007a) Zinc isotope fractionation during high-affinity and low-affinity ziinc transport by the marine diatom *Thalassiosira oceanica*. Limnol Oceanogr 52:2710–2714

John SG, Park JG, Zhang Z, Boyle EA (2007b) The isotopic composition of some common forms of anthropogenic zinc. Chem Geol 245:61–69

John SG, Rouxel OJ, Craddock PR, Engwall AM, Boyle EA (2008) Zinc stable isotopes in seafloor hydrothermal vent fluids and chimneys. Earth Planet Sci Lett 269:17–28

John SG, Adkins J (2012) The vertical distribution of iron stable isotopes in the North Atlantic near Bermuda. Global Biogeochem Cycles 26:GB2034

John SG, Helgoe J, Townsend E (2018) Biogeochemical cycling of Zn and Cd and their stable isotopes in the Eastern Tropical South Pacific. Mar Chem 201:256–262

John T, Layne GD, Haase KM, Barnes JD (2010) Chlorine isotope evidence for crustal recycling into the Earth's mantle. Earth Planet Sci Lett 298:175–182

Johnson AC, Aarons SM, Dauphas N, Nie NX, Zeng H, Helz RT, Romaniello SJ, Anbar AD (2019) Titanium isotope fractionation in Kilauea Iki lava lake driven by oxide crystallization. Geochim Cosmochim Acta 264:180–190

Johnson TM (2004) A review of mass-dependent fractionation of selenium isotopes and implications for other heavy stable isotopes. Chem Geol 204:201–214

Johnson CM, Beard BL (1999) Correction of instrumentally produced mass fractionation during isotopic analysis of Fe by thermal ionization mass spectrometry. Int J Mass Spectr 193:87–99

Johnson CM, Skulan JL, Beard BL, Sun H, Nealson KH, Braterman PS (2002) Isotopic fractionation between Fe(III) and Fe(II) in aqueous solutions. Earth Planet Sci Lett 195:141–153

Johnson CM, Beard BL, Roden EE (2008) The iron isotope fingerprints of redox and biogeochemical cycling in modern and ancient Earth. Ann Rev Earth Planet Sci 36:457–493

Johnson CM, Beard B, Weyer S (2020) Iron geochemistry: an isotopic perspective. Springer (2020)

Johnson TM, Bullen TD (2003) Selenium isotope fractionation during reduction by Fe(II)-Fe(III) hydroxide-sulfate (green rust). Geochim Cosmochim Acta 67:413–419

Johnson TM, Herbel MJ, Bullen TD, Zawislanski PT (1999) Selenium isotope ratios as indicators of selenium sources and oxyanion reduction. Geochim Cosmochim Acta 63:2775–2783

Johnston DT (2011) Multiple sulphur isotopes and the evolution of the Earth's sulphur cycle. Earth Sci Rev 106:161–183

Johnston DT, Farquhar J, Wing BA, Kaufman AJ, Canfield DE, Habicht KS (2005) Multiple sulphur isotope fractionations in biological systems: a case study with sulphate reducers and sulphur disproportionators. Am J Sci 305:645–660

Jouvin D, Louvat P, Juillot F, Marechal CN, Benedetti MF (2009) Zinc isotopic fractionation: why organic matters. Environ Sci Tech 43:5747–5754

Jouvin D, Weiss DJ, Mason TF, Bravin MN, Louvat P, Zhao F, Ferec F, Hinsinger P, Benedetti MF (2012) Stable isotopes of Cu and Zn in higher plants: evidence for Cu reduction at the root surface and two conceptional models for isotopic fractionation processes. Environ Sci Technol 46:2652–2660

Juillot F, Marechal C, Ponthieu M, Cacaly S, Morin G, Benedetti M, Hazemann JL, Proux O, Guyot F (2008) Zn isotopic fractionation caused by sorption on goethite and 2-Lines ferrihydrite. Geochim Cosmochim Acta 72:4886–4900

Jørgensen BB, Böttcher MA, Lüschen H, Neretin LN, Volkov II (2004) Anaerobic methane oxidation and a deep $H_2S$ sink generate isotopically heavy sulfides in Black Sea sediments. Geochim Cosmochim Acta 68:2095–2118

Kaplan IR, Rittenberg SC (1964) Microbiological fractionation of sulphur isotopes. J Gen Microbiol 34:195–212
Kang JT, Zhu HL, Liu YF, Liu F, Wu F, Hao YT, Zhi XC, Zhang ZF, Huang F (2016) Calcium isotopic composition of mantle xenoliths and minerals from Eastern China. Geochim Cosmochim Acta 174:334–335
Kang JT, Ionov DA, Liu F, Zhang CL, Golovin AV, Qin LP, Zhang ZF, Huang F (2017) Calcium isotopic fractionation in mantle peridotites by melting and metasomatism and Ca isotope composition of the bulk silicate earth. Earth Planet Sci Lett 474:128–137
Kang JT, Ionov DA, Zhu HL, Liu F, Zhang ZF, Liu Z, Huang F (2019) Calcium isotope sources and fractionation during melt-rock interaction in the lithospheric mantle: evidence from pyroxenites, wehrlites, and eclogites. Chem Geol 524:272–282
Kasemann SA, Jeffcoate AB, Elliott T (2005a) Lithium isotope composition of basalt glass reference material. Ann Chem 77:5251–5257
Kasemann SA, Hawkesworth CJ, Prave AR, Fallick AE, Pearson PN (2005b) Boron and calcium isotope composition in Neproterozoic carbonate rocks from Namibia: evidence for extreme environmental change. Earth Planet Sci Lett 231:73–86
Kasemann SA, Schmidt D, Pearson P et al (2008) Biological and ecological insights into Ca isotopes in planktic foraminifera as a paleotemperature proxy. Earth Planet Sci Lett 271:292–302
Kashiwabara T, Kubo S, Tanaka M, Senda R, Iizuka T, Tanimizu M, Takahashi Y (2017) Stable isotope fractionation of tungsten during sdsorption on Fe and Mn (oxyhydr)oxides. Geochim Cosmochim Acta 204:52–67
Kast ER, Stolper DA, Auderset A, Higgins JA, Ren H, Wang XT, Martinez-Garcia A, Haug GH, Sigman DM (2019) Nitrogen isotope evidence for expanded ocean suboxia in the early Cenozoic. Science 364:386–389
Kato C, Moynier F, Foriel J (2015a) Ga isotopes in terrestrial and meteoritical samples. Goldschmidt 2015, abstract
Kato C, Moynier F, Foriel J, Teng FZ, Puchtel IS (2017) The gallium isotopic composition of the bulk silicate earth. Chem Geol 448:164–172
Kato C, Moynier F, Valdes MC, Dhaliwal JK, Day JM (2015b) Extensive volatile loss during formation and differentiation of the Moon. Nat Commun 6:7617. https://doi.org/10.1038/ncomms8617
Kaufmann RS, Long A, Bentley H, Davis S (1984) Natural chlorine isotope variations. Nature 309:338–340
Kaufmann RS, Long A, Bentley H, Campbell DJ (1986) Chlorine isotope distribution of formation water in Texas and Louisiana. Bull Am Assoc Petrol Geol 72:839–844
Kelley SP, Fallick AE (1990) High precision spatially resolved analysis of $\delta^{34}S$ in sulphides using a laser extraction technique. Geochim Cosmochim Acta 54:883–888
Kelley KD, Wilkinson JJ, Chapman JB, Crowther HL, Weiss DJ (2009) Zinc isotopes in sphalerite from base metal deposits in the Red Dog district, northern Alaska. Econ Geol 104:767–773
Kemp ALW, Thode HG (1968) The mechanism of the bacterial reduction of sulphate and of sulphite from isotopic fractionation studies. Geochim Cosmochim Acta 32:71–91
Kendall B, Dahl TW, Anbar AD (2017) The stable isotope geochemistry of molybdenum. Rev Mineral Geochem 82:683–732
Kendall C, Grim E (1990) Combustion tube method for measurement of nitrogen isotope ratios using calcium oxide for total removal of carbon dioxide and water. Anal Chem 62:526–529
Kendall C (1998) Tracing nitrogen sources and cycling in catchments. In: Kendall C, McDonnell JJ (eds) Isotope tracers in catchment hydrology. Elsevier Science, Amsterdam, pp 519–576
Kerstel ER, Gagliardi G, Gianfrani L, Meijer HA, van Trigt R, Ramaker R (2002) Determination of the $^{2}H/^{1}H$, $^{17}O/^{16}O$ and $^{18}O/^{16}O$ isotope ratios in water by means of tunable diode laser spectroscopy at 1.39 μ. Spectrochim Acta A 58:2389–2396

Kiczka M, Wiederhold JG, Kraemer SM, Bourdon B, Kretzschmar R (2010) Iron isotope fractionation during Fe uptake and translocation in Alpine plants. Environ Sci Techn 44:6144–6150

Kieffer SW (1982) Thermodynamic and lattice vibrations of minerals: 5. Application to phase equilibria, isotopic fractionation and high-pressure thermodynamic properties. Rev Geophys Space Phys 20:827–849

Kim ST, O'Neil JR (1997) Equilibrium and nonequilibrium oxygen isotope effects in synthetic carbonates. Geochim Cosmochim Acta 61:3461–3475

Kim ST, Mucci A, Taylor BE (2007) Phosphoric acid fractionation factors for calcite and aragonite between 25 and 75 °C. Chem Geol 246:135–146

Kimball BE, Mathur R, Dohnalkova AC, Wall AJ, Runkel RL, Brantley SL (2009) Copper isotope fractionation in acid mine drainage. Geochim Cosmochim Acta 73:1247–1263

Kipp MA, Stüeken EE, Bekker A, Buick R (2017) Selenium isotopes record expensive marine suboxia during the Great Oxidation Event. PNAS 114:875–880

Kirshenbaum I, Smith JS, Crowell T, Graff J, McKee R (1947) Separation of the nitrogen isotopes by the exchange reaction between ammonia and solutions of ammonium nitrate. J Chem Phys 15:440–446

Kita NT, Ushikubo T, Fu B, Valley JW (2009) High precision SIMS oxygen isotope analysis and the effect of sample topography. Chem Geol 264:43–57

Kitchen JW, Johnson TM, Bullen TD, Zhu J, Raddatz A (2012) Chromium isotope fractionation factors for reduction of Cr(VI) by aqueous Fe(II) and organic molecules. Geochim Cosmochim Acta 89:190–201

Kitchen NE, Valley JW (1995) Carbon isotope thermometry in marbles of the Adirondack Mountains, New York. J Metamorph Geol 13:577–594

Kiyosu Y, Krouse HR (1990) The role of organic acid in the abiogenic reduction of sulfate and the sulfur isotope effect. Geochemical J 24:21–27

Kleine T, Touboul M, Bourdon B, Nimmo F, Mezger K, Palme H, Jacobsen S, Yin Q-Z, Halliday AN (2009) Hf-W chronology of the accretion and early evolution of asteroids and terrestrial planets. Geochim Cosmochim Acta 73:5150–5188

Klochko K, Kaufman AJ, Yao W, Byrne RH, Tossell JA (2006) Experimental measurement of boron isotope fractionation in seawater. Earth Planet Sci Lett 248:276–285

Kohn MJ, Schoeninger MJ, Valley JW (1996) Herbivore tooth oxygen isotope compositions: effects of diet and physiology. Geochim Cosmochim Acta 60:3889–3896

Kohn MJ, Valley JW (1998a) Oxygen isotope geochemistry of amphiboles: isotope effects of cation substitutions in minerals. Geochim Cosmochim Acta 62:1947–1958

Kohn MJ, Valley JW (1998b) Effects of cation substitutions in garnet and pyroxene on equilibrium oxygen isotope fractionations. J Metam Geol 16:625–639

Kohn MJ, Valley JW (1998c) Obtaining equilibrium oxygen isotope fractionations from rocks: theory and examples. Contr Mineral Petrol 132:209–224

Kolodny Y, Luz B, Navon O (1983) Oxygen isotope variations in phosphate of biogenic apatites, I. Fish bone apatite—rechecking the rules of the game. Earth Planet Sci Lett 64:393–404

Kolodny Y, Torfstein A, Weiss-Sarusi K, Zakon Y, Halicz L (2017) $^{238}U$–$^{235}U$–$^{234}U$ fractionation between tetravalent and hexavalent uranium in seafloor phosphorites. Chem Geol (in press)

Konter JG, Pietruszka AJ, Hanan BB, Finlayson VA, Craddock PR, Jackson MG, Dauphas N (2017) Unusual $\delta^{56}Fe$ values in Samoan rejuvenated lavas generated in the mantle. Earth Planet Sci Lett 450:221–232

König S, Wille M, Voegelin A, Schoenberg R (2016) Molybdenum isotope systematics in subduction zones. Earth Planet Sci Lett 447:95–102

Kowalski PM, Jahn S (2011) Prediction of equilibrium Li isotope fractionation between minerals and aqueous solutions at high p and T: an efficient ab initio approach. Geochim Cosmochim Acta 75:6112–6123

Kowalski N, Dellwig O et al (2013a) Pelagic molybdenum concentration anomalies and the impact of sediment resuspension on the molybdenum budget in two tidal systems of the North Sea. Geochim Cosmochim Acta 119:198–211

Kowalski PM, Wunder B, Jahn S (2013b) Ab initio prediction of equilibrium boron isotope fractionation between minerals and aqueous fluids at high p and T. Geochim Cosmochim Acta 101:285–301

Kozdon R, Kita RN, Huberty JM, Fournelle JH, Johnson CA, Valley JW (2010) In situ sulfur isotope analysis of sulfide minerals by SIMS: precision and accuracy with application to thermometry of similar to 3.5 Ga Pilbara cherts. Chem Geol 275:243–253

Krabbe N, Kruijer TS, Kleine T (2017) Tungsten stable isotope compositions of terrestrial samples and meteorites determined by double spike MC-ICPMS. Chem Geol 450:135–144

Krabbenhöft A, Eisenhauer A et al (2010) Constraining the marine strontium budget with natural isotope fractionations ($^{87}Sr/^{86}Sr$, $\delta^{88/86}Sr$) of carbonates, hydrothermal solutions and river waters. Geochim Cosmochim Acta 74:4097–4109

Krabbenhöft A, Fietzke J, Eisenhauer A, Liebetrau V, Böhm F, Vollstaedt H (2009) Determination of radiogenic and stable strontium isotope ratios ($^{87}Sr/^{86}Sr$; $\delta^{88/86}Sr$) by thermal ionization mass spectrometry applying an $^{87}Sr/^{84}Sr$ double spike. J Anal at Spectr 24:1267–1271

Kritee K, Barkay T, Blum JD (2009) Mass-dependent stable isotope fractionation of mercury during *mer* mediated microbial degradation of monoethylmercury. Geochim Cosmochim Acta 73:1285–1296

Kritee K, Blum JD, Johnson MW, Bergquist BA, Barkay T (2007) Mercury stable isotope fractionation during reduction of Hg(II) to Hg(0) by mercury resistant microorganisms. Environ Sci Technol 41:1889–1895

Krouse HR, Thode HG (1962) Thermodynamic properties and geochemistry of isotopic compounds of selenium. Can J Chem 40:367–375

Krouse HR, Viau CA, Eliuk LS, Ueda A, Halas S (1988) Chemical and isotopic evidence of thermochemical sulfate reduction by light hydrocarbon gases in deep carbonate reservoirs. Nature 333:415–419

Ku TCW, Walter LM, Coleman ML, Blake RE, Martini AM (1999) Coupling between sulfur recycling and syndepositional carbonate dissolution: evidence from oxygen and sulfur isotope composition of pore water sulfate, South Florida Platform, USA. Geochim Cosmochim Acta 63:2529–2546

Kueter N, Schmidt MW, Lilley MD, Bernasconi SM (2019) Experimental determination of equilibrium $CH_4$–$CO_2$–CO carbon isotope fractionation factors (300–1200 °C). Earth Planet Sci Lett 506:64–75

Kunzmann M, Halverson GP, Sossi PA, Raub TD, Payne JL, Kirby J (2013) Zn isotope evidence for immediate resumption of primary productivity after snowball Earth. Geology 41:27–30

Kurzawa T, König S, Labidi J, Yierpan A, Schoenberg R (2017) A method for Se isotope analysisof low ng-level geological samples via double spike and hydride generation MC-ICP-MS. Chem Geol 466:219–228

Kurzawa T, König S, Alt JC, Yierpan A, Schoenberg R (2019) The role of subduction recycling on the selenium isotope signature of the mantle: constraints from Mariana arc lavas. Chem Geol 513:239–249

Kurzweil F, Wille M, Schoenberg R, Taubald H, Van Kranendonk MJ (2015) Continuously increasing $\delta^{98}Mo$ values in Neoarchean black shales and iron formations from the Hamersley Basin. Geochim Cosmochim 164:523–542

Kurzweil F, Münker C, Tusch J, Schoenberg R (2018) Accurate stable tungsten isotope measurements of natural samples using a $^{180}W$–$^{183}W$ double spike. Chem Geol 476:407–417

Kurzweil F, Münker C, Grupp M, Braukmüller N, Fechtner L, Christian M, Hohl SV, Schoenberg R (2019) The stable tungsten isotope composition of modern igneous reservoirs. Geochim Cosmochim Acta 251:176–191

Kurzweil F, Münker C, Hoffmann JE, Tusch J, Schoenberg R (2020) Stable W isotope evidence for redistribution of homogeneous $^{182}W$ anomalies in SW Greenland. Geochem Perspec Lett 14:53–57

Lacan F, Francois R, Ji Y, Sherrell RM (2006) Cadmium isotopic composition in the ocean. Geochim Cosmochim Acta 70:5104–5118

Lai YJ, Pogge von Strandmann P, Dohmen R, Takazawa E, Elliott T (2015) The influence of melt infiltration on the Li and Mg isotopic composition of the Horoman peridotite massif. Geochim Cosmochim Acta 164:318–332

Lam P, Lavik G, Jensen MM, van de Vossenberg J, Schmid M, Woebken D, Gutierrez D, Amann R, Jetten MS, Kuypers MM (2009) Revising the nitrogen cycle in the Peruvian oxygen minimum zone. PNAS 106:4752–4757

Lambelet M, Rehkämper M, van de Flierdt T, Xue Z, Kreissig K, Coles B, Porecelli D, Andersson P (2013) Isotopic analysis of Cd in the mixing zone of Siberian rivers with the Arctic Ocean—new constraints on marine Cd cycling and the isotopic composition of riverine Cd. Earth Planet Sci Lett 361:64–73

Land LS (1980) The isotopic and trace element geochemistry of dolomite: the state of the art. In: Concepts and models of dolomitization. Soc Econ Paleontol Min Spec Publ 28:87–110

Larson PB, Maher K, Ramos FC, Chang Z, Gaspar M, Meinert LD (2003) Copper isotope ratios in magmatic and hydrothermal ore-forming processes. Chem Geol 201:337–350

Lau KV, Maher K, Altiner D, Kelley BM, Kump LR, Lehrmann DJ, Silva-Tamayo JC, Weaver KL, Yu M, Payne JL (2016) Marine anoxia delayed Earth system recovery after the end-Permian extinction. PNAS 113:2360–2365

Lau KV and 12 others (2017) The influence of seawater carbonate chemistry, mineralogy and diagenesis on calcium isotope variations in Lower-Middle Triassic carbonate rocks. Chem Geol 471:13-37

Lauretta DS, Klaue B, Blum JD, Buseck PR (2001) Mercury abundances and isotopic compositions in the Murchison (CM) and Allende (CV) carbonaceous chondrites. Geochim Cosmochim Acta 65:2807–2816

Laws EA, Popp BN, Bidigare RR, Kennicutt MC, Macko SA (1995) Dependence of phytoplankton carbon isotopic composition on growth rate and $CO_{2aq}$: theoretical considerations and experimental results. Geochim Cosmochim Acta 59:1131–1138

Laws EA, Bidigare RR, Popp BN (1997) Effect of growth rate and $CO_2$ concentration on carbon isotope by the marine diatom *Phaeodactylum tricornutum*. Limnol Oceanogr 42:1552–1560

Laycock A, Coles B, Kreissig K, Rehkämper M (2016) High precision $^{142}Ce/^{140}Ce$ stable isotope measurements of purified materials with a focus on $CeO_2$ nanoparticles. J Anal at Spectr 31:297–302

Layne G, Godon A, Webster J, Bach W (2004) Secondary ion mass spectrometry for the determination of $\delta^{37}Cl$. Part I: Ion microprobe analyses of glasses and fluids. Chem Geol 207:277–289

Layton-Matthews D, Leybourne M, Peter JM, Scott SD, Cousens B, Eglington BM (2013) Multiple sources of selenium in ancient seafloor hydrothermal systems: compositional and Se, S and Pb isotopic evidence from volcanic-hosted and volcanic-sediment hosted massive sulphide deposits of the Finlayson Lake district, Yukon, Canada. Geochim Cosmochim Acta 117:313–331

Lazar C, Young ED, Manning CE (2012) Experimental determination of equilibrium nickel isotope fractionation between metal and silicate from 500 °C to 950 °C. Geochim Cosmochim Acta 86:276–295

Le Roux PJ, Shirey SB, Benton L, Hauri EH, Mock TD (2004) In situ, multiple-multiplier, laser ablation ICP-MS measurement of boron isotopic composition ($\delta^{11}B$) at the nanogram level. Chem Geol 203:123–138

Lee DC, Halliday AN (1996) Hf-W isotopic evidence for rapid accretion and differentiation in the early solar system. Science 274:1876–1879

Leeman WP, Tonarini S, Chan LH, Borg LE (2004) Boron and lithium isotopic variations in a hot subduction zone—the southern Washington cascades. Chem Geol 212:101–124
Lehmann M, Siegenthaler U (1991) Equilibrium oxygen- and hydrogen-isotope fractionation between ice and water. J Glaciol 37:23–26
Lemarchand D, Gaillardet J, Lewin E, Allègre CJ (2002) Boron isotope systematics in large rivers: implications for the marine boron budget and paleo-pH reconstruction over the Cenozoic. Chem Geol 190:123–140
Lemarchand E, Schott J, Gaillardet J (2005) Boron isotopic fractionation related to boron sorption on humic acid and the structure of surface complexes formed. Geochim Cosmochim Acta 69:3519–3533
Lemarchand D, Wasserburg GJ, Papanastassiou DA (2004) Rate-controlled calcium isotope fractionation in synthetic calcite. Geochim Cosmochim Acta 68:4665–4678
Lepak R, Yin R, Krabbenhoft DP, Ogorek JM, DeWild JF, Holsen TM, Hurley JP (2015) Use of stable isotope signatures to determine mercury sources in the Great Lakes. Environ Sci Techn Lett 2:335–341
Letolle R (1980) Nitrogen-15 in the natural environment. In: Fritz P, Fontes JCh (eds) Handbook of environmental isotope geochemistry. Elsevier, Amsterdam, pp 407–433
Li D, Li ML, Liu WR, Qin ZZ, Liu SA (2018) Cadmium isotope ratios as standard solutions and geological reference materials measured by MC-ICP-MS. Geostand Geoanal Res 42:593–605
Li S, Li W, Beard BL, Ramo ME, Wang X, Chen Y, Chen J (2019a) K isotopes as a tracer for continental weathering and geological K cycling. PNAS 116:8740–8745
Li SG, Wei Y, 15 others (2016a) Deep carbon cycles constrained by a large-scale mantle Mg isotope anomaly in eastern China. Nat Sci Rev https://doi.org/10.1093/nsr/nww070
Li SZ, Zhu XK, Wu LH, Luo YM (2016b) Cu isotope compositions in *Elsholtzia splendens*: Influence of soil condition and growth period on Cu isotopic fractionation in plant tissue. Chem Geol 444:49–58
Li W, Chakraborty S, Beard BL, Romanek CS, Johnson CM (2012) Magnesium isotope fractionation during precipitation of inorganic calcite under laboratory conditions. Earth Planet Sci Lett 333–314:304–316
Li W, Jackson SE, Pearson NJ, Alard O, Chappell BW (2009a) The Cu isotope signature of granites from the Lachlan Fold Belt, SE Australia. Chem Geol 258:38–49
Li W, Beard BL, Li S (2016c) Precise measurement of stable potassium isotope ratios using a single focusing collision cell multi-collector ICP-MS. JAAS 31:1023–1029
Li W, Kwon KD, Li S, Beard BL (2017) Potassium isotope fractionation between K-salts and saturated aqueous solutions at room temperature: laboratory experiments and theoretical calculations. Geochim Cosmochim Acta 214:1–13
Li WY, Teng FZ, Ke S, Rudnick R, Gao S, Wu FY, Chappell B (2010) Heterogeneous magnesium isotopic composition of the upper continental crust. Geochim Cosmochim Acta 74:6867–6884
Li WY, Teng FZ, Wing BA, Xiao Y (2014) Limited magnesium isotope fractionation during metamorphic dehydration in metapelites from the Onawa contact aureole, Maine. Geochem Geophys Geosys 15(10). https://doi.org/10.1002/2013GC004992
Li Y, Wang W, Wu Z, Huang S (2019) First-principles investigation of equilibrium K isotope fractionation among K-bearing minerals. Geochim Cosmochim Acta 264:30–42
Li X, Zhao H, Tang M, Liu Y (2009) Theoretical prediction for several important equilibrium Ge isotope fractionation factors and geological implications. Earth Planet Sci Lett 287:1–11
Li X, Liu Y (2010) First principles study of Ge isotopic fractionation during adsorption onto Fe (III)-oxyhydroxidessurfaces. Chem Geol 278:15–22
Li X, Liu Y (2011) Equilibrium Se isotope fractionation parameters: a first principle study. Earth Planet Sci Lett 304:113–120
Liang YH, Halliday AN, Siebert C, Fitton JG, Burton KW, Wang KL, Harvey J (2017) Molybdenum isotope fractionation in the mantle. Geochim Cosmochim Acta 199:91–111

Liebscher A, Meixner A, Romer R, Heinrich W (2005) Liquid-vapor fractionation of boron and boron isotopes: experimental calibration at 400 °C/23 Mpa to 450 °C/42 Mpa. Geochim Cosmochim Acta 69:5693–5704

Lin M, Zhang X, Li M, Xu Y, Zhang Z, Tao J, Su B, Liu L, Shen Y, Thiemens M (2018) Five-S-isotope evidence of two distict mass-independent sulfur isotope effects and implications for the modern and Archean atmospheres. PNAS 115(34):8541–8546

Little SH, Vance D, Walker-Brown C, Landing WM (2014a) The oceanic mass balance of copper and zinc isotopes, investigated by analysis of their inputs, and outputs to ferromanganese oxide sediments. Geochim Cosmochim Acta 125:653–672

Little SH, Sherman DM, Vance D, Hein JR (2014b) Molecular controls on Cu and Zn isotopic fractionation in Fe-Mn crusts. Earth Planet Sci Lett 396:213–222

Little SH, Vance D, McManus J, Severmann S (2016) Key role of continental margin sediments in the oceanic mass balance of Zn and Zn isotopes. Geology 44:207–210

Little SH, Archer C, Milne A, Schlosser C, Achterberg EP, Lohan MC, Vance D (2018) Paired dissolved and particulate phase Cu isotope distributions in the South Atlantic. Chem Geol 502:29–43

Little SH, Archer C, McManus J, Najorka J, Wegorzewski AV, Vance D (2020) Towards balancing the oceanic Ni budget. Earth Planet Sci Lett 547:116461

Liu H, Wang K, Sun WD, Xiao Y, Xue YY, Tuller-Ross B (2020) Extremely light K in subducted low-T altered oceanic crust: implications for K-recycling in subduction zone. Geochim Cosmochim Acta 277:206–223

Liu R, Hu L, Humayun M (2017) Natural variations in the rhenium isotopic composition of meteorites. Meteoritics Planet Sci 52:479–492

Liu SA, Teng FZ, Yang W, Wu FY (2011) High-temperature inter-mineral magnesium isotope fractionation in mantle xenoliths from the North China craton. Earth Planet Sci Lett 308:131–140

Liu SA, Huang J, Liu J, Wörner G, Yang W, Tang YJ, Chen Y, Tang L, Zheng J, Li S (2015) Copper isotope composition of the silicate Earth. Earth Planet Sci Lett 427:95–103

Liu SA, Wang ZZ, Li SG, Huang J, Yang W (2016) Zinc isotopic evidence for a large-scale carbonated mantle beneath eatern China. Earth Planet Sci Lett 444:169–178

Liu S, Li Y, Ju Y, Liu J, Liu J, Shi Y (2018) Equilibrium nickel isotope fractionation in nickel sulfide minerals. Geochim Cosmochim Acta 222:1–16

Liu YF, Qi HW, Bi XW, Hu RZ, Qi LK, Yin RS, Tang YY (2021) Mercury and sulfur isotopic composition of sulfides from sediment-hosted lead-zinc deposits in Lanping basin, Southwestern China. Chem Geol 559:119910

Livermore B, Dahl TW, Bizarro M, Connelly JN (2020) Uranium isotope compositions of biogenic carbonates—implications for the U uptake in shells and the application of the paleo-ocean oxygenation proxy. Geochim Cosmochim Acta 287:50–64

Lobo L, Degryse P, Shortland A, Vanhaeke F (2013) Isotopic analysis of antimony using multi-collector ICP-mass spectrometry for provenance determination of Roman glass. J Anal at Spectrom 28:1213–2129

Long A, Eastoe CJ, Kaufmann RS, Martin JG, Wirt L, Fincey JB (1993) High precision measurement of chlorine stable isotope ratios. Geochim Cosmochim Acta 57:2907–2912

Longinelli A, Craig H (1967) Oxygen-18 variations in sulfate ions in sea-water and saline lakes. Science 156:56–59

Louvat P, Bouchez J, Paris G (2011) MC-ICP-MS isotope measurements with direct injection nebulisation (d-DIHEN): optimisation and application to boron in seawater and carbonate samples. Geostand Geoanal Res 35:75–88

Lu D, Liu Q, Zhang T, Cai Y, Yin Y, Jiang G (2016) Stable silver isotope fractionation in the natural transformation process of silver nanoparticles. Nat Nanotechnol 11:682–687

Luais B (2012) Germanium chemistry and MC-ICPMS isotopic measurements of Fe–Ni, Zn alloys and silicate matrices: insights into deep Earth processes. Chem Geol 334:295–311

Luck JM, Ben Othman D, Albarede F (2005) Zn and Cu isotopic variations in chondrites and iron meteorites: early solar nebula reservoirs and parent-body processes. Geochim Cosmochim Acta 69:5351–5363

Lugmair GW, Shukolyukov A (1998) Early solar system timescales according to $^{53}Mn$–$^{53}Cr$ systematics. Geochim Cosmochim Acta 62:2863–2886

Lundstrom CC, Chaussidon M, Hsui AT, Keleman P, Zimmermann M (2005) Observations of Li isotope variations in the Trinity ophiolite: evidence for isotope fractionation by diffusion during mantle melting. Geochim Cosmochim Acta 69:735–751

Luo Y, Dabek-Zlotorzynska E, Celo V, Muir D, Yang L (2010) Accurate and precise determination of silver isotope fractionation in environmental samples by multicollector-ICPMS. Anal Chem 82:3922–3928

Luz B, Barkan E (2010) Variations of $^{17}O/^{16}O$ and $^{18}O/^{16}O$ in meteoric waters. Geochim Cosmochim Acta 74:6276–6286

Lyons TW, Reinhard CT, Planavsky NJ (2014) The rise of oxygen in Earth' s early ocean and atmosphere. Nature 506:307–315

Lécuyer C, Grandjean P, Reynard B, Albarede F, Telouk P (2002) $^{11}B/^{10}B$ analysis of geological materials by ICP-MS Plasma 54: application to bron fractionation between brachiopod calcite and seawater. Chem Geol 186:45–55

Ma J, Hintelmann H, Kirk JL, Muir DC (2013a) Mercury concentrations and mercury isotope compositionin lake sediment cores Chem Geol 336:96–102

Ma J, Wei G, Liu Y, Ren Z, Xu Y, Yang Y (2013b) Precise measurement of stable neodymium isotopes of geological materials by using MC-ICP-MS. JAAS 28:1926–1931

Machel HG, Krouse HR, Sassen P (1995) Products and distinguishing criteria of bacterial and thermochemical sulfate reduction. Appl Geochemistry 10:373–389

Macris CA, Young ED, Manning CE (2013) Experimental determination of equilibrium magnesium isotope fractionation between spinel, forsterite and magnesite from 600 to 800 ° C. Geochim Cosmochim Acta 118:18–32

Macris CA, Manning CE, Young ED (2015) Crystal chemical constraints on inter-mineral Fe isotope fractionation and implications for Fe isotope disequilibrium in San Carlos mantle xenoliths. Geochim Cosmochim Acta 154:168–185

Magenheim AJ, Spivack AJ, Michael PJ, Gieskes JM (1995) Chlorine stable isotope composition of the oceanic crust: implications for earth's distribution of chlorine. Earth Planet Sci Lett 131:427–432

Magenheim AJ, Spivack AJ, Volpe C, Ranson B (1994) Precise determination of stable chlorine isotope ratios in low-concentration natural samples. Geochim Cosmochim Acta 58:3117–3121

Mahata S, Bhattacharya SK, Wang CH, Liang MC (2013) Oxygen isotope exchange between $O_2$ and $CO_2$ over hot platinum: an innovative technique for measuring $\Delta^{17}O$ in $CO_2$. Anal Chem 85:6894–6901

Maher K, Larson P (2007) Variation in copper isotope ratios and controls on fractionation in hypogene skarn mineralization at Coroccohuayco and Tintaya, Peru. Econ Geol 102:225–237

Maher K, Jackson S, Mountain B (2011) Experimental evaluation of the fluid-mineral fractionation of Cu isotopes at 250 °C and 300 °C. Chem Geol 286:229–239

Malinovskiy D, Moens L, Vanhaecke F (2009) Isotopic fractionation of Sn during methylation and demethylation in aqueous solution. Environ Sci Tech 43:4399–4404

Manaka T, Araoka D, Yoshimura T, Hossain HM, Nishio Y, Suzuki A, Kawahata H (2017) Downstream and Seasonal Changes of Lithium Isotope Ratios in the Ganges-Brahmaputra river systèm. Geochem Geophys Geosys 18:3003–3015

Manzini M, Bouvier AS, 11 others (2017) SIMS chlorine isotope analyses in melt inclusions from arc settings. Chem Geol 449:112–122

Maréchal CN, Télouk P, Albarède F (1999) Precise analysis of copper and zinc isotopic compositions by plasma-source mass spectrometry. Chem Geol 156:251–273

Maréchal CN, Albarede F (2002) Ion-exchange fractionation of copper and zinc isotopes. Geochim Cosmochim Acta 66:1499–1509

Marin-Carbonne J, Robert F, Chaussidon M (2014) The silicon and oxygen isotope compositions of Precambrian cherts: a record of oceanic paleo-temperatures? Precambrian Res 247:223–234
Mariotti A, Germon JC, Hubert P, Kaiser P, Letolle R, Tardieux P (1981) Experimental determination of nitrogen kinetic isotope fractionation: some principles, illustration for the denitrification and nitrification processes. Plant Soil 62:413–430
Markl G, Lahaye Y, Schwinn G (2006a) Copper isotopes as monitors of redox processes in hydrothermal mineralization. Geochim Cosmochim Acta 70:4215–4228
Markl G, von Blanckenburg F, Wagner T (2006b) Iron isotope fractionation during hydrothermal ore deposition and alteration. Geochim Cosmochim Acta 70:3011–3030
Marriott CS, Henderson GM, Belshaw NS, Tudhope AW (2004) Temperature dependence of $\delta^7$Li, $\delta^{44}$Ca and Li/Ca during growth of calcium carbonate. Earth Planet Sci Lett 222:615–624
Marschall HR, Altherr R, Kalt A, Ludwig T (2008) Detrital, metamorphic and metasomatic tourmaline in high-pressure metasediments from Syros (Greece): intra-grain boron isotope patterns determined by secondary-ion mass spectrometry. Contr Mineral Petrol 155:703–717
Marschall HR, Jiang SY (2011) Tourmaline isotopes: no element left behind. Elements 7:313–319
Marschall HR, Wanless VD, Shimizu N, Pogge von Strandmann PA, Elliott T, Monteleone BD (2017) The boron and lithium isotopic composition of mid-ocean ridge basalts and the mantle. Geochim Cosmochim Acta 207:102–138
Marschall H (2018) Foster G (2018) Boron Isotopes. Springer, The fifth element
Marschall HR, Tang M (2020) High-temperature processes: is it time for lithium isotopes? Elements 16:247–252
Martin JE, Vance D, Balter V (2014) Natural variation of magnesium isotopes in mammal bones and teeth from two South African trophic chains. Geochim Cosmochim Acta 130:12–20
Martin JE, Vance D, Balter V (2015) Magnesium stable isotope ecology using mammal tooth enamel. PNAS 112:430–435
Martin E, Bindeman I, Balan E, Palandri J (2017) Hydrogen isotope determination by TC/EA technique in application to volcanic glass as a window into secondary hydration. J Volcan Geother Res 348:49–61
Marty B, Humbert F (1997) Nitrogen and argon isotopes in oceanic basalts. Earth Planet Sci Lett 152:101–112
Marty B, Zimmermann L (1999) Volatiles (He, C, N, Ar) in mid-ocean ridge basalts: assesment of shallow-level fractionation and characterization of source composition. Geochim Cosmochim Acta 63:3619–3633
Mason TFD et al (2005) Zn and Cu isotopic variability in the Alexandrinka volcanic-hosted massive sulphide (VHMS) ore deposit, Urals, Russia. Chem Geol 221:170–187
Mason AH, Powell WG, Bankoff HA, Mathur R, Bulatovic A, Filipovic V, Ruiz J (2016) Tin isotope characterization of bronze artefacts of the central Balkans. J Archaeolog Sci 69:110–117
Mason AH, Powell WG, Bankoff HA, Mathur R, Price M, Bulatovic A, Filipovic V (2020) Provenance of tin in the Late Bronze Age Balkans based on probabilistic and spatial analysis of tin isotopes. J Archaeolog Sci 122:105181
Mathur R, Ruiz J, Titley S, Liermann L, Buss H, Brantley S (2005) Cu isotopic fractionation in the supergene environment with and without bacteria. Geochim Cosmochim Acta 69:5233–5246
Mathur R, Titley S, Barra F, Brantley S, Wilson M, Phillips A, Munizaga F, Maksaev V, Vervoort J, Hart G (2009) Exploration potential of Cu isotope fractionation in porphyry copper deposits. J Geochem Explor 102:1–6
Mathur R, Brantley S, Anbar A, Munizaga F, Maksaev R, Vervoort J, Hart G (2010a) Variations of Mo isotopes from molybdenite in high-temperature hydrothermal ore deposits. Mineral Deposita 45:43–50
Mathur R, Dendas M, Titley S, Phillips A (2010b) Patterns in the copper isotope composition of minerals in porphyry copper deposits in southwestern United States. Econ Geol 105:1457–1467

Mathur R, Arribas A, Megaw P, Wilson M, Stroup S, Meyer-Arrivillaga D, Arribas I (2018) Fractionation of silver isotopes in native silver explained by redox reactions. Geochim Cosmochim Acta 224:313–326

Matt P, Powell W, Mathur R, deLorraine WF (2020) Zn-isotopic evidence for fluid assisted ore remobilization at the Balmat zinc mine, NY. Ore Geology Rev 116:103227

Matthews A, Goldsmith JR, Clayton RN (1983a) Oxygen isotope fractionation between zoisite and water. Geochim Cosmochim Acta 47:645–654

Matthews A, Goldsmith JR, Clayton RN (1983b) On the mechanics and kinetics of oxygen isotope exchange in quartz and feldspars at elevated temperatures and pressures. Geol Soc Am Bull 94:396–412

Matthews DE, Hayes JM (1978) Isotope-ratio-monitoring gas chromatography-mass spectrometry. Anal Chem 50:1465–1473

Mavromatis V, van Zuilen K, Purgstaller B, Baldermann A, Nägler TF, Dietzel M (2016) Barium isotope fractionation during witherite ($BaCO_3$) dissolution, precipitation and at equilibrium. Geochim Cosmochim Acta 190:72–78

Mavromatis V, Gonzalez AG, Dietzel M, Schott J (2019) Zinc isotope fractionation during inorganic precipitation of calcite—towards a new pH proxy. Geochim Cosmochim Acta 244:99–112

Mazza SE, Stracke A, Gill JB, Kimura JI, Kleine T (2020) Tracing dehydration and melting of the subducted slab with tungsten isotopes in arc lavas. Earth Planet Sci Lett 530:115942

McClelland JW, Montoya JP (2002) Trophic relationships and the nitrogen isotope composition of amino acids in plankton. Ecology 83:2173–2180

McCoy-West AJ, Millet MA, Burton KW (2017) The neodymium stable isotope composition of the silicate Earth and chondrites. Earth Planet Sci Lett 480:121–132

McCoy-West AJ, Millet MA, Nowell GM, Burton KW (2018) The neodymium stable isotope composition of oceanic basalts. Goldschmidt 2018, Abstract

McCoy-West AJ, Millet MA, Nowell GM, Nebel O, Burton KW (2020a) Simultaneous measurement of neodymium stable and radiogenic isotopes from a single aliqot using a double spike. JAAS 35:388–402

McCoy-West A, Millet MA, Burton KW (2020b) The neodymium stable isotope composition of the oceanic crust: reconciling the mismatch between erupted mid-ocean ridge basalts and lower crustal gabbros. Frontiers Earth Sci 8:25

McCrea JM (1950) On the isotopic chemistry of carbonates and a paleotemperature scale. J Chem Phys 18:849–857

McCready RGL (1975) Sulphur isotope fractionation by Desulfovibrio and Desulfotomaculum species. Geochim Cosmochim Acta 39:1395–1401

McCready RGL, Kaplan IR, Din GA (1974) Fractionation of sulfur isotopes by the yeast Saccharomyces cerevisiae. Geochim Cosmochim Acta 38:1239–1253

McIlivin MR, Altabet MA (2005) Chemical conversion of nitrate and nitrite to nitrous oxide for nitrogen and oxygen isotopic analysis in freshwater and seawater. Anal Chem 77:5589–5595

McKibben MA, Riciputi LR (1998) Sulfur isotopes by ion microprobe. In: Applications of microanalytical techniques to understanding mineralizing processes. Rev Econ Geol 7:121–140

McMahon KW, McCarthy MD (2016) Embracing variability in amino acid $\delta^{15}N$ fractionation: mechanisms, implications and applications for trophic ecology. Ecosphere 7:e01511

McManus J et al (2006) Molybdenum and uranium geochemistry in continental margin sediments: palaeoproxy potential. Geochim Cosmochim Acta 70:4643–4662

McManus J, Nägler T, Siebert C, Wheat CG, Hammond D (2002) Oceanic molybdenum isotope fractionation: diagenesis and hydrothermal ridge flank alteration. Geochem Geophys Geosyst 3:1078. https://doi.org/10.1029/2002GC000356

McMullen CC, Cragg CG, Thode HG (1961) Absolute ratio of $^{11}B/^{10}B$ in Searles Lake borax. Geochim Cosmochim Acta 23:147

Mead C, Herckes P, Majestic BJ, Anbar AD (2013) Source apportionment of aerosol iron in the marine environment using iron isotope analysis. Geophys Res Lett 40:5722–5727

Méheut M, Lazzeri M, Balan E, Mauri F (2007) Equilibrium isotopic fractionation in the kaolinite, quartz, water system: prediction from first-principles density functional theory. Geochim Cosmochim Acta 71:3170–3181

Méheut M, Lazzeri M, Balan E, Mauri F (2009) Structural control over equilibrium silicon and oxygen isotope fractionation: a first principles density-functional theory study. Chem Geol 258:28–37

Méheut M, Lazzeri M, Balan E, Mauri F (2010) First-principles calculation of H/D isotopic fractionation between hydrous minerals and water. Geochim Cosmochim Acta 74:3874–3882

Méheut M, Schauble EA (2014) Silicon isotope fractionation in silicate minerals: insights from first-principles models of phyllosilicate, albite and pyrope. Geochim Cosmochim Acta 134:137–154

Méheut M, Ibanez-Meija M, Tissot LH (2021) Drivers of zirconium isotope fractionation in Zr-bearing phases and melts: the role of vibrational, nuclear field shift and diffusive effects. Geochim Cosmochim Acta 292:203–216

Meier M, Cloquet C, Marty R (2016) Mercury (Hg) in meteorites: variations in abundance, thermal release profile, mass-dependent and mass-independent isotopic fractionation. Geochim Cosmochim Acta 182:55–72

Meng YM, Qi HW, Hu RZ (2015) Determination of germanium isotopic compositions of sulfides by hydride generation MC-ICP-MS and its application to the Pb–Zn deposits in SW China. Ore Geology Rev 65:1095–1109

Miller MF, Franchi IA, Sexton AS, Pillinger CT (1999) High precision $\delta^{17}O$ isotope measurements of oxygen from silicates and other oxides: method and applications. Rapid Commun Mass Spect 13:1211–1217

Miller CA, Peucker-Ehrenbrink B, Ball L (2009) Precise determination of rhenium isotope composition by multi-collector inductively-coupled plasma mass spectrometry. JAAS 24:1069–1078

Miller CA, Peucker-Ehrenbrink B, Schauble EA (2015) Theoretical modeling of rhenium isotope fractionation, natural variations across a black shale weathering profile, and potential as a paleoredox proxy. Earth Planetary Sci Lett 430:339–348

Millet MA, Baker JA, Payne CE (2012) Ultra-precise stable Fe isotope measurements by high resolution multi-collector inductively coupled mass spectrometry with a $^{57}Fe$–$^{58}Fe$ double spike. Chem Geol 304–305:18–25

Millet MA, Dauphas N (2014) Ultra-precise titanium stable isotope measurements by double-spike high resolution MC-ICP-MS. JAAS 29:1444–1458

Millet MA, Dauphas N, Greber ND, Burton KW, Dale CW, Debret B, Macpherson CG, Nowell GM, Williams HM (2016) Titanium stable isotope investigation of magmatic processes on the Earth and Moon. Earth Planet Sci Lett 449:197–205

Millot R, Guerrot C, Vigier N (2004) Accurate and high-precision measurement of lithium isotopes in two reference materials by MC-ICP-MS. Geostand Geoanal Res 28:153–159

Millot R, Vigier N, Gaillardet J (2010a) Behaviour of lithium and its isotopes during weathering in the Mackenzie Basin, Canada. Geochim Cosmochim Acta 74:3897–3912

Millot R, Petelet-Giraud E, Guerrot C, Negrel P (2010b) Multi-isotopic composition ($\delta^7Li$-$\delta^{11}B$-$\delta D$-$\delta^{18}O$) of rainwaters in France: origin and spatio-temporal characterization. Appl Geochem 25:1510–1524

Misra S, Froelich PN (2012) Lithium isotope history of Cenozoic seawater: changes in silicate weathering and reverse weathering. Science 335:818–823

Mitchell K, Mason P, Van Cappellen P, Johnson TM, Gill BC, Owens JD, Diaz J, Ingall E, Reichart GJ, Lyons T (2012) Selenium as paleo-oceanographic proxy: a first assessment. Geochim Cosmochim Acta 89:302–317

Mitchell K, Couture RM, Johnson TM, Mason PRD, Van Cappellen P (2013) Selenium sorption and isotope fractionation: iron(III) oxides versus iron(II) sulfides. Chem Geol 342:21–28

Miyazaki T, Kimura JI, Chang Q (2014) Analysis of stable isotope ratios of Ba by double-spike standard-sample bracketing using multiple-collector inductively coupled plasma mass spectrometry. J Anal at Spectrom 29:483–490

Mizutani Y, Rafter TA (1973) Isotopic behavior of sulfate oxygen in the bacterial reduction of sulfate. Geochem J 6:183–191

Möller K, Schoenberg R, Pedersen RB, Weiss D, Dong S (2012) Calibration of the new certified reference materials ERM-AE633 and ERM-AE647 for copper and IRMM-3702 for zinc isotope amount ratio determinations. Geostand Geoanal Res 36:177–199

Mondillo N, Wilikinson JJ, Boni M, Weiss DJ, Mathur R (2018) A global assessment of Zn isotope fractionation in secondary Zn minerals from sulfide and non-sulfide ore deposits and a model for fractionation control. Chem Geol 500:182–193

Monson KD, Hayes JM (1982) Carbon isotopic fractionation in the biosynthesis of bacterial fatty acids. Ozonolysis of unsaturated fatty acids as a means of determining the intramolecular distribution of carbon isotopes. Geochim Cosmochim Acta 46:139–149

Montoya-Pino C, Weyer S, Anbar AD, Pross J, Oschmann J, van de Schootbrugge B, Arz HW (2010) Global enhancement of ocean anoxia during Oceanic Anoxic Event 2: a quantitative approach using U isotopes. Geology 38:315–318

Mook WG, Bommerson JC, Stavermann WH (1974) Carbon isotope fractionation between dissolved bicarbonate and gaseous carbon dioxide. Earth Planet Sci Lett 22:169–174

Morgan JL, Skulan JL, Gordon GW, Romaniello SJ, Smith SM, Anbar AD (2012) Rapidly assessing changes in bone mineral balance using stable calcium isotopes. PNAS 109:9989–9994

Morgan LE, Santiago Ramos DP, Davidheiser-Kroll B, Faithfull J, Lloyd NS, Ellam RM, Higgins JA (2018) High-precision $^{41}K/^{39}K$ measurements by MC-ICP-MS indicate terrestrial variability of $\delta^{41}K$. JAAS 33:175–186

Moriguti T, Nakamura E (1998) Across-arc variation of Li-isotopes in lavas and implications for crust /mantle recycling at subduction zones. Earth Planet Sci Lett 163:167–174

Moynier F, Bouvier A, Blichert-Toft J, Telouk P, Gasperini D, Albarede F (2006) Europium isotopic variations in Allende CAIs and the nature of mass-dependent fractionation in the solar nebula. Geochim Cosmochim Acta 70:4287–4294

Moynier F, Pichat S, Pons ML, Fike D, Balter V, Albarède F (2008) Isotope fractionation and transport mechanisms of Zn in plants. Chem Geol 267:125–130

Moynier F, Beck P, Jourdan F, Qin QZ, Reimold U, Koeberl C (2009) Isotopic fractionation of Zn in tektites. Earth Planet Sci Lett 277:482–489

Moynier F, Agranier A, Hezel DC, Bouvier A (2010) Sr stable isotope composition of Earth, the Moon, Mars, Vesta and meteorites. Earth Planet Sci Lett 300:359–366

Moynier F, Yin QZ, Schauble E (2011) Isotopic evidence of Cr partitioning into Earth's core. Science 331:1417–1420

Moynier F, Paniello RC, Gounelle M, Albarede F, Beck P, Podosek F, Zanda B (2011) Nature of volatile depletion and genetic relationships in enstatite chondrites and aubrites inferred from Zn isotopes. Geochim Cosmochim Acta 75:297–307

Moynier F, Fujii T (2017) Theoretical isotopic fractionation of magnesium between chlorophylls. Sci Rep 7:6973

Moynier F, Vance D, Fujii T, Savage P (2017) The isotope geochemistry of zinc and copper. Rev Mineral Geochem 82:543–600

Moynier F, Chen J, Zhang K, Cai H, Wang Z, Jackson MG, Day JM (2020) Chondritic mercury isotopic composition of earth and evidence for evaporative equilibrium degassing during the formation of eucrites. Earth Planet Sci Lett 551:116544

Murphy MJ, Stirling CH, Kaltenbach A, Turner SP, Schaefer BF (2014) Fractionation of $^{238}U/^{235}U$ by reduction during low temperature uranium mineralization processes. Earth Planet Sci Lett 388:306–317

Nabelek PI (1991) Stable isotope monitors. In: Contact metamorphism. Rev Mineral 26:395–435

Nabelek PI, Labotka TC (1993) Implications of geochemical fronts in the Notch Peak contact-metamorphic aureole, Utah, USA. Earth Planet Sci Lett 119:539–559
Nägler TF et al (2014) Proposal for an international molybdenum isotope reference standard and data representation. Geostand Geoanal Res 38:149–151
Nägler TF, Eisenhauer A, Müller A, Hemleben C, Kramers J (2000) The $\delta^{44}$Ca-temperature calibration on fossil and cultured *Globigerinoides sacculifer*: new tool for reconstruction of past sea surface temperatures. Geochem Geophys Geosyst G3 1(2000GC000091)
Nägler TF, Siebert C, Lüschen H, Böttcher ME (2005) Sedimentary Mo isotope records across the Holocene fresh-brackish water transition of the Black Sea. Chem Geol 219:283–295
Nägler TF, Neubert N, Böttcher ME, Dellwig O, Schnetger B (2011) Molybedenum isotope fractionation in pelagic euxinia: evidence from the modern Black and Baltic Seas. Chem Geol 289:1–11
Nakada R, Takahashi Y, Tanimizu M (2013a) Isotopic and speciation study of cerium during its solid-water distribution with implication for Ce stable isotope as a paleo-redox proxy. Geochim Cosmochim Acta 103:49–62
Nakada R, Tanimizu M, Takahashi Y (2013b) Difference in the stable isotopic fractionations of Ce, Nd and Sm during adsorption on iron and manganese oxides and its interpretation based on their local structures. Geochim Cosmochim Acta 121:105–119
Nakada R, Takahashi Y, Tanimizu M (2016) Cerium stable isotope ratios in ferromanganese deposits and their potential as a paleo-redox proxy. Geochim Cosmochim Acta 181:89–100
Nakada R, Tanaka M, Tanimizu M, Takahashi Y (2017) Aqueous speciation is likely to control stable isotope fractionation of cerium at varying pH. Geochim Cosmochim Acta 218:273–290
Nakada R, Asakura N, Nagaishi K (2019) Examination of analytical conditions of cerium (Ce) isotope and stable isotope ratio of Ce in geochemical standards. Geochem J 53:293–304
Nakano T, Nakamura E (2001) Boron isotope geochemistry of metasedimentary rocks and tourmalines in a subduction zone metamorphic suite. Phys Earth Planet Inter 127:233–252
Nan X, Wu F, Zhang Z, Hou Z, Huang F, Yu H (2015) High-precision barium isotope measurements by MC-ICP-MS. JAAS 30:2307–2315
Nan XY, Yu HM, Rudnick RL, Gaschnig RM, Xu J, Li WY, Zhang Q, Jin ZD, Li XH, Huang F (2018) Barium isotope composition of the upper continental crust. Geochim Cosmochim Acta 233:33–49
Nanne JA, Millet MA, Burton KW, Dale CW, Nowell GM, Williams HM (2017) High precision osmium stable isotope measurements by double spike MC-ICP-MS and N-TIMS. JAAS 32:749–765
Navarette JU, Borrok DM, Viveros M, Elzey JT (2011) Copper isotope fractionation during surface adsorption and intracellular incorporation by bacteria. Geochim Cosmochim Acta 75:784–799
Nebel O, Mezger K, van Westrenen W (2011) Rubidium isotopes in primitive chondrites: constraints on Earth' s volatile element depletion and lead isotope evolution. Earth Planet Sci Lett 301:1–8
Neddermeyer A (2019) The development and application of uranium, molybdenum, and vanadium stable isotope ratios as redox-proxies in samples from modern times and the early earth. Dissertation, Hannover (2019)
Needham AW, Porcelli D, Russell SS (2009) An Fe isotope study of ordinary chondrites. Geochim Cosmochim Acta 73:7399–7413
Neretin LN, Böttcher ME, Grinenko VA (2003) Sulfur isotope geochemistry of the Black Sea water column. Chem Geol 200:59–69
Neubert N, Heri AR, Voegelin AR, Schlunegger F, Villa IM (2011) The molybdenum isotopic composition in river water; constraints from small catchments. Earth Planet Sci Lett 304:180–190
Neubert N, Nägler TF, Böttcher ME (2008) Sulfidity controls molybdenum isotope fractionation into euxinic sediments: evidence from the modern Black Sea. Geology 36:775–778

Neymark LA, Premo WR, Mel'nikov NN, Emsbo P (2014) Precise determination of $\delta^{88}Sr$ in rocks, minerals and waters by double-spike TIMS: a powerful tool in the study of geological, hydrological and biological processes. J Anal At Spectr 29:65–75
Ni P, Chabot NL, Ryan CJ, Shahar A (2020) Heavy iron isotope composition of iron meteorites explained by core crystallization. Nat Geosci 13:611–615
Nielsen LC, Druhan JL, Yang W, Brown ST, DePaolo DJ (2011a) Calcium isotopes as tracers of biogeochemical processes. In: Handbook of environmental isotope geochemistry. Springer, New York, pp 105–124
Nielsen SG et al (2005) Thallium isotope composition of the upper continental crust and rivers—an investigation of the continental sources of dissolved marine thallium. Geochim Cosmochim Acta 69:2007–2019
Nielsen SG, Rehkämper M, Norman MD, Halliday AN, Harrison D (2006) Thallium isotopic evidence for ferromanganese sediments in the mantle source of Hawaiian basalts. Nature 439:314–317
Nielsen SG, Rehkämper M, Brandon AD, Norman MD, Turner S, O'Reilly SY (2007) Thallium isotopes in Iceland and Azores lavas—implications for the role of altered crust and mantle geochemistry. Earth Planet Sci Lett 264:332–345
Nielsen SG, Mar-Gerrison S, Gannoun A, LaRowe D, Klemm V, Halliday A, Burton KW, Hein JR (2009) Thallium isotope evidence for a permanent increase in marine organic carbon export in the early Eocene. Earth Planet Sci Lett 278:297–307
Nielsen SG, Rehkämper M (2011a) Thallium isotopes and their application to problems in earth and environmental science. In: Baskaran M (ed) Handbook of environmental isotope geochemistry, vol 1. Springer, New York, pp 247–269
Nielsen SG, Goff M, Hesselbo SP, Jenkyns HC, LaRowe DE, Lee CT (2011b) Thallium isotopes in early diagentic pyrite—a paleoredox proxy? Geochim Cosmochim Acta 75:6690–6704
Nielsen SG, Prytulak J, Halliday AN (2011c) Determination of precise and accurate $^{51}V/^{50}V$ isotope ratios by MC-ICP-MS, Part 1: Chemical separation of vanadium and mass spectrometric protocols. Geostand Geoanal Res 35:293–306
Nielsen SG, Prytulak J, Wood BJ, Halliday AN (2014) Vanadium isotopic difference between the silicate Earth and meteorites. Earth Planet Sci Lett 389:167–175
Nielsen SG, Yogodzinski GM, Prytulak J, Plank T, Kay SM, Kay RW, Blusztain J, Owens JD, Auro M, Kading T (2016) Tracking along-arc sediment inputs to the Aleutian arc using thallium isotopes. Geochim Cosmochim Acta 181:217–237
Nielsen SG, Rehkämper M, Prytulak J (2017) Investigation and application of thallium isotope fractionation. Rev Mineral Petrol 82:759–798
Nielsen SG, Horner TJ, Pryer HV, Blusztain J, Shu Y, Kurz MD, Le Roux V (2018) Barium isotope evidence for pervasive sediment recycling in the upper mantle. Sci Adv 4:eaas8675
Nielsen SG, Shu Y, Auro M, Yogodzinski G, Shinjo R, Plank T, Kay SM, Horner TJ (2020) Barium isotope systematics of subduction zones. Geochim Cosmochim Acta 275:1–18
Nishio Y, Nakai S, Yamamoto J, Sumino H, Matsumoto T, Prikhod'ko VS, Arai S (2004) Lithium isotope systematics of the mantle derived ultramafic xenoliths: implications for EM1 origin. Earth Planet Sci Lett 217:245–261
Nitzsche HM, Stiehl G (1984) Untersuchungen zur Isotopenfraktionierung des Stickstoffs in den Systemen Ammonium/Ammoniak und Nitrid/Stickstoff. ZFI Mitt 84:283–291
Noordmann J, Weyer S, Montoya-Pino C, Dellwig O, Neubert N, Eckert S, Paetzel M, Böttcher ME (2015) Uranium and molybdenum isotope systematics in modern euxinic basins: case studies from the central Baltic Sea and the Kyllaren fjord (Norway). Chem Geol 396:182–195
Noordmann J, Weyer S, Georg RB, Jöns S, Sharma M (2016) $^{238}U/^{235}U$ isotope ratios of crustal material, rivers and products of hydrothermal alteration: new insights on the oceanic U isotope mass balance. Isotope Environm Health Stud 52:141–163
Ockert C, Gussone N, Kaufhold S, Teichert BM (2013) Isotope fractionation during Ca exchange on clay minerals in a marine environment. Geochim Cosmochim Acta 112:374–388

O'Connell TC (2017) "Trophic" and "source" amino acids in trophic estimation: a likely metabolic explanation. Oecologia 184:317–326

Oduro H, Harms B, Sintim HO, Kaufman AJ, Cody G, Farquhar J (2011) Evidence of magnetic isotope effects during thermochemical sulfate reduction. PNAS 108:17635–17638

Oelze M, von Blanckenburg F, Hoellen D, Dietzel M, Bouchez J (2014) Si stable isotope fractionation during adsorption and the competition between kinetic and equilibrium isotope fractionation: implications for weathering systems. Chem Geol 380:161–171

Oeser M, Weyer S, Horn I, Schuth S (2014) High-precision Fe and Mg isotope ratios of silicate reference glasses determined in situ by femtosecond LA-MC-ICP-MS and by solution nebulisation MC-ICP-MS. Geostand Geoanal Res 38:311–328

Oeser RA, von Blanckenburg F (2020) Strontium isotopes trace biological activity in the Critical Zone along a climate and vegetation gradient. Chem Geol 558:119861

Oeser M, Dohmen R, Horn I, Schuth S, Weyer S (2015) Processes and time scales of magmatic evolution as revealed by Fe–Mg chemical and isotopic zoning in natural olivines. Geochim Cosmochim Acta 154:130–150

Ohmoto H (1986) Stable isotope geochemistry of ore deposits. Rev Mineral 16:491–559

Ohmoto H, Goldhaber MB (1997) Sulfur and carbon isotopes. In: Barnes HL (ed) Geochemistry of hydrothermal ore deposits, 3rd edn. Wiley Interscience, New York, pp 435–486

Ohmoto H, Rye RO (1979) Isotopes of sulfur and carbon. In: Geochemistry of hydrothermal ore deposits, 2nd edn. Holt Rinehart and Winston, New York

Ohno T, Hirata T (2013) Determination of mass-dependent isotopic fractionation of cerium and neodymium in geochemical samples by MC-ICPMS. Anal Sci 29:47–53

Ono S, Shanks WC, Rouxel OJ, Rumble D (2007) S-33 constraints on the seawater sulphate contribution in modern seafloor hydrothermal vent sulfides. Geochim Cosmochim Acta 71:1170–1182

Ono S, Wing BA, Johnston D, Farquhar J, Rumble D (2006) Mass-dependent fractionation of quadruple sulphur isotope system as a new tracer of sulphur biogeochemical cycles. Geochim Cosmochim Acta 70:2238–2252

Opfergelt S, Georg RB, Delvaux B, Cabidoche YM, Burton KW, Halliday AN (2012) Mechanism of magnesium isotope fractionation in volcanicsoil weathering sequences, Guadeloupe. Earth Planet Sci Lett 341–344:176–185

Owens NJP (1987) Natural variations in $^{15}N$ in the marine environment. Adv Mar Biol 24:390–451

Owens NJP (2019) Application of thallium isotopes. Cambridge University Press

Owens JD, Nielsen SG, Horner TJ, Ostrander CM, Peterson LC (2017) Thallium-isotopic compositions of euxinic sediments as a proxy for global manganese-oxide burial. Geochim Cosmochim Acta 213:291–307

O'Leary MH (1981) Carbon isotope fractionation in plants. Phytochemistry 20:553–567

O'Neil JR, Epstein S (1966) A method for oxygen isotope analysis of milligram quantities of water and some of its applications. J Geophys Res 71:4955–4961

O'Neil JR, Roe LJ, Reinhard E, Blake RE (1994) A rapid and precise method of oxygen isotope analysis of biogenic phosphate. Isr J Earth Sci 43:203–212

O'Neil JR, Taylor HP (1967) The oxygen isotope and cation exchange chemistry of feldspars. Am Mineral 52:1414–1437

O'Neil JR, Truesdell AH (1991) Oxygen isotope fractionation studies of solute-water interactions. In: Stable isotope geochemistry: a tribute to Samuel Epstein. Geochemical Soc Spec Publ 3:17–25

Pack A (2021) Isotopic traces of atmospheric $O_2$ in rocks, minerals and melts. Rev Mineral Geochem 86:217–240

Pack A, Herwartz D (2014) The triple oxygen isotope composition of the Earth mantle and understanding $\Delta^{17}O$ variations in terrestrial rocks and minerals. Earth Planet Sci Lett 390:138–145

Pagani M, Lemarchand D, Spivack A, Gaillardet J (2005) A critical evaluation of the boron isotope-$p_H$ proxy: the accuracy of ancient ocean $p_H$ estimates. Geochim Cosmochim Acta 69:953–961

Page B, Bullen T, Mitchell M (2008) Influences of calcium availability and tree species on Ca isotope fractionation in soil and vegetation. Biogeochemistry 88:1–13

Palmer MR (2017) Boron cycling in subduction zones. Elements 13:237–242

Palmer MR, Slack JF (1989) Boron isotopic composition of tourmaline from massive sulfide deposits and tourmalinites. Contr Mineral Petrol 103:434–451

Palmer MR, Spivack AJ, Edmond JM (1987) Temperature and pH controls over isotopic fractionation during the absorption of boron on marine clays. Geochim Cosmochim Acta 51:2319–2323

Paniello RC, Day JM, Moynier F (2012) Zinc isotopic evidence for the origin of the Moon. Nature 490:376–379

Parendo CA, Jacobsen SB, Wang K (2017) K isotopes as a tracer of seafloor hydrothermal alteration. PNAS 114:1827–1831

Paris G, Sessions A, Subhas AV, Adkins JF (2013) MC-ICP-MS measurement of $\delta^{34}$S and $\Delta^{33}$S in small amounts of dissolved sulphate. Chem Geol 345:50–61

Park R, Epstein S (1960) Carbon isotope fractionation during photosynthesis. Geochim Cosmochim Acta 21:110–126

Parkinson IJ, Hammond SJ, James RH, Rogers NW (2007) High-temperature lithium isotope fractionation: insights from lithium isotope diffusion in magmatic systems. Earth Planet Sci Lett 257:609–621

Payne JL, Turchyn AV, Paytan A, DePaolo DJ, Lehrmann DJ, Yu M, Wei J (2010) Calcium isotope constraints on the end-Permian mass exttinction. PNAS 107:8543–8548

Pearce CR, Parkinson IJ, Gaillardet J, Charlier BL, Mokadem F, Burton KW (2015) Reassessing the stable ($\delta^{88/86}$Sr) and radiogenic ($^{87}$Sr/$^{86}$Sr) strontium isotopic composition of marine inputs. Geochim Cosmochim Acta 157:125–146

Pearson PN, Palmer MR (1999) Middle Eocene seawater pH and atmospheric carbon dioxide. Science 284:1824–1826

Pearson PN, Palmer MR (2000) Atmospheric carbon dioxide concentrations over the past 60 million years. Nature 406:695–699

Pellerin A, Antler G, Holm SA, Findlay AJ, Crockford PW, Turchyn AV, Jorgensen BB, Finster K (2019) Large sulfur isotope fractionation by bacterial sulfide oxidation. Sci Adv 5:eaaw1480

Penniston-Dorland S, Liu XM, Rudnick RL (2017) Lithium isotope geochemistry. Rev Mineral Geochem 82:165–217

Pereira NS, Voegelin AR, Paulukat C, Sial AN, Ferreira VP, Frei R (2016) Chromium-isotope signatures in scleractinian corals from the Rocas Atoll, Tropical South Atlantic. Geobiology 14:54–67

Peterson BJ, Fry B (1987) Stable isotopes in ecosystem studies. Ann Rev Ecol Syst 18:293–320

Piasecki A, Sessions A, Lawson M, Ferreira AA, Neto EVS, Eiler JM (2016) Analysis of the site-specific carbon isotope composition of propane by gas source isotope ratio mass spectrometry. Geochim Cosmochim Acta 188:58–72

Piasecki A, Sessions A, Lawson M, Ferreira AA, Neto EVS, Ellis GS, LewanMD EJM (2018) Position-specific $^{13}$C distributions within propane from experiments and natural gas samples. Geochim Cosmochim Acta 220:110–124

Pichat S, Douchet C, Albarede F (2003) Zinc isotope variations in deep-sea carbonates from the eastern equatorial Pacific over the last 175 ka. Earth Planet Sci Lett 210:167–178

Pistiner JS, Henderson GM (2003) Lithium-isotope fractionation during continental weathering processes. Earth Planet Sci Lett 214:327–339

Planavsky NJ, Reinhard CT, Wang X, Thomson D, McGoldrick P, Rainbird RH, Johnson T, Fischer WW, Lyons TW (2014) Low Mid-Proterozoic atmospheric oxygen levels and the delayed rise of animals. Science 346:635–638

Plessen B, Harlov DE, Henry D, Guidotti CV (2010) Ammonium loss and nitrogen isotopic fractionation in biotite as a function of metamorphic grade in metapelites from western Maine, USA. Geochim Cosmochim Acta 74:4759–4771

Pogge von Strandmann PA, Burton KW, James RH, van Calsteren P, Gislason SR, Sigfusson B (2008) The influence of weathering processes on riverine magnesium isotopes in a basaltic terrain. Earth Planet Sci Lett 276:187–197

Pogge von Strandmann PA, Elliott T, Marschall HR, Coath C, Lai YJ, Jeffcoate AB, Ionov DA (2011) Variations of Li and Mg isotope ratios in bulk chondrites and mantle xenoliths. Geochim Cosmochim Acta 75:5247–5268

Pogge von Strandmann PA, Forshaw J, Schmidt DN (2014) Modern and Cenozoic records of magnesium behaviour from foraminiferal Mg isotopes. Biogeosci Discuss 11:7451–7464

Pogge von Strandmann PA, Stüeken EE, Elliott T, Poulton SW, Dehler CM, Canfield DE, Catling DC (2015) Selenium isotope evidence for progressive oxidation oft he Neoproterozoic biosphere. Nature Commun 6:10157

Pogge von Strandmann PA, Frings PJ, Murphy MJ (2017) Lithium isotope behavior during weathering in the Ganges Alluvial Plain. Geochim Cosmochim Acta 198:17–31

Pogge von Strandmann PA, Kasemann SA, Wimpenny JB (2020) Lithium and lithium isotopes in Earth's surface cyles. Elements 16:253–258

Poitrasson F (2017) Silicon Isotope Geochemistry. Rev Mineral Geochem 82:289–344

Poitrasson F, Freydier R (2005) Heavy iron isotope composition of granites determined by high resolution MC-ICP-MS. Chem Geol 222:132–147

Poitrasson F, Levasseue S, Teutsch N (2005) Significance of iron isotope mineral fractionation in pallasites and iron meteorites for the core-mantle differentiation of terrestrial planets. Earth Planet Sci Lett 234:151–164

Poitrasson F, Roskosz M, Corgne A (2009) No iron isotope fractionation between molten alloys and silicate melt to 2000 °C and 7.7 GPa: experimental evidence and implications for planery differentiation and accretion. Earth Planet Sci Lett 278:376–385

Poitrasson F, Halliday AN, Lee DC, Levasseur S, Teutsch N (2004) Iron isotope differences between Earth, Moon, Mars and Vesta as possible records of contrasted accretion mechanisms. Earth Planet Sci Lett 223:253–266

Poitrasson F, Cruz Vieira L et al (2014) Iron isotope composition of the bulk waters and sediments from the Amazon River basin. Chem Geol 377:1–11

Poitrasson F, Zambardi T, Magna T, Neal CR (2019) A reassessment of the iron isotope composition of the Moon and its implications for the accretion and differentiation of terrestrial planets. Geochim Cosmochim Acta 267:257–274

Pokrovsky OS, Viers J, Emnova EE, Kompantseva EI, Freydier R (2008) Copper isotope fractionation during its interaction with soil and aquatic microorganisms and metal oxy(hydr) oxides: possible structural control. Geochim Cosmochim Acta 72:1742–1757

Pokrovsky OS, Galy A, Schott J, Propovsky GS, Mantoura S (2014) Germanium isotope fractionation during Ge adsorption on goethite and its coprecipitation with Feoxy(hydr)oxides. Geochim Cosmochim Acta 131:138–149

Polyakov VB, Mineev SD, Clayton RN, Hu G, Mineev KS (2005) Determination of tin equilibrium fractionation factors from synchrotron radiation experiments. Geochim Cosmochim Acta 69:5531–5536

Polyakov VB, Clayton RN, Horita J, Mineev SD (2007) Equilibrium iron isotope fractionation factors of minerals: reevaluation from the data of nuclear inelastic resonant X-ray scattering and Mossbauer spectroscopy. Geochim Cosmochim Acta 71:3833–3846

Popp BN, Laws EA, Bidigare RR, Dore JE, Hanson KL, Wakeham SG (1998) Effect of phytoplankton cell geometry on carbon isotope fractionation. Geochim Cosmochim Acta 62:69–77

Porter SJ, Selby D, Cameron V (2014) Characterising the nickel isotopic composition of organic-rich marine sediments. Chem Geol 387:12–21

Post DM (2002) Using stable isotopes to estimate trophic position: models, methods, and assumptions. Ecology 83:703–718
Poulson RL, Siebert C, McManus J, Berelson WM (2006) Authigenic molybdenum isotope signatures in marine sediments. Geology 34:617–620
Poulson-Brucker RL, McManus J, Severmann S, Berelson WM (2009) Molybdenum behaviour during early diagenesis: insights from Mo isotopes. Geochem Geophys Geosys 10(Q06010):1–25
Prentice AJ, Webb EA (2016) The effect of progressive dissolution on the oxygen and silicon isotope composition of opal-A phytoliths: implications for palaeoenvironmental reconstruction. Palaeogeogr Palaeoclimatol Palaeoecol 453:42–51
Pretet C, van Zuilen K, Nägler TF, Reynaud S, Immenhauser A, Böttcher ME, Samankassou E (2015) Constraints on barium isotope fractionation during aragonite precipitation by corals. Depositional Record 1:118–129
Pringle EA, Moynier F, Savage PS, Jackson MG, Moreira M, Day JM (2016) Silicon isotopes reveal recycled oceanic crust in the mantle sources of ocean island basalts. Geochim Cosmochim Acta 189:282–295
Pringle EA, Moynier F (2017) Rubidium isotopic composition of the Earth, meteorites and the Moon: evidence for the origin of volatile loss during planetary accretion. Earth Planet Sci Lett 473:62–70
Prytulak J, Nielsen RG, Halliday AN (2011) Determination of precise and accurate $^{51}V/^{50}V$ isotope ratios by multi-collector ICP-MS, Part 2: isotope composition of six reference materials plus the Allende chondrite and verification tests. Geostand Geoanal Res 35:307–318
Prytulak J, Nielsen SG, Plank T, Barker M, Elliott T (2013a) Assessing the utility of thallium and thallium isotopes for tracing subduction zone inputs to the Mariana arc. Chem Geol 345:139–149
Prytulak J, Nielsen SG et al (2013b) The stable vanadium isotope composition of the mantle and mafic lavas. Earth Planet Sci Lett 365:177–189
Prytulak J, Sossi PA, Halliday AN, Plank T, Savage PS, Woodhead JD (2017a) Stable vanadium isotopes as a redox proxy in magmatic systems? Geochem Perspect Lett 3:75–84
Prytulak J, Brett A, Webb M, Plank T, Rehkämper M, Savage PS, Woodhead J (2017b) Thallium elemental behavior and stable isotope fractionation during magmatic processes. Chem Geol 448:71–83
Puchelt H, Sabels BR, Hoering TC (1971) Preparation of sulfur hexafluoride for isotope geochemical analysis. Geochim Cosmochim Acta 35:625–628
Qi HW, Rouxel O, Hu RZ, Bi XW, Wen HJ (2011) Germanium isotopic systematics in Ge-rich coal from the Lincang Ge deposit, Yunnan. Southwestern China. Chem Geol. 286:252–265
Qi YH, Wu F, Ionov DA, Puchtel IS, Carlson RW, Nicklas RW, Yu HM, Kang JT, Li CH, Huang F (2019) Vanadium isotope composition of the bulk silicate earth: constraints from peridotites and komatiites. Geochim Cosmochim Acta 259:288–301
Qin L, Xia J, Carlson RW, Zhang Q (2015) Chromium stable isotope composition of meteorites. 46thLunar Planet Sci Conference (2015) (abstr)
Qin L, Wang X (2017) Chromium isotope geochemistry. Rev Mineral Geochem 82:379–414
Qin T, Wu F, Wu Z, Huang F (2016) First-principles calculations of equilibrium fractionation of O and Si isotopes in quartz, albite, anorthite, and zircon. Contr Mineral Petrol 171:91
Ra K (2010) Determination of Mg isotopes in chlorophyll *a* for marine bulk phytoplankton from the northwestern Pacific Ocean. Geochem Geophys Geosys 11(12):Q12011. https://doi.org/10.1029/2010GC003350
Ra K, Kitagawa H (2007) Magnesium isotope analysis of different chlorophyll forms in marine phytoplankton using multi-collector ICP-MS. J Anal at Spectrom 22:817–821
Raddatz J, Liebetrau V et al (2013) Stable Sr-isotope, Sr/Ca, Mg/Ca, Li/Ca and Mg/Li ratios in the scleractinian cold-water coral *Lophelia pertusa*. Chem Geol 352:143–152

Rader ST, Mazdab FK, Barton MD (2018) Mineralogical thallium geochemistry and isotope variations from igneous, metamorphic, and metasomatic systems. Geochim Cosmochim Acta 243:42–65

Radic A, Lacan F, Murray JW (2011) Iron isotopes in the seawater of the equatorial Pacific Ocean: new constraints for the oceanic iron cycle. Earth Planet Sci Lett 306:1–10

Rae J (2017) Boron isotopes in foraminifera: systematics, biomineralisation and $CO_2$ reconstruction. In: Boron, the fifth element. Advances in Isotope Geochemistry. Springer (2017)

Ransom B, Spivack AJ, Kastner M (1995) Stable Cl isotopes in subduction-zone pore waters: implications for fluid-rock reactions and the cycling of chlorine. Geology 23:715–718

Rasbury ET, Hemming NG (2017) Boron isotopes: A "Paleo-pH Meter" for tracking ancient atmospheric $CO_2$. Elements 13:243–248

Ratié G, Jouvin D and 10 others (2015) Nickel isotope fractionation during tropical weathering of ultramafic rocks. Chem Geol 402:68–76

Rees CE (1978) Sulphur isotope measurements using $SO_2$ and $SF_6$. Geochim Cosmochim Acta 42:383–389

Rehkämper M, Frank M, Hein JR, Halliday A (2004) Cenozoic marine geochemistry of thallium deduced from isotopic studies of feromanganese crusts and pelagic sediments. Earth Planet Sci Lett 219:77–91

Rehkämper M, Frank M, Hein JR, Porcelli D, Halliday A, Ingri J, Libetrau V (2002) Thallium isotope variations in seawater and hydrogenetic, diagenetic and hydrothermal ferromanganese deposits. Earth Planet Sci Lett 197:65–81

Rehkämper M, Wombacher F, Horner TJ, Xue Z (2011) Natural and anthropogenic Cd isotope variations. In: Baskaran M (ed) Handbook of environmental isotope geochemistry. Springer, New York, pp 125–154

Resongles E, Freydier R, Casiot C, Viers J, Chmeleff J, Elbaz-Poulichet F (2015) Antimony isotopic composition in river waters affected by ancient mining activity. Talanta 144:851–861

Revels BN, Rickli J, Moura CA, Vance D (2021) The riverine flux of molybdenum and its isotopes to the ocean: weathering processes and dissolved-particulate partitioning in the Amazon basin. Earth Planet Sci Lett 559:116773

Revesz K, Böhlke JK, Yoshinari T (1997) Determination of $^{18}O$ and $^{15}N$ in nitrate. Anal Chem 69:4375–4380

Reinhard CT, Planavsky NJ, Wang X, Fischer WW, Johnson TM, Lyons TW (2014) The isotopic composition of authigenic chromium in anoxic sediments: a case study from the Cariaco Basin. Earth Planet Sci Lett 407:9–18

Richter F, Watson EB, Mendybaev RA, Teng FZ (2008) Magnesium isotope fractionation in silcate melts by chemical and thermal diffusion. Geochim Cosmochim Acta 72:206–220

Rickli J, Janssen DJ, Hassler C, Ellwood MJ, Jaccard SL (2019) Chromium biogeochemistry and stable isotope distribution in the Southern Ocean. Geochim Cosmochim Acta 262:188–206

Ripley EM, Dong SF, Li CS, Wasylenki LE (2015) Cu isotope variations between conduit and sheet-style, Ni-Cu-PGE sulfide mineralization in the Midcontinent Rift System, North America. Chem Geol 414:59–68

Rippberger S, Rehkämper VM, Porcelli D, Halliday AN (2007) Cadmium isotope fractionation in seawater—a signature of biological activity. Earth Planet Sci Lett 261:670–684

Robert F, Chaussidon M (2006) A paleotemperature curve for the Precambrian oceans based on silicon isotopes in cherts. Nature 443:969–972

Rodler A, Sanchez-Pastor N, Fernandez-Diaz L, Frei R (2015) Fractionation behavior of chromium isotopes during coprecipitation with calcium carbonate: implications for their use as paleoclimatic proxy. Geochim Cosmochim Acta 164:221–235

Rolison JM, Landing WM, Luke W, Cohen M, Salters VJ (2013) Isotopic composition of species-specific atmospheric Hg in a coastal environment. Chem Geol 336:37–49

Rollion-Bard C, Blamart D, Trebosc J, Tricot G, Mussi A, Cuif JP (2011) Boron isotopes as pH proxy: a new look at boron speciation in deep-sea corals using $^{11}B$ MAS NMR and EELS. Geochim Cosmochim Acta 75:1003–1012

Rollion-Bard C, Erez J (2010) Intra-shell boron isotope ratios in the symbiont-bearing benthic foraminifera *Amphistegina lobifera*: implications for $\delta^{11}B$ vital effects and paleo-pH reconstructions. Geochim Cosmochim Acta 74:1530–1536

Rollion-Bard C, Vigier N, Spezzaferri S (2007) In-situ measurements of calcium isotopes by ion microprobe in carbonates and application to foraminifera. Chem Geol 244:679–690

Romanek CS, Grossman EL, Morse JW (1992) Carbon isotope fractionation in synthetic aragonite and calcite: effects of temperature and precipitation rate. Geochim Cosmochim Acta 56:419–430

Romaniello SJ, Herrmann AD, Anbar AD (2013) Uranium concentrations and $^{238}U/^{235}U$ isotope ratios in modern carbonates from the Bahamas: assessing a novel paleoredox proxy. Chem Geol 362:305–316

Romaniello SJ, Herrmann AD, Anbar AD (2016) Syndepositional diagenetic control of molybdenum isotope variations in carbonate sediments from the Bahamas. Chem Geol 438:84–90

Romer RL, Meixner A (2014) Lithium and boron isotopic fractionation in sedimentary rocks during metamorphism—the role of rock composition and protolith mineralogy. Geochim Cosmochim Acta 128:158–177

Rose EF, Chaussidon M, France-Lanord C (2000) Fractionation of boron isotopes during erosion processes: the example of Himalayan rivers. Geochim Cosmochim Acta 64:397–408

Rosenbaum J, Sheppard SMF (1986) An isotopic study of siderites, dolomites and ankerites at high temperatures. Geochim Cosmochim Acta 50:1147–1150

Roskosz M, Amet Q, Fitoussi C, Dauphas N, Bourdon B, Tissandier L, Hu MY, Said A, Alatas A, Alp EE (2020) Redox and structural controls on tin isotopic fractionations among magmas. Geochim Cosmochim Acta 268:42–55

Rosman JR, Taylor PD (1998) Isotopic compositions of the elements (technical report): commission on atomic weights and isotopic abundances. Pure Appl Chem 70:217–235

Rouxel O, Ludden J, Carignan J, Marin L, Fouquet Y (2002) Natural variations in Se isotopic composition detemined by hydride generation multiple collector inductively coupled plasma mass spectrometry. Geochim Cosmochim Acta 66:3191–3199

Rouxel O, Ludden J, Fouquet Y (2003) Antimony isotope variations in natural sytems and implications for their use as geochemical tracers. Chem Geol 200:25–40

Rouxel O, Fouquet Y, Ludden JN (2004a) Copper isotope systematics of the Lucky Strike, Rainbow and Logatschev seafloor hydrothermal fields on the Mi-Atlantic Ridge. Econ Geol 99:585–600

Rouxel O, Fouquet Y, Ludden JN (2004b) Subsurface processes at the Lucky Strike hydrothermal field, Mid-Atlantic Ridge: evidence from sulfur, selenium and iron isotopes. Geochim Cosmochim Acta 68:2295–2311

Rouxel O, Bekker A, Edwards KJ (2005) Iron isotope constraints on the Archean and Proterozoic ocean redox state. Science 307:1088–1091

Rouxel O, Galy A, Elderfield H (2006) Germanium isotope variations in igneous rocks and marine sediments. Geochim Cosmochim Acta 70:3387–3400

Rouxel O, Toner B, Manganini S, German C (2016) Geochemistry and iron isotope systematics of hydrothermal plume fall-out at EPR 9°50′N. Chem Geol 441:212–234

Rouxel O, Luais B (2017) Germanium isotope geochemistry. Rev Mineral Geochem 82:601–656

Rozanski K, Araguas-Araguas L, Gonfiantini R (1993) Isotopic patterns in modern global precipitation. In: Climate change in continental isotopic records. Geophys Monograph 78:1–36

Rubinson M, Clayton RN (1969) Carbon-13 fractionation between aragonite and calcite. Geochim Cosmochim Acta 33:997–1002

Rudnick RL, Ionov DA (2007) Lithium elemental and isotopic disequilibrium in minerals from peridotite xenoliths from far-east Russia: product of recent melt/fluid-rock interaction. Earth Planet Sci Lett 256:278–293

Rudnicki MD, Elderfield H, Spiro B (2001) Fractionation of sulfur isotopes during bacterial sulfate reduction in deep ocean sediments at elevated temperatures. Geochim Cosmochim Acta 65:777–789

Ruiz J, Mathur R, Young S, Brantley S (2002) Controls of copper isotope fractionation. Geochim Cosmochim Acta Spec Suppl 66:A654

Rumble D, Miller MF, Franchi IA, Greenwood RC (2007) Oxygen three-isotope fractionation lines in terrestrial silicate minerals: an inter-laboratory comparison of hydrothermal quartz and eclogitic garnet. Geochim Cosmochim Acta 71:3592–3600

Russell WA, Papanastassiou DA, Tombrello TA (1978) Ca isotope fractionation on the Earth and other solar system materials. Geochim Cosmochim Acta 42:1075–1090

Rustad JR, Dixon DA (2009) Prediction of iron-isotope fractionation between nematite ($\alpha$-$Fe^2O^3$) and ferric and ferrous iron in aqueous solution from density functional theory. J Phys Chem 113:12249–12255

Rustad JR, Casey WH, Yin QZ, Bylaska EJ, Felmy AR, Bogatko SA, Jackson VE, Dixon DA (2010) Isotopic fractionation of $Mg^{2+}$(aq), $Ca^{2+}$(aq) and $Fe^{2+}$(aq) with carbonate minerals. Geochim Cosmochim Acta 74:6301–6323

Ryan BM, Kirby JK, Degryse F, Harris H, McLaughlin MJ, Scheiderich K (2013) Copper speciation and isotopic fractionation in plants: uptake and translocation mechanism. New Phytol 199:367–378

Rye RO (1974) A comparison of sphalerite-galena sulfur isotope temperatures with filling-temperatures of fluid inclusions. Econ Geol 69:26–32

Ryu JS, Jacobson AD, Holmden C, Lundstrom C, Zhang Z (2011) The major ion, $\delta^{44/40}$Ca, $\delta^{44/42}$Ca and $\delta^{26/24}$Mg geochemistry of granite weathering at pH = 1 and T = 25 °C: power-law processes and the relative reactivity of minerals. Geochim Cosmochim Acta 75:6004–6602

Ryu JS, Vigier N, Derry L, Chadwick OA (2021) Variations of Mg isotope geochemistry in soils over a Hawaiian 4 Myr chronosequence. Geochim Cosmochim Acta 292:94–114

Rüggeburg A, Fietzke J, Liebetrau V, Eisenhauer A, Dullo WC, Freiwald A (2008) Stable strontium isotopes ($\delta^{88/86}$Sr) in cold-water corals—a new proxy for reconstruction of intermediate ocean water temperatures. Earth Planet Sci Lett 269:570–575

Saccocia PJ, Seewald JS, Shanks WC (2009) Oxygen and hydrogen isotope fractionation in serpentine-water and talc-water systems from 250 to 450 °C, 50 MPa. Geochim Cosmochim Acta 73:6789–6804

Sachse D, Billault I et al (2012) Molecular paleohydrology: interpreting the hydrogen-isotopic composition of lipid biomarkers from photosynthesizing organisms. Ann Rev Earth Planet Sci 40:221–249

Sadofsky SJ, Bebout GE (2000) Ammonium partitioning and nitrogen isotope fractionation among coexisting micas during high-temperature fluid-rock interaction. Examples from the New England Appalachians. Geochim Cosmochim Acta 64:2835–2849

Saenger C, Wang Z (2014) Magnesium isotope fractionation in biogenic and abiogenic carbonates: implications for paleoenvironmental proxies. Quart Sci Rev 90:1–21

Saji NS, Wielandt D, Paton C, Bizarro M (2016) Ultra-high precision Nd-isotope measurements of geological materials by MC-ICPMS. JAAS 31:1490–1504

Sakai H (1968) Isotopic properties of sulfur compounds in hydrothermal processes. Geochem J 2:29–49

Santiago-Ramos DS, Morgan LE, Lloyd NS, Higgins JA (2018) Reverse weathering in marine sediments and the geochemical cycle of potassium in seawater: insights from the K isotopic composition ($^{41}K/^{39}K$) of deep-sea pore-fluids. Geochim Cosmochim Acta 236:99–120

Sanyal A, Nugent M, Reeder RJ, Bijma J (2000) Seawater $p_H$ control on the boron isotopic composition of calcite: evidence from inorganic calcite precipitation experiments. Geochim Cosmochim Acta 64:1551–1555

Saunders NJ, Barling J, Harvey J, Halliday AN (2020) Heterogeneous nickel isotope compositions in the terrestrial mantle—Part 1. Geochim Cosmochim Acta 285:129–149

Sauer PE, Eglinton TI, Hayes JM, Schimmelmann A, Sessions AL (2001) Compound-specific D/H ratios of lipid biomarkers from sediments as a proxy for environmental and climatic conditions. Geochim Cosmochim Acta 65:213–222

Saunier G, Pokrovski GS, Poitrasson F (2011) First experimental determination of iron isotope fractionation between hematite and aqueous solution at hydrothermal conditions. Geochim Cosmochim Acta 75:6629–6654

Savage PS, Georg RB, Williams HM, Burton KW, Halliday AN (2011) Silicon isotope fractionation during magmatic differentiation. Geochim Cosmochim Acta 75:6124–6139

Savage PS, Georg RB, Williams HM, Turner S, Halliday AN, Chappell BW (2012) The silicon isotope composition of granites. Geochim Cosmochim Acta 92:184–202

Savage PS, Georg RB, Williams HM, Halliday AN (2013a) The silicon isotope composition of the upper continental crust. Geochim Cosmochim Acta 109:384–399

Savage PS, Georg RB, Williams HM, Halliday AN (2013b) silicon isotopes in granulite xenoliths: insights into isotopic fractionation during igneous processes and the composition of the deep continental crust. Earth Planet Sci Lett 365:221–231

Savage PS, Armytage R, Georg RB, Halliday AN (2014) High temperature silicon isotope geochemistry. Lithos 190–191:500–519

Savage PS, Moynier F, Chen H, Shofner G, Siebert J, Badro J, Puchtel IS (2015) Copper isotope evidence for large-scale sulphide fractionation during Earth's differentiation.Geochem Perspect Lett 1:53–64

Savin SM, Lee M (1988) Isotopic studies of phyllosilicates. Rev Mineral 19:189–223

Schauble EA (2004) Applying stable isotope fractionation theory to new systems. Rev Mineral Geochem 55:65–111

Schauble EA (2007) Role of nuclear volume in driving equilibrium stable isotope fractionation of mercury, thallium and other very heavy elements. Geochim Cosmochim Acta 71:2170–2189

Schauble EA (2011) First principles estimates of equilibrium magnesium isotope fractionation in silicate, oxide, carbonate and hexa-aquamagnesium(2+) crystals. Geochim Cosmochim Acta 75:844–869

Schauble EA, Rossman GR, Taylor HP (2001) Theoretical estimates of equilibrium Fe isotope fractionations from vibrational spectroscopy. Geochim Cosmochim Acta 65:2487–2498

Schauble ES, Rossman GR, Taylor HP (2003) Theoretical estimates of equilibrium chlorine-isotope fractionations. Geochim Cosmochim Acta 67:3267–3281

Schauble EA, Rossman GR, Taylor HP (2004) Theoretical estimates of equilibrium chromium isotope fractionations. Chem Geol 205:99–114

Scheele N, Hoefs J (1992) Carbon isotope fractionation between calcite, graphite and $CO_2$. Contr Mineral Petrol 112:35–45

Scheiderich K, Amini M, Holmden C, Francois R (2015) Global variability of chromium isotopes in seawater demonstrated by Pacific, Atlantic and Arctic Ocean samples. Earth Planet Sci Lett 423:87–97

Schiller M, Paton C, Bizarro M (2012) Calcium isotope measurement by combined HR-MC-ICPMS and TIMS. JAAS 27:38–49

Schilling K, Johnson TM, Wilcke W (2011) isotope fractionation of selenium during fungal biomethylation by *Alternaria alternata*. Environ Sci Technol 45:2670–2676

Schilling K, Basu A, Wanner RR, Sanford RA, Pallud C, Johnson TM, Mason RD (2020) Mass-dependent selenium isotopic fractionation during microbial reduction of seleno-oxyanions by phylogenetically diverse bacteria. Geochim Cosmochim Acta 276:274–288

Schimmelmann A, Sessions AL, Mastalerz M (2006) Hydrogen isotopic (D/H) composition of organic matter during diagenesis and thermal maturation. Ann Rev Earth Planet Sci 34:501–533

Schmidt M, Maseyk K, Lett C, Biron P, Richard P, Bariac T, Seibt U (2010) Concentration effects on laser based $\delta^{18}O$ and $\delta^{2}H$ measurements and implications for the calibration of vapour measurements with liquid standards. Rapid Comm Mass Spectrom 24:3553–3561

Schmitt AD, Galer SJ, Abouchami W (2009) Mass-dependent cadmium isotopic variations in nature with emphasis on the marine environment. Earth Planet Sci Lett 277:262–272
Schmitt AD, Cobert F, Bourgeade P et al (2013) Calcium isotope fractionation during plant growth under a limited nutrient supply. Geochim Cosmochim Acta 110:70–83
Schoenberg R, von Blanckenburg F (2005) An assessment of the accuracy of stable Fe isotope ratio measurements on samples with organic and inorganic matrices by high-resolution multicollector ICP-MS. Int J Mass Spectr 242:257–272
Schoenberg R, von Blanckenburg F (2006) Modes of planetary-scale Fe isotope fractionation. Earth Planet Sci Lett 252:342–359
Schoenberg R, Zink S, Staubwasser M, von Blanckenburg F (2008) The stable Cr isotope inventory of solid Earth reservoirs determined by double-spike MC-ICP-MS. Chem Geol 249:294–306
Schoenberg R, Merdian A, Holmden C, Kleinhanns IC, Haßler K, Wille M, Reitter E (2016) The stable Cr isotope compositions of chondrites and silicate planetary reservoirs. Geochim Cosmochim Acta 183:14–30
Schoenheimer R, Rittenberg D (1939) Studies in protein metabolism: I. General considerations in the application of isotopes to the study of protein metabolism. The normal abundance of nitrogen isotopes in amino acids. J Biol Chem 127:285–290
Schönbächler M, Carlson RW, Horan MF, Mock TD, Hauri EH (2007) High precision Ag isotope measurements in geologic materials by multiple-collector ICPMS: an evaluation of dry versus wet plasma. Inter J Mass Spectrom 261:183–191
Schulze M, Ziegerick M, Horn I, Weyer S, Vogt C (2017) Determination of tin isotope ratios in cassiterite by femtosecond laser ablation multicollector inductively coupled mass spectrometry. Spectrochimica Acta Part B 130:26–34
Schüßler JA, Schoenberg R, Behrens H, von Blanckenburg F (2007) The experimental calibration of iron isotope fractionation factor between pyrrhotite and peralkaline rhyolitic melt. Geochim Cosmochim Acta 71:417–433
Schuth S, Horn I, Brüske A, Wolf PE, Weyer S (2017) First vanadium isotope analyses of V-rich minerals by femtosecond laser ablation and solution-nebulization MC-ICP-MS. Ore Geol Rev 81:1271–1286
Schuth S, Brüske A, Hohl SV, Jiang SY, Gregory DD, Viehmann S, Weyer S (2019) Vanadium isotopes of the Yangtze river system, China. Goldschmidt 2019 Abstr
Scott C, Lyons TW (2012) Contrasting molybdenum cycling and isotopic properties in euxinix versus non-euxinic sediments and sedimentary rocks: refining the paleoproxies. Chem Geol 324–325:19–27
Seal RR (2006) Sulfur isotope geochemistry of sulfide minerals. Rev Mineral Geochem 61:633–677
Seal RR, Alpers CN, Rye RO (2000) Stable isotope systematics of sulfate minerals. Rev Mineral 40:541–602
Sedaghatpour F, Teng FZ, Liu Y, Sears DW, Taylor LA (2013) Magnesium isotope composition of the Moon. Geochim Cosmochim Acta 120:1–16
Sedaghatour F, Jacobsen SB (2019) Magnesium stable isotopes support the lunar magma ocean cumulate remelting model for mare basalts. PNAS 116:73–78
Seitz HM, Brey GP, Zipfel J, Ott U, Weyer S, Durali S, Weinbruch S (2007) Lithium isotope composition of ordinary and carbonaceous chondrites and differentiated planetary bodies: bulk solar system and solar reservoirs. Earth Planet Sci Lett 260:582–596
Sessions AL, Burgoyne TW, Schimmelmann A, Hayes JM (1999) Fractionation of hydrogen isotopes in lipid biosynthesis. Org Geochem 30:1193–1200
Severmann S, Johnson CM, Beard BL, German CR, Edmonds HN, Chiba H, Green DR (2004) The effect of plume processes on the Fe isotope composition of hydrothermally derived Fe in the deep ocean as inferred from the Rainbow vent site, Mid-Atlantic Ridge, 36°14′N. Earth Planet Sci Lett 225:63–76

Severmann S, Johnson CM, Beard BL, McManus J (2006) The effect of early diagenesis on the Fe isotope composition of porewaters and authigenic minerals in continental margin sediments. Geochim Cosmochim Acta 70:2006–2022

Severmann S, Lyons TW, Anbar A, McManus J, Gordon G (2008) Modern iron isotope perspective on the benthic iron shuttle and the redox evolution of ancient oceans. Geology 36:487–490

Severmann S, McManus J, Berelson WM, Hammond DE (2010) The continental shelf benthic flux and its isotope composition. Geochim Cosmochim Acta 74:3984–4004

Shahar A, Young ED, Manning CE (2008) Equilibrium high-temperature Fe isotope fractionation between fayalite and magnetite: an experimental calibration. Earth Planet Sci Lett 268:330–338

Shahar A, Ziegler K, Young ED, Ricollaeu A, Schauble E, Fei Y (2009) Experimentally determined Si isotope fractionation between silicate and Fe metal and implications for the Earth's core formation. Earth Planet Sci Lett 288:228–234

Shalev N, Gavrieli I, Halicz L, Sandler A, Stein M, Lazar B (2017) Enrichment of $^{88}$Sr in continental waters due to calcium carbonate precipitation. Earth Planet Sci Lett 459:381–393

Shao J, Stott LD, Gray WR, Greenop R, Pecher I, Neil HL, Coffin RB, Davy B, Rae JW (2019) Atmosphere-ocean $CO_2$ exchange across the last deglaciation from the boron isotope proxy. Paleoceanography and Paleoclimatology 34:1650–1670

Sharma T, Clayton RN (1965) Measurement of 18O/16O ratios of total oxygen of carbonates. Geochim Cosmochim Acta 29:1347–1353

Sharma M, Polizzotto M, Anbar AD (2001) Iron isotopes in hot springs along the Juan de Fuca Ridge. Earth Planet Sci Lett 194:39–51

Sharp ZD (1990) A laser-based microanalytical method for the in-situ determination of oxygen isotope ratios of silicates and oxides. Geochim Cosmochim Acta 54:1353–1357

Sharp ZD, Atudorei V, Durakiewicz T (2001) A rapid method for determination of hydrogen and oxygen isotope ratios from water and hydrous minerals. Chem Geol 178:197–210

Sharp ZD, Barnes JD, Brearley AJ, Chaussidon M, Fischer TP, Kamenetsky VS (2007) Chlorine isotope homogeneity of the mantle, crust and carbonaceous chondrites. Nature 446:1062–1065

Sharp ZD, Shearer CK, McKeegan KD, Barnes JD, Wang YD (2010) The chlorine isotope composition of the Moon and implications for an anhydrous mantle. Science 239:1050–1053

Sharp ZD, Gibbons JA, Maltsev O, Atudorei V, Pack A, Sengupta S, Shock EL, Knauth LP (2016) A calibration of the triple oxygen isotope fractionation in the $SiO_2$–$H_2O$ system and applications to natural samples. Geochim Cosmochim Acta 186:105–119

Sharp ZD, Wostbrock JA, Pack A (2018) Mass-dependent triple oxygen isotope variations in terrestrial materials. Geochem Persp Lett 7:27–31

Shen B, Jacobson B, Lee CT, Yin QZ, Mourton DM (2009) The Mg isotopic systematics of granitoids in continental arcs and implications for the role of chemical weathering in crust formation. PNAS 106:20652–20657

Shen J, Qin L, Fang Z, Zhang Y, Liu J, Liu W, Wang FY, XiaoY YuH, Wei S (2018) High-temperature inter-mineral Cr isotope fractionation: a comparison of ionic model predictions and experimental investigations of mantle xenoliths from the North China craton. Earth Planet Sci Lett 499:278–290

Shen J, Yu J, Chen J, Algeo TJ, Xu G, Feng Q, Shi X, Planavsky NJ, Shu W, Xie S (2019) Mercury evidence of intense volcanic effects on land during the Permian-Triassic transition. Geology 47:1117–1121

Sheppard SMF, Nielsen RL, Taylor HP (1971) Hydrogen and oxygen isotope ratios in minerals from Porphyry Copper Deposits. Econ Geol 66:515–542

Sherman DM (2013) Equilibrium isotope fractionation of copper during oxidation/reduction, aqueous complexation and ore-forming processes: prediction from hybrid density functional theory. Geochim Cosmochim Acta 118:85–97

Sherman LS, Blum JD, Johnson KP, Keeler GJ, Barres JA, Douglas TA (2010) Mass-independent fractionation of mercury isotopes in Arctic snow driven by sunlight. Nat Geosci 3:173–177

Sherman LS, Blum JD, Nordstrom DK, McCleskey RB, Barkay T, Vetriani C (2009) Mercury isotope composition of hydrothermal systems in the Yellowstone Plateau volcanic field and Guaymas Basin sea-floor rift. Earth Planet Sci Lett 279:86–96

Shiel AE, Weis D, Orians KJ (2010) Evaluation of zinc, cadmium and lead isotope fractionation during smelting and refining. Sci Tot Environ 408:2357–2368

Shields WR, Goldich SS, Garner EI, Murphy TJ (1965) Natural variations in the abundance ratio and the atomic weight of copper. J Geophys Res 70:479–491

Shimizu K, Ushikubo T, Murai T, Matsu'ura F, Ueno Y (2019) In situ analyses of hydrogen and sulfur isotope ratios in basaltic glass using SIMS. Geochem J 53:195–207

Shollenberger QR, Brennecka GA (2020) Dy, Er, and Yb isotope compositions of meteorites and their components: Constraints on presolar carriers of the rare earth elements. Earth Planet Sci Lett 529:115866

Shouakar-Stash O, Alexeev SV, Frape SK, Alexeeva LP, Drimmie RJ (2007) Geochemistry and stable isotope signatures including chlorine and bromine isotopes of the deep groundwaters of the Siberian Platform, Russia. Appl Geochem 22:589–605

Shouakar-Stash O, Drimmie RJ, Frape SK (2005) Determination of inorganic chlorine stable isotopes by continuous flow isotope mass spectrometry. Rapid Commun Mass Spectr 19:121–127

Shu Y, Nielsen SG, Marschall HR, John T, Blusztajn J, Auro M (2019) Closing the loop: subducted eclogites match the thallium isotope compositions of ocean island basalts. Geochim Cosmochim Acta 250:130–148

Siebert C, Nägler TF, von Blanckenburg F, Kramers JD (2003) Molybdenum isotope records as potential proxy for paleoceanography. Earth Planet Sci Lett 211:159–171

Siebert C, Ross A, McManus J (2006a) Germanium isotope measurements of high-temperature geothermal fluids using double-spike hydride generation MC-ICP-MS. Geochim Cosmochim Acta 70:3986–3995

Siebert C, McManus J, Bice A, Poulson R, Berelson WM (2006b) Molybdenum isotope signatures in continental margin sediments. Earth Planet Sci Lett 241:723–733

Sigman DM, Casciotti KL, Andreani M, Barford C, Galanter M, Böhlke JK (2001) A bacterial method for the nitrogen isotopic analysis of nitrate in seawater and freshwater. Anal Chem 73:4145–4153

Sikora ER, Johnson TM, Bullen TD (2008) Microbial mass-dependent fractionation of chromium isotopes. Geochim Cosmochim Acta 72:3631–3641

Sim MS, Bosak T, Ono S (2011) Large sulfur isotope fractionation does not require disproportionnation. Science 333:74–77

Sime NG, De la Rocha C, Galy A (2005) Negligible temperature dependence of calcium isotope fractionation in 12 species of planktonic foraminifera. Earth Planet Sci Lett 232:51–66

Simon JI, dePaolo DJ (2010) Stable calcium isotopic composition of meteorites and rocky planets. Earth Planet Sci Lett 289:457–466

Sio CK, Dauphas N, Teng NZ, Chaussidon M, Helz RT, Roskosz M (2013) Discerning crystal growth from diffusion profiles in zoned olivine by in-situ Mg-Fe isotopic analyses. Geochim Cosmochim Acta 123:302–321

Skulan JL, Beard BL, Johnson CM (2002) Kinetic and equilibrium isotope fractionation between aqueous Fe(III) and hematite. Geochim Cosmochim Acta 66:2505–2510

Skulan JL, DePaolo DJ, Owens TL (1997) Biological control of calcium isotopic abundances in the global calcium cycle. Geochim Cosmochim Acta 61:2505–2510

SkulanJL DePaolo DJ (1999) Calcium isotope fractionation between soft and mineralised tissues as a monitor of calcium use in vertebrates. PNAS 96:13709–13713

Slack JF, Palmer MR, Stevens BPJ, Barnes RG (1993) Origin and significance of tourmaline-rich rocks in the Broken Hill district, Australia. Econ Geol 88:505–541

Smith CN, Kesler SE, Blum JD, Rytuba JR (2008) Isotope geochemistry of mercury in source rocks, mineral deposits and spring deposits of the California Coast Ranges, USA. Earth Planet Sci Lett 269:398–406

Smith MP, Yardley BWD (1996) The boron isotopic composition of tourmaline as a guide to fluid processes in the southwestern England orefield: an ion microprobe study. Geochim Cosmochim Acta 60:1415–1427

Smithers RM, Krouse HR (1968) Tellurium isotope fractionation study. Can J Chem 46:583–591

Soderman CR, 11 others (2021) Heavy $\delta^{57}$Fe values in ocean island basalts: a non-unique signature of processes and source lithologies in the mantle. Geochim Cosmochim Acta 292:309-332

Sonke JE, Sivry Y et al (2008) Historical variations in the isotopic composition of atmospheric zinc deposition from a zinc smelter. Chem Geol 252:145–157

Sonke JE, Schäfer J et al (2010) Sedimentary mercury stable isotope records of atmospheric and riverine pollution from two major European heavy metal refineries. Chem Geol 279:90–100

Sonke JE (2011) A global model of mass independent mercury stable isotope fractionation. Geochim Cosmochim Acta 75:4577–4590

Sossi P, O'Neill HS (2017) The effect of bonding environment on iron isotope fractionation between minerals at high temperature. Geochim Cosmochim Acta 196:121–143

Sossi P, Prytulak J, O'Neill HS (2018a) Experimental calibration of vanadium partitioning and stable isotope fractionation between hydrous granitic melt and magnetite at 800 °C and 0.5 GPa. Contr Mineral Petrol 173:27

Sossi P, Moynier F, van Zuilen K (2018b) Volatile loss following cooling and accretion of the Moon revealed by chromium isotopes. PNAS 115:10920–10925

Sossi P, Nebel O, O'Neill HS, Moynier F (2018c) Zinc isotope composition of the earth and its behavior during planetary accretion. Chem Geol 477:73–84

Spivack AJ, Edmond JM (1986) Determination of boron isotope ratios by thermal ionization mass spectrometry of the dicesium metaborate cation. Anal Chem 58:31–35

Spivack AJ, Kastner M, Ransom B (2002) Elemental and isotopic chloride geochemistry in the Nankai trough. Geophysical Res Lett 29:1661. https://doi.org/10.1029/2001GL014122

Spivak-Birndorf LJ, Wang SJ, Bish DL, Wasylenki LE (2018) Nickel isotope fractionation during continental weathering. Chem Geol 476:316–326

Stefurak EJ, Fischer WW, Lowe DR (2015) Texture-specific Si isotope variations in Barberton Greenstone Belts cherts record low temperature fractionations in early Archean seawater. Geochim Cosmochim Acta 150:26–52

Steig EJ, Gkinis V, Schauer AJ, Samek K, Hoffnagle J, Dennis KT, Tan SM (2014) Calibrated high-precision $^{17}$O excess measurements using cavity ring-down spectroscvopy with laser-current-tuned cavity resonance. Atmos Meas Tech 7:2421–2435

Stephant A, Anand M, Zhao X, Chan Q H, Bonifacie M, Franchi IA (2019) The chlorine isotopic composition of the Moon: insights from melt inclusions. Earth Planet Sci Lett 523:115715

Stetson SJ, Gray JE, Wanty RB, MacLady DL (2009) Isotope variability of mercury in ore, mine-waste calcine, and leachates of mine-waste calcine from areas mined for mercury. Environ Sci Technol 43:7331–7336

Steuber T, Buhl D (2006) Calcium-isotope fractionation in selected modern and ancient marine carbonates. Geochim Cosmochim Acta 70:5507–5521

Stevenson EI, Hermoso M, Rickaby RE, Tyler JJ, Minoletti F, Parkinson IJ, Mokadem F, Burton KW (2014) Controls on stable strontium isotope fractionation in coccolithophores with implications for the marine Sr cycle. Geochim Cosmochim Acta 128:225–235

Stirling CH, Andersen MB, Potter EK, Halliday AN (2007) Low-temperature isotopic fractionation of uranium. Earth Planet Sci Lett 264:208–225

Stirling CH, Andersen MB, Warthmann R, Halliday RN (2015) Isotope fractionation of $^{238}$U and $^{235}$U during biologically-mediated uranium reduction. Geochim Cosmochim Acta 163:200–218

Stotler RL, Frape SK, Shouakar-Stash O (2010) An isotopic survey of $\delta^{81}$Br and $\delta^{37}$Cl of dissolved halides in the Canadian and Fennoscandian shields. Chem Geol 274:38–55

Stracke A, Tipper ET, Klemme S, Bizimis M (2018) Mg isotope systematics during magmatic processes: inter-mineral fractionation in mafic to ultramafic Hawaiian xenoliths. Geochim Cosmochim Acta 226:192–205

Stüeken EE (2017) Selenium isotopes as a biogeochemical proxy in deep time. Rev Mineral Petrol 82:657–682

Stüeken EE, Foriel J, Nelson BK, Buick R, Catling DC (2013) Selenium isotope analysis of organic-rich shales: advances in sample preparation and isobaric interference correction. J Anal Atom Spectr 28:1734–1749

Stüeken EE, Foriel J, Buick R, Schoepfer SD (2015a) Selenium isotope ratios, redox changes and biological productivity across the end-Permian mass extinction. Chem Geol 410:28–39

Stüeken EE, Buick R, 8 others (2015b) The evolution of the global selenium cycle: secular trends in Se isotopes and abundances. Geochim Cosmochim Acta 162:109–125

Stüeken EE, Buick R, Anbar AD (2015c) Selenium isotopes support free $O_2$ in the latest Archean. Geology 43:259–262

Stüeken EE, Kipp MA, Koehler MC, Buick R (2016) The evolution of Earth' s biogeochemical nitrogen cycle. Earth Sci Rev 160:220–239

Stüeken EE, Kipp MA (2020) Selenium isotope paleobiogeochemistry. Cambridge University Press

Sturchio NC, Hatzinger PB, Atkins MD, Suh C, Heraty LJ (2003) Chlorine isotope fractionation during microbial reduction of perchlorate. Environ Sci Technol 37:3859–3863

Stylo M, Neubert N, Wang Y, Monga N, Romaniello SJ, Weyer S, Bernier-Latmani R (2015) Uranium isotopes fingerprint biotic reduction. PNAS 112:5619–5624

Sun J, 12 others (2021) Ca isotope systematics of carbonatites: insight into carbonatiote source and evolution. Geochem Perspective Lett 17. https://doi.org/10.7185/geochemlet.2107

Sun R, Sonke JE, Liu G (2016a) Biogeochemical controls on mercury stable isotope compositions of world coal deposits: a review. Earth-Sci Rev 152:1–13

Sun Y, Wu LH, Li XY (2016b) Experimental determination of of silicon isotope fractionation in rice. PLos One 11(12):e0168970

Sun Y, Teng FZ, Xin YH, Kwan YC, Pang N (2019) Tracing subducted oceanic slabs in the mantle by using potassium isotopes. Geochim Cosmochim Acta 278:353–360

Sutton JN, 5 others (2018) A review of the stable isotope bio-geochemistry of the global silicon cycle and its associated trace elements. Front Earth Sci 5:112

Suzuoki T, Epstein S (1976) Hydrogen isotope fractionation between OH-bearing minerals and water. Geochim Cosmochim Acta 40:1229–1240

Swart PK, Burns SJ, Leder JJ (1991) Fractionation of the stable isotopes of oxygen and carbon in carbon dioxide during the reaction of calcite with phosphoric acid as a function of temperature and technique. Chem Geol 86:89–96

Swihart GH, Moore PB (1989) A reconnaissance of the boron isotopic composition of tourmaline. Geochim Cosmochim Acta 53:911–916

Tanimizu M, Araki Y, Asaoka S, Takahashi Y (2011) Determination of natural isotopic variation in antimony using inductively coupled plasma mass spectrometry for an uncertainty estimation of the standard atomic weight of antimony. Geochem J 45:27–32

Tang M, Rudnick RL, Chauvel C (2014) Sedimentary input to the source of Lesser Antilles lavas: a Li perspective. Geochim Cosmochim Acta 144:43–58

Tarutani T, Clayton RN, Mayeda TK (1969) The effect of polymorphism and magnesium substitution on oxygen isotope fractionation between calcium carbonate and water. Geochim Cosmochim Acta 33:987–996

Tatzel M, von Blanckenburg F, Oelze M, Schuessler JA, Bohrmann G (2015) The silicon isotope record of early silica diagenesis. Earth Planet Sci Lett 428:293–303

Taube H (1954) Use of oxygen isotope effects in the study of hydration ions. J Phys Chem 58:523

Taylor HP (1968) The oxygen isotope geochemistry of igneous rocks. Contr Mineral Petrol 19:1–71

Taylor HP (1974) The application of oxygen and hydrogen isotope studies to problems of hydrothermal alteration and ore deposition. Econ Geol 69:843–883

Taylor HP, Epstein S (1962) Relation between $^{18}O/^{16}O$ ratios in coexisting minerals of igneous and metamorphic rocks. I Principles and experimental results. Geol Soc Am Bull 73:461–480

Taylor TI, Urey HC (1938) Fractionation of the lithium and potassium isotopes by chemical exchange with zeolites. J Chem Phys 6:429–438

Telus M, Dauphas N, Moynier F, Tissot F, Teng FZ, Nabelek PI, Craddock PR, Groat LA (2012) Iron, zinc, magnesium and uranium isotopic fractionation during continental crust differentiation: the tale from migmatites, granitoids and pegmatites. Geochim Cosmochim Acta 97:247–265

Teng FZ (2017) Magnesium isotope geochemistry. Rev Mineral Geochem 82:219–287

Teng FZ et al (2004) Lithium isotope composition and concentration of the upper continental crust. Geochim Cosmochim Acta 68:4167–4178

Teng FZ, McDonough WF, Rudnick RL, Walker RJ (2006) Diffusion-driven extreme lithium isotopic fractionation in country rocks of the Tin Mountain pegmatite. Earth Planet Sci Lett 243:701–710

Teng FZ, McDonough WF, Rudnick RL, Wing BA (2007a) Limited lithium isotopic fractionation during progressive metamorphic dehydration in metapelites: a case study from the Onawa contact aureole, Maine. Chem Geol 239:1–12

Teng FZ, Wadhwa M, Helz RT (2007b) Investigation of magnesium isotope fractionation during basalt differentiation: implications for a chondritic composition of the terrestrial mantle. Earth Planet Sci Lett 261:84–92

Teng FZ, Dauphas N, Helz R (2008) Iron isotope fractionation during magmatic differentiation in Kilauea Iki lava lake. Science 320:1620–1622

Teng FZ, Rudnick RL, McDonough WF, Wu FY (2009) Lithium isotope systematics of A-type granites and their mafic enclaves: further constraints on the Li isotopic composition of the continental crust. Chem Geol 262:415–424

Teng FZ, Dauphas N, Helz RT, Gao S, Huang S (2011) Diffusion-driven magnesium and iron isotope fractionation in Hawaiian olivine. Earth Planet Sci Lett 308:317–324

Teng FZ, Yang W (2013) Comparison of factors affecting the accuracy of high-precision magnesium isotope analysis by multi-collector inductively coupled plasma mass spectrometry. Rapid Commun Mass Spectrom 28:19–24

Teng FZ, Dauphas N, Huang S, Marty B (2013) Iron isotope systematics of oceanic basalts. Geochim Cosmochim Acta 107:12–26

Teng FZ and 20 others (2015a) Magnesium isotopic compositions of international geological reference materials. Geostand Geoanal Res 39:329–339

Teng FZ and 12 others (2015b) Interlaboratory comparison of magnesium isotope composition of 12 felsic to ultramafic igneous rock standards analyzed by MC-ICPMS Geochem Geophys Geosyst 16:3197–3209

Teng FZ, Hu Y, Ma ZL, Wei GJ, Rudnick RL (2020) Potassium isotope fractionation during continental weathering and implications for global K isotopic balance. Geochim Cosmochim Acta 278:261–272

Tesdal JE, Galbraith ED, Kienast M (2013) Nitrogen isotopes in bulk marine sediments: linking seafloor observations with subseafloor records. Biogeosciences 10:101–118

Tessalina S (2015) Silver isotope systematics in native Ag from hydrothermal Mo–Cu, Cu–Pb–Zn and Ag–Au deposits. Abstr Goldschmidt-Conference 2015

Teutsch N, Schmid M, Muller B, Halliday AN, Burgmann H, Wehrli B (2009) Large iron isotope fractionation at the oxic-anoxic boundary in lake Nyos. Earth Planet Sci Lett 285:52–60

Thamdrup B, Dalsgaard T (2002) Production of $N_2$ through anaerobic ammonium oxidation coupled to nitrate reduction in marine sediments. Appl Environ Microbiol 68:1312–1318

Them TR, Gill BC, Caruthers AH, Gerhardt AM, Gröcke DR, Lyons TW, Marroquin SM, Nielsen SG, AlexandreJP OJD (2018) Thallium isotopes reveal protracted anoxia during the

Toarcian (Early Jurassic) associated with volcanism, carbon burial, and mass extinction. PNAS 115:6596–6601
Them TR, Jagoe CH, Caruthers AH, Gill BC, Grasby SE, Gröcke DR, Yin R, Owens JD (2019) Terrestrial sources as the primary delivery mechanism of mercury to the oceans across the Toarcian Oceanic Anoxic Event (Early Jurassic). Earth Planet Sci Letters 507:62–72
Thibodeau AM, Berqquist BA (2017) Do mercury isotopes record the signature of massive volcanism in marine sedimentary records? Geology 45:95–96
Thiemens MH, Lin M (2021) Discoveries of mass independent isotope effects in the solar system: past, present and future. Rev Mineral Geochem 86:35–95
Thoby M, Konhauser KO, Fralick PW, AltermannW VPT, Lalonde SV (2019) Global importance of oxic molybdenum sinks prior to 2.6 Ga revealed by the Mo isotope composition of Precambrian carbonates. Geology 47:559–562
Thode HG, Macnamara J, Collins CB (1949) Natural variations in the isotopic content of sulphur and their significance. Can J Res 27B:361
Tian H, Yang W, Li SG, Ke S, Chu ZY (2016) origin of low $\delta^{26}$Mg basalts with EM-1 component: evidence for interaction between enriched lithosphere and carbonate asthenosphere. Geochim Cosmochim Acta 188:93–105
Tian S, Inglis E, Creech J, Zhang W, Wang Z, Hu Z, Liu Y, Moynier F (2020a) The zirconium stable isotope compositions of 22 geological reference materials, 4 zircons and 3 standard solutions. Chem Geol 555:119791
Tian S, Moynier F, Inglis EC, Creech J, Bizarro M, Siebert J, Day JM, Puchtel IS (2020b) Zirconium isotopic composition of the mantle through time. Geochem Persp Lett 15:40–43
Tipper ET, Galy A, Bickle MJ (2006a) Riverine evidence for a fractionated reservoir of Ca and Mg on the continents: implications for the oceanic Ca cycle. Earth Planet Sci Lett 247:267–279
Tipper ET, Galy A, Gaillardet J, Bickle MJ, Elderfield H, Carder EA (2006b) The magnesium isotope budget of the modern ocean: constraints from riverine magnesium isotope ratios. Earth Planet Sci Lett 250:241–253
Tipper ET, Galy A, Bickle MJ (2008) Calcium and magnesium isotope systematics in rivers draining the Himalaya-Tibetan–Plateau region: lithological or fractionation control? Geochim Cosmochim Acta 72:1057–1075
Tipper ET, Gaillardet J, Galy A, Louvat P, Bickle MJ, Capmas F (2010) Calcium isotope ratios in the world's largest rivers: a constraint on the maximum imbalance of oceanic calcium fluxes. Global Biogeochem Cyles 24. https://doi.org/10.1029/2009GB003574
Tissot FL, Dauphas N (2015) Uranium isotopic composition of the crust and ocean: age corrections, U budget and global extent of modern anoxia. Geochim Cosmochim Acta 167:113–143
Tomascak PB, Tera F, Helz RT, Walker RJ (1999) The absence of lithium isotope fractionation during basalt differentiation: new measurements by multicollector sector ICP-MS. Geochim Cosmochim Acta 63:907–910
Tomascak PB, Ryan JG, Defant MJ (2000) Lithium isotope evidence for light element decoupling in the Panama subarc mantle. Geology 28:507–510
Tomascak PB, Widom E, Benton LD, Goldstein SL, Ryan JG (2002) The control of lithium budgets in island arcs. Earth Planet Sci Lett 196:227–238
Tomascak PB (2004) Lithium isotopes in earth and planetary sciences. Rev Mineral Geochem 55:153–195
Tomascak PB, Magna T, Dohmen R (2016) Advances in lithium isotope geochemistry. Springer
Tompkins HG, Zieman LJ, Ibanez-Mejia M, Tissot FL (2020) Zirconium stable isotope analysis of zircon by MC-ICP-MS: methods and application to evaluating intra-crystalline zonation in a zircon megacryst. JAAS 35:1157–1186

Tostevin R, Turchyn AV, Farquhar J, Johnston DT, Eldrige DL, Bishop JK, McIlvin M (2014) Multiple sulfur isotope constraints on the modern sulfur cycle. Earth Planet Sci Lett 396:14–21

Toutain JP, Sonke J et al (2008) Evidence for Zn isotopic fractionation at Merapi volcano. Chem Geol 253:74–82

Trail D, Boehnke P, Savage PS, Liu MC, Miller ML, Bindeman I (2018) Origin and significance of Si and O isotope heterogeneities in Phanerozoic, Archean and Hadean zircon. PNAS 115:10287–10292

Trail D, Savage PS, Moynier F (2019) Experimentally determined Si isotope fractionation between zircon and quartz. Geochim Cosmochim Acta 260:257–274

Trofimov A (1949) Isotopic constitution of sulfur in meteorites and in terrestrial objects. Dokl Akad Nauk SSSR 66:181 (in Russian)

Trudinger PA, Chambers LA, Smith JW (1985) Low temperature sulphate reduction: biological versus abiological. Can J Earth Sci 22:1910–1918

Truesdell AH (1974) Oxygen isotope activities and concentrations in aqueous salt solution at elevated temperatures: Consequences for isotope geochemistry. Earth Planet Sci Lett 23:387–396

Trumbull RB, Slack JF (2017) Boron isotopes in the continental crust: granites, pegmatites, felsic volcanic rocks, and related ore deposits. In: Boron isotopes—the fifth element. Advances in isotope geochemistry. Springer (2017)

Trumbull RB, CodecoMS, Jiang SY, Palmer MR, Slack JF (2020) Boron isotope variations in tourmaline from hydrothermal ore deposits: a review of controlling factors and insights for mineralizing systems. Ore Geology Rev 125:103682

Tuller-Ross B, Savage PS, Chen H, Wang K (2019) Potassium isotope fractionation during magmatic differentiation of basalt to rhyolite. Chem Geol 525:37–45

Tuller-Ross B, Marty B, Chen H, Kelley KA, Lee H, Wang K (2019) Potassium isotope systematics of oceanic basalts. Geochim Cosmochim Acta 259:144–154

Turner JV (1982) Kinetic fractionation of carbon-13 during calcium carbonate precipitation. Geochim Cosmochim Acta 46:1183–1192

Urey HC, Brickwedde FG, Murphy GM (1932) A hydrogen isotope of mass 2 and its concentration. Phys Rev 40:1

Usdowski E, Hoefs J (1993) Oxygen isotope exchange between carbonic acid, bicarbonate, carbonate, and water: a re-examination of the data of McCrea (1950) and an expression for the overall partitioning of oxygen isotopes between the carbonate species and water. Geochim Cosmochim Acta 57:3815–3818

Usdowski E, Michaelis J, Böttcher MB, Hoefs J (1991) Factors for the oxygen isotope equilibrium fractionation between aqueous $CO^2$, carbonic acid, bicarbonate, carbonate, and water. Z Phys Chem 170:237–249

Ushikobo T, Kita N, Cavoisie AJ, Wilde SA, Rudnick RL, Valley JW (2008) Lithium in Jack Hill zircons: evidence for extensive weathering of Earth's earliest crust. Earth Planet Sci Lett 272:666–676

Ushikubo T, Williford KW, Farquhar J, Johnson DT, van Kranendonk J, Valley JW (2014) Development of an in-situ four-sulfur isotope analysis with multiple Faraday cup detectors by SIMS and applications to pyrite grains in a Paleoproterozoic glaciogenic sandstone. Chem Geol 383:86–99

Uvarova YA, Kyser TK, Geagea ML, Chipley D (2014) Variations in the uranium isotopic composition of uranium ores from different types of uranium deposits. Geochim Cosmochim Acta 146:1–17

Valdes MC, Moreira M, Foriel J, Moynier F (2014) The nature of Earth's building blocks as revealed by calcium isotopes. Earth Planet Sci Lett 394:135–145

Valdes MC, Debaille V, Berger J, Armytage RM (2019) The effects of high-temperature fractional crystallization on calcium isotopic composition. Chem Geol 509:77–91

Valley JW, O'Neil JR (1981) $^{13}C/^{12}C$ exchange between calcite and graphite: a possible thermometer in Greville marbles. Geochim Cosmochim Acta 45:411–419

Van Acker M, Shahar A, Young ED, Cöleman ML (2006) GC/Multiple Collector-ICPMS method for chlorine stable isotope analysis of chlorinated aliphatic hydrocarbons. Anal Chem 78:4663–4667

Van den Boorn SH, van Bergen MJ, Vroon PZ, de Vries ST, Nijman W (2010) Silicon isotope and trace element constraints on the origin of $\approx$ 3.5 Ga cherts: implications for early Archaean marine environments. Geochim Cosmochim Acta 74:1077–1103

Van Warmerdam EM, Frape SK, Aravena R, Drimmie RJ, Flatt H, Cherry JA (1995) Stable chlorine and carbon isotope measurements of selected chlorinated organic solvents. Appl Geochem 10:547–552

Van Zuilen K, Nägler TF, Bullen TD (2016) Barium isotopic compositions of geological reference materials. Geostand Geoanal Res 40:543–558

Van Zuilen K, Müller T, Nägler TF, Dietzel M, Küsters T (2016) Experimental determination of barium isotope fractionation during diffusion and adsorption processes at low temperatures. Geochim Cosmochim Acta 186:226–241

Vance D, Archer C, Bermin J, Perkins J, Statham PC, Lohan MC, Ellwood MJ, Mills RA (2008) The copper isotope geochemistry of rivers and oceans. Earth Planet Sci Lett 274:204–213

Vance D, Little SH, Archer C, Cameron V, Andersen MB, Rijkenberg MJ, Lyons TW (2016) The oceanic budgets of nickel and zinc isotopes: the importance of sulfidic environments as illustrated by the Black Sea. Phil Trans R Soc A 374:20150294

Varas-Reus MI, König S, Yierpan A, Lorand JP, Schoenberg R (2019) Selenium isotopes as tracers of a late volatile contribution to earth from the outer Solar System. Nat Geosci 12:779–782

Varela DE, Pride CJ, Brzezinski MA (2004) Biological fractionation of silicon isotopes in southern-ocean surface waters. Global Biogeochem Cycles 18. https://doi.org/10.1029/2003GB002140

Velinsky DJ, Pennock JR, Sharp JH, Cifuentes LA, Fogel ML (1989) Determination of the isotopic composition of ammonium-nitrogen at the natural abundance level from estuarine waters. Mar Chem 26:351–361

Vengosh A, Chivas AR, McCulloch M, Starinsly A, Kolodny Y (1991a) Boron isotope geochemistry of Australian salt lakes. Geochim Cosmochim Acta 55:2591–2606

Vengosh A, Heumann KG, Juraske S, Kasher R (1994) Boron isotope application for tracing sources of contamination in groundwater. Environ Sci Tech 28:1968–1974

Vengosh A, Starinsky A, Kolodny Y, Chivas AR (1991b) Boron isotope geochemistry as a tracer for the evolution of brines and associated hot springs from the Dead Sea, Israel. Geochim Cosmochim Acta 55:1689–1695

Vennemann TW, O'Neil JR (1993) A simple and inexpensive method of hydrogen isotope and water analyses of minerals and rocks based on zinc reagent. Chem Geol 103:227–234

Vennemann TW, Fricke HC, Blake RE, O'Neil JR, Colman A (2002) Oxygen isotope analysis of phosphates: a comparison of techniques for analysis of $Ag_3PO_4$. Chem Geol 185:321–336

Vennemann T, O'Neil JR (1996) Hydrogen isotope exchange reactions between hydrous minerals and hydrogen: I. A new approach for the determination of hydrogen isotope fractionation at moderate temperatures. Geochim Cosmochim Acta 60:2437–2451

Ventura GT, Gall L, Siebert C, Prytulak J, Szatmari P, Hürlimann M, Halliday AN (2015) The stable isotope composition of vanadium, nickel and molybdenum in crude oils. Appl Geochem 59:104–117

Viers J et al (2007) Evidence of Zn isotope fractionation in a soil-plant system of a pristine tropical watershed (Nsimi, Cameroon). Chem Geol 239:124–137

Vho A, Lanari P, Rubatto D (2019) An internally-consistent database for oxygen isotope fractionation between minerals. J Petrol 60:21012130

Voegelin AR, Nägler TF, Beukes NJ, Lacassie JP (2010) Molybdenum isotopes in late Archean carbonate rocks: implications for early Earth oxygenation. Precambr Res 182:70–82

Voegelin AR, Nägler TF, Samankassou E, Villa IM (2009) Molybdenum isotopic composition of modern and Carboniferous carbonates. Chem Geol 265:488–498

Voegelin AR, Pettke T, Greber ND, von Niederhäusern B, Nägler TF (2014) Magma differentiation fractionates Mo isotope ratios: Evidence from the Kos Plateau Tuff (Aegean Arc). Lithos 190–191:440–448

Vogel JC, Grootes PM, Mook WG (1970) Isotopic fractionation between gaseous and dissolved carbon dioxide. Z Physik 230:225–238

Voigt J, Ec H, Frank M, Vollstaedt H, Eisenhauer A (2015) Variability of carbonate diagenesis in equatorial Pacific sediments deduced from radiogenic and stable Sr isotopes. Geochim Cosmochim Acta 148:360–377

Voisey CR, Maas R, Tomkins AG, Brauns M, Brügmann G (2019) Extreme silver isotope variation in orogenic gold systems implies multistaged metal remobilization during ore genesis. Econ Geol 114:233–242

Vollstädt H, Eisenhauer A et al (2014) The Phanerozoic $\delta^{88/86}$Sr record of seawater: new constraints on past changes in oceanic carbonate fluxes. Geochim Cosmochim Acta 128:249–265

Von Allmen K, Böttcher ME, Samankassou E, Nägler TF (2010) Barium isotope fractionation in the global barium cycle: first evidence from barium minerals and precipitation experiments. Chem Geol 277:70–77

Wacey D, Noffke N, Cliff J, Barley ME, Farquhar J (2015) Micro-scale quadruple sulfur isotope analysis of pyrite from the 3480 Ma dresser formation: new insights into sulfur cycling on the early Earth. Precambrian Res 258:24–35

Wachter EA, Hayes JM (1985) Exchange of oxygen isotopes in carbon dioxide-phosphoric acid systems. Chem Geol 52:365–374

Wakaki S, Tanaka T (2012) Stable isotope analysis of Nd by double spike thermal ionization mass spectrometry. Int J Mass Spectrometry 323–324:45–54

Wang C, 10 others (2021) Chromium isotope systematics and the diagenesis of marine carbonates. Earth Planet Sci Lett 562:116824

Wang D, Mathur R, Powell W, Godfrey L, Zheng Y (2019a) Experimental evidence for fractionation of tin chlorides by redox and vapor mechanisms. Geochim Cosmochim Acta 250:209–218

Wang K, Jacobsen SB (2016a) An estimate of the bulk silicate earth potassium isotopic composition based on MC-ICP-MS measurements of basalts. Geochim Cosmochim Acta 178:223–232

Wang K, Jacobsen (2016b) Potassium isotopic evidence for a high-energy giant impact origin of the Moon. Nature 538:487–489

Wang K, Peucker-Ehrenbrink B, Chen H, Lee H, Hasenmueller EA (2020) Dissolved potassium isotopic composition of major world rivers. Geochim Cosmochim Acta 294:145–159

Wang W, Zhou C, Liu Y, Wu Z, Huang F (2019b) Equilibrium Mg isotope fractionation among aqueous $Mg^{2+}$, carbonates, brucite and lizardite: insights from first-principles molecular dynamics simulations. Geochim Cosmochim Acta 250:117–129

Wang QL, Chetelat B, Zhao ZQ, Ding H, Li SL, Wang BL, Li J, Liu XL (2015a) Behavior of Lithium isotopes in the Changjiang river system: source effects and response to weathering and erosion. Geochim Cosmochim Acta 151:117–132

Wang Y, Sessions AL, Nielsen JR, Goddard WA (2009a) Equilibrium $^{2}H/^{1}H$ fractionations in organic molecules. I. Calibration of ab initio calculations. Geochim Cosmochim Acta 73:7060–7075

Wang Y, Sessions AL, Nielsen RJ, Goddard WA (2009b) Equilibrium $^{2}H/^{1}H$ fractionations in organic molecules. II: Linear alkanes, alkenes, ketones, carboxylic acids, esters, alcohols and ethers. Geochim Cosmochim Acta 73:7076–7086

Wang Y, Sessions AL, Nielsen RJ, Goddard WA (2013a) Equilibrium $^{2}H/^{1}H$ fractionation in organic molecules. III Cyclic ketones and hydrocarbons. Geochim Cosmochim Acta 107:82–95

Wang X, Johnson TM, Lundstrom CC (2014) Isotope fractionation during oxidation of tetravalent uranium by dissolved oxygen. Geochim Cosmochim Acta 150:160–170

Wang X, Johnson TM, Lundstrom CC (2015b) Low temperature equilibrium isotope fractionation and isotope exchange kinetics between U(IV) and U(VI). Geochim Cosmochim Acta 158:262–275

Wang X, Planavsky NJ, Reinhard CT, Hein JR, Johnson TM (2016a) A cenozoic seawater redox record derived from $^{238}U/^{235}U$ in ferromanganese crusts. Am J Sci 316:64–83

Wang X, Planavsky NJ, Reinhard CT, Zou H, Ague JJ, Wu Y, Gill BC, Schwarzenbach EM, Peucker-Ehrenbrink B (2016b) Chromium isotope fractionation during subduction-related metamorphism, black shale weathering and hydrothermal alteration. Chem Geol 423:19–33

Wang X, Fitoussi C, Bourdon B, Amet Q (2018) Tin isotope fractionation during magmatic processes and the isotope composition of the bulk silicate earth. Geochim Cosmochim Acta 228:320–335

Wang XL, Planavsky NJ, Hull PM, Tripati AE, Zou HJ, Elder L, Henehan M (2017a) Chromium isotopic composition of core-top planktonic foraminifera. Geobiology 15:51–64

Wang Z, Hu P, Gaetani G, Liu C, Saenger C, Cohen A, Hart S (2013b) Experimental calibration of Mg isotope fractionation between aragonite and seawater. Geochim Cosmochim Acta 102:113–123

Wang Z, Ma J, Li J, Wei G, Chen X, Deng W, Xie L, Lu W, Zou L (2015c) Chemical weathering controls on variations in the molybdenum isotopic composition of river water: evidence from large rivers in China. Chem Geol 410:201–212

Wang ZZ, Liu SA, Liu J, Huang J, Xiao Y, Chu ZY, Zhao XM, Tang L (2017b) Zinc isotope fractionation during mantle melting and constraints on the Zn isotope composition of Earth's upper mantle. Geochim Cosmochim Acta 198:151–167

Wang Z, Park JW, Wang X, Zou Z, Kim J, Zhang P, Li M (2019c) Evolution of copper isotopes in arc systems: insight from lavas and molten sulfur in Niuatahi Volcano, Tonga rear arc. Geochim Cosmochim Acta 250:18–33

Wanner C, Sonnenthal EL, Liu XM (2014) Seawater $\delta^{7}Li$: a direct proxy for global $CO_2$ consumption by continental silicate weathering? Chem Geol 381:154–167

Wasserman NL, Johnson MT (2020) Measurements of mass-dependent Te isotopic variation by hydride generation MC-ICP-MS. JAAS 35:307–319

Wasylenki LE, Rolfe BA, Weeks CL, Spiro TB, Anbar AD (2008) Experimental investigation of the effects of temperature and ionic strength on Mo isotope fractionation during adsorption to manganese oxides. Geochim Cosmochim Acta 72:5997–6005

Wasylenki LE, Swihart JW, Romaniello SJ (2014) Cadmium isotope fractionation during adsorption to Mn oxyhydroxide at low and high ionic strength. Geochim Cosmochim Acta 140:212–226

Wasylenki LE, Howe HD, Spivak-Birndorf LJ, Bish DL (2015) Ni isotope fractionation during sorption to ferrihydrite: implications for Ni in banded iron formations. Chem Geol 400:56–64

Wasylenki LE, Schaefer AT, Chanda P, Farmer JC (2020) Differential behavior of tungsten stable isotopes during sorption to Fe versus Mn oxyhydroxides at low ionic strength. Chem Geol 558:119836

Watanabe Y, Farquhar J, Ohmoto H (2009) Anomalous fractionation of sulfur isotopes during thermochemical sulfate reduction. Science 324:370–373

Wei R, Guo Q, Wen H, Liu C, Yang J, Peters M, Hu J, Zhu G, Zhang H, Tian L, Han X, Ma J, Zhu C, Wan Y (2016) Fractionation of stable cadmium isotopes in the cadmium tolerant *ricinus communis* and hyperaccumulator *solanum nigrum*. Sci Rep 6:24309. https://doi.org/10.1038/srep24309

Wei W, Klaebe R, Ling HF, Huang F, Frei R (2020) Biogeochemical cycle of chromium isotopes at the Earth's surface and its applications as a paleo-environment proxy. Chem Geol 541:119570

Weinstein C, Moynier F, Wang K, Paniello R, Foriel J, Catalano J, Pichat S (2011) Isotopic fractionation of Cu in plants. Chem Geol 286:266–271

Weiss DJ, Mason TFD, Zhao FJ, Kirk GJD, Coles BJ, Horstwood MSA (2005) Isotopic discrimination of zinc in higher plants. New Phytol 165:703–710

Weiss DJ, Rausch N, Mason TFD, Coles BJ, Wilkinson JJ, Ukonmaanaho L, Arnold T, Nieminen TM (2007) Atmospheric deposition and isotope biogeochemistry of zinc in ombrotrophic peat. Geochim Cosmochim Acta 71:3498–3517

Welch SA, Beard BL, Johnson CM, Braterman PS (2003) Kinetic and equilibrium Fe isotope fractionation between aqueous Fe(II) and Fe(III). Geochim Cosmochim Acta 67:4231–4250

Wen H, Carignan J (2011) Selenium isotopes trace the source and redox processes in the black shale-hosted Se-rich deposits in China. Geochim Cosmochim Acta 75:1411–1427

Wen H, Carignan J, Chu X, Fan H, Cloquet C, Huang J, Zhang X, Chang H (2014) Selenium isotopes trace anoxic and ferruginous seawater conditions in the early Cambrian. Chem Geol 390:164–172

Wen H, Zhang Y, Cloquet C, Zhu C, Fan H, Luo C (2015) Tracing sources of pollution in soils from the Jinding Pb–Zn mining district in China using cadmium and lead isotopes. Appl Geochem 52:147–154

Wen H, Zhu C, Zhang Y, Cloquet C, Fan H, Fu S (2016) Zn/Cd ratios and cadmium isotope evidence for the classification of lead-zinc deposits. Sci Rep 6:25273. https://doi.org/10.1038/srep25273

Wen B, Zhou J, Zhou A, Liu C, Li L (2018) A review of antimony (Sb) isotopes analytical methods and application in environmental systems. Inter Biodet Biodegr 128:109–116

Wenzel B, Lecuyer C, Joachimski MM (2000) Comparing oxygen isotope records of Silurian calcite and phosphate—$\delta^{18}O$ composition of brachiopods and conodonts. Geochim Cosmochim Acta 69:1859–1872

Weyer S, Schwieters JB (2003) High precision Fe isotope measurements with high mass resolution MC-ICPMS. Inter J Mass Spectr 226:355–368

Weyer S, Anbar AD, Brey GP, Münker C, Mezger K (2005) Iron isotope fractionation during planetary differentiation. Earth Planet Sci Lett 240:251–264

Weyer S, Ionov D (2007) Partial melting and melt percolation in the mantle: the message from Fe isotopes. Earth Planet Sci Lett 259:119–133

Weyer S, Anbar AD, Gerdes A, Gordon GW, Algeo TJ, Boyle EA (2008) Natural fractionation of $^{238}U/^{235}U$. Geochim Cosmochim Acta 72:345–3359

Weyrauch M, Oeser M, Brüske A, Weyer S (2017) In situ high-precision Ni isotope analysis of metals by femtosecond-LA-MC-ICP-MS. JAAS 32:1312–1319

Whiteman JP, Elliott Smith EA, Besser AC, Newsome SD (2019) A guide to using compound-specific stable isotope analysis to study the fate of molecules in organisms and ecosystems. Diversity 11:8. https://doi.org/10.3390/d11010008

Widanagamage IH, Schauble EA, Scher HD, Griffith EM (2014) Stable strontium isotope fractionation in synthetic barite. Geochim Cosmochim Acta 147:58–75

Widanagamage IH, Griffith EM, Singer DM, Scher HD, Buckley WP, Senko JM (2015) Controls on stable Sr-isotope fractionation in continental barite. Chem Geol 411:215–227

Wiechert U, Fiebig J, Przybilla R, Xiao Y, Hoefs J (2002) Excimer laser isotope-ratio-monitoring mass spectrometry for in situ oxygen isotope analysis. Chem Geol 182:179–194

Wiechert U, Hoefs J (1995) An excimer laser-based microanalytical preparation technique for in-situ oxygen isotope analysis of silicate and oxide minerals. Geochim Cosmochim Acta 59:4093–4101

Wiechert U, Halliday AN (2007) Non-chondritic magnesium and the origin of the inner terrestrial planets. Earth Planet Sci Lett 256:360–371
Wiederhold JG (2015) Metal stable isotope signatures as tracers in environmental geochemistry. Environ Sci Tech 49:2606–2614
Wiederhold JG, Kraemer SM, Teutsch N, Borer PM, Halliday AN, Kretzschmar R (2006) Iron isotope fractionation during proton-promoted, ligand-controlled and reductive dissolution of goethite. Environ Sci Tech 40:3787–3793
Wiegand BA, Chadwick OA, Vitousek PM, Wooden JH (2005) Ca cycling and isotopic fluxes in forested ecosystems in Hawaii. Geophys Res Lett 32:L11404
Wiggenhauser M, Bigalke M, Imseng M, Müller M, Keller A, Murphy K, Kreissig K, Rehkämper M, Wilcke W, Frossard E (2016) Cadmium isotope fractionation in soil-wheat systems. Environm Sci Tech 50:9223–9231
Wilkinson JJ, Weiss DJ, Mason TF, Coles BJ (2005) Zinc isotope variation in hydrothermal systems: preliminary evidence from the Irish Midlands ore field. Econ Geol 100:583–590
Willbold M, Elliott T (2017) Molybdenum isotope variations in magmatic rocks. Chem Geol 449:253–268
Wille M, Kramers JD, Nägler TF, Beukes NJ, Schroder S, Meiser T, Lacassie JP, Voegelin AR (2007) Evidence for a gradual rise of oxygen between 2.6 and 2.5 Ga from Mo isotopes and Re-PGE signatures in shales. Geochim Cosmochim Acta 71:2417–2435
Wille M, Sutton J, Ellwood MJ, Sambridge M, Maher W, Eggins S, Kelly M (2010) Silicon isotopic fractionation in marine sponges: a new model for understanding silicon isotope variations in sponges. Earth Planet Sci Lett 292:281–289
Williams HM, Peslier AH, McCammon C, Halliday AN, Levasseur S, Teutsch N, Burg JP (2005) Systematic iron isotope variations in mantle rocks and minerals: the effects of partial melting and oxygen fugacity. Earth Planet Sci Lett 235:435–452
Williams HM, Markowski A, Quitte G, Halliday AN, Teutsch N, Levasseur S (2006) Fe isotope fractionations in iron meteorites: new insight into metal-sulphide segregation and planetary accretion. Earth Planet Sci Lett 250:486–500
Williams HM, Bizimis M (2014) Iron isotope tracing of mantle heterogeneity within the source regions of oceanic basalts. Earth Planet Sci Lett 404:396–407
Williams LB, Ferrell RE, Hutcheon I, Bakel AJ, Walsh MM, Krouse HR (1995) Nitrogen isotope geochemistry of organic matter and minerals during diagenesis and hydrocarbon migration. Geochim Cosmochim Acta 59:765–779
Williams LB, Hervig RL, Holloway JR, Hutcheon I (2001) Boron isotope geochemistry during diagenesis. Part I. Experimental determination of fractionationduring illitization of smectite. Geochim Cosmochim Acta 65:1769–1782
Williams LB, Hervig RL (2004) Boron isotopic composition of coals: a potential tracer of organic contaminated fluids. Appl Geochem 19:1625–1636
Wimpenny J, Gislason SR, James RH, Gannoun A, Pogge von Strandmann P, Burton KW (2010) The behavior of Li and Mg isotopes during primary phase dissolution and secondary mineral formation in basalt. Geochim Cosmochim Acta 74:5259–5279
Wimpenny J, Marks N, Knight K, Borg L, Badro J, Ryerson R (2020) Constraining the behavior of gallium isotopes during evaporation at extreme temperatures. Geochim Cosmochim Acta 286:54–71
Wombacher F, Rehkämper M, Mezger K, Münker C (2003) Stable isotope composition of cadmium in geological materials and meteorites determined by multiple-collector ICPMS. Geochim Cosmochim Acta 67:4639–4654
Wombacher F, Rehkämper M, Mezger K (2004) Dependence of the mass-dependence in cadmium isotope fractionation during evaporation. Geochim Cosmochim Acta 68:2349–2357
Wombacher F, Rehkämper M, Mezger K (2008) Cadmium stable isotope cosmochemistry. Geochim Cosmochim Acta 72:646–667

Woodland SJ, Rehkämper M, Halliday AN, Lee DC, Hattendorf B, Günther D (2005) Accurate measurement of silver isotopic compositions in geological materials including low Pd/Ag meteorites. Geochim Cosmochim Acta 69:2153–2163

Wortmann UG, Bernasconi SM, Böttcher ME (2001) Hypersulfidic deep biosphere indicates extreme sulfur isotope fractionation during single-step microbial sulfate reduction. Geology 29:647–650

Wu F, Qin T, Li X, Liu Y, Huang JH, Wu Z, Huang F (2015) First-principles investigation of vanadium isotope fractionation in solution and during adsorption. Earth Planet Sci Lett 426:216–224

Wu F, Qi Y, Perfit MR, Gao Y, Langmuir CH, Wanless VD, Yu H, Huang F (2018) Vandium isotope compositions of mid-ocean ridge lavas and altered oceanic crust. Earth Planet Sci Lett 493:128–139

Wu F, Owens JD, Huang T, Sarafian A, Huang KF, Sen IS, Horner TJ, Blusztain J, Morton P, Nielsen SG (2019) Vanadium isotope composition of sea water. Geochim Cosmochim Acta 244:403–415

Wu F, Owens JD, Scholz F, Huang L, Li S, Riedinger N, Petersen LC, German CR, Nielsen SG (2020b) Sedimentary vanadium isotope signatures in low oxygen marine conditions. Geochim Cosmochim Acta 284:134–155

Wu F, Turner S, Schaefer BF (2020b) Melange versus fluid and melt enrichment of subarc mantle: a novel test using barium isotopes in the Tonga-Kermadec arc. Geology 48:1053–1057

Wu H, He Y, Bao L, Zhu C, Li S (2017a) Mineral composition control on inter-mineral iron isotopic fractionation in granitoids. Geochim Cosmochim Acta 198:208–217

Wu L, Beard BL, Roden EE, Johnson CM (2011) Stable iron isotope fractionation between aqueous Fe (II) and hydrous ferric oxide. Environ Sci Technol 45:1845–1852

Wu W, Wang X, Reinhard CT, Planavsky NJ (2017b) Cr isotope systematics in the Connecticut River. Chem Geol 456:98–111

Wunder B, Meixner A, Romer R, Wirth R, Heinrich W (2005) The geochemical cycle of boron: constraints from boron isotope partitioning experiments between mica and fluid. Lithos 84:206–216

Wunder B, Meixner A, Romer R, Heinrich W (2006) Tempearature-dependent isotopic fractionation of lithium between clinopyroxene and high-pressure hydrous fluids. Contr Mineral Petrol 151:112–120

Wunder B, Meixner A, Romer RL, Feenstra A, Schettler G, Heinrich W (2007) Lithium isotope fractionation between Li-bearing staurolite, Li-mica and aqueous fluids: an experimental study. Chem Geol 238:277–290

Xia J, Qin L, Shen J, Carlson R, Ionov DA, Mock TD (2017) Chromium isotope heterogeneity in the mantle. Earth Planet Sci Lett 464:103–115

Xiao YK, Liu WG, Qi HP, Zhang CG (1993) A new method for the high-precision isotopic measurement of bromine by thermal ionization mass spectrometry. Int J Mass Spectrom Ion Proc 123:117–123

Xiao Y, Teng FZ, Zhang HF, Yang W (2013) Large magnesium isotope fractionation in peridotite xenoliths from eastern North China craton: product of melt-rock interaction. Geochim Cosmochim Acta 115:241–261

Xie H, Ponton C, Formolo MJ, Lawson M, Peterson BK, Lloyd MK, Sessions AL, Eiler JM (2018) Position-specific hydrogen isotope equilibrium in propane. Geochim Cosmochim Acta 238:193–207

Xie RC, Galer SJ, Abouchami W, Rijkenberg MJ, de Baar HJ, De Jong J, Andreae MO (2017) Non-Rayleigh control of upper-ocean Cd isotope fractionation in the western South Atlantic. Earth Planet Sci Lett 417:94–103

Xie RC, Rehkämper M, Grasse P, de Flierdt T, Frank M, Xue Z (2019) Isotopic evidence for complex biogeochemical cycling of Cd in the eastern tropical South Pacific. Earth Planet Sci Lett 512:134–146

Xu CX, Yin RS, Peng JT, Hurley JP, Lepak RF, Gao JF, Feng XB, Hu RZ, Bi XW (2018) Mercury isotope constraints on the source for sediment-hosted lead-zinc deposits in the Chengdu area, southwestern China. Mineral Deposita 53:339–352
Xu YK, Xin YH, Tian YC, Huang Y, Sletten RS, Zhu D, Teng FZ (2019) Potassium isotopic compositions of international geological reference materials. Chem Geol 513:101–107
Xu W, Zhu JM, Johnson TM, Wang X, Lin ZQ, Tan D, Qin H (2020) Selenium isotope fractionation during adsorption by Fe, Mn and Al oxides. Geochim Cosmochim Acta (in press)
Xue Z, Rehkämper M, Horner TJ, Abouchami W, Middag R, van de Flierd T, de Baar HJ (2013) Cadmium isotope variations in the Southern Ocean. Earth Planet Sci Lett 382:161–172
Yamaguchi KE, Johnson CM, Beard BL, Ohmoto H (2005) Biogeochemical cycling of iron in the Archean-Paleoproterozoic earth: constraints from iron isotope variations in sedimentary rocks from the Kapvaal and Pilbara cratons. Chem Geol 218:135–169
Yamazaki E, Nakai S, Yokoyama T, Ishihara S, Tang HF (2013) Tin isotopic analysis of cassiterites from southeastern and eastern Asia. Geochem J 47:21–35
Yang C, Liu SA (2019) Zinc isotope constraints on recycled oceanic crust in the mantle sources of the Emeishan Large Igneous province. J Geophys Res (solid Earth) 124:12537–12555
Yang J, Siebert C, Barling J, Savage P, Liang YH, Halliday A (2015a) Absence of molybdenum isotope fractionation during magmatic differentiation at Hekla volcano, Iceland. Geochim Cosmochim Acta 162:126–136
Yang J, Li Y, Liu S, Tian H, Chen C, Liu J, Shi Y (2015b) Theoretical calculations of Cd isotope fractionation in hydrothermal fluids. Chem Geol 391:74–82
Yang J, Barling J, Siebert C, Fietzke J, Stephens E, Halliday AN (2017) The molybedenum isotopic compositions of I-, S- and A-type granitic suites. Geochim Cosmochim Acta 205:168–186
Yang S-C, Lee D-C, Ho L-Y (2012a) The isotopic composition of cadmium in the water column of the South China Sea. Geochim Cosmochim Acta 98:66–77
Yang W, Teng FZ, Zhang HF (2009) Chondritic magnesium isotopic composition of the terrestrial mantle: a case study of peridotite xenoliths from the North China craton. Earth Planet Sci Lett 288:475–482
Yang W, Teng FZ, Zhang HF, Li SG (2012b) Magnesium isotopic systematics of continental basalts from the North China craton: implications for tracing subducted carbonate in the mantle. Chem Geol 328:185–194
Yang W, Teng FZ, Li WY, Liu SA, Ke S, Liu YS, Zhang HF, Gao S (2016) Magnesium isotope composition of the deep continental crust. Am Mineral 101:243–252
Yao J, Mathur R, Powell W, Lehmann B, Tornos F, Wilson M, Ruiz J (2018) Sn-isotope fractionation as a record of hydrothermal redox reactions. Am Mineralogist 103:1591–1598
Yesavage T, Fantle MS, Vervoort J, Mathur R, Jin L, Liermann LJ, Brantley SL (2012) Fe cycling in the Shale Hills critical zone observatory, Pennsylvania: an analysis of biogeochemical weathering and Fe isotope fractionation. Geochim Cosmochim Acta 99:18–38
Yierpan A, König S, Labidi J, Kurzawa T, Babechuk MG, Schoenberg (2018) Chemical sample processing for combinedselenium isotope and selenium-tellurium elemental investigation of the Earth' s igneous reservoirs. Geochem Geophys Geosyst 19:516–533
Yierpan A, König S, Labidi J, Schoenberg R (2019) Selenium isotope and S-Se-Te elemental systematics along the Pacific-Antarctic ridge: role of mantle processes. Geochim Cosmochim Acta 249:199–224
Yin NH, Sivry Y, Benedetti MF, Lens PN, van Hullebusch ED (2016) Application of Zn isotopes in environmental impact assessment of Zn–Pb metallurgical industries: a mini review. Appl Geochem 64:128–135
Yin R, Feng X, Shi W (2010) Application of the stable isotope system to the study of sources and fate of Hg in the environment: a review. Appl Geochem 25:1467–1477

Yin R, Feng X, Wang J, Li P, Liu J, Zhang Y, Chen J, Zheng L, Hu T (2013) Mercury speciation and mercury isotope fractionation during ore roasting process and their implication to source identification of downstream sediment in the Wanshan mercury mining area, SW China. Chem Geol 366:39–46

Yin R, Feng X, Chen B, Zhang J, Wang W, Li X (2015) Identifying the sources and processes of mercury in subtropical estuarine and ocean sediments using Hg isotopic composition. Environ Sci Technol 49:1347–1355

Yin R, Feng X, Hurley JP, Krabbenhoft DP, Lepak RF, Hu R, Zhang Q, Li Z, Bi X (2016b) Mercury isotopes as proxies to identify sources and environmental impacts of mercury in sphalerites. Sci Rep 6:18686. https://doi.org/10.1038/srep18686

Yokochi R, Marty B, Chazot G, Burnard P (2009) Nitrogen in perigotite xenoliths: lithophile behaviour and magmatic isotope fractionation. Geochim Cosmochim Acta 73:4843–4861

Young ED, Galy A (2004) The isotope geochemistry and cosmochemistry of magnesium. Rev Mineral Geochem 55:197–230

Young ED, Galy A, Nagahara H (2002) Kinetic and equilibrium mass-dependent isotope fractionation laws in nature and their geochemical and cosmochemical significance. Geochim Cosmochim Acta 66:1095–1104

Young ED, 23 others (2017) The relative abundances of resolved $^{12}CH_2D_2$ and $^{13}CH_3D$ and mechanisms controlling isotopic bond ordering in abiotic and biotic methane gases. Geochim Cosmochim Acta 203:235–264

Young MB, McLaughlin K, Kendall C, Stringfellow W, Rollow M, Elsbury K, Donald E, Payton A (2009) Characterizing the oxygen isotopic composition of phosphate sources to aquatic ecosystems. Environ Sci Techn 43:5190–5196

Yuan W, Chen JB, Birck JL, Yin ZY, Yuan SL, Cai HM, Wang ZW, Huang Q, Wang ZH (2016) Precise analysis of gallium isotopic composition by MC-ICP-MS. Anal Chem 88:9606–9613

Yuan W, Saldi GD, Chen JB, Zuccolini MV, Birck JL, Liu Y, Schott J (2018) Gallium isotope fractionation during Ga adsorption on calcite and goethite. Geochim Cosmochim Acta 223:350–363

Zakharov DO, Marin-Carbonne J, Alleon J, Bindeman IN (2021) Triple oxygen isotope trend recorded by Precambrian cherts: a perspective from combined bulk and in situ Secondary Ion Probe measurements. Rev Mineral Geochem 86:323–365

Zambardi T, Poitrasson F, Corgne A, Meheut M, Quitte Anand M (2013) Silicon isotope variations in the inner solar system: implications for planetary formation, differentiation and composition. Geochim Cosmochim Acta 121:67–83

Zambardi T, Sonke JE, Toutain JP, Sortino F, Shinohara H (2009) Mercury emissions and stable isotope compositions at Vulcano Island (Italy). Earth Planet Sci Let 277:236–243

Zeebe RE (2007) An expression for the overall oxygen isotope fractionation between the sum of dissolved inorganic carbon and water. Geochem Geophys Geosys 8. https://doi.org/10.1029/2007GC001663

Zerkle AL, Schneiderich K, Maresca JA, Liermann LJ, Brantley SL (2011) Molybdenum isotope fractionation by cyanobacterial assimilation during nitrate utilization and $N_2$ fixation. Geobiology 9:94–106

Zhang L, Chan LH, Gieskes JM (1998) Lithium isotope geochemistry of pore waters from Ocean Drilling Program Sites 918 and 919, Irminger Basin. Geochim Cosmochim Acta 62:2437–2450

Zhang Z, Ma J, Wang Z, Zhang L, He X, Zhu G, Zeng T, Wei G (2021) Rubidium isotope fractionation during chemical weathering of granite. Geochim Cosmochim Acta (in press)

Zhang J, Quay PD, Wilbur DO (1995) Carbon isotope fractionation during gas-water exchange and dissoltion of $CO_2$. Geochim Cosmochim Acta 59:107–114

Zhang H, Wang Y, He Y, Teng FZ, Jacobsen SB, Helz RT, Marsh BD, Huang S (2018a) No measurable calcium isotopic fractionation during crystallization of Kilauea Iki lava lake. Geochem Geophys Geosys 19:3128–3139

Zhang T, Zhou L, Yang L, Wang Q, Feng LP, Liu YS (2016) High precision measurements of gallium isotopic compositions in geological materials by MC-ICP-MS. JAAS. https://doi.org/10.1039/c6ja00202a
Zhang F, Xiao S, Kendall B, Romaniello SJ, Cui H, Meyer M, Anbar AD (2018b) Extensive marine anoxia during the terminal Ediacaran Period. Sci Adv 4:aan8983
Zhang F, 15 others (2020) Uranium isotopes in marine carbonates as a global ocean paleoredox proxy: a critical review. Geochim Cosmochim Acta 287:27–49
Zhang Q, Amor K, Galer SJ, Thompson I, Porcelli D (2018c) Variations of stable isotope fractionation during bacterial chromium reduction processes and their implications. Chem Geol 481:155–164
Zhang W, Wang Z, Moynier F, Inglis E, Tian S, Li M, Liu Y, Hu Z (2019) Determination of Zr isotopic ratios in zircons using laser-ablation multiple-collector inductively coupled plasma mass-spectrometry. JAAS 34:1800–1809
Zhao M, Tarhan LG, Zhang Y, Hood A, Asael D, Reid RP, Planavsky NJ (2021) Evaluation of shallow-water carbonates as a seawater zinc isotope archive. Earth Planet Sci Lett 553:116599
Zhao Y, Vance D, Abouchami W, de Baar HJ (2014) Biogeochemical cycling of zinc and its isotopes in the Southern Ocean. Geochim Cosmochim Acta 125:653–672
Zhao Y, Xue C, Liu SA, Symons DT, Zhao X, Yang Y, Ke JJ (2017a) Copper isotope fractionation during sulfide-magma differentiation in the Tulaergen magmatic Ni–Cu deposit, NW China. Lithos 286–287:206–215
Zhao X, Zhang ZF, Huang S, Liu Y, Li X, Zhang H (2017b) Coupled extremely light Ca and Fe isotopes in peridotites. Geochim Cosmochim Acta 208:368–380
Zheng W, Hintelmann H (2010) Nuclear field shift effects in isotope fractionation of mercury during abiotic reduction in the absence of light. J Phys Chem A 114:4238–4245
Zheng W, Gilleaudeau GJ, Kah LC, Anbar AD (2018) Mercury isotope signatures record photic zone euxinia in the Mesoproterozoic ocean. PNAS 115:10594–10599
Zhou JX, Huang ZL, Zhou MF, Zhu XK, Muchez P (2014) Zinc, sulphur and lead isotopic variations in carbonate-hosted Pb–Zn sulfide deposits, Southwest China. Ore Geol Rev 58:41–54
Zhu C, Wen H, Zhang Y, Fu S, Fan H, Cloquet C (2017a) Cadmium isotope fractionation in the Fule Mississippi Valley-type deposit, Southwest China. Min Deposita 52:675–686
Zhu XK, O'Nions RK, Guo Y, Belshaw NS, Rickard D (2000) Determination of natural Cu-isotope variations by plasma-source mass spectrometry: implications for use as geochemical tracers. Chem Geol 163:139–149
Zhu XK et al (2002) Mass fractionation processes of transition metal isotopes. Earth Planet Sci Lett 200:47–62
Zhu JM, Johnson TM, Clark SK, Zhu XK, Wang XL (2014) Selenium redox cycling during weathering of Se-rich shales: a selenium isotope study. Geochim Cosmochim Acta 126:228–249
Zhu P, MacDougall JD (1998) Calcium isotopes in the marine environment and the oceanic calcium cycle. Geochim Cosmochim Acta 62:1691–1698
Zhu Z, Meija J, Zheng A, Mester Z, Yang L (2017b) Determination of the isotopic composition of iridium using Multicollector-ICPMS. Anal Chem 89:9375–9382
Ziegler K, Chadwick OA, Brzezinski MA, Kelly EF (2005a) Natural variations of $\delta^{30}$Si ratios during progressive basalt weathering. Geochim Cosmochim Acta 69:4597–4610
Ziegler K, Chadwick OA, White AF, Brzezinski MA (2005b) $\delta^{30}$Si systematics in a granitic saprolite, Puerto Rico. Geology 33:817–820
Ziegler K, Young ED, Schauble E, Wasson JT (2010) Metal-silicate silicon isotope fractionationin enstatite meteorites and constraints on Earth's core formation. Earth Planet Sci Lett 295:487–496

Zink S, Schoenberg R, Staubwasser M (2010) Isotopic fractionation and reaction kinetics between Cr(III) and Cr(VI) in aqueous media. Geochim Cosmochim Acta 74:5729–5745
Zou Z, Wang Z, Li M, Becker H, Geng X, Hu Z, Lazarov M (2019) Copper isotope variations during magmatic migration in the mantle: insights from mantle pyroxenites in Balmuccia peridotite massif. J Geophys Res (solid Earth) 124(11):11130–11149

# 3 Variations of Stable Isotope Ratios in Nature

## 3.1 Extraterrestrial Materials

Extraterrestrial materials consist of samples from the moon, Mars, Vesta and a variety of smaller bodies such as asteroids and comets. These planetary samples have been used to deduce the evolution of our solar system. A major difference between extraterrestrial and terrestrial materials is the existence of primordial isotopic heterogeneities in the early solar system. These heterogeneities are not observed on Earth, because they have become obliterated during high-temperature processes over geologic time. Nevertheless, isotopes have been used as a genetic link between meteorites and the Earth (i.e. Clayton 2004). Small differences in isotope composition between the Earth and meteorite groups may identify the type of meteorites that are representative of precursor material that formed the early Earth (Simon and DePaolo 2010; Valdes et al. 2014; Ireland et al. 2020).

Heterogeneities in isotope composition indicate incomplete mixing of distinct presolar materials during formation of the solar system. Such isotope anomalies have been documented on all scales, from microscopic zoning in meteoritic minerals to bulk asteroids. The most extreme examples, however, have been documented from minute presolar grains extracted from primitive meteorites and measured with the ion microprobe. These high-temperature grains of silicon carbide, graphite, diamond etc. have been formed by condensation in cooling gases and show isotope variations that may vary by several orders of magnitude, too large to be explained by chemical or physical fractionation, but pointing to nuclear reactions. They have acquired their isotope characteristics before the solar system has been formed. The implications of these variations for models of stellar formation have been summarized by Zinner (1998), Hoppe and Zinner (2000), Clayton and Nittler (2004) and others.

As shown by high precision MC-ICP-MS measurements, many elements display small nucleosynthetic isotope variations in meteorites originating from the heterogeneous distribution of presolar material in the solar system. A number of studies have reported correlated mass-independent isotope variations in meteorites.



J. Hoefs, *Stable Isotope Geochemistry*, Springer Textbooks in Earth Sciences, Geography and Environment, https://doi.org/10.1007/978-3-030-77692-3_3

Examples have been published by Trinquier et al. (2009) for titanium, Steele et al. (2012) for nickel, Akram et al. (2015) for zirconium, Ek et al. (2017) for palladium. The small variations are useful for the identification of meteorite classes and provide indicators of mixing processes.

### 3.1.1 Chondrites

Primitive meteorites of chondritic composition are stony undifferentiated bodies that have formed from the primitive solar material during the formation of the solar system. Chondritic meteorites can be divided into different classes on the basis of their volatile contents and of their total iron content distributed between Fe in silicates and Fe in metal.

Most chondrites have experienced a complex history, which includes primary formation processes and secondary processes that include thermal metamorphism and aqueous alteration. It is generally very difficult to distinguish between the effects of primary and secondary processes on the basis of isotope composition.

#### 3.1.1.1 Oxygen

It is generally agreed that variations in the oxygen isotope composition within the solar system result from mixing of two distinct reservoirs: an $^{16}O$-rich and an $^{17}O$, $^{18}O$-rich reservoir relative to Earth. The first observation, that clearly demonstrated isotopic inhomogeneities in the early solar system, was made by Clayton et al. (1973a). Previously, it had been thought that in a plot of $^{17}O/^{16}O$ versus $^{18}O/^{16}O$, all physical and chemical processes must produce mass–dependent O-isotope fractionations yielding a straight line with a slope of 0.52. This line has been called the "Terrestrial Fractionation Line". Figure 3.1 shows that O-isotope data from terrestrial and lunar samples fall along the predicted mass–dependent fractionation line. Bulk meteorites, the Moon and Mars lie within a few ‰ above or below the terrestrial fractionation line. Thus, triple oxygen isotopes became a tool to classify bulk meteorites, the Moon, Mars and Vesta.

Selected anhydrous high-temperature minerals in carbonaceous chondrites, do not fall along the chemical fractionation trend, but instead define another trend with a slope of 1. The first evidence for oxygen isotope anomalies was found in Ca–Al-rich refractory inclusions (CAI) in the Allende carbonaceous chondrite, which are composed predominantly of melilite, pyroxene, and spinel.

The origin of the line with a slope of 1 was first interpreted to mean an injection of supernova material into an early forming disk. However, no similar "anomalies" have been later found in isotopes of other elements. Thiemens and Heinderich (1983) observed that UV photolysis and formation of ozone fractionates oxygen along the slope of 1, potentially explaining the trend.

The carbonaceous chondrites display the widest range in oxygen isotope composition of any meteorite group (Clayton and Mayeda 1999). The evolution of these meteorites can be interpreted as a progression of interactions between dust and gas components in the solar nebula followed by solid/fluid interactions within parent

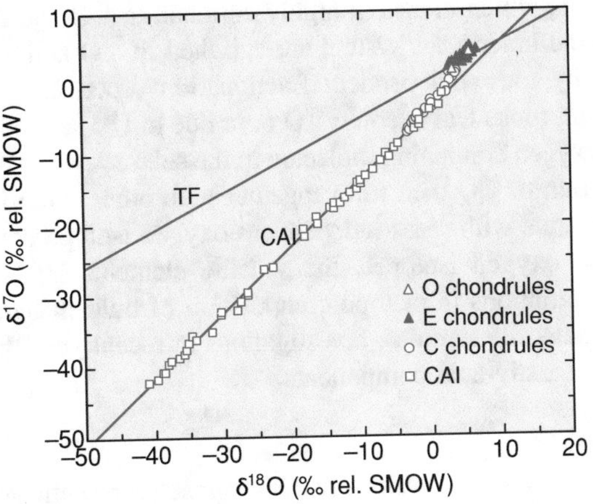


**Fig. 3.1** $^{17}O$ versus $^{18}O$ isotopic composition of Ca–Al rich inclusions (CAI) from chondrites (Clayton 1993)

bodies. Young et al. (1999) have shown that reactions between rock and water inside a carbonaceous chondrite parent body could have produced groups of different carbonaceous chondrite types having different paragenesis of secondary minerals. The analysis of the isotope compositions of phyllosilicates, carbonates etc. provide evidence under which conditions aqueous alteration took place. Clumped isotope temperatures between 20 and 70 °C allow the reconstruction of aqueous alterations in carbonaceous chondrites (Guo and Eiler 2007).

Yurimoto et al. (2008) have summarized the oxygen isotope composition of the chondrite components (refractory inclusions, chondrules and matrix) and concluded that O isotope variations within a chondrite are typically larger than O isotope variations among bulk chondrites. The question remains as to where, when, and how the isotopic anomalies were originally produced (Thiemens 1988). Even without full understanding of the causes of isotope variations in meteorites, oxygen isotopes are very useful in classifying meteorites and in relating meteorites to their precursor asteroids and planets (Clayton 2004). Oxygen isotope signatures have confirmed that eucrites, diogenites, howardites and mesosiderites originate from one single parent body probably derived from the asteroid 4 Vesta, as shergottites, nakhlites and chassignites come from another parent body, most probably Mars (Clayton and Mayeda 1996). The main group of pallasites represent intermixed core-mantle material from a single disrupted asteroid with no equivalent known (Greenwood et al. 2006).

In the past it had been assumed that the oxygen isotope composition of the Sun is the same as that of the Earth. This view has changed with the suggestion of Clayton (2002) that the Sun and the initial composition of the solar system is $^{16}O$ rich. The O-isotope composition of the Sun is reflected in the composition of the solar wind. McKeegan et al. (2011) measured the solar wind collected during the Genesis

Discovery mission, which indeed is highly enriched in $^{16}O$ and they demonstrated that rocks from the inner solar system are enriched in $^{17}O$ and $^{18}O$ by about 70‰ relative to $^{16}O$ by mass-independent fractionation processes. According to this model, solar sytem rocks had become $^{16}O$ poor due to UV self shielding of CO, the most abundant oxygen containing molecule in the solar system. Oxygen released by the UV dissociation of CO then form together with other components of the solar system solid minerals with mass-independent oxygen isotope compositions.

In addition to oxygen isotopes, the volatile elements H, C, N and S show extremely large variations in isotope composition of bulk meteorites. Rather than analyzing bulk meteorite samples, investigations in recent years have concentrated on the analyses of individual components.

#### 3.1.1.2 Hydrogen

The solar system consists of water containing reservoirs with very different D/H isotope compositions that can be used as fingerprints for the origin of water in planetary bodies (Saal et al. 2013; Sarafian et al. 2014). Hydrogen isotopes indicate a gradient through the solar system as a function of distance from the Sun: the protosolar nebula is very D-depleted whereas ice in the outer solar system is very D enriched. Because retention of hydrogen is a function of the gravitational field, lighter planets such as Mars loose $H_2$ prefererentially to D-H becoming progressively heavier due to overall hydrogen escape.

Similar ranges of D/H ratios among carbonaceous chondrites, Earth, Mars and Moon suggest a common source region for water in these planetary bodies. Alexander et al. (2012) compared D-isotope ratios of chondritic meteorites with those in comets and demonstrated that they are distinct from one another with comets being highly enriched in D relative to chondrites. Since the various types of chondrites have D-contents being similar to Earth, the dominant source of volatiles on Earth appear to be from asteroids (Sarafian et al. 2014). In specific, enstatite chondrites have similar isotopic composition to terrestrial rocks (Javoy et al. 2010). Piani et al. (2020) concluded that enstatite chondrites not only have similar hydrogen and nitrogen isotope composition, but have sufficient hydrogen to be delivered to Earth at least 3 times the mass of ocean water.

In extraterrestrial material, hydrogen is bound in hydrated minerals and in organic matter. Hydrogen isotopes, thus, may provide insight not only into the origin of water in planetary material (Robert 2001; Alexander et al. 2012; Marty 2012; Saal et al. 2013), but also in the origin of organic molecules (Deloule and Robert 1995; Deloule et al. 1998).

Bulk D/H ratios of meteorites give a relatively homogeneous composition with a mean $\delta^2$H-value of −100‰ (Robert et al. 2000). This relatively homogeneous composition masks the very heterogeneous distribution of individual components. Considerable efforts have been undertaken to analyze D/H ratios of the different compounds (Robert et al. 1978; Kolodny et al. 1980; Robert and Epstein 1982; Becker and Epstein 1982; Yang and Epstein 1984; Kerridge 1983; Kerridge et al. 1987; Halbout et al. 1990; Krishnamurthy et al. 1992). Eiler and Kitchen (2004) have evaluated the hydrogen isotope composition of water-rich carbonaceous chondrites by stepwise-heating analysis of very small amounts of separated water-rich material. They observed a decrease in $\delta^2H$ with increasing extent of aqueous alteration from 0‰ (least altered, most volatile rich) to −200‰ (most altered, least volatile rich).

Hydrogen in organic matter reveals a $\delta^2H$-variation from −500 to +6000‰ whereas water in silicates gives a variation from −400 to +3700‰ (Deloule and Robert 1995; Deloule et al. 1998). Two mechanisms have been proposed to account for the deuterium enrichment: (i) for organic molecules, high D/H ratios can be explained by ion molecule reactions that occur in interstellar space and (ii) for phyllosilicates the enrichment can be produced via isotope exchange between water and hydrogen (Robert et al. 2000).

Alexander et al. (2010) reported even larger D-enrichment up to almost +12,000‰ in insoluble organic material. They suggested that such large enrichments may be produced in the meteorite parent body through the loss of isotopically very light $H_2$ generated through Fe oxidation by water at temperatures below 200 °C.

#### 3.1.1.3 Carbon

Besides the bulk carbon isotopic composition, the various carbon phases occurring in carbonaceous chondrites (kerogen, carbonates, graphite, diamond, silicon carbide) have been individually analyzed. The $\delta^{13}C$–values of the total carbon fall into a narrow range, whereas $\delta^{13}C$–values for different carbon compounds in single meteorites show extremely different $^{13}C$-contents. Figure 3.2 shows one such example, the Murray meteorite after Ming et al. (1989).

Of special interest are the minute grains of silicon carbide and graphite in primitive carbonaceous chondrites, which obviously carry the chemical signature of the pre–solar environment (Ott 1993). The SiC grains, present at a level of a few ppm, have a wide range in silicon and carbon isotope composition, with accompanying nitrogen also being isotopically highly variable. The $^{12}C/^{13}C$ ratio ranges

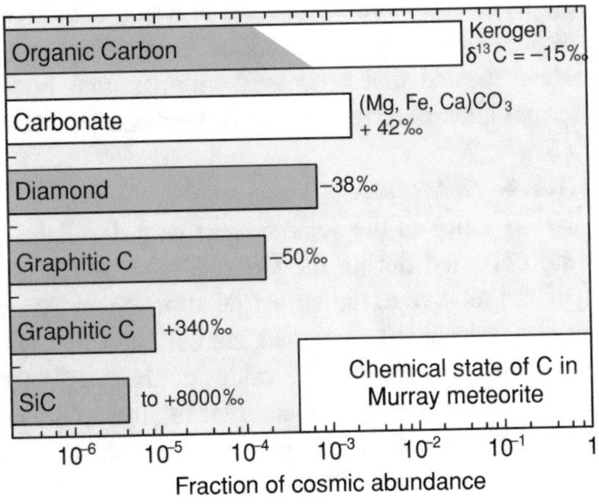


**Fig. 3.2** Carbon compounds in primitive meteorites. Species classified as interstellar on the basis of C-isotopes are *coloured*. Only a minor fraction of organic carbon is interstellar (after Ming et al. 1989)

from 2 to 2500, whereas it is 89 for the bulk Earth. According to Ott (1993), the SiC grains can be regarded as "star dust", probably from carbon stars that existed long before our solar system. Amari et al. (1993) presented ion microprobe data of individual micrometer sized graphite grains in the Murchison meteorite, that also deviate from values typical for the solar system. These authors interpreted the isotope variability as indicating at least three different types of stellar sources.

The analysis of meteoritic organic matter may provide information about the origin of prebiotic organic matter in the early solar system. Carbonaceous chondrites contain organic carbon in solvent-insoluble form (about 70%) and a mixture of solvent-soluble organic compounds (about 30%). The organic carbon is substantially enriched in $^{13}C$ and $^{15}N$, indicating that the material is not a terrestrial contaminant.

Two hypotheses have dominated the debate over formation mechanisms for the organic matter: I. formation by a Fischer-Tropsch type process (the synthesis of hydrocarbons from carbon monoxide and hydrogen) promoted by catalytic mineral grains and II. formation by Miller–Urey type reactions (the production of organic compounds by radiation or electric discharge) in an atmosphere in contact with an aqueous phase. However, the isotopic variability exhibited by the volatile elements in different phases in carbonaceous chondrites is not readily compatible with abiotic syntheses. Either complex variants of these reactions must be invoked, or totally different types of reactions need to be considered. $\delta^{13}C$-values reported for amino acids in the Murchison meteorite vary between +23 and +44‰ (Epstein et al. 1987). Engel et al. (1990) analyzed individual amino acids in the Murchison meteorite and also confirmed a strong $^{13}C$ enrichment. Of particular importance is the discovery of a distinct $\delta^{13}C$ difference between D- and L-alanine, which suggests that optically active forms of material were present in the early solar system.

Compound specific C- and H-isotope compositions of carboxylic acids, the most abundant class of soluble organic compounds in carbonaceous chondrites show a large range in $\delta^{13}C$-values (from −31 to +32‰) and positive $\delta^{2}H$-values, that give evidence that these materials are not terrestrial contaminants (Huang et al. 2005a, b).

In the insoluble macromolecular organic matter Alexander et al. (2007) observed very large variations within and between chondrite classes. These authors excluded Fischer–Tropsch type reactions being responsible for the large variations but instead argued that processes within parent bodies, such as different degrees of thermal alteration, may cause differences in $\delta^{2}H$-values.

#### 3.1.1.4 Nitrogen

$^{15}N/^{14}N$ ratios in the solar system vary dramatically (Hüri and Marty 2015). Solar wind collected during the Genesis mission has a $^{15}N$-content that is about 400‰ depleted relative to the terrestrial atmosphere (Marty et al. 2011), whereas the inner planets, asteroids and comets are enriched in $^{15}N$. Organic matter in carbonaceous chondrites may reach $\delta^{15}N$-values of about +5000‰ (Chakraborty et al. 2014). The large $^{15}N$-enrichment in meteorites relative to the protosolar gas cannot be explained by isotope fractionation processes in planetary environments, but requires the existence of especially enriching $^{15}N$-reactions. Chakraborty et al. (2014) observed extreme N-isotope fractionations during vacuum UV photodissociation of $N_2$.

#### 3.1.1.5 Sulfur

There are many sulfur components in meteorites which may occur in all possible valence states (−2 to +6). Troilite is the most abundant sulfur compound of iron meteorites and has a relatively constant S-isotope composition (recall that troilite from the Canyon Diablo iron meteorite is the international sulfur standard). Carbonaceous chondrites contain sulfur in the form of sulfates, sulfides, elemental sulfur and complex organic sulfur-containing molecules. Monster et al. (1965), Kaplan and Hulston (1966) and Gao and Thiemens (1993a, b) separated the various sulfur components and demonstrated that sulfides are characterized by the highest $\delta^{34}S$-values, whereas sulfates have the lowest $\delta^{34}S$-values, just the opposite from what is generally observed in terrestrial samples. This is strong evidence against any microbiological activity and instead favors a kinetic isotope fractionation in a sulfur–water reaction (Monster et al. 1965). The largest internal isotope fractionation (7‰) is found in the Orgueil carbonaceous chondrite (Gao and Thiemens 1993a). Orgueil and Murchison have internal isotopic variations between different specimens, which may indicate that sulfur isotope heterogeneity existed in meteorite parent bodies.

Quadruple sulfur isotope measurements potentially may help in identifying genetic relationships between meteorites in a similar way to oxygen isotopes. Early measurements by Hulston and Thode (1965) and Kaplan and Hulston (1966), and those by Gao and Thiemens (1993a, b), did not indicate any isotope anomaly. However, more recent measurements by Rai et al. (2005) on achondrites and by Rai and Thiemens (2007) on chondrites did show the presence of mass independent sulfur isotope fractionations indicating photochemical reactions of gaseous sulfur species in the early solar nebula.

Antonelli et al. (2014) observed anomaleous $^{33}S$ depletions in differentiated iron meteorites along with $^{33}S$ enrichments in several other groups. The complementary positive and negative $^{33}S$ compositions are explained by photolysis of gaseous $H_2S$ in the solar nebula. Photochemically predicted $^{33}S$ depletions imply that the starting composition of inner solar system sulfur was chondritic.

#### 3.1.1.6 Metals

Mass-dependent and mass-independent metal isotope fractionations in meteorites may be due to (i) initial heterogeneities of the solar nebula, (ii) fractionation processes during condensation and planetary accretion and (iii) differentiation processes after planet formation. Metal isotope studies of meteorites have been used in particular to characterize the conditions of planetary accretion, including core formation and the loss of volatiles. Mass dependent Mg, Si and Fe isotope fractionations among different extraterrestrial objects may, in principle, result from loss of planetary material to space through vaporization or from loss of material to a planet's core.

Of special interest are iron meteorites, generally used as analogues of planetary core formation. As shown by Williams et al. (2006) Fe isotope differences between metal and troilite in the range of 0.5‰—the metal phase being heavier than the sulfide phase troilite—may be interpreted as equilibrium fractionations. On the other hand, metal-sulfide fractionations for Cu isotopes are very variable, being one order of magnitude greater than for Fe isotopes and thus cannot represent equilibrium conditions (Williams and Archer 2011). $\delta^{66}Zn$ values in iron meteorites are indistinguishable from the bulk silicate earth (Bridgestock et al. 2014).

Differences in Fe isotope composition between chondrites, iron meteorites and terrestrial basalts may indicate isotope fractionations between metallic and ferrous Fe during segregation of planetary objects into a metallic core and a silicate mantle (Poitrasson et al. 2005; Weyer et al. 2005; Schoenberg and von Blanckenburg 2006; Williams et al. 2012; Craddock et al. 2013). Temperatures above 1000 °C necessary for core segregation were long considered too high to record any detectable isotope fractionation. Fe, Mg and Si isotope investigations have shown, however, that this is not the case (Georg et al. 2007; Weyer and Ionov 2007; Wiechert and Halliday 2007; Fitoussi et al. 2009; Ziegler et al. 2010). Whether core-mantle segregation fractionates Fe isotopes on Earth is a matter of debate. At temperatures between 1750 and 2000 °C, Poitrasson et al. (2009) observed no Fe isotope fractionation between Fe–Ni alloy and ultramafic melt (see also discussion on p. xxx). Recently, Ni et al. (2020) demonstrated that during core crystallization solid iron metal becomes enriched in heavy Fe isotopes relative to liquid iron. Ni isotope compositions have been measured for the search of extinct $^{60}Fe$ (Moynier et al. 2007; Steele et al. 2011).

Depletion of volatile elements during planet formation is usually associated with stable isotope fractionation. A classic example are potassium isotopes, first demonstrated by Humayun and Clayton (1985), later investigated in detail for the Moon by Wang and Jacobsen (2016) and Tian et al. (2020). Due to its high volatility, Zn isotopes as another example may be used to explore variations in the impact history of planets. Large isotope variations (over 6‰) in $\delta^{66}Zn$ values have been explained by impact induced volatilization with preferential loss of the light isotopes in the gas phase (Moynier et al. 2007; Chen et al. 2013).

Calcium is a pure lithophile element which does not partition into planetary cores and is not affected by evaporation. Therefore, Ca isotopes may indicate genetic links between Earth and different classes of meteorites. Earth, Moon, Mars and differentiated asteroids are indistinguishable from primitive chondrites in Ca isotope composition (Simon and DePaolo 2010; Valdes et al. 2014; Huang and Jacobsen 2017).

Metal isotopes can be also used to investigate the formation of chondrules and calcium-aluminium inclusions (CAI). Chondrules have the same Mg and Si isotope composition as most other components of the solar system, but CAIs generally have higher Mg and Si isotope compositions. The systematic enrichment of heavy isotopes in CAIs has been interpreted as being due to evaporation of molten CAIs at low pressures (Shahar and Young 2007; Rumble et al. 2011). Chondrules show a large variation in iron isotope compositions, whereas chondrites, samples from the Moon and Mars have nearly indistinguishable Fe isotope ratios (Craddock and Dauphas 2010).

#### 3.1.1.7 Meteorite-Earth Relationship

It is generally accepted that chondritic meteorites represent the building blocks of the Earth. Which type of meteorites best reflect the composition of the Earth has been a matter of debate. Since the early finding of Clayton and Mayda (1984) demonstrating that the oxygen isotope composition of enstatite chondrite is indistinguishable from the Earth's mantle and later findings that more or less all isotope systems investigated so far show no difference between the bulk silicate earth and enstatite chondrites, it is reasonable to assume that the composition of enstatite chondrites implies a genetic link to the composition of the silicate earth (Javoy et al. 2010). This relationship has been confirmed for many elements such as hydrogen (Piani et al. 2020), nitrogen (Javoy et al. 1986; Piani et al. 2020), sulfur (Defouilloy et al. 2016), titanium (Trinquier et al. 2009), calcium (Valdes et al. 2014), strontium (Moynier et al. 2010), iron (Wang et al. 2014), zinc (Moynier et al. 2011), nickel (Regelous et al. 2008), ruthenium (Fischer-Gödde and Kleine 2017). Silicon seems to be an exception, because of its relatively large difference in isotope composition between enstatite chondrites and bulk silicate earth. However, as shown by Sikdar and Rai (2020), Si isotopes possess a bimodal distribution among silicate and metallic fractions in enstatite chondrites. The latter is highly depleted in heavy Si isotopes, but the former is close to the Si isotope composition of the bulk silicate earth value.

### 3.1.2 The Moon

The classic giant-impact theory for the origin of the Moon requires the collision between a proto-Earth and an impactor often referred to as Theia. Size and composition of the impactor continue to be debated. A comparison of isotope compositions of the Earth–Moon system plays an important role in the discussion about the origin of the Moon.

#### 3.1.2.1 Oxygen

Since the early days of the Apollo missions it has been shown that the oxygen isotope composition of the common lunar igneous minerals is very constant, with very little variation from one sampled locality to another (Onuma et al. 1970; Clayton et al. 1973b). Small $^{18}O$ differences between low-Ti and high Ti-basalts are obviously due to modal mineralogical differences (Spicuzza et al. 2007; Liu et al. 2010). This constancy implies that the lunar interior should have a $\delta^{18}O$–value of about 5.5‰, essentially identical to terrestrial mantle rocks. The fractionations observed among coexisting minerals indicate temperatures of crystallization of about 1000 °C or higher, similar to values observed in terrestrial basalts (Onuma et al. 1970).

Theoretical considerations suggest that the moon forming material should mainly derive from the impacting body. This means that even very small differences in $^{17}O$ and $^{18}O$ content between the impacting body and the proto-Earth should leave a detectable difference in lunar rocks. Precise $^{17}O$ and $^{18}O$-isotope measurements by Wiechert et al. (2001), Liu et al. (2010), Hallis et al. (2010), Herwartz et al. (2014), Young et al. (2016), Greenwood et al. (2018a, b), Cano et al. (2020) and Fischer et al. (in press) revealed no or only very small differences between the Earth and the Moon. The different studies demonstrate that the classic Giant impact model is not entirely accurate unless Theia and the proto-Earth had been composed of the same building blocks.

#### 3.1.2.2 Hydrogen

For years it was thought that the Moon is very dry and therefore very low in volatiles. Early studies of lunar samples (soils and breccia) reported variable $H_2O$ concentrations and $\delta^2H$ compositions, which were interpreted as hydrogen being implanted on the lunar surface due to the interaction with solar wind. Water extracted from basalts has been interpreted as being terrestrial water that has contaminated samples.

This picture has changed as recent progress in SIMS techniques has enabled measurement of very low OH concentrations in volcanic glass, olivine hosted melt inclusions and apatite. Hauri et al. (2011, 2017) demonstrated by studying melt inclusions in rare tephra that some parts of the Moon contain as much water as the Earth's upper mantle.

The interpretation of lunar hydrogen isotope data is complicated, because water may originate from the lunar mantle, from solar wind protons and/or comets. Greenwood et al. (2011) and Barnes et al. (2013) reported $\delta^2H$-values in apatite from +600 to +1100‰ and postulated that a significant portion of the water originated from comets. On the other hand, Saal et al. (2013) and Safarian et al. (2014) concluded that lunar water is indistinguishable from bulk water in carbonaceous chondrites and similar to terrestrial water implying an asteroidal origin for the Earth and the Moon. By summarizing the existing data, Hui et al. (2017) concluded that the lunar magma ocean has an initial $\delta^2H$-value of about −280‰ which was increased to a $\delta^2H$-value of about +310‰ by losing more than 95% of the initial hydrogen content through preferential degassing of molecular $H_2$ vs D-H under reducing conditions.

#### 3.1.2.3 Other Volatile Elements

The Moon is depleted in volatile elements relative to Earth. Volatile element depletions may be inherited from the early stages of accretion, but may also have occurred during the impact and post impact processes that formed the Moon (Righter 2019). The chronology of volatile depletion suggests both inheritance of volatile-depleted materials to form the Moon as well as later volatile depletion events (Day and Moynier 2014). The most probable causes appear to be global scale depletion from a giant impact or inefficient volatile loss from a magma ocean phase.

Moon and Earth show little or no isotope variations for many elements (O, Si, Mg, Ca, Fe Ti), but show variations for a number of volatile elements (S: Wing and Farquhar 2015; Cl: Sharp et al. 2010; Stephant et al. 2019; Gargano et al. 2020; K:

Wang et al. 2016; Rb; Pringle et al. 2017; Zn: Paniello et al. 2012; Ga: Kato et al. 2015), which are characterized by isotope enrichments relative to Earth.

Chlorine isotopes, for instance, analyzed in lunar whole rocks, olivine-hosted melt inclusions and in apatite revealed large enrichments in $\delta^{37}$Cl values (up to +80‰ in apatites). Several hypotheses have been proposed to explain the $^{37}$Cl enrichment. Most commonly chlorine loss and isotope enrichment have been explained by degassing during the Giant impact. Boyce et al. (2015), Stephant et al. (2019) and Barnes et al. (2019) linked Cl isotope fractionation to the differentiation of the Moon via the Lunar Magma Ocean.

For sulfur, Wing and Farquhar (2015) have shown that lunar basalts are enriched in heavy sulfur isotopes compared to fresh, uncontaminated terrestrial basalts and estimated that <1–10% of lunar sulfur was lost after a moon-forming impact event.

Enrichments of the heavy isotopes on the surfaces of the lunar fines are most probably due to the influence of the solar wind. Detailed interpretation of their isotopic variations is difficult due to both, the lack of knowledge of the isotopic composition of the solar wind, and uncertainties of the mechanisms for trapping. Kerridge (1983) demonstrated that nitrogen trapped in lunar surface rocks consists of at least two components differing in release characteristics during experimental heating and isotopic composition: the low-temperature component is consistent with solar wind nitrogen, whereas the high-temperature component consists of solar energetic particles.

### 3.1.3 Mars

In the late 1970s and early 1980s it was realized that differentiated meteorites referred to as the SNC (Shergottites, Nakhlites, Chassignites) group were samples from Mars (McSween et al. 1979; Bogard and Johnson 1983, besides others). This conclusion is based on young crystallization ages compared to that of other meteorites and compositions of trapped volatiles that match those of the martian atmosphere.

#### 3.1.3.1 Oxygen

SNC-meteorites have an average $\delta^{18}$O-value of 4.3‰, which is distinctly lower than the 5.5‰ value for the Earth–Moon system (Clayton and Mayeda 1996; Franchi et al. 1999). Small $^{18}$O-variations among the different SNC-meteorites result primarily from different modal abundances of the major minerals. On a three-isotope plot the $\delta^{17}$O offset between Mars and Earth is 0.3‰ (see Fig. 3.3). In this connection it is interesting to note that the so-called HED (howardites, eucrites and diogenites) meteorites, possibly reflecting material from the asteroid Vesta, have an oxygen isotope composition of 3.3‰ (Clayton and Mayeda 1996). The $\delta^{17}$O-offset to the Earth is about −0.3‰ (Fig. 3.3). These differences in O-isotope composition among the terrestrial planets must reflect differences in the raw material from which the planets were formed.

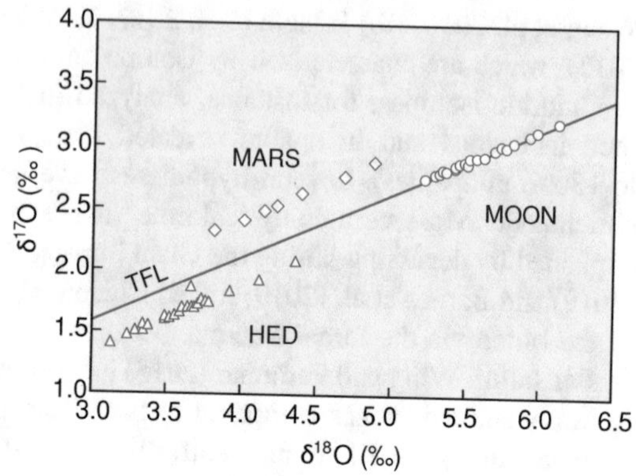


**Fig. 3.3** Three oxygen isotope plot of lunar, Martian rocks and HED meteorites supposed to be fragments of asteroid Vesta (after Wiechert et al. 2001)

### 3.1.3.2 Hydrogen

Volatiles, especially water, on Mars are of special relevance to reveal the geological and geochemical evolution of the planet. The hydrogen isotope composition of Mars can be estimated from two sources: (i) in situ measurements of the present-day Martian atmosphere (Webster et al. 2013) and (ii) hydrogen containing compounds in Martian meteorites (Usui et al. 2012 and others). (i) D/H ratios of the atmosphere are strongly enriched showing large latitudinal gradients. Near polar regions $\delta^2$H-values are around 3000‰, whereas near equatorial regions $\delta^2$H-values are around 6000‰. This enrichment is thought to result from preferential loss of H relative to D from the Martian atmosphere over time (Owen et al. 1988; Villanueva et al. 2015). (ii) Ion microprobe studies of amphibole, biotite, and apatite in SNC meteorites by Watson et al. (1994) and stepwise heating studies by Leshin et al. (1996) reported large variations in $\delta^2$H-values. These authors observed that water in the samples originated from two sources: a terrestrial contaminant released largely at low temperatures and an extraterrestrial component at high temperatures showing extreme D-enrichments. Studies by Boctor et al. (2003), Greenwood et al. (2008), Hu et al. (2014), Usui et al. (2012, 2015) and Mane et al. (2016) revealed a very large range in deuterium isotope composition with values from −111 to 6034‰. From olivine hosted melt inclusions Usui et al. (2012) and Mane et al. (2016) argue that the most depleted $\delta^2$H-values represent the primordial Martian mantle. Thus, it appears that Martian and Earth's mantle have similar hydrogen isotope compositions, indicating similar sources.

### 3.1.3.3 Carbon

As is the case for hydrogen, carbon isotope signatures in Martian meteorites present evidence for different carbon reservoirs. Wright et al. (1990) and Romanek et al. (1994) distinguished three carbon compounds: one component released at temperatures below ≈500 °C, mostly derived from terrestrial contamination, a second component, released between 400 and 700 °C in heating experiments or by reaction with acid, originates mostly from breakdown of carbonates and gives $\delta^{13}$C-values up to +40‰ and the third component, released at temperatures above 700 °C, has

$\delta^{13}C$-values between −20 and −30‰ reflecting the isotope composition of magmatic carbon on Mars.

Carbonates in Martian meteorites have been especially well studied due to the hypothesis that they might indicate past life on Mars (McKay et al. 1996). Understanding the conditions of formation of the carbonates is thus crucial to the whole debate. Despite extensive chemical and mineralogical studies, the environment of carbonate formation has remained unclear. $\delta^{18}O$-values of the carbonates are highly variable ranging from about 5–25‰ depending on different investigators and the carbonate investigated (Romanek et al. 1994; Valley et al. 1997; Leshin et al. 1998). In situ C isotope analysis by Niles et al. (2005) gave highly zoned $\delta^{13}C$-values from ≈ +30 to +60‰ consistent with a derivation from the Martian atmosphere and suggesting abiotic formation.

Further evidence about a nonbiogenic origin of Martian carbonates have been presented by Farquhar et al. (1998) and Farquhar and Thiemens (2000). By measuring $\delta^{17}O$ and $\delta^{18}O$-values Farquhar et al. (1998) observed an $^{17}O$ anomaly in carbonates relative to silicates which they interpreted as being produced by photochemical decomposition of ozone just as in the Earth's stratosphere.

McKay et al. (1996) furthermore suggested on the basis of morphology that tiny sulfide grains inside the carbonates may have formed by sulfate-reducing bacteria. $\delta^{34}S$-values of sulfides range from 2.0 to 7.3‰ (Greenwood et al. 1997), which is similar to values from terrestrial basalts and probably not the result of bacterial reduction of sulfate.

The isotopic results are, therefore, not in favor of a microbiological activity on Mars, but the discussion will certainly continue on this exciting topic.

Finally, it should be mentioned that in situ isotope measurements of the Martian atmosphere from the Curiosity Rover indicate large enrichments of the heavy isotopes, which might reflect substantial atmospheric loss (Webster et al. 2013; Mahaffy et al. 2013, 2015) (see Table 3.1).

#### 3.1.3.4 Sulfur

Mars seems to be rich in sulfur (King and McLennan 2009). Sulfur has been observed as primary igneous sulfides and most importantly as secondary sulfates near the surface of Mars. Mass-independent $^{33}S$ anomalies have been identified in both sulfides and sulfates (Farquhar et al. 2007; Franz et al. 2014), which obviously result from photochemical reactions in the Martian atmosphere favoring a surficial

**Table 3.1** Isotopic composition of atmospheric gases in ‰ measured by Curiosity Rover's Instruments (Mahaffy et al. 2015). δ-values clearly indicate atmospheric degassing that has taken place probably over billion of years

| | |
|---|---|
| $\delta^{13}C$ of $CO_2$ | 46 ± 4 |
| $\delta^{18}O$ of $CO_2$ | 48 ± 5 |
| $\delta^{17}O$ of $CO_2$ | 25 ± 5 |
| $\delta^{2}H$ of $H_2O$ | 4950 ± 1080 |
| $\delta^{15}N$ of $N_2$ | 572 ± 82 |

sulfur cycle for at least 3.5 billion years. Variations observed in $\Delta^{33}S$, but the absence of $\Delta^{36}S$ anomalies implies MIF production by different pathways from those on Earth.

### 3.1.4 Venus

Venus is Earth's closest planetary neighbor with similar size and mass, but very different surface conditions. The Pioneer mission in 1978 obtained data on gases, temperatures and pressures in the atmosphere of Venus. The mass spectrometer on the spacecraft measured the isotopic atmospheric composition relative to $CO_2$, the dominant atmospheric constituent. The $^{13}C/^{12}C$ and $^{18}O/^{16}O$ ratios were observed to be close to the Earth value, whereas the $^{15}N/^{14}N$ ratio is within 20% of that of the Earth (Hoffman et al. 1979). One of the major problems related to the origin and evolution of Venus is that of its "missing water". There is no liquid water on the surface of Venus today and the water vapor content in the atmosphere is probably not more than 220 ppm (Hoffman et al. 1979). This means that either Venus was formed from material very poor in water or whatever water was originally present has disappeared, possibly as the result of escape of hydrogen into space. And indeed Donahue et al. (1982) measured a 100-fold enrichment of deuterium relative to Earth, which is consistent with such an outgassing process. The magnitude of this process is, however, difficult to quantify. In summary, the isotope composition of Venus is poorly known, but future robotic sample missions will hopefully fil the gap of knowledge.

## 3.2 Mantle

Considerable geochemical and isotopic evidence has accumulated supporting the concept that many parts of the mantle have experienced a complex history of partial melting, melt emplacement, crystallization, recrystallization, deformation and metasomatism. A result of this complex history is that the mantle is chemically and isotopically heterogeneous with a complex pattern of depletion and fertilisation signatures. A major goal of isotope mantle geochemistry is the characterization of distinct mantle reservoirs and the processes for their evolution.

Heterogeneities in stable isotopes are difficult to detect, because stable isotope ratios are affected by various partial melting-crystal fractionation processes that are governed by temperature-dependent fractionation factors between residual crystals and partial melt and between cumulate crystals and residual liquid. Unlike radiogenic isotopes, stable isotopes are also fractionated by low temperature surface processes. Therefore, they offer a potentially important means by which recycled crustal material can be distinguished from intra-mantle fractionation processes.

O, H, C, S, and N isotope compositions of mantle derived rocks are substantially more variable than expected from the small fractionations at high temperatures. The most plausible process that may result in variable isotope ratios in the mantle is the input of subducted oceanic crust, and less frequent sedimens of continental crust, into some portions of the mantle. Because different parts of subducted slabs have different isotopic compositions, the released fluids may also differ in the O, H, C, N and S isotope composition. In this context, the process of mantle metasomatism is of special significance. Metasomatic fluids rich in $Fe^{3+}$, Ti, K, LREE, P and other LIL (Large Ion Lithophile) elements tend to react with peridotite mantle and form secondary micas, amphiboles and other accessory minerals. The origin of metasomatic fluids is likely to be either (i) exsolved fluids from an ascending magma or (ii) fluids or melts derived from subducted, hydrothermally altered crust and its overlying sediments.

With respect to the volatile behavior during partial melting, it should be noted that volatiles will be enriched in the melt and depleted in the parent material. During ascent of melts, volatiles will be degassed preferentially, and this degassing will be accompanied by isotopic fractionation (see discussion under "magmatic volatiles").

Sources of information about the isotopic composition of the upper portion of the lithospheric mantle come from the direct analysis of unaltered ultramafic xenoliths brought rapidly to the surface in explosive volcanic vents. Due to rapid transport, these peridotite nodules are in many cases chemically fresh and considered by most workers to be the best samples available from the mantle. The other primary source of information is from basalts, which represent partial melts of the mantle. The problem with basalts is that they do not necessarily represent the mantle composition because partial melting processes may have caused an isotopic fractionation relative to the precursor material. Partial melting of peridotites would result in the preferential melting of Ca–Al-rich minerals leaving behind refractory residues dominated by olivine and orthopyroxene which may differ slightly in the isotopic composition from the original materials. Also, basaltic melts may interact with the crustal lithosphere through which the magmas pass on their way to the Earth's surface. The following section will focus on ultramafic xenoliths, the isotopic characteristics of basalts is discussed under "magmatic rocks".

### 3.2.1 Oxygen

The $\delta^{18}O$ value of the bulk Earth is constrained by the composition of lunar basalts and bulk chondritic meteorites to be close to 6‰. Insight into the detailed oxygen isotope composition of the subcontinental lithospheric mantle come from the analysis of peridotitic xenoliths entrained in alkali basalts and kimberlites. The first oxygen isotope studies of such ultramafic nodules by Kyser et al. (1981, 1982) created much debate. Gregory and Taylor (1986) suggested that the large fractionations in the peridotite xenoliths analyzed by Kyser et al. (1981, 1982) arose through open-system exchange with fluids having variable oxygen isotope compositions and with olivine exchanging $^{18}O$ more rapidly than pyroxene.

It should be recognized, however, that olivine is a very refractory mineral and, as a result, quantitative reaction yields are generally not achieved, when analyzed by convential fluorination techniques. Mattey et al. (1994) analysed 76 samples of olivine in spinel-, garnet- and diamond-facies peridotites using laser fluorination techniques and observed an almost invariant O-isotope composition around 5.2‰. Assuming modal proportions of olivine, orthopyroxene and clinopyroxene of 50:40:10, the calculated bulk mantle $\delta^{18}O$-value would be 5.5‰. Such a mantle source could generate liquids, depending on melting temperatures and degree of partial melting, with O-isotope ratios equivalent to those observed for MORB and many ocean island basalts.

Although the results of Mattey et al. (1994) have been confirmed by Chazot et al. (1997), it should be kept in mind that most of the mantle peridotites that have been analyzed for $\delta^{18}O$ originate from the continental lithospheric mantle and not from the mantle as a whole. More recently there have been several indications that the O-isotope composition of mantle xenoliths from certain exotic settings can be more variable than indicated by Mattey et al. (1994) and Chazot et al. (1997). Zhang et al. (2000) and Deines and Haggerty (2000) documented complex disequilibrium features among peridotitic minerals and intra-crystalline isotope zonations, which presumably result from metasomatic fluid/rock interactions.

Eclogite xenoliths from diamondiferous kimberlites constitute an important suite of xenoliths because they may represent the deepest samples of the continental lithospheric mantle that were originally basaltic crust. Eclogite xenoliths have the most diverse range in $\delta^{18}O$-values between 2.2 and 7.9‰ (McGregor and Manton 1986; Ongley et al. 1987). This large range of $^{18}O$-variation indicates that the oxygen isotope composition of the continental lithosphere varies substantially, at least in any region where eclogite survives and is the most compelling evidence that some nodules represent metamorphic equivalents of hydrothermally altered and subducted oceanic crust.

As a new tool to describe the oxygen isotope composition of mantle rocks, the determination of precise $\Delta^{17}O$-values has been introduced, which reflect the departure in $^{17}O/^{16}O$ relative to $^{18}O/^{16}O$ assuming solely mass-dependent isotope fractionations (Pack and Herwartz 2014; Bao et al. 2016; Sharp et al. 2018).

### 3.2.2 Hydrogen

The origin of the water on Earth is a controversial topic with very different schools of thought. One view postulates, that water was delivered to Earth before and after the Giant Impact from exogeneous sources such as comets and/or meteorites; the other holds that Earth's water has an indigeneous origin (Drake and Righter 2002). Delivery of water from comets and meteorites can be evaluated in the light of their D/H ratios, suggesting that comets cannot be the **major** source of water on Earth, but should be <10% (Marty 2012). Estimates of the D/H ratio of the bulk Earth are uncertain, because volatiles derived from mantle-derived rocks may have been lost and fractionated during magma degassing.

In this connection the concept of "juvenile water" has to be introduced, which has influenced thinking in various fields of igneous petrology and ore genesis. Juvenile water is defined as water that originates from degassing of the mantle and that has never been part of the surficial hydrologic cycle. The analysis of OH-bearing minerals such as micas and amphiboles of deep-seated origin has been considered to be a source of information for juvenile water (e.g. Sheppard and Epstein 1970). Because knowledge about fractionation factors is limited and temperatures of final isotope equilibration between the minerals and water not known, calculations of the H-isotope composition of water in equilibrium with the mantle is rather crude.

Figure 3.4 gives $\delta^2$H-data on phlogopites and amphiboles, indicating that the hydrogen isotope composition of mantle water should lie in general between −80 and −50‰, the range first proposed by Sheppard and Epstein (1970) and subsequently supported by several other authors. Also shown in Fig. 3.4 are a considerable number of phlogopites and amphiboles which have $\delta^2$H-values higher than −50‰. Such elevated $\delta^2$H-values may indicate that water from subducted oceanic crust has played a role in the genesis of these minerals. Similar conclusions have been reached as a result of the analysis of water of submarine basalts from the Mariana arc (Poreda 1985) and from estimates of the original $\delta^2$H-values in boninites from Bonin Island (Dobson and O'Neil 1987).

Water in the mantle is found in different states: as a fluid especially near subduction zones, as a hydrous phase and as a hydroxyl point defect in nominally anhydrous minerals. $\delta^2$H-values between −90 and −110‰ have been obtained by

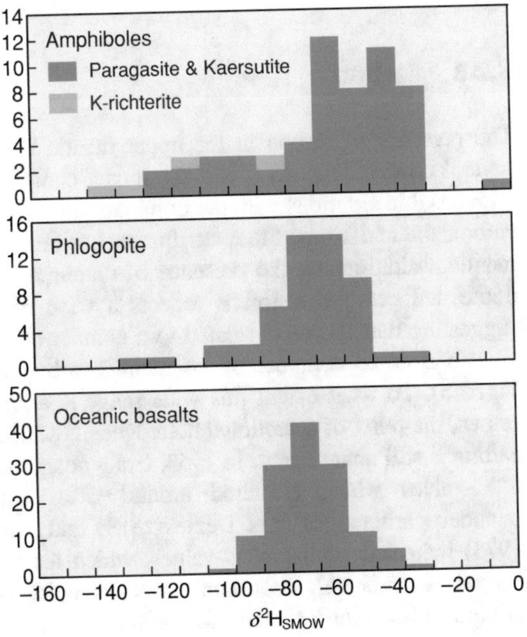


Fig. 3.4 Hydrogen isotope variations in mantle-derived minerals and rocks (modified after Bell and Ihinger 2000)

Bell and Ihinger (2000) analyzing nominally anhydrous mantle minerals (garnet, pyroxene) containing trace quantities of OH. Moine et al. (2020) presented evidence for the presence of molecular $H_2$ in omphacite from eclogites. $\delta^2H$-values increase with dehydration suggesting a positive H-isotope fractionation between minerals and $H_2$-bearing fluid contrary to what is expected by isotope exchange between minerals and $H_2O$-fluids.

Water is carried into the mantle at subduction zones in the form of hydrous minerals. The average $\delta^2H$-composition of the oceanic crust is estimated to be −50‰ (Agrinier et al. 1995; Shaw et al. 2008). As hydrous minerals break down with increasing pT-conditions, the D/H ratio is fractionated producing D-enriched fluids and a D-depleted residual slab. Analyzing olivine-hosted melt inclusions from a subduction zone-setting, Shaw et al. (2008) determined $\delta^2H$-values from −12 to −55‰. Continuous losses of D-enriched fluids lead to a depletion in remaining water containing phases. Thus, slab-bound water will evolve to progressively lower D-values as D-enriched waters are released to the mantle wedge.

For long, the generally accepted $\delta^2H$-values of MORB and its mantle source has been −80 ± 10‰ (Kyser and O'Neil 1984; Poreda et al. 1986; Chaussidon et al. 1991 and others). As shown more recently by Dixon et al. (2017) the $\delta^2H$-value of MORB is heterogeneous as a result of variable contributions from recycled material. They showed for instance that the average $\delta^2H$-value in the North Atlantic is −90, wheras it is −60‰ in the Pacific. As argued by Loewen et al. (2019) high $\delta^2H$-values result from water recycled during subduction whereas low values may reflect multi-stage melting histories. Shaw et al. (2012) presented hydrogen isotope data from volcanic glasses in the Manus back-arc that span a wide range in $\delta^2H$-values from −33 to −126‰.

### 3.2.3 Carbon

The presence of carbon in the upper mantle has been well documented through several observations: $CO_2$ is a significant constituent in volcanic gases associated with basaltic eruptions with the dominant flux at mid-ocean ridges. The eruption of carbonatite and kimberlite rocks further testifies to the storage of $CO_2$ in the upper mantle. Additionally, the presence of diamond and graphite in kimberlites, peridotite and eclogite xenoliths reflects a wide range of mantle redox conditions, suggesting that carbon is related to a number of different processes in the mantle.

The isotopic composition of mantle carbon varies by more than 30‰ (see Fig. 3.5). To what extent this wide range is a result of mantle fractionation processes, the relict of accretional heterogeneities, or a product of recycling of crustal carbon is still unanswered. In 1953, Craig noted that diamonds exhibited a range of $\delta^{13}C$-values which clustered around −5‰. Subsequent investigations which included carbonatites (e.g. Deines 1989) and kimberlites (e.g. Deines and Gold 1973) indicated similar $\delta^{13}C$-values, which led to the concept that upper mantle carbon is relatively constant in C-isotopic composition, with $\delta^{13}C$-values around −5‰. During the formation of a carbonatite magma, carbon is

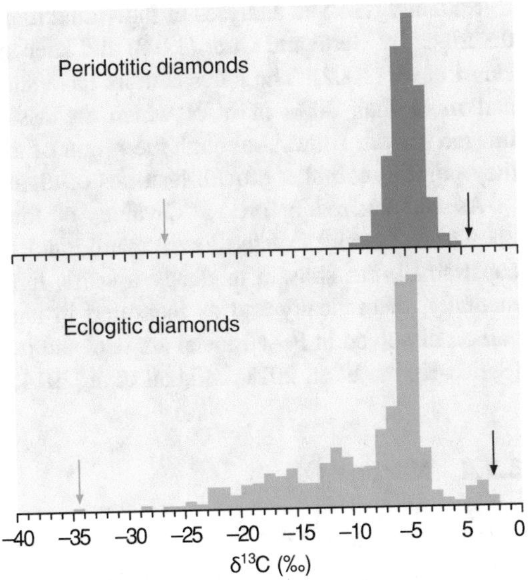


**Fig. 3.5** Carbon isotope variations of diamonds (*arrows* indicate highest and lowest $\delta^{13}$C-values (modified after Cartigny 2005)

concentrated in the melt and is almost quantitatively extracted from its source reservoir. Since the carbon content of the mantle is low, the high carbon concentration of carbonatite melts requires extraction over volumes up to 10,000 times higher than the volume of a carbonatite magma (Deines 1989). Thus, the mean $\delta^{13}$C-value of a carbonatite magma should represent the average carbon isotope composition of a relatively large volume of the mantle.

The C-isotope distribution of diamonds is in contrast to that for carbonatites. As more and more data for diamonds became available (at present more than 4000 C-isotope data) (Deines et al. 1984; Galimov 1985; Cartigny 2005; Cartigny et al. 2014), the range of C-isotope variation broadened to more than 40‰. (from −41 to +5‰ (Galimov 1991; Kirkley et al. 1991; Cartigny 2005; Stachel et al. 2009). More than 70% of the data vary in the narrower range from −8 to −2‰ with a mean of −5‰, being similar to the range of carbon in other mantle derived rocks. The large $^{13}$C variability is not random but restricted to certain genetic classes: Common "peridotitic diamonds" (diamonds associated with peridotitic xenoliths) have less variable carbon isotope compositions than "eclogitic diamonds", which span the entire range of $^{13}$C/$^{12}$C variations (see Fig. 3.5; Cartigny 2005). Diamond formed in metamorphic rocks subducted to ultrahigh pressures have $\delta^{13}$C-values from −30 to −3‰, whereas carbonados, a unique type of polycrystalline diamond (Cartigny 2010) have C-isotope values around −25‰. Current debate centers on whether the more extreme values are characteristic of the mantle source regions or whether they have resulted from isotope fractionation processes linked to diamond formation. What appears to be obvious: the observed ranges cannot be assigned to a single process or to variations in the carbon source alone (Stachel et al. 2009). A combination of processes and multiple carbon sources are required.

Spatially resolved analyses of individual diamonds by SIMS measurements first described by Harte and Otter (1992) and later by others have been summarized by Hauri et al. (2002). The latter authors have shown $\delta^{13}C$ variations of about 10‰ and more than 20‰ in $\delta^{15}N$ which are associated with cathodoluminescence-imaged growth zones. Although the origin of these large variations is still unclear, they point to complex growth histories of diamonds.

As summarized before, $\delta^{13}C$-values of the upper mantle are relatively well known; $\delta^{13}C$-values of the lower mantle and the bulk earth, in contrast, are less constrained, but show a tendency towards lighter $^{13}C$ values. As shown experimentally, theoretically and as measured in natural samples, Fe-, Si-carbides and carbon dissolved in Fe–Ni metal are depleted in $^{13}C$ relative to other carbon phases (Satish-Kumar et al. 2011; Mikhail et al. 2014; Horita and Polyakov 2015).

### 3.2.4 Nitrogen

A large fraction of Earth's total nitrogen resides in the mantle, either being primordial or being recycled crustal nitrogen. In silicates nitrogen as $NH_4^+$ replaces $K^+$, in melts and fluids nitrogen speciation depends on redox conditions. Nitrogen trapped in MORB and OIB glasses has been analyzed by Marty and Humbert (1997) and Marty and Zimmermann (1999) (see Fig. 3.6). By analysing separate minerals in peridotite xenoliths Yokochi et al. (2009) observed large N-isotope disequilibria. $\delta^{15}N$-values as low as −17.3‰ have been measured for phlogopite, whereas clinopyroxene and olivine show positive $^{15}N$ values. Positive δ-values of about +3‰ have been found also in deep mantle material sampled by mantle plumes which may suggest that recycling of oceanic crust may account for heavy nitrogen in the deep mantle (Dauphas and Marty 1999).

Nitrogen is the main trace component in diamonds. Nitrogen isotopes have been measured in over 700 diamond samples with $\delta^{15}N$-values ranging from +13 to −23‰. Despite this broad distribution, the majority varies around −5‰ (Javoy et al. 1986; Boyd et al. 1992; Boyd and Pillinger 1994; Hauri et al. 2002; Cartigny 2005; Cartigny et al. 1997, 1998, 2014). Nitrogen in diamonds, thus, is depleted in $^{15}N$ compared to atmospheric nitrogen (0‰) and sedimentary nitrogen which is enriched in $^{15}N$ (Cartigny and Marty 2013). The negative δ-values in diamonds clearly indicate that the mantle contains non-atmospheric nitrogen.

Nitrogen may have been segregated into the core. Li et al. (2016) have investigated experimentally nitrogen isotope fractionations between liquid Fe-rich metal and silicate melts. They have shown that core-mantle segregation may have enriched the silicate mantle in $^{15}N$.

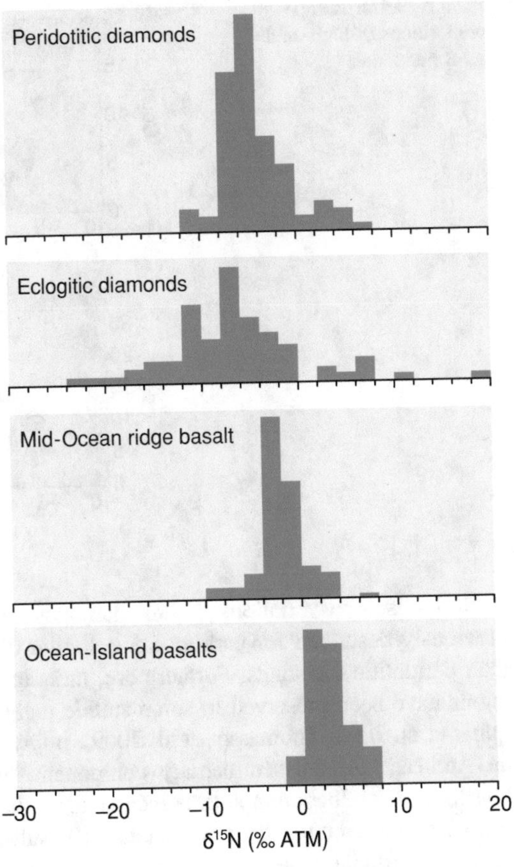


**Fig. 3.6** Nitrogen isotope variations in mantle derived materials (modified after Marty and Zimmermann 1999)

### 3.2.5 Sulfur

Sulfur occurs in a variety of forms in the mantle, the major sulfur phase is monosulfide solid solution between Fe, Ni and Cu. Ion microprobe measurements on sulfide inclusions from megacrysts and pyroxenite xenoliths from alkali basalts and kimberlites and in diamonds gave $\delta^{34}S$-values from −11 to +14‰ (Chaussidon et al. 1987, 1989; Eldridge et al. 1991).

Interesting differences in sulfur isotope compositions are observed when comparing high-S peridotitic tectonites with low-S peridotite xenoliths (Fig. 3.7). Tectonites from the Pyrenees predominantly have negative $\delta^{34}S$-values of around −5‰, whereas low-S xenoliths from Mongolia have largely positive $\delta^{34}S$-values of up to +7‰. Ionov et al. (1992) determined sulfur contents and isotopic compositions in some 90 garnet and spinel lherzolites from six regions in southern Siberia and Mongolia for which the range of $\delta^{34}S$ values is from −7 to +7‰. Ionov et al. (1992) concluded that low sulfur concentrations (<50 ppm) and largely positive $\delta^{34}S$-values predominate in the lithospheric continental mantle worldwide.

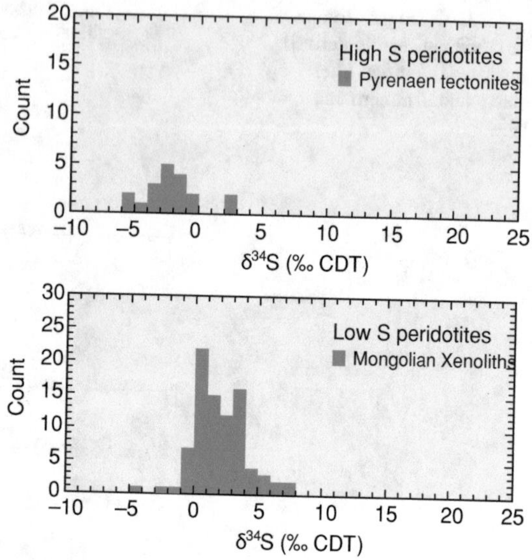


**Fig. 3.7** Sulfur isotope compositions of high- and low-S peridotites

Sulfur isotope variations within diamonds exhibit the same characteristics as previously described for carbon: i.e. eclogitic diamonds are much more variable than peridotitic diamonds. Furthermore, mass independent sulfur isotope fractionations have been preserved in some sulfide inclusions in eclogitic diamonds (Farquhar et al. 2002; Thomassot et al. 2009), implying that sulfide inclusions contain an Archean surficial sedimentary component. Smit et al. (2019) observed two episodes of subduction in sulfide inclusions of diamonds. Palaeoarchean diamonds from the Slave craton do not contain MIF sulfur suggesting no involvement of recycled surficial material, whereas younger diamonds from the West African craton contain MIF sulfur suggesting construction of the cratonic mantle through subduction-style horizontal processes.

### 3.2.6 Stable Isotope Composition of the Core

Although the composition of the core remains largely unknown, cosmochemical and geophysical arguments indicate that the core must contain lighter elements in addition to Fe and Ni to account for its lower density. One plausible candidate is carbon. Satish-Kumar et al. (2011) demonstrated that temperature dependent carbon isotope fractionations occur between reduced carbon and iron carbide, the latter being depleted in $^{13}C$ relative to carbon, suggesting that the core is depleted in $^{13}C$ relative to the mantle. The earth's core may be regarded also as a major nitrogen sink. Li et al. (2016) showed experimentally that metal is depleted in $^{15}N$ relative to silicate.

Silicon is another element of interest, because liquid iron reacts with silicate to form an iron-silicon alloy at the relevant pT conditions of the core. Determinations on silicon isotope fractionations between metal and silicate yielded controversial results (Georg et al. 2007; Shahar et al. 2009, 2011; Zambardi et al. 2013; Hin et al. 2014). As demonstrated by Shahar et al. (2016) pressure effects cannot be ignored when discussing the isotope composition of the core. Because of the different bonding environments of the two phases, silicon in silicate should be enriched in $^{28}Si$ relative to the alloy phase (Schauble 2004; Georg et al. 2007). As shown experimentally by Shahar et al. (2009, 2011), at temperatures between 1800 and 2200 °C, silicates are distinctly enriched in $^{28}Si$ relative to metal. At the even higher temperatures of the core, a 1.2‰ depletion of the metal phase in the core relative to silicates in the mantle can be anticipated. Somewhat smaller Si isotope fractionations have been determined by Hin et al. (2014). Helffrich et al. (2018) estimated that the $\delta^{30}Si$ of the core is between −0.92 and −1.36‰ compared to −0.29‰ of the bulk earth.

Iron and nickel isotopes may fractionate during core formation as discussed by Bourdon et al. (2018). With respect to iron isotopes, an enrichment of the heavy Fe isotopes in the metal phase relative to Fe-oxides should be expected (Young et al. 2015). Experiments on the iron isotope distribution between metal and silicates have yielded, however, contrasting results (Poitrasson et al. 2009; Hin et al. 2012; Shahar et al. 2016). Experiments by Elardo and Shahar (2017) showed that Ni plays a very important role in controlling iron isotope fractionation during core formation.

Lazar et al. (2012) observed Ni isotope fractionations between metal and silicate, the Ni in metal being isotopically enriched relative to Ni in silicate. Molybdenum as another interesting element of core formation shows opposite signs of fractionations with Mo in metal being isotopically depleted relative to silicate (Hin et al. 2013).

## 3.3 Magmatic Rocks

On the basis of their high temperature of formation, it could be expected that magmatic rocks exhibit relatively small differences in isotopic composition. However, as a result of secondary alteration processes and the fact, that magmas can have a crustal and a mantle origin, the variation observed in isotopic composition of magmatic rocks can actually be quite large.

Provided an igneous rock has not been affected by subsolidus isotope exchange or hydrothermal alteration, its isotope composition will be determined by:

(i) the isotope composition of the source region in which the magma was generated,
(ii) the temperature of magma generation and crystallization,
(iii) the mineralogical composition of the rock, and
(iv) the evolutionary history of the magma including processes of isotope exchange, assimilation of country rocks, magma mixing, etc.

In the following sections, which concentrate on $^{18}O/^{16}O$ measurements, some of these points are discussed in more detail (see also Taylor 1968, 1986a, b; Taylor and Sheppard 1986). Isotope variations of metal isotopes reported for magmatic rocks are discussed briefly in Chap. 2 under the specific elements.

### 3.3.1 Fractional Crystallization

Because fractionation factors between melt and solid are small at magmatic temperatures, fractional crystallization is expected to play only a minor role in influencing the isotopic composition of magmatic rocks. Matsuhisa (1979), for example, reported that $\delta^{18}O$ values increased by approximately 1‰ from basalt to dacite within a lava sequence from Japan. Muehlenbachs and Byerly (1982) analyzed an extremely differentiated suite of volcanic rocks at the Galapagos spreading center and showed that 90% fractionation only enriched the residual melt by about 1.2‰. On Ascension Island Sheppard and Harris (1985) measured a difference of nearly 1‰ in a volcanic suite ranging from basalt to obsidian. Furthermore, modelling closed-system crystal fractionation, an $^{18}O$ enrichment of about 0.4‰ per 10 wt% increase in $SiO_2$ content can be predicted.

Fractional crystallization may affect silicon isotopes: $\delta^{30}Si$-values become enriched with increasing $SiO_2$ contents (Douthitt 1982; Savage et al. 2011). In several metal isotope systems fractional crystallization may also cause measurable isotope fractionations, which in specific is relevant for iron isotopes due to potential redox changes (Poitrasson and Freydier 2005; Teng et al. 2008; Schuessler et al. 2009 and others).

### 3.3.2 Differences Between Volcanic and Plutonic Rocks

Systematic differences in O-isotope composition are observed between fine-grained, rapidly quenched volcanic rocks and their coarse-grained plutonic equivalents (Taylor 1968; Anderson et al. 1971). Fractionations among minerals in plutonic mafic rocks are on average about twice as great as for the corresponding fractionations observed in equivalent extrusive mafic rocks. This difference may result from retrograde exchange between minerals or post-crystallization exchange reactions of the plutonic rocks with a fluid phase. This interpretation is supported by the fact that basaltic and gabbroic rocks from the lunar surface yield the same "isotopic temperatures" corresponding to their initial temperatures of crystallization. Due to the low water concentration on the Moon, retrograde exchange is very limited.

### 3.3.3 Low Temperature Alteration Processes

Because of their high glass contents and very fine grain size, volcanic rocks are very susceptible to low-temperature processes such as hydration and weathering, which are characterized by large $^{18}O$-enrichment effects in the altered rocks.

In general, it is probable that Tertiary and older volcanic rocks will exhibit O-isotope compositions that have been modified to higher $^{18}O/^{16}O$ ratios from their primary state (Taylor 1968; Muehlenbachs and Clayton 1972; Cerling et al. 1985; Harmon et al. 1987). Although there is no way to ascertain the magnitude of these $^{18}O$-enrichments on a sample by sample basis, a crude estimate can be made by determining the water (and carbon dioxide) content and "correcting" to what are considered primary values of the suite of rocks to be analyzed (Taylor et al. 1984; Harmon et al. 1987). Thus, any water content >1% could be of secondary origin and the $\delta^{18}O$-value for such samples should be corrected before such $^{18}O$-measurements are to be used for primary, magmatic interpretations.

### 3.3.4 Assimilation of Crustal Rocks

Because the various surface and crustal environments are characterized by different and distinctive isotope compositions, stable isotopes provide a powerful tool for discriminating between the relative role of mantle and crust in magma genesis. This is especially true when stable isotopes are considered together with radiogenic isotopes, because variations within these independent isotopic systems may arise from unrelated geologic causes. For instance, a mantle melt that has been affected by contamination processes within the upper crust will exhibit increases in $^{18}O/^{16}O$ and $^{87}Sr/^{86}Sr$ ratios that correlate with an increase in $SiO_2$ and decrease in Sr content. In contrast, a mantle melt, which evolves only through differentiation unaccompanied by interaction with crustal material, will have an O-isotope composition that mainly reflects that of its source region, independent of variations in chemical composition. In this latter case, correlated stable and radiogenic isotope variations would be an indication of variable crustal contamination of the source region, (i.e. crustal material that has been recycled into the mantle via subduction).

Modelling by Taylor (1980) and James (1981) has demonstrated that it is possible to distinguish between the effects of source contamination as well as crustal contamination. Magma mixing and source contamination are two-component mixing processes which obey two-component hyperbolic mixing relations, whereas crustal contamination is a three-component mixing process, involving the magma, the crustal contaminant, and the cumulates, that results in more complex mixing trajectories on an oxygen—radiogenic isotope plot. Finally, it has to be mentioned, that in contrast to the radiogenic isotopes, oxygen is the major component in rocks, implying that modification of the $\delta^{18}O$-value by several tenths of 1‰ requires uptake of volumetrically significant sediment masses, that may cause a space problem.

## 3.3.5 Glasses from Different Tectonic Settings

### 3.3.5.1 Oxygen

Early investigations of oxygen isotopes in igneous rocks relied on whole rock data analyzed by the classical reaction with fluorine compounds. Relatively large oxygen isotope variations can be due to secondary alteration effects. Correcting for these low-temperature effects, Harmon and Hoefs (1995) assembled a database consisting of 2855 O-isotope analyses of Neogene volcanic rocks worldwide. They observed a 5‰ variation in the $\delta^{18}$O-values of fresh basalts and glasses, which they have taken as evidence of significant oxygen isotope heterogeneities in the mantle sources of the basalts. This is documented in Fig. 3.8, which plots $\delta^{18}$O-values versus Mg-numbers (Harmon and Hoefs 1995).

The usage of whole rock data has, however, its ambiguities. Estimates of original magmatic $\delta^{18}$O values are best achieved through analysis of unaltered phenocrysts within rocks in particular refractory phenocrysts such as olivine and zircon. Laser-based extraction methods on small amounts of separated mineral phases have documented subtle, but resolvable differences among different types of basaltic lavas (Eiler et al. 1996, 2000, 2011; Dorendorf et al. 2000; Cooper et al. 2004; Bindeman et al. 2004, 2005, 2008 and others).

MORB has a rather uniform O-isotope composition of all basalt types (5.7 ± 0.2‰) and can be used as a reference against which basalts erupted in other tectonic settings can be compared. By performing high precision laser isotope

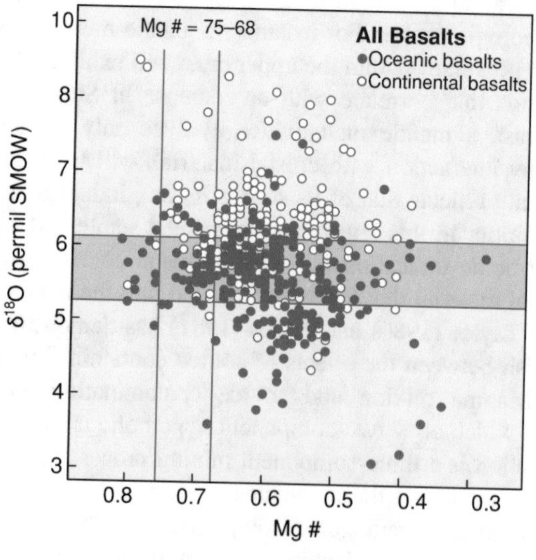


**Fig. 3.8** Plot of $\delta^{18}$O-values versus Mg numbers for oceanic basalts (*filled circles*) and continental basalts (*open circles*). The *shaded field* denotes the ±2 σ range of a MORB mean value of +5.7‰. The *clear vertical field* denotes the range for primary basaltic partial melts in equilibrium with a peridotitic source (Harmon and Hoefs 1995)

analyses on MORB glasses from the North Atlantic, Cooper et al. (2004) observed a $\delta^{18}O$ variation range of about 0.5‰, which is larger than originally thought by Harmon and Hoefs (1995). $^{18}O$ variations correlate with geochemical parameters of mantle enrichment such as high $^{87}Sr/^{86}Sr$ and low $^{143}Nd/^{144}Nd$ ratios. According to Cooper et al. (2004) the enriched material reflects subducted altered dehydrated oceanic crust.

The largest variability in oxygen isotope composition has been found in subduction related basalts. Bindeman et al. (2005) observed a $\delta^{18}O$ range in olivine phenocrysts between 4.9 and 6.8‰. Oxygen isotope variations in arc-related lavas can constrain the contributions of subducted sediments and fluids to the sub-arc mantle assuming the $\delta^{18}O$ of the subducted component is known (Eiler et al. 2000; Dorendorf et al. 2000). These authors demonstrated that crustal assimilation or a contribution of oceanic sediments is negligible (<1–2%). Instead, the observed $^{18}O$-enrichment in olivines and clinopyroxenes may result from exchange with high $^{18}O$ fluids derived from subducted altered oceanic crust.

Continental basalts tend to be enriched in $^{18}O$ relative to oceanic basalts and exhibit considerably more variability in O-isotope composition, a feature attributed to interaction with $^{18}O$-enriched continental crust during magma ascent (Harmon and Hoefs 1995; Baker et al. 2000).

#### 3.3.5.2 Hydrogen

Water dissolves in silicate melts and glasses in at least two distinct forms: water molecules and hydroxyl groups. Because the proportions of these two species change with total water content, temperature and chemical composition of the melt, the bulk partitioning of hydrogen isotopes between vapor and melt is a complex function of these variables. Dobson et al. (1989) determined the fractionation between water vapor and water dissolved in felsic glasses in the temperature range from 530 to 850 °C. Under these conditions, the total dissolved water content of the glasses is below 0.2%, with all water present as hydroxyl groups. The measured hydrogen fractionation factors vary from 1.051 to 1.035 and are greater than those observed for most hydrous mineral—water systems, perhaps reflecting the strong hydrogen bonding of hydroxyl groups in glasses.

Hydrogen isotope and water content data for MORB, OIB and BAB glasses have been determined by Kyser and O'Neil (1984), Poreda (1985), and Poreda et al. (1986). The range of $\delta^2H$-values for MORB glasses is from −90 to −40‰ and is indistinguishable from that reported for phlogopites and amphiboles from kimberlites and peridotites (see Fig. 3.4).

D/H ratios and water content in fresh submarine basalt glasses can be altered by (i) degassing, (ii) addition of seawater at magmatic temperature and (iii) low-temperature hydration. Clog et al. (2013) have reinvestigated the processes of potential contamination and argued that previous measurements on MORB glasses may have suffered from analytical artefacts concluding that the upper depleted mantle has a $\delta^2H$-value close to −60‰ that has been also reported by Dixon et al. (2017).

The process of degassing has been documented best for rhyolitic magmas where water-rich magmas (about 2%) have a $\delta^2$H-value of −50‰. At very late eruption stages with remaining water contents of around 0.1% the $\delta^2$H-value is around −120‰ (Taylor et al. 1983; Taylor 1986a, b). For this process the decisive parameter is the isotopic fractionation between the vapor and the melt, which can be between 15 and 35‰ (Taylor 1986a, b) and the amount of water lost from the system (Rayleigh fractionation). De Hoog et al. (2009) modeled hydrogen isotope fractionation during degassing taking the variation of water species with water content and temperature into account. Progressively increasing OH/$H_2O$ ratios during degassing in melts lead to increasing H fractionation factors.

The degassing process produces an opposite trend to meteoric water hydrothermal alteration, showing decreasing $\delta^2$H-values with increasing water content (Hudak and Bindeman 2018). Hudak and Bindeman (2018) investigated glass hydration in ignimbrite sheets and demonstrated that the hydration formed late in the ignimbrite cooling history when temperature dropped below at 100 °C and meteoric water invaded from below.

#### 3.3.5.3 Carbon

Isotopic fractionation between $CO_2$ and dissolved carbon in melts has been estimated by various authors to vary between 2 and 4‰ (as summarized by Holloway and Blank 1994), the vapor being enriched in $^{13}$C relative to the melt. This fractionation can be used to interpret the carbon isotope composition of glasses and $CO_2$ in volcanic gases and to estimate the initial carbon concentration of undegassed basaltic melts.

Reported $\delta^{13}$C-values for basaltic glass vary from −30 to about −3‰ that represent isotopically distinct carbon extracted at different temperatures by stepwise heating (Pineau et al. 1976; Pineau and Javoy 1983; Des Marais and Moore 1984; Mattey et al. 1984). A “low-temperature” component of carbon is extractable below 600 °C, whereas a “high-temperature” fraction of carbon is liberated above 600 °C. There are two different interpretations regarding the origins of these two different types of carbon. While Pineau et al. (1976) and Pineau and Javoy (1983) consider that the whole range of carbon isotope variation observed to represent primary dissolved carbon, which becomes increasingly $^{13}$C depleted during multi-stage degassing of $CO_2$, Des Marais and Moore (1984) and Mattey et al. (1984) suggest that the “low-temperature” carbon originates from surface contamination. For MORB glasses, the “high-temperature” carbon has an isotopic composition typical for that of mantle values. Island arc glasses have lower $\delta^{13}$C-values, which might be explained by mixing two different carbon compounds in the source regions: a MORB—like carbon and an organic carbon component from subducted pelagic sediments (Mattey et al. 1984).

#### 3.3.5.4 Nitrogen

The determination of nitrogen isotopes in basaltic glasses is severely complicated by its low concentration, which makes nitrogen sensitive to atmospheric contamination and to addition of surface-derived materials i.e. organic matter. Nitrogen in basaltic glasses has been determined by Exley et al. (1987), Marty and Humbert (1997) and Marty and Zimmermann (1999). Marty and coworkers reported that nitrogen in MORB and OIB glasses has an average $\delta^{15}$N-value of around −4 ± 1‰ (see Fig. 3.6). The major factors affecting its isotopic composition appear to be magma degassing and assimilation of surface-derived matter.

#### 3.3.5.5 Sulfur

The behavior of sulfur in magmatic systems is particularly complex: sulfur can exist as both sulfate and sulfide species in four different forms: dissolved in the melt, as an immiscible sulfide melt, in a separate gas phase, and in various sulfide and sulfate minerals. To determine the source of sulfur in magmatic rocks requires knowledge of complex parameters such as oxygen fugacity, speciation of dissolved sulfur in melt and what is most important the degassing history. Mandeville et al. (2009) have demonstrated that magmatic degassing can modify the initial sulfur isotope composition by up to 14‰ On the other hand de Moor et al. (2010) demonstrated that degassing of a magma body resulted in a slight $^{34}$S enrichment only.

Early measurements on MORB glasses and submarine Hawaiian basalts indicated a very narrow range in sulfur isotope composition, with $\delta^{34}$S-values clustering around zero (Sakai et al. 1982, 1984). More recent measurements by Labidi et al. (2012) showed that published MORB data are affected by incomplete sulfur recovery during analytical extraction. Labidi et al. (2012, 2014) argued that the sulfur isotope composition of the depleted mantle is more negative than previously thought and has a $\delta^{34}$S-values of − 1.4‰. Negative $\delta^{34}$S-values for the mantle could result from a low-$^{34}$S oceanic crust recycled within the MORB mantle source (Cabral et al. 2013) or from sulfur isotope fractionation during core-mantle segregation leading to a $^{34}$S enriched core and a $^{34}$S depleted mantle (Labidi et al. 2013). Noteworthy is the discovery that OIB's may carry negative $\Delta^{33}$S values (Mangaia; Cabral et al. 2013; Pitcairn; Delavault et al. 2016), indicating assimilation of Archean sedimentary material that has remained in the mantle for billion of years before returning to the Earth's surface. Hutchison et al. (2019) investigated the S-isotope characteristics of alkaline magmatic rocks that span an age range of two billion years, thus providing a time record of S-isotope magma sources. They observed a $\delta^{34}$S-range from −5 to +5‰ with Proterozoic rocks restricted to positive values and Phanerozoic rocks showing greater diversity.

In subaerial basalts, the variation of $\delta^{34}$S-values is larger and generally shifted towards positive values. One reason for this larger variation is the loss of a sulfur-bearing phase during magmatic degassing. Isotopic shifts that accompany degassing depend on temperature and speciation, the latter is directly proportional to the fugacity of oxygen (Sakai et al. 1982) and on open-system conditions (immediate removal from the magma) or closed-system conditions (vapor exsolved remains in equilibrium with the magma) (Taylor 1986a, b).

### 3.3.6 Magnesium and Iron

Magnesium and iron are closely linked in their geochemical behavior by their mutual substitution in common mantle minerals, but show different behavior during partial melting and fractional crystallization as Mg preferentially partitions into the solid whereas Fe partitions into the melt. A primary difference of the Mg and Fe isotopic behavior is that magnesium occurs in only one valence state whereas iron occurs in two. Similar Mg isotope compositions among MORB's, OIB's and peridotites suggest a lack of significant Mg isotope fractionations during partial melting of peridotites and differentiation of basaltic magma. By contrast, since $Fe^{3+}$ is more incompatible than $Fe^{2+}$ during partial melting, heavy Fe isotopes are slightly enriched in basalts relative to peridotites.

Mg isotope compositions in mantle-derived rocks are relatively uniform, wheras Fe isotopes show considerable variability (Liu et al. 2011; Williams and Bizimis 2014; An et al. 2017). Williams and Bizimis (2014) explored Fe isotopes as tracer of peridotitic and pyroxenetic components in the mantle. They showed that pyroxenites are enriched in $\delta^{56}$Fe-values relative to peridotites being consistent with Fe isotope fractionations during partial melting, where isotopically heavy Fe is extracted in the melt phase leaving behind depleted $\delta^{56}$Fe-values in peridotites. This relationship favors a heterogeneous mantle containing portions of light and heavy $\delta^{56}$Fe-values.

As demonstrated by Zhao et al. (2010) and Poitrasson et al. (2013), partial melting accounts for small Fe isotope variations only in the mantle; the main cause for Fe isotope heterogeneity is metasomatism by melts and by fluids, which seems to be also true for Mg isotope variations (Hu et al. 2016). Zhao et al. (2017) observed in reaction products between peridotites and silicate melts large Fe isotope variations indicating disequilibrium fractionations.

Chemical diffusion may generate Fe and Mg isotope fractionations that exceed equilibrium isotope fractionations by an order of magnitude (Dauphas et al. 2010; Sio et al. 2013; Zhao et al. 2017). Fe and Mg isotope fractionations during Fe–Mg exchange between olivine and melt has been first investigated by Teng et al. (2011). Olivines are particularly useful to investigate diffusion effects because olivines are commonly zoned in Mg and Fe. As olivine crystallizes, the resulting disequilibrium between melt and olivine causes diffusion of Mg from olivine into the melt and Fe from the melt into olivine. This results in olivine grains that are isotopically zoned showing negatively correlated Mg and Fe isotope compositions (see Fig. 3.9). Thermo (or Soret) diffusion causes a positively correlated Mg–Fe isotope fractionation. This allows the distinction of the two processes.

### 3.3.7 Lithium and Boron

Since lithium and boron isotope variations mainly occur during low temperature processes, Li and B isotopes may provide a good tracer of surface material that is recycled to the mantle. Heterogeneous distributions of subducted oceanic and

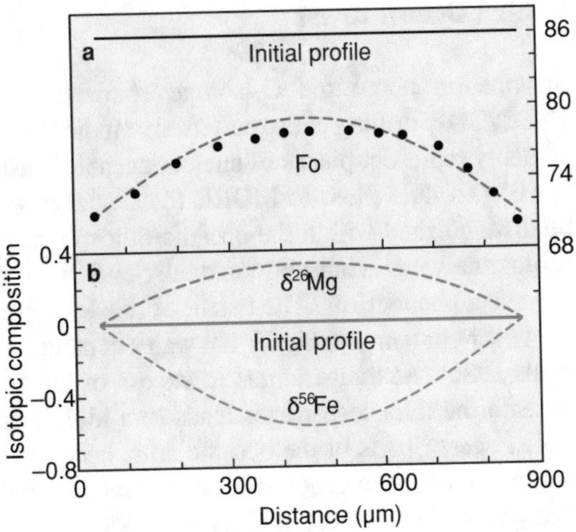


**Fig. 3.9** **a** Forsterite variations in an olivine phenocryst from the Kilauea lava lake. **b** Predicted Mg and Fe isotope fractionation induced by interdiffusion. Solid lines represent initial conditions, dotted curves are calculated diffusion profiles (after Teng et al. 2011)

continental crust in the mantle should thus result in variations of Li and B isotope ratios of mantle rocks. Dehydration processes in subduction zones deliver additional processes fractionating Li and B isotopes.

Seitz et al. (2004), Magna et al. (2006) and Jeffcoate et al. (2007) reported significant Li isotope fractionation among mantle minerals. Olivines are about 1.5‰ lighter than coexisting orthopyroxenes, clinopyroxenes and phlogopites are, in contrast, highly variable, which might indicate isotope disequilibrium. In situ SIMS analyses show Li isotope zonations in peridotite minerals. Jeffcoate et al. (2007) report a 40‰ variation in a single orthopyroxene crystal from San Carlos, which is attributed to diffusive fractionation during ascent and cooling. Metasomatic overprinting by fluids or melts is another process that can be responsible for the great variability of $\delta^7$Li-values observed in mantle peridotites (Tang et al. 2007). Thus, in the last years, it became evident that Li isotope variations in mantle rocks are primarily controlled by diffusion that exceeds the effects of equilibrium processes (Marschall and Tang 2020).

Since boron concentrations in mantle minerals are exceedingly low, boron isotope analysis of mantle minerals are analytically demanding. On the basis of a boron budget between mantle and crust, Chaussidon and Marty (1995) estimated that the primitive mantle had a $\delta^{11}$B value of $-10 \pm 2$‰. Marschall et al. (2017) concluded that MORB glasses have a homogeneous mean B isotope composition of −7.1‰. Since B isotope fractionation during mantle melting and crystal fractionation appears to be small, the average MORB glass value may reflect the B isotope composition of the depleted mantle and the bulk silicate Earth. Potential variations among these reservoirs are hard to be resolved at present (Marschall et al. 2017).

### 3.3.8 Ocean Crust

Information about the O-isotope character of the oceanic crust comes from DSDP/ODP drilling sites and from studies of ophiolite complexes, which presumably represent pieces of ancient oceanic crust. Primary, unaltered oceanic crust has $\delta^{18}$O-values close to MORB ($\delta^{18}$O: 5.7 ± 0.2‰). Two types of alteration can be distinguished within the oceanic lithosphere: at low temperatures (< ~100 °C) submarine weathering may markedly enrich the groundmass of basalts in $^{18}$O, but not affect phenocrysts. The extent of this low temperature alteration correlates with the water content: the higher the water content, the higher the $\delta^{18}$O-values (e.g. Alt et al. 1986). At temperatures in excess of about 300 °C hydrothermal circulation beneath the midocean ridges leads to a high-temperature water/rock interaction in which deeper parts of the oceanic crust become depleted in $^{18}$O by 1–2‰. Similar findings have been reported from ophiolite complexes, the most cited example is that of Oman (Gregory and Taylor 1981). Maximum $^{18}$O contents occur in the uppermost part of the pillow lava sequence and decrease through the sheeted dike complex. Below the base of the dike complex down to the Moho, $\delta^{18}$O-values are lower than typical MORB values by about 1–2‰.

Thus, separate levels of the oceanic crust are simultaneously enriched and depleted in $^{18}$O relative to "normal" mantle values because of reaction with sea water at different temperatures. Muehlenbachs and Clayton (1976) and Gregory and Taylor (1981) concluded that the $^{18}$O enrichments are balanced by the $^{18}$O depletions which acts like a buffer for the oxygen isotope composition of ocean water.

Gao et al. (2006) evaluated the existing data base and concluded that apparent differences in mass-weighed $\delta^{18}$O-values exist among profiles through the recent and the fossil oceanic crust depending on differences in spreading rates. Oceanic crust formed under fast spreading ridges usually have depleted or balanced $\delta^{18}$O-values, whereas oceanic crust formed under slow spreading ridges is characterized by enriched $\delta^{18}$O-values. This difference might be due to different depths of seawater penetration in fast and slow spreading ridges.

Of special significance in the oceanic crust are serpentinites that are formed by the hydration of olivine-rich ultramafic rocks, because they play an important role in the recycling of water and other volatiles from the surface to the deep lithosphere and back to the surface via mantle wedges and arc magmas (Evans et al. 2013). Serpentinization, thus, may take place over a range of temperatures and in a variety of geologic settings. Experimentally determined H isotope fractionation factors (Saccocia et al. 2009) allow to constrain fluid sources. Serpentine from mid-ocean ridge environments, for instance, has been formed by interaction with hot ocean water.

## 3.3.9 Granitic Rocks

### 3.3.9.1 Whole-Rock Oxygen

On the basis of $^{18}O/^{16}O$ ratios, Taylor (1977, 1978) subdivided granitic rocks into three groups: (i) normal $^{18}O$-granitic rocks with $\delta^{18}O$-values between 6 and 10‰, (ii) high $^{18}O$ granitic rocks with $\delta^{18}O$-values >10‰ and (iii) low $^{18}O$ granitic rocks with $\delta^{18}O$-values <6‰. Although this is a somewhat arbitrary grouping it nevertheless turns out to be a useful geochemical classification. Bindeman (2008) suggested that the term "normal-$\delta^{18}O$" be used more restrictively for mante-derived rocks and products of their differentiates.

Isotopic compositions of granites depend on their origin; A-type (anorogenic, intraplate, including rapakivi) are normal $\delta^{18}O$ granites. Many granitic plutonic rocks throughout the world have relatively uniform $^{18}O$-contents with $\delta^{18}O$-values between 6 and 8‰. Normal $\delta^{18}O$ granites have been described from oceanic island—arc areas where continental crust is absent (e.g., Chivas et al. 1982). Such plutons are considered to be entirely derived by differentiation of mantle magmas. Granites at the high end of the normal $^{18}O$-group most commonly have formed by partial melting of crust that contained both a sedimentary and a magmatic fraction. It is interesting to note that many of the normal $^{18}O$-granites are of Precambrian age and that metasediments of this age quite often have $\delta^{18}O$-values below 10‰ (Longstaffe and Schwarcz 1977), because Archean sedimentary rocks are lower in $\delta^{18}O$ (Bindeman 2021).

Granitic rocks with $\delta^{18}O$-values higher than 10‰ require derivation from some type of $^{18}O$-enriched sedimentary or metasedimentary protolith. For instance, such high $\delta^{18}O$-values are observed in many Hercynian granites of western Europe (Hoefs and Emmermann 1983), in Damaran granites of Africa (Haack et al. 1982) and in leucogranites from the Himalayas of Central Asia (Blattner et al. 1983). All these granites are easily attributed to anatexis within a heterogeneous crustal source, containing a large high-$\delta^{18}O$ metasedimentary component.

Granitic rocks with $\delta^{18}O$-values lower than 6‰ cannot be derived by any known differentiation process from basaltic magmas. Whole rock $\delta^{18}O$-values below mantle–like values require direct or indirect oxygen isotope exchange with low-$^{18}O$ surface waters at high temperatures (Taylor 1987a, b; Fu et al. 2012; Blum et al. 2016). The largest low-$^{18}O$ magmatic province is the Snake River Plain—Yellowstone igneous province which contains more than 10,000 km$^3$ of felsic volcanic rocks derived from low-$^{18}O$ magma (Blum et al. 2016; Colon et al. 2018). As shown by these authors, low-$\delta^{18}O$ values and isotope variability in minerals result from remelting of hydrothermally altered rocks with low $\delta^{18}O$ controlled by the timing and depth of water–rock interactions.

### 3.3.9.2 Non-traditional Isotopes

Si isotopes have been also used to distinguish among different granite types (Savage et al. 2012). Because weathering leads to the formation of $^{30}Si$ depleted clay minerals, granites derived from sedimentary rocks (S-type granites) are isotopically more variable and on average more depleted than I- and A-type granites. However, the relatively small variations indicate that Si-isotopes are less sensitive than O-isotopes (Trail et al. 2018).

Comparable results have been obtained from iron isotope investigations showing that $\delta^{56}$Fe values are generally positively correlated with $SiO_2$ contents. Particular heavy Fe isotope values have been observed in granites with $SiO_2$ contents >70% (Poitrasson and Freydier 2005; Heimann et al. 2008). These authors suggested that exsolution of fluids has removed light Fe isotopes causing the enrichment of $SiO_2$-rich granitoids.

Since Li and B show large isotope variations in surface materials and more or less do not fractionate during melting, Li and B isotope compositions of granites, in principle, may be suitable as a source indicator. However, because potential source lithologies generally show small ranges in Li and B isotopes, both elements are not very sensitive fingerprints as source indicators, except for a few source lithologies with extreme Li and B isotope values (Romer et al. 2014).

#### 3.3.9.3 Zircon

Zircon ($ZrSiO_4$) contains 3 isotope systems with potential isotope variations, all of which have been investigated separately or in conjunction (Zr: Ibanez-Mejia and Tissot 2019; Si and O: Trail et al. 2018).

In the past years, the preferred interest has lain on the analyses of oxygen isotopes (i.e. Valley et al. 2005; Spencer et al. 2019 and others) demonstrating that the combination of radiogenic, especially hafnium, and stable isotope measurements in zircons allow a better understanding of the petrogenesis of granites and the evolution of the continental crust (Hawkesworth and Kemp 2006). Non-metamict zircons preserve their $\delta^{18}$O-value from the time of crystallization because of their refractory and robust nature (Valley 2003). The $\delta^{18}$O-value of zircons, thus, can be used to trace relative contributions of mantle-derived crust and of crust derived by reworking of pre-existing igneous or (meta)-sedimentary crust. Magmas in equilibrium with the mantle crystallize zircon that have a narrow range in $\delta^{18}$O-values of 5.3 ± 0.3‰. Zircons from plutonic oceanic crust have average $\delta^{18}$O-values of 5.2 ± 0.5‰, thus indicating that plagiogranites and differentiated gabbros do not carry a significant seawater signature (Grimes et al. 2011).

$\delta^{18}$O-variations towards higher values result if the parental magma incorporates higher $^{18}$O material (supracrustal rocks through melting or assimilation). Zircons with $\delta^{18}$O-values lower than 5.3‰ indicate an origin of low $^{18}$O magmas pointing to meteoric water–rock interaction. Valley et al. (2005) have analyzed 1200 dated zircons representing the whole spectrum of geologic ages. Uniformly low $\delta^{18}$O-values are found in the first half of Earth history, but much more varied values are observed in younger rocks. In contrast to the Archean, $^{18}$O-values during the Proterozoic gradually increase possibly indicating a maturation of the crust (see Fig. 3.10). After 1.5 Ga high $\delta^{18}$O-values above 8‰ reflect gradual changes in the composition of sediments and the rate and style of recycling of surface-derived material into magmas (Valley et al. 2005).

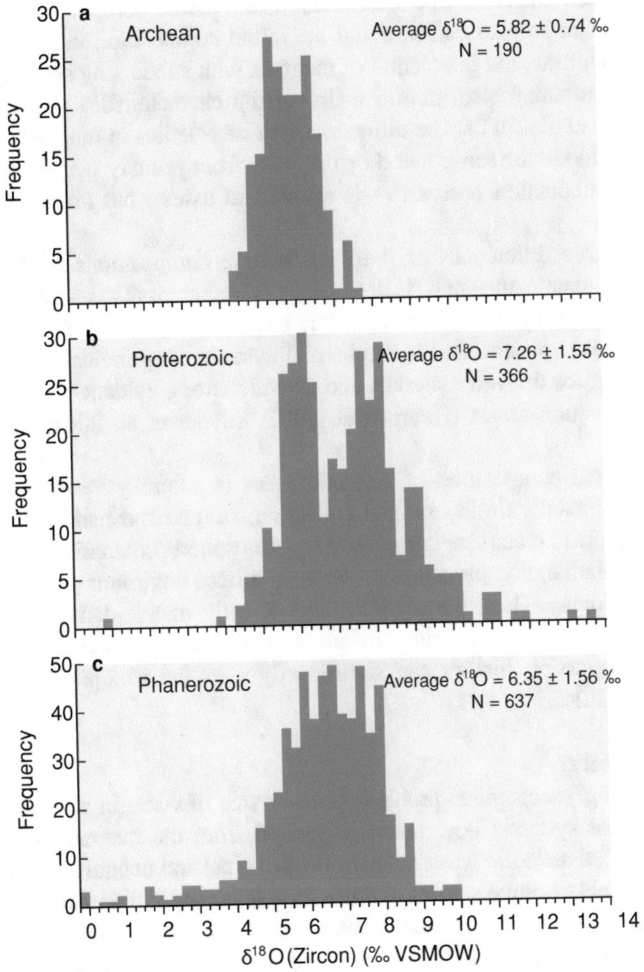


**Fig. 3.10** Histogram of $\delta^{18}$O-values for igneous zircons (**a** *Archean*, **b** *Proterozoic*, **c** *Phanerozoic*) (after Valley et al. 2005)

Including Si isotopes, the combined Si and O-isotope analysis provides more insights in the identity of material involved in zircon formation. Trail et al. (2018) presented evidence for the existence of cherts and banded iron formations >4 Ga.

### 3.3.10 Volatiles in Magmatic Systems

The isotope composition of magmatic volatiles and related isotope fractionation processes can be deduced by analyses of volcanic gases and hot springs. The main process that can cause isotope fractionation of volatile compounds is degassing. Informations can be also gained about the initial composition in the melt prior to gas loss. In addition, the interaction of magmas with subducting slabs, oceanic and continental crust may also imprint their volatile characteristics onto those of the source (Hahm et al. 2012). The ultimate origin of volatiles in magmatic systems—whether juvenile in the sense that they originate from primary mantle degassing, or recycled by subduction processes—is difficult to assess, but may be deduced in some cases.

Because large differences exist in the isotope compositions of surface rocks relative to the mantle, the analysis of volatiles is important in assessing the extent of volatile transfer from the surface reservoirs to the mantle via subduction. Volatiles from arc related volcanic and hydrothermal systems may indicate an appreciable amount of surface derived materials and provide strong evidence of volatile recycling in subduction zones (Hauri et al. 2002; Snyder et al. 2001; Fischer et al. 2002).

The chemical composition of volcanic gases is naturally variable and can be modified significantly during sample collection, storage and handling. While it is relatively simple to recognize and correct for atmospheric contamination, the effects of natural contamination processes in the near-surface environment are much more difficult to address. Thus, the identification of truly mantle-derived gases except helium remains very challenging. In addition to assimilation/contamination processes, the degassing history can significantly alter the isotopic composition of magmatic volatiles.

#### 3.3.10.1 Water

A long-standing geochemical problem is the source of water in volcanic eruptions and geothermal systems: how much is derived from the magma itself and how much is recycled meteoric water? One of the principal and unequivocal conclusions drawn from stable isotope studies of fluids in volcanic hydrothermal systems is that most hot spring waters are meteoric waters derived from local precipitation (Craig et al. 1956; Clayton et al. 1968; Clayton and Steiner 1975; Truesdell and Hulston 1980; Taylor 1986a, b and others).

Most hot spring waters have deuterium contents similar to those of local precipitation, but are usually enriched in $^{18}O$ as a result of isotopic exchange with the country rock at elevated temperatures. The magnitude of the oxygen isotope shift depends on the original O-isotope composition of both water and rock, the mineralogy of the rock, temperature, water/rock ratio, and the time of interaction.

There is increasing evidence, however, that a magmatic water component cannot be excluded in some volcanic systems. As more and more data have become available from volcanoes around the world, especially from those at very high latitudes, Giggenbach (1992) demonstrated that "horizontal" $^{18}O$ shifts are actually the exception rather than the rule: shifts in oxygen isotope composition are also accompanied by a change in the deuterium content (Fig. 3.11). Giggenbach (1992) argued that these waters all followed similar trends corresponding to mixing of local ground waters with a water having a rather uniform isotopic composition with a

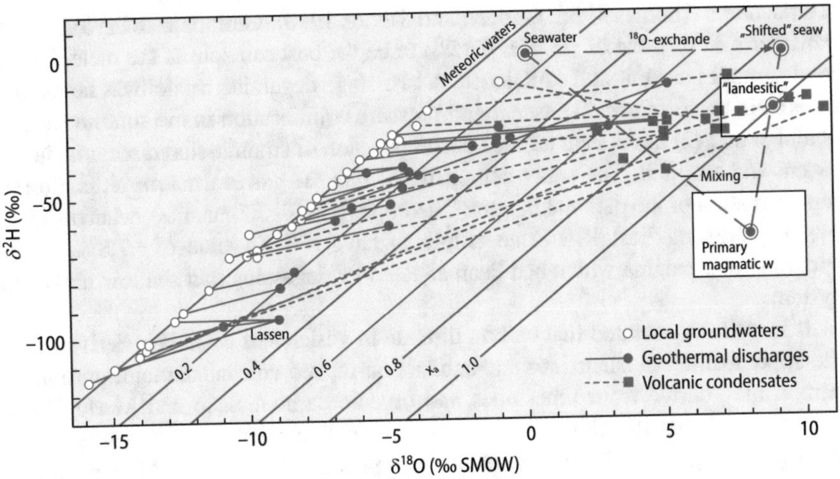


Fig. 3.11 Isotopic composition of thermal waters and associated local ground waters. Lines connect corresponding thermal waters to local groundwaters (Giggenbach 1992)

$\delta^{18}$O-value of about 10‰ and a $\delta^{2}$H-value of about −20‰. He postulated the existence of a common magmatic component in andesite volcanoes having a $\delta^{2}$H of −20‰ which is much higher than the generally assumed mantle water composition. The most likely source would be recycled seawater carried to zones of arc magma generation by the subducted slab.

What is sometimes neglected in the interpretation of isotope data in volcanic degassing products are the effects of boiling. Loss of steam from a geothermal fluid can cause isotopic fractionations. Quantitative estimates of the effects of boiling on the isotopic composition of water can be made using known temperature-dependent fractionation coefficients and estimates of the period of contact between the steam and liquid water during the boiling process (Truesdell and Hulston 1980). As shown by Zakharov et al. (2019), the measurement of the $^{17}$O composition in geothermal fluids can provide additional information on the conditions of water/rock interaction.

### 3.3.10.2 Carbon

$CO_2$ is the second most abundant gas species in magmatic systems. In a survey of $CO_2$ emanations from tectonically active areas worldwide, Barnes et al. (1978) attributed $\delta^{13}$C-values between −8 and −4‰ to a mantle source. This is, however, problematic, because average crustal and mantle isotope compositions are more or less identical and surficial processes that can modify the carbon isotope composition are numerous. A more promising approach may be to analyze the $^{13}$C-content of $CO_2$ collected directly from magmas at high temperatures.

The volcano where gases have been collected and analyzed for the longest time is Kilauea in Hawaii, the data base covering a period from about 1960 to 1985

(Gerlach and Thomas 1986; Gerlach and Taylor 1990). Gerlach and Taylor (1990) consider a $\delta^{13}$C-value of −3.4 ± 0.05‰ to be the best estimate of the mean for the total summit gas emission of Kilauea. A two-stage degassing model was developed to explain these values: (1) ascent and pressure equilibration in the summit magma chamber and (2) rapid, near surface decompression of summit-stored magma during ascent and eruption. The study demonstrated that the gas at the summit is a direct representation of the parental magma C-isotope ratio ($\delta^{13}$C: −3.4‰), whereas gases given off during East Rift Zone eruptions have a $\delta^{13}$C-value of −7.8‰, corresponding to a magma which had been affected by degassing in a shallow magmatic system.

It is well documented that carbon dioxide in vesicles of MORB is derived from the upper mantle. In island arcs and subduction-related volcanism major portions of carbon may derive from limestones and organic carbon. Sano and Marty (1995) demonstrated that the $CO_2/^3He$ ratio in combination with the $\delta^{13}$C-value can be used to distinguish between sedimentary organic, limestone and MORB carbon. Using this approach Nishio et al. (1998) and Fischer et al. (1998) concluded that about two-thirds of the carbon in a subduction zone originates from carbonates, whereas up to one third is derived from organic carbon. Even larger portions (>80%) of $CO_2$ derived from marine carbonates have been found by Shaw et al. (2003) in volcanoes from the Central American arc. Mason et al. (2017) compiled a global data set for carbon and helium isotopes from volcanic arcs and demonstrated that $CO_2$ emitted from arc volcanoes ($\delta^{13}$C-values from −3.8 to −4.6‰) is considerably heavier than $CO_2$ from MORB (−6‰), indicating that limestones are an important $CO_2$ source of arc volcanoes.

Besides $CO_2$, methane has been reported in high-temperature hydrothermal vent fluids (Welhan 1988; Ishibashi et al. 1995). The origin of this methane is somewhat unclear, even in systems which are associated with $^3$He anomalies. Whereas a non-biogenic magmatic origin of methane has been assumed for the East Pacific Rise (Welhan 1988), a thermogenic origin has been proposed for the Okinawa trough (Ishibashi et al. 1995).

In recent years there is growing evidence that methane can be produced abiogenic during a Fischer–Tropsch type synthesis (reduction of CO or $CO_2$ by $H_2$ in the presence of a catalyst) (Sherwood-Lollar et al. 2006; McCollom and Seewald 2006 and others). Hydrocarbons ($C_1$–$C_4$) synthesized under abiogenic hydrothermal conditions are significantly depleted in $^{13}$C relative to their $CO_2$ source. The magnitude of $^{13}$C depletion may be similar to C isotope fractionations during biological processes making it impossible to distinguish between biogenic and abiogenic sources of reduced carbon. This finding has important implications for the discussion of the Earth earliest biosphere. Sherwood-Lollar et al. (2002) observed a trend of decreasing $^{13}$C contents with increasing carbon numbers $C_1$–$C_4$ just opposite to gases derived from biologic sources. Experiments by Fu et al. (2007), however, could not confirm the trend observed by Sherwood-Lollar et al. (2002).

### 3.3.10.3 Nitrogen

Nitrogen in particular is a potential tracer of volatile recycling between the surface and the mantle, because of the large differences in N-isotope composition of MORB ($\delta^{15}N$: −5‰), the atmosphere (0‰) and sediments (6–7‰). As demonstrated by Zimmer et al. (2004), Clor et al. (2005) and Elkins et al. (2006), nitrogen isotopes are very well suited for determining the fate of organic matter in subduction zones. These authors have demonstrated variable contributions of organic matter-derived nitrogen along arcs in Costa Rica, Nicaragua and Indonesia. For instance, Elkins et al. (2006) estimated that sediment contributions to volcanic and geothermal gases in the Nicaraguan volcanic front are around 70%.

### 3.3.10.4 Sulfur

Elucidation of the origin of sulfur in volcanic systems is complicated by the fact that next to $SO_2$, significant amounts of $H_2S$, sulfate and elemental sulfur can also be present. The bulk sulfur isotope composition must be calculated using mass balance constraints. The principal sulfur gas in equilibrium with basaltic melts at low pressure and high temperature is $SO_2$. With decreasing temperature and/or increasing water fugacity, $H_2S$ becomes more stable. $\delta^{34}S$-values of $SO_2$ sampled at very high temperatures provide the best estimate of the $^{34}S$-content of magmas (Taylor 1986a, b). Sakai et al. (1982) reported $\delta^{34}S$-values of 0.7–1‰ in the solfataric gases of Kilauea which compare well with the $\delta^{34}S$-values of 0.9–2.6‰ for Mount Etna gases, measured by Allard (1983) and Liotta et al. (2012). De Moor et al. (2013) investigated sulfur isotope systematics in gases and rocks from a relative reduced volcanic system (Erta Ale in Ethopia) and a relative oxidized system (Masaya in Nicaragua). $\delta^{34}S$-values in Erta Ale ($\delta^{34}S_{(gas)}$ −0.5‰,

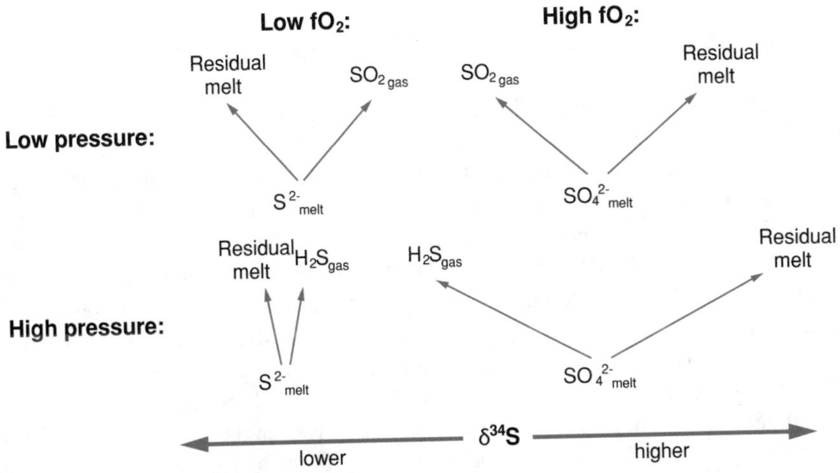


Fig. 3.12 S-isotope degassing scenarios at high and low pressures and at high and low oxygen fugacities (De Moor et al. 2013)

$\delta^{34}S_{(rock)}$ +0.9‰) are considerably more depleted than $\delta^{34}$Svalues from the arc volcano Masaya ($\delta^{34}S_{(gas)}$ +4.8‰, $\delta^{34}S_{(rock)}$ +7.4‰). High values in Masaya obviously reflect recycling of subducted sulfate. Figure 3.12 shows schematically sulfur isotope degassing scenarios at high and low pressures one one hand and high and low oxygen fugacities on the other.

Volcanic sulfur compounds play a key role for the monitoring of volcanoes, because $SO_2$ may convert to submicron particles of sulfate aerosol that may lead to a cooling of the atmosphere for months or even years. The injection of large quantities of volcanic $SO_2$ from explosive eruptions into the stratosphere, thus, may have a significant impact on global climate. Bindeman et al. (2007) and Martin and Bindeman (2009) investigated the sulfur and oxygen isotope composition of sulfate in volcanic ash. They observed a large range in $\delta^{34}S$, $\delta^{18}O$ and $\Delta^{33}S$- and $\Delta^{17}O$-isotope values. The existence of mass-independent S-isotope fractionations demonstrates that the chemistry required for MIF generation occurs in the stratosphere. MIF of oxygen occurs through oxidation of $SO_2$ in the upper atmosphere by interaction with mass-independent ozone. $\Delta^{33}S$- and $\Delta^{17}O$-isotope signatures of ice core sulfate from Antarctica allowed a 2600-year chronology of stratospheric volcanic events (Gautier et al. 2019).

In summary, stable isotope analysis (H, C, S) of volcanic gases and hot springs allow for estimates of the isotopic composition of the mantle source. However, it must be kept in mind that numerous possibilities for contamination, assimilation, and gas phase isotopic fractionation, especially in the surficial environment, make such deductions challenging. In cases where it is possible to "see through" these secondary effects, small differences in H, C, N and S isotope compositions of volcanic gases and hot springs might be characteristic of different geotectonic settings.

### 3.3.11 Isotope Thermometers in Geothermal Systems

Although there are many isotope exchange processes occurring within a geothermal fluid, many of which have the potential to provide thermometric information, only a few have generally been applied, because of suitable exchange rates for achieving isotope equilibrium (Hulston 1977; Truesdell and Hulston 1980; Giggenbach 1992). Temperatures are determined on the basis of calculated fractionation factors of Richet et al. (1977). Differences among geothermometers in the C–O–H–S system are generally ascribed to differences in exchange rates in the decreasing order $CO_2$–$H_2O$ (oxygen) > $H_2O$–$H_2$ (hydrogen) > $SO_2$–$H_2S$ (sulfur) > $CO_2$–$CH_4$ (carbon). Especially pronounced are the differences for the $CO_2$–$CH_4$ thermometer which are often higher than the actual measured temperatures. Investigations on Nisyros volcano, Greece, however, suggest that chemical and isotopic equilibrium between $CO_2$ and $CH_4$ may occur to temperatures as low as 320 °C (Fiebig et al. 2004).

## 3.4 Metamorphic Rocks

The isotope composition of metamorphic rocks is mainly controlled by three factors, besides the temperature of exchange (i) the composition of the pre-metamorphic protolith, (ii) the effects of volatilization with increasing temperatures and (iii) an exchange with infiltrating fluids or melts. The relative importance of these three factors can vary extremely from area to area and from rock type to rock type; and the accurate interpretation of the causes of isotope variations in metamorphic rocks requires knowledge of the reaction history of the respective metamorphic rocks.

(i) The isotope composition of the precursor rock—either sedimentary or magmatic—is usually difficult to estimate. Only in relatively dry non-volatile-bearing precursor rocks do retain metamorphic rocks their original composition.
(ii) Prograde metamorphism of sediments causes the liberation of volatiles, which can be described by two end-member processes (Valley 1986):
(a) Batch volatilization, where all fluid is evolved before any is permitted to escape and (b) Rayleigh volatilization, which requires that once fluid is generated it is isolated immediately from the rock. Natural processes seem to fall between both end-member processes, nevertheless they describe useful limits. Metamorphic volatilization reactions generally reduce the $\delta^{18}O$-value of a rock because $CO_2$ and, in most cases, $H_2O$ lost are enriched in $^{18}O$ compared to the bulk rock. The magnitude of $^{18}O$ depletion can be estimated by considering the relevant fractionations at the respective temperatures. In most cases the effect on the $\delta^{18}O$-value should be small (around 1‰), because the amount of oxygen liberated is small compared to the remaining oxygen in the rock and isotope fractionations at these rather high temperatures are small and, in some cases, may even reverse sign.
(iii) The infiltration of externally derived fluids is a controversal idea, but has gained much support in recent years. Many studies have convincingly demonstrated that a fluid phase plays a far more active role than was previously envisaged, although it is often not clear that the isotopic shifts observed are metamorphic rather than diagenetic (see also Kohn and Valley 1994).

A critical issue is the extent to which the isotope composition of a metamorphic rock is modified by a fluid phase. Volatilization reactions leave an isotope signature greatly different from that produced when fluid-rock interaction accompanies mineral-fluid reaction. Changes of 5–10‰ are a strong indication that fluid-rock interaction rather than volatilization reactions occurred during the metamorphic event. Coupled O–C depletions are seen in many metamorphic systems involving carbonate rocks. Figure 3.13 summarizes results from 28 studies of marble mostly in contact metamorphic settings. In each of the localities shown in Fig. 3.13, the O–C trend has a negative slope, qualitatively similar to the effects of devolatilization.

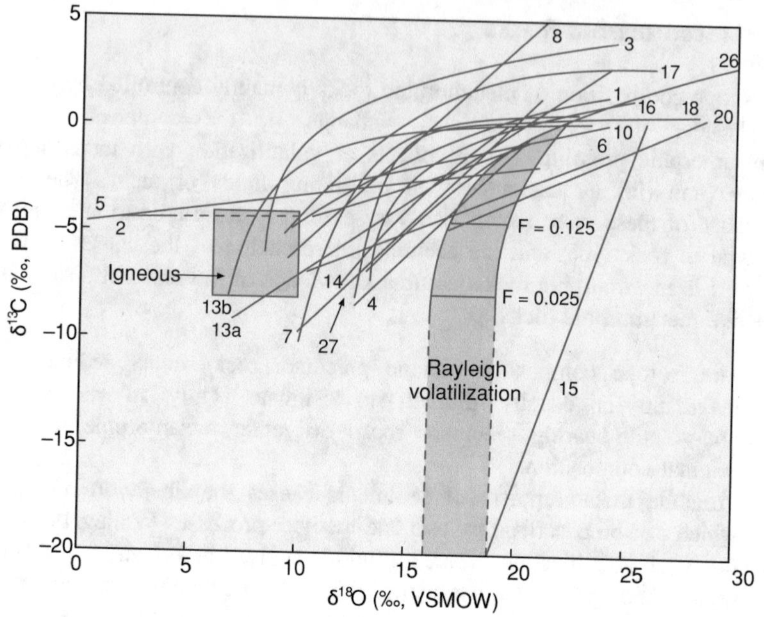


Fig. 3.13 Coupled C–O trends showing decreasing values of $\delta^{13}C$ and $\delta^{18}O$ with increasing metamorphic grade from contact metamorphic localities (Baumgartner and Valley 2001)

However, in each area the magnitude of depletions is too large to be explained by closed-system devolatilization processes, but fluid infiltration and exchange with low $^{18}O$ and $^{13}C$ fluids is indicated (Valley 1986; Baumgartner and Valley 2001).

Two end-member situations can be postulated in which coexisting minerals would change their isotopic composition during fluid-rock interaction (Kohn and Valley 1994): (i) A pervasive fluid moves independently of structural and lithologic control through a rock and leads to a homogenization of whatever differences in isotopic composition may have existed prior to metamorphism. (ii) A channelized fluid leads to local equilibration on the scale of individual beds or units, but does not result in isotopic homogenization of all rocks or units. Channelized flow favors chemical heterogeneity, allowing some rocks to remain unaffected. Although both types of fluid flow appear to be manifest in nature, the latter type appears to be more common.

Numerical modeling of isotope exchange amongst minerals has provided a detailed view of how fluid flow occurs during metamorphism. Stable isotope fronts similar to chromatographic fronts will develop when fluids infiltrate rocks that are not in equilibrium with the infiltrating fluid composition. Isotope ratios increase or decrease abruptly at the front depending on the initial ratio in the rock and infiltrating fluid. Taylor and Bucher-Nurminen (1986), for instance, report sharp isotopic gradients of up to 17‰ in $\delta^{18}O$ and 7‰ in $\delta^{13}C$ over distances of a few mm in calcite around veins in the contact aureole of the Bergell granite. Similar sharp gradients have been also observed in other metasomatic zones but are often unrecognized because an unusually detailed mm-scale sampling is required.

Well defined stable isotope profiles may be used to provide quantitative information on fluid fluxes such as the direction of fluid flow and the duration of infiltration events (Baumgartner and Rumble 1988; Bickle and Baker 1990; Cartwright and Valley 1991; Dipple and Ferry 1992; Baumgartner and Valley 2001). In well constrained situations, fluid flow modeling permits estimation of fluid fluxes that are far more realistic than fluid/rock ratios calculated from a zero-dimensional model.

Due to the invention of micro-analytical techniques (laser sampling and ion microprobe), it has become possible to document small-scale isotope gradients within single mineral grains. Oxygen isotope zoning may develop at a variety of scales, from outcrop scale to the grain scale. In a detailed ion microprobe study, Ferry et al. (2014) observed in a large number of different minerals large intercrystalline and intracrystalline $^{18}O$ variability. Patterns of zoning may reflect multiple processes including diffusive oxygen isotope exchange and infiltration of external fluids. For garnets, zoning has been observed in several cases with increases or decreases from core to rim (Kohn et al. 1993; Young and Rumble 1993; Xiao et al. 2002; Errico et al. 2012; Russell et al. 2013). The shape of the isotopic gradient across a grain will allow distinction among processes controlled by open-system fluid migration or closed-system diffusion.

### 3.4.1 Contact Metamorphism

Because the isotopic composition of igneous rocks is quite different from those of sedimentary rocks, studies of the isotope variations in the vicinity of an intrusive contact offer the possibility of investigating the role of fluids interacting with rocks around cooling plutons. Two types of aureole can be distinguished (Nabelek 1991): (a) "closed" aureoles where fluids are derived from the pluton or the wall-rock and (b) "open" aureoles that for at least part of their metamorphic history have been infiltrated by fluids of external origin. Some aureoles will be dominated by magmatic or metamorphic fluids, whereas others by surface-derived fluids. The occurrence of meteoric-hydrothermal systems around many plutonic complexes has been documented by Taylor and his coworkers and has been described in more detail on p.. The depth to which surface-derived fluids can penetrate is likely that of the brittle-ductile transition and at temperatures close to ~400 °C but depths are still under debate, but most meteoric-hydrothermal systems appear to have developed at depths <~6 km (Criss and Taylor 1986). However, Wickham and Taylor (1985) suggested that seawater infiltration has been observed to a depth of 12 km in the Trois Seigneur Massif, Pyrenees.

In many contact aureoles combined petrologic and isotope studies have provided evidence that fluids were primarily locally derived. Oxygen isotope compositions of calc-silicates from many contact aureoles have revealed that the $^{18}O$-contents of the calc-silicate hornfelses approach those of the respective intrusions. This, together with characteristic hydrogen and carbon isotope ratios, has led many workers to conclude that magmatic fluids were dominant during contact metamorphism with meteoric fluids becoming important during subsequent cooling only (Taylor and O'Neil 1977; Nabelek et al. 1984; Bowman et al. 1985; Valley 1986). Ferry and Dipple (1992) developed different models to simulate fluid-rock interaction on the Notch Peak aureole, Utah. Their preferred model assumes fluid flow in the direction of increasing temperature, thus arguing against magmatic fluids, but instead proposing fluids derived from volatilization reactions. Nabelek (1991) calculated model $\delta^{18}$O-profiles which should result from both "down-temperature" and "up-temperature" flow in a contact aureole. He demonstrated that the presence of complex isotopic profiles can be used to get information about fluid fluxes. Gerdes et al. (1995) have examined meter-scale $^{13}$C and $^{18}$O transport in a thin marble layer near a dike in the Adamello contact aureole, Southern Alps. They observed systematic stable isotope changes in the marble over <1 m as the dike is approached with $\delta^{13}$C-values ranging from 0 to −7‰ and $\delta^{18}$O values from 22.5 to 12.5‰. These authors have compared the isotope profiles to one- and two-dimensional models of advective–dispersive isotope transport. Best agreement is obtained using a two-dimensional model that specifies (i) a high permeability zone flow and (ii) a lower permeability zone in marble away from the dike.

### 3.4.2 Regional Metamorphism

It is a general observation that low-grade metamorphic pelites have $\delta^{18}$O-values between 15 and 18‰ whereas high-grade gneisses have $\delta^{18}$O-values between 6 and 10‰ (Garlick and Epstein 1967; Shieh and Schwarcz 1974; Longstaffe and Schwarcz 1977; Rye et al. 1976; Wickham and Taylor 1985; Peters and Wickham 1995). In the absence of infiltration of a fluid phase, isotopic shifts resulting from net transfer reactions in typical amphibolite or lower granulite facies metapelites and metabasites are about 1‰ or less for about 150 °C of heating (Kohn et al. 1993; Young 1993). Thus, the processes responsible for this decrease in $^{18}$O must be linked to large-scale fluid transport in the crust.

There are several factors which control fluid transport. One is the lithology of a metamorphic sequence. Marbles, in particular, are relatively impermeable during metamorphism (Nabelek et al. 1984) and, therefore, may act as barriers to fluid flow, limiting the scale of homogenization and preferentially channeling fluids through silicate layers. Marbles may act as local high-$^{18}$O reservoirs and may even increase the $^{18}$O content of adjacent lithologies (Peters and Wickham 1995). Therefore, massive marbles generally preserve their sedimentary isotope signatures, even up to the highest metamorphic grades (Valley et al. 1990).

Sedimentary sequences undergoing a low-grade metamorphism initially may contain abundant connate pore fluids which provide a substantial low-$^{18}$O reservoir and a medium for isotopic homogenization. An additional important fluid source is provided by metamorphic dehydration reactions at higher grades of metamorphism (e.g. Ferry 1992). In some areas, petrological and stable isotope studies suggest that metamorphic fluid compositions were predominantly internally buffered by devolatilization reactions and that large amounts of fluid did not interact with the rocks during regional metamorphism (e.g. Valley et al. 1990). In a high-grade poly-metamorphic terrane, later metamorphic events are likely to be dominated by magmatic fluid sources since previous events would have caused extensive dehydration, thereby limiting potential fluid sources (Peters and Wickham 1995). A detailed study of the O-isotope composition of pelites, amphibolites and marbles from the island of Naxos, Greece demonstrates that the isotopic pattern observed today is the result of at least three processes: two fluid flow events and a pre-existing isotopic gradient (Baker and Matthews 1995).

Shear zones are particularly good environments to investigate fluid flow at various depths within the crust (Kerrich et al. 1984; Kerrich and Rehrig 1987; McCaig et al. 1990; Fricke et al. 1992). During retrograde metamorphism aqueous fluids react with dehydrated rocks and fluid flow is concentrated within relatively narrow zones. By analyzing quartzite mylonites in Nevada, Fricke et al. (1992) demonstrated that significant amounts of meteoric waters must have infiltrated the shear zone during mylonitization to depths of at least 5–10 km. Similarly, McCaig et al. (1990) showed that formation waters were involved in shear zones in the Pyrenees and that the mylonitization process occurred at a depth of about 10 km.

Unusually low $\delta^{18}$O-values—as light as −5 to −10‰—have been observed in ultra-high pressure (UHP)-rocks from Dabie Shan and Sulu, China (Rumble and Yui 1998; Zheng et al. 1998; Xiao et al. 2006 besides others). UHP-rocks are characterized by coesite and microdiamond in eclogite and other crustal rocks, which is strong evidence that a sizable segment of ancient continental crust was subducted to mantle depths. The extremely low $\delta^{18}$O-values result from meteoric water interaction prior to UHP metamorphism. Surprisingly, these rocks have preserved their extremely low $\delta^{18}$O-values indicating a short residence time at mantle depth followed by a rapid uplift. Quartz-garnet oxygen isotope temperatures in the range 700–900 °C are consistent with an approach to grain-scale oxygen isotope equilibrium under UHP conditions (Rumble and Yui 1998; Xiao et al. 2006).

The Dabie-Sulu terrain is the largest among the UHP belts worldwide and covers an area of 5000 km$^2$ in Dabie and >10,000 km$^2$ in Sulu. The huge amounts of meteoric water necessary to cause the $^{18}$O-depletions probably originate from the deglaciation of the Neoproterozoic snowball earth.

More recently, even larger oxygen isotope depletions ($\delta^{18}$O-values as low as −27.3‰) have been reported in 2.3–2.4 Ga old rocks from Karelia, Russia. Very low $\delta^{18}$O hydrothermally-altered rocks have been discovered over 500 km along the Baltic Shield which are related to Paleoproterozoic Snowball earth glacial episodes (Bindeman and Serebryakov 2011). Triple oxygen isotope systematics suggest that infiltrating waters had $\delta^{18}$O-values below −40‰ corresponding to glacial ice from Paleoproterozoic Snowball Earth events (Herwartz et al. 2015; Zakharov et al. 2017, 2019; Herwartz 2021).

### 3.4.3 Subduction Zone Metamorphism

In recent years numerous studies have investigated isotope fractionation processes during subduction zone metamorphism. The subducting oceanic lithosphere, typically composed of a section of igneous oceanic crust with a sedimentary cover and an underlying ultramafic lithospheric mantle section, transports material to the deep mantle via the descending slab and returns material to the surface via arc magmatims. Dehydration and decarbonation accompanied by the release of oxidizing species play a key role in controlling redox changes in the subducting slab and overlying mantle wedge which result in the oxidation of the subarc mantle. The primary sources of water and $CO_2$ have been linked to the breakdown of hydrous and carbonate-containing minerals in the oceanic crust. Volatilization of serpentinite representing the largest potential source of fluids has been considered to play a decisive role (Alt et al. 2013; Spandler and Pirard 2013).

The most prominent isotope systems that have been studied in this connection are the light isotopes, tracing fluid sources and pathways. Oxygen isotope investigations have modeled the effect of dehydrations reactions on the bulk rock $\delta^{18}O$-value and the effect of an external fluid influx (Vho et al. 2020). In a closed system, dehydration produces minor to negligible $^{18}O$ variations; in an open system, fluid-rock interactions produce larger $^{18}O$ shifts. With respect to hydrogen, slab-derived fluids are generally heavy in deuterium (Shaw 2008, 2012). As dehydration progresses and D is increasingly removed from the slab, the released fluids and the slab itself becomes isotopically lighter. Nitrogen in subduction zones is relative stable in the form of $NH_4$ in potassic minerals (phengite). Busigny et al. (2011) and Halama et al. (2011) estimated that the majority of nitrogen (80%?) buried in subduction zones is recycled to the deep mantle with an estimated $\delta^{15}N$-values between 3 and 5‰.

Of particular importance are the redox sensitive elements C, S, Fe (Evans 2012 and others). The behavior of carbon in subduction zones has been summarized by Plank and Manning (2019). They showed that pT-paths experienced in each subduction zone separate two carbon reservoirs with distinct isotope compositions: (i) carbon that returns to the surface and (ii) carbon that is carried deep into the mantle either as molten carbonate or as diamond.

In contrast to carbon, reduced sulfur species dominate in slab fluids (Walters et al. 2019; Li et al. 2020). Variations in sulfur isotope composition depend on their origins from (i) metasediments, (ii) altered oceanic crust or (iii) serpentinite. Li et al. (2020) concluded that slab fluids provide negligible sulfate to enrich arc mantle in heavy S isotopes.

Debret et al. (2018) have documented iron isotope fractionations in serpentinite due to interaction with low $\delta^{56}Fe$ sediment derived fluids. Insoluble iron can be mobilized by complexation with carbonate and sulphate bearing fluids during the early stages of subduction. Pons et al. (2016) used Zn isotopes in subducted serpentinites to decipher the chemical nature of slab derived fluids.

Lithium and boron isotopes also provide robust tracers of surface material that is recycled to the mantle (Elliott et al. 2004; De Hoog and Savov 2018). Preferential loss of the heavier isotopes $^{7}Li$ and $^{11}B$ during dehydration results in decreasing $\delta^{7}Li$ and $\delta^{11}B$-values with slab depths. As the slab becomes isotopically lighter, transfer of lithium and boron in the fluid or melt from the subducting slab into the overlying mantle has a strong impact on the isotopic composition of arc rocks.

Other fluid mobile elements showing characteristic isotope fractionations during dehydration processes of subducting slabs are molybdenum (Gaschnig et al. 2017; Willbold and Elliott 2017), uranium (Freymuth et al. 2019) and tungsten (Mazza et al. 2020).

### 3.4.4 Lower Crustal Rocks

Granulites constitute the dominant rock type in the lower crust. Granulites may be found at the Earth's surface in two different settings: (i) exposed in high grade regional metamorphic belts and (ii) found as small xenoliths in basaltic pipes. Both types of granulites suggest a compositionally diverse lower crust ranging in composition from mafic to felsic.

Stable isotope studies of granulite terranes (Sri Lanka—Fiorentini et al. 1990; South India—Jiang et al. 1988; Limpopo Belt—Hoernes and Van Reenen 1992; Venneman and Smith 1992; Adirondacks—Valley and coworkers) have shown that terranes are isotopically heterogeneous and are characterized by $\delta^{18}O$-values that range from "mantle-like" values to typical metasedimentary values above 10‰. Investigations of amphibolite/granulite transitions have shown little evidence for a pervasive fluid flux as a major factor in granulite facies metamorphism (Valley et al. 1990; Cartwright and Valley 1991; Todd and Evans 1993).

Similar results have been obtained from lower crustal granulite xenoliths, which also exhibit a large range in $\delta^{18}O$-values from 5.4 to 13.5‰ (Mengel and Hoefs 1990; Kempton and Harmon 1992). Mafic granulites are characterized by the lowest $\delta^{18}O$-values and range of $^{18}O$-contents. By contrast, silicic meta-igneous and meta-sedimentary granulites are significantly enriched in $^{18}O$ with an average $\delta^{18}O$-value around 10‰. The overall variation of 8‰ emphasizes the O-isotope heterogeneity of the lower crust and demonstrates that pervasive deep crustal fluid flow and isotopic homogenization is not a major process.

### 3.4.5 Thermometry

Oxygen isotope thermometry is widely used to determine temperatures of metamorphic rocks. The principal concern in isotope thermometry continues to be the preservation of peak metamorphic temperatures during cooling. It has long been recognized that oxygen isotope thermometers often record discordant temperatures in slowly cooled metamorphic rocks. Figure 3.14 gives a compilation of literature data (Kohn 1999) showing $\delta^{18}O$ values and calculated temperature ranges for quartz-magnetite and muscovite-biotite. Muscovite-biotite pairs from rocks whose metamorphic conditions range from greenschist to granulite facies cluster around an

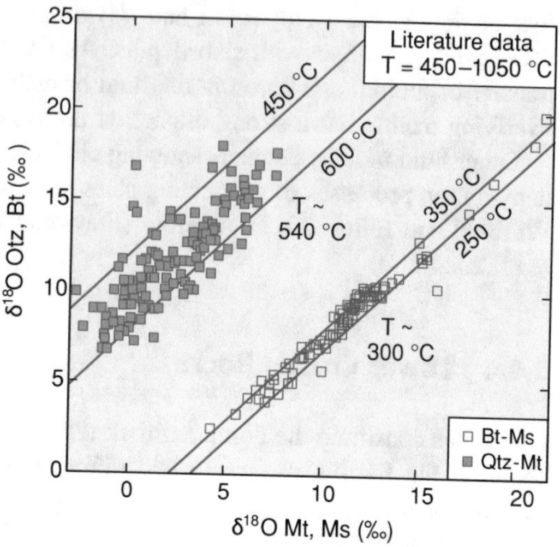


**Fig. 3.14** Plot of $\delta^{18}O$ of quartz versus $\delta^{18}O$ magnetite (*solid squares*) and of biotite versus muscovite (*open squares*) from rocks whose peak metamorphic conditions range from greenschist through granulite facies (after Kohn 1999)

apparent temperature of ~300 °C, whereas quartz-magnetite pairs have an apparent temperature of ~540 °C. These data demonstrate substantial diffusional resetting, which is consistent with relatively high water fugacities during cooling (Kohn 1999).

Assuming that a rock behaves as a closed system and consists of the three mineral assemblage quartz, feldspar and horblende, then hornblende will be the slowest diffusing phase and feldspar the fastest diffusing phase. Using the formulation of Dodson (1973) for closure temperature and a given set of parameters (diffusion constants, cooling rate and grain size), Giletti (1986) calculated apparent temperatures that would be obtained in rocks with different modal proportions of the three minerals once all isotope exchange had ceased in the rock. In the Giletti model, the apparent quartz—hornblende temperature is dependent only on the quartz/feldspar ratio and is independent of the amount of hornblende in the rock, since hornblende is the first phase to reach its closure temperature. Eiler et al. (1992, 1993), however, demonstrated that the abundance of the slow diffusing phase (e.g. hornblende) can affect apparent equilibrium temperatures because of continued exchange between the grain boundaries of this phase and fast diffusing phases. Thus, retrograde diffusion related oxygen isotope exchange makes the calculation of peak metamorphic temperatures impossible, but can be used to estimate cooling rates.

Diffusion modelling, on the other hand, also predicts that accurate temperatures can be obtained from refractory accessory minerals, if they occur in a rock that is modally dominated by a readily exchangeable mineral (Valley 2001). The basis of this approach is that the accessory mineral preserves the isotope composition from crystallization because of slow diffusion while the dominant mineral preserves its isotope composition by mass balance because there are no other sufficiently abundant exchangeable phases.

Several refractory accessory mineral thermometers have been applied, including aluminosilicate, magnetite, garnet and rutile in quartz-rich rocks and magnetite, titanite or diopside in marble. Refractory minerals are defined based on their relative diffusion rates relative to the matrix of the total rock. Thus plagioclase—magnetite or plagioclase—rutile may be good thermometers in amphibolite or eclogite-facies basic rocks, but fail in the granulite facies.

Other suitable phases for the preservation of peak metamorphic temperatures are the $Al_2SiO_5$ polymorphs kyanite and sillimanite, having both slow oxygen diffusion rates. By analyzing the alumosilicate polymorphs from a variety of rocks with different temperature histories, Sharp (1995) could derive empirical equilibrium fractionation factors for kyanite and sillimanite. In some rocks oxygen isotope temperatures are far higher than the regional metamorphic temperatures, possibly reflecting early high-temperature contact metamorphic effects that are preserved only in the most refractory phases.

Despite extensive diffusional resetting under water-buffered conditions, some rocks clearly retain oxygen isotope fractionations that are not reset by diffusion during cooling. Farquhar et al. (1996) have investigated two granulite terrains from NW Canada and Antarctica. Quartz-garnet temperatures of around 1000 °C are in good agreement with a variety of independent temperature estimations. Quartz-pyroxene temperatures are significantly lower and still lower quartz-magnetite temperatures of around 670 °C are attributed to a combination of faster oxygen diffusion in quartz and magnetite and recrystallization during late-stage deformation. The "dry" nature of granulites is obviously critical for preservation of high-temperature records. Cooler and more hydrous rocks seem to be less capable of retaining a record of peak temperatures.

Carbon isotope partitioning between calcite and graphite is another example of a favorable thermometer to record peak metamorphic temperatures in marbles because calcite is the abundant phase with relatively high carbon diffusivities whereas graphite is of minor abundance and has a very slow diffusion rate. Figure 3.15 shows the decrease of fractionation of calcite and graphite ($\Delta$) with increasing metamorphic grade. The narrow range of graphite $\delta$-values associated with granulite facies rocks indicates isotope equilibrium between carbonate and graphite at high temperatures. Figure 3.15 also indicates that under granulite-facies conditions the original carbon isotope composition has been obliterated due to exchange between carbonate and reduced carbon. As shown by Kueter et al. (2019) the best calibration of the carbonate-graphite thermometer is the combination high temperature experiments with empirical metamorphic calibrations.

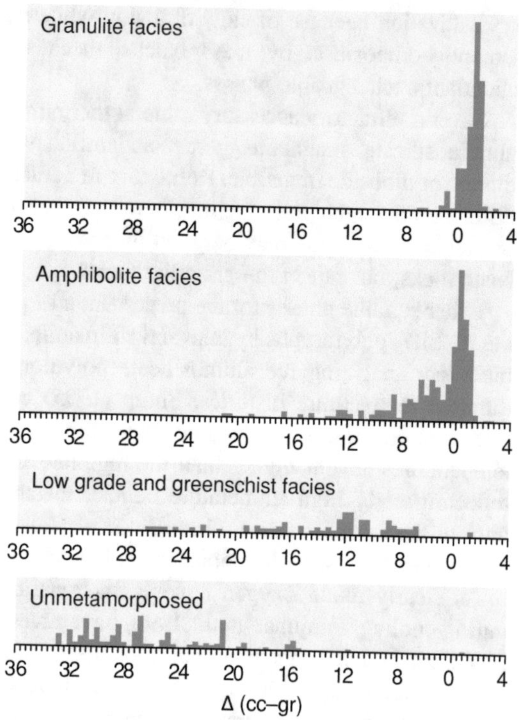


**Fig. 3.15** Frequency distribution of calcite-graphite fractionations (Δ) with increasing metamorphic grade (after Des Marais 2001)

## 3.5 Ore Deposits and Hydrothermal Systems

Stable isotopes have become an integral part of ore deposits studies. The determination of light isotopes of H, C, O and S can provide information about the diverse origins of ore fluids, about temperatures of mineralization and about physico-chemical conditions of mineral deposition. The investigation of the metal isotope composition adds important constraints on formation processes, in particular low temperature redox conditions. In contrast to early views, which assumed that almost all metal deposits owed their genesis to magmas, stable isotope investigations have convincingly demonstrated that ore formation has taken place in the Earths near-surface environment by recycling processes of fluids, metals, sulfur, and carbon. Reviews of the application of the light stable isotopes to the genesis of ore deposits have been given by Ohmoto (1986), Taylor (1987a, b, 1997).

Inasmuch as water is the dominant constituent of ore-forming fluids, knowledge of its origin is fundamental to any theory of ore genesis. There are two ways for determining $\delta^2$H- and $\delta^{18}$O-values of ore fluids: (i) by direct measurement of fluid inclusions contained within hydrothermal minerals, or (ii) by analysis of hydroxyl-bearing minerals and calculation of the isotopic composition of fluids from known temperature-dependent mineral–water fractionations, assuming that minerals were precipitated from solutions under conditions of isotope equilibrium.

(I) Fluids and gases may be extracted from rocks by (i) thermal decrepitation by heating in vacuum and by (ii) crushing and grinding in vacuum. Serious analytical difficulties may be associated with both techniques. The major disadvantage of the thermal decrepitation technique is that, although the amount of gas liberated is higher than by crushing, compounds present in the inclusions may exchange isotopically with each other and with the host mineral at the high temperatures necessary for decrepitation. Crushing in vacuum largely avoids isotope exchange processes. However, during crushing large new surfaces are created which easily adsorb some of the liberated gases and that, in turn, might be associated with fractionation effects. Both techniques preclude separating the different generations of inclusions in a sample and, therefore, the results obtained represent an average isotopic composition of all generations of inclusions.

Numerous studies have used the $\delta^2$H-value of the extracted water to deduce the origin of the hydrothermal fluid. However, without knowledge of the internal distribution of hydrogen in quartz, such a deduction can be misleading (Simon 2001). Hydrogen in quartz mainly occurs in two reservoirs: (i) in trapped fluid inclusions and (ii) in small clusters of structurally bound molecular water. Because of hydrogen isotope fractionation between the hydrothermal fluid and the structurally bound water, the total hydrogen extracted from quartz does not necessarily reflect the original hydrogen isotope composition. This finding may explain why $\delta^2$H-values from fluid inclusions often tend to be lower than $\delta^2$H-values from associated minerals (Simon 2001).

(II) The indirect method of deducing the isotope composition of ore fluids is more frequently used, because it is technically easier. Uncertainties arise from several sources: uncertainty in the temperature of deposition, and uncertainty in the equations for isotope fractionation factors. Another source of error is an imprecise knowledge of the effects of fluid chemistry ("salt effect") on mineral–water fractionation factors.

Several studies (e.g. Berndt et al. 1996; Driesner and Seward 2000; Horita et al. 1995; Shmulovich et al. 1999) have demonstrated that the approach of using mineral—pure water fractionation factors to deduce the origin of the water is incorrect. Isotope fractionations involving aqueous solutions depend not only on temperature and fluid composition, but also on the presence or absence of phase separation ("boiling"). Phase separation is an important process causing potentially isotope fractionation. Hydrogen isotope studies (Berndt et al. 1996; Shmulovich et al. 1999) indicate that high temperature phase separation produces D-enrichment in the vapor and D-depletion in the conjugate fluid. If the fractionation effect inherent in a boiling fluid system is disregarded, one may easily misinterpret the isotope composition of hydrothermal minerals, since boiling may mask the source of the parent fluids. In addition, for hydrogen isotope fractionations, pressure may have some control on mineral–water fractionations (Driesner 1997; Horita et al. 1999).

The mineral alunite, and its iron equivalent jarosite, are a special case. Alunite ($KAl_3(SO_4)_2(OH)_6$) contains four sites where elements containing stable isotopes are found and both the sulfate and hydroxyl anionic groups may provide information on fluid source and condition of formation.

### 3.5.1 Origin of Ore Fluids

Ore fluids may be generated in a variety of ways. The principal types include (i) sea water, (ii) meteoric waters and (iii) juvenile water, all of which have a strictly defined isotopic composition. All other possible types of ore fluids such as formation, metamorphic and magmatic waters can be considered recycled derivatives or mixtures from one or more of the three reference waters (see Fig. 3.16).

(i) Sea water

The oxygen isotopic composition of present-day ocean water is more or less constant with δ-values close to zero permil. The isotopic composition of ancient ocean water, however, is less well constrained (see Sect. 3.7), but still should not be removed from zero by more than 1 or 2‰. Many volcanogenic massive sulfide deposits are formed in submarine environments from heated oceanic waters. This concept gains support from the recently observed hydrothermal systems at ocean

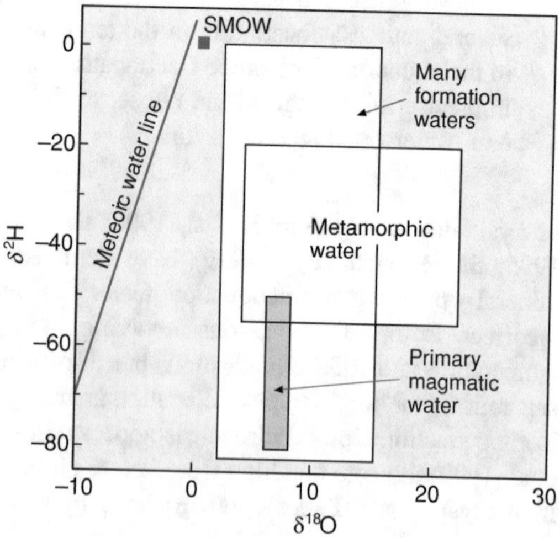


Fig. 3.16 Plot of $\delta^2H$ versus $\delta^{18}O$ of waters of different origin

ridges, where measured isotopic compositions of fluids are only slightly modified relative to 0‰. $\delta^{18}O$ and $\delta^{2}H$-values of vent fluids are best understood in terms of seawater interaction with the ocean crust (Shanks 2001).

Bowers and Taylor (1985) have modelled the isotopic composition of an evolving seawater hydrothermal system. At low temperatures, the $\delta^{18}O$-value of the fluid decreases relative to ocean water because the alteration products in the oceanic crust are $^{18}O$ rich. At around 250 °C, the solution returns to its initial seawater isotopic composition. Further reaction with basalt at 350 °C increases the $\delta^{18}O$ value of modified seawater to ~2‰. The $\delta^{2}H$-value of the solution increases slightly at all temperatures because mineral–water fractionations are generally all less than zero. At 350 °C, the $\delta^{2}H$-value of the solution is 2.5‰. The best documented example for the role of ocean water during ore deposition is for the Kuroko-type deposits (see the extensive monograph by Ohmoto and Skinner 1983).

(ii) Meteoric waters

Heated meteoric waters are a major constituent of ore-forming fluids in many ore deposits and may become dominant during the latest stages of ore deposition. The latter has been documented for many porphyry skarn-type deposits. The isotopic variations observed for several Tertiary North American deposits vary systematic with latitude and, hence, palaeo-meteoric water composition (Sheppard et al. 1971). The ore-forming fluid has commonly been shifted in O-isotope composition from its meteoric $\delta^{18}O$-value to higher $^{18}O$ contents through water–rock interaction. Meteoric waters may become dominant in epithermal gold deposits and other vein and replacement deposits.

(iii) Juvenile water

The concept of juvenile water has influenced early discussions about ore genesis tremendously. The terms "juvenile water" and "magmatic water" have been used synonymously sometimes, but they are not exactly the same. Juvenile water originates from degassing of the mantle and has never existed as surface water. Magmatic water is a non-genetic term and simply means a water that has equilibrated with a magma.

It is difficult to prove that juvenile water has ever been sampled. One way to search for juvenile water is by analyzing hydroxyl-bearing minerals of mantle origin (Sheppard and Epstein 1970). The estimated isotopic composition of juvenile water from such an approach is $\delta^{2}H$: − 60 ± 20‰ and $\delta^{18}O$: + 6 ± 1‰ (Ohmoto 1986).

#### 3.5.1.1 Magmatic Water

Despite the close association of intrusions with many ore deposits, there is still debate about the extent to which magmas contribute water and metals to ore-forming fluids. Many early studies of the stable isotope composition of hydrothermal minerals indicated a dominance of meteoric water (Taylor 1974), more recent studies show that magmatic fluids are commonly present, but that their isotopic compositions may be masked or erased during later events such as the influx of meteoric waters.

The $\delta^2H$-value of magmatic water changes progressively during degassing, resulting in a positive correlation between $\delta^2H$ and the residual water content of an igneous body. Thus, late-formed hydroxyl-bearing minerals represent the isotopic composition of a degassed melt rather than that of the initial magmatic water. The $\delta^2H$ values of most of the water exsolved from many felsic melts is in the range of −60 to −30‰, whereas the associated magmatic rocks may be significantly depleted in deuterium.

The calculated range of isotopic composition for magmatic waters is commonly 6–10‰ for $\delta^{18}O$-values and − 50 to − 80‰ for $\delta^2H$-values. Magmatic fluids may change their isotopic composition during cooling through isotope exchange with country rocks and mixing with fluids entrained within the country rocks. Thus, the participation of a magmatic water component during an ore-forming process is generally not easily detected.

#### 3.5.1.2 Metamorphic Water

Metamorphic water is defined as water associated with metamorphic rocks during metamorphism. Thus, it is a descriptive, non-genetic term and may include waters of different ultimate origins. In a narrower sense, metamorphic water refers to the fluids generated by dehydration of minerals during metamorphism. The isotopic composition of metamorphic water may be highly variable, depending on the respective rock types and their history of fluid/rock interaction. A wide range of $\delta^{18}O$-values (5–25‰) and $\delta^2H$-values (−70 to −20‰) is generally attributed to metamorphic waters (Taylor 1974).

#### 3.5.1.3 Formation Waters

The changes in the D- and $^{18}O$-contents of pore fluids depend on the origin of initial fluid (ocean water, meteoric water), temperature and the lithology of rocks with which the fluids are or have been associated. Generally, formation waters with the lowest temperature and salinity have the lowest $\delta^2H$- and $\delta^{18}O$-values, approaching those of meteoric waters. Brines of the highest salinities are generally more restricted in isotopic composition. It is still an unanswered question though whether meteoric water was the only source of water to these brines. The final isotope composition of brines can be produced by reactions between meteoric water and sediments, or result from mixtures of fossil ocean water trapped in the sediments and meteoric water.

### 3.5.2 Wall-Rock Alteration

Information about the origin and genesis of ore deposits can also be obtained by analyzing the alteration products in wall-rocks. Hydrogen and oxygen isotope zonation in wall-rocks around hydrothermal systems can be used to define the size and the conduit zones of a hydrothermal system. The fossil conduit is a zone of large water fluxes, generally causing a strong alteration in the rocks and lowering the $\delta^{18}O$-values. Thus, fossil hydrothermal conduits can be outlined by following the zones of $^{18}O$-depletion. Oxygen isotope data are especially valuable in rock types that do not show diagnostic alteration mineral assemblages as well as those in which the assemblages have been obliterated by subsequent metamorphism (e.g. Beaty and Taylor 1982; Green et al. 1983). Criss et al. (1985, 1991) found excellent spatial correlations between low $\delta^{18}O$-values and economic mineralization in siliceous rocks. Similar zonation around ore deposits in carbonate rocks have also been observed (e.g. Vazquez et al. 1998). Thus, zones having anomalously low $\delta^{18}O$-values may be a useful guide for exploration of hydrothermal ore deposits.

### 3.5.3 Fossil Hydrothermal Systems

Mainly through the work of Taylor and coworkers, it has become well established that many epizonal igneous intrusions have interacted with meteoric groundwaters on a very large scale. The interaction and transport of large amounts of meteoric water through hot igneous rocks produces a depletion in $^{18}O$ in the igneous rocks by up to 10–15‰ and a corresponding shift in the $^{18}O$ content of the water. About 60 of such systems have been observed to date (Criss and Taylor 1986). They exhibit great variations in size from relatively small intrusions (<100 km$^2$) to large plutonic complexes (>1000 km$^2$). Amongst the best documented examples are the Skaergaard intrusion in Greenland, the Tertiary intrusions of the Scottish Hebrides, and the Tertiary epizonal intrusions of the northwestern United States and southern British Columbia, where 5% of the land surface has been altered by meteoric hydrothermal water (Criss et al. 1991).

The best-studied example of a hydrothermal system associated with a gabbro is the Skaergaard intrusion (Taylor and Forester 1979; Norton and Taylor 1979). The latter authors carried out a computer simulation of the Skaergaard hydrothermal system and found a good match between calculated and measured $\delta^{18}O$-values. They further demonstrated that most of the sub-solidus hydrothermal exchange took place at very high temperatures (400–800 °C), which is compatible with the general absence of hydrous alteration products in the mineral assemblages and with the presence of clinopyroxene.

In granitic hydrothermal systems, temperatures of alteration are significantly lower because of differences in the intrusion temperatures. The most conspicious petrographic changes are chloritization of mafic minerals, particularly of biotite, and a major increase in the turbidity of feldspars. Large non-equilibrium quartz—feldspar oxygen isotope fractionations are typical. Steep linear trajectories on plots of $\delta^{18}O_{(feldspar)}$ versus $\delta^{18}O_{(quartz)}$ are a characteristic feature of these hydrothermally altered rocks (see Fig. 2.17). The trajectories result from the fact that feldspar exchanges $^{18}O$ with hydrothermal fluids much faster than coexisting quartz and from the fact that the fluids entering the rock system have $\delta^{18}O$-values which are out of equilibrium with the mineral assemblage. The process seldom goes to completion, so the final mineral assemblage is in isotope disequilibrium, which is the most obvious fingerprint of the hydrothermal event.

Taylor (1988) distinguished three types of fossil hydrothermal systems on the basis of varying water/rock ratios, temperatures, and the length of time that fluid/rock interaction proceeds.

Epizonal systems with a wide variation in whole rock $^{18}O$-contents and extreme oxygen isotope disequilibrium among coexisting minerals. These systems typically have temperatures between 200 and 600 °C and life-times <$10^6$ y.

Deeper-seated and/or longer-lived systems, also with a wide spectrum of whole rock $^{18}O/^{16}O$ ratios, but with equilibrated $^{18}O/^{16}O$ ratios among coexisting minerals. Temperatures are between 400 and 700 °C and life-times >$10^6$y.

Equilibrated systems with a relatively uniform oxygen isotope composition in all lithologies. These systems require a large water/rock ratio, temperatures between 500 and 800 °C, and life times around $5 \times 10^6$ y.

These types are not mutually exclusive, Type III systems for example may have been subjected to Type I or Type II conditions at an earlier stage of their hydrothermal history.

### 3.5.4 Hydrothermal Carbonates

The measured $\delta^{13}C$- and $\delta^{18}O$-values of carbonates can be used to estimate the carbon and oxygen isotope composition of the fluid in the same way as has been discussed before for oxygen and hydrogen. The isotopic composition of carbon and oxygen in any carbonate precipitated in isotopic equilibrium with a fluid depends on the isotopic composition of carbon and oxygen in the fluid, the temperature of formation, and the relative proportions of dissolved carbon species ($CO_2$, $H_2CO_3$, $HCO_3^-$, and/or $CO_3^{2-}$). To determine carbonate speciation, pH and temperature must be known; however, in most geologic fluids with temperatures above about 100 °C, the content of $HCO_3^-$ and $CO_3^{2-}$ is negligible compared to $CO_2$ and $H_2CO_3$.

Experimental investigations have shown that the solubility of carbonate increases with decreasing temperature. Thus, carbonate cannot be precipitated from a hydrothermal fluid due to simple cooling in a closed system. Instead, an open system is required in which processes such as $CO_2$ degassing, fluid-rock interaction or fluid mixing can cause the precipitation of carbonate. These processes result in correlation trends in $\delta^{13}C$ versus $\delta^{18}O$ space for hydrothermal carbonates as often observed in nature and theoretically modeled by Zheng and Hoefs (1993).

Figure 3.17 presents $\delta^{13}C$ and $\delta^{18}O$-values of hydrothermal carbonates from the Pb–Zn deposits of Bad Grund and Lautenthal, Germany. The positive correlation between $^{13}C/^{12}C$- and $^{18}O/^{16}O$-ratios can be explained either by calcite precipitation due to the mixing of two fluids with different salt contents or by calcite precipitation from a $H_2CO_3$-dominant fluid due to a temperature effect coupled with either $CO_2$ degassing or with fluid-rock interaction.

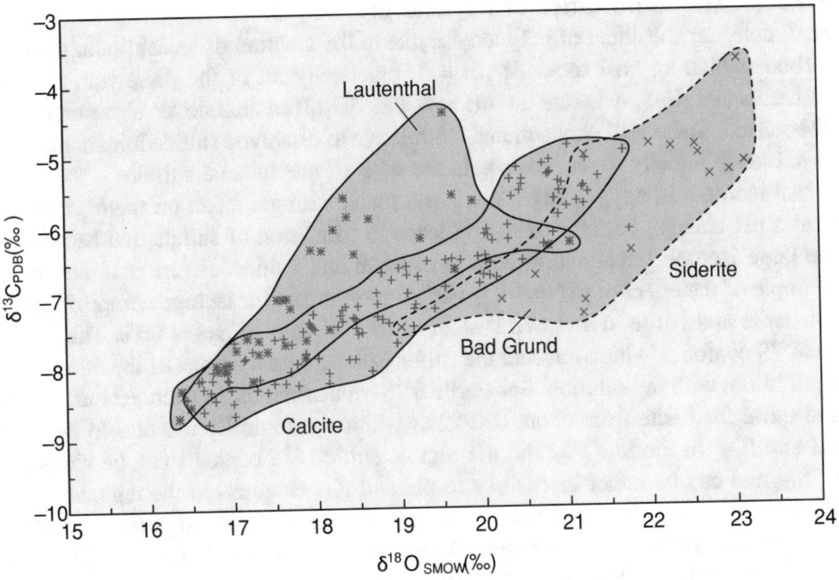


Fig. 3.17 C- and O-isotope compositions of calcites and siderites from the Bad Grund and Lautenthal deposits, Harz (after Zheng and Hoefs 1993)

### 3.5.5 Sulfur Isotope Composition of Ore Deposits

A huge amount of literature exists about the sulfur isotope composition in hydrothermal ore deposits. Out of the numerous papers on the subject the reader is referred to comprehensive reviews by Ohmoto and Rye (1979), Ohmoto (1986), Taylor (1987a, b) and Ohmoto and Goldhaber (1997). The basic principles to be followed in the interpretation of $\delta^{34}S$ values in sulfidic ores were elucidated by Sakai (1968), Ohmoto (1972) and recently by Hutchinson et al. (2020). The development of analytical techniques to perform multiple S-isotope analysis with high spatial resolution has enhanced our knowledge about sulfur sources and ore-forming fractionation processes. For example, MIF sulfur fractionations in ore deposits indicate the input of Archean rocks or in recent hydrothermal seafloor systems the input of biogenic sulfur (see p. 116).

The isotopic composition of a hydrothermal sulfide is determined by a number of the factors such as (1) isotopic composition of the hydrothermal fluid from which the mineral is deposited, (2) temperature of deposition, (3) chemical composition of the dissolved element species including pH and $fO_2$ at the time of mineralization, and (4) relative amount of mineral deposited from the fluid. The first parameter is characteristic of the source of sulfur, the three others relate to the conditions of deposition.

#### 3.5.5.1 The Importance of $fO_2$ and pH

First, consider the effect of pH-increase due to the reaction of an acidic fluid with a carbonate-bearing host rock. At pH = 5, practically all of the dissolved sulfur is undissociated $H_2S$, whereas at pH = 9 the dissolved sulfide is almost entirely dissociated. Since $H_2S$ concentrates $^{34}S$ relative to dissolved sulfide ion, an increase in pH leads directly to an increase in the $\delta^{34}S$ of precipitated sulfides.

An increase in oxygen fugacities has a much stronger effect on the $\delta^{34}S$-values than a pH change, because oxidation leads to formation of sulfate and because of the large isotope fractionation between sulfate and sulfide. Figure 3.18 shows an example of the effect of pH and $fO_2$ variation on the sulfur isotope compositions of sphalerite and barite in a closed system at 250 °C with $\delta^{34}S_{\Sigma S}$ = O‰. The curves are $\delta^{34}S$ contours, which indicate the sulfur isotope compositions of the minerals in equilibrium with the solution. Sphalerite $\delta^{34}S$-values can range from −24 to +5.8‰ and those for barite from about 0–24.2‰ within geologically reasonable limits of pH and $fO_2$. In the low $fO_2$ and pH region, sulfide $^{34}S$ contents can be similar to $\delta^{34}S_{\Sigma S}$ and can be rather insensitive to pH and $fO_2$ changes. In the region of high $fO_2$ values where the proportion of sulfate species becomes significant, mineral $\delta^{34}S$ values can be greatly different from $\delta^{34}S_{\Sigma S}$ and small changes in pH or $fO_2$ may result in large changes in the sulfur isotope composition of either sulfide or sulfate. Such a change must, however, be balanced by a significant change in the ratio of sulfate to sulfide.

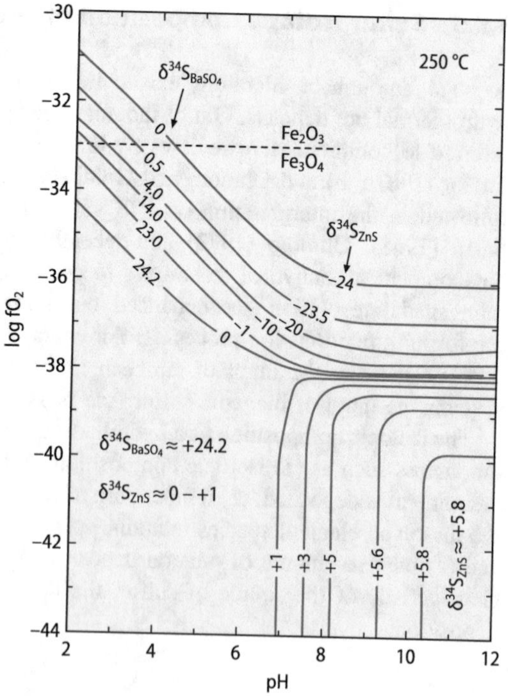


**Fig. 3.18** Influence of $fO_2$ and pH on the sulfur isotope composition of sphalerite and barite at 250 °C and $\delta^{34}S_{\Sigma S}$ = 0‰ (modified after Ohmoto 1972)

In summary, interpretation of the distribution of $\delta^{34}S$-values relies on information about the source of sulfur and on a knowledge of the mineral parageneses that constrain the ambient temperature, Eh and pH. If the oxidation state of the fluid is below the sulfate/$H_2S$ boundary, then the $^{34}S/^{32}S$ ratios of sulfides will be insensitive to redox shifts.

In the following section different classes of ore deposits are discussed.

#### 3.5.5.2 Magmatic Ore Deposits

Magmatic deposits are characterized by sulfides which precipitate from mafic silicate melts rather than hydrothermal fluids. They can be divided into S-poor (deposits of platinum group elements) and S-rich magmatic sulfide systems (Ni–Cu deposits) (Ripley and Li 2003). The majority of this type of deposits are hosted within sedimentary country rocks in which the sulfur is assimilated or volatized during magma emplacement. Typical examples are the deposits of Duluth, Stillwater, Bushveld, Sudbury and Norils'k. In many of these deposits relatively large deviations in $\delta^{34}S$-values from the presumed mantle melt value near zero are observed, indicating magma contamination by interactions with country rocks. The large spread in $\delta^{34}S$ is generally attributed to assimilation of sulfur from the wall rocks, provided that the sulfur isotope composition of the country rocks is significantly different from the magma.

#### 3.5.5.3 Porphyry Copper Deposits

This group of deposits is closely associated in space and time with magmatic intrusions that were emplaced at relatively shallow depths. They have been developed in hydrothermal systems driven by the cooling of magma. From $\delta^{2}H$- and $\delta^{18}O$-measurements, it has been concluded that porphyry copper deposits show the clearest affinity of a magmatic water imprint (Taylor 1974) with variable involvement of meteoric water generally at late stages of ore formation.

The majority of $\delta^{34}S$-values of sulfides fall between −3 and 1‰ and of sulfates between 8 and 15‰ (Field and Gustafson 1976; Shelton and Rye 1982; Rye 2005). Sulfate-sulfide isotope date suggest a general approach to isotope equilibrium. Calculated sulfate-sulfide temperatures, for conditions of complete isotope equilibrium, are typically between 450 and 600 °C and agree well with temperatures estimated from other methods. Thus, the sulfur isotope data and temperatures support the magmatic origin of the sulfur in porphyry deposits. $Cu^{-}$ (Gregory and Mathur 2017) and Fe-isotope (He et al. 2020) data also indicate a magmatic origin in the mineralized core of porphyry copper deposits.

#### 3.5.5.4 Recent and Fossil Sulfide Deposits at Mid-Ocean Ridges

Numerous sulfide deposits have been discovered on the seafloor along the East Pacific Rise, Juan de Fuca Ridge, Explorer Ridge and Mid-Atlantic Ridge (Shanks 2001). These deposits are formed from hydrothermal solutions which result from the interaction of circulating hot seawater with oceanic crust. Sulfides are derived mainly from two sources: (i) leaching from igneous and sedimentary wall rocks and

(ii) thermochemical sulfate reduction due to interaction with ferrous silicates and oxides or with organic matter.

The role of sulfur in these vents is complex and often obscured by its multiple redox states and by uncertainties in the degree of equilibration. Studies by Styrt et al. (1981), Arnold and Sheppard (1981), Skirrow and Coleman (1982), Kerridge et al. (1983), Zierenberg et al. (1984), and others have shown that the sulfur in these deposits is enriched in $^{34}S$ relative to a mantle source (typical $\delta^{34}S$ ranges are between 1 and 5‰), implying small additions of sulfide derived from seawater.

Vent sulfides at sediment covered hydrothermal systems may carry, in addition, signatures of sulfides derived from bacterial reduction. $\delta^{34}S$-values alone may be unable to distinguish between the different sulfur sources. High precision measurements of $\delta^{33}S$, $\delta^{34}S$ and $\delta^{36}S$ allow, however, the distinction of biological isotope fractionation from abiological fractionation (Ono et al. 2007; Rouxel et al. 2008a, b). Biogenic sulfides are characterized by relatively high $\Delta^{33}S$ values compared to hydrothermal sulfides. Sulfides from the East Pacific Rise and the Mid-Atlantic Ridge, analyzed by Ono et al. (2007), gave low $\Delta^{33}S$ values compared to biogenic sulfides suggesting no contribution of biogenic sulfides. In altered oceanic basalts at ODP Site 801, however, Rouxel et al. (2008a, b) provided evidence for secondary biogenic pyrite. These authors estimated that at least 17% of pyrite sulfur was derived from bacterial reduction.

Like sulfur, Fe isotopes show complex patterns (Severmann et al. 2004; Rouxel et al. 2004a, b, 2008a, b; Bennett et al. 2009). High temperature vent fluids are depleted in $^{56}Fe$ relative to their basaltic and ultramafic source rocks. Precipitation of $Fe^{2+}_{aq}$ primarily produces iron sulfides and $Fe^{3+}$ hydroxides; the proportion varies depending on the Fe/S ratio of the hydrothermal fluid and the amount of dissolved oxygen in seawater. Precipitating marcasite and pyrite from active vents are isotopically lighter than chalcopyrite. Further Fe isotope fractionation occurs, when vent fluids enter oxygen-rich ocean water and isotopically light polymetallic sulfides and isotopically heavy Fe hydroxides precipitate (Bennett et al. 2009).

#### 3.5.5.5 Biogenic Deposits

The discrimination between bacterial sulfate and thermal sulfate reduction in ore deposits on the basis of $\delta^{34}S$-values is rather complex. The best criterion to distinguish between both types is the internal spread of δ-values. If individual sulfide grains at a distance of only a few millimeters exhibit large and nonsystematic differences in $\delta^{34}S$-values, then it seems reasonable to assume an origin involving bacterial sulfate reduction. Irregular variations in $^{34}S$-contents are attributed to bacteria growing in reducing microenvironments around individual particles of organic matter. In contrast, thermal sulfate reduction requires higher temperatures supplied by external fluids, which is not consistent with the closed system environment of bacterial reduction.

Two types of deposits, where the internal S-isotope variations fit the expected scheme of bacterial reduction, but where the biogenic nature was already known from other geological observations, are the "sandstone-type" uranium mineralization in the Colorado Plateau (Warren 1972) and the Kupferschiefer in Central

Europe (Marowsky 1969), although thermal sulfate reduction may have occurred at the base of the Kupferschiefer (Bechtel et al. 2001). In addition, sandstone-type uranium deposits show characteristic uranium isotope compositions which are distinctly different from magmatic uranium ores (Bhattacharyya et al. 2017).

#### 3.5.5.6 Metamorphosed Deposits

It is generally assumed that metamorphism reduces the isotopic variations in a sulfide ore deposit. Recrystallization, liberation of sulfur from fluid and vapor phases, such as the breakdown of pyrite into pyrrhotite and sulfur, and diffusion at elevated temperatures should tend to reduce initial isotopic heterogeneities.

Studies of regionally metamorphosed sulfide deposits (Seccombe et al. 1985; Skauli et al. 1992) indicate, however, little evidence of homogenisation on the deposit scale. Significant changes may take place in certain restricted parts of the deposit as a result of special local conditions, controlled by factors such as fluid flow regimes and tectonics. Thus, a very limited degree of homogenisation takes place during metamorphism (Cook and Hoefs 1997). The extent of this is obscured by primary distribution and zonation patterns.

### 3.5.6 Metal Isotopes

One of the most important questions in the genesis of ore deposits is the origin of the metals. MC-ICP-MS techniques have provided a new tool for the precise analysis of metal isotopes (Cu, Fe, Zn, Mo, Ni). Since the silicate earth (crust + mantle) shows a limited range of metal isotope compositions, different metal reservoirs with distinct isotopic compositions are not easily recognizable. It is therefore necessary to determine the ranges of metal isotope compositions in different ore deposit types and to investigate the mechanism that fractionate metal isotopes. Variations in metal isotope ratios depend on known parameters such as formation temperatures, abiotic or biotic processes and redox state during ore formation making interpretation of metal isotope ratios in ore deposits complex.

Like sulfur, mass balance among reduced and oxidized species controls the isotopic composition of metal sulfides (Asael et al. 2009). Thus far, Cu and Fe have received the greatest attention in applying metal stable isotopes to ore deposits (see the summary of Li et al. 2010a, b and Mathur and Wang 2019).

#### 3.5.6.1 Copper

By far the largest variations of metal isotopes in ore deposits have been measured for Cu isotopes in a wide spectrum of ore deposits, including black smokers (Zhu et al. 2000a, b; Rouxel et al. 2004a, b), massive sulfide deposits (Mason et al. 2005; Ikehate et al. 2011), porphyry deposits (Graham et al. 2004; Mathur et al. 2010; Li et al. 2010a, Zheng et al. 2019), skarn (Maher and Larson 2007) and other hydrothermal deposits (Markl et al. 2006a, b). A common feature of these investigations is that copper mineralizations affected by low temperature redox processes show larger variations than high temperature mineralizations.

Native copper found in the Horoman peridotite, Japan shows very little Cu isotope fractionation (Ikehata and Hirata 2012). The primary native copper and the copper in whole rocks yield homogeneous Cu isotope values around 0‰, implying no significant copper isotope fractionation during high-temperature magmatic processes. As argued by Mathur and Wang (2019), without sulfide segregation no Cu isotope fractionation seems to occur.

Porphyry copper deposits have received the most attention in Cu isotope ore deposit studies. Although the copper source is still a matter of debate, Cu isotope distributions in porphyry copper deposits are characterized by light Cu isotope values in the central parts of the mineralization dominated by high-temperature potassic alteration and heavy Cu isotope ratios in the peripheral sections dominated by low-temperature phyllic and argillic alteration. Low values in the center of the mineralization may reflect the Cu isotope composition of the source. Copper isotopes, therefore, may be an effective tool for mineral exploration.

The magnitude of Cu isotope fractionations increases with increasing degrees of secondary alterations. A range of more than 5‰ in $\delta^{65}Cu$ has been interpreted by Markl et al. (2006a) as being due to redox processes among dissolved Cu-species and to fractionations during precipitation of Cu minerals. Thus, an important aspect of metal isotope investigations is the identifaction of low-temperature alteration processes, where biogenic and abiogenic redox processes potentially lead to significant isotope fractionations as already has been demonstrated in Sects. 2.13, 2.14 and 2.18, for Fe, Cu and Mo isotopes.

### 3.5.6.2 Iron

Iron isotopes have been investigated in a variety of different types of ore deposits including banded iron formations (Johnson et al. 2003, 2008 and others), modern seafloor hydrothermal deposits (Rouxel et al. 2004a, 2008a, b) and magmatic-hydrothermal deposits (Horn et al. 2006; Markl et al. 2006a, b; Wang et al. 2011a, b; Wawryck and Foden 2015; Li et al. 2018). Iron in hydrothermal mineralizations may encompass a considerable range in isotopic composition that may occur at very small spatial and temporal scales. Of special importance are redox processes. Progressive precipitation of Fe(II) minerals, being depleted in heavy Fe isotopes, will enrich the remaining fluid, whereas precipitation of Fe(III) minersals, enriched in heavy isotopes will deplete the ore fluid. Thus, a 2.5‰ variation of iron minerals in $\delta^{56}Fe$ can be explained by mixing either through mixing with oxygen-rich surface waters or through mixing with $CO_2$-rich fluids (Markl et al. 2006b).

Since Cu and Fe are both sensitive to redox processes, it might be expected that Cu and Fe isotope variations in a specific ore deposit are coupled, which, however seems to be not the case. One reason for a decoupling might be that the redox potential of $Cu^{2+}/Cu^{+}$ is much lower than for $Fe^{3+}/Fe^{2+}$ making Cu isotopes more sensitive to redox processes.

The combination of Fe- with O-isotopes may give insight into the origin of a special class of ore deposits, namely the Kiruna-type apatite-iron oxide deposits. Various ore-forming processes have been controversially discussed, including a low-temperature hydrothermal and a high-temperature magmatic origin. Studies by Bilenker et al. (2016) and Troll et al. (2019) have shown that Fe- and O-isotopes in apatite-iron oxide deposits argue for a high-temperature magmatic origin.

#### 3.5.6.3 Zinc

Zn isotopes in ore deposits—generally measured in sphalerites—show much smaller isotope variations than Fe and in specific Cu. A characteristic feature is that Zn isotopes become enriched in heavier isotopes from early to late stages of precipitation recording a temporal evolution of the ore forming fluid. Secondary non-sulfide minerals such as willemite ($Zn_2SiO_4$) and smithsonite ($ZnCO_3$) formed during supergene oxidation show larger isotope fractionations in positive and negative directions relative to the precursor mineral, generally sphalerite (Mondillo et al. 2018).

## 3.6 Hydrosphere

First, some definitions concerning water of different origin are given. The term "**meteoric**" applies to water that has been part of the meteorological cycle, and participated in processes such as evaporation, condensation, and precipitation. All continental surface waters, such as rivers, lakes, and glaciers, fall into this general category. Because meteoric water may seep into the underlying rock strata, it will also be found at various depths within the lithosphere dominating all types of continental ground waters. The **ocean**, although it continuously receives the continental run-off of meteoric waters as well as rain, is not regarded as being meteoric in nature. **Connate** water is water, which has been trapped in sediments at the time of burial. **Formation** water is present in sedimentary rocks and may be a useful nongenetic term for waters of unknown origin and age within these rocks.

### 3.6.1 Meteoric Water–General Considerations

When water evaporates from the surface of the ocean, the water vapor is enriched in H and $^{16}O$ because $H_2{}^{16}O$ has a higher vapor pressure than HDO and $H_2{}^{18}O$ (Table 1.1). Under equilibrium conditions at 25 °C, the fractionation factors for evaporating water are 1.0092 for $^{18}O$ and 1.074 for D (Craig and Gordon 1965). However, under natural conditions, the actual isotopic composition of water is more negative than the predicted equilibrium values due to kinetic isotope effects (Craig and Gordon 1965). Vapor leaving the surface of the ocean cools as it rises and rain forms when the dew point is reached. During removal of rain from a moist air mass, the residual vapor is continuously depleted in the heavy isotopes, because the rain leaving the system is enriched in $^{18}O$ and D. If the air mass moves poleward and becomes cooler, additional rain formed will contain less $^{18}O$ than the initial rain. This relationship is schematically shown in Fig. 3.19. The isotope composition of

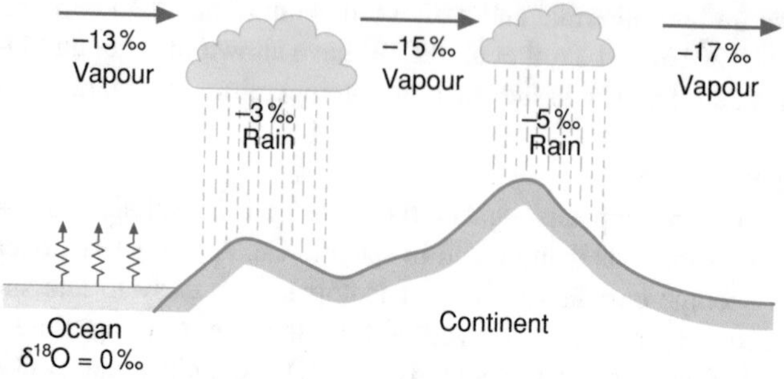


**Fig. 3.19** Schematic O-isotope fractionation of water in the atmosphere (after Siegenthaler 1979)

mean world-wide precipitation is estimated to be $\delta^2H = -22$ and $\delta^{18}O = -4‰$ (Craig and Gordon 1965).

The theoretical approaches to explain isotope variations in meteoric waters evolved from the "isolated air mass" models, which are based on Rayleigh condensation, with immediate removal of precipitation and with a part of the condensate being kept in the cloud during the rain-out process. Isotope studies of individual rain events have revealed that successive portions of single events may vary drastically (Rindsberger et al. 1990). Quite often the pattern is "V-shaped", a sharp decrease of δ-values is usually observed at the beginning of a storm with a minimum somewhere in the middle of the event. The most depleted isotope values usually correspond to the period of most intense rain with little evaporation experienced by individual rain drops. It has also been observed that convective clouds produce precipitation with higher δ-values than stratiform clouds. Thus, the isotope composition of precipitation from a given rain event depends on meteorological history of the air mass in which the precipitation is produced and the type of cloud through which it falls. Liquid precipitation (rain) and solid precipitation (snow, hail) may differ in their isotope composition insofar as rain drops may undergo evaporation and isotope exchange with atmospheric vapor on their descent to the surface. By analyzing hailstones, discrete meteorological events can be studied because hailstones keep a record on the internal structure of a cloud. Jouzel et al. (1975) concluded that hailstones grow during a succession of upward and downward movements in a cloud.

The International Atomic Energy Agency (IAEA) conducts a world-wide survey of the isotope composition of monthly precipitation for more than 50 years. The global distribution of D and $^{18}O$ in rain has been monitored since 1961 through a network of stations (Yurtsever 1975). From this extensive data base, it can be deduced how geographic and meteorological factors (rainout, temperature, humidity) influence the isotopic composition of precipitation.

The first detailed evaluation of the equilibrium and non-equilibrium factors that determine the isotopic composition of precipitation was published by Dansgaard (1964). He demonstrated that the observed geographic distribution in isotope

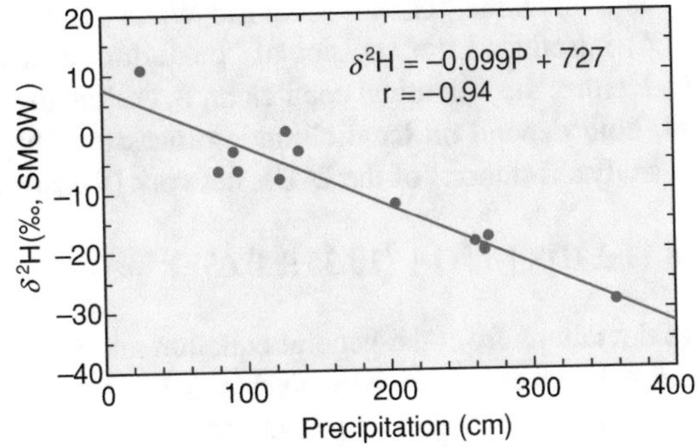


**Fig. 3.20** Average $\delta^2$H-values of the annual precipitation from oceanic islands as a function of the amount of annual rainfall. The island stations are distant from continents, within 30° of the equator and at elevations <120 m (after Lawrence and White 1991)

composition is related to a number of environmental parameters that characterize a given sampling site, such as latitude, altitude, distance to the coast, amount of precipitation, and surface air temperature. Out of these, two factors are of special significance: temperature and the amount of precipitation. As shown in Fig. 3.20, the best temperature correlation is observed in continental regions nearer to the poles, whereas the correlation with amount of rainfall is most pronouced in tropical regions (Lawrence and White 1991). The apparent link between local surface air temperature and the isotope composition of precipitation is of special interest mainly because of the potential importance of stable isotopes as palaeoclimatic indicators. The amount effect is ascribed to gradual saturation of air below the cloud, which diminishes any shift to higher $\delta^{18}$O-values caused by evaporation during precipitation (Fricke and O'Neil 1999).

A compilation of studies throughout the world's mountain belts has revealed a consistent and linear relationship between change in the isotopic composition of precipitation and change in elevation (Poage and Chamberlain 2001). The isotopic composition of precipitation decreases linearly with increasing elevation by about 0.28‰/100 m in most regions of the world except in the Himalayas and at elevations above 5000 m.

#### 3.6.1.1 $\delta^2$H–$\delta^{18}$O Relationship, Deuterium (D)—Excess

In all processes concerning evaporation and condensation, hydrogen isotopes are fractionated in proportion to oxygen isotopes, because a corresponding difference in vapor pressures exists between $H_2O$ and HDO in one case and $H_2{}^{16}O$ and $H_2{}^{18}O$ in the other. Therefore, hydrogen and oxygen isotope distributions are correlated in meteoric waters. Craig (1961) first defined the following relationship:

$$\delta^2H = 8\delta^{18}O + 10$$

which is generally known as the "Global Meteoric Water Line".

Dansgaard (1964) introduced the concept of "deuterium excess", d defined as $d = \delta^2H - 8\,\delta^{18}O$. Neither the numerical coefficient, 8, nor the deuterium excess, d, are really constant, both depend on local climatic processes. The long-term arithmetic mean for all analyzed stations of the IAEA network (Rozanski et al. 1993) is:

$$\delta^2H = (8.17 \pm 0.06)\,\delta^{18}O + (10.35 \pm 0.65)\ r^2 = 0.99,\ n = 206$$

Relatively large deviations from the general equation are evident when monthly data for individual stations are considered (Table 3.2). In an extreme situation, represented by the St. Helena station, a very poor correlation between $\delta^2H$ and $\delta^{18}O$ exists. At this station, it appears that all precipitation comes from nearby sources and represents the first stage of the rain-out process. Thus, the generally weaker correlations for the marine stations (Table 3.2) may reflect varying contributions of air masses with different source characteristics and a low degree of rain-out.

The imprint of local conditions can also be seen at other coastal and continental stations. The examples in Table 3.2 demonstrate that varying influences of different sources of vapor with different isotope characteristics, different air mass trajectories, or evaporation and isotope exchange processes below the cloud base, may often lead to much more complex relationships at the local level between $\delta^2H$ and $\delta^{18}O$ than suggested for the regional or continental scale by the global "Meteoric Water Line" equation.

Knowledge about the isotopic variations in precipitation is increased when single rain events are analyzed from local stations. Especially under mid-latitude weather

**Table 3.2** Variations in the numerical constant and the deuterium excess for selected stations of the IAEA global network (Rozanski et al. 1993)

| Station | Numerical constant | Deuterium excess | $r^2$ |
|---|---|---|---|
| *Continental and coastal stations* | | | |
| Vienna | 7.07 | −1.38 | 0.961 |
| Ottawa | 7.44 | +5.01 | 0.973 |
| Addis Ababa | 6.95 | +11.51 | 0.918 |
| Bet Dagan, Israel | 5.48 | +6.87 | 0.695 |
| Izobamba (Ecuador) | 8.01 | +10.09 | 0.984 |
| Tokyo | 6.87 | +4.70 | 0.835 |
| *Marine stations* | | | |
| Weathership E (N.Atlantic) | 5.96 | +2.99 | 0.738 |
| Weathership V (N.Pacific) | 5.51 | −1.10 | 0.737 |
| St.Helena (S.Atlantic) | 2.80 | +6.61 | 0.158 |
| Diego Garcia Isl. (Indian Oc.) | 6.93 | +4.66 | 0.880 |
| Midway Isl. (N.Pacific) | 6.80 | +6.15 | 0.840 |
| Truk Isl. (N.Pacific) | 7.07 | +5.05 | 0.940 |

conditions, such short-term variations arise from varying contributions of tropical, polar, marine, and continental air masses.

The d-excess in oceanic water vapour is determined by evaporative conditions (surface temperature, relative humidity, wind speed) (e.g. Merlivat and Jouzel 1979). Deuterium excess over the oceans increases when humidity over the ocean decreases. Thus, reduced d-excess values in Antartic ice cores have been interpreted as indicators of higher relative humidity in the oceanic source area providing the moisture for Antarctic precipitation (Jouzel et al. 1982). Later Johnsen et al. (1989), followed by others, showed that besides humidity temperatures in the source regions also have an effect on the size of the d-excess.

Deuterium excess profiles from Greenland and Antarctic ice cores show well defined climatic changes being negatively correlated with $\delta^{18}O$-values. Combining $\delta^{18}O$-values with deuterium excess values, temperature estimates at the site of precipitation and at the source region of the moisture can be achieved (Masson-Delmotte et al. 2005). A recent helpful online resource has been presented by Bowen and Good (2015), that permits users getting information on water isotopes at any coordinates on earth's surface.

#### 3.6.1.2 $\delta^{17}O$–$\delta^{18}O$ Relationships, $^{17}O$ Excess

Although mass-independent fractionations are not known to occur in water, $^{17}O$ is becoming as another useful tracer within the hydrologic cycle (Angert et al. 2004; Surma et al. 2021). Improvements in analytical techniques allow to measure $\delta^{17}O$ and $\delta^{18}O$ with a precision of a few 0.01‰ which permits calculation of $\Delta^{17}O$ with similar precision and thus the detection of very small $\delta^{17}O$ variations.

As already demonstrated the isotopic composition of water is controlled by two mass-dependent processes. (i) the equilibrium fractionation that is caused by the different vapour pressures of $^{17}O$ and $^{18}O$ and (ii) the kinetic fractionation that is caused by the different diffusivities of $^{17}O$ and $^{18}O$ during transport in air. Angert et al. (2004) have demonstrated that for kinetic water transport in air, the slope in a $\delta^{17}O$–$\delta^{18}O$ diagram is 0.511, whereas it is 0.526 for equilibrium effects. Similar values have been given by Barkan and Luz (2007).

Similar to the deuterium excess, the deviation from an expected $^{17}O/^{16}O$–$^{18}O/^{16}O$ relationship has been defined as $^{17}O$ excess (Barkan and Luz 2007) that can be described as the triple oxygen meteoric water line (Luz and Barkan 2010; Surma et al. 2021).

$$\delta^{17}O = 0.528\delta^{18}O + 0.033$$

Atmospheric vapor collected above the ocean shows the existence of a small $^{17}O$ excess and a negative correlation between $^{17}O$ excess and relative humidity. The $^{17}O$ excess originates from evaporation of sea water into marine air that is under-saturated in water vapor and from the transfer of vapor to liquid water or snow (Luz and Barkan 2010).

$^{17}O$-excess is thus a unique tracer, which is, in contrast to the deuterium excess, temperature independent and which may give additional informations on humidity relations. Steep spatial gradients of $^{17}$O-excess in precipitation across Antarctica indicate higher values in marine influenced regions and lower values in the Antarctic interior (Schoenemann et al. 2014). Glacial-interglacial $^{17}$O records (Landais et al. 2008; Uemura et al. 2010) reveal small shifts in $^{17}$O excess from low values in glacial periods to high values in interglacial periods. According to Schoenemann et al. (2014) fractionations during snow formation control the $^{17}$O excess in Antarctic precipitation. Variations in moisture source relative humidity play a minor role in determining $^{17}$O excess changes.

Last not least, it should be noted that d-excess and $^{17}$O excess have different definitions: while d-excess is defined in a linear scale, $^{17}$O excess is in a logarithm scale.

#### 3.6.1.3 Meteoric Waters in the Past

Assuming that the H- and O-isotope compositions and temperatures of ancient ocean waters are comparable to present-day values, the isotopic composition of ancient meteoric waters may have been governed by relations similar to those existing presently. However, given the local complexities, the application of this relationship back through time should be treated with caution. To date, however, there is no compelling evidence that the overall systematics of ancient meteoric waters were very different from the present meteoric water relationship (Sheppard 1986). If the isotope composition of ocean water has changed with time, but global circulation patterns were like today, the "meteoric water line" at a specific time would be parallel to the modern meteoric water line, that is the slope would remain at a value of 8, but the intercept would be different.

The systematic behavior of stable isotopes in precipitation as a function of altitude can be used to provide estimates of paleoaltitude. For paleoelevation reconstruction the isotope relationship between precipitation and elevation must be quantitatively known or assumed. In this approach the isotopic composition of paleoprecipitation is determined from the analysis of in situ formed authigenic minerals (Chamberlain and Poage 2000; Blisnink and Stern 2005 and others). The effect of topography on the isotopic composition of precipitation is most straightforward in temperate mid-latitude regions and in topographically and climatically simple settings and varies generally between 2 and 5‰ per 1 km. Paleoelevation can be also reconstructed by using clumped isotope thermometry (Huntington et al. 2010; Quade et al. 2011).

### 3.6.2 Ice Cores

The isotopic composition of snow and ice deposited in polar regions and at high elevations in mountains depend primarily on temperature. Snow deposited during the summer has less negative $\delta^{18}$O and $\delta^{2}$H-values than snow deposited during the winter. A good example of the seasonal dependence has been given by Deutsch et al. (1966) on an Austrian glacier, where the mean $\delta^{2}$H-difference between winter and summer snow was observed to be −14‰. This seasonal cycle has been used to determine the annual stratigraphy of glaciers and to provide short-term climatic records. However, alteration of the snow and ice by seasonal meltwater can result in changes of the isotopic composition of the ice, thus biasing the historical climate record. Systematic isotope studies also have been used to study the flow patterns of glaciers. Profiles through a glacier should exhibit lower isotope ratios at depth than nearer the surface, because deep ice may have originated from locations upstream of the ice-core site, where temperatures should be colder.

In the last decades, several ice cores over 1000 m depth have been recovered from Greenland and Antarctica. In these cores, seasonal variations are generally observed only for the uppermost portions. After a certain depth, which depends on accumulation rates, seasonal variations disappear completely and isotopic changes reflect long-term climatic variations. No matter how thin a sample one cuts from the ice core, its isotope composition will represent a mean value of several years of snow deposition.

The ice cores—investigated in great detail by large groups of researchers—are the Vostok core from East Antarctica (Lorius et al. 1985; Jouzel et al. 1987), the GRIP and GISP 2 cores from Greenland (Dansgaard et al. 1993; Grootes et al. 1993) and more recently the EPICA Antarctic core penetrating almost 800,000 years of ice and 6 glacial cycles (Augustin et al. 2004). In the Vostok core, the low accumulation rate of snow in Antarctica results in very thin annual layers, which means that climate changes of a century or less are difficult to resolve. The newer Greenland ice cores GRIP and GISP 2 were drilled in regions with high snow accumulation near the centre of the Greenland ice sheet. In these cores it is possible to resolve climate changes on the timescale of decades or less, even though they occurred a hundred thousand years ago. The GRIP and GISP 2 data indicate a dramatic difference between our present climate and the climate of the last interglacial period. Whereas the present interglacial climate seems to have been very stable over the last 10,000 years, the early and late parts of the last interglacial (c.135,000 and c.115,000 years before present, respectively) were characterized by rapid fluctuations between temperatures, both warmer and very much colder than the present. It apparently took only a decade or two to shift between these very different climatic regimes.

Figure 3.21 compares $\delta^{18}O$ profiles from Antarctica and Greenland. The dramatic δ-shifts observed in Greenland cores are less pronounced in the δ-record along the Vostok core, probably because the shifts in Greenland are connected to rapid ocean/atmosphere circulation changes in the North Atlantic (for more details, see Sect. 3.12.1).

### 3.6.3 Groundwater

In temperate and humid climates, the isotopic composition of groundwater is similar to that of the precipitation in the area of recharge (Gat 1971). This is strong evidence for direct meteoric recharge to an aquifer. The seasonal variation of all meteoric water is strongly attenuated during transit and storage in the ground. The

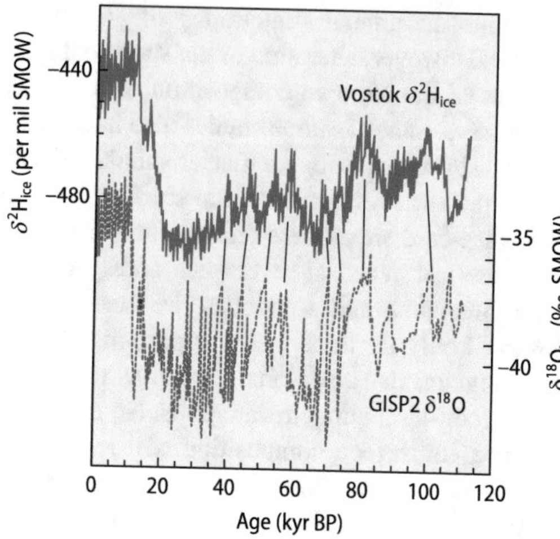


**Fig. 3.21** Correlations of $\delta^2H$ and $\delta^{18}O$ values of Greenland (GISP-2) and Antarctic (Vostok) ice cores covering the last glacial-interglacial cycles (http://www.gisp2.sr.unh.edu/GISP2/DATA/Bender.html)

degree of attenuation varies with depth and with surface and bedrock geologic characteristics, but in general deep groundwaters show no seasonal variation in $\delta^2H$ and $\delta^{18}O$ values and have an isotopic composition close to amount-weighted mean annual precipitation values.

The characteristic isotope fingerprint of precipitation provides an effective means for identifying possible groundwater recharge areas and hence subsurface flow paths. For example, in areas close to rivers fed from high altitudes, groundwaters represent a mixture of local precipitation and high-altitude low-$^{18}O$ waters. In suitable cases, quantitative estimates about the fraction of low-$^{18}O$ river water in the groundwater can be carried out as a function of the distance from the river.

The main mechanisms that can cause variations between precipitation and recharged groundwater are (Gat 1971):

(1) recharge from partially evaporated surface water bodies,
(2) recharge that occurred in past periods of different climate when the isotopic composition of precipitation was different from that at present,
(3) isotope fractionation processes resulting from differential water movement through the soil or the aquifer or due to kinetic or exchange reactions within geologic formations.

In semi-arid or arid regions, evaporative losses before and during recharge shift the isotopic composition of groundwater towards higher δ-values. Furthermore, transpiration of shallow groundwater through plant leaves, may also be an important evaporation process. Detailed studies of soil moisture evaporation have shown that evaporation loss and isotopic enrichment are greatest in the upper part of the soil profile and are most pronounced in unvegetated soils (Welhan 1987). In some arid regions, groundwater may be classified as paleowaters, which were recharged under different meteorological conditions than present in a region today and which imply ages of water of several thousand years. Gat and Issar (1974) have demonstrated that the isotopic composition of such paleowaters can be distinguished from more recently recharged groundwaters, which have experienced some evaporation.

In summary, the application of stable isotopes to groundwater studies is based on the fact that the isotopic composition of water behaves conservatively in low-temperature environments where water–rock contact times are short relative to the kinetics of mineral–water isotope exchange reactions.

### 3.6.4 Rivers

Isotopic relationships observed for groundwater also are valid for rivers. A global database GNIR (Global Network of Isotopes in Rivers) has been introduced by Halder et al. (2015) collecting data on 218 river stations. O- and H-isotopes in rivers follow the Meteoric Water Line, but hydrological processes, such as catchment storage and reservoir mixing (lakes or dams) may modify the isotope composition of rivers. As shown by Dutton et al. (2005)—comparing oxygen isotope ratios of precipitation and river water across the United States—regions with river water being more positive than local precipitation (the Great Plains) can be divided from regions with river water being more negative than local precitation (the western USA). The difference mainly results from the "catchment" effect and evaporation effects. Oxygen isotope ratios of rivers are more negative than precipitation in regions with high elevations of the recharge area and more positive in regions with relatively high rates of evaporation. Furthermore, as shown by Halder et al. (2015), seasonal variations in rivers do not increase with altitude as observed for precipitation. The average annual seasonal $\delta^{18}O$ amplitude is 2.5‰ for rivers compared to 7.5‰ for precipitation. Low seasonal variations in rivers suggest the prevalence of mixing and storage such as occurs in lakes and due to groundwater influx. An additional factor affecting the isotopic composition of rivers is anthropogenic activity.

### 3.6.5 Isotope Fractionations During Evaporation

In an evaporative environment, one could expect to find extreme enrichments in the heavy isotopes D and $^{18}O$. However, this is generally not the case. Taking the Dead Sea as the typical example of an evaporative system, Fig. 3.22 shows only moderately enriched $\delta^{18}O$-values and even to an even lesser degree $\delta^{2}H$-values (Gat 1984). Isotope fractionations accompanying evaporation are rather complex and can be best described by subdividing the evaporation process into several steps (Craig and Gordon 1965):

(a) the presence of a saturated sublayer of water vapor at the water-atmosphere interface, which is depleted in the heavy isotopes,

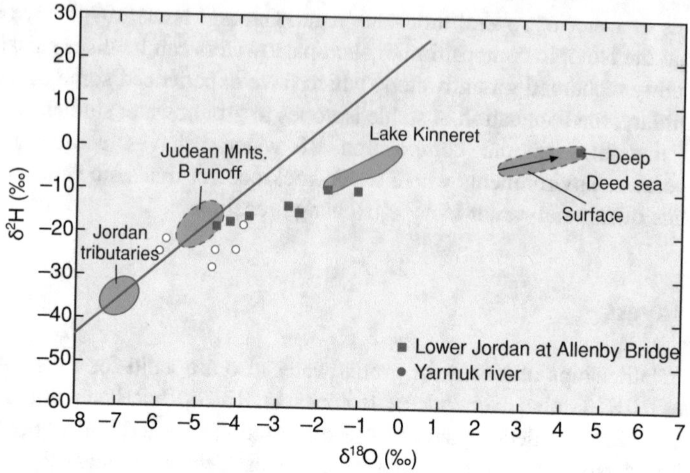


Fig. 3.22 $\delta^2H$ versus $\delta^{18}O$ values of the Dead Sea and its water sources as an example of an evaporative environment (after GAT 1984)

(b) the migration of vapor away from the boundary layer, which results in further depletion of heavy isotopes in the vapor due to different diffusion rates,
(c) the vapor reaching a turbulent region where mixing with vapor from other sources occurs, and
(d) the vapor of the turbulent zone then condensing and back-reacting with the water surface.

This model qualitatively explains the deviation of isotopic compositions away from the "Meteoric Water Line" because molecular diffusion adds a non-equilibrium fractionation term and the limited isotopic enrichment occurs as a consequence of molecular exchange with atmospheric vapor. It is mainly the humidity which controls the degree of isotope enrichment. Only under very arid conditions, and only in small water bodies, really large enrichments in D and $^{18}O$ are observed. For example, Gonfiantini (1986) reported a $\delta^{18}O$-value of +31.3‰ and a $\delta^2H$-value of +129‰ for a small, shallow lake in the western Sahara.

### 3.6.6 Ocean Water

#### 3.6.6.1 Oxygen and Hydrogen Isotopes

The isotopic composition of ocean water has been discussed in detail by Craig and Gordon (1965), and Broecker (1974). It is governed by fractionation during evaporation and sea-ice formation and by the isotope content of precipitation and runoff entering the ocean.

Ocean water with 3.5% salinity is well mixed and exhibits a very narrow range in isotopic composition. There is, however, a strong correlation with salinity because evaporation, which increases salinity, also concentrates $^{18}O$ and D. Low salinities, which are caused by freshwater and meltwater dilution, correlate with low D and $^{18}O$ concentrations. As a consequence, modern ocean waters plot along two trends that meet at an inflection point where salinity is 3.55% and $\delta^{18}O$ is 0.5‰ (Fig. 3.22).

The high-salinity trend represents areas where evaporation exceeds precipitation and its slope is determined by the volume and isotopic composition of the local precipitation and the evaporating water vapor. However, isotope enrichments due to evaporation are limited in extent, because of back-exchange of atmospheric moisture with the evaporating fluid. The slope of the low salinity trend (see Fig. 3.23) extrapolates to a freshwater input of about −21‰ for $\delta^{18}O$ at zero salinity, reflecting the influx of high-latitude precipitation and glacial meltwater. This δ-value is, in all probability, not typical of freshwater influx in non-glacial periods. Thus, the slope of the low salinity trend may have changed through geologic time.

Delaygue et al. (2000) have modeled the present day $^{18}O$ distribution in the Atlantic and Pacific Ocean and its relationship with salinity (see Fig. 3.23). A good agreement is found between observed and simulated $\delta^{18}O$ values using an oceanic circulation model. As shown in Fig. 3.24 the Atlantic Ocean is enriched by more than 0.5‰ relative to the Pacific Ocean, but both ocean basins show the same general patterns with high $^{18}O$-values in the sub-tropics and lower values at high latitudes. Deep ocean waters are isotopically more stable.

Short-term fluctuations in the isotope composition of sea water must arise during glacial periods. If all the present ice sheets in the world were melted, the $\delta^{18}O$-value

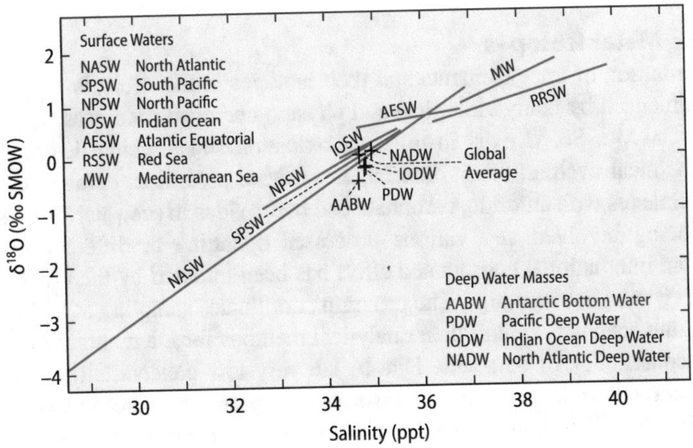


**Fig. 3.23** Salinity versus $\delta^{18}O$ relationships in modern ocean surface and deep waters (after Railsback et al. 1989)

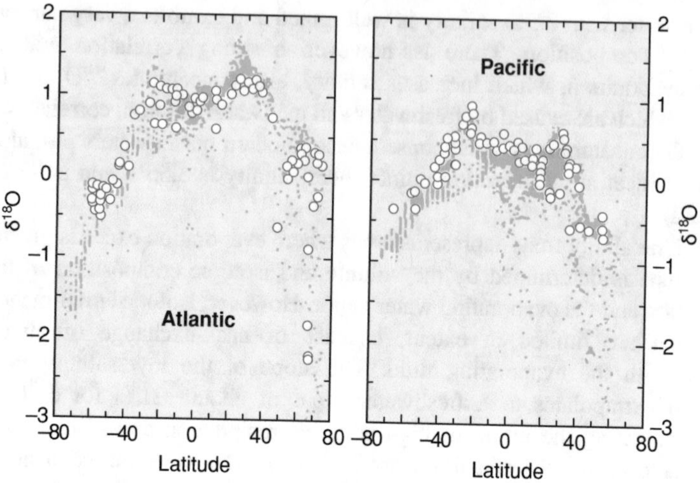


Fig. 3.24 Comparison of measured and modeled $\delta^{18}O$ values of surface ocean waters. Characteristic features are: tropical maxima, equatorial low- and high-latitude minima, enrichment of the Atlantic relative to the Pacific (after Delaygue et al. 2000)

of the ocean would be lowered by about 1‰. By contrast, Fairbanks (1989) has calculated an $^{18}O$-enrichment of 1.25‰ for ocean water during the last maximum glaciation.

Another important question concerning the isotopic composition of ocean water is how constant its isotopic composition has been throughout geological history. This remains an area of ongoing controversy in stable isotope geochemistry (see Sect. 3.8).

#### 3.6.6.2 Metal Isotopes

The distribution of trace elements and their isotopes in the ocean is very complex, being influenced by many biological and physico-chemical processes. Some metals like Fe, Cr, Mo, Se, U exist in different oxidation states which determines their biogeochemical cycling; others like Ni, Zn, Cd are present as divalent cation and form complexes with chloride, carbonate and hydroxides in seawater. With so many factors being involved and various processes operating at different scales and regions, an international coordinated effort has been initiated by the GEOTRACES program (www.geotraces.org). The program coordinates sampling at a global scale; participating scientists submit their analytical methods for intercalibration to ensure internal consistency of data sets. Due to the very low concentration of metals in ocean water, determinations of metal isotope compositions in ocean water are very demanding. The challenge is to extract and purify the metal from large volumes of seawater. A recent issue of the journal "Elements" summarizes important aspects of the GEOTRACES program (Jeandel, Chase and Hatje, editors, 2018: Elements 14, No. 6).

Metals in ocean water can be classified on the basis of their mean residence times relative to the mean residence time of water molecules. Metals with residence times longer than the mixing time of ocean water (about 1000 years) have a homogeneous isotope composition at all water depths. Examples are Li, Mg, Mo, Sr, Tl and U. Metals with shorter residence times vary with water depths having homogeneous deep-water compositions, but variable nutrient-dependent surface water compositions. Like the major nutrients C, P, N, Si, trace metals, such as Fe, Zn, Ni, Mo, are essential micronutrients. They display nutrient-like depth profiles: low concentrations and variable isotope compositions in surface waters; higher concentrations and relatively constant isotope compositions in deep waters.

Combined with metal concentration data, metal isotopes may constrain marine sources and sinks. Sources to the ocean include continental weathering products transported via rivers and winds and hydrothermal fluxes from midocean ridges. Of special importance in this connection are Zn, Cd and Fe isotopes (Conway and John 2015). Zn isotopes indicate the importance of scavenging and Zn uptake (Little et al. 2014; Zhao et al. 2014); Cd isotopes provide data about biological and physical cycling of Cd (Ripperger et al. 2007). Phytoplankton preferentially incorporates light Zn and Cd isotopes, while scavenging to organic matter preferentially absorbs heavy Zn isotopes. Biological uptake of Zn and Cd therefore leads to an enrichment of surface waters, whereas scavenging of Zn has the opposite effect.

Another important isotope system is Fe (Dauphas et al. 2017). Due to the low concentration and the short residence time of Fe in the ocean, significant interoceanic differences in Fe isotopes may provide information on the dominant Fe sources: (i) oxic and reducing sediments, (ii) hydrothermal input and (iii) atmospheric dust.

(i) Fe released from oxic sediments appears to be isotopically similar to continental material (Homoky et al. 2013). Fe released from reducing sediments is isotopically light compared to the other Fe sources. Insoluble Fe(III) is reduced to soluble Fe(II) introducing large Fe isotope fractionations (Anbar and Rouxel 2007 and others).

(ii) Hydrothermal vents may represent an important Fe source. Although most Fe is precipitated near hydrothermal vents, a small portion of hydrothermal Fe may be transported over great distances. As shown by Conway and John (2014), Fe isotope values in the North Atlantic range from −0.1 to −1.35‰, indicating that hydrothermal vents may be a source of isotopically light Fe.

(iii) The importance of atmospheric dust is reflected in the global distribution of iron concentration. Higher concentrations in the North Atlantic can be related to the input from the Sahara, even to individual storm events. Natural aerosols should have $\delta^{56}$Fe-values similar to the continental crust, however, surface waters in the North Atlantic showing obvious signs of dust input are about 0.6‰ heavier than total dust (Conway and John 2014).

In comparison to ocean water, large isotope variations of dissolved metals are observed in rivers, reflecting isotope variations of catchment rocks and differences in weathering processes. In general, suspended particles display less isotope variations than dissolved metal compounds.

### 3.6.7 Pore Waters

In the marine environment oxygen and hydrogen isotope compositions of pore waters may be inherited from ocean water or influenced by diagenetic reactions in the sediment or underlying basement. Knowledge of the chemical composition of sedimentary pore waters has increased considerably since the beginning of the Deep-Sea-Drilling-Project. From numerous drill sites, similar depth-dependent trends in the isotopic composition have been observed.

For oxygen this means a decrease in $^{18}O$ from an initial δ-value very near 0‰ (ocean water) to about −2‰ at depths around 200 m (Perry et al. 1976; Lawrence and Gieskes 1981; Brumsack et al. 1992). Even lower $\delta^{18}O$-values of about −4‰ at depths of around 400 m have been observed by Matsumoto (1992). This decrease in $^{18}O$ is mainly due to the formation of authigenic $^{18}O$-enriched clay minerals such as smectite from alteration of basaltic material and volcanic ash. Other diagenetic reactions include recrystallization of biogenic carbonates, precipitation of authigenic carbonates and transformation of biogenic silica (opal-A) through opal-CT to quartz. The latter process, however, tends to increase $\delta^{18}O$-values of the water. Material balance calculations by Matsumoto (1992) have indicated that the $^{18}O$-shift towards negative δ-values is primarily controlled by low-temperature alteration of basement basalts, which is slightly compensated by the transformation of biogenic opal to quartz.

D/H ratios may also serve as tracers of alteration reactions. Alteration of basaltic material and volcanic ash should increase $\delta^{2}H$-values of pore waters because the hydroxyl groups in clay minerals incorporate the light hydrogen isotope relative to water. However, measured $\delta^{2}H$-values of pore waters generally decrease from seawater values around 0‰ at the core tops to values that are 15–25‰ lower, with a good correlation between $\delta^{2}H$ and $\delta^{18}O$. This strong covariation suggests that the same process is responsible for the D and $^{18}O$ depletion observed in many cores recovered during DSDP/ODP drilling. Quite a different process has been suggested by Lawrence and Taviani (1988) to explain the depth-dependent decrease in porewater $\delta^{2}H$-values. They proposed oxidation of local organic matter or oxidation of biogenic or mantle methane. Lawrence and Taviani (1988) favored the oxidation of mantle methane, or even hydrogen, noting that oxidation of locally-derived organic compounds may not be feasible because of the excessive quantity of organic material required. In conclusion, the depletion of deuterium in porewaters is not clearly understood.

Additional informations about processes altering the isotopic composition of pore waters may be obtained through the analysis of Fe, Ca and Mg isotope compositions. To interpret pore water profiles, it is necessary to understand fractionation processes during precipitation and dissolution and cation exchange (Teichert et al. 2009; Ockert et al. 2013).

Porewaters typically have light $\delta^{56}$Fe-values (Severmann et al. 2006, 2010), which is attributed to dissimilatary iron reduction during bacterial decomposition of organic matter. In continental shelf settings $\delta^{56}$Fe-values of pore fluids are more depleted in heavy Fe isotopes than in deep sea sediments (Homoky et al. 2009) which reflects the differences in supply of organic carbon and iron.

$\delta^{44/40}$Ca-values of pore waters in carbonates equilibrate within the upper tens of meters (Fantle and de Paolo 2007), in siliciclastic, organic rich sediments equilibration occurs at greater depth due to reduced carbonate dissolution (Turchyn and de Paolo 2011). Mg isotope composition of pore fluids show large variations and display different trends with depth depending on the type of minerals formed (Higgins and Schrag 2012; Geske et al. 2015a, b). In some ODP sites Mg isotope ratios increase with depth due to precipitation of dolomite, whereas in other ODP sites Mg isotope ratios decrease with depth due to Mg incorporation into clay minerals.

### 3.6.8 Formation Water

Formation waters are saline with salt contents ranging from ocean water to very dense Ca–Na–Cl brines. Their origin and evolution are still controversial, because the processes involved in the development of saline formation waters are complicated by the extensive changes that have taken place in the brines after sediment deposition.

Oxygen and hydrogen isotopes are a powerful tool in the study of the origin of subsurface waters. Prior to the use of isotopes, it was generally assumed that most of the formation waters in marine sedimentary rocks were of connate marine origin. This widely held view was challenged by Clayton et al. (1966), who demonstrated that waters from several sedimentary basins were predominantly of local meteoric origin.

Although formation waters show a wide range in isotopic composition, waters within a sedimentary basin are usually isotopically distinct. As is the case with surface meteoric waters, there is a general decrease in isotopic composition from low to high latitude settings (Fig. 3.25). Displacements of $\delta^2$H and $\delta^{18}$O-values from the Meteoric Water Line (MWL) are very often correlated with salinity: the most depleted waters in D and$^{18}$O are usually the least saline, fluids most distant from the MWL tend to be the most saline.

Presently, in the view of numerous subsequent studies, (i.e. Hitchon and Friedman 1969; Kharaka et al. 1974; Banner et al. 1989; Connolly et al. 1990; Stueber and Walter 1991), it is obvious that basin subsurface waters have complicated histories and frequently are mixtures of waters with different origins. As was proposed by Knauth and Beeunas (1986) and Knauth (1988), formation waters in sedimentary basins may not require complete flushing by meteoric water, but instead can result from mixing between meteoric water and the remnants of original connate waters.

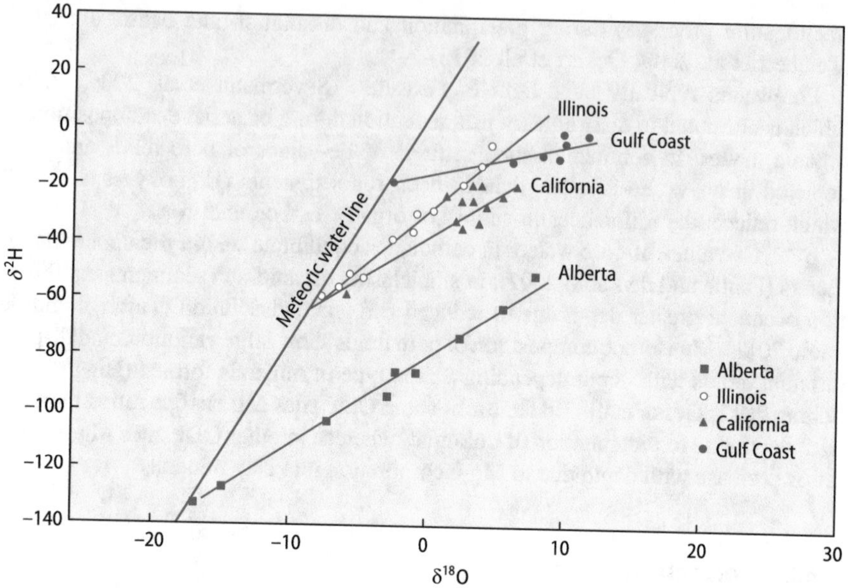


Fig. 3.25 $\delta^2H$ versus $\delta^{18}O$ values for formation waters from the midcontinental region of the United States (after Taylor 1974)

The characteristic $\delta^{18}O$ shift observed in formation waters may be due to isotopic exchange with $^{18}O$-rich sedimentary minerals, particularly carbonates. The $\delta^2H$-shift is less well understood, possible mechanisms for D-enrichment are (i) fractionation during membrane filtration, and/or (ii) exchange with $H_2S$, hydrocarbons and hydrous minerals. (i) It is well known that shales and compacted clays can act as semipermeable membranes which prevent passage of ions in solution while allowing passage of water (ultrafiltration). Coplen and Hanshaw (1973) have shown experimentally that ultrafiltration may be accompanied by hydrogen and oxygen isotope fractionation. However, the mechanism responsible for isotopic fractionation is poorly understood. Phillips and Bentley (1987) proposed that fractionation may result from increased activity of the heavy isotopes in the membrane solution, because high cation concentrations increase hydration sphere fractionation effects. (ii) Hydrogen isotope exchange between $H_2S$ and water will occur in nature, but probably will not be quantitatively important. Due to the large fractionation factor between $H_2S$ and $H_2O$, this process might be significant on a local scale. Isotope exchange with methane or higher hydrocarbons will probably not be important, because exchange rates are extremely low at sedimentary temperatures.

Somewhat unusual isotopic compositions have been observed in highly saline deep waters from Precambrian crystalline rocks as well as in deep drill holes, which plot above or to the left of the Meteoric Water Line (Frape et al. 1984; Kelly et al. 1986; Frape and Fritz 1987). There are two major theories about the origin of these Ca-rich brines.

(a) the brines represent modified Paleozoic seawater or basinal brines (Kelly et al. 1986),
(b) the brines are produced by leaching of saline fluid inclusions in crystalline rocks or by intense water/rock interactions (Frape and Fritz 1987).

Since then quite a number of studies have indicated that the unusual composition is a wide-spread phenomenon in low-permeability fractured rocks with slow water movement and not too high temperatures. Kloppman et al. (2002) summarized the existing data base of 1300 oxygen and hydrogen isotope analyses from crystalline rocks and suggested that the isotope shift to the left side can be explained by seawater which has dissolved and precipitated fracture minerals and subsequently been diluted by meteoric waters. Bottomley et al. (1999) argued that the extremely high concentrations of chloride and bromide in the brines make crystalline host rocks a less likely source for the high salinities. By measuring Li-isotopes these authors postulated that the brines in crystalline rocks share a common marine origin.

### 3.6.9 Water in Hydrated Salt Minerals

Many salt minerals have water of crystallization in their crystal structure. Such water of hydration can provide information on the isotope compositions and/or temperatures of brines from which the minerals were deposited. To interpret such isotope data, it is necessary to know the fractionation factors between the hydration water and the solution from which they are deposited. Several experimental studies have been made to determine these fractionation factors (Matsuo et al. 1972; Matsubaya and Sakai 1973; Stewart 1974; Horita 1989). Because most saline minerals equilibrate only with highly saline solutions, the isotopic activity and isotopic concentration ratio of water in the solution are not the same (Sofer and Gat 1972). Most studies determined the isotopic concentration ratios of the source solution and as Horita (1989) demonstrated, these fractionation factors have to be corrected using the "salt effect" coefficients when applied to natural settings (Table 3.3).

For the water-gypsum system, fractionation factors have been redetermined by Gazquez et al. (2017) and Herwatz et al. (2017). Hydrogen fractionations agree with previous estimates, but for oxygen they found a somewhat lower value of 1.0035 and 1.0037. The additionally determined $^{17}O$ values in combination with the deuterium excess may provide informations about the relative effects of humidity at the time of gypsum formation. Gazquez et al. (2018) estimated changes in relative humidity over the last 15,000 years in Lake Estanya (Spain). They reported the driest conditions for the Younger Dryas 12–13,000 years ago.

**Table 3.3** Experimentally determined fractionation factors of salt minerals and their corrections using "salt effect" coefficients (after Horita 1989)

| Mineral | Chemical formula | T°C | αD | $\alpha D_{(corr)}$ | $\alpha^{18}O$ | $\alpha^{18}O_{(corr)}$ |
|---|---|---|---|---|---|---|
| Borax | $Na_2B_4O_7 \times 10\ H_2O$ | 25 | 1.005 | 1.005 | – | – |
| Epsomite | $MgSO_4 \times 7\ H_2O$ | 25 | 0.999 | 0.982 | – | – |
| Gaylussite | $Na_2CO_3 \times CaCO_3 \times 5\ H_2O$ | 25 | 0.987 | 0.966 | – | – |
| Gypsum | $CaSO_4 \times 2\ H_2O$ | 25 | 0.980 | 0.980 | 1.0041 | 1.0041 |
| Mirabilite | $Na_2SO_4 \times 10\ H_2O$ | 25 | 1.017 | 1.018 | 1.0014 | 1.0014 |
| Natron | $Na_2CO_3 \times 10\ H_2O$ | 10 | 1.017 | 1.012 | – | – |
| Trona | $Na_2CO_3 \times NaHCO_3 \times 2\ H_2O$ | 25 | 0.921 | 0.905 | – | – |

## 3.7 The Isotopic Composition of Dissolved and Particulate Compounds in Ocean and Fresh Waters

The following chapter discusses the carbon, nitrogen, oxygen and sulfur isotope composition of dissolved and particulate compounds in ocean and fresh waters. The isotopic compositions of released components in waters of different origins depend on a variety of processes such as the composition of the minerals which have been weathered, the inorganic or organic nature of the precipitation process, and exchange with atmospheric gases. Investigations of non-traditional isotope systems in recent years have demonstrated that chemical weathering is a complex process that may induce large isotope fractionations. The weathering of silicates rarely results in the dissolution of the initial mineral, but instead in the formation of secondary minerals with isotopic compositions that differ from the initial mineral. Of special importance are biological processes acting mainly in surface waters, which tend to deplete certain elements such as carbon, nitrogen and silicon in surface waters by biological uptake, and which subsequently are returned at depth by oxidation and dissolution processes.

### 3.7.1 Carbon Species in Water

#### 3.7.1.1 Bicarbonate in Ocean Water

In addition to organic carbon, four other carbon species exist in natural water: dissolved $CO_2$, $H_2CO_3$, $HCO_3^-$ and $CO_3^{2-}$ all of which tend to equilibrate as a function of temperature and pH. $HCO_3^-$ is the dominant C-bearing species in ocean water. The first global $\delta^{13}C$ measurements of dissolved inorganic carbon (DIC) were published by Kroopnick et al. (1972) and Kroopnick (1985) within the geochemical ocean sections study (GEOSECS). These studies have yielded a global average $\delta^{13}C$-value of 1.5‰ with a variation range of ±0.8‰ with the least variations at equatorial regions and greater variability at higher latitudes.

The distribution of $\delta^{13}C$-values with water depth is mainly controlled by biological processes: Conversion of $CO_2$ into organic matter removes $^{12}C$ resulting in a $^{13}C$ enrichment of the residual DIC. In turn, the oxidation of organic matter releases $^{12}C$-enriched carbon back into the inorganic reservoir, which results into a depth-dependent isotope profile. A typical example is shown in Fig. 3.26.

North Atlantic Deep Water (NADW), which is formed with an initial $\delta^{13}C$-value between 1.0 and 1.5‰, becomes gradually depleted in $^{13}C$ as it moves southward and mixes with Antarctic bottom water, which has an average $\delta^{13}C$-value of 0.3‰ (Kroopnick 1985). As deep-water moves to the Pacific Ocean, its $^{13}C/^{12}C$ ratio is further reduced by 0.5‰ by the continuous flux and oxidation of organic matter in the water column. This is the basis for using $\delta^{13}C$-values as a tracer of paleo-oceanographic changes in deep water circulation (e.g. Curry et al. 1988).

The uptake of anthropogenic $CO_2$ by the ocean is a crucial process for the carbon cycle, resulting in changes of the $\delta^{13}C$-value of dissolved oceanic bicarbonate (Quay et al. 1992; Bacastow et al. 1996; Gruber 1998; Gruber et al. 1999; Sonnerup et al. 1999). Quay et al. (1992) first demonstrated that the $\delta^{13}C$-value of dissolved bicarbonate in the surface waters of the Pacific has decreased by about 0.4‰ between 1970 and 1990. If this number is valid for the ocean as a whole, it would

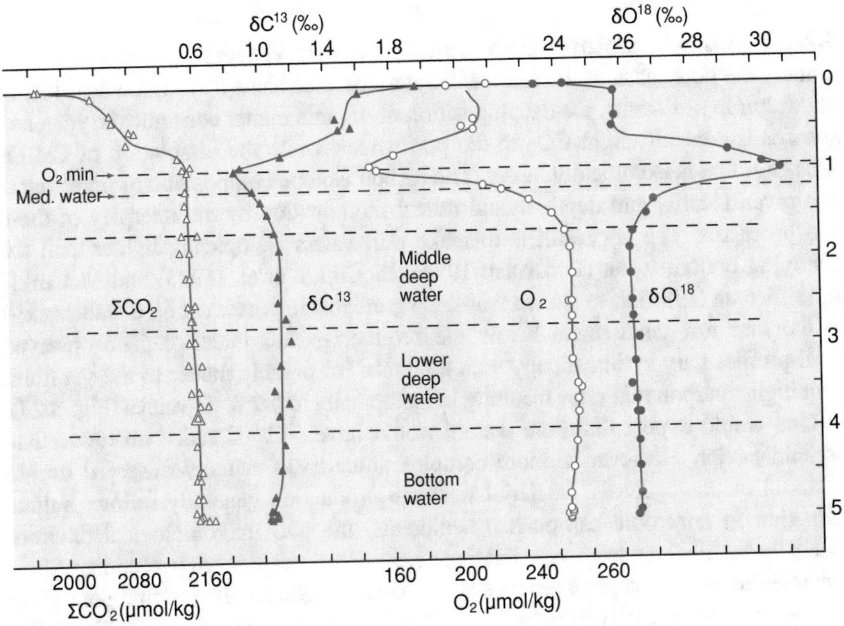


**Fig. 3.26** Vertical profiles of dissolved $CO_2$, $\delta^{13}C$, dissolved $O_2$ and $\delta^{18}O$ in the North Atlantic (Kroopnick et al. 1972)

allow a quantitative estimate for the net sink of anthropogenically produced $CO_2$. Recent studies estimate that the Earth's ocean has absorbed around 50% of the $CO_2$ emitted over the industrial period (Mikaloff-Fletcher et al. 2006).

#### 3.7.1.2 Particulate Organic Matter (POM)

POM in the ocean originates largely from plankton in the euphotic zone and reflects living plankton populations. Between 40°N and 40°S $\delta^{13}C$ of POM varies between −18.5 and −22‰. In cold Arctic waters $\delta^{13}C$-values are on average −23.4‰ and in high latitude southern ocean $\delta^{13}C$ are even lower with values between −24 and −36‰ (Goericke and Fry 1994). As POM sinks, biological reworking changes its chemical composition, the extent of this reworking depends on the residence time in the water column. Most POM profiles described in the literature exhibit a general trend of surface isotopic values comparable to those for living plankton, with $\delta^{13}C$-values becoming increasingly lower with depth. Jeffrey et al. (1983) interpreted this trend as the loss of labile, $^{13}C$-enriched amino acids and sugars through biological reworking which leaves behind the more refractory, isotopically light lipid components.

C/N ratios of POM increase with depth of the water column consistent with preferential loss of amino acids. This implies that nitrogen is more rapidly lost than carbon during degradation of POM, which is the reason for the much greater variation in $\delta^{15}N$-values than in $\delta^{13}C$-values (Saino and Hattori 1980; Altabet and McCarthy 1985).

#### 3.7.1.3 Carbon Isotope Composition of Pore Waters

Initially the pore water at the sediment/water interface has a $\delta^{13}C$-value near that of sea water. In sediments, the decomposition of organic matter consumes oxygen and releases isotopically light $CO_2$ to the pore water, while the dissolution of $CaCO_3$ adds $CO_2$ that is isotopically heavy. The carbon isotope composition of pore waters at a given locality and depth should reflect modification by the interplay of these two processes. The net result is to make porewaters isotopically lighter than the overlying bottom water (Grossman 1984). McCorkle et al. (1985) and McCorkle and Emerson (1988) have shown that steep gradients in porewater $\delta^{13}C$-values exist in the first few centimeters below the sediment–water interface. The observed $\delta^{13}C$-profiles vary systematically with the "rain" of organic matter to the sea floor, with higher carbon rain rates resulting in isotopically lower $\delta^{13}C$-values (Fig. 3.27).

One would expect that pore waters would have $^{13}C/^{12}C$ ratios no lower than organic matter. However, a more complex situation is actually observed due to bacterial methanogenesis. Bacterial methane production generally follows sulfate reduction in anaerobic carbon-rich sediments, the two microbiological environments being distinct from one another, except for substrate-rich sections. Since methane-producing bacteria produce very $^{12}C$-rich methane, the residual pore water can become significantly enriched in $^{13}C$ as shown in some profiles in Fig. 3.27.

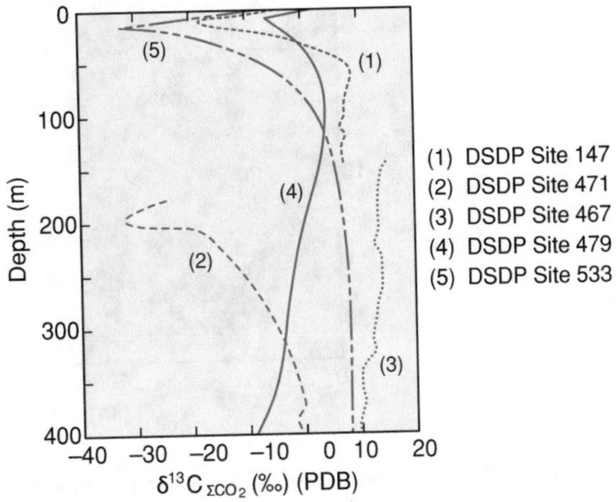


**Fig. 3.27** $\delta^{13}C$ records of total dissolved carbon from pore waters of anoxic sediments recovered in various DSDP sites (after Anderson and Arthus 1983)

#### 3.7.1.4 Carbon in Fresh Waters

Chemical weathering consumes atmospheric $CO_2$ through two pathways. (I) atmospheric $CO_2$ dissolves in rain and surface waters and reacts with rock forming minerals generating $HCO_3^-$ and (2) atmospheric $CO_2$ is converted to plant organic matter and subsequently released as soil $CO_2$. The mixing proportion from the two different sources determine the carbon isotope composition of fresh waters resulting in extremely variable isotopic composition, because varying mixtures of carbonate species derived from weathering of carbonates and of $CO_2$ originating from biogenic sources in soils are isotopically different (Hitchon and Krouse 1972; Longinelli and Edmond 1983; Pawellek and Veizer 1994; Cameron et al. 1995).

Although the $CO_2$ partial pressures in rivers vary widely, studies of major rivers often show that $CO_2$ concentrations are about 10–15 times greater than expected for equilibrium conditions with the atmosphere. Rivers thus are actively degassing $CO_2$ into the atmosphere, affecting the natural carbon cycle. This explains an increased interest in analyzing river systems for their carbon isotope composition. Despite the fact that the carbon isotopic composition of carbonate minerals and of soil-$CO_2$ are distinctive, the observed $\delta^{13}C$-variations of dissolved inorganic carbon are often not easy to interpret, because riverine respiration and exchange processes with atmospheric $CO_2$ play a role. Figure 3.28 gives some examples where carbon sources can be clearly identified. In the Amazon dissolved $CO_2$ originates from decomposition of organic matter (Longinelli and Edmond 1983), whereas in the St. Lawrence river system $CO_2$ originates from the dissolution of carbonates and equilibration with the atmosphere (Yang et al. 1996). The Rhine represents a mixture of both sources (Buhl et al. 1991).

In river systems often a $^{13}C$ enrichment is observed from upstream to downstream due to enhanced isotopic exchange with atmospheric $CO_2$ and/or in situ photosynthetic activity (Telmer and Veizer 1999). Variable seasonal signals can be explained by changes in the oxidation rate of $^{13}C$-depleted organic matter from the soils in watersheds. Rivers that are characterized by the presence of large lakes at

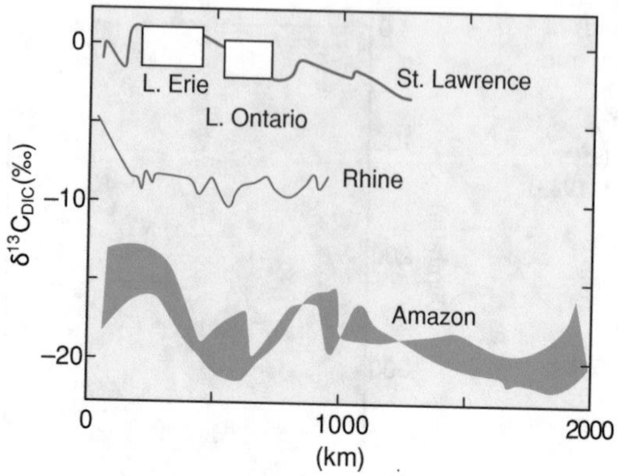


**Fig. 3.28** Carbon isotopic composition of total dissolved carbon in large river systems. *Data source* Amazon: Longinelli and Edmond (1983), Rhine: Buhl et al. (1991), St. Lawrence: Yang et al. (1996)

their head—like the Rhone and St. Lawrence—show heavy $^{13}C$-values at their head (Ancour et al. 1999; Yang et al. 1996). Due to the long residence time of dissolved carbon in lakes, the bicarbonate is in near equilibrium with atmospheric $CO_2$.

### 3.7.2 Silicon

Silicon isotope variations in the ocean are caused by biological Si-uptake through siliceous organisms like diatoms. Insofar strong similarities exist with C-isotope variations. Diatoms preferentially incorporate $^{28}Si$ as they form biogenic silica. Thus, high $\delta^{30}Si$ values in surface waters go parallel with low Si-concentrations and depend on differences in silicon surface water productivity. In deeper waters dissolution of sinking silica particles causes an increase in Si concentration and a decrease of $\delta^{30}Si$-values. Therefore, in ocean water distinct $^{30}Si$ gradients with depth exist (Georg et al. 2006; Beucher et al. 2008). Surface waters may show a large variation from +2.2 to +4.4‰ (Grasse et al. 2013). Deep water masses have on the other hand more $^{30}Si$ depleted values with regional variations indicating mixing of different water masses (Ehlert et al. 2013).

Vertical and horizontal gradients of Si isotopes have been observed in ocean water profiles, preferentially in the Southern Ocean having higher Si concentrations than the Northern Ocean (Beucher et al. 2008; De Souza et al. 2012; Fripiat et al. 2012). Dissolved silica in North Atlantic Deep Water has a $\delta^{30}Si$-value being 0.5‰ higher than deep water of the Southern Ocean which suggests export of Si from surface waters of the Southern Ocean (De Souza et al. 2012). The clearest gradient in the $^{30}Si$ distribution with water depth has been observed in the North Atlantic Ocean (de Souza et al. 2012, 2015; Brzezinski and Jones 2015). De Souza et al. (2015) estimated that the high $^{30}Si$-values in the deep water of the North Atlantic are affected up to 60% by southern-ocean surface waters.

### 3.7.3 Nitrogen

Nitrogen is one of the limiting nutrients in the ocean. Apparently, the rate of nitrate formation is so slow, and marine denitrification so rapid, that nitrate is in short supply. Dissolved nitrogen is subject to isotope fractionation during microbial processes and during biological uptake. Nitrate dissolved in oceanic deep waters has a $\delta^{15}N$-value of 6–8‰ (Cline and Kaplan 1975; Wada and Hattori 1976). Denitrification seems to be the principal mechanism that keeps marine nitrogen at higher $\delta^{15}N$-values than atmospheric nitrogen. Measurement of the O isotope composition of nitrate provides an additional signature for interpreting processes of nitrate production and consumption (Casciotti 2016 and others).

The $\delta^{15}N$-value of particulate material was originally thought to be determined by the relative quantities of marine and terrestrial organic matter. However, temporal variations in the $^{15}N$-content of particulate matter predominate and obscure N-isotopic differences previously used to distinguish terrestrial from marine organic matter. Altabet and Deuser (1985) observed seasonal variations in particles sinking to the ocean bottom and suggested that $\delta^{15}N$-values of sinking particles represent a monitor for nitrate flux in the euphotic zone. Natural $^{15}N$-variations can thus provide information about the vertical structure of nitrogen cycling in the ocean.

Saino and Hattori (1980) first observed distinct vertical changes in the $^{15}N$ content of suspended particulate nitrogen and related these changes to particle diagenesis. A sharp increase in $^{15}N$ below the base of the euphotic zone has been ubiquitously observed (Altabet and McCarthy 1985; Saino and Hattori 1987; Altabet 1988). These findings imply that the vertical transport of organic matter is mediated primarily by rapidly sinking particles and that most of the decomposition of organic matter takes place in the shallow layer beneath the bottom of the euphotic zone.

### 3.7.4 Oxygen

As early as 1951, Rakestraw et al. demonstrated that dissolved $O_2$ in the oceans is enriched in $^{18}O$ relative to atmospheric oxygen. Like its concentration, the $\delta^{18}O$ of dissolved oxygen is affected by three processes: air–water gas exchange, respiration and photosynthesis (see recent review by Mader et al. 2017). When gas exchange dominates over photosynthesis and respiration as in the surface ocean dissolved oxygen is close to saturation and the $\delta^{18}O$ is ~24.2‰, because there is a 0.7‰ equilibrium fractionation during gas dissolution (Quay et al. 1993).

Extreme enrichments up to 14‰ (Kroopnick and Craig 1972) occur in the oxygen minimum region of the deep ocean due to preferential consumption of $^{16}O$ by bacteria in abyssal ocean waters, which is evidence for a deep metabolism (see Fig. 3.22).

Precise measurements of the $^{17}O$ content of dissolved oxygen in seawater indicate a small $^{17}O$ anomaly that can be used to estimate overall photosynthetic oxygen production in seawater (Luz and Barkan 2000, 2005; Juranek and Quay 2010).

Quay et al. (1995) measured $^{18}O/^{16}O$ ratios of dissolved oxygen in rivers and lakes of the Amazon Basin. They observed a large $\delta^{18}O$ range from 15 to 30‰. In fresh waters, when respiration dominates over photosynthesis, dissolved $O_2$ will be undersaturated and $\delta^{18}O$ is >24.2‰; when photosynthesis exceeds respiration, dissolved $O_2$ will be supersaturated and $\delta^{18}O$ will be <24.2‰.

### 3.7.5 Sulfate

Modern ocean water sulfate has a fairly constant $\delta^{34}S$-value of 21‰ (Rees et al. 1978), more recently Tostevin et al. (2014) specified this value to 21.24‰. The $\delta^{34}S$-value depends on the river input thought to be 5 and 15‰ and on the fraction and sulfur isotope fractionation associated with pyrite burial. Additional constraints on S isotope fractionations during biological reductions may be placed from the minor S isotopes $^{33}S$ and $^{36}S$ (Farquhar et al. 2003; Johnston 2011).

The $\delta^{18}O$-value of ocean water is 9.3‰ (Lloyd 1967, 1968; Longinelli and Craig 1967). From theoretical calculations of Urey (1947), it is quite clear that the $\delta^{18}O$-value of dissolved sulfate does not represent equilibrium with $\delta^{18}O$-value of the water, because under surface conditions oxygen isotope exchange of sulfate with ambient water is extremely slow (Chiba and Sakai 1985). By using quantum-chemical calculations Zeebe (2010) estimated the equilibrium fractionation between dissolved sulfate and water to be 23‰ at 25 °C.

Lloyd (1967, 1968) proposed a model in which the fast-bacterial turnover of sulfate at the sea bottom determines the oxygen isotope composition of dissolved sulfate. Böttcher et al. (2001), Aharon and Fu (2000, 2003) and others demonstrated that the $\delta^{18}O$ of sulfate is not only influenced by microbial sulfate reduction, but also by disproportionation and reoxidation of reduced sulfur compounds. In marine pore waters, $^{18}O$ enrichments up to 30‰ have been observed, generally associated with strong $^{34}S$ enrichments. By plotting $\delta^{18}O_{(SO4)}$ vs $\delta^{34}S_{(SO4)}$ two different slopes can be distinguished: in some cases $\delta^{18}O$ increases linearly with $\delta^{34}S$ in residual sulfate (slope 1), whereas in most cases the $\delta^{18}O$ increases initially until it reaches a constant value with no further increase while $\delta^{34}S$ may continue to increase (slope 2). Böttcher et al. (1998), Brunner et al. (2005) and Antler et al. (2013) discussed models to explain the different slopes of $\delta^{18}O$–$\delta^{34}S$ plots: (i) a model that postulates the predominance of kinetic oxygen isotope fractionation steps linked to different sulfate reduction steps and (ii) a model postulating a predominance of oxygen isotope exchange between cell-internal sulfur compounds and ambient water (Brunner et al. 2005; Wortmann et al. 2007).

In freshwater environments the sulfur and oxygen isotope composition of dissolved sulfate is much more variable and potentially the isotope ratios can be used to identify the sources: (i) oxidation of sedimentary and magmatic sulfides, (ii) dissolution of evaporates, (iii) atmospheric aerosols, (iv) anthropogenic input. However, such attempts have been only partially successful because of the variable composition of the different sources. $\delta^{34}S$-values of dissolved sulfate of different rivers and lakes show a rather large spread as is demonstrated in Fig. 3.29. The data

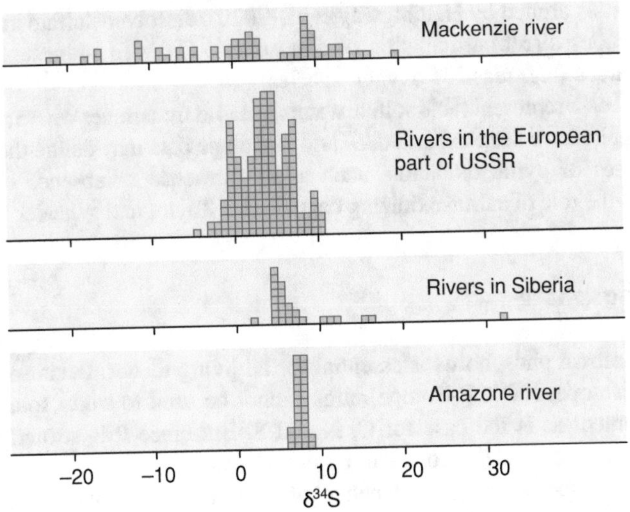


**Fig. 3.29** Frequency distribution of $\delta^{34}$S-values in river sulfate

of Hitchon and Krouse (1972) for water samples from the MacKenzie River drainage system exhibit a wide range of $\delta^{34}$S-values reflecting contributions from marine evaporites and shales. Calmels et al. (2007) argue that around 85% of the sulfate in the MacKenzie river is derived from pyrite oxidation and not from sedimentary sulfate. For the Amazon River, Longinelli and Edmond (1983) found a very narrow range in $\delta^{34}$S-values which they interpreted as representing a dominant Andean source for sulfate from the dissolution of Permian evaporites with a lesser admixture of sulfide sulfur. Rabinovich and Grinenko (1979) reported time-series measurements for the large European and Asian rivers in Russia. The sulfur in the European river systems should be dominated by anthropogenically derived sources, which in general have $\delta^{34}$S-values between 2 and 6‰. On the basis of sulfate measurements from more than 160 rivers, accounting for close to 50% of the world's freshwater flux to the ocean, Burke et al. (2018) estimated the global riverine S-isotope as 4.4 and 4.8‰ when the most polluted rivers are excluded.

Additional constraints on the origin and formation of sulfate have been presented by the analysis of $^{17}$O (Balci et al. 2007; Kohl and Bao 2011; Waldeck et al. 2019). Weathering of sulfides to sulfate has been postulated to directly trace past oxygen isotope compositions. Tropospheric $O_2$ carries a negative $^{17}$O anomaly (see p. "atmospheric oxygen"). The magnitude of the $^{17}$O anomaly depends on the ratio of atmospheric $O_2$ and $CO_2$ partial pressures (Waldeck et al. 2019; Hemingway et al. 2020). Thus, the $^{17}$O content in tropospheric oxygen can be regarded as a quantitative $pCO_2$ tracer.

However, as argued by Hemingway et al. (2020), dissolved sulfate in Himalayan rivers precludes direct participation of atmospheric $O_2$, who suggested alternative oxygen sources (e.g. reactive oxygen species).

A special case represents acid sulfate waters released from mines where metal sulfide ores and lignite have been exploited. S- and O-isotope data may define the conditions and processes of pyrite oxidation, such as the presence or absence of dissolved oxygen and the role of sulfur-oxidizing bacteria (i.e. Taylor and Wheeler 1994).

### 3.7.6 Phosphate

As is well known phosphorus is essential for all living matter. Because P has only one stable isotope, stable P-isotope ratios cannot be used to study sources of P in the environment as is the case for C, N and S. But since P is strongly bound to oxygen, O isotope investigations can be used instead.

Oxygen isotope exchange between phosphate and water under purely abiotic conditions is negligible (Tudge 1960; Blake et al. 1997 and others), but is fast in biologically mediated systems (Luz and Kolodny 1985; Blake et al. 1997, 2005). Experiments with microbiological cultures as well as with enzymes indicate that oxygen isotope fractionations depend on growth conditions, phosphate concentrations and sources (Blake et al. 2005). Thus, the $\delta^{18}$O-value of phosphate in fresh and ocean water can be used to distinguish different P sources and biological pathways.

Phosphate depth profiles in the Atlantic and Pacific showed that $^{18}$O is near equilibrium with water (Colman et al. 2005), whereas it is not in near coastal shallow waters (McLaughlin et al. 2006). In a 2 years time series experiment these authors observed seasonal $^{18}$O variations up to 6‰ in the Monterey Bay. Isotope equilibrium is approached during episodic upwelling events when phosphate is extensively cycled by the biological community; lower non-equilibrium values have been observed when phosphate is not extensively used. Even larger $^{18}$O variations have been observed in pore waters (Goldhammer et al. 2011).

Identification of phosphate sources is important to reduce anthropogenic inputs of phosphorus to the environment. Young et al. (2009a, b) measured the $\delta^{18}$O-value of different phosphate sources such as fertilizers, detergents, animal feces. They observed a large range in $\delta^{18}$O-values from 8 to 25‰. Although $\delta^{18}$O-values overlap, Young et al. (2009a, b) concluded that in suitable cases some phosphate sources are distinct and can be identified.

## 3.8 Isotopic Composition of the Ocean During Geologic History

The growing concern with respect to "global change" brings with it the obvious need to document and understand the geologic history of sea water. From paleo-ecological studies it can be deduced that ocean water should not have changed its chemical composition very drastically, since marine organisms can only tolerate relatively small chemical changes in their marine environment. The similarity of the mineralogy and to some extent paleontology of sedimentary rocks during the Earth's history strengthens the conclusion that the chemical composition of ocean water has not varied substantially. This was the general view for many years. More recently, however, fluid inclusions in evaporite minerals have indicated that the chemical concentrations of major ions in ocean water such as Ca, Mg and $SO_4$ have changed substantially over the Phanerozoic (Horita et al. 2002a, b and others). It is thus likely that steady state conditions of input fluxes to and output fluxes from the oceans are not always equal during earth's history. The rate of these changes in ocean chemistry is dictated by the residence time of ions in the ocean.

One of the most sensitive tracers recording the composition of ancient sea water is the isotopic composition of chemical sediments precipitated from sea water. The following discussion concentrates on the stable isotope composition of oxygen, carbon, and sulphur. More recently, other isotope systems have been investigated such as Ca (De La Rocha and DePaolo 2000b; Schmitt et al. 2003; Fantle and DePaolo 2005; Farkas et al. 2007), B (Lemarchand et al. 2000, 2002; Joachimski et al. 2005) and Li (Hoefs and Sywall 1997; Misra and Froelich 2012; Wanner et al. 2014). One of the fundamental questions in all these approaches is which kind of sample provides the necessary information, in the sense that it represents the coexisting ocean water composition during the time of sediment formation and has not been modified subsequently by diagenetic reactions. Furthermore, since most chemical sediments are deposited close to the continental margins, they are not necessarily representative for the whole ocean.

### 3.8.1 Oxygen

It is generally agreed that continental glaciation and deglaciation induce changes in the $\delta^{18}O$-value of the ocean on short time scales. There is, however, considerable debate about long-term changes.

The present ocean is depleted in $^{18}O$ by at least 6‰ relative to the total reservoir of oxygen in the crust and mantle. Muehlenbachs and Clayton (1976) presented a model in which the isotopic composition of ocean water is held constant by two different processes: (i) low temperature weathering of oceanic crust which depletes ocean water in $^{18}O$, because $^{18}O$ is preferentially bound in weathering products and (ii) high-temperature hydrothermal alteration of ocean ridge basalts which enriches ocean water in $^{18}O$, because $^{16}O$ is preferentially incorporated into the solid phase during the hydrothermal alteration of oceanic crust. If sea floor-spreading ceased, or its rate were to decline, the $\delta^{18}O$-value of the oceans would slowly change to lower values because of continued continental and submarine weathering. Gregory and Taylor (1981) and Muehlenbachs (1998) presented further evidence for this rock/water buffering and argued that the $\delta^{18}O$ of sea water should be invariant within about ±1‰, as long as sea-floor spreading was operating at a rate of at least 50% of its modern value.

The sedimentary record, however, is not in accord with this model for constant oxygen isotope compositions because in a general way carbonates, cherts, shales and phosphates show a decrease in $\delta^{18}O$ in progressively older samples (Veizer and Hoefs 1976; Knauth and Lowe 1978; Shemesh et al. 1983; Bindeman 2021).

Figure 3.30 summarizes age-dependent trends in $\delta^{18}O$-values for cherts, carbonates and shales as compiled by Bindeman et al. (2016) and Bindeman (2021).

The prime issue arising from these trends is whether they are of primary or secondary (post-depositional) origin. Veizer et al. (1997, 1999) presented strong evidence that they are, at least partly, of primary origin. Based on well-selected Phanerozoic low-Mg calcite shells (mostly brachiopods), they observed a 5‰ decline from the Quaternary to the Cambrian. Because well preserved textures and trace element contents are comparable to modern low-Mg calcitic shells, Veizer and coworkers argued that the shells preserved the primary oxygen isotope composition and can be used to deduce the past ocean composition. Prokoph et al. (2008) provided on updated compilation of 39,000 $\delta^{18}O$ and $\delta^{13}C$ isotope data for the entire earth history confirming earlier observation of Veizer and coworkers.

Jaffrés et al. (2007) reviewed models about the potential influence of varying chemical weathering and hydrothermal circulation rates. The most likely explanation for the long-term trend in seawater $\delta^{18}O$ involves stepwise increases in the ratio of high- to low-temperature fluid/rock interactions, a conclusion favored by Kasting

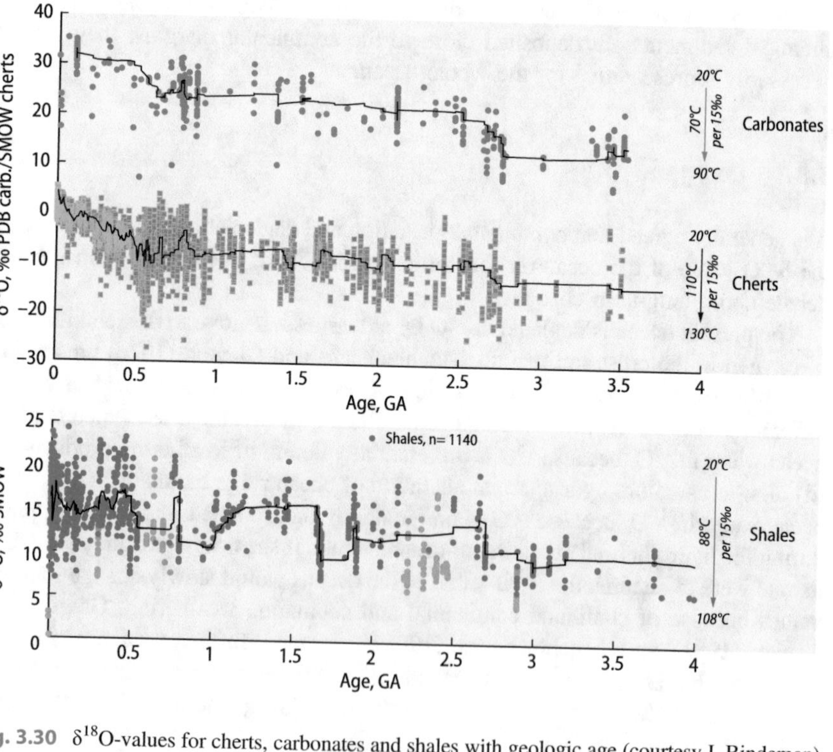


Fig. 3.30 $\delta^{18}O$-values for cherts, carbonates and shales with geologic age (courtesy I. Bindeman)

et al. (2006). Model calculations on the geological water cycle by Wallman (2001) support the idea that seawater $\delta^{18}O$-values were not constant through time, but evolved from an $^{18}O$ depleted state to the current value.

By analyzing marine iron oxides covering an age span of 2 billion years, Galili et al. (2019) observed an $^{18}O$ increase of about 10‰ over the last 2 billion years. Since the temperature dependence of iron oxide–water fractionation is weak, Galili et al. (2019) concluded that the $^{18}O$ content of the ocean has increased since the Archean. A different approach has been used by Pope et al. (2012), who measured the O- and H-isotopic composition of serpentine from Isua (Greenland) to elucidate the isotope composition of the Archean ocean. They concluded that the oxygen isotope composition was comparable to today, but that the hydrogen isotope composition was depleted by about 25‰.

New insights into the unsolved question of constant or variable oxygen isotope composition during geologic history can be obtained through the analysis of clumped isotopes (Cummins et al. 2014). As demonstrated by these authors, clumped isotopes may detect diagenetic alteration at high temperature, because carbonate that is recrystallized at high temperature will reflect an elevated clumped isotope temperature. Thus, clumped isotopes provide a more sensitive indicator of diagenetic recrystallization than $^{18}O$ or trace element content. Cummins et al. (2014) measured clumped isotope compositions of very well-preserved Silurian brachiopods and corals from Gotland, Sweden and demonstrated that out of a large sample set only a small subset of samples retained primary oxygen water temperatures. These results along with those of Dennis et al. (2013), Bergmann et al. (2018) and Henkes et al. (2018) suggest that the oxygen isotope composition of seawater has remained constant over Phanerozoic time.

Triple oxygen isotope determinations further allow to constrain temperatures of formation and conditions of diagenesis for carbonates and silicates (Wostbrock and Sharp 2021) indicating that neither increasing ocean temperatures nor increasing depletions of heavy oxygen isotopes can explain the observed trend in $\delta^{18}O$-values, but favor a complex diagenesis as decisive mechanism.

### 3.8.2 Carbon

The $^{13}C$ content of a marine carbonate is closely related to that of the dissolved marine bicarbonate from which the carbonate precipitated. For a long time, the $\delta^{13}C$-value of ancient oceans was regarded as essentially constant around 0‰. It was in the 1980s when it was first realized that the observed fluctuations represent regular secular variations. Shifts in the carbon isotopic composition of marine carbonates may be interpreted as representing shifts in the amount of organic carbon being buried. An increase in the amount of buried organic carbon means that $^{12}C$ would be preferentially removed from seawater, so that the ocean reservoir would become isotopically heavier. Negative $\delta^{13}C$-shifts accordingly may indicate a decrease in the rate of carbon burial and/or enhanced oxidative weathering of once buried organic matter.

$\delta^{13}C$-values of limestones vary mostly within a band of 0 ± 3‰ since at least 3.5 Ga (Veizer and Hoefs 1976). The long-term C-isotope trend for carbonates has been punctuated by sudden shifts over short time intervals named "carbon isotope events", which are considered to represent characteristic features, and have been used as time markers for stratigraphic correlations. Characteristic carbon isotope events are the Paleocene-Eocene Thermal Maximum (Cohen et al. 2007), the Jurassic-Cretaceous Oceanic Anoxic events (Jenkyns 2010) and the Permian–Triassic extinction (Payne and Kump 2007) (see also Carbon isotope stratigraphy, p.).

Especially noteworthy are very high $\delta^{13}C$-values of up to 16‰ and higher for 2.2–2.0 Ga old carbonates—the socalled Lomagundi-Jatuli event in the Palaeoproterozoic—and at the end of the Proterozoic with both periods representing periods of increased burial of organic carbon (Knoll et al. 1986; Baker and Fallick 1989; Derry et al. 1992 and others). As summarized by Martin et al. (2013) the Lomagundi-Jatuli event follows the oxygenation of the Earth's atmosphere and lasts for more than 100 million years. By compiling the data base for the whole Proterozoic, Shields and Veizer (2002) (Fig. 3.31) demonstrated $^{13}C$ fluctuations of at least 15‰, coincident with wide spread glaciations (see also Special Issue of Chemical Geology 237, 2007). Highly $^{13}C$ enriched intervals are related to interglacial times, where the $^{13}C$ enrichment appears to be the result of unusually efficient burial of organic carbon. Hayes and Waldbauer (2006), on the other hand, interpreted the unusual $^{13}C$-enrichment as indication for the importance of methanogenic bacteria in sediments.

Negative $\delta^{13}C$ intervals are generally associated with glaciations (Kaufmann and Knoll 1995). The most negative $^{13}C$-values have been found in massive carbonates that cap glaciogenic sequences ("cap" carbonates), which record the most profound

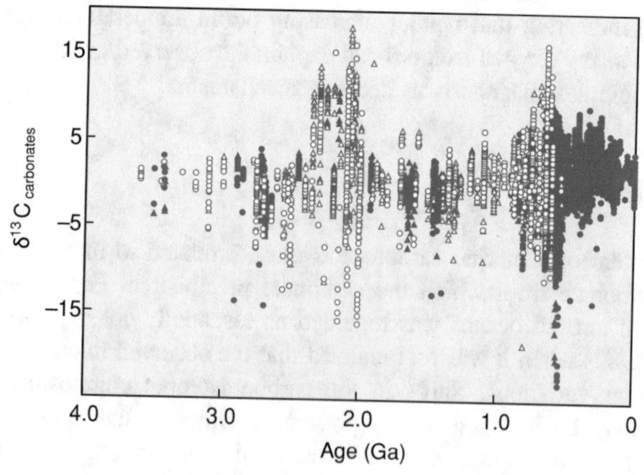


Fig. 3.31 $\delta^{13}C$-values for marine carbonates over time. Note persistent values of 0–3‰ for the last 600 Ma, anomaleous variability at 0.6–0.8 Ga and 2.0–2.3 Ga correlative with snowball earth episodes (Shields and Veizer 2002)

carbon isotope variations on Earth. The change from very heavy to very light $\delta^{13}C$-values has been interpreted as a collapse of biological productivity for millions of years due to global glaciations (Hoffmann et al. 1998) representing one the central arguments of the "snowball Earth" hypothesis. Glaciations ended abruptly when subaerial volcanic outgassing raised atmospheric $CO_2$ to very high levels shifting the $^{13}C$ of carbonates to values around −5‰.

Because of the relationship between carbonate and organic carbon, a parallel shift in the isotope composition of both carbon reservoirs should be observed. Unfortunately, very often carbonate-carbon and organic carbon have not been investigated together. Hayes et al. (1999) have compiled the existing data base on both reservoirs. In contrast to previous assumptions, the long-term fractionation is invariant and its average close to 30‰ rather than 25‰. Variations in the fractionations between the two reservoirs can, in principle, be interpreted as reflecting variations in the $pCO_2$ content of the atmosphere (Kump and Arthur 1999). By employing a simple model which is subjected to different perturbations each lasting 500,000 years, Kump and Arthur (1999) demonstrated that increased burial of organic carbon leads to a fall in atmospheric $pCO_2$ and to positive $^{13}C$-shifts in both carbonate and organic carbon. Lately, shifts in $^{13}C$ have been correlated to variations in the $O_2/CO_2$ ratio of the ambient atmosphere (Strauß and Peters-Kottig 2003).

### 3.8.3 Sulfur

Because isotope fractionation between dissolved sulfate in ocean water and gypsum/anhydrite is small (Raab and Spiro 1991), evaporite sulfates should closely reflect the sulfur isotope composition of marine sulfate through time. The first S-isotope "age curves" were published by Nielsen and Ricke (1964) and Thode and Monster (1964). Since then, this curve has been updated by many more analyses (Holser and Kaplan 1966; Holser 1977; Claypool et al. 1980). The sulfur isotope curve varies from a maximum of $\delta^{34}S$ = +30‰ in early Paleozoic time, to a minimum of +10‰ in Permian time. These shifts are considered to reflect net fluxes of isotopically light sulfur generated during bacterial reduction of oceanic sulfate to the reservoir of reduced sulfide in sediments, thus increasing the $^{34}S$-content in the remaining oceanic sulfate reservoir. Conversely, a net return flux of the light sulfide to the ocean during weathering or enhanced hydrothermal sulfide input lead to a decrease of marine sulfate $\delta^{34}S$-values. Modeling by Kump (1989) has indicated that pyrite burial was twice as large as today during most of the early Paleozoic followed by a decrease to values that are about half of today's rate during the Carboniferous and Permian and by approximately constant rates for the last 180 Ma (Kump 1989).

Since evaporites through geologic time contain large gaps and considerable scatter in sulfur isotope composition, two alternative approaches for the reconstruction of seawater $\delta^{34}S$ values through time have been utilized: (I) Structurally substituted sulfate in marine carbonates (Burdett et al. 1989; Kampschulte and Strauss 2004). This approach avoids apparent disadvantages of the evaporite record namely that evaporites are discontinuous with a poor age resolution representing continental margin formations with potential influence from nearby continents. Hence, a much better temporal resolution from structural sulfate records has been obtained. (II) Marine barite in pelagic sediments. Paytan et al. (1998, 2004) generated a seawater sulfur curve for the Cenozoic and for the Cretaceous with a resolution of $\sim 1$ million years. Barite has advantages over the other two sulfate proxies, because of its resistance to diagenesis as long as dissolved pore water is present to prevent barite dissolution (see Fig. 3.32). Since pelagic sediments are restricted to the modern ocean, the barite record lasts for the last 150 million years only.

The oxygen isotope composition of marine barite might be also a useful tracer for the sulfate cycle in the past. Turchyn and Schrag (2004, 2006) observed a 5‰ variability in $\delta^{18}O$ over the past 10 million years. Oxygen is incorporated into sulfate through sulfide oxidation and released through sulfate reduction. Turchin and Schrag (2004) suggested that sea level fluctuations reducing the area of continental shelves and increasing sulfide weathering may be responsible for the observed variations.

It might be expected that a parallel age curve to that for sulfates should exist for sedimentary sulfides. However, the available S-isotope data for sulfides range widely and seem to depend strongly on the degree to which the reduction system is "open" and on the sedimentation rate so that age trends are obscured (Strauß 1997,

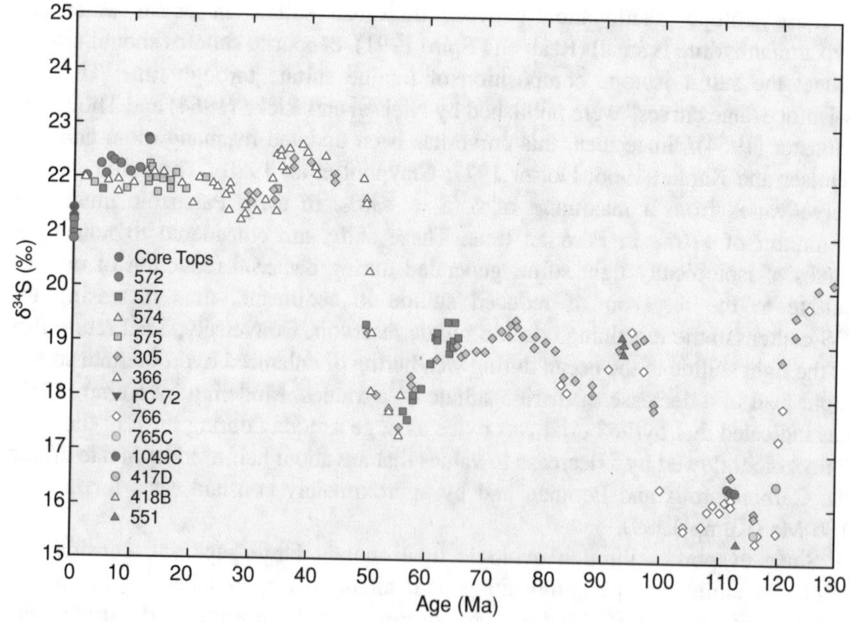


**Fig. 3.32** Marine sulfate $\delta^{34}S$ curve of marine barite for 130 Ma to present (Paytan et al. 2004)

1999). Changes in the maximum sulphur isotope fractionation between sulphides and coexisting sulfates were used to propose changes in the complexity of the sulphur cycle (Canfield and Teske 1996). The large variability in $\delta^{34}S_{sulfide}$ values within age-equivalent strata might be best explained by time-dependent steps of pyrite formation during progressive diagnesis.

Considering a difference in $\delta^{34}$S-values of 40–60‰ between bacteriogenic sulfide and marine sulfate in present-day sedimentary environments, similar fractionations in ancient sedimentary rocks may be interpreted as evidence for the activity of sulfate-reducing bacteria. The presence or absence of such fractionations in sedimentary rocks thus may constrain the time of emergence of sulfate-reducing bacteria. In early Archean sedimentary rocks most sulfides and the rare sulfates have $\delta^{34}$S-values near 0‰ (Monster et al. 1979; Cameron 1982). The lack of substantial isotope fractionation between sulfate and sulfide has been interpreted initially as indicating an absence of bacterial reduction in the Archean, but could also indicate complete sulfate reduction. Ohmoto et al. (1993) employed a laser microprobe approach to analyze single pyrite grains from the ca 3.4 Ga Barberton greenstone belt and observed a variation of up to 10‰ among pyrites from a single rock specimen, which could imply that bacterial reduction has occurred since at least 3.4 Ga. Shen and Buick (2004) argued that the large spread in $\delta^{34}$S values of microscopic pyrites aligned along growth faces of former gypsum in the 3.47 Ga North Pole barite deposit, Australia represents the oldest evidence for microbial sulfate reduction.

### 3.8.4 Lithium

The two major sources of Li to the ocean are rivers and hydrothermal input at spreading centers, major sinks are Li-incorporation into marine sediment and low temperature oceanic crust. By analyzing well-dated planktonic foraminifera, Misra and Froelich (2012) presented a Li-isotope curve for the last 68 Ma. They observed a Li isotope increase of 9‰ over the past 50 million years, which they interpreted to reflect a general increase in continental weathering rates. As reported by Washington et al. (2020), brachiopods record a similar magnitude of change in Li isotopes as foraminifers. Since Li–in contrast to Ca and Sr—is preferentially incorporated in silicate minerals, the Li isotope record is sensitive to changes in the weathering of silicate rocks. Modelling by Wanner et al. (2014) revealed a correlation between $\delta^{7}$Li-values and $CO_2$ consumption rates by silicate weathering. Thus, the Li isotope record may quantify atmospheric $CO_2$ consumption and, therefore, may give informations about the evolution of the global climate.

### 3.8.5 Boron

Geochemical modelling has indicated significant variations in the boron isotope composition of sea water with geologic time (Lemarchand et al. 2002; Simon et al.

2006). $\delta^{11}$B-values of past seawater depend, like Li and other elements, on continental erosion rates and rates of chemical exchange at oceanic ridges. What makes B unique is its dependence on the pH of ocean water (see p. 127). Studies on foraminifera to reconstruct the $\delta^{11}$B-value of past ocean water have been used either to determine the pH-value (Pearson and Palmer 2000; Pearson et al. 2009), or to determine changes in the boron isotope composition of ocean water (Raitzsch and Hönisch 2014). By using independent estimates of past deep-ocean pH, benthic foraminifera, being less pH affected than planktonic foraminifera, demonstrate oscillations of 2‰ in $^{11}$B-values with a striking $^{11}$B increase of about 3‰ since the Eocene (Raitzsch and Hönisch 2014). Greenop et al. (2017) reached similar conclusions by measuring the difference of $\delta^{11}$B-values between paired planktic and benthic foraminifera.

### 3.8.6 Calcium

Several studies have documented secular changes in the Ca isotope composition of the ocean (De La Rocha and De Paolo 2000; Griffith et al. 2008a, b, c; Steuber and Buhl 2006; Farkas et al. 2007; Fantle 2010; Lau et al. 2017), that indicate a dynamic Ca cycle during earth's history and suggest feedbacks between the Ca and the C-cycle to buffer the oceanic carbon reservoir. Besides changes in the input and output fluxes to the ocean, additional processes might change the Ca isotope composition such as a shift from Early Paleozoic calcitic oceans to late Paleozoic aragonitic oceans or changes in the magnitude of dolomite formation. Similarities in the pattern of the lithium (Misra and Froelich, 2012), boron (Greenop et al. 2017) and calcium (Griffith et al. 2008a; b, c) isotope composition in the Neogene suggest a common mechanism such as an increase in continental weathering. For carbonate-apatites in phosphorites of Cretaceous age, Soudry et al. (2006) interpreted an increase in $\delta^{44}$Ca values over 90 Myr as being due to a decrease in weathering fluxes and an expansion of carbonate sedimentation.

## 3.9 Atmosphere

The basic chemical composition of the atmosphere is quite simple, being made up almost entirely of three elements: nitrogen, oxygen and argon. Other elements and compounds are present in amounts that although small are nevertheless significant. A mixture of gases with different molecular weights should partially segregate and fractionate in a gravity field. However, the lower atmosphere—the troposphere—is much too turbulent for gravitational fractionation to be observed. While it appears possible that certain gases in the upper atmosphere—the stratosphere—could be affected by this process, isotopic evidence for this has not been found so far (Thiemens et al. 1995). (Gravitational fractionation can however be observed in air trapped in ice cores and in sand dunes (Sowers et al. 1992) see p. 44).

In recent years, tremendous progress has been achieved in the analysis of the isotope composition of important trace compounds in the atmosphere, mainly through the introduction of the GC-IRMS technique allowing the precise analysis of nanomole quantities of $O_3$, $CH_4$, $N_2O$, CO, $H_2$ and also sulfate and nitrate. Of special importance is the isotope composition of ozone having a unique composition that affects other trace components.

Trace gases continually break apart and recombine in a multitude of photochemical reactions, which may produce isotope fractionations (Kaye 1987; Brenninkmeijer et al. 2003). Isotope analysis is increasingly employed in studies of the cycles of atmospheric trace gases e.g. $CH_4$ and $N_2O$, giving insights into sources and sinks and transport processes of these compounds. The rationale is that various sources have characteristic isotope ratios and that sink processes are accompanied by isotope fractionation.

Many of the processes responsible for isotope fractionations in the Earth's atmosphere may also occur in the atmospheres of other planetary systems, such as the atmospheric escape of atoms and molecules to outer space. Likely unique to Earth are isotope fractionations related to biological processes or to interactions with the ocean. One aspect of atmospheric reserach which has great potential for the application of stable isotope investigations is the study of anthropogenic pollution.

Compared to the troposphere very different fractionation effects and reactions can be observed in the stratosphere. Due to photochemical isotope fractionations, gases like $CO_2$, $N_2O$, $CH_4$, $H_2$ become highly enriched in heavy isotopes. Of special importance is the isotope composition of stratospheric ozone. In situ mass-spectrometric measurements by Mauersberger (1981, 1987) demonstrated that an equal enrichment in $^{17}O$ and $^{18}O$ exists in the stratosphere, with mass 49 ($^{17}O$-bearing) and mass 50 (mostly $^{18}O$-bearing) roughly following a 1:1 relationship. The isotopic signature of ozone is transferred through exchange reactions to other molecules in the stratosphere, leading to mass-independent signatures in tropospheric gases such as $O_2$ and $CO_2$. Similar effects have also been observed in stratospheric nitrous oxide (Cliff and Thiemens 1997).

Figure 3.33 summarizes mass independent isotope compositions of a number of atmospheric molecules such as ozone, $CO_2$, $N_2O$, and CO (Thiemens 1999, 2006).

### 3.9.1 Atmospheric Water Vapour

While the major compounds nitrogen, oxygen and argon have a constant concentration in the lower part of the atmosphere, water vapour concentrations are highly variable: Craig and Gordon (1965) first measured the isotopic composition of atmospheric water vapour over the North Pacific. Later Rozanski and Sonntag (1982) and Johnson et al. (2001) observed in vertical profiles of troposheric and stratospheric water vapour a gradual depletion of $\delta^2H$ (and $\delta^{18}O$) with increasing altitude up to the tropopause with a reversal in the stratosphere. The depletion trend in the troposphere can be explained by isotope fractionation associated with cloud formation and rainout processes leading to preferential removal of heavy isotopes

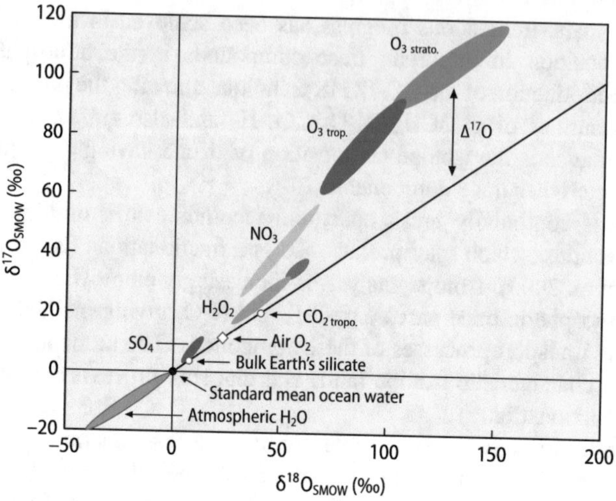


**Fig. 3.33** $\delta^{17}O$ versus $\delta^{18}O$ plot of atmospheric oxygen species (Thiemens 2006)

from water vapour. In the stratosphere photochemical oxidation of methane might be responsible for the observed increase in $\delta^2H$.

In the past, isotope measurements of atmospheric vapour were collected by traditional mass-spectrometric methods involving time-consuming collection of vapour samples by cryogenic technique. Due to the development of infrared laser spectroscopy, isotope compositions of water vapour can be now measured in-situ with high precisions and high temporal resolution (Gupta et al. 2009 and others). Wei et al. (2019) published a global database of water vapour isotopes from 35 worldwide sites presenting hourly $\delta^{18}O$- and $\delta^2H$-values. Such large data sets help to understand moisture paths and convective mixing processes at multiple heights in the atmosphere (i.e. Galewsky et al. 2016). The largest meteorological factor influencing isotope ratios of water vapor is specific humidity. As shown by Guilpart et al. (2017) and others, significant diurnal variations in specific humidity and isotope composition is caused by large-scale atmospheric transport in horizontal and vertical direction.

### 3.9.2 Nitrogen

Nearly 80% of the atmosphere consists of elemental nitrogen. This nitrogen, collected from different altitudes, exhibits a constant isotopic composition (Dole et al. 1954; Sweeney et al. 1978) and represents the "zero-point" of the naturally occuring isotope variations. Besides the overwhelming predominance of elemental nitrogen, there are various other nitrogen compounds in the atmosphere, which play a key role in atmospheric pollution and determining the acidity of precipitation.

Combustion of fossil fuels and biomass form reactive $NO_x$ ($NO + NO_2$); other sources of $NO_x$ include microbial processes in soils, but anthropogenic activities currently dominate NOx production.

Fractionations during the conversion of $NO_x$ to nitrate appear to be small, therefore $\delta^{15}N$-values should reflect sources of $NO_x$. Heaton (1986) has discussed the possibility of isotopically differentiating between naturally produced and anthropogenic $NO_x$. Since very little isotope fractionation is expected at the high temperatures of combustion in power plants and vehicles, the $\delta^{15}N$-value of pollution nitrate is expected to be similar to that of the nitrogen which is oxidized.

In soils, $NO_x$ is produced by nitrification and denitrification processes which are kinetically controlled. This, in principle, should lead to more negative $\delta^{15}N$-values in natural nitrate compared to anthropogenic nitrate. However, Heaton (1986) concluded that this distinction cannot be made on the basis of $^{15}N$-contents, which has been confirmed by Durka et al. (1994).

$^{18}O$ variations in atmospheric nitrate are very large (ranging from +25 to +115‰, Morin et al. 2008; Michalski et al. 2011), and vary during a yearly cycle. Higher $\delta^{18}O$-values are found in wintertime, lower values in summer time. High latitude nitrate has higher $^{18}O$-values than mid-latitude nitrate. Similar trends are observed in $\Delta^{17}O$-values, which indicate a strong mass-independent anomaly derived from exchange with ozone.

#### 3.9.2.1 Nitrous Oxide

Besides $NO_x$ oxides, there is nitrous oxide ($N_2O$), which is of special interest in isotope geochemistry. $N_2O$ is present in air at around 300 ppb and increases by about 0.2% per year. Nitrous oxide is an important greenhouse gas that is, on a molecular basis, a much more effective contributor to global warming than $CO_2$ and has also a major chemical control on stratospheric ozone budgets.

$N_2O$ forms during microbial nitrification and denitrifaction processes in soils and water, the global budget is, however poorly constrained. Nitrification is the main source under aerobic conditions; denitrification under anaerobic conditions. The main sink of $N_2O$ is destruction in the stratosphere by UV photolysis.

The first $\delta^{15}N$-values for $N_2O$ were determined by Yoshida et al. (1984), the first $\delta^{18}O$-values were published by Kim and Craig (1990) and the first dual isotope determinations have been presented by Kim and Craig (1993). $\delta^{15}N$ and $\delta^{18}O$ values of atmospheric $N_2O$ today range from 6.4 to 7.0‰ and 43–45.5‰ (Sowers 2001). First isotope measurements of $N_2O$ from the Vostok ice core by Sowers (2001) indicate large $^{15}N$ and $^{18}O$ variations with time ($\delta^{15}N$ from 10 to 25‰ and $\delta^{18}O$ from 30 to 50‰), which have been interpreted to result from in situ $N_2O$ production via nitrification.

Terrestrial emissions, mainly from soils, have generally lower δ-values than marine sources. As shown by Kool et al. (2009), $\delta^{18}O$ signatures in $N_2O$ are determined by oxygen isotope exchange with ambient water. Due to kinetic effects, production of $N_2O$ from both nitrification and denitrification yields $N_2O$ which is isotopically light relative to its precursors whereas reduction during denitrification results in a $^{15}N$ and $^{18}O$ enrichment in the residual $N_2O$ (Well and Flessa 2009).

Atmospheric nitrous oxide exhibits a small mass-independent $^{17}O$ component (Cliff and Thiemens 1997; Cliff et al. 1999), providing a characteristic isotope signature from $^{17}O$ enriched ozone. $\delta^{15}N$ and $\delta^{18}O$-values of stratospheric $N_2O$ gradually increase with altitude due to preferential photodissociation of the lighter isotopes (Rahn and Wahlen 1997).

There is another aspect that makes $N_2O$ a very interesting compound for isotope geochemists. $N_2O$ is a linear molecule in which one nitrogen atom is at the centre and one at the end. The center site is called α-position, the end site β-position. Yoshida and Toyoda (2000) and Röckmann et al. (2003) showed that the $^{15}N$ content in the two N-positions varies, reactions involved in denitrification result in strong $^{15}N$ enrichment at the central position, while reactions during nitrification result in smaller enrichments (Perez et al. 2006; Park et al. 2011). In contrast to $^{18}O$ and mean $^{15}N$-values, the difference between the $N^{\alpha}$ and the $N^{\beta}$ position is independent of the isotope composition of the precursor (Popp et al. 2002). The uneven intramolecular distribution, thus, may help to identify the sources and sinks of $N_2O$ (see p. on site specific isotope composition).

Because of its asymmetric structure, $N_2O$ reflects a large number of different isotope processes. Magyar et al. (2016) described a method to measure six singly and doubly substituted isotope species of $N_2O$ constraining the values of $\delta^{15}N$, $\delta^{18}O$, $\Delta^{17}O$, $^{15}N$ site preference and the clumped isotopomers $^{14}N^{15}N^{18}O$ and $^{15}N^{14}N^{18}O$. The 6 isotopic variants on a single sample provide additional constraints on the sources and sinks of $N_2O$.

By measuring $N_2O$ extracted from ice cores covering the last 3000 years, Prokopiou et al. (2018) showed that isotopic signatures decreased with age. The $\delta^{18}O$ decrease is larger during the early part, $\delta^{15}N$ decrease is larger during the late part of the industrial period, implying a decoupling of sources over the industrial period.

### 3.9.3 Oxygen

Atmospheric oxygen has a rather constant isotopic composition (Dole et al. 1954; Kroopnick and Craig 1972; Bender et al. 1994) with a $\delta^{18}O$-value of 23.5‰, which, more recently, has been re-determined to be 23.88‰ (Barkan and Luz 2005). Oxygen is produced by photosynthesis without fractionation with respect to the substrate water (Helman et al. 2005). Because the ocean is the largest water reservoir on Earth, the $\delta^{18}O$-value of atmospheric oxygen, therefore, is linked to the seawater composition.

Urey (1947) calculated that under equilibrium conditions atmospheric oxygen should be enriched in $^{18}O$ relative to water by 6‰ at 25 °C. This means atmospheric oxygen cannot be in equilibrium with the hydrosphere and thus the $^{18}O$-enrichment of atmospheric oxygen, the so-called "Dole" effect, must have another explanation. It is generally agreed that the $^{18}O$-enrichment is of biological and kinetic origin and results from the fact that during respiration most species preferentially use $^{16}O$ (Lane and Dole 1956). Oxygen consumed during respiration has a $^{18}O$-content that is about 20‰ lower than the intake of $O_2$ (Guy et al. 1993).

The Dole effect can be separated into terrestrial and oceanic contributions. Bender et al. (1994) estimated that the terrestrial contribution should be 22.4‰ whereas the marine contribution should be 18.9‰. The δ-value should be thus very sensitive to changes in the ratio of marine to terrestrial photosynthesis when the climate shifted from glacial to interglacial periods. As has been shown by the analysis of molecular oxygen trapped in ice cores, the $\delta^{18}O$-value of atmospheric oxygen has indeed varied with geologic time. Sowers et al. (1991), Bender et al. (1994) and Severinghaus et al. (2009) have pioneered the analysis of $\delta^{18}O$ of $O_2$ in air bubbles trapped in ice cores by measuring the difference between the $\delta^{18}O$-value of atmospheric oxygen and ocean water. $\delta^{18}O$-values within glacial-interglacial cycles vary within 1.5‰ and follow the $\delta^{18}O$ value of sea water (Severinhaus et al. 2009).

Further insight into the isotopic composition of atmospheric oxygen comes from the measurement of the $^{17}O$ content having a $\delta^{17}O$-value of 12.03‰ (Luz et al. 1999; Luz and Barkan 2000, 2005; Barkan and Luz 2011). As proposed by Luz et al. (1999) and Luz and Barkan (2000) the $^{17}O$ anomaly can be used as a tracer of global biosphere production rates.

Triple oxygen isotope studies have been successfully applied in two areas: photochemical stratospheric reactions among $O_3$, $O_2$ and $CO_2$ (Thiemens et al. 1995) lead to a mass independent isotope fractionation of tropospheric oxygen, whereas reactions during photosynthesis and respiration fractionate $^{17}O$ and $^{18}O$ in a mass dependent way. As a result, tropospheric oxygen is depleted in $^{17}O$ by about 0.3‰ relative to oxygen affected by photosynthesis and respiration alone. The magnitude of the $^{17}O$ depletion depends on the relative proportions of biological productivity and stratospheric mixing. Young et al. (2014) have considered a multitude of atmospheric reactions that involve large fractionations and mass independent effects related to high $\Delta^{17}O$ ozone formation and concluded that a negative −0.3‰ $\Delta^{17}O$ value has mass-dependent and mass independent contributions.

The $^{17}O$ signal of atmospheric oxygen may be transferred to crustal minerals such as gypsum and barite through oxidative weathering of continental sulfides. Thus, by analysing terrestrial sulfates, a record of $^{17}O$ anomalies through geological history may be obtained (Bao et al. 2000, 2001; Bao 2015; Cao and Bao 2021). Pack (2021) has summarized the list of materials that may carry information about the isotope anomaly of air oxygen such as cosmic spherules, tektites and ferro-manganese nodule.

Additional informations about atmospheric oxygen may be obtained by the measurement of the very rare $^{18}O^{18}O$ and $^{17}O^{18}O$ proportions in tropospheric $O_2$ (Yeung et al. 2012, 2014), showing that clumped isotopes of oxygen do not reflect isotope equilibrium in the troposphere. More recent measurements by Yeung et al. (2015) demonstrated that photosynthetic oxygen is depleted in $^{18}O^{18}O$ and $^{17}O^{18}O$ relative to its stochastic distribution which these authors interpreted as unique biological signatures.

#### 3.9.3.1 Evolution of Atmospheric Oxygen

Geological, mineralogical and geochemical indicators have been used to deduce oxygen levels of past atmospheres. For half of Earth history, oxygen contents probably have been less than 0.001% of the present atmospheric level (PAL). Stable isotope proxies of redox sensitive metals document the oxygenation of the Earth's atmosphere and oceans. The increase in oxygen concentration seems to have occurred in several steps. The first major step occurred at about 2.4 Ga (Farquhar et al. 2000; Farquhar and Wing 2003 and others), the socalled "Great Oxidation Event (GOE)", which is characterized by oxidative weathering. The most convincing argument for the existence of the GOE are mass-independent sulfur isotope fractionations that have persisted until the onset of the GOE.

Recent studies have indicated that the evolution of atmospheric oxygen is more complex than a single stage transition from anoxic in the Archean to oxic in the Paleoproterozoic (Anbar and Rouxel 2007; Wille et al. 2007, 2013; Frei et al. 2009; Voegelin et al. 2010). The Great Oxidation Event seems to be a protracted process rather than a discrete event or with other words a transitional interval with ups and downs of atmospheric oxygen concentrations (Lyons et al. 2014 and others).

The evolution of atmospheric oxygen has been also deduced from redox sensitive isotope systems, i.e. S, Cr, Fe, Se, Mo, U, that have been used to constrain the transition from an anoxic to an oxygenated atmosphere and ocean. Of special importance are iron isotopes. Besides carbon and sulfur, iron as a third element controls the redox chemistry of the ocean. Rouxel et al. (2005) demonstrated a progressive change in iron cycling from 3.5 to 0.5 Ga that was associated with the oxygenation of the ocean (see Fig. 3.34). According to Rouxel et al. (2005) the iron isotope distribution during earth's history can be divided into 3 stages: stage I (2.8–2.3 Ga) is characterized by highly variable and negative $\delta^{56}$Fe values of pyrite, stage II (2.3–1.6 Ga) is characterized by unusually high δ-values and stage III (from 1.6 Ga till today) is characterized by pyrite having a small $\delta^{56}$Fe range from about 0 to −1‰. These different stages might reflect changes in the redox state of the earth. In stage I (older than 2.3 Ga), iron was removed from the ocean as iron oxides and as pyrite. Iron oxides enriched in $^{56}$Fe were precipitated by anaerobic oxidation, which drove the ocean toward lower $\delta^{56}$Fe-values (Kump 2005). In stage II from 2.3 to 1.8 Ga the atmosphere became oxidized, but the ocean remained more or less anoxic. In stage III atmosphere and ocean were oxygenated, ensuring that iron did not accumulate in the ocean, but was removed as insoluble $Fe^{3+}$ that retained the iron isotope composition of the iron inputs to the ocean which are close to the crustal average.

Supporting results have been achieved from other redox-sensitive trace elements and isotope system. Using a range of elemental and isotope proxies (S, Mo, Se, U) the 2.5 Ga Mount McRae shale drillcore—as an example—indicates episodic increases of $O_2$ levels from an anoxic to an oxygenated atmosphere (Anbar et al. 2007; Lyons et al. 2014; Kendall et al. 2017). In the Phanerozoic redox sensitive isotope proxies have been used to trace the extent of oceanic euxinia. A particular promising approach appears the combination of Mo and U isotopes taking advantage of their differing redox response (Kendall et al. 2017; Andersen et al. 2017).

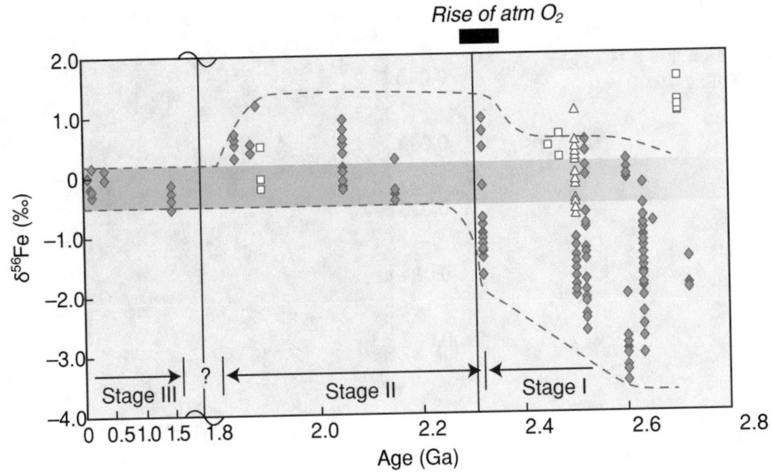


Fig. 3.34 $\delta^{56}Fe$ values of pyrite and iron oxides versus time showing three evolutionary stages of the ocean (Anbar and Rouxel 2007)

## 3.9.4 Carbon Dioxide

### 3.9.4.1 Carbon

The increasing $CO_2$-content of the atmosphere is a problem of world-wide concern. By measuring both the concentration and isotope composition of $CO_2$ on the same samples of air, it is possible to determine whether variations are of anthropogenic, oceanic or biologic origin. Carbon dioxide sequestration is considered an important option to reduce greenhouse-gas emissions. Sedimentary basins in general and deep saline aquifers are regarded as possible repositories for anthropogenic $CO_2$ and the isotopic composition of injected $CO_2$ may provide an ideal tracer for the fate of injected $CO_2$ in the reservoir (Kharaka et al. 2006 and others).

The first extensive measurements of the carbon isotope ratio of $CO_2$ were made in 1955/56 by Keeling (1958, 1961). He noted daily, seasonal, secular, local and regional variations as regular fluctuations. Daily variations exist over continents, which depend on plant respiration and reach a distinct maximum around midnight or in the early morning hours. At night there is a measurable contribution of respiratory $CO_2$, which shifts $\delta^{13}C$-values towards lower values (see Fig. 3.35). Seasonal variations in $^{13}C$ are very similar to $CO_2$-concentrations and result from terrestrial plant activity. As shown in Fig. 3.36 the seasonal cycle diminishes from north to south, as expected from the greater seasonality of plant activity at high latitude and the larger amount of land area in the northern hemisphere. This effect is hardly discernible in the southern hemisphere (Keeling et al. 1989).

Long-term measurements of atmospheric $CO_2$ are available for a few clean-air locations on an almost continuous basis since 1978 (Keeling et al. 1979,1984,1989,1995,2017; Mook et al. 1983; Ciais et al. 1995). These

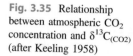

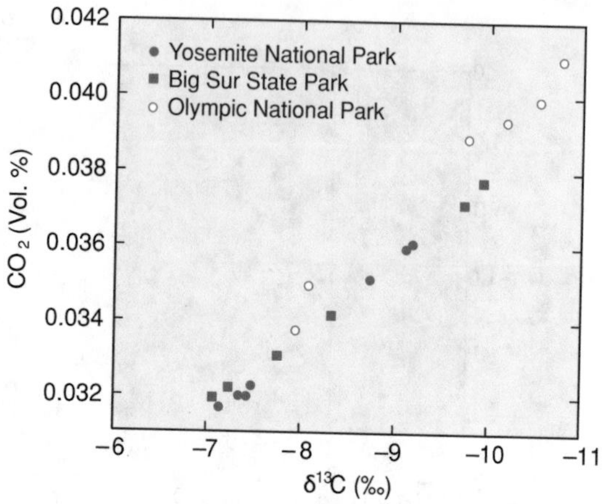


**Fig. 3.35** Relationship between atmospheric $CO_2$ concentration and $\delta^{13}C_{(CO2)}$ (after Keeling 1958)

measurements clearly demonstrate that on average atmospheric $CO_2$ increases by about 1.5 ppm per year while the isotope ratio shifts towards lower $^{13}C/^{12}C$ ratios. The annual combustion of $10^{15}$ g of fossil fuel with an average $\delta^{13}C$-value of −27‰ would change the $^{13}C$-content of atmospheric $CO_2$ by −0.02‰ per year. The observed change is, however, much smaller. Of the $CO_2$ emitted into the atmosphere roughly half remains in the atmosphere and the other half is absorbed into the oceans and the terrestrial biosphere. The partitioning between these two sinks is a matter of debate. Whereas most oceanographers argue that the oceanic sink is not large enough to account for the entire absorption, terrestrial ecologists doubt that the terrestrial biosphere can be a large carbon sink. Including more recent $CO_2$ records, Keeling et al. (2017) postulated that no combination of sources and sinks of $CO_2$ from fossil fuels, land and oceans can explain the observed decrease in $^{13}C/^{12}C$ ratios of atmospheric $CO_2$, unless the isotope discrimination of land plants has increased.

#### 3.9.4.2 Oxygen

Atmospheric $CO_2$ has a $\delta^{18}O$-value of about +41‰, which means that atmospheric $CO_2$ is in approximate isotope equilibrium with ocean water, but not with atmospheric oxygen (Keeling 1961; Bottinga and Craig 1969). Measurements by Mook et al. (1983) and Francey and Tans (1987) have revealed large-scale seasonal and regional variations. There is a North–South shift in $^{18}O$-contents of almost 2‰ increasing towards the south, about ten times larger than for $^{13}C$. Seasonal cycles are similar in magnitude to those of $\delta^{13}C$ (see Fig. 3.37). This north–south gradient is caused by the unequal distribution of ocean and land between the two hemispheres and by the very different oxygen isotope composition of ocean and meteoric water.

Farquhar et al. (1993) demonstrated that much more $CO_2$ comes into contact with leaf water than is actually taken up by plants during photosynthesis. For every $CO_2$ molecule that is taken up by photosynthesis, two others enter the leaf through

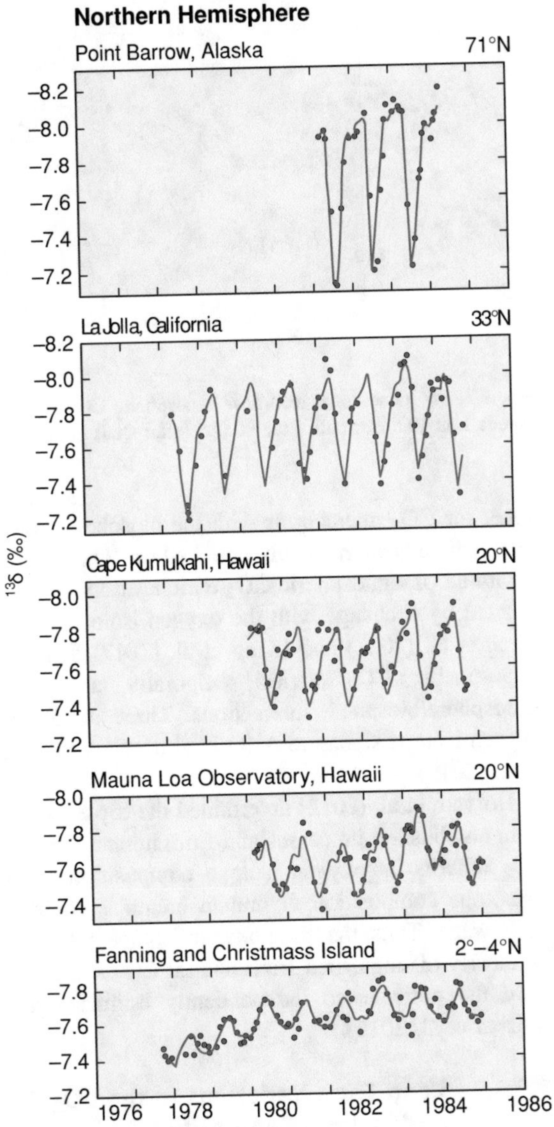


**Fig. 3.36** Seasonal $\delta^{13}C$ variations of atmospheric $CO_2$ from five stations in the Northern Hemisphere. *Dots* denote monthly averages, *oscillating curves* are fits of daily averages (after Keeling et al. 1989)

the stomata. They rapidly equilibrate with the leaf water and then diffuse back to the atmosphere without having been incorporated by the plant. This large flux therefore only influences the $^{18}O$ content of atmospheric $CO_2$, but has no influence on the $\delta^{13}C$-value.

Additional insight into the cycling of atmospheric $CO_2$ can be gained by the analysis of triple oxygen isotopes. Hoag et al. (2005) suggested that the determination of the $^{17}O$ content besides the $^{18}O$ content of tropospheric $CO_2$ might be a

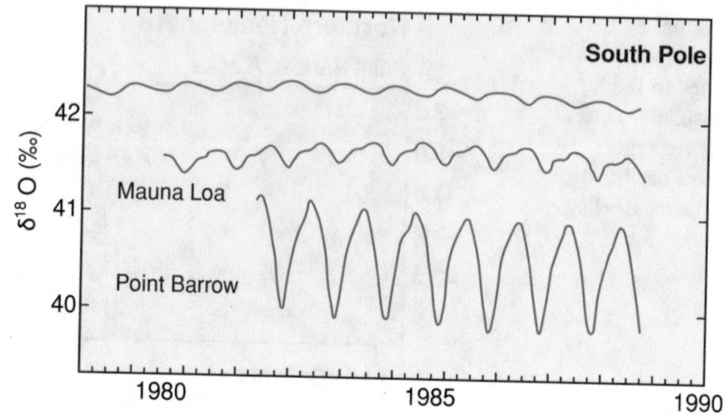


Fig. 3.37 $\delta^{18}O$ seasonal record of atmospheric $CO_2$ from three stations: Point Barrow 71.3°N, Mauna Loa 19.5°S, South Pole 90.0°S (after Ciais et al. 1998)

tracer for $CO_2$ interactions with the biosphere and the hydrophere. As $CO_2$ in the stratosphere is anomaleouly enriched in $^{17}O$ and $^{18}O$ through exchange with ozone, the influx of stratospheric $CO_2$ enriches tropospheric $CO_2$ in $^{17}O$ and $^{18}O$ which is resetted by exchange with the oxygen isotope composition of surface water. Hofmann et al. (2017) and Liang et al. (2017) demonstrated that the $\Delta^{17}O$ signal of tropospheric $CO_2$ varies temporally and spatially being controlled by atmosphere/biosphere interactions. Three dimensionsal simulations of the triple oxygen isotope signature $\Delta^{17}O$ in atmospheric $CO_2$ have been presented by Koren et al. (2019).

Horvath et al. (2012) determined the triple oxygen isotope composition of $CO_2$ from combustion processes and from human breath. High-temperature combustion $CO_2$ inherits its oxygen isotope composition from ambient air $O_2$, wheras the O-isotope composition of human breath is controlled by isotope exchange with body water. Thus, the triple oxygen isotope composition of anthropogenic $CO_2$ can be clearly distinguished from natural $CO_2$ sources. Clumped isotopes, in addition have the potential to independently estimate anthropogenic $CO_2$ contributions (Laskar et al. 2016).

#### 3.9.4.3 Long Term Variations in the $CO_2$ Concentration and Isotope Composition

There is increasing awareness that the $CO_2$ content of the Earth's atmosphere has varied considerably over the last 500 Ma. The clearest evidence comes from measurements of $CO_2$ from ice cores, which have yielded an impressive record of $CO_2$ variations over the past 800,000 years.

In a much broader context, Berner (1990) has modeled how long-term changes in $CO_2$ concentrations can result from the shifting balance of processes that deliver $CO_2$ to the atmosphere (such as volcanic activity) and processes that extract $CO_2$ (such as weathering and the deposition of organic material). The carbon dioxide curve calculated for the past 500 Ma matches the climate record at several key points: it is low during the ice age of the Carboniferous and Permian and rises to a maximum in the Cretaceous. Although the exact curve is far from being known, it is clear that fluctuations in the $CO_2$ content of the ancient atmosphere may have played a critical role in determining global surface paleotemperatures. To elucidate these short- and long-term $CO_2$-fluctuations, several promising "$CO_2$-paleobarometers" use variations of carbon isotopes in different materials.

Short-term carbon isotope variations in tree rings have been interpreted as indicators of anthropogenic $CO_2$ combustion (Freyer 1979; Freyer and Belacy 1983). While different trees show wide variability in their isotope records due to climatic and physiological factors, many tree-ring records indicate a 1.5‰ decrease in $\delta^{13}$C-values from 1750 to 1980. Freyer and Belacy (1983) reported C-isotope data for the past 500 years on two sets of European oak trees: forest trees exhibit large non-systematic $^{13}$C variations over the 500 years, whereas free-standing trees show smaller $^{13}$C fluctuations, which can be correlated to climatic changes. For the free-standing trees, the $\delta^{13}$C record is characterized by a systematic decrease of about 2‰ since the industrialization around 1850.

The most convincing evidence for changes in atmospheric $CO_2$-concentrations and $\delta^{13}$C-values comes from air trapped in ice cores in Antarctica. Figure 3.38 shows a high time-resolution record for the last 1000 years from analysis of the Law Dome, Antarctica ice core (Trudinger et al. 1999). Changes in $CO_2$ concentration and in $\delta^{13}$C values during the last 150 years are clearly related to the increase of anthropogenic fossil fuel burning. During the last ice age with low $CO_2$-concentrations, atmospheric $CO_2$ was isotopically lighter by about 0.3‰ relative to interglacial periods (Leuenberger et al. 1992). Schmitt et al. (2012) presented $\delta^{13}$C-data for the past 24,000 years from two Antarctic ice cores and observed a 0.3‰ decrease from about 17,500–14,000, a time where $CO_2$ concentrations rose, which they interpreted as being due to upwelling of $CO_2$-enriched waters in the southern ocean.

A high-resolution high precision deglacial record from Taylor Glacier, Antarctica spanning the age interval from 22,000 to 11,000 years has been presented by Bauska et al. (2016). An initial increase in $CO_2$ concentrations is marked by a decrease in $\delta^{13}$C-values which might be due to a weakened biological pump whereas the continuing increase in $CO_2$ concentrations is associated with small changes in $^{13}$C-contents suggesting a combination of sources and processes (Bauska et al. 2016).

Two different classes of approaches have been used in the study of long-term atmospheric $CO_2$ change: (i) one utilizing deep-sea sediments, (ii) the other studying continental sediments. (I) This approach uses the relationship between the concentration of molecular $CO_2$ and the $\delta^{13}$C-value of marine organic plankton (Rau et al. 1992). Attempts to quantify the relationship between $CO_{2(aq)}$ and $\delta^{13}C_{org}$ have resulted in several empirically derived calibrations (Jasper and Hayes 1990; Jasper et al. 1994; Freeman and Hayes 1992 and others). Theoretical considerations and experimental work demonstrated that cellular growth rate (Laws et al. 1995; Bidigare et al. 1997) and cell geometry (Popp et al. 1998) also exert considerable

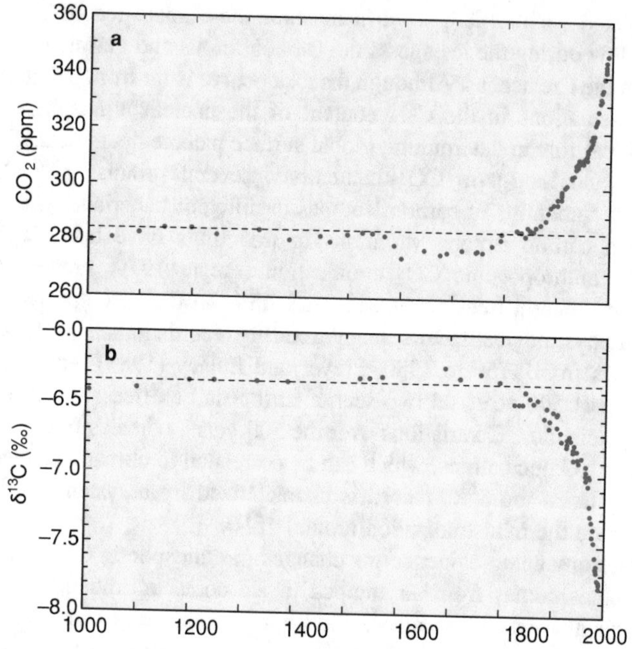


Fig. 3.38 Law Dome ice core $CO_2$ and $\delta^{13}C$ record for the last 1000 years (after Trudinger et al. 1999)

control on $\delta^{13}C_{org}$, insofar as they influence the intracellular $CO_2$ concentration. Other complicating factors are potential contamination of terrestrial organic matter and marine photosynthesizers with varying carbon fixation pathways that are integrated in bulk organic matter. Therefore, it is preferable to use specific biomarkers, such as alkenones. Alkenones are long-chain ($C_{36}$–$C_{39}$) unsaturated ketones, produced by a few taxa of phytoplankton such as the common *Emiliani huxleyi*, in which the number of double bonds is correlated with the water temperature at the time of synthesis. Palaeo-$CO_2$ levels can be estimated from the carbon isotope composition of alkenones and coeval carbonates (Jasper and Hayes 1990; Pagani et al. 1999a, b). (II) On the continents, a comparable approach analyzes the bulk $\delta^{13}C$-value of C3 land plants or the $\delta^{13}C$-value of selected plant lipids (Schubert and Jaren 2012; Chapman et al. 2019). This approach assumes that $CO_2$ concentrations affect the C isotope composition of plant tissues via photorespiration. Cui et al. (2020) presented, as an example, a $CO_2$ record across the past 23 m.y., but this method is applicable, in principle, to the entire 400 m.y. history of C3-photosynthesis.

Another alternative to reconstruct the $CO_2$ content of the ancient atmosphere is by analyzing fossil soil carbonate that formed from $CO_2$ diffusion from the atmosphere or plant roots (Cerling 1991). This method also relies on certain assumptions and prerequisites. One, for instance, is the necessity of differentiating pedogenic calcretes from those formed in equilibrium with groundwater, which cannot be used for $p_{CO2}$ determinations (Quast et al. 2006).

The boron isotope approach (see Sect. 2.3.2) to estimate $pCO_2$ relies on the fact that a rise in the atmospheric $CO_2$ concentration will increase $pCO_2$ of the surface ocean which in turn causes a reduction of pH. By measuring the boron isotope composition of planktonic foraminifera Palmer et al. (1998) and Pearson and Palmer (2000) have reconstructed the pH-profile of Eocene seawater and estimated past atmospheric $CO_2$ concentrations. However, Lemarchand et al. (2000) argued that $\delta^{11}B$ records of planktonic foraminifera partly reflect changes in the marine boron isotope budget rather than changes in ocean pH.

### 3.9.5 Carbon Monoxide

Carbon monoxide is an important trace gas, which has a mean residence time of about 2 months and a mean concentration of the order of 0.1 ppm. The principal sources of CO are (i) oxidation of methane and other higher hydrocarbons, (ii) biomass burning, (iii) traffic, industry and domestic heating, (iv) oceans and (v) vegetation. The dominant sinks are (i) in situ oxidation by hydroxyl radical (OH), which is responsible for the removal of contaminant gases from the troposphere and (ii) uptake by soils. The first isotope data on CO have been presented by Stevens et al. (1972), which have later been confirmed by Brenninkmeijer (1993) and Brenninkmeijer et al. (1995). Seasonal variations in $\delta^{13}C$ values appear to reflect a shift in the relative contributions from two major sources, biomass burning and atmospheric oxidation of methane. $\delta^{18}O$-values are even more variable than $\delta^{13}C$ due to a kinetic isotope effect accompanying the removal of CO from the atmosphere. Oxygen in CO also exhibits a mass independent fractionation with a pronounced $^{17}O$ excess of up to 7.5‰, which must be related to the removal reaction with OH (Röckmann et al. 1998).

Röckmann et al. (2002) measured the complete isotope composition of CO from high northern latitude stations. $\delta^{13}C$, $\delta^{17}O$ and $\delta^{18}O$ values show strong seasonal variations and indicate mixing between mid and high northern latitude CO. In winter high amounts of combustion CO from industrial regions are transported to high latitudes. Large variations in C- and O- isotope compositions have been observed in CO from ice cores, which have been interpreted as being due to changes in biomass burning (Wang et al. 2011a, b).

### 3.9.6 Methane

Based on the concentration measured in air contained in polar ice cores, methane concentrations have more than doubled over the past several hundred years (Stevens 1988). Concentrations were increasing at almost 1% per year in the late 70 s and early 80 s, the growth rate has slowed down then for some years, but more recently increases again. This has lead to controversial discussions about the causes (i.e. Turner et al. 2019).

Methane enters the atmosphere from biological and anthropogenic sources and is destroyed by reaction with the hydroxyl radical. Thus, a mass-weighted average composition of all $CH_4$ sources is equal to the mean $\delta^{13}$C-value of atmospheric methane, corrected for any isotope fractionation effects in $CH_4$ sink reactions. Atmospheric methane has a mean $\delta^{13}$C-value of around −47‰ (Stevens 1988). Methane extracted from air bubbles in polar ice up to 350 years in age has a $\delta^{13}$C-value which is 2‰ lower than at present (Craig et al. 1988). Sowers (2010) presented a dual record covering the Holocene. $\delta^{13}$C-values decrease from −46.4‰ at 11,000 to −48.4‰ at 1000, δD-values shift by 20‰ between 4000 and 1000.

Quay et al. (1999) presented global time series records between 1988 and 1995 on the carbon and hydrogen isotope composition of atmospheric methane. They measured spatial and temporal variation in $^{13}$C and D with a slight enrichment observed for the southern hemisphere relative to the northern hemisphere. Since 2007 the $\delta^{13}$C-value of methane is changing slightly to lighter values (from −47.1 to −47.3‰) which indicates a change in the relative proportions of emissions from biogenic, thermogenic and pyrogenic sources and/or a decline in the sinks (Nisbet et al. 2019).

Sherwood et al. (2017) presented a global database of carbon and hydrogen isotope compositions of methane liberated from fossil fuel, microbial and biomass burning sources. Fossil fuel sources have been divided into conventional gas, shale gas and coal; microbial sources have been subdivided into wetlands, rice paddies, ruminants, termites, landfills and wastes, summing up in an estimated mean $\delta^{13}$C-value of −53.6‰, which differs from the actual measured value of −47.3‰. The difference is due to isotope fractionations during photochemical destruction of methane. For hydrogen, estimates are much more uncertain, with mean $\delta^{2}$H-values between −245 and −415‰ compared to a measured value of −95‰. Again, the difference is due to photochemical isotope fractionation.

Stratospheric methane collected over Japan gave a $\delta^{13}$C-value of −47.5‰ at the tropopause and increased to −38.9‰ at around 35 km (Sugawara et al. 1998). These authors suggested that reaction with Cl in the stratosphere might be responsible for the $^{13}$C-enrichment.

### 3.9.7 Hydrogen

Molecular hydrogen ($H_2$) is after methane the second most abundant reduced gas in the atmosphere with an average concentration of 0.53 ppm (Ehhalt and Rohrer 2009). Although hydrogen distribution is rather uniform, the concentration in the southern hemisphere is around 3% higher than in the northern hemisphere. The isotope geochemistry of hydrogen in the atmosphere is very complex, because there are numerous hydrogen-containing compounds undergoing continuous chemical and physical transformations.

Hydrogen sources are the photo-oxidation of methane and other hydrocarbons and combustion processes (biomass and fossil fuel burning), sinks are soil uptake and oxidation by hydroxyradicals. Due to the large mass difference between H and D, large isotope fractionations occur in the processes that produce or remove hydrogen. Of special importance are kinetic isotope effects during soil uptake of atmospheric hydrogen (Rice et al. 2011). Photochemical sources of $H_2$ lead to $\delta^2$H-values between +100 to +200‰ (Rahn et al. 2003). Hydrogen from fossil fuel combustion and biomass burning yields $\delta^2$H-values between −200 and −300‰, even more depleted $\delta^2$H-values have been observed in oceanic derived dissolved hydrogen which depend on water temperature (Walter et al. 2016). The major result from these studies is that there are large seasonal and latitudinal variations in deuterium content with higher $\delta^2$H-values in the southern hemisphere than in the northern hemisphere (Batenburg et al. 2011).

Summarizing, source and sink processes lead to $\delta^2$H-values for tropospheric hydrogen of +130‰ (Gerst and Quay 2001; Batenburg et al. 2011). Considering low $\delta^2$H-values of $H_2$ sources during bacterial processes, fossil fuel combustion and biomass burning (−250‰ and lower), the D-enrichment of atmospheric hydrogen is difficult to explain. One way is to attribute the enrichment with a kinetic fractionation during reaction with OH, the other is photochemical production of $H_2$ from methane and higher hydrocarbons.

Extreme D enrichments in $H_2$ have been found in stratospheric air samples (Rahn et al. 2003). $\delta^2$H-values vary up to +440‰, representing the most D-enriched natural material on Earth.

### 3.9.8 Sulfur

Sulfur is found in trace compounds in the atmosphere, where it occurs in aerosols as sulfate and in the gaseous state. Sulfur can orginate naturally (volcanic, sea spray, aeolian weathering, biogenic) or anthropogenically (combustion and refining of fossil fuels, ore smelting, gypsum processing). These different sources differ greatly in their isotopic composition as shown in Fig. 3.39. The complexities involved in the isotopic composition of atmospheric sulfur have been discussed in the SCOPE 43 report, edited by Krouse and Grinenko (1991). The isotopic compositions of the industrial sulfur sources are generally so variable, that the assessment of anthropogenic contributions to the atmosphere is extremely difficult. Krouse and Case (1983) were able to give semiquantitative estimates for a unique situation in Alberta where the industrial $SO_2$ had a constant $\delta^{34}$S-value near 20‰. Generally, situations are much more complicated which limits the "fingerprint" character of the sulfur isotope composition of atmospheric sulfur to such rare cases.

Seasonal dependencies for sulfur in precipitation and in aerosol samples have been observed by Nriagu et al. (1991). $\delta^{34}$S-data for aerosol samples of the Canadian arctic show pronounced $^{34}$S enrichments in summer compared to winter. This situation is quite different from that observed for airborne sulfur in southern Canada. In rural and remote areas of southern Canada, the $\delta^{34}$S-values of

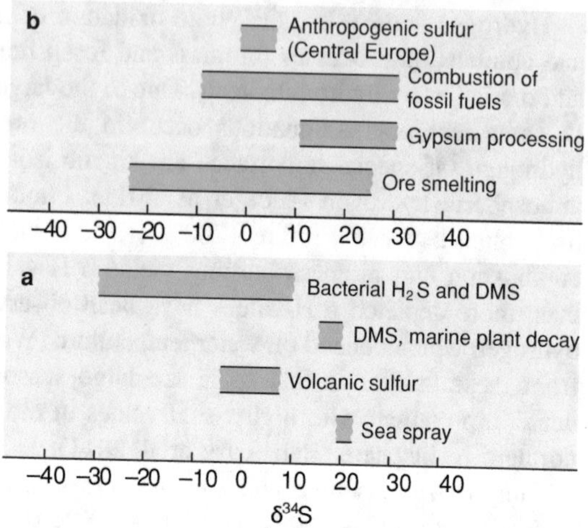


**Fig. 3.39** S-isotope composition of **a** natural and **b** anthropogenic sulfur sources in the atmosphere. *DMS* Dimethylsulfide

atmospheric samples are higher in winter and lower in summer. While during the winter sulfur is mainly derived from sources used for heating and industrial sources, in summer the large emission of $^{34}$S-depleted biogenic sulfur from soils, vegetation, marshes, and wetlands results in the lowering of the $\delta^{34}$S-values of airborne sulfur. The opposite trend observed for aerosol sulfur in the Arctic suggests a different origin of the sulfur in these high latitude areas.

The major sulfur gas in the atmosphere is carbonyl sulfide (COS) with a life time of a few years. Sulfur isotope measurements of atmospheric COS is challenging, because atmospheric concentrations are around 0.5 ppb. Angert et al. (2019) described a method to measure COS in air. Davidson et al. (2021) reported a $\delta^{34}$S-value of 13.9 ± 0.1‰ for tropospheric COS. Air samples influenced by anthropogenic activity show a $\delta^{34}$S-value of 8 ± 1‰.

The main natural source of COS is the ocean, either as COS emission or as dimethyl sulfide (DMS) emissions that rapidly oxidize to COS. DMS represents the largest natural source of oceanic biogenic sulfur to the atmosphere. Sulfur isotope ratios of DMS and its precursor dimethylsulfoniopropionate (DMSP) collected from different oceanic regions revealed a rather homogeneous range from +18.9 to +20.3‰ (Amrani et al. 2013).

In aerosols, multiple sulfur isotope compositions are very suitable to constrain potential sulfur sources and formation processes of sulfate aerosols. As shown by Han et al. (2017) $\Delta^{33}$S-values in aresole sulfate from Beijing, China show a pronounced seasonality with positive values in spring, summer and autumn and negative values in winter. Positive values may reflect air mass transport between the troposphere and the stratosphere; negative values in winter are probably related to incomplete combustion of coal during the heating season (Han et al. 2017).

Shaheen et al. (2014) measured large S-isotope mass-independent anomalies in Antarctic snow unaffected by volcanic activity, which indicate that tropospheric sulfate produced during fossil-fuel and biomass burning contributes to the stratospheric sulfate aerosol layer. Large sulfur isotope anomalies have been also found in sulfate aerosols of volcanic origin extracted from snow by Baroni et al. (2007). By extracting age-related sulfate from the Antartic ice sheet, Baroni et al. (2007) demonstrated that sulfate from the Agung and Pinatubo eruptions exhibit large mass-independent sulfur isotope fractionations. The sign of the $\Delta^{33}S$ changed over time from an initial positive component to a negative component, which indicates a fast process during photochemical oxidation of $SO_2$ to sulfuric acid on a time scale of months. Martin (2018) demonstrated that multi-isotope S- and O-isotope analyses of volcanic sulfate allow the distinction of different sulfur oxidation pathways in the atmosphere. Burke et al. (2019) observed distinct differences between the S-isotope signatures of tropical versus non-tropical eruptions.

### 3.9.9 Perchlorate

Perchlorate has been detected in soils, waters, plants and food in concentrations that may cause health problems. $\delta^{37}Cl$-, $\delta^{18}O$- and $\Delta^{17}O$-values vary for individual chlorate sources (Bao and Gu 2004; Böhlke et al. 2005). Perchlorate in the environment is either of man-made or of natural origin. Synthetic perchlorate is used as a constituent of explosives, missiles and rockets or in car airbags. Natural perchlorate is a minor component in hyperarid salt deposit, such as the Atacama dersert. Synthetic perchlorate is produced by electrolytic reactions from aqueous chloride. Its $\delta^{37}Cl$-values vary around zero ‰ like natural Cl-sources. $\delta^{18}O$-values range from about −25 to −12‰ and are much lower than presumed water sources indicating isotope fractionations during synthesis. Natural perchlorate shows the lowest $\delta^{37}Cl$-values reported so far (Bao and Gu 2004; Böhlke et al. 2005), indicating fractionations during formation in the atmosphere. $\delta^{18}O$-values vary too, but the most prominent property are large positive $^{17}O$-anomalies reflecting photochemical reactions of atmospheric Cl-species with ozone. In contrast to natural perchlorates, synthetic perchlorates show no $\Delta^{17}O$-anomalies. Thus, the combination of $\delta^{37}Cl$- and $\Delta^{17}O$-values allow a clear distinction between natural and anthropogenic perchlorates.

Sturchio et al. (2007, 2014) have applied the characteristic isotopic differences to trace the origin of perchlorate in groundwater. Abnormally high concentrations of perchlorate in groundwaters of the Pomona district, southern California can be traced to fertilizer shipped from the Atacama desert by a mining company that sold nitrate fertilizer containing perchlorate as an impurity to citrus farmers. The case has lead to a lawsuit in which the City of Pomona wanted the costs for a cleanup from the mining company. The case was brought to court in 2010, in 2015 the court rejected the claim of Pomona despite convincing scientific proof.

### 3.9.10 Metal Isotopes

Trace metals in aerosols originate from natural and anthropogenic sources, the latter is more abundant than the former. Source identification of metals in atmospheric particulates is thus of prime importance for air quality programmes. Schleicher et al. (2020) determined Zn–Cu isotope compositions in urban aerosols and natural dust materials. They summarized natural mineral dust data from deserts in Asia and Africa and from anthropogenic sources in Beijing and Xian. The main sources of particulate matter in big cities are emissions from fossil fuel combustion, exhaust and non-exhaust traffic sources (brake and tire abrasion), industrial processes, domestic heating and last but not least biomass burning. The variability of sources of urban aerosols in cities results in a characteristic signature. Airborne particles collected at different times in London and Barcelona showed characteristic isotope patterns (Ochoa Gonzalez et al. 2016). For Zn, non-exhaust emissions from verhicles are the major source of pollution; for Cu, fossil fuel combustion is a very important source.

## 3.10 Biosphere

As used here, the term "biosphere" includes the total sum of living matter—plants, animals, and microbial biomass and the residues of the living matter in the geological environment such as coal and petroleum. A fairly close balance exists between photosynthesis and respiration, although over the whole of geological time respiration has been exceeded by photosynthesis, and the energy derived from this is stored mostly in disseminated organic matter, and, of course, in coal and petroleum.

Photosynthesis is responsible for isotope fractionations in the biosphere, not only for carbon, but also for hydrogen and oxygen. Nevertheless, the transformation of biogenic matter to organic matter in sediments also involves isotope fractionations, occurring in two stages: a biochemical and a geochemical stage. During the biochemical stage microorganisms play the major role in reconstituting the organic matter. During the geochemical stage, increasing temperature and to a much lesser extent pressure are responsible for the further transformation of organic matter (Galimov 2006 and others).

### 3.10.1 Living Organic Matter

#### 3.10.1.1 Bulk Carbon

Wickman (1952) and Craig (1953) were the first to demonstrate that marine plants are about 10‰ enriched in $^{13}C$ relative to terrestrial plants. Since that time numerous studies have broadened this view and provided a much more detailed account of isotope variations in the biosphere. The reason for the large C-isotope differences found in plants were only satisfactorily explained after the discovery of new photosynthetic pathways in the 1960s. The majority of land plants (80–90%) employ the C3 (or Calvin) photosynthetic pathway which results in organic carbon approximately 18‰ depleted in $^{13}C$ with respect to atmospheric $CO_2$. $\delta^{13}C$-values of $C_3$ plants vary between −22 and −36‰ with a mean value around −27‰ (Kohn 2010). Heavy values are preferentially found in arid regions whereas light values are found in tropical forests. Although $\delta^{13}C$-values depend on many environmental factors, the most important is the annual precipitation rate showing an increase in $\delta^{13}C$-values with decreasing precipitation rates.

$\delta^{13}C$-values in plants also depend, as shown above, on the $\delta^{13}C$ of atmospheric $CO_2$. There is an ongoing debate whether $\delta^{13}C$-values depend on $CO_2$ concentration (Hare et al. 2018). While Feng and Epstein (1995) and Schubert and Jahren (2012) besides others have shown that increasing $pCO_2$ increases carbon isotope fractionation, Kohn (2016) observed no $pCO_2$ effect.

Around 10–20% of carbon uptake by modern land plants is via C4 (or Hatch-Slack) photosynthesis with a carbon isotope fractionation of only 6‰ on average. The C4 pathway is thought to represent an adaptation to $CO_2$ limited photosynthesis, which developed relatively late in the Earth's history. It is advantageous under warm, dry and saline environmental conditions. Differences in the isotope composition of C3 and C4 plants are widely used as a palaeoenvironmental indicator to trace climatic changes or changes in the diet of animals and humans. Ecosystems with abundant C4 biomass have been documented only from the late Neogene to the present (Cerling et al. 1993, 1997). In South Asia, isotopic records from soil carbonates and tooth enamel reveal a dramatic increase in the abundance of C4 plants at 7 ± 1 million years ago (Quade et al. 1992; Quade and Cerling 1995 and others).

One of the most important groups of all living matter is marine phytoplankton. Natural oceanic phytoplankton populations vary in $\delta^{13}C$-value by about 15‰ (Sackett et al. 1973; Wong and Sackett 1978). Rau et al. (1982) demonstrated that different latitudinal trends in the carbon isotope composition of plankton exist between the northern and the southern oceans: south of the equator the correlation between latitude and plankton $^{13}C$-content is significant, whereas a much weaker relationship exists in the northern oceans.

The unusual $^{13}C$ depletion in high latitude Southern Ocean plankton has been puzzling for years. Rau et al. (1989, 1992) found a significant inverse relationship between high-latitude $^{13}C$-depletion in plankton and the concentration of molecular $CO_2$ in surface waters. Thus, it has been assumed that the major factor controlling the C isotope composition of phytoplankton is the availability of aqueous dissolved $CO_2$. However, as has been shown in culture experiments with marine microalgae (Laws et al. 1995; Bidigare et al. 1997; Popp et al. 1998) the carbon isotope composition of phytoplankton depends on many more factors including cell wall permeability, growth rate, cell size, the ability of the cell to actively assimilate inorganic carbon and the influence of nutrients on cell growth. Therefore, estimates of paleo-$CO_2$ concentrations based on the C-isotope composition of marine organic matter need to consider the paleoenvironmental conditions at the time of phytoplankton production, which are difficult to constrain for the geologic past.

Organic material that comprises living matter consists of carbohydrates (saccharides, "Sacc")—the first product of carbon fixation—and proteins ("Prot"), nucleic acids ("NA") and lipids ("Lip") with prevailing regularities within these compound classes:

$\delta$NA ~ $\delta$Prot,
$\delta$Prot − $\delta$Sacc ~ −1‰ and
$\delta$Lip − $\delta$Sacc ~ −6‰ (Hayes 2001).

What is known for a long time is that lipids are depleted in $^{13}C$ by 5–8‰ relative to the bulk biomass. On the other hand, the carbohydrate fraction of various organisms is on average enriched in $^{13}C$ by 4.6‰ relative to the bulk (Teece and Fogel 2007). Even larger variations are observed for individual amino acids (Abelson and Hoering 1961) and individual carbohydrates (Teece and Fogel 2007), where variations are associated with different metabolic pathways during their synthesis.

The $\delta^{13}C$-value of the total marine organic matter represents a mixed isotope signal derived from land plant detritus, primary production by aquatic organisms and microbial biomass. The possibility of analyzing individual components has refined the interpretation of bulk $\delta^{13}C$-data. Compound-specific isotope analyses allow the resolution of the isotopic composition of material derived from primary sources from that of secondary inputs. These source-specific molecules have become known as biomarkers, which are complex organic compounds derived from living organisms, showing little structural difference from their parent biomolecules, being not affected by diagenesis as long as the basic biological structure is preserved. Due to the specificity of their origin, biomarkers allow for an investigation of the extent to which various organisms contribute organic materials to complex mixtures. In the Messel Shale, Freeman et al. (1990) observed C-isotope variations of individual compounds between −73.4 and −20.9‰ (see Table 3.4). This large range can be interpreted as representing a mixture of secondary, bacterially mediated processes and primary producers. While the major portion of the analyzed hydrocarbons reflect the primary biological source material, some hydrocarbons having low concentrations are extremely $^{13}C$ depleted indicating their secondary microbial origin in a methane-rich environment. Later studies by Summons et al. (1994), Thiel et al. (1999), Hinrichs et al. (1999) and Peckmann and Thiel (2005) clearly suggested that fermentative and chemoautotrophic organisms must have made significant contributions to total sedimentary organic matter. For example, extremely depleted $\delta^{13}C$-values as low as −120‰ of specific biomarkers indicate that $^{13}C$-depleted methane must be the carbon source for the respective archaea rather than the metabolic product. In another example Schoell et al. (1994) demonstrated that steranes and hopanes can be used as a monitor of water depth. These authors showed that $\delta^{13}C$-values of $C_{35}$ hopanes and the $\Delta$-difference between steranes and hopanes follow the climatic evolution of the Miocene in the Pacific Ocean.

**Table 3.4** $\delta^{13}C$-values of separated individual hydrocarbons from the Messel shale (Freeman et al.1990)

| Peak | $\delta^{13}C$ | Compound |
|---|---|---|
| 1 | −22.7 | Norpristane |
| 2 | −30.2 | C19 acyclic isoprenoid |
| 3 | −25.4 | Pristane |
| 4 | −31.8 | Phytane |
| 5 | −29.1 | C23 acyclic isoprenoid |
| 8 | −73.4 | C32 acyclic isoprenoid |
| 9 | −24.2 | Isoprenoid alkane |
| 10 | −49.9 | 22, 29, 30-trisnorhopane |
| 11 | −60.4 | Isoprenoid alkane |
| 15 | −65.3 | 30-norhopane |
| 19 | −20.9 | Lycophane |

#### 3.10.1.2 Position Specific Isotope Composition

As has been demonstrated by Abelson and Hoering (1961), DeNiro and Epstein (1977) and Monson Hayes (1980) adjacent carbon positions within fatty and amino acids can differ by up to 30‰. Blair et al. (1985) showed that the methyl ($CH_3$ group) and the carboxyl group (COO) in acetate differ in $\delta^{13}C$ values by up to 20‰. Such differences obviously reflect isotope effects associated with biosynthetic pathways. In recent years it has become possible to measure the C and H isotope composition at the site-specific level (Eiler 2013; Eiler et al. 2014) using a high resolution IRMS instrument. Piasecki et al. (2016, 2018) analyzed propane which is the simplest organic molecule that could record site-specific carbon isotope variations. They demonstrated that propane inherits a site-specific structure from its precursors and records the mechanisms of cracking reactions. In propane there are two singly substituted carbon groups, a terminal ($^{13}CH_3–CH_2–CH_3$) and a central group ($CH_3–^{13}CH_2–CH_3$). As Gilbert et al. (2016, 2019) showed the difference in carbon isotope composition between terminal and central group range from −1.8 to −12.9‰, varying widely for different propane sources. Anaerobic bacterial oxidation leads to a pronounced $^{13}C$ enrichment in the central position of propane which contrasts with the isotope signature of thermogenic produced propane (Gilbert et al. 2019). Anaerobic bacterial degradation of propane thus can be detected in natural gas reservoirs.

Xie et al. (2018) investigated the D/H fractionation between the central and the terminal position of propane. They observed large temperature dependent isotope differences between the two positions which indicate that intramolecular hydrogen exchange under anhydrous conditions occurs much faster than exchange between propane and water.

#### 3.10.1.3 Hydrogen

During photosynthesis plants remove hydrogen from water and transfer it to organic compounds. Because plants utilize environmental water during photosynthesis,

$\delta^2$H-values of plants are primarily determined by the $\delta^2$H-value of the water available for plant growth. Hydrogen enters the plant as water from roots in the case of terrestrial plants or via diffusion in the case of aquatic plants. In both cases, the water enters the organisms without any apparent fractionation. In higher terrestrial plants water transpires from the leaf due to evaporation, which is associated with a H-isotope fractionation of up to 40–50‰ (White 1989).

Large negative isotope fractionations occur in biochemical reactions during the synthesis of organic compounds (Schiegl and Vogel 1970). A generalized picture of the hydrogen isotope fractionations in the metabolic pathway of plants is shown in Fig. 3.40 (after Sachse et al. 2012).

There are systematic differences in the D/H ratios among classes of compounds in plants: lipids usually contain less deuterium than the protein and the carbohydrate fractions (Hoering 1975; Estep and Hoering 1980). Lipids can be divided into two groups: straight-chain lipids are depleted in D by 150–200‰ relative to water whereas isoprenoid lipids are depleted by about 200–300‰.

The component typically analyzed in plants is cellulose, which is the major structural carbohydrate in plants (Epstein et al. 1976, 1977). Cellulose contains 70% carbon-bound hydrogen, which is isotopically non-exchangeable and 30% of exchangeable hydrogen in the form of hydroxyl groups (Epstein et al. 1976; Yapp and Epstein 1982). The hydroxyl-hydrogen readily exchanges with the environmental water and its D/H ratio is not a useful indicator of the D/H ratio of the water used by the plants.

Compound-specific analysis of individual lipids have revealed a large range of $\delta^2$H-values from about –400 to +200‰ (Sachse et al. 2012 besides others) that can be related to isotope fractionations associated with different biosynthetic pathways and secondary hydrogenation and dehydrogenation exchange reactions. These

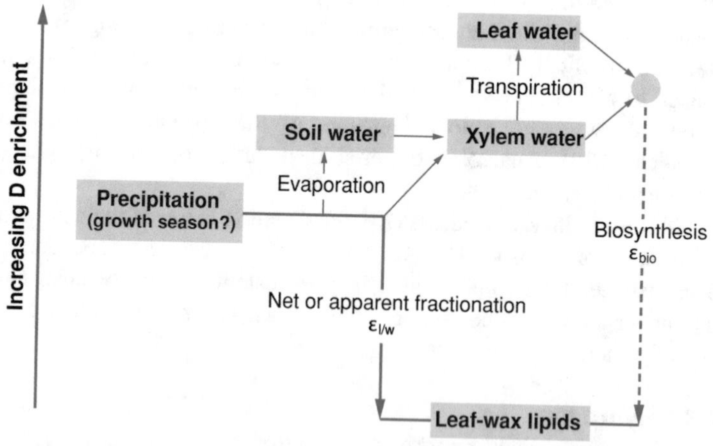


Fig. 3.40 Generalized scheme of hydrogen isotope changes in plants (Sachse et al. 2012)

effects have to be known when interpreting $\delta^2H$-values of lipid biomarkers as paleoclimate indicators.

In amino acids, hydrogen isotopes can provide a tracer for food and water. Fogel et al. (2016) demonstrated that 46% of hydrogen in some nonessential amino acids originated from water, whereas much less water (12%) can be traced in essential amino acids.

Hydrogen and carbon in organic matter, although both of biological origin, undergo very different changes during diagenesis and maturation. Whereas carbon tends to be preserved, hydrogen is exchanged during various diagenetic reactions with environmental water. The timescale for H-isotope exchange depends on the structure of the organic molecule and can reach millions of years. Schimmelmann et al. (2006) demonstrated as thermal maturation increases $\delta^2H$-values of individual hydrocarbons steadily increase while the 100‰ biosynthetic difference between linear and isoprenoid structures disappear, which, as shown by Wang et al. (2009), can be attributed to hydrogen isotope exchange towards an equilibrium state. The endpoint of isotope exchange results in fractionations between hydrocarbons and pore water to be in the range of −80 to −110‰.

As summarized by Sessions (2016) four major factors control the hydrogen isotope composition of hydrocarbons: (i) the D/H isotope composition of environmental water, (ii) metabolic processes that fix water hydrogen into organic molecules, (iii) hydrogen exchange processes that alter D/H ratios over geologic time and (iv) kinetic fractionations during thermal cracking of sedimentary organic matter. Since these 4 parameters overlap in nature, deuterium contents of organic molecules reflect a complex mixture of environmental, biologic and diagenetic signals, which are difficult to be disentangled.

#### 3.10.1.4 Oxygen

Due to the rapid exchange between organically bound oxygen, in particular the oxygen of carbonyl and carboxyl functional groups, with water, studies on the oxygen isotope fractionation within living systems have concentrated on cellulose, the oxygen of which is only very slowly exchangeable (Epstein et al. 1977; DeNiro and Epstein 1979, 1981). Oxygen potentially may enter organic matter from three different sources: $CO_2$, $H_2O$ and $O_2$. DeNiro and Epstein (1979) have shown that $^{18}O$-contents of cellulose for two sets of plants grown with water having similar oxygen isotope ratios, but with $CO_2$ initially having different oxygen isotope ratios, did not differ significantly. This means that $CO_2$ is in oxygen isotope equilibrium with the water. Therefore, the isotopic composition of water determines the oxygen isotope composition of organically bound oxygen. Similar to hydrogen, oxygen isotope fractionation does not occur during uptake of soil water through the root, but rather in the leaf because of evapotranspiration causing isotope enrichment, the extent depends on the ratio of external to internal water vapor pressures. A high $\delta^{18}O$-value in cellulose can thus reflect an increase in temperature or a decrease in relative humidity, making the interpretation of $\delta^{18}O$-values ambiguous (Sternberg et al. 2002).

Current methods of $^{18}O$ analysis give a mean value of the individual positions in the cellulose molecule. Waterhouse et al. (2013) describe a method for the measurement of different oxygen positions in cellulose by demonstrating that different oxygen positions undergo variable degrees of O-isotope exchange. The method potentially enables a separate determination of temperatures and humidities of the past.

#### 3.10.1.5 Nitrogen

There are various pathways by which inorganic nitrogen can be fixed into organic matter during photosynthesis. N-autotrophs can utilize a variety of materials and thus can have a wide range of $\delta^{15}N$-values depending on environmental conditions. However, most plants have $\delta^{15}N$-values between −5 and +2‰. Plants fixing atmospheric nitrogen have δ-values between 0 and +2‰. Isotope fractionation will occur when the inorganic nitrogen source is in excess (Fogel and Cifuentes 1993). Isotope fractionations during assimilation of $NH_4$ by algae varied extensively from −27 to 0‰ (Fogel and Cifuentes 1993). A similar range of fractionations has been observed with algae grown on nitrate as the source of nitrogen.

A large fraction of organic nitrogen is comprised by amino acids. As first shown by McClelland and Montoya (2002) internal differences in $^{15}N$ contents among different amino acids are due to differences in metabolic pathways. McClelland and Montoya (2002) distinguished two groups of amino acids: a "source" group reflecting the $^{15}N$ composition of the system and a "trophic level" group showing $^{15}N$ enrichments relative to the source. "Source" amino acids" (i.e. phenylalanine) more or less do not fractionate $^{15}N$ during trophic transfer, whereas "trophic" amino acids (i.e. glutamic acid) are enriched 6–8‰ during each trophic step (Ohkouchi et al. 2017). Compound-specific nitrogen isotope analysis of the two amino acids thus gives insight into the trophic position of organisms.

#### 3.10.1.6 Sulfur

The processes responsible for the direct primary production of organically bound sulfur are the direct assimilation of sulfate by living plants and microbiological assimilatory processes in which organic sulfur compounds are synthesized. Generally inorganic sulfate and atmospheric $SO_2$ serve as the major sulfur sources in plants. Typically, plants have $\delta^{34}S$-values that are about 1‰ depleted relative to environmental sulfate (Trust and Fry 1992).

Since biosynthetic organic sulfur occurs in chemically labile forms, such as amino acids, sulfur contents in organic matter should decrease during diagenesis. However, this is not the case, generally S-contents increase. Sulfurization of organic matter increases its stability by replacing alcohols, aldehydes and conjugated double bonds. Most organic sulfur contained in humic and fulvic acids results from secondary sulfurization during early diagenesis being considerably depleted in $^{34}S$ relative to the original plant material. This indicates the addition of isotopically depleted sulfides from bacterially reduced sulfate (for more details see Sect. 3.11.12).

#### 3.10.1.7 Metals in Plants

Metals play an essential role in plant nutrition. The amounts required for plant growth vary by orders of magnitude depending on plant species. Metals such as Zn and Cu are important cofactors of enzymes; Fe plays a vital role in various redox reactions and the biosynthesis of chlorophyll; Zn is important for carbohydrate and protein metabolism; Mo and Cu are important micronutrients; excesses of these elements can lead to toxic effects.

Bioessential metals are extracted from soils and cycled through living organic matter. During these cycling processes various fractionation processes do occur and metal isotopes, thus, can be used to study the transfer processes from soils to plants and within plants (Von Blanckenburg et al. 2009; Caldelas and Weiss 2017 and others).

The range of isotope variations of metal isotopes in plants and animals is of similar magnitude as those reported for geological materials (Jaouen et al. 2014). Metal isotope compositions vary between seeds, stem and leaves, all being isotopically different from the growth media.

Examples of metal isotope fractionations in plants have been presented for Fe (Guelke and von Blanckenburg 2007; Kiczka et al. 2010), for Zn and Mg (Moynier et al. 2008; Viers et al. 2007; Black et al. 2008) and for Ca (Page et al. 2008; Cobert et al. 2011). Fe isotope studies demonstrated that the uptake of Fe by plants at the plasma membrane creates a Fe pool that is depleted in heavy isotopes. Studies on Zn and Mg isotopes have demonstrated the complex chemistry in plants. Ca isotope investigations have identified 3 isotope fractionation steps in higher plants that may allow the study of Ca transfer mechanisms within plants.

In summary, the extent and direction of metal isotope fractionations in plants are metal dependent and still unknown in many cases. Potentially, like the light elements, metals may create isotope signatures characteristic for life.

### 3.10.2 Indicators of Diet and Metabolism

A similarity in $\delta^{13}C$-values between animals and plants from the same environment was first noted by Craig (1953). Later, many field and laboratory studies have documented small shifts of 1–2‰ in $^{13}C$ and even smaller shifts in $^{34}S$ between an organism and its food source (DeNiro and Epstein 1978; Peterson and Fry 1987; Fry 1988).

This technique has been widely used in tracing the origin of carbon, sulfur and nitrogen in modern and prehistoric food webs (e.g. DeNiro and Epstein 1978) and culminates in the classic statement "You are what you eat plus/minus a few permil". The precise magnitude of the isotopic difference between diet and a particular tissue depends on the extent to which the heavy isotope is incorporated or lost during synthesis. In contrast to carbon and sulfur, nitrogen shows a 3–4‰ enrichment in $^{15}N$ in the muscle tissue, bone collagen or whole organism relative to the food source (Minigawa and Wada 1984; Schoeninger and DeNiro 1984). Considering the 3–4‰ fractionation, nitrogen isotopes are also a good indicator of the dietary source. Due to the preferential excretion of $^{14}N$, the 3–4‰ shift in $\delta^{15}N$ values occurs with each trophic level along the food chain and thus provides a basis for establishing a trophic structure.

Multi-element isotope analysis on the origin of organic compounds reveal new applications of stable isotope investigations termed stable isotope forensics (e.g. Meier-Augustein 2010). In this branch of research not only the origin of food such as honey, wine or whisky are traced, but also attempts have been carried out distinguishing sources for drugs and elucidating explosives (for more details, see p. 462).

Archaeological studies have used the stable isotope analysis of collagen extracted from fossil bones to reconstruct the diet of prehistoric human populations (e.g. Schwarcz et al. 1985). Metal isotopes may provide additional constraints to disentangle complex diets (Martin et al. 2015; Jaouen et al. 2016). Martin et al. (2015) investigated Mg isotope variations in tooth enamel from modern mammals. $\delta^{26}Mg$-values become enriched from herbivores to omnivores which may result from $^{26}Mg$ enrichment in muscle relative to bone. For zinc, Jaouen et al. (2016) observed that herbivores are isotopically enriched in Zn isotopes relative to carnivores, which they explained by consumption of isotopically enriched leaves relative to other parts of plants. More details to metal isotope fractionations in animals: see section "Medical applications", p. 463.

### 3.10.3 Tracing Anthropogenic Organic Contaminant Sources

The identification of organic compounds polluting the environment is a problem of worldwide concern. Compound-specific stable isotope analysis has become a powerful tool to study the sources of organic contaminants and their transformation reactions in the environment. The first studies on degradation of groundwater pollutants were published in the late 1990. Since then the field has rapidly grown resulting in many articles that monitor natural attenuation of contaminated sites (i.e. Schmidt et al. 2004; Philp 2007; Hofstetter et al. 2008 and others).

Types of contaminants in the environment are manifold and include natural seepage of crude oils, leaking tanks and pipelines, polychlorinated biphenyls, and other types of chemicals. The ultimate goal of many of these studies is the question who was responsible for the contamination and will have to pay for the cleanup.

Temporal and spatial isotope variations of individual organic contaminants may reveal by which pathway contaminants may degrade or even in some cases to which degree a reaction has progressed. When a biotic or abiotic transfer reaction process takes place, a kinetic isotope effect usually occurs making the reaction products initially lighter than their parent products.

Natural attenuation processes may preclude easy application of the isotope ratios as a tracer of pollution. Besides bacterial degradation, isotope fractionations during evaporation and migration of organic contaminants may affect the isotope composition. Of special concern are chlorinated hydrocarbons; by coupling C- with Cl-isotopes, sources, pathways and degradation of chlorinated hydrocarbons can be traced (Heraty et al. 1999; Huang et al. 1999; Jendrzejewski et al. 2001). The use of

C- and Cl-isotopes requires the isotope ratios of the polluting product to be significantly different from the natural abundance. Jendrzejewski et al. (2001) demonstrated on a set of chlorinated hydrocarbons from various manufacturers that both carbon ($\delta^{13}C$ from –24 to –51‰) and chlorine ($\delta^{37}Cl$ from –2.7 to +3.4‰) had a large compositional range. The range for chlorine is especially significant, because it is much larger than that of inorganic chlorine. Sullivan-Ojeda et al. (2020) extended the analytical approach by investigating C, H, Cl and Br isotopes in individual halogenated hydrocarbons.

The nitrogen cycle has been also influenced considerably by human activities including agriculture and fossil fuel burning, adding reactive nitrogen to the environment on a local and a global scale. As demonstrated by Hastings et al. (2009, 2013), nitrogen isotopes of reactive nitrogen can be used to trace its origin. For example, Hastings et al. (2009) analysed N isotopes in a 100 m long ice core and observed a decrease from pre-industrial $\delta^{15}N$-values of +11‰ to present day values of −1‰. Other studies have shown that fertilizer, animal wastes or sewage are the main sources of nitrate pollution in the hydrosphere. Under favorable conditions, these N-bearing compounds can be isotopically distinguished from each other (Heaton 1986). Anthropogenic fertilizers have $\delta^{15}N$-values in the range −4 to +4‰ reflecting their atmospheric source, whereas animal waste typically has $\delta^{15}N$-values >5‰. Soil-derived nitrate and fertilizer nitrate commonly have overlapping $\delta^{15}N$-values.

### 3.10.4 Marine Versus Terrestrial Organic Matter

The commonly observed difference in $\delta^{13}C$ of about 7‰ between organic matter of marine primary producers and land plants has been successfully used to trace the origin of recent organic matter in coastal oceanic sediments (e.g. Westerhausen et al. 1993). Samples collected along riverine-offshore transects reveal very consistent and similar patterns of isotopic change from terrestrial to marine values (for instance Sackett and Thompson 1963; Kennicutt et al. 1987 and others). It is evident that the decreasing contribution of terrestrial organic matter to distal marine sediments is reflected in the C-isotope composition of the marine sedimentary organic matter. But even deep-sea sediments deposited in areas remote from continents may contain a mixture of marine and continental organic matter.

The C-isotope difference between terrestrial and marine organic matter cannot, however, be used as a facies indicator as originally thought. Carbon isotope fractionation associated with the production of marine organic matter has changed with geologic time, while that associated with the production of terrestrial organic matter has been nearly constant (Arthur et al. 1985; Hayes et al. 1989; Popp et al. 1989; Whittacker and Kyser 1990). Particularly intriguing has been the unusually $^{13}C$-depleted organic matter in Cretaceous marine sediments, which has been interpreted as resulting from elevated aqueous $CO_2$ concentrations allowing for greater discrimination during algal photosynthesis.

Hayes et al. (1999) systematically evaluated the carbon isotope fractionation between carbonates and coeval organic matter for the past 800 Ma. They concluded that earlier assumptions of a constant fractionation between carbonate and organic matter is untenable and that fractionations may vary by about 10‰ depending on the dominant biogeochemical pathway as well on environmental conditions.

Not only carbon, but the nitrogen isotope composition of sediments also is primarely determined by the source organic matter. Source studies have been undertaken to trace the contribution of terrestrial organic matter to ocean water and to sediments (i.e. Sweeney et al. 1978; Sweeney and Kaplan 1980). Such studies are based, however, on the assumption that the $^{15}N$ content remains unchanged in the water column. Investigations by Cifuentes et al. (1989), Altabet et al. (1991), and Montoya et al. (1991) have demonstrated that there may be rapid temporal (even on a time scale of days) and spatial changes in the nitrogen isotope composition of the water column due to biogeochemical processes. This complicates a clear distinction between terrestrial and marine organic matter, although marine organic matter generally has a higher $^{15}N/^{14}N$ ratio than terrestrial organic matter.

### 3.10.5 Fossil Organic Matter

Similar to living organisms, organic matter in the geosphere is a complex mixture of particulate organic remains and living bacterial organisms. This complexity results from the multitude of source organisms, variable biosynthetic pathways, and transformations that occur during diagenesis and catagenesis. Of special importance are different stabilities of organic compounds in biological and inorganic degradation processes during diagenesis and subsequent metamorphism.

Immediately after burial of the biological organic material into sediments, complex diagenetic changes occur. Two processes have been proposed to explain the observed changes in carbon isotope composition: (i) preferential degradation of organic compounds which have different isotope composition compared to the preserved organic compounds. Since easily degradable organic compounds like amino acids are enriched in $^{13}C$ compared to the more resistant compounds like lipids, this causes a shift to slightly more negative δ-values. (ii) Isotope fractionations due to metabolism of microorganisms. Early diagenesis does not only encompass degradation of organic matter, but also production of new compounds that potentially have different isotopic compositions than the original source material. A classic example has been presented by Freeman et al. (1990) analyzing hydrocarbons from the Messel shale in Germany (see Table 3.3). Considered as a whole, recent marine sediments show a mean $\delta^{13}C$-value of −25‰ (Deines 1980). Some $^{13}C$ loss occurs with transformation to kerogen, leading to an average $\delta^{13}C$-value of −27.5‰ (Hayes et al. 1983). This $^{13}C$ depletion might be best explained by the large losses of $CO_2$ that occur during the transformation to kerogen and which are especially pronounced during the decarboxylation of some $^{13}C$-rich carboxyl groups. With further thermal maturation the opposite effect (a $^{13}C$ enrichment) is observed. Experimental studies of Peters et al. (1981) and Lewan

(1983) indicate that thermal alteration produces a maximum $^{13}C$ change of about +2‰ in kerogens. Changes of more than 2‰ are most probably not due to isotope fractionation during thermal degradation of kerogen, but rather to isotope exchange reactions between kerogen and carbonates.

Whereas carbon tends to be preserved during diagenesis and maturation, hydrogen is exchanged during various diagenetic reactions with environmental water. $\delta^2H$-values of organic compounds, therefore, can be regarded as a continuously evolving system that can provide information about processes during burial of sedimentary rocks (Sessions et al. 2004). Radke et al. (2005) examined how maturation processes alter the $\delta^2H$-value of individual compounds. They demonstrated that aliphatic hydrocarbons are most favourable to record the primary composition because they resist hydrogen exchange. Pedentschouk et al. (2006) concluded that n-alkanes and isoprenoids have the potential to preserve the original biological signal till the onset of oil generation.

The isotopic compositions of the end products of organic matter diagenesis—carbon dioxide, methane and insoluble complex kerogen- may record the primary depositional environment. Boehme et al. (1996) determined the C-isotope budget in a well-defined coastal site. These authors demonstrated that the degradation of biogenic carbon proceeds via sulfate reduction and methanogenesis. The dominant carbon isotope effect during diagenesis is associated with methanogenesis, which shifts the carbon isotope value of the carbon being buried towards higher $^{13}C$-contents.

### 3.10.6 Oil

Questions concerning the origins of coal and petroleum center on three topics: the nature and composition of the parent organisms, the mode of accumulation of the organic material, and the reactions whereby this material was transformed into the end products.

Petroleum or crude oil is a naturally occurring complex mixture, composed mainly of hydrocarbons. Although there are, without any doubt, numerous compounds that have been formed directly from biologically produced molecules, the majority of petroleum components are of secondary origin, either decomposition products or products of condensation and polymerization reactions.

Combined stable isotope analysis ($^{13}C$, D, $^{15}N$, $^{34}S$) has been used successfully in petroleum exploration (Stahl 1977; Schoell 1984; Sofer 1984). The isotopic composition of crude oil is mainly determined by the isotopic composition of its source material, more specifically, the type of kerogen and the sedimentary environment in which it has been formed and by its degree of thermal alteration (Tang et al. 2005). Other secondary effects like biodegradation, water washing, and migration distances appear to have only minor effects on its isotopic composition.

Variations in $^{13}C$ have been the most widely used parameter. Generally, oils are depleted by 1–3‰ compared to the carbon in their source rocks. The various chemical compounds within crude oils show small, but characteristic

$\delta^{13}C$-differences. With increasing polarity the $^{13}C$-content increases from the saturated to aromatic hydrocarbons to the heterocomponents (N, S, O compounds) and to the asphaltene fraction. These characteristic differences in $^{13}C$ have been used for correlation purposes. Sofer (1984) plotted the $^{13}C$-contents of the saturated and aromatic fractions against each other. Oils and suspected source rock extracts that are derived from similar types of source materials will plot together in such a graph whereas those derived from different types of source material will plot in other regions of the graph. The approach of Stahl (1977) and Schoell (1984) is somewhat different: the $^{13}C$-contents of the different fractions are plotted as shown in Fig. 3.41. In this situation, oils derived from the same source rock will define a near linear relationship in the plot. Figure 3.41 illustrates a positive oil-oil correlation and a negative oil-source rock correlation.

Combined compound specific $^{13}C$ and D-analyses have been applied in a number of areas of petroleum geochemistry. Tang et al. (2005) demonstrated that variations in $\delta^2H$-values of long chain hydrocarbons provide a sensitive measure of the extent of thermal maturation. Such studies have demonstrated that thermal maturation processes tend to alter the shape of the curves, particularly the curves for the saturate fraction, making correlations more difficult. Furthermore, oil migration might affect the isotope composition. Generally, a slight $^{13}C$ depletion is observed

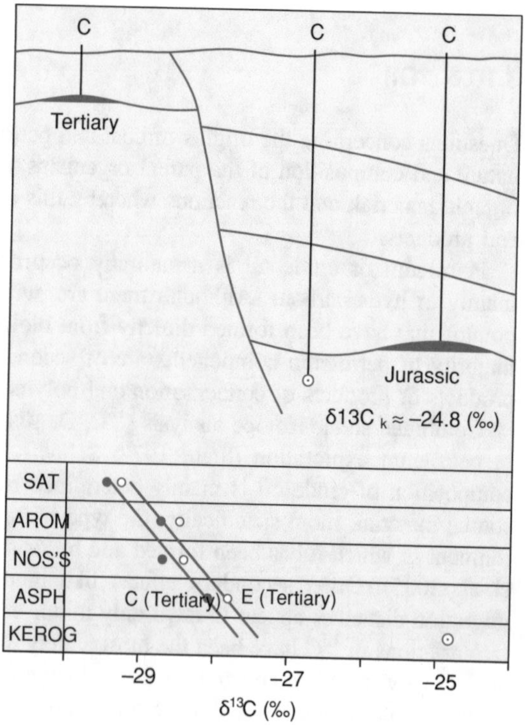


**Fig. 3.41** Petroleum-type curves of different oil components from the North Sea showing a positive oil-oil correlation and a negative source rock—oil correlation (*SAT* saturated hydrocarbons, *AROM* aromatic hydrocarbons, *NOS'S* heterocompounds, *ASPH* asphaltenes (Stahl 1977))

with migration distance, which is caused by a relative increase in the saturate fraction and a loss in the more $^{13}C$-enriched aromatic and asphaltene fraction.

Compound-specific analyses also indicate that $^{13}C$ differences between the isoprenoid-hydrocarbons, pristane, and phytane, for which a common origin from chlorophyll is generally assumed, point to different origins of these two components (Freeman et al. 1990). Other classes of biomarkers, such as the hopanes, are also not always derived from a common precursor. Schoell et al. (1992) have demonstrated that hopanes from an immature oil can be divided into two groups: one that is $^{13}C$ depleted by 2–4‰ relative to the whole oil, whereas the other is depleted by 9‰, which suggests that the latter group is derived from chemoautotrophic bacteria which utilize a $^{13}C$-depleted source. These results indicate that the origin and fate of organic compounds are far more complicated than was previously assumed.

Compound-specific sulfur analysis of individual organosulfur compounds may reveal $^{34}S$-variations of more than 60‰ in oils of different ages and petroleum provinces (Amrani et al. 2012; Amrani 2014; Greenwood et al. 2018; Cai et al. 2015, 2016 and others). $^{34}S$-variations between individual sulfur compounds may reflect different stages of thermal and microbial sulfate reductions as well as migration and secondary alteration processes.

Crude oils often contain high concentrations of metals, in specific V and Ni and to a lesser extent Mo. Ventura et al. (2015) presented a first data set of V, Ni and Mo isotope compositions and observed a large variation in V- and Mo-isotopes, but a narrower range for Ni isotopes. The major factor governing the metal isotope composition appears to be the isotope composition of the primary source rocks.

### 3.10.7 Coal

Carbon and hydrogen isotope compositions of coals are rather variable (Schiegl and Vogel 1970; Redding et al. 1980; Smith et al. 1982; Schimmelmann et al. 1999; Mastalerz and Schimmelmann 2002). Different plant communities and climates may account for these variations. Due to the fact that during coalification, the amount of methane and other higher hydrocarbons liberated is small compared to the total carbon reservoir, very little change in the carbon isotope composition seems to occur with increasing grade of coalification.

The D/H ratio in coals is usually measured on total hydrogen, although it consists of two portions: exchangeable and non-exchangeable hydrogen. In lignite up to 20% of hydrogen consists of isotopically labile hydrogen that exchanges fast and reversibly with ambient water. With increasing temperature (maturity) the exchangeable portion decreases to about 2% (Schimmelmann et al. 1999; Mastalerz and Schimmelmann 2002). Non-exchangeable organic hydrogen may have preserved original biochemical D/H ratios. $\delta^2H$-values in coals typically become isotopically heavier with increasing maturity, which suggests that exchange between organic hydrogen and formation water occurs during thermal maturation.

The origin and distribution of sulfur in coals is of special significance, because of the problems associated with the combustion of coals. Sulfur in coals usually occurs in different forms, as pyrite, organic sulfur, sulfates, and elemental sulfur. Pyrite and organic sulfur are the most abundant forms. Organic sulfur is primarily derived from two sources: the originally assimilated organically- bound plant sulfur preserved during the coalification process, and biogenic sulfides which reacted with organic compounds during the biochemical alteration of plant debris.

Studies by Smith and Batts (1974), Smith et al. (1982), Price and Shieh (1979) and Hackley and Anderson (1986) have shown that organic sulfur exhibits rather characteristic S-isotope variations, which correlate with sulfur contents. In low-sulfur coals $\delta^{34}S$-values of organic sulfur are rather homogeneous and reflect the primary plant sulfur. By contrast, high-sulfur coals are isotopically more variable and typically have more negative $\delta^{34}S$-values, suggesting a significant contribution of sulfur formed during bacterial processes.

#### 3.10.7.1 Black Carbon

The incomplete combustion of organic material under restricted oxygen concentration produces carbon-rich materials such as charcoal and soot. This black carbon is a common minor component in many recent and palaeo-environments, because it is resistant to decay and thus the carbon isotope composition may allow deductions about its origin (Bird and Ascough 2012 and others). Charcoal and soot may change slightly its isotope composition during combustion depending on temperature and pyrolysis conditions.

Forest fires are the predominant source of terrestrial black carbon, a smaller portion is emitted to the atmosphere as aerosols primarely derived from fossil fuel combustion. Black carbon deposited on continents is incorporated into soils, where it may be mobilized entering aquatic systems as dissolved black carbon. Wagner et al. (2019) used compound-specific carbon isotope analyses to reveal that dissolved black carbon in the ocean is 6‰ enriched in $^{13}C$ compared to dissolved black carbon exported by rivers. Wagner et al. (2019) suggested that riverine dissolved black carbon is rapidly degraded before it reaches the open ocean and that marine dissolved black carbon originates from a source with an isotopic composition similar to that of marine phytoplankton.

### 3.10.8 Natural Gas

Natural gases are dominated by a few simple hydrocarbons, which may form in a wide variety of environments. While methane is always the major constituent of the gas, other components may be higher hydrocarbons (ethane, propane, butane), $CO_2$, $H_2S$, $N_2$ and rare gases. Two different types of gas occurrences can be distinguished—biogenic and thermogenic gas—the most useful parameters in distinguishing both types are their $^{13}C/^{12}C$ and D/H ratios. Complications in assessing sources of natural gases are introduced by mixing, migration and oxidative alteration processes. For practical application an accurate assessment of the origin of a gas, the maturity of the source rock and the timing of gas formation would be desirable. A variety of models has been published that describes the carbon and hydrogen isotope variations of natural gases (Berner et al. 1995; Galimov 1988; James 1983, 1990; Rooney et al. 1995; Schoell 1983, 1988).

Rather than using the isotopic composition of methane alone James (1983, 1990) and others have demonstrated that carbon isotope fractionations between the hydrocarbon components (particularly propane, iso-butane and normal butane) within a natural gas can be used with distinct advantages to determine maturity, gas-source rock and gas–gas correlations. With increasing molecular weight, from $C_1$ to $C_4$, a $^{13}C$ enrichment is observed which approaches the carbon isotope composition of the source.

Genetic models for natural gases were based in the past primarily on field data and on empirical models. More recently, mathematical modeling based on Rayleigh distillation theory and kinetic isotopic theory (Rooney et al. 1995; Tang et al. 2000) may explain why, in a single gas $\delta^{13}C$ values increase from $C_1$ to $C_4$ and why in different gases $\delta^{13}C$ values of a given hydrocarbon increase with increasing thermal maturity. Such models may provide information on the isotope composition of each gas at any stage of generation. By re-assessing carbon isotope systematics of methane, ethane and propane during thermal maturity, Cesar et al. (2020) demonstrated that carbon isotope distributions in low-permeability reservoirs differ from conventional hydrocarbon accumulations by driving the carbon isotopes to an even isotope distribution of 6‰ between methane and ethane, and ethane and propane.

Although most natural gas occurences yield the sequence $\delta^{13}C_1$ (methane) ≤ $\delta^{13}C_2$ (ethane) ≤ $\delta^{13}C_3$ (propane), an increasing number of studies (Jenden et al. 1993; Burruss and Laughrey 2010; Tilley and Muehlenbachs 2013; Xia et al. 2013 and others) have described reversed isotope trends with $\delta^{13}C_1 \geq \delta^{13}C_2 \geq \delta^{13}C_3$. Gases with reversed trends can be explained by mixing of primary gas (methane from kerogen cracking) and secondary gas ("wetter" gas from intermediate products of kerogen with a higher proportion of higher alkanes).

Apart from gas sources and formation mechanisms, isotope effects during migration might affect the isotope composition of natural gas. Early experimental work has indicated that migrating methane could be enriched in $^{12}C$ or $^{13}C$ depending on the mechanism of migration and on the properties of the medium through which the gas is moving. Experiments by Zhang and Kroos (2001) on natural shales with different organic matter contents demonstrate variable $^{13}C$ depletions (1–3‰) during migration, which depend on the amount of organic matter in shales.

Of special interest in recent years has been the analysis of natural gas hydrates that form in marine sediments and polar rocks when saline pore waters are saturated with gas at high pressure and low temperature. Large $\delta^{13}C$ and $\delta^2H$-variations of hydrate bound methane, summarized by Kvenvolden (1995) and Milkov (2005), suggest that gas hydrates represent complex mixtures of gases of both microbial and thermogenic origin. The proportions of both gas types can vary significantly even between proximal sites.

As has been proposed by numerous studies (e.g. Röhl et al. 2000; Dickens 2003) the massive release of gas hydrates could modify climate. The best example for this hypothesis are sedimentary rocks deposited at around 55 Ma during the Paleocene-Eocene thermal maximum, where a $\delta^{13}C$ decrease of 2–3‰ in carbonate-carbon is interpreted as a consequence of an abrupt thermal release of gas-hydrate methane and its subsequent incorporation into the carbonate pool.

#### 3.10.8.1 Biogenic Gas

According to Rice and Claypool (1981), over 20% of the world's natural gas accumulations are of biogenic origin. Biogenic methane commonly occurs in recent anoxic sediments and is well documented in both freshwater environments, such as lakes and swamps, and in marine environments, such as estuaries and shelf regions. Two primary metabolic pathways are generally recognized for methanogenesis: fermentation of acetate and reduction of $CO_2$. Although both pathways may occur in marine and freshwater environments, $CO_2$-reduction is dominant in the sulfate-free zone of marine sediments, while acetate fermentation is dominant in freshwater sediments.

During microbial action, kinetic isotope fractionations on the organic material by methanogenic bacteria result in methane that is highly depleted in $^{13}C$, typically with $\delta^{13}C$-values between −110 and −50‰ (Schoell 1984, 1988; Rice and Claypool 1981; Whiticar et al. 1986). In marine sediments, methane formed by $CO_2$ reduction is often more depleted in $^{13}C$ than methane formed by acetate fermentation in freshwater sediments. Thus, typical $\delta^{13}C$ ranges for marine sediments are between −110 and −60‰, while those for methane from freshwater sediments are from −65 to −50‰ (Whiticar et al. 1986; Whiticar 1999).

The difference in composition between methane of freshwater and of marine origin is even more pronounced on the basis of hydrogen isotopes. Marine bacterial methane has $\delta^2H$-values between −250 and −170‰ while biogenic methane in freshwater sediments is strongly depleted in D with $\delta^2H$-values between −400 and −250‰ (Whiticar et al. 1986; Whiticar 1999). Different hydrogen sources may account for these large differences: formation waters supply the hydrogen during $CO_2$ reduction, whereas during fermentation up to three quarters of the hydrogen come directly from the methyl group, which is extremely depleted in D.

#### 3.10.8.2 Thermogenic Gas

Thermogenic gas is produced when organic matter is deeply buried and—as a consequence—temperature rises. Thereby, increasing temperatures modify the organic matter due to various chemical reactions, such as cracking and hydrogen diproportionation in the kerogen. $^{12}C$–$^{12}C$ bonds are preferentially broken during the first stages of organic matter maturation. As this results in a $^{13}C$-enrichment of the residue, more $^{13}C$–$^{12}C$ bonds are broken with increasing temperatures which produces higher $\delta^{13}C$-values. Thermal cracking experiments carried out by Sackett (1988) have confirmed this process and demonstrated that the resulting methane is depleted in $^{13}C$ by some 4–25‰ relative to the parent material. As shown by Thiagarajan et al. (2020), thermogenic gas is initially produced by irreversible cracking reactions; during thermal maturation, C- and H-isotopic compositions within and among light n-alkanes approach thermodynamic equilibrium under the conditions of gas formations or later during reservoir storage.

Thermogenic gas typically has $\delta^{13}C$-values between −50 and −20‰ (Schoell 1980, 1988). Gases generated from non-marine (humic) source rocks are isotopically enriched relative to those generated from marine (sapropelic) source rocks at equivalent levels of maturity. In contrast to $\delta^{13}C$-values, $\delta^{2}H$-values are independent of the composition of the precursor material, but solely depend on the maturity of kerogen.

In conclusion, the combination of carbon and hydrogen isotope analysis of natural gases is a powerful tool to discriminate different origins of gases. In a plot of $\delta^{13}C$ versus $\delta^{2}H$ (see Fig. 3.42) not only is a distinction of biogenic and thermogenic gases from different environments clear, but it is also possible to delineate mixtures between the different types.

#### 3.10.8.3 Abiogenic Methane

To distinguish abiogenic from biogenic organic compounds on the basis of their $\delta^{13}C$ and $\delta^{2}H$ signatures is demanding (Taran et al. 2007; Bradley and Summons 2010; Etiope and Sherwood-Lollar 2013; Zwicker et al. 2018). Abiogenic methane is defined as methane that does not involve biogenic organic precursors (Welhan 1988). Etiope and Sherwood-Lollar (2013) and Etiope and Schoell (2014) listed nine specific mechanisms of $CH_4$ production in two main environments: (i) high-temperature magmatic processes and (ii) low-temperature (below 150 °C) serpentinization processes of ultramafic rocks. The isotopic composition may be

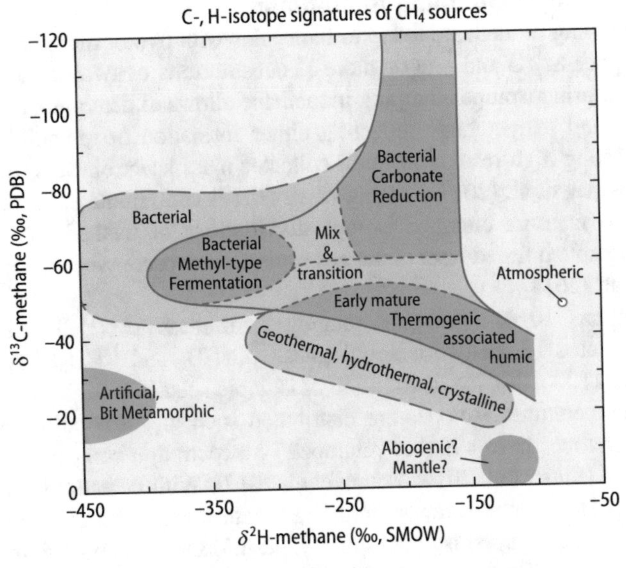


Fig. 3.42 $\delta^{13}C$ and $\delta^{2}H$ variations of natural gases of different origins (after Whiticar 1999)

divided into two groups: $^{13}C$ and D enriched ($\delta^{13}C$-values ≥ −20 and $\delta^{2}H$ ≥ −200) and $^{13}C$ and D depleted ($\delta^{13}C$ ≤ −30 and $\delta^{2}H$ ≤ −200). The two groups may reflect variable mantle or crustal sources and/or variable degrees of $CO_2$ and $CH_4$ isotope exchange.

Methane emanating in mid-ocean ridge hydrothermal systems is one of the occurrences for which an abiogenic formation can be postulated with some confidence. Considerably higher $\delta^{13}C$-values than biogenic methanes (up to −7‰; Abrajano et al. 1988) were thought to be the characteristic feature of abiogenic methane. Horita and Berndt (1999) demonstrated that abiogenic methane can be formed under hydrothermal conditions in the presence of a nickel–iron catalyst. Isotope fractionations induced by the catalyst, however, result in very low $\delta^{13}C$-values.

From a large C- and H-isotope data set of n-alkanes in volcanic-hydrothermal fluids, Fiebig et al. (2019a, b) concluded that the bulk isotope compositions of the gases follow trends of high-temperature organic matter degradation. In sediment-free volcanic systems organic matter may be supplied by surface waters. In serpentinites, the circulation of water induces mineral reactions, which may release hydrogen ($H_2$) that under Fischer–Tropsch type reactions form methane.

Another important source of abiogenic methanogenesis has been found in crystalline rocks from the Canadian and Ferroscandian shield areas (Sherwood Lollar et al. 1993, 2002; Warr et al. (2021), ; presented a nine-year record of methane from the Kidd Creek observatory in Ontario. On the basis of clumped isotopologue studies, they identified multiple processes of methane production including abiotic and biotic methane production.

#### 3.10.8.4 Isotope Clumping in Methane

Isotope clumping in methane refers to molecules with two or more heavy isotopes, meaning either a $^{13}C$ and one or more D substitutions or two or more D substitutions. Measuring isotope clumping in methane allows to determine bond ordering in methane and temperatures reflecting either formation or re-equilibration temperatures. Using high resolution multi-collector mass spectrometers, Stolper et al. (2014), Douglas et al. (2017), Young et al. (2017) and Giunta et al. (2019, 2021) were able to measure clumped isotope distributions of methane. An alternative method—tunable infrared laser direct absorption spectroscopy—has been described by Ono et al. (2014).

Methane has 10 isotopologues ranging from mass 16 ($^{12}CH_4$) to mass 21 ($^{13}CD_4$). At mass 18 are two isotopologues, $^{12}CH_2D_2$ and $^{13}CH_3D$. At very high temperatures (1000 °K), the distribution of isotopic species is random or stochastic. At lower temperatures, isotopes are distributed such that isotopologues with two heavy rare isotopes have a higher "clumped" concentration compared to a random concentration (Eiler et al. 2014; Young et al. 2017). With decreasing temperatures, the effects of two heavy isotopes clumping together increases. In internal equilibrium, excesses of clumped isotopologues yield informations about temperatures of methane formation (Douglas et al. 2017; Young et al. 2017; Labidi et al. 2020). By analyzing methane in deep-sea hydrothermal systems, Labidi et al. (2020)

suggested that $^{13}CH_3D$ isotopologue data agree with methane being formed at around 350 °C, but $^{12}CH_2D_2$ data show post formation residence temperatures.

As shown by Young et al. (2017) (see their Fig. 1) at 100 °C for example, the abundance of $^{13}CH_3D$ is about 4‰ enriched relative to the random distribution and the abundance of $^{12}CH_2D_2$ is about 12‰ enriched; at room temperature the values increase to 6‰ and 20‰ respectively. Thus, when methane forms in internal equilibrium, the paired determination of $^{12}CH_2D_2$ and $^{13}CH_3D$ can provide formation temperatures for abiogenic, thermogenic and biogenic methane. Deviations from isotope equilibrium have been described by Douglas et al. (2017) and others, which can be due to mixing of two gases, kinetic and diffusion processes.

Giuntas et al. (2019) reported clumping data of methane from two sedimentary basins. In one basin, clumping indicated thermodynamic equilibrium while in the other large offsets from thermodynamic equilibrium have been found which may be associated with microbial methanogenesis.

#### 3.10.8.5 Nitrogen in Natural Gas

Nitrogen is sometimes a major constituent of natural gases, but the origin of this nitrogen is still enigmatic. As shown by Zhu et al. (2000a, b, c), Huang et al. (2005a, b), Li et al. (2009) and others, nitrogen concentrations and isotope compositions in natural gas can be very variable. While a certain fraction is released from degrading sedimentary organic matter during burial, several non-sedimentary sources of nitrogen may also contribute to the natural gas. Natural gases from California's Great Valley had a complex origin involving mixing of multiple sources (Jenden et al. 1988). These authors interpreted relatively constant $\delta^{15}N$-values between 0.9 and 3.5‰ as indicating a deep-crustal metasedimentary origin. Hydrocarbon-rich and nitrogen-rich gases can thus be genetically unrelated.

#### 3.10.8.6 Isotope Signatures of Early Life on Earth

Carbon and sulfur isotopes ratios changing in characteristic manner by biological processes have been regarded as being most suitable to detect early life forms on Earth. Detection of biosignatures requires deconvolution of potential secondary alterations that increase with increasing geologic age. Thus, any evidence of life preserved in the most ancient rocks are likely to be ambiguous. Therefore, isotope data alone cannot be taken as proof, but must be accompanied by geological and petrographic data, that specify the stratigraphic context, metamorphic grade and diagenetic and metasomatic overprint. Furthermore, carbon and sulfur isotope fractionation analogous to that of living organic matter can result from abiotic processes.

Early investigators, i.e. Schidlowski (2001) have claimed that carbon isotope ratios from Isua (Greenland) that have an age of 3.8–3.9 Ga represent biosignatures, however, carbonaceous graphitic material from Isua have been metamorphosed to 400–500 °C allowing carbon isotope exchange with carbonates. Graphitization processes may have degraded any original organic molecules leaving graphite with little information about its origin.

In recent years, in-situ SIMS techniques have been developed to measure C- and S-isotopes of organic matter, carbonates and pyrite on microscales with high spatial resolution and accuracy (Lepot et al. 2013; Williford et al. 2016). According to these studies, the 3.45 Ga old Strelley Pool Formation in Western Australia can be regarded as the oldest sediments indicating biogenecity.

## 3.11 Sedimentary Rocks

Sediments are the weathering products and residues of magmatic, metamorphic, and sedimentary rocks and reflect weathering, erosion, transport and accumulation in water and air. As a result, sediments may be complex mixtures of material that has been derived from multiple sources. It is convenient to consider sedimentary rocks, and the components of sedimentary rocks, in two categories: clastic and chemical. Transported fragmental debris of all kinds makes up the clastic component of the rock. Inorganic and organic precipitates from water belong to the chemical constituents. According to their very different constituents and low temperatures of formation, sedimentary rocks may be extremely variable in isotopic composition. For example, the $\delta^{18}$O-values of sedimentary rocks span a large range from about +10 (certain sandstones) to about +44‰ (certain cherts).

Shales represent the most abundant sedimentary rock type consisting of a mixture of authigenic clays and detrital quartz with other authigenic and detrital minerals being minor constituents. Early studies by Savin and Epstein (1970a, b), Sheppard and Gilg (1996) and others have demonstrated that due to low porosities shales more or less preserve their O- and H isotope compositions. More recent studies by Bindeman et al. (2016) and Bindeman et al. (2018) have reported a shift of several‰ to depleted $^{18}$O values in shales that are associated with Proterozoic galaciations providing evidence of diagenetic alterations in contact with low $^{18}$O meltwaters. Triple oxygen isotope measurements of shales by Bindeman et al. (2018) demonstrated a stepwise decrease of 0.08‰ across the Archean-Proterozoic boundary. Bindeman et al. (2018) suggested that the observed trend is best explained by a shift of water–rock interactions from near-coastal in the Archean to continental in the Proterozoic (see also p.).

### 3.11.1 Fractionations During Weathering

Chemical weathering converts rocks/minerals into soluble and insoluble secondary products, thereby causing large isotope fractionations. Bouchez et al. (2014) presented a framework to model metal isotope fractionations during weathering. In specific, metal isotope fractionations may occur during mineral dissolution, formation of secondary minerals, ab- and desorption and very importantly during biological activities. Of special relevance are redox reactions, either inorganic or microbiologically mediated. As for the light elements, the reduced species of the metal is generally isotopically lighter than the oxidized species.

Studies on isotope fractionation during weathering have used 3 different approaches (i) investigating weathering profiles, (ii) investigating the isotope composition of river water and sediments and iii) experimental studies.

(I) The analysis of weathering profiles can potentially provide insight into the continental climate during their formation. Despite this potential, only few studies (Bird and Chivas 1989; Bird et al. 1992) have used this approach because of the (i) imprecise knowledge of mineral–water fractionations at surficial temperatures and (ii) the difficulty of obtaining pure phases from complex, very fine-grained rocks. Bird et al. (1992) developed partial dissolution techniques and used this methodology to separate nine pure minerals from a lateritic soil in Haiti (see Fig. 3.43). The measured $\delta^{18}$O-values for some minerals agree with $^{18}$O/$^{16}$O ratios predicted from available fractionation factors, whereas other do not. Discrepancies might be due to incorrect fractionation factors for the respective minerals or to processes that may have influenced the formation of particular minerals (e.g. evaporation) (Bird et al. 1992). Metal isotope fractionations during weathering has been reviewed by Bullen (2014). Weathering profiles and soils have revealed that the extent of metal isotope fractionations primarily depends on the type of secondary minerals formed.

(II) Measuring the isotope ratios of the dissolved and the particulate phase in rivers generally display significant isotope variations between rivers and rocks. Since numerous kinetic and equilibrium processes potentially affect

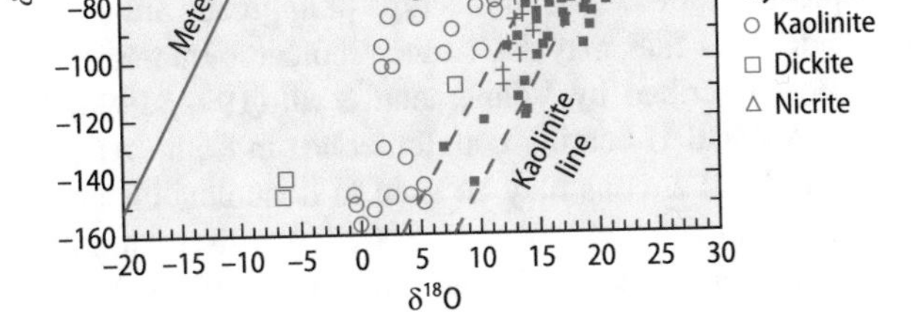

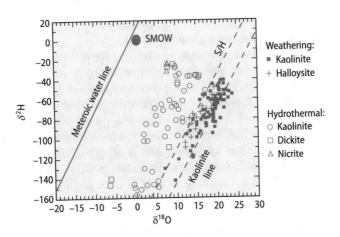

**Fig. 3.43** Predicted (*bars*) and measured (*crosses*) oxygen isotope composition of separated minerals from Haitian weathering profiles. The range of predicted $\delta^{18}$O-values are calculated assuming a temperature of 25 °C and a meteoric water $\delta^{18}$O-value of −3.1‰ (after Bird et al. 1992)

the isotope composition of rivers, the magnitude of metal isotope fractionation may vary depending on the specific weathering process. Light isotopes preferentially partition into secondary phases, whereas heavy isotopes preferentially partition into the associated fluid leading to river water generally enriched in heavy isotopes relative to weathered bedrocks. Suspended river sediments provide watershed-averaged pictures of weathering depending on different climate zones.

(III) Laboratory dissolution experiments to simulate weathering reactions help to determine the magnitude of isotope fractionations of specific weathering reactions and to interpret field data. It has to be kept in mind, however, that natural weathering in rocks and soils is more complex than simplified laboratory conditions and proceed on far longer timescales.

### 3.11.2 Clastic Sediments

Clastic sedimentary rocks are composed of detrital grains that normally retain the oxygen isotope composition of their source and of authigenic minerals formed during weathering and diagenesis, whose isotopic composition is determined by the physicochemical environment in which they formed. This means authigenic minerals formed at low temperatures will be enriched in $^{18}O$ compared to detrital minerals of igneous origin (Savin and Epstein 1970b). Due to the difficulty of separating authigenic overgrowths from detrital cores in quartz, few studies of this kind have been reported in the literature. However, recent improvements in the precision of ion microbe analysis with high spatial resolution (1–10 μm) both types of quartz can be clearly distinguished (see Fig. 3.44, Kelly et al. 2007). These authors suggested that the homogeneous $\delta^{18}O$ values of quartz overgrowth formed from meteoric waters at low temperatures (10–30 °C).

Detrital minerals in clastic sediments can be used for provenance studies. If not recrystallized, many common rock-forming minerals, such as quartz, muscovite, garnets etc. can retain their original source rock compositions up to medium-grade metamorphic conditions. Hence, they can potentially be used as tracers of provenance to the sediments. Applications of this type of approach are useful, particularly for siliciclastic sediments that may lack other indicator minerals of provenance; examples have been described by Vennemann et al. (1992, 1996) for the provenance of Archean Au- and U-bearing conglomerates in South Africa and Canada. $\delta^{18}O$-values of well dated zircons may be used to document changes with time in the composition of sediments (Valley et al. 2005) (see discussion on p. 219).

$^{18}O$ enrichments of authigenic minerals are controlled by fluid composition, temperature, and the effective mineral/water ratio. Is the fluid a low-$^{18}O$ meteoric water, the oxygen isotope composition of the precipitating mineral will have a low-$^{18}O$ signature, assuming no change in temperature (Longstaffe 1989). Thus, the changes that occur in sedimentary rocks during diagenesis are largely a function of fluid composition, fluid/rock ratio and temperature.

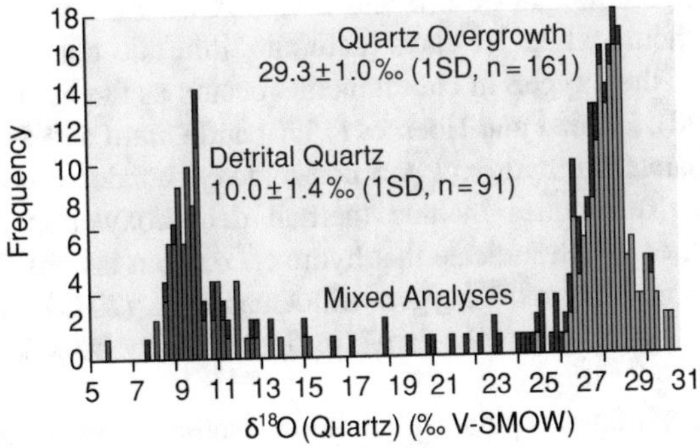


**Fig. 3.44** Histogram of $\delta^{18}O$-values of quartz in sandstone from 6 to 10 μm spots by ion microprobe. Mixed analyses are on the boundary of detrital quartz and quartz overgrowth (Kelly et al. 2007)

One way to estimate temperatures employs the oxygen isotope composition of diagenetic assemblages. For example, using quartz–illite pairs from the Precambrian Belt Supergroup, Eslinger and Savin (1973) calculated temperatures that range from 225 to 310 °C, with increasing depth. In this case the $\delta^{18}O$-values were consistent with the observed mineralogy and fractionations between minerals are reasonable for the grade of burial metamorphism. This approach assumes that the diagenetic minerals used have equilibrated their O-isotopes with each other and that no retrograde re-equilibration occurred following maximum burial.

### 3.11.3 Clay Minerals

Savin and Epstein (1970a, b) and Lawrence and Taylor (1971) established a general isotope systematics of clay minerals from continental and oceanic environments. Subsequent reviews by Savin and Lee (1988) and Sheppard and Gilg (1996) have summarized the isotope studies of clay minerals applied to a wide range of geological problems. All applications depend on the knowledge of isotope fractionation factors between clay minerals and water, the temperature, and the time when isotopic exchange with the clay ceased. Because clay minerals may be composed of a mixture of detrital and authigenic components, and because particles of different ages may have exchanged to varying degrees, the interpretation of isotopic variations of clay minerals requires a firm understanding of the clay mineralogy of a given sediment.

By comparison with many other silicate minerals, isotope studies of natural clays are complicated by a number of special problems related to their small particle size and, hence, much larger specific surface area and the presence of interlayer water in certain clays. Surfaces of clays are characterized by 1 or 2 layers of adsorbed water. Savin and Epstein (1970a) demonstrated that adsorbed and interlayer water in smectites and serpentine can absorb and exchange its isotopes with atmospheric water vapour in hours while 1:1 clays (kaolinite, illite) do not.

One portion of the oxygen in clay minerals occurs as the hydroxyl ion. Hamza and Epstein (1980), Bechtel and Hoernes (1990) and Girard and Savin (1996) have attempted to separate the hydroxyl and non-hydoxyl bonded oxygen for separate isotope analysis. Techniques include thermal dehydroxylation and incomplete fluorination, both of which indicate that hydroxyl oxygen is considerably depleted in $^{18}O$ relative to non-hydroxyl oxygen. Bindeman et al. (2019) demonstrated that water in clays is depleted in $^{18}O$ relative to the bulk clay by 8‰ across different climate zones.

Do natural clay minerals retain their initial isotopic compositions? Evidence concerning the extent of isotopic exchange for natural systems is contradictory (Sheppard and Gilg 1996). Many clay minerals such as kaolinite, smectite and illite are often out of equilibrium with present-day local waters. This is not to imply that these clay minerals never underwent any post-formational or retrograde exchange. Sheppard and Gilg (1996) concluded that convincing evidence for complete O- and/or H-isotope exchange without recrystallization is usually lacking, unless the clay has been subjected to either higher temperatures or an unusual set of geological circumstances. Thus, isotopic compositions of clay minerals that formed in contact with meteoric waters should have isotopic compositions that plot on sub-parallel lines to the Meteoric Water Line, the offset being related to their respective fractionation factor (see Fig. 3.45). This implies that some information of past environments is usually recorded in clay minerals and in suitable cases, especially

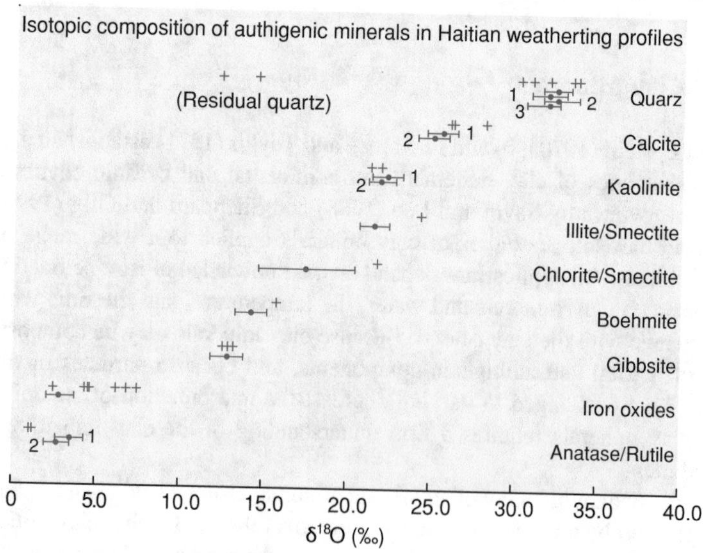


Fig. 3.45 $\delta^2H$ and $\delta^{18}O$ values of kaolinites and related minerals from weathering and hydrothermal environments. The Meteoric Water Line, kaolinite weathering and supergene/hypogene (S/H) lines are given for reference (after Sheppard and Gilg 1996)

involving monomineralic clays, can be used as a paleoclimate indicator (Stern et al. 1997; Chamberlain and Poage 2000; Gilg 2000). By analysing a large number of smectites from the Basin and Range Province and the Great Plains in North America, Mix and Chamberlain (2014) concluded that in some localities temperature change is the decisive factor in controlling the D and $^{18}O$ isotope composition, while in other localities the change in meteoric water composition is responsible for the variations in isotope composition.

## 3.11.4 Biogenic Silica and Cherts

### 3.11.4.1 Biogenic Silica

Due to the large oxygen isotope fractionation between $SiO_2$ and water at low temperatures, biogenic silica and cherts represent the "heaviest" oxygen isotope components in nature. Just as is the case for carbonates, the oxygen isotope composition of biogenic silica such as diatoms and radiolarians is potentially a paleoclimate indicator, which would enable the extension of climate records into oceanic regions depleted in $CaCO_3$ such as high latitude regions. Thus, a variety of techniques have been developed for the extraction of biogenic silica oxygen. The presence of loosely bound water within cherts and biogenic silica precipitates complicates measurements of the O-isotope composition of biogenic silica. Biogenic silica has an amorphous structure containing not only Si–O–Si bonds, but also Si–OH bonds and crystallization water which easily can exchange with environmental water and making it imperative to be removed before isotope analysis. At present 3 techniques exist (Chapligin et al. 2011):

(i) Controlled isotope exchange. Using controlled exchange with waters of different isotope composition, Labeyrie and Juillet (1982) and Leclerc and Labeyrie (1987) were able to estimate the isotope ratio of both exchanged and unexchanged silica-bound oxygen.

(ii) Stepwise fluorination.
Haimson and Knauth (1983), Matheney and Knauth (1989) and Dodd and Sharp (2010) noted that the first fractions of oxygen were $^{18}O$ depleted compared with oxygen recovered in later fractions, suggesting that the water-rich components of hydrous silica react preferentially in the early steps of fluorination.

(iii) High temperature carbon reduction (Lücke et al. 2005).
The technique is based on inductive high temperature heating (>1500 °C) leading to carbon monoxide. It enables complete dehydration and decomposition in a single continuous process.
Silica-water oxygen isotope fractionation factors differ considerably: Diatoms from sediment cores (Matheney and Knauth 1989) are up to 8‰ higher than living fresh water diatoms (Brandriss et al. 1998; Dodd and Sharp 2010) or diatoms from sediment traps (Moschen et al. 2006; Schmidt et al. 2001).

Schmidt et al. (2001) demonstrated that the enrichment in sedimentary diatoms can be correlated with structural and compositional changes arising from the in-situ condensation of Si–OH groups during silica maturation in surface sediments. Dodd et al. (2013) argued that the $^{18}O$ enrichment in sedimentary diatoms is due to post-mortem alteration. They demonstrated that diatoms can reach near equilibrium silica water compositions within half a year after diatom death.

#### 3.11.4.2 Cherts

In general, modern cherts form via biological precipitation of siliceous organisms while Precambrian cherts form by inorganic precipitation. Cherts may also form by silification of precursor material. Whether O and Si isotopes record primary environmental conditions or diagenetic dissolution/reprecipitation processes is a matter of debate.

As was shown from the early studies of Degens and Epstein (1962), cherts exhibit temporal isotopic variations like carbonates: the older cherts having lower $^{18}O$ contents. Thus, cherts of different geological ages may contain a record of temperature, isotopic composition of ocean water, and diagenetic history. However, because cherts may have formed by sedimentary, hydrothermal and volcanic silification and may have been altered by metamorphic fluids, the reconstruction of ocean water temperatures on the basis of $^{18}O$-values remain a matter of debate. Based on the study of triple O isotopes of Archean cherts, Sengupta and Pack (2018), Sengupta et al. (2020) and Zakharov et al. (2021) argued that O isotope compositions do not support "hot" nor "low $^{18}O$" oceans, but can be explained by precipitation from hydrothermal fluids.

High resolution in situ O- and Si-isotope SIMS analysis of cherts (Marin et al. 2010; Marin-Carbonne et al. 2011, 2012, 2014a, b; Steinhoefel et al. 2010; Chakrabarti et al. 2012; Stefurak et al. 2015) reveal very large O- and Si-isotope variations on the micrometer-scale indicating oxygen and silicon isotope exchange during burial diagenesis and the formation of microquartz from diagenetic or metamorphic fluids.

Low $\delta^{30}Si$-values in Archean cherts favor a hydrothermal source of silica; increasing $\delta^{30}Si$-values during the Proterozoic may reflect an increase in continental Si-sources relative to hydrothermal ones (Chakrabarti et al. 2012).

### 3.11.5 Marine Carbonates

#### 3.11.5.1 Oxygen

In 1947, Urey discussed the thermodynamics of isotopic systems and suggested that variations in the temperature of precipitation of calcium carbonate from water should lead to measurable variations in the $^{18}O/^{16}O$ ratio of the calcium carbonate. He postulated that the determination of temperatures of the ancient oceans should be possible, in principle, by measuring the $^{18}O$ content of fossil shell calcite. The first paleotemperature "scale" was introduced by Mc Crea (1950). Subsequently this scale has been refined several times. Through experiments which compare the actual growth temperatures of foraminifera with calculated isotope temperatures Erez and Luz (1983) determined the following temperature equation

$$T^{\circ}C = 17.0 - 4.52(\delta^{18}O_c - \delta^{18}O_w) + 0.03(\delta^{18}O_c - \delta^{18}O_w)^2$$

where $\delta^{18}O_c$ is the O-isotope composition of $CO_2$ derived from carbonate and $\delta^{18}O_w$ is the O-isotope composition of $CO_2$ in equilibrium with water at 25 °C.

According to this equation an $^{18}O$ increase of 0.26‰ in carbonate represents a 1 °C temperature decrease. Bemis et al. (1998) have re-evaluated the different temperature equations and demonstrated that they can differ as much as 2 °C in the temperature range between 5 and 25 °C. The reason for these differences is that in addition to temperature and water isotopic composition, the $\delta^{18}O$ of a shell may be affected by species specific kinetic effects (so called vital effects, see below) and by photosynthetic activity of algal symbionts.

Laboratory experiments and field studies on biogenic and inorganic $CaCO_3$ have demonstrated that nonequilibrium effects such as carbonate concentration, pH and precipitation rates may also affect measured $CaCO_3$ compositions. As Dietzel et al. (2009) argued there is no definite proof that spontaneously precipitated calcite from aqueous solution is in true oxygen isotope equilibrium and it can well be that currently adapted equilibrium values for calcite-water fractionations are actually too low (Coplen 2007).

Before a meaningful temperature calculation of a fossil organism can be carried out several assumptions have to be fulfilled. The isotopic composition of an aragonite or calcite shell will remain unchanged until the shell material dissolves and recrystallizes during diagenesis. In most shallow depositional systems, C- and O-isotope ratios of calcitic shells are fairly resistant to diagenetic changes, but many organisms have a hollow structure allowing diagenetic carbonate to be added. With increasing depths of burial and time the chances of diagenetic effects generally increase. Because fluids contain much less carbon than oxygen, $\delta^{13}C$-values are thought to be less affected by diagenesis than $\delta^{18}O$-values. Criteria of how to prove primary preservation are not always clearly resolved (see discussion "diagenesis of limestones"). Schrag (1999) argued that carbonates formed in warm tropical surface oceans are particularly sensitive to the effects of diagenesis, because pore waters—having much lower temperatures than tropical surface waters—could shift the primary composition to higher δ-values. This is not the case for high latitude carbonates, where surface and pore fluids are quite similar in their average temperature.

Shell-secreting organisms to be used for paleotemperature studies must have been precipitated in isotope equilibrium with ocean water. As was shown by studies of Weber and Raup (1966a, b), some organisms precipitate their skeletal carbonate in equilibrium with the water in which they live, but others do not. Wefer and Berger (1991) summarized the importance of the so-called "vital effect" on a broad range of organisms (see Fig. 3.46). For oxygen isotopes, most organisms precipitate

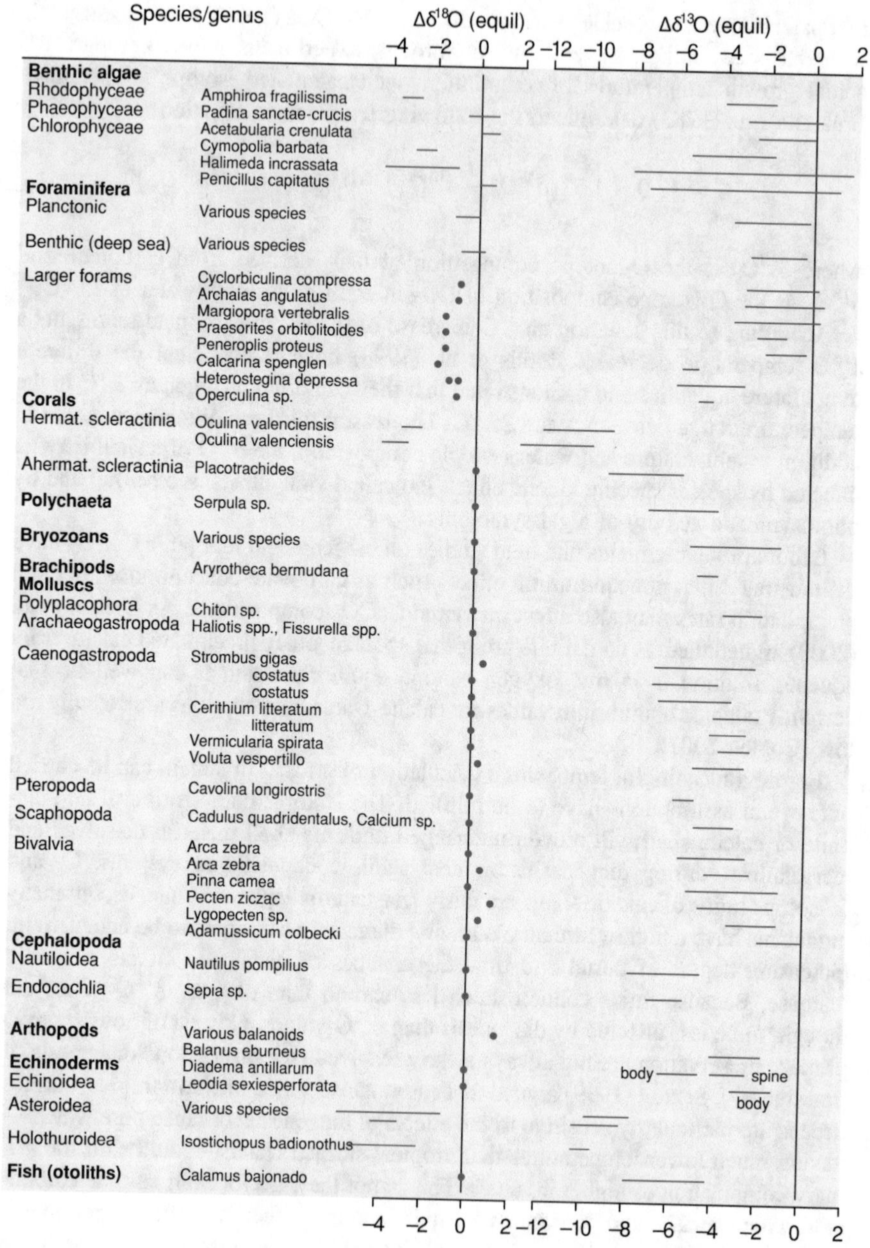


Fig. 3.46 $\Delta^{18}O$ and $\Delta^{13}C$ differences from equilibrium isotope composition of extant calcareous species (after Wefer and Berger 1991)

$CaCO_3$ with isotope compositions close to predicted equilibrium values; if disequilibrium prevails, the isotopic difference from equilibrium is rather small. For carbon, disequilibrium is the rule, with $\delta^{13}C$-values being more negative than expected at equilibrium. As discussed below, this does not preclude the reconstruction of the $^{13}C/^{12}C$ ratio of the palaeo-ocean waters.

Isotopic disequilibria effects can be classified as either metabolic or kinetic (McConnaughey 1989a, b). Metabolic isotope effects apparently result from changes in the isotopic composition of dissolved inorganic carbon in the neighborhood of the precipitating carbonate caused by photosynthesis and respiration. Kinetic isotope effects result from discrimination against $^{13}C$ and $^{18}O$ either during hydration ($CO_2 + H_2O \rightarrow H_2CO_3$ and hydroxylation of $CO_2$ ($CO_2 + OH^- \rightarrow HCO_3^-$) or at the solid-water interface of the growing carbonate. Strong kinetic disequilibrium fractionation often is associated with high calcification rates (McConnaughey 1989a, b).

Besides temperature, a variable isotopic composition of the ocean is another factor responsible for $^{18}O$ variations in foraminifera. A crucial control is salinity: ocean waters with salinities greater than 3.5% have a higher $^{18}O$ content, because $^{18}O$ is preferentially depleted in the vapor phase during evaporation, whereas waters with salinities lower than 3.5% have a lower $^{18}O$ content due to dilution by fresh waters, especially meltwaters. The other factor which causes variations in the isotopic composition of ocean water is the volume of low-$^{18}O$ ice present on the continents. As water is removed from the ocean during glacial periods, and temporarily stored on the continents as $^{18}O$-depleted ice, the $^{18}O/^{16}O$ ratio of the global ocean increases in direct proportion to the volume of continental and polar glaciers. The magnitude of the temperature effect versus the ice volume effect can be largely resolved by separately analyzing planktonic and benthic foraminifera. Planktonic foraminifera live vertically dispersed in the upper water column of the ocean recording the temperature and the isotopic composition of the water. The $^{18}O$ difference between shallow and deep-living planktonic foraminifera increases from nearly 0‰ in subpolar regions to ~3‰ in the tropics. The difference between shallow and deep-calcifying taxa can be used to calculate the vertical temperature gradient in the upper 250 m of the oceans.

It is expected that the temperature of deep-water masses is more or less constant, as long as ice caps exist at the poles. Thus, the oxygen isotope composition of benthic organisms should preferentially reflect the change in the isotopic composition of the water (ice-volume effect), while the $\delta^{18}O$-values of planktonic foraminifera are affected by both temperature and isotopic water composition.

The best approach to disentangle the effect of ice volume and temperature is to study shell material from areas where constant temperatures have prevailed for long periods of time, such as the western tropical Pacific Ocean or the tropical Indian Ocean. On the other end of the temperature spectrum is the Norwegian Sea, where deep water temperatures are near the freezing point today and, therefore, cannot have been significantly lower during glacial time, particularly as the salinities are also already high in this sea. Within the framework of this set of limited assumptions, a reference record of the $^{18}O$ variations of a water mass which has experienced no temperature variations during the last climatic cycle can be obtained (Labeyrie et al. 1987).

A direct approach to measuring the $\delta^{18}O$-value of seawater during the Last Glacial Maximum (LGM) is based on the isotopic composition of pore fluids (Schrag et al. 1996). Variations in deep water $\delta^{18}O$ caused by changes in continental ice volume diffuse down from the seafloor leaving a profile of $\delta^{18}O$ versus depth in the pore fluid. Using this approach Schrag et al. (1996) estimated that the global $\delta^{18}O$ change of ocean water during LGM is $1.0 \pm 0.1$‰.

In addition to these variables, the interpretation of $^{18}O$-values in carbonate shells is complicated by the seawater carbonate chemistry. In culture experiments with living foraminifera Spero et al. (1997) demonstrated that higher pH-values or increasing $CO_3^{2-}$ concentrations result in isotopically lighter shells, which is due to changing sea water chemistry. As shown by Zeebe (1999) an increase of seawater pH by 0.2–0.3 units causes a decrease in $^{18}O$ of about 0.2–0.3‰ in the shell. This effect has to be considered for instance when samples from the last glacial maximum are analyzed.

Another approach to distinguish between the temperature effect and the unknown water composition is the clumped isotope thermometer (Eiler 2007; Ghosh et al. 2006a, b; Tripati et al. 2010; Thiagarajan et al. 2011) that has the potential to circumvent the ambiguities of the classic carbonate thermometer of Urey (1947). Clumping of $^{13}C$ and $^{18}O$ into carbonate structures is independent of the $\delta^{18}O$ of the water from which the mineral is formed. Under isotope equilibrium the clumped isotope composition ($\Delta_{47}$) is solely a function of the temperature of formation. Calibrations of $\Delta_{47}$ for inorganic and biogenic calcite result in a sensitivity of about 0.004–0.005‰/°C (Huntington et al. 2010; Tripati et al. 2010; Dennis and Schrag 2010; Wacker et al. 2014; Petersen et al. 2019) (see Fig. 1.5). As has been discussed above, it is not certain, however, whether organisms precipitate carbonate under thermodynamic equilibrium. Bajnai et al. (2020) showed that dual clumped isotope analysis (simultaneous $\Delta_{47}$ and $\Delta_{48}$ measurements) can identify the origin and quantify the extent of kinetic effects.

#### 3.11.5.2 Carbon

A large number of studies have investigated the use of $^{13}C$-contents of foraminifera as a paleo-oceanographic tracer. As previously noted, $\delta^{13}C$-values are not in equilibrium with sea water. However, by assuming that disequilibrium $^{13}C/^{12}C$ ratios are, on average, invariant with time, systematic variations in C-isotope composition may reflect variations in $^{13}C$ content of ocean water. The first record of carbon isotope compositions in Cenozoic deep-sea carbonates was given by Shackleton and Kennett (1975). They clearly demonstrated that planktonic and benthic foraminifera yield consistent differences in $\delta^{13}C$-values, the former being enriched in $^{13}C$ by about 1‰ relative to the latter. This $^{13}C$-enrichment in planktonic foraminifera is due to photosynthesis which preferentially incorporates $^{12}C$ in organic carbon thereby depleting surface waters in $^{12}C$. A portion of the organic matter is transferred to deep waters, where it is reoxidized, which causes a $^{12}C$

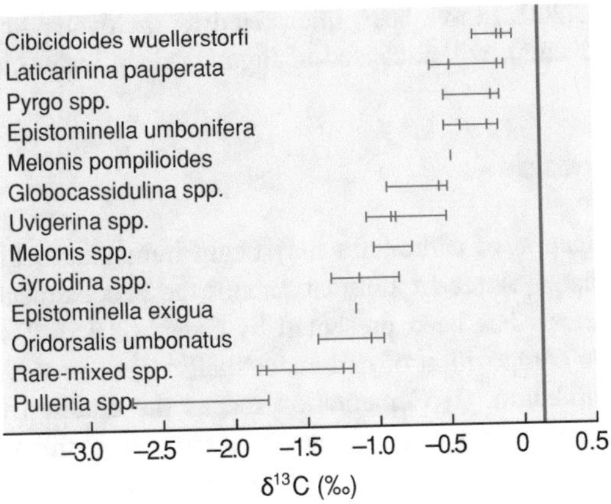


**Fig. 3.47** $\delta^{13}C$-values of benthic foraminifera species. The $\delta^{13}C$-value for the dissolved bicarbonate in deep equatorial water is shown by the vertical line (after Wefer and Berger 1991)

enrichment in the deeper water masses. Figure 3.47 presents $\delta^{13}C$-values of benthic foraminifera ranked according to their relative tendency to concentrate $^{13}C$.

$\delta^{13}C$-values in planktonic and benthic foraminifera can be used to monitor $CO_2$ variations in the atmosphere by measuring the vertical carbon isotope gradient, which is a function of the biological carbon pump. This approach was pioneered by Shackleton et al. (1983), who showed that enhanced contrast between surface waters and deeper waters was correlated with intervals of reduced atmospheric $CO_2$ contents. Increased organic carbon production in surface waters (possibly caused by enhanced nutrient availabilty) leads to removal of carbon from surface waters, which in turn draws down $CO_2$ from the atmospheric reservoir through re-equilibration.

Another application of carbon isotopes in foraminifera is to distinguish distinct water masses and to trace deep water circulation (Bender and Keigwin 1979; Duplessy et al. 1988). Since dissolved carbonate in the deeper waters becomes isotopically lighter with time and depths in the area of their formation due to the increasing oxidation of organic material, comparison of sites of similar paleodepth in different areas can be used to trace the circulation of deep waters as they move from their sources. Such a reconstruction can be carried out by analyzing $\delta^{13}C$-values of well-dated foraminifera.

Reconstructions of pathways of deep-water masses in the North Atlantic during the last 60,000 years have been performed by analyzing high resolution records of benthic foraminifera *Cibicides wuellerstorfi*, as this species best reflects changes in the chemistry of bottom waters (Duplessy et al. 1988; Sarntheim et al. 2001). The initial $\delta^{13}C$-signature of North Atlantic Deep Water (NADW) is ~1.3–1.5‰. As NADW flows southward the ongoing oxidation of organic matter results in a progressive $^{13}C$-depletion down to less than 0.4‰ in the Southern Ocean. Reductions in $^{13}C$ observed in many cores from the North-Atlantic (Sarntheim et al.

2001; Elliot et al. 2002) have been interpreted as meltwater input to the surface ocean (Heinrich events), which caused changes in deep water circulation.

### 3.11.6 Diagenesis

Diagenetic modification of carbonates may begin immediately after the formation of primary carbonates. A recent summary about the geochemical reactions occurring during diagenesis has been presented by Swart (2015). Two processes may change the isotope composition of carbonate shells: (i) cementation and (ii) dissolution and reprecipitation. (i) Cementation means the addition of abiogenic carbonate from ambient pore waters. Cements added early after primary formation may be in equilibrium with ocean water, whereas late cements depend on the isotope composition of pore waters and temperature. (ii) Dissolution and reprecipitation occurs in the presence of a bicarbonate containing pore fluid and represents the solution of an unstable carbonate phase such as aragonite and the reprecipitation of a stable carbonate phase, mostly low Mg-calcite.

Diagenetic modification may occur in two subsequent pathways, often termed as burial and meteoric diagenesis. In recent years, metal isotopes, in specific Ca and Mg isotopes have become valuable tools to indicate the diagenetic history and to constrain the extent and rate of carbonate recrystallization (Fantle et al. 2020 and others).

#### 3.11.6.1 Burial Pathway

This type of diagenetic stabilization is best documented in deep sea environments. Entrapped pore waters are of marine origin and in equilibrium with the assemblage of carbonate minerals. The conversion of sediment into limestone is not achieved by a chemical potential gradient, but rather through a rise in pressure and temperature due to deposition of additional sediments. In contrast to the meteoric pathway, fluid flow is confined to squeezing off pore waters upwards into the overlying sedimentary column. Theoretically, O-isotope ratios should not change appreciably with burial, because the $\delta^{18}O$ is of sea water origin. Yet, with increasing depth, the deep-sea sediments and often also the pore waters exhibit $^{18}O$ depletions by several permil (Lawrence 1989). The major reason for this $^{18}O$ depletion seems to be a low-temperature exchange with the oceanic crust in the underlying rock sequence. The $^{18}O$ shift in the solid phases is mostly due to an increase in temperature with increasing burial. Independent estimates of diagenetic temperatures may be obtained by clumped isotope thermometry (Huntington et al. 2011; Ferry et al. 2011).

The other important diagenetic process is the oxidation of organic matter. With increasing burial, organic matter in sediments passes successively through different zones which are characterized by distinct redox reactions that are mediated by assemblages of specific bacteria. The usual isotopic changes of these processes will result in a shift towards lighter C-isotope values, the degree of $^{13}C$-depletion being proportional to the relative contribution of carbon from the oxidation of organic matter. Under special conditions of fermentation, the $CO_2$ released may be isotopically heavy, which may cause a shift in the opposite direction.

#### 3.11.6.2 Meteoric Pathway

Carbonate sediments deposited in shallow marine environments are often exposed to the influence of meteoric waters during their diagenetic history. Meteoric diagenesis lowers $\delta^{18}O$- and $\delta^{13}C$-values, because meteoric waters have lower $\delta^{18}O$-values than seawater. For example, Hays and Grossman (1991) demonstrated that oxygen isotope compositions of carbonate cements depend on the magnitude of $^{18}O$ depletion of respective meteoric waters. $\delta^{13}C$-values are lowered because soil bicarbonate is $^{13}C$-depleted relative to ocean water bicarbonate.

A more unusual effect of diagenesis is the formation of carbonate concretions in argillaceous sediments. Isotope studies by Hoefs (1970), Sass and Kolodny (1972), and Irwin et al. (1977) suggest that micobiological activity created localized supersaturation of calcite in which dissolved carbonate species were produced more rapidly than they could be dispersed by diffusion. Extremely variable $\delta^{13}C$-values in these concretions indicate that different microbiological processes participated in concretionary growth. Irwin et al. (1977) presented a model in which organic matter is diagenetically modified in a sequence by (a) sulfate reduction, (b) fermentation and (c) thermally induced abiotic $CO_2$ formation which can be distinguished on the basis of their $\delta^{13}C$-values, (a) −25‰, (b) +15‰ and (c) −20‰.

### 3.11.7 Limestones

Early limestone studies utilized whole-rock samples. In later studies, individual components, such as different generations of cements, have been analyzed (Hudson 1977; Dickson and Coleman 1980; Moldovany and Lohmann 1984; Given and Lohmann 1985; Dickson et al. 1990). These studies suggest that early cements exhibit higher $\delta^{18}O$ and $\delta^{13}C$ values with successive cements becoming progressively depleted in both $^{13}C$ and $^{18}O$. The $^{18}O$ trend may be due to increasing temperatures and to isotopic evolution of pore waters. Employing a laser ablation technique, Dickson et al. (1990) identified a very fine-scale O-isotope zonation in calcite cements, which they interpreted as indicating changes in the isotope composition of the pore fluids.

#### 3.11.7.1 Carbon Isotope Stratigraphy

Carbon isotopes in sediments have, in general, a relatively uniform composition with $\delta^{13}C_{(carb)}$ around 0‰ and $\delta^{13}C_{(org)}$ around −25‰. However, at certain periods throughout Earth's history, systematic carbon isotope excursions in carbonates and organic matter may occur over relatively short time scales. These carbon isotope events may be used as stratigraphic markers, being indicative of environmental changes. During the past decades numerous carbon isotope excursions have been described aiming for regional or global stratigraphic correlations (i.e. Saltzman and Thomas 2012).

Carbon isotope events have been observed in all geological time periods. However, not all carbon isotope excursions should be classified as isotope events, especially not those containing only minor $\delta^{13}$C-shifts and those with low carbonate contents. In many of these excursions $\delta^{13}C_{(carb)}$ and $\delta^{13}C_{(org)}$ are covariant indicating that carbonate and organic matter has been produced in the surface ocean, while decoupled $\delta^{13}C_{(carb)}$ and $\delta^{13}C_{(org)}$ shifts have been interpreted as evidence for diagenetic alteration (i.e. Jiang et al. 2012).

Well-known carbon isotope events are oceanic anoxic events (OAEs) which are characterized by coeval globally distributed organic rich sediments (i.e. Jenkyns 2010). Typical examples of OAEs are the Jurassic-Cretaceous OAEs and the Paleocene-Eocene Thermal Maximum. Carbon isotope events associated with OEAs are characterized by an initial negative carbon isotope excursion which may be attributed to the release of $^{13}$C depleted submarine methane hydrates. The following positive carbon isotope excursion is attributed to an increase of burial rates of $^{12}$C-enriched organic matter.

Another characteristic carbon isotope event has been observed at the Permian/Triassic boundary, in which a negative shift in $\delta^{13}C_{(carb)}$-values is associated with a mass extinction event taking place within less than 200 kyr (Corsetti et al. 2005 and others). Other mass extinction events associated with carbon isotope excursions have been described for the late Devonian (Joachimski et al. 2002). Very pronounced events (i.e. Shuram excursion) have been found in carbonates deposited during the Precambrian-Cambrian transition, which show the lightest $\delta^{13}C_{(carb)}$-values and the largest net shifts of the whole carbon isotope record (Corsetti 2017).

### 3.11.8 Dolomites

Dolomite is found abundantly in Paleozoic and older strata, but is rare in younger rocks. Two reqirements are necessary for dolomite formation: (i) the presence of a high Mg/Ca fluid and (ii) large volumes of fluid that are pumped through limestones. There are only few locations where dolomite is forming today. In laboratory experiments, researchers have struggled to produce dolomite at temperatures and pressures realistic for its sedimentary formation (Horita 2014). This is the crux of the “dolomite problem”.

Since dolomitization takes place in the presence of water, oxygen isotope compositions are controlled by the pore fluid composition, the temperature of formation and to a lesser extent by the salt content. Carbon isotope compositions, in contrast, are determined by the precursor carbonate composition, because pore fluids generally have low carbon contents, so that the $\delta^{13}$C-value of the precursor is generally retained. Two problems complicate the interpretation of isotope data to delineate the origin and diagenesis of dolomites: (i) extrapolations of high-temperature experimental dolomite-water fractionations to low temperatures suggest that at 25 °C dolomite should be enriched in $^{18}$O relative to calcite by 4–7‰ (e.g. Sheppard and Schwarcz 1970). On the other hand, the oxygen isotope fractionation observed between Holocene calcite and dolomite is somewhat lower, namely in the range between 2 and 4‰ (Land 1980; McKenzie 1984), in agreement with recent theoretical predictions (Zheng and Böttcher 2016). The fractionation also may depend partly on the crystal structure, more specifically on the composition and the degree of crystalline order. (ii) For many years it has not been possible to determine the equilibrium oxygen isotope fractionations between dolomite and water at sedimentary temperatures directly, because the synthesis of dolomite at these low temperatures is problematic. With the discovery, that bacteria mediate the precipitation of dolomite, Vasconcelos et al. (2005) presented a new paleothermometer enabling the reconstruction of temperature conditions of ancient dolomite deposits.

Horita (2014) determined experimentally C- and O-isotope fractionations by precipitation of dolomite at 80 °C and by dolomitization of $CaCO_3$ in the temperature range 100–350 °C. In this temperature range dolomite is enriched relative to calcite by 0.7–2.6‰. As postulated by Horita (2014) fractionations can be extrapolated to lower temperatures. By applying clumped isotopes, Ferry et al. (2011) demonstrated that dolomite in the Italian dolomites formed at temperatures between 40 and 80 °C suggesting temperatures of formation, but not burial temperatures of recrystallization.

Figure 3.48 summarizes oxygen and carbon isotope compositions of some recent and Pleistocene dolomite occurences (after Tucker and Wright 1990). Variations in oxygen isotope composition reflect the involvement of different types of waters (from marine to fresh waters) and varying ranges of temperatures. With respect to carbon, $\delta^{13}$C-values between 0 and 3‰ are typical of marine compositions. In the presence of abundant organic matter, negative $\delta^{13}$C-values in excess of −20‰ indicate that carbon is derived from the decomposition of organic matter. Very positive $\delta^{13}$C-values up to +15‰ result from fermentation of organic matter (Kelts and McKenzie 1982). Such isotopically heavy dolomites have been described, for example, from the Guaymas Basin, where dolomite formation has taken place in the zone of active methanogenesis.

Besides C- and O-isotope compositions, Ca isotopes (Holmden 2009; Blättler et al. 2015) and Mg isotopes (Geske et al. 2015a, b; Blättler et al. 2015; Huang et al. 2015) have been investigated in a wide range of dolomite types. Ca and Mg isotope ratios of dolomites are affected by various factors including Ca and Mg sources and precipitation/dissolution processes complicating its application as a proxy for dolomite formation models. But, combined Ca and Mg isotope studies in dolomites may help to constrain the diagenetic history of dolomites (Fantle and Higgins 2014; Blättler et al. 2015). Given different residence times of C, Ca and Mg in the ocean, coherent isotope variations of these elements argue for depth changes in the isotopic composition of pore fluids (Blättler et al. 2015). Mg and Ca isotopes that do not correlate with geologic age or dolomite type may reflect multiple parameters such as changes in diagenetic history.

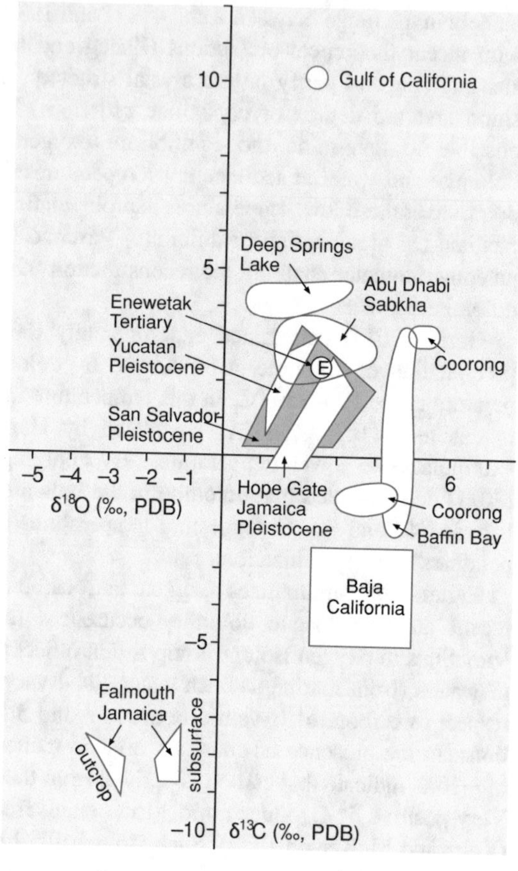


Fig. 3.48 Carbon and oxygen isotope composition of some recent and Pleistocene dolomite occurences (after Tucker and Wright 1990)

### 3.11.9 Freshwater Carbonates

Carbonates deposited in freshwater lakes exhibit a wide range in isotopic composition, depending upon the isotopic composition of the rainfall in the catchment area, its amount and seasonality, the temperature, the rate of evaporation, the relative humidity, and the biological productivity. Lake carbonates typically consist of a matrix of discrete components, such as detrital components, authigenic precipitates, neritic and benthic organisms. The separate analysis of such components has the potential to permit investigation of the entire water column. For example, the oxygen isotopic composition of authigenic carbonates and diatoms can be used to obtain a surface water signal of changes in temperature and meteoric conditions, while the composition of bottom dwellers can be used as a monitor of the water composition, assuming that the bottom water temperatures remained constant.

The carbon and oxygen isotope compositions of carbonate precipitated from many lakes show a strong covariance with time, typically in those lakes which represent closed systems or water bodies with long residence times (Talbot 1990).

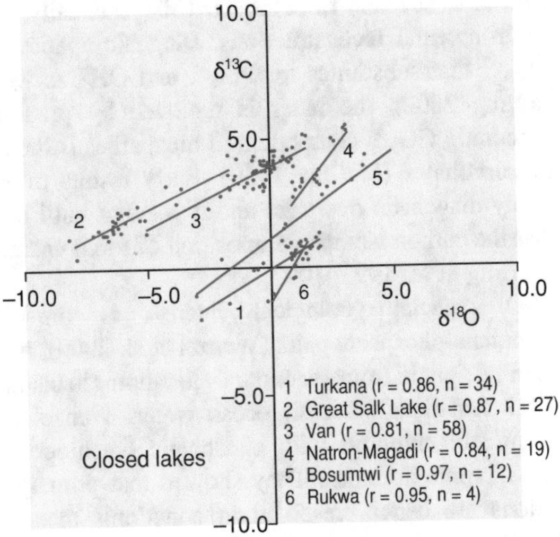


**Fig. 3.49** Carbon and oxygen isotope compositions of freshwater carbonates from recently closed lakes (after Talbot 1990)

In contrast, weak or no temporal covariance is typical of lakes which represent open systems with short residence times. Figure 3.49 gives examples of such covariant trends. Each closed lake appears to have a unique isotopic identity defined by its covariant trend, which depends on the geographical and climatic setting of a basin, its hydrology and the history of the water body (Talbot 1990).

### 3.11.10 Phosphates

The stable isotope composition of biogenic phosphates records a combination of environmental parameters and biological processes. Biogenic phosphate, $Ca_5(PO_4, CO_3)_3(F, OH)$, for paleoenvironmental reconstructions were first used by Longinelli (e.g. Longinelli 1966, 1984; Longinelli and Nuti 1973), and later by Kolodny and his coworkers (Kolodny et al. 1983; Luz and Kolodny 1985). However, the use was rather limited for many years, because of analytical difficulties. More recently these problems have been overcome by refinements in analytical techniques (Crowson et al. 1991; O'Neil et al. 1994; Cerling and Sharp 1996; Vennemann et al. 2002; Lecuyer et al. 2002), so the isotope analyses of phosphates for paleoenvironmental reconstruction has been used much more widely.

Under abiotic surface conditions phosphate is resistant to oxygen isotope exchange. During biological reactions, however, phosphate-water oxygen isotope exchange is rapid due to enzymatic catalysis (Kolodny et al. 1996; Blake et al. 1997, 2005; Paytan et al. 2002). O'Neil et al. (1994) have shown the importance of phosphate speciation in determining O isotope fractionation among different $PO_4$(aq) species and between $PO_4$(aq) species and water.

Phosphate materials that may be analyzed are bone, dentine, enamel, fish scales and invertebrate shells. In contrast to bone and dentine, enamel is extremely dense, so it is least likely to be affected diagenetically and the prime candidate for paleoevironmental reconstructions. Biogenic apatites contain besides the $PO_4$ group $CO_3^{2-}$ that substitutes for $PO_4^{3-}$ and $OH^-$ as well as "labile" $CO_3^{2-}$ (Kohn and Cerling 2002), the latter is removed by pretreatment with a weak acid. The remaining $CO_3^{2-}$ component in bioapatites is then analyzed similar to the analysis of carbonates (McCrea 1950). Early results of the carbonate-carbon seemed to imply diagenetic overprint and it was not until the 1990s that it became accepted that the carbon isotope composition of tooth enamel carbonate is a recorder of diet (Cerling et al. 1993, 1997).

Of special geological interest is the isotopic analyses of coeval carbonate-phosphate pairs (Wenzel et al. 2000), which helps to distinguish primary marine signals from secondary alteration effects and sheds light on the causes for $\delta^{18}O$ variations of fossil ocean water. Wenzel et al. (2000) compared Silurian calcitic brachiopods with phosphatic brachiopods and conodonts from identical stratigraphic horizons. They showed that primary marine oxygen isotope compositions are better preserved in conodonts than in brachiopod shell apatite and suggested that conodonts record paleotemperature and $^{18}O/^{16}O$ ratios of Silurian sea water. Joachimski et al. (2004) reached similar conclusions for Devonian seawater.

Studies on mammals, invertebrates and fishes clearly indicate that the oxygen isotope composition of biogenic apatite varies systematically with the isotope composition of the body water that depends on local drinking water (Longinelli 1984; Luz et al. 1984; Luz and Kolodny; 1985). For mammals, there is a constant offset between the $\delta^{18}O$ of body water and $PO_4$ (~18‰, Kohn and Cerling 2002) and between $PO_4$ and $CO_3$ components of bioapatite of ~8‰ (Bryant et al. 1996; Iacumin et al. 1996). Studies by Luz et al. (1990), and Ayliffe and Chivas (1990) demonstrated that $\delta^{18}O$ of biogenic apatite can also depend on humidity and on diet.

A different approach to get informations about the earth's climate has been used by Pack et al. (2013) by measuring the triple oxygen isotope composition of bone and teeth apatite of small mammals. The approach relies on the fact that atmospheric oxygen along with drinking water and water in food is one of the oxygen sources for mammals. Pack et al. (2013) used the relation between the $^{17}O$ anomaly of air oxygen and atmospheric $CO_2$, which is transferred to bone apatite thereby giving hints to atmospheric $CO_2$ concentrations during the animal's lifetime.

### 3.11.11 Iron Oxides

#### 3.11.11.1 Oxygen

Iron oxides/hydroxides are ubiquitous in soils and sediments and are common precursors to goethite and hematite. The initial precipitates in natural settings are water-rich ferric oxide gels and poorly ordered ferrihydrite, which are later slowly aged to goethite and hematite. The determination of oxygen isotope fractionations in the iron oxide—water system has led to controversial results (Yapp 1983, 1987,

2007; Bao and Koch 1999), yet oxygen isotope fractionations in bulk minerals are small and relatively insensitive to changes in temperatures. Miller et al. (2020) described a method to assess the oxygen isotope difference between the two sites of oxygen in goethite (FeOOH), (i) the Fe–O bond site and (ii) the site with bonds to iron and hydrogen. Miller et al. (2020) found temperature dependent O-isotope fractionations between these two sites, resulting in a single-mineral geothermometer. Natural goethites analyzed by Miller et al. (2020) gave reasonable formation temperatures between 15 and 41 °C.

Bao and Koch (1999), on the other hand, showed that iron oxides may record the oxygen isotope composition of ambient waters. They argued that the isotopic composition of original ferric oxide gels and ferrihydrite are erased by later exchange with ambient water during the ageing process. Thus, $\delta^{18}O$-values of natural crystalline iron oxides may monitor the long-term average $\delta^{18}O$-value of waters. In iron oxides covering the past 2 billion years, Galili et al. (2019) observed an increase in $^{18}O$, potentially reflecting an $^{18}O$ increase of seawater with time. Galili et al. (2019) suggested, however, that the $^{18}O$ enrichment was driven by an increase in terrestrial sediment cover or by a change in the proportion of high to low temperature hydrothermal water/rock interactions.

During conversion of goethite to hematite only small fractionation effects seem to occur, because most of the oxygen remains in the solid (Yapp 1987). Thus, in principle it should be possible to reconstruct the sedimentary environment of iron oxides from Precambrian Banded Iron Formations (BIF). By analyzing the least metamorphosed BIFs, Hoefs (1992) concluded, however, that the situation is not so simple. Infiltration of external fluids during diagenesis and/or low temperature metamorphism appears to have erased the primary isotope record in these ancient sediments.

#### 3.11.11.2 Iron

Due to the poorly crystalline state of ferric hydrous oxides and due to their fast transformations to stable minerals, Fe isotope fractionations between iron hydroxides and stable Fe phases are not well known. Approaches to determine the equilibrium fractionation factor between $FeII_{(aq)}$ and Fe hydroxides yield Fe fractionations around −3.2‰, making Fe(III) minerals the most enriched in $^{56}Fe$ (Johnson et al. 2002; Welch et al. 2003; Wu et al. 2011). Since fractionations between $FeII_{aq}$ and Fe hydroxides are similar to fractionations between $FeII_{aq}$ and $FeIII_{aq}$ (Johnson et al. 2002; Welch et al. 2003), Fe isotope fractionations between $Fe(III)_{aq}$ and Fe-hydroxides should be close to zero.

Special attention has been given to banded iron formations (BIFs), in which one of the largest ranges of Fe-isotope compositions on Earth are observed (Johnson et al. 2003, 2008; Steinhöfel et al. 2009, 2010; Halverson et al. 2011). Although models of BIF formation are still under debate, there is however general agreement, that the large Fe-isotope variations result from reduction and oxidation of iron in the water column, in the sedimentary environment and during diagenetic overprint (Steinhöfel et al. 2009, 2010). It is noteworthy that small-scale heterogeneities in iron oxides remain preserved to very high metamorphic stages (Frost et al. 2007).

#### 3.11.11.3 Fe–Mn Crusts

Ferromanganese oxide crusts are potential archives of the isotope composition of ocean water through geologic time. During their slow growth (1–15 mm/Ma) they incorporate a wide range of trace elements either by structural incorporation or surface complexation. Fe-Mn crusts are generally classified into three types depending on the predominant genetic origin during their growth: (i) hydrogenetic (seawater-derived), (ii) hydrothermal and (iii) diagenetic. The temporal record found in Fe–Mn crusts depends on the residence time of specific elements in seawater and their response to changes in relative fluxes. Fe isotope records of Fe–Mn crusts, for instance, do not reflect the Fe isotope composition of the global ocean, but that of the local ocean in which the crust has been formed, since the residence time of dissolved iron (200–500 years) is less than the mixing time of average water molecules (around 1000 years). Comparison of the Fe isotope composition of modern Fe–Mn crust growth surfaces and ambient seawater indicates that the Fe bound in the crust is isotopically lighter than that of dissolved seawater (Horner et al. 2015). A determination of the metal isotope composition of seawater in the past thus requires that the fractionation factor during sorption is known.

As shown by Horner et al. (2015) a reconstruction of deep-sea Fe isotope compositions for the last 76 Ma in the Central Pacific yield large systematic changes in the Fe isotope composition indicating several distinct Fe sources. Metals having longer mean residence times like Mo (Siebert et al. 2003), U (Goto et al. 2014; Wang et al. 2016), and Ni (Gueguen et al. 2016), indicate more or less constant isotope compositions during the Cenozoic.

### 3.11.12 Sedimentary Sulfur and Pyrite

#### 3.11.12.1 Sulfur

Analysis of the sulfur isotope composition of sediments may yield important information about the origin and further transformations of sulfur compounds. There has been much progress to identify and measure the isotopic composition of different forms of sulfur in sediments (e.g. Mossmann et al. 1991; Zaback and Pratt 1992; Brüchert and Pratt 1996; Neretin et al. 2004). Pyrite is generally considered to be the end product of sulfur diagenesis in anoxic marine sediments. Acid-volatile sulfides (AVS), which include "amorphous" FeS, mackinawite, greigite and pyrrhotite, are considered to be transient early species, but investigations by Mossmann et al. (1991) have demonstrated that AVS can form before, during and after precipitation of pyrite within the upper tens of centimeters of sediment.

Up to six or even seven sulfur species have been separated and analyzed for their isotope composition by Zaback and Pratt (1992), Brüchert and Pratt (1996) and Neretin et al. (2004). Their data provides information regarding the relative timing of sulfur incorporation and the sources of the individual sulfur species. Pyrite exhibits the greatest $^{34}S$ depletion relative to sea water. Acid-volatile sulfur and sulfur in organic compounds are generally enriched in $^{34}S$ relative to pyrite. This indicates that pyrite is precipitated nearest to the sediment–water interface under mildly reducing conditions, while AVS and kerogen sulfur resulted from formation at greater depth under more reducing conditions with low concentrations of pore water sulfate. Elemental sulfur is most abundant in surface sediments and, probably, formed by oxidation of sulfide diffusing across the sediment–water interface. In iron depleted systems reduced sulfur species also react with organic matter to form organic sulfur compounds. Relative to pyrite organic sulfur compounds are generally enriched in $^{34}S$, which can be explained by the relative timing of pyrite (early) and organic sulfur (late) formation (Shawar et al. 2018). By using a GC–MC-ICP-MS technique, Raven et al. (2015) were able to measure the compound-specific S isotope composition of organic sulfur compounds. In contrast to earlier findings, extractable organic S-compounds are $^{34}S$ depleted relative to kerogen and porewater sulfide providing additional informations about organic matter sulfurization.

Pyrite is the end product of sedimentary S- and Fe-cycling and their stable isotopes record variations of redox changes. Bacterial sulfate reduction is accomplished by the oxidation of organic matter:

$$2\,CH_2O + SO_4^{2-} \rightarrow H_2S + 2\,HCO_3^-$$

the resulting $H_2S$ reacting with available iron, which is in the reactive non-silicate bound form (oxy-hydroxides). Thus, the amount of pyrite formed in sediments may be limited by (i) the amount of sulfate, (ii) the amount of organic matter and (iii) the amount of reactive iron. Based upon the relationships between these three reservoirs different scenarios for pyrite formation in anoxic environments can be envisaged (Raiswell and Berner 1985). In normal marine sediments, where oxygen is present in the overlying water body, the formation of pyrite appears to be limited by the supply of organic matter.

Due to the activity of anaerobic sulfate reducing bacteria, most sulfur isotope fractionation takes place in the uppermost mud layers in shallow seas and tidal flats. As a result, sedimentary sulfides are depleted in $^{34}S$ relative to ocean water sulfate. The depletion is usually in the order of 20–60‰ (Hartmann and Nielsen 1969; Goldhaber and Kaplan 1974), although bacteria in pure cultures have been observed to produce fractionations up to a maximum reported value of 47‰ (Kaplan and Rittenberg 1964; Bolliger et al. 2001). Therefore, sedimentary sulfides depleted in $^{34}S$ by more than the apparent limit of 47‰ suggest additional fractionations that probably accompany sulfide oxidation and formation of sulfur intermediates and further metabolism. To explain the discrepancy between culture experiments and natural environments the bacterial disproportionation of intermediate sulfur compounds has been proposed (Canfield and Thamdrup 1994; Cypionka et al. 1998; Böttcher et al. 2001).

Sulfur isotope variations in sediments reflect a record of primary syngenetic as well as secondary diagenetic processes (Jorgenson et al. 2004). For a given range of sulfur isotope values the most negative value should represent the least affected, most primary signal or the one that is most affected by the oxidative part of the sulfur cycle. In a few cases pyrite sulfur with higher $\delta^{34}S$-values than coexisting seawater has been found in the fossil record, which has been attributed to post-depositional diagenetic overprint by anaerobic methane oxidation (Jorgensen et al. 2004).

$\delta^{34}S$-values of bulk pyrite integrate over the time, in which pyrite has formed by different processes and in different environments. Microanalytical techniques has opened the possibility to determine intra-grain and inter-grain variability of bulk pyrite. Generally, pyrite grains become enriched in $^{34}S$ towards the margin of grains which has been interpreted as evidence for microbial sulfate reduction in closed systems.

Investigating pyrites from Devonian carbonates with the ionprobe, Riciputi et al. (1996) observed a bimodal distribution of sulfides that are very heterogeneous on a thin section scale varying by as much as 25‰. The predominantly low δ-values indicate bacterial sulfate reduction, whereas the higher values reflect formation at much greater depths by thermochemical sulfate reduction. Correlations between pyrite morphology and isotope values suggest that sulfate reduction was a very localized process, which varied considerably on a small scale. McKibben and Riciputi (1998) reported $\delta^{34}S$-variations of about 105‰ over 200 μm in single pyrite grains.

Besides bacterial sulfate reduction, thermochemical sulfate reduction (TSR) in the presence of organic matter is another process which can produce large quantities of $H_2S$. The crucial question is whether abiological sulfate reduction can occur at temperatures as low as 100 °C, which is just above the limit of microbiological reduction. Experimental and field evidence indicates that TSR starts at temperatures between 100 and 140 °C (Goldstein and Aizenshat 1994) and that the presence of reduced sulfur is essential for initiating thermochemical sulfater reduction (Zhang et al. 2008a, b). TSR-generated $H_2S$ typically has $\delta^{34}S$-values close to the source sulfate (Krouse 1977; Cai et al. 2003 and others). Sulfur isotope fractionations associated with the formation of organic sulfur compounds during TSR by Meshoulam et al. (2016) have demonstrated that individual organic sulfur compounds may give informations that cannot be discerned from the $\delta^{34}S$-values of the bulk phases, $H_2S$ and sulfates.

#### 3.11.12.2 Iron

Like sulfur, the interpretation of the large range of Fe-isotope variations in pyrite is demanding (Severmann et al. 2006; Guilbaud et al. 2011; Rolison et al. 2018; Mansor and Fantle 2019 and others). As argued by Mansor and Fantle (2019) predominantly negative, but highly variable $\delta^{56}Fe$ values suggest a predominance of kinetic isotope fractionations during pyrite formation; positive values, on the other hand, suggest predominantly equilibrium isotope fractionations.

Mackinawite ($FeS_x$) is regarded as a precursor mineral for pyrite formation and Fe isotope fractionations in the $Fe^{2+}$(aq)—FeS system have been taken as decisive for pyrite Fe isotope signatures. Butler et al. (2005) and Guilbaud et al. (2011) demonstrated experimentally that FeS is depleted in $^{56}Fe$ relative to $Fe^{2+}$. Johnson et al. (2008) argued that $\delta^{56}$Fe values of pyrite reflect a mixture of FeS compounds formed during bacterial reduction and Fe that is produced by dissimilatory iron reduction. According to Marin-Carbonne et al. (2014a, b) coupled Fe and S isotope variations in pyrite indicate different mineral precursors: (i) mackinawite that is precipitated in the water column and (ii) greigite that is formed in the sediment.

In summary, detailed investigations on sedimentary pyrite have revealed large variations in sulfur and iron isotope compositions that potentially may allow the distinction between biogenic and abiogenic processes of formation and even may indicate different metabolic processes (i.e. Archer and Vance 2006; Marin-Carbonne et al. 2014a, b, 2020).

## 3.12 Palaeoclimatology

Past climates leave their imprint in the geologic record in many ways. For temperature reconstructions the most widely used geochemical method is the measurement of stable isotope ratios. Samples for climate reconstruction have in common that their isotope composition depends in a sensitive way on the temperature at the time of their formation.

Climatic records can be divided into (i) marine and (ii) continental records. Because the ocean system is very large and well-mixed, the oceanic record carries a global signal, while continental records are affected by regional factors. One restriction in reconstructing climates is the temporal resolution. This is especially true for marine sediments. Sedimentation rates in the deep-ocean generally are between 1 and 5 cm/$10^3$ y; highly productive areas have 20 cm/$10^3$y, which limits the temporal resolution to 50 years for productive areas and to 200 years for the other areas. Furthermore, benthic organisms can mix the top 20 cm of marine sediments, which further reduces temporal resolutions.

### 3.12.1 Continental Records

Isotopic reconstruction of climatic conditions on the continents is difficult, because land ecosystems and climates exhibit great spatial and temporal heterogeneity. The most readily determined terrestrial climatic parameter is the isotopic composition of precipitation, which is in turn dependent largely but not exclusively on temperature. Relevant climatic information from meteoric precipitation is preserved in a variety of natural archives, such as (i) tree rings, (ii) organic matter and (iii) hydroxyl-bearing minerals.

#### 3.12.1.1 Tree Rings

Tree rings offer an absolute chronology with annual resolution, but the scarcity of suitable old material and uncertainties about the preservation of original isotope ratios are major restrictions in the application of tree rings. The cellulose component of plant material is generally used for isotope studies because of its stability and its well-defined composition. Numerous studies have investigated the stable isotope composition of tree rings. However, in many respects climatic applications are limited.

An increasing number of studies have investigated the complex processes that transfer the climatic signal in meteoric water to tree cellulose (for instance White et al. 1994; Tang et al. 2000). Although there are strong correlations of $\delta^2H$ and $\delta^{18}O$ with source water, there are variable fractionations between water and cellulose. The complexities result from the interplay of various factors such as humidity, amount of precipitation, topography, biological isotope fractionation, root structure, ageing of late-wood. Tang et al. (2000) assessed both systematic (variations of temperature, humidity, precipitation etc.) and random isotopic variations in tree rings from a well characterized area in the northwestern United States, and demonstrated for instance that temperature only explains up to 26% of the total variance of $\delta^2H$ values of cellulose nitrate.

#### 3.12.1.2 Organic Matter

The utility of D/H ratios in organic matter as paleoclimatic proxies relies on the preservation of its primary biosynthetic signal. In recent years the D/H analysis of compound-specific sedimentary biomarkers have been increasingly used. $\delta^2H$-values of lipid biomarkers from aquatic organism and terrestrial plants, for instance, can be used as palaeohydrological proxies (Sachse et al. 2012 and others).

$\delta^2H$-values of leaf waxes, being stable over long time periods, have been increasingly used in terrestrial paleoclimate research (Niedermeyer et al. 2016; Daniels et al. 2017). Leaf waxes reflect the hydrogen isotope composition of precipitation with a large offset relative to the $\delta^2H$-values of precipitation. Accurate knowledge of the hydrogen isotope fractionation is necessary for a climatic interpretation.

The question arises at what point paleoclimatic information is lost during diagenesis and thermal maturation. Schimmelmann et al. (2006) argued that in the earliest stages of diagenesis $\delta^2H$-values of most lipid biomarkers are unaffected. With the onset of catagenesis quantitative information diminishes, but qualitative information may be still preserved. At the highest levels of maturity, biomarkers become thermally unstable and can undergo degradation leading to extensive hydrogen isotope exchange (Sessions et al. 2004) and therefore limiting paleoclimate information.

#### 3.12.1.3 Hydroxyl-Bearing Minerals

Hydoxyl bearing minerals might be regarded as another tool to reconstruct climatic changes. Again, there are major difficulties that restrict a general application. Fractionation factors of clay minerals and hydroxides are not well constrained, especially at low temperatures and meaningful $\delta^2H$ and $\delta^{18}O$ measurements require pure mineral separates, which are extremely difficult to achieve due to their small particle size and because these phases are often intergrown. Furthermore, there is a concern that some clays are detrital, whereas others are authigenic; thus, mixtures may be difficult to interpret.

#### 3.12.1.4 Lake Sediments

The isotope composition of biogenic and authigenic mineral precipitates from lake sediments can be used to infer changes in either temperature or the isotope composition of lake water. Knowledge of the factors that may have influenced the isotope composition of the lake water is essential for the interpretation of the precipitated phases (Leng and Marshall 2004). In many lakes the combined analysis of different types of authigenic components (precipitated calcite, ostracodes, bivalves, diatoms, gypsum etc.) may offer the possibility of obtaining seasonally specific informations. Combining hydrogen and triple oxygen isotopes in gypsum hydration water, Gazquez et al. (2018) presented a method to quantitatively estimate the paleo-humidity in lakes.

One of the most useful components for estimating past climate variations are non-marine ostracodes (small bivalved crustaceans), which can live in most types of fresh-water and can be regarded as the "foraminifera of the continent". An increasing number of studies have demonstrated the potentials of ostracodes to reconstruct changes in temperatures of mean annual precipitation, changes in paleohydrology and evaporation histories (Lister et al. 1991; Xia et al. 1997a, b; Von Grafenstein et al. 1999; Schwalb et al. 1999). A number of authors have demonstrated systematic differences in $\delta^{18}O$ of up to 2‰ between ostracodes and calcite precipitated under equilibrium conditions and even larger differences for $\delta^{13}C$. These differences have not been explained satisfactorily, because the knowledge about life cycles, habitat preferences and valve formation mechanisms of ostracodes is still limited.

#### 3.12.1.5 Speleothems

Speleothems have been frequently used in recent years for climate reconstructions, because of their wide occurrence and their precisely dated highly resolvable records. Two features in caves facilitate the use of stable isotopes as a palaeoarchive: (i) cave air temperatures remain relatively constant throughout the year and are similar to the mean annual temperature above the cave. (ii) In cool temperate climate regions, cave air is characterized by very high humidity that minimizes evaporation effects.

Interest in speleothems as recorders of continental palaeo-environments has increased considerably in recent years. The potential of speleothems as climate indicators was first discussed by Hendy and Wilson (1968) followed by Thompson et al. (1974). These early investigators already recognized the complexity of cave carbonate isotope compositions (see reviews by McDermott 2004 and Lachniet 2009). A working group called SISAL (Speleothem Isotope Synthesis and Analysis) compiled a database for carbon and oxygen isotopes. As shown by Baker et al. (2019), dripwater $\delta^{18}O$ closely reflects amount-weighted precipitation $\delta^{18}O$ in cool climates. In warm and dry environments speleothems have a seasonal bias towards the precipitation of recharge periods.

Most isotope studies on speleothems have concentrated on $\delta^{18}O_{calcite}$ as the principal paleoclimatic indicator. Fewer studies have discussed the potential use of $\delta^{2}H$ and $\delta^{18}O$ of fluid inclusions in speleothems (Dennis et al. 2001; McGarry et al. 2004; Zhang et al. 2008a, b). With an improved crushing technique for the liberation of the fluid inclusion water, Zhang et al. (2008a, b) were able to recover the water without isotopic fractionation. Affolter et al. (2014) described a cavity ring spectroscopy method to measure simultaneously $\delta^{2}H$, $\delta^{17}O$ and $\delta^{18}O$.

With respect to water oxygen in fluid inclusions, post-depositional isotope exchange may complicate $\delta^{18}O$ interpretation, leaving hydrogen to be the reliable climate indicator. Complications may also arise from kinetic isotope effects during rapid degassing of $CO_2$. As shown by Affek et al. (2008), Daeron et al. (2011) and others, clumped isotopes may provide a sensitive indicator for disequilibrium effects. In such cases decreased $\Delta_{47}$ values correlate with increased $\delta^{18}O$-values corresponding to higher apparent temperatures. Using hydrogen isotopes of fluid inclusion water from the Milandre Cave in Switzerland, Affolter et al. (2019) established a mid-latitude European mean annual temperature reconstruction for the last 14,000 years.

#### 3.12.1.6 Phosphates

Oxygen isotope compositions of phosphates have also been used as a paleotemperature indicator. Since the body temperature of mammals is constant at around 37 °C, $\delta^{18}O$ values in either bones or teeth depend only on the $\delta^{18}O$ value of the body water, which in turn depends on drinking water (Kohn 1996). Thus, phosphates from continental environments are an indirect proxy of ancient meteoric waters.

The best proxy appears to be mammalian tooth enamel (Ayliffe et al. 1994; Fricke et al. 1998a, b), which forms incrementally from the crown to the base of the tooth. Enamel, therefore, preserves a time series of $\delta^{18}O$ values of precipitation along the direction of growth that reflect only $^{18}O$-changes of ingested water. Oxygen isotope data for teeth of mammal herbivores that lived over a wide range of climatic conditions demonstrate that intra tooth $\delta^{18}O$-values mirror both seasonal and mean annual differences in the $^{18}O$ content of local precipitation (Fricke et al. 1998a). Records going back to glacial-interglacial transitions have been described by Ayliffe et al. (1992). Fricke et al. (1998b) even postulated that tooth enamel may provide a temperature record as far back as the Early Cenozoic.

### 3.12.2 Ice Cores

Ice cores from polar regions represent prime recorders of past climates. They have revolutionized our understanding of Quaternary climates by providing high resolution records of changing isotope compositions of snow or ice and of changing air compositions from air bubbles occluded in the ice. The best documented ice-core record from Greenland is a pair of 3 km long ice cores from the summit of Greenland. These cores provide a record of climate as far back as 110,000 years ago. Precise counting of individual summer and winter layers extends back to at least 45,000 years ago.

The Antarctic ice sheet also has provided numerous ice cores for paleoclimate research. Antarctica is colder and its ice sheet is larger and thicker than that on Greenland. It accumulates more slowly than in Greenland, such that its temporal resolution is not as good. The Vostok ice core has provided strong evidence of the nature of climate changes over the past 420 ky. More recently, a core from Dome C, Antarctica has almost doubled the age range to the past 740 ky (Epica 2004). Besides, good agreement with the Vostok core was observed for the 4 most recent glacial cycles, the Dome C core extends back to 8 glacial cycles.

High elevation ice cores from low latitudes, that have been drilled in Africa (Kilimanjaro), South America and Asian Himalayas (e.g. Thompson et al. 2006) represent an important addition to the polar region ice cores. Some of these high altitude, low latitude ice cores span the last 25,000 years, representing a high-resolution record of the late glacial stage and the Holocene (Thompson et al. 2000). The interpretation of δ-values is, however, challenging, because of large seasonal differences in precipitation regimes (amount effect) in the tropics.

Oxygen and hydrogen isotope ratios and various atmospheric constituents in ice cores have revealed a detailed climatic record for the past 700 ky. To convert isotopic changes to temperatures, temperature—$\delta^{18}O$ correlations must be known. In early work, Dansgaard et al. (1993) proposed a relationship of 0.63‰ per 1 °C, whereas Johnsen et al. 1995) have used 0.33‰ per 1 °C (but see the remarks of caution by Allen and Cuffey 2001). The δ-T relationship varies with climatic conditions, especially between interglacial and glacial periods, because a more extensive sea-ice cover increases the distance to moisture sources and the isotopic composition of oceans changed during glacial periods.

Figure 3.50 compares $\delta^{18}O$ ice core data from GRIP and NGRIP in Greenland for the time period 50,000–30,000 years with significantly colder temperatures during the Last Glacial Maximum (LGM) than the time period for the last 10,000 years. Characteristic features of Fig. 3.50 are fast changes in $\delta^{18}O$-values fluctuating between −37 and −45‰. These socalled Dansgaard-Oeschger events (Dansgaard et al. 1993; Grootes et al. 1993) are characterized by rapid warming episodes within decades leading to higher $\delta^{18}O$ followed by gradual cooling over a longer period. 23 Dansgaard-Oeschger events have been identified between 110,000 and 23,000 years before present, the causes for these sawtooth patterns are still unclear.

Besides O- and H-isotopes, nitrogen isotope investigations in ice core have been carried out to trace climatic (Hastings et al. 2005) and anthropogenic-driven (Geng et al. 2014) changes in nitrate sources as nitrate is one of the most abundant ions in ice cores. Geng et al. (2014) demonstrated the N-isotope composition of nitrate has undergone a marked change during the past 200 years by burning fossil fuels and making synthetic fertilizers.

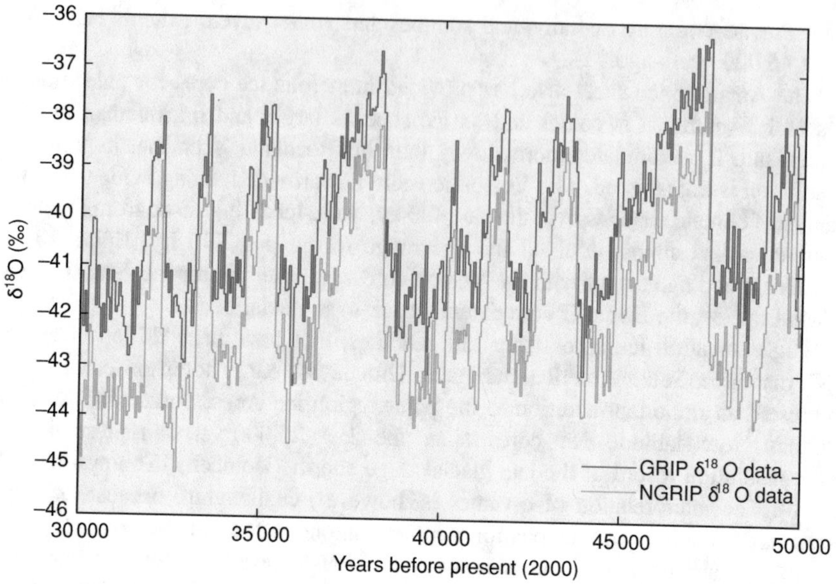


**Fig. 3.50** Dansgaard-Oeschger events in the time period from 45,000 to 30,000 years before present from GRIP and NGRIP ice core data (http://en.wikipedia.org/wiki/Image:Grip-ngrip-do18-closeup.png)

### 3.12.2.1 Correlations of Ice-Core Records

Ice-core isotope stratigraphy represents a major advance in paleoclimatology because it enables the correlation of climate records from the two poles with each other and with the high-resolution deep-sea marine climate records over the past 100 ka (Bender et al. 1994), allowing the study of phasing between the ocean and the atmosphere. One of the most difficult problems in correlating ice-cores is determining the age-depth relationship. If accumulation rates are high enough, accurate timescales have been achieved for the last 10,000 years. Prior to that there is increasing uncertainty, but in recent years new approaches have been developed, improving age determinations and allowing age correlations between different ice cores (see Fig. 3.50).

A very promising method for correlation purposes relies on changes in atmospheric gas composition. As the mixing time of the atmosphere is on the order of 1–2 years, changes in gas composition should be synchronous. Bender et al. (1994) have used variations of $\delta^{18}O$ in gas inclusions from ice-cores correlating the Vostok and GISP-2 ice cores. Similar $^{18}O$-variations in both cores makes an aligment of the two records possible (Sowers et al. 1991, 1993), which then allows the comparison of other parameters such as $CO_2$ and $CH_4$ with temperature changes as deduced from the isotopic composition of the ice.

### 3.12.2.2 Gas-Inclusions in Ice Cores

Atmospheric trace gas chemistry is a rapidly growing field of paleo-atmospheric research, because the radiative properties of $CO_2$, $CH_4$ and $N_2O$ make them potential indicators of climate change. A fundamental problem in constructing a record of trace gas concentrations from ice-cores is the fact that the air in bubbles is always younger than the age of the surrounding ice. This is because as snow is buried by later snowfalls and slowly becomes transformed to firn and ice, the air between the snow crystals remains in contact with the atmosphere until the air bubbles become sealed at the firn/ice transition, when density increases to about 0.83 g/cm$^3$. The trapped air is thus younger than the matrix, with the age difference depending mainly on accumulation rate and temperature. In Greenland, for instance the age difference varies between 200 and 900 years.

Sowers et al. (1993) and Bender et al. (1994) showed that it is possible to construct an oxygen isotope curve similar to that derived from deep-sea foraminifera from molecular $O_2$ trapped in ice. These authors argued that $\delta^{18}O$ (atm) can serve as a proxy for ice volume just as $\delta^{18}O$ values in foraminifera. The isotope signal of atmospheric oxygen can be converted from seawater via photosynthetic marine organisms according to the following scheme

$$\delta^{18}O(\text{seawater}) \rightarrow \text{photosynthesis} \rightarrow \delta^{18}O(\text{atm}) \rightarrow \text{polar ice} \rightarrow \delta^{18}O(\text{ice})$$

This conversion scheme is, however, complex and several hydrological and ecological factors have to be considered. Sowers et al. (1993) argued that these factors remained near constant over the last glacial-interglacial cycle, so that the dominant signal in the atmospheric oxygen isotope record represents an ice-volume signal.

Air composition in ice cores is slightly modified by physical processes, such as gravitational and thermal fractionation. A gas mixture in ice cores with different molecular weights will partially segregate due to thermal diffusion and gravitational fractionation. Generally, the species with greater mass will migrate towards the bottom and/or the cold end of a column of air. By slow diffusion, air trapped in ice-cores can develop slight changes in atmospheric ratios such as the Ar/$N_2$ ratio as well as fractionate the nitrogen and oxygen isotope composition of air molecules. This approach was pioneered by Severinghaus et al. (1996), who first showed that thermal diffusion can be observed in sand dunes. Later Severinghaus et al. (1998), Severinghaus and Brook (1999) and Grachev and Severinghaus (2003) demonstrated that thermally driven isotopic anomalies are detectable in ice core air bubbles. Since gases diffuse about 50 times faster than heat, rapid climatic temperature changes will cause an isotope anomaly. Nitrogen in bubbles in snow thus may serve as a tracer for palaeoclimatic reconstructions because the $^{15}N/^{14}N$ ratio of atmospheric $N_2$ has stayed constant in the atmosphere. The measurement of nitrogen isotope ratios can, therefore, supplement the oxygen isotope record and can be used to determine the rapidity and scale of climate change. By measuring the thickness of ice separating nitrogen and oxygen isotope anomalies at the end of Younger Dryas 11,500 years ago, Severinghaus et al. (1998) estimated that the rate of temperature change to be <50–100 years and suggested that the Younger Dryas was about 15 °C colder than today which is about twice as large as estimated from Dansgaard-Oeschger events.

### 3.12.3 Marine Records

Most oceanic paleoclimate studies have concentrated on foraminifera. In many cases analyses have been made both of planktonic and benthonic species. Since the first pioneering paper of Emiliani (1955), numerous cores from various sites of the DSDP and ODP program have been analyzed and, when correlated accurately, have produced a well-established oxygen isotope curve for the Pleistocene and Tertiary. These core studies have demonstrated that similar $\delta^{18}O$-variations are observed in all areas. With independently dated time scales on hand, these systematic $\delta^{18}O$ variations result in synchronous isotope signals in the sedimentary record because the mixing time of the oceans is relatively short ($10^3$ years). These signals provide stratigraphic markers enabling correlations between cores which may be thousands of kilometers apart. Several Pleistocene biostratigraphic data have been calibrated with oxygen isotope stratigraphy, which helps to confirm their synchrony. This correlation has greatly facilitated the recognition of both short and long time periods of characteristic isotopic compositions, and times of rapid change from one period with characteristic composition to another, thus, making oxygen isotope stratigraphy a practical tool in modern paleoceanographic studies. Figure 3.51 shows the oxygen isotope curve for the Pleistocene. This diagram exhibits several striking features: the most obvious one is the cyclicity. Furthermore, fluctuations never go beyond a certain maximum value on either side of the range. This seems to imply that very effective feedback mechanisms are at work stopping the cooling and warming trends at some maximum level. The "sawtooth"-like curve as in Fig. 3.51 is characterized by very steep gradients: maximum cold periods are immediately followed by maximum warm periods, although the trend of $\delta^{18}O$ variation as a function of cooling is opposite in foraminifera compared to ice.

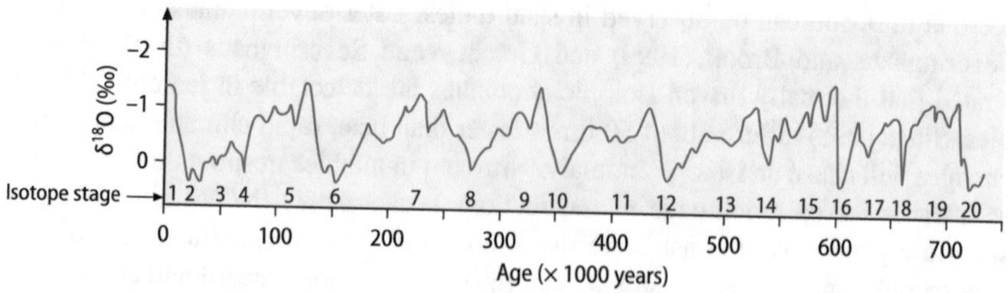


Fig. 3.51 Composite $\delta^{18}O$ fluctuation in the foraminifera species G saculifer from Caribbean cores (Emiliani 1978)

Emiliani (1955) introduced the concept of "isotopic stages" by designating stage numbers for identifiable events in the marine foraminiferal oxygen isotope record for the Pleistocene. Odd numbers identify interglacial or interstadial (warm) stages, whereas even numbers define $^{18}O$ enriched glacial (cold) stages. A second terminology used for subdividing isotope records is the concept of terminations labeled with Roman numbers I, II, III etc. which describe rapid transitions from peak glacial to peak interglacial values. This approach was used by Martinson et al. (1987) to produce a high-resolution chronology, called the Specmap time scale which is used when plotting different isotope records on a common time scale. With these different techniques a rather detailed chronology can be worked out.

A careful examination of the curve shown in Fig. 3.51 shows a periodicity of approximately 100,000 years. Hays et al. (1976) argued that the main structure of the oxygen isotope record is caused by variations in solar insolation, promoted by variations in the Earth's orbital parameters. Thus, isotope data have played a capital role in the confirmation of the "Milankovitch Theory" which argues that the isotope and paleoclimate record is a response to the forcing of the orbital parameters operating at specific frequencies. Lisiecki and Raymo (2005) presented an updated compilation of global $\delta^{18}O$ time series for benthic foraminifera for the past several million years that is widely used for astrochronologic correlations.

### 3.12.3.1 Corals

Reef-building corals provide high-resolution records up to several centuries that potentially are ideal tools for the reconstruction of tropical climate. Annual banding provides chronological control and high year-round growth rates allows annual to subannual resolution. Coral skeletons are well known for strong vital effects, their oxygen isotope composition is generally depleted relative to equilibrium by 1–6‰. Because of this strong non-equilibrium fractionation early workers were highly skeptical about the usefulness of $\delta^{18}O$-values as climate indicators. Later workers, however, realized that the $\delta^{18}O$ records reveal subseasonal variations in seawater temperature and salinity. Most climate studies circumvent the problem of equilibrium offsets by assuming a time independent constant offset and interpret relative changes only. Thus, $\delta^{18}O$ values of corals generally are not interpreted as temperature records, but as records reflecting combinations of temperature and salinity changes. $\delta^{18}O$ values in corals may record anomalies associated with El Nino (Cole et al. 1993; Dunbar et al. 1994), including the dilution effect on $\delta^{18}O$ by high amounts of precipitation (Cole and Fairbanks 1990).

Coral growth rates vary over the course of a year, which is expressed in an annual banding. Leder et al. (1996) demonstrated that a special microsampling technique (fifty samples a year) is necessary to accurately reproduce annual sea surface conditions. Generally, $\delta^{18}O$ records show a long-term warming and/or decrease in salinity throughout the tropical oceans (Gagan et al. 2000; Grottoli and Eakin 2007). Fossil coral samples imply an additional problem. Since corals dominantly are composed of aragonite, subaereal exposure of fossil corals will easily change oxygen isotope values due to diagenetic recrystallization to calcite.

#### 3.12.3.2 Conodonts

Conodonts are tooth-like phosphatic microfossils that are widespread in both space and time. Since the early work of Longinelli (1966), Longinelli and Nuti (1973) and Kolodny et al. (1983) phosphates have been used to reconstruct temperatures. Although being more difficult to analyze, phosphates are advantageous over carbonates because they are more resistant towards isotope exchange. Puceat et al. (2010) redetermined the phosphate-water oxygen isotope fractionation on fish raised under controlled conditions and observed a similar slope to earlier equations, but an offset of about +2‰, shifting calculated temperatures to 4–8 °C higher temperatures. With this temperature calibration, reasonable temperatures can be obtained for the Devonian (Joachimski et al. 2009); at the Permian/Triassic boundary a large temperature increase has been observed (Joachimski et al. 2012). For the early Paleozoic Goldberg et al. (2021) observed good agreement of bulk rock, brachiopod and conodonts $\delta^{18}O$ records. Diagenetic overprint has been assessed by pairing clumped isotopes and $\delta^{18}O$ data.

#### 3.12.3.3 Characteristic Climatic Events

During the last two decades a rapid growth of high-resolution isotope records across the Cenozoic has taken place. Zachos et al. (2001) have summarized 40 DSDP and ODP sites representing various intervals in the Cenozoic. Their compilation of benthic foraminifera shows a $\delta^{18}O$ range of 5.4‰ over the course of the Cenozoic. This variation provides constraints on the evolution of deep-sea temperature and continental ice volume. Because deep ocean waters are derived primarily from cooling and sinking of water in polar regions, the deep-sea temperature data also reflect high-latitude sea-surface temperatures.

One of the most dramatic climatic events during the Cenozoic is the Paleocene-Eocene-Thermal-Maximum (PETM) at about 56 Ma lasting less than 200,000 years (McInerney and Wing 2011). The PETM is characterized by an abrupt temperature increase of about 5 °C or even up to 8 °C in conjunction with a large negative carbon isotope anomaly.

For the period prior to the first onset of Antarctic glaciation (around 33 Ma), oxygen isotope variations in global benthic foraminifera records reflect temperature changes only. Oxygen isotope data suggest the deep oceans of Cretaceous and Paleocene age may have been as warm as 10–15 °C, which is very different from today's conditions, when deep waters vary from about + 4 to −1 °C. The compilation of Zachos et al. (2001) indicates a bottom water temperature increase of about 5 °C over 5 million years during the Paleocene to the early Eocene.

Variations in the benthic foraminifera record after 33 Ma indicate fluctuations in global ice volume in addition to temperature changes. Since then the majority of the $\delta^{18}O$ variations can be attributed to fluctuations in the global ice volume. Thus, Tiedemann et al. (1994) demonstrated the presence of at least 45 glacial-interglacial cycles over the last 2.5 Ma.

Zachos et al. (2001) discussed the Cenozoic climatic history in respect to three different time frames: (i) long-term variations driven mainly by tectonic processes on time scales of $10^5$–$10^7$ years, (ii) rhythmic and periodic cycles driven by orbital processes with characteristic frequencies of roughly 100, 40 and 23 kyr. (These orbitally driven variations in the spatial and seasonal distribution of solar radiation are thought to be the fundamental drivers of glacial and interglacial oscillations), (iii) brief, aberrant events with durations of $10^3$–$10^5$ years. These events are usually accompanied by a major perturbation in the global carbon cycle; the 3 largest occurred at 55, 34 and 23 Ma.

Figure 3.52 summarizes the oxygen isotope curve for the last 65 Ma. The most pronounced warming trend is expressed by a 1.5‰ decrease in $\delta^{18}O$ and occurred early in the Cenozoic from 59 to 52 Ma, with a peak in Early Eocene. Coinciding with this event is a brief negative carbon isotope excursion, explained as a massive

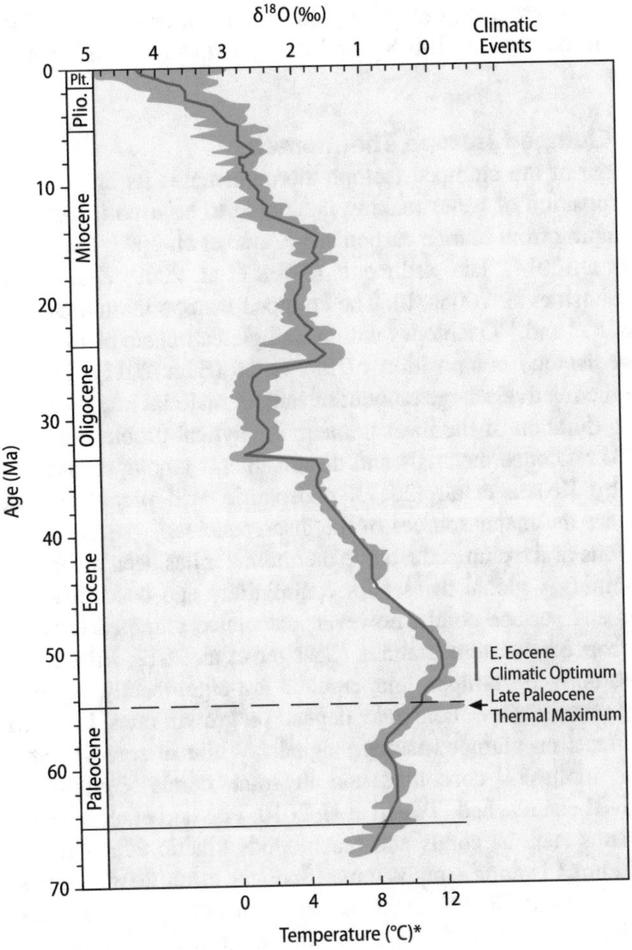


Fig. 3.52 Global deep-sea isotope record from numerous DSDP and ODP cores (Zachos et al. 2001)

release of methane into the atmosphere (Norris and Röhl 1999). These authors used high resolution analysis of sedimentary cores to show that two thirds of the carbon shift occured just in a few thousand years, indicating a catastrophic release of carbon from methane clathrates into the ocean and atmosphere.

A 17 Ma trend toward cooler conditions followed, as expressed by a 3‰ rise in $\delta^{18}O$, which can be attributed to a 7 °C decline in deep-sea temperatures. All subsequent changes reflect a combined effect of ice-volume and temperature.

To investigate the rhythmic scales, Zachos et al. (2001) looked in detail to four-time intervals (0–4.0; 12.5–16.5; 20.5–24.5; 31–35 Ma) each representing an interval of major continental ice-sheet growth or decay. These intervals demonstrate that climate varies in a quasi-periodic fashion. In terms of frequency, Zachos et al. (2001) concluded that much of the power in the climate spectrum appears to be related with changes in the obliquity (40 ky). This inference of a 40 ky periodicity contrasts with the obvious 100 Ky periodicity indicated by isotope curves for the last 1–2 Ma.

#### 3.12.3.4 Clumped Isotope Thermometry

The advantage of the clumped isotope thermometer is its independence from the isotope composition of water making it suitable to be used in very different environments ranging from marine carbonates (Came et al. 2007; Finnegan et al. 2011; Cummins et al. 2014), lake sediments (Affek et al. 2008; Zaarur et al. 2016), to paleosols (Ghosh et al. 2006a, b). The clumped isotope thermometer measures the clumping of $^{13}C$ and $^{18}O$ isotopes within a single carbonate phase without the need to know the isotope composition of the water (Eiler 2011). Limitations of the method are the relatively large amount of sample material necessary for an analysis and the long duration of the measurement. Analytical problems may arise by discrepancies of reference materials and disagreements among empirical calibrations. As argued by Kelson et al. (2017), phosphoric acid preparation and $CO_2$ gas purification are the major sources of the discrepancies.

Applications of the clumped isotope thermometer has been given by Tripati et al. (2010) presenting a global dataset for foraminifera and coccoliths. In the case of speleothems and surface corals, however, calculated clumped temperatures markedly differ from known temperatures (Saenger et al. 2012; Eiler et al. 2014). For speleothems, derived clumped temperatures are significantly higher than known growth temperatures. Deviations may depend on growth rates, implying the impact of kinetic effects on clumped isotope signals. While in some organisms such as foraminifera, mollusks, coccoliths and in some corals, equilibrium conditions appear to have been reached (Tripati et al. 2010; Piasecki et al. 2019 and others), in other organisms such as corals and brachiopods kinetic effects disturb the application of clumped isotope temperatures (Spooner et al. 2016; Bajnai et al. 2018; Guo 2020 and others). Simultaneous measurements of $\Delta_{47}$ and $\Delta_{48}$ in $CO_2$ evolved from phosphoric acid digestion can identify disequilibrium effects and thus can achieve reasonable paleotemperatures corrected for kinetic effects (Fiebig et al. 2019a, b, Bajnai et al. 2020).

Finally, it should be noted that clumped isotope studies clearly have indicated that the oxygen isotope composition of the ocean has not changed during the Phanerozoic (Bergmann et al. 2018; Henkes et al. 2018 and others).

## 3.13 Additional Applications

### 3.13.1 Forensic Isotope Geochemistry

Forensic isotope geochemistry may be defined as the application of stable isotope compositions in criminal investigations. The field is expanding rapidly; a book by Meier-Augenstein (Stable isotope forensics, 2nd edition, 2018) and several review articles (Chesson et al. 2014; Cerling et al. 2016; Chesson et al. 2018; Bartelink and Chesson 2019) have presented overviews of recent developments and applications. Applications range from (i) analysis of explosives, (ii) sourcing and authentication of food, beverages, drugs, art and money, iii) provenancing unidentified human remains and poached wildlife. For the determination of sources and provenance of inorganic and organic materials the concept of isotope landscapes (isoscapes) has been developed, representing maps of the spatial and temporal distribution of H, C, N and O isotopes across geographic areas (i.e. Bowen 2010).

Examples of forensic applications are:

(i) Ammonium nitrate ($NH_4NO_3$) and urea nitrate ($CH_5N_3O_4$) (the latter has been used by the terrorists of the 9/11 attack) are common explosives made from readily available materials. Although ammonium nitrate is a widely used fertilizer, it is considered as an explosive ingredient and its storage and transport being highly regulated. Explosives are generally easily identified, but it is very difficult to distinguish between the source materials used, because they are chemically identical, but may differ in isotope composition. By using a multi element approach, Widory et al. (2009), Aranda et al. (2011) and others demonstrated that distinct explosive families (PETN, TNT, ANFO) can be clearly differentiated by their specific isotope signature.

(ii) Stable isotopes help to understand trade patterns of illegal materials. Examples are illegal wildlife (elephant ivory, rhino horn), drugs (marijuana, heroin, cocaine) and counterfeits or mislabeled products (honey, wine). A long list of a variety of foods, drugs, synthetic material, art and antiquities, in which multi isotope analysis as well as "isotopic landscapes" (isoscapes) have been used to elucidate possible regions of origin. As documented for many food products, stable isotopes convincingly show how "authencity" can be distinguished from "adulteration". Adulteration is a legal term and means that food products do not meet legal standards and have become poorer in quality by adding extraneous sources. A commonly used adulteration approach is adding sugars from corn or sugar cane (C4 plants) to fruit juices, wines, honey etc. (C3 plants) to increase the volume.

H- and O-isotopes analysis allowed a distinction between authentic and fraudulent US$ 100 notes. As shown by Cerling et al. (2016) authentic bank notes were homogeneous in H and O-isotope composition, whereas fraudulent notes were variable in composition due to the use of papers from different areas. Another example to document illegal trade patterns is the determination of source-areas of ivory. Van der Merwe et al. (1990) and Ziegler et al. (2016) demonstrated that multi-isotope analysis of ivory from different regions in Africa allow the identification of the area in which an elephant lived.

(iii) The identication of human bodies is one of the central themes of forensic sciences (Chesson et al. 2018; Bartelink and Chesson 2019). Major goals of isotope analysis are whether an unidentified decedent was a resident of the region in which the remains were found or of likely regions from which the unknown decedent could have been.

Variations in the H, O, C, N, S isotope composition of the human body reflect the isotope composition of water consumed either directly as drinking water and other beverages or indirectly by food. Since the H- and O-isotope composition of water varies geographically (Meteoric Water Line relationship), it is possible to predict the source region in which the tissue (i.e. hair, nail) has been formed. Differences in the diet among populations of Europe, Asia, North- and South America lead to differences in C, N, S isotopes of body tissues. Hair and nail provide a record of diet and water, whereas apatite in bones and teeth reflect the isotope composition of components in the drinking water. These relationships help to identify human remains.

### 3.13.2 Medical Studies

The classic stable isotope systems (CHON) have been rarely used for medical purposes, because they are too abundant and too unspecific in tracing biological processes. Metal isotopes, in contrast, play a number of specific key roles in biological reactions. Thus, metal isotope fractionations induced by physiological processes in the human body have great potential to be used in medical research (Albarede 2015). In specific, this has been shown in studies using Ca, Fe, Zn and Cu isotopes. In recent years these isotope systems have been increasingly used to develop methods for medical diagnosis (see reviews by Albarede et al. 2016, 2017; Tanaka and Hirata 2018). Fe and Cu isotopes may change in oxidation states which may generate large isotope fractionations in human tissues, Zn isotope fractionations occur among tissues without changes in oxidation states. Ca isotopes are a powerful tracer for the calcium metabolism of bones.

(i) Ca

Calcium is one of the most important elements in human physiology. The average Ca content of adults is 1000–1200 g with 99% stored in bones. Bones are continuously formed and resorbed during life-time. With increasing age, bone resorption may exceed bone formation resulting in bone loss, known as osteoporosis. First applications have used Ca isotopes in urine and blood as an indicator of Ca metabolism (Skulan et al. 2007; Heuser and Eisenhauer 2009; Morgan et al. 2012). These studies have shown that Ca isotopes can be used to quantify the fluxes in and out of the bone, thus being an important tool to study osteoporosis and bone-involved cancers. Skulan and de Paolo (1999) first demonstrated that during bone formation lighter Ca isotopes are preferentially incorporated into bones whereas during resorption of bone no Ca isotope fractionation occurs. Thus, when bone formation exceeds resorption Ca isotope compositions of soft tissues (blood and urine) shift towards heavier values, but shift to lighter values when resorption exceeds bone formation (Morgan et al. 2012; Channon et al. 2015). As shown by Eisenhauer et al. (2019), Ca isotopes can be used as a new biomarker for early diagnosis of osteoporosis. Gordon et al. (2014) have demonstrated how Ca isotopes can be used to predict multiple myeloma disease activity, a type of blood cancer that causes bone destruction.

(ii) Zn

Zinc is the second most abundant trace element in the human body. It is essential for the functioning of many enzymatic reactions. Zinc deficiency causes serious health problems and is one of the most prevalent micronutrient deficiencies. The Zn isotope composition in the human body is controlled by the isotope composition of the ingested food and by an individual-dependent isotope fractionation during absorption in the intestine (Moynier et al. 2013). Studies of sheep and mice showed that different organs differ in their Zn isotope composition (Balter et al. 2010; Moynier et al. 2013). For instance, red blood cells and bones are isotopically enriched in heavy Zn isotopes relative to serum, brain and liver (Moynier et al. 2013).

As shown by Larner et al. (2015) and Schilling et al. (2020) Zn isotopes may be used as an early biomarker for breast, pancreatic and prostate cancer. Larner et al. (2015) demonstrated that breast cancer tumours are significantly depleted in heavy Zn isotopes relative to blood, serum and healthy breast tissue. Schilling et al. (2020) showed that changes in the Zn isotope composition of urine may be indicative for early pancreatic cancer detection. Moynier et al. (2017) investigated the distribution of Zn isotopes during Alzheimer'sdisease's. Due to the formation of Zn-rich amyloid plaques in the brain, homeostasis of Zn is dysregulated in Alzheimer patients. As shown by Moynier et al. (2017) brains from transgenic mice with Alzheimer symptoms are enriched in heavy Zn isotopes compared to a wild-type control group. The results were subsequently confirmed in humans by Moynier et al. (2020).

(iii) Fe

Iron is essential for humans, because Fe(II)-bearing hemoglobin carries oxygen in blood. Walczyk and von Blankenburg (2002) first reported Fe isotope variations in blood and among organs. Blood and muscle tissues are 1–2‰ depleted in heavy Fe isotopes relative to the diet. Although the exact mechanisms of Fe isotope fractionation during intestinal uptake is not well known, the iron isotope composition of whole blood is an indicator of the efficiency of dietary Fe uptake (Hotz and Walczyk 2013). Impaired regulation of iron absorption may lead to (i) iron deficiency (anemia) or (ii) iron overload (hemochromatosis). (i) Iron deficiency is the most common nutritional deficiency disorder caused by inadequate absorption and/or excessive iron loss. Anoshkina et al. (2017) have applied Fe isotopes to distinguish between iron deficient anemia and EPO-related anemia. While iron deficiency anemia leads to a change in the serum Fe isotope composition, EPO-related anemia does not show this effect. (ii) Hereditary hemochromatosis may be indicated by heavier Fe isotope values compared to healthy individuals (Hotz et al. 2012). Cikomola et al. (2017) showed that Fe isotope values of blood in diabetic patients were higher compared to non-diabetic patients correlating with the body mass index value.

(iv) Cu

Copper is an essential trace element playing a key role in the function of many enzymes. Cu is required for metabolic function and a potential toxin to the cell. Therefore, a delicate balance has to be maintained to achieve homoestatic regulation. As shown by Balter et al. (2013) Cu isotope ratios in organs and body fluids vary considerably. Kidneys in specific are enriched in heavy Cu isotopes which may reflect fractionations during redox processes.

The liver is the main storage site of Cu and plays a key role in homoestatic regulation. Cu isotope values of blood serum in patients with liver cirrhosis are depleted relative to a healthy control group increasing with the severity of the disease (Costas-Rodriguez et al. 2015).

Copper isotopes may indicate rapidly evolving cancer (Telouk et al. 2015; Balter et al. 2015; Albarede et al. 2016). As shown in these studies Cu isotope ratios of patients with colon, breast and liver cancer are depleted in heavy isotopes relative to healthy individuals. Telouk et al. (2015) suggested that low Cu isotope ratios not only can be used for prognosis in end-stage cancer, but may predict mortality earlier than molecular biomarkers. Copper is also implicated in neurodegenerative diseases. Sauzeat et al. (2018) observed higher Cu isotope ratios in patients with amythrophic lateral sclerosis (ALS) compared to a control group. Thus, it appears that Cu isotope studies may be applicable to the diagnostics of neurodegenerative malfunctions such as ALS and Parkinson.

Summarizing, the papers cited above demonstrate the great potential of metal isotope studies to a wide variety of medical problems. Metal isotope compositions may differentiate between healthy and diseased states, sometimes even before conventional methods allow a distinction. In conjunction with metal concentrations, metal isotopes, thus, may become an important diagnostic tool in medical sciences.

## References


Abelson PH, Hoering TC (1961) Carbon isotope fractionation in formation of amino acids by photosynthetic organisms. PNAS 47:623

Abrajano TA, Sturchio NB, Bohlke JH, Lyon GJ, Poreda RJ, Stevens MJ (1988) Methane—hydrogen gas seeps Zambales ophiolite, Phillippines: deep or shallow origin. Chem Geol 71:211–222

Affek HP, Bar-Matthews M, Ayalon A, Matthews A, Eiler JM (2008) Glacial/interglacial temperature variations in Soreq cave speleothems as recorded by "clumped isotope" thermometry. Geochim Cosmochim Acta 72:5351–5360

Affolter S, Fleitmann D, Leuenberger M (2014) New online method for water isotope analysis of speleothem fluid inclusions using laser absorption spectroscopy (WS-CRDS). Clim past 10:1291–1304

Affolter S, Häuselmann A, Fleitmann D, Edwards RL, Cheng H, Leuenberger M (2019) Central Europe temperature constrained by speleothem fluid inclusion water isotopes over the past 14000 years. Sci Adv 5:eaav3809

Agrinier P, Hekinian R, Bideau D, Javoy M (1995) O and H stable isotope compositions of oceanic crust and upper mantle rocks exposed in the Hess Deep near the Galapagos Triple Junction. Earth Planet Sci Lett 136:183–196

Aharon P, Fu B (2000) Microbial sulfate reduction rates and sulfur and oxygen isotope fractionation at oil and gas seeps in deepwater Gulf of Mexico. Geochim Cosmochim Acta 64:233–246

Aharon P, Fu B (2003) Sulfur and oxygen isotopes of coeval sulphate-sulfide in pore fluids of cold seep sediments with sharp redox gradients. Chem Geol 195:201–218

Akram W, Schönbächler M, Bisterzo S, Gallino (2015) Zirconium isotope evidence for the heterogeneous distribution of s-process materials in the solar system. Geochim Cosmochim Acta 165:484–500

Albarede F (2015) Metal stable isotopes in the human body: a tribute of geochemistry to medicine. Elements 11:265–269

Albarede F, Telouk P, Balter V, Bondanese VP, Albalat E, Oger P, Bonaventura P, Miossec P, Fujii T (2016) Medical applications of Cu, Zn, and S isotope effects. Metallomics 8:1056–1070

Albarede F, Telouk P, Balter V (2017) Medical applications of isotope metallomics. Rev Mineral Geochem 82:851–887

Alexander CM, Fogel M, Yabuta H, Cody GD (2007) The origin and evolution of chondrites recorded in the elemental and isotopic compositions of their macromolecular organic matter. Geochim Cosmochim Acta 71:4380–4403

Alexander CM, Newsome SD, Fogel ML, Nittler LR, Busemann H, Cody GR (2010) Deuterium enrichments in chondritic macromolecular material—implications for the origin and evolution of organics, water and asteroids. Geochim Cosmochim Acta 74:4417–4437

Alexander CM, Bowden R, Fogel ML, Howard KT, Herd CD, Nittler LR (2012) The provenances of asteroids and their contributions to the volatile inventories of the terrestrila planets. Science 337:721–723

Allard P (1983) The origin of hydrogen, carbon, sulphur, nitrogen and rare gases in volcanic exhalations: evidence from isotope geochemistry. In: Tazieff H, Sabroux JC (eds) Forecasting volcanic events. Elsevier Publlishing Co., pp 337–386

Allen RB, Cuffey KM (2001) Oxygen- and hydrogen-isotopic ratios of water in precipitation: beyond paleothermometry. Rev Mineral Geochem 43:527–553

Alt JC, Muehlenbachs K, Honnorez J (1986) An oxygen isotopic profile through the upper kilometer of the oceanic crust, DSDP hole 504 B. Earth Planet Sci Lett 80:217–229
Alt JC, Schwarzenbach EM, Früh-Green GL, Shanks WC, Bernasconi SM, Garrido CJ et al (2013) The role of serpentinites in cycling of carbon and sulfur: Seafloor serpentinization and subduction metamorphism. Lithos 178:40–54
Altabet MA, Deuser WC (1985) Seasonal variations in natural abundance of $^{15}N$ in particles sinking to the deep Sargasso Sea. Nature 315:218–219
Altabet MA, McCarthy JJ (1985) Temporal and spatial variations in the natural abundance of $^{15}N$ in POM from a warm-core ring. Deep Sea Res 32:755–772
Altabet MA, Deuser WG, Honjo S, Stienen C (1991) Seasonal and depth related changes in the source of sinking particles in the North Atlantic. Nature 354:136–139
Amari S, Hoppe P, Zinner E, Lewis RS (1993) The isotopic compositions of stellar sources of meteoritic graphite grains. Nature 365:806–809
Amrani A (2014) Organosulfur compounds: molecular and isotopic evolution from biota to oil and gas. Ann Rev Earth Planet Sci 42:733–768
Amrani A, Deev A, Sessions AL, Tang Y, Adkins JF, Hill RL, Moldowan JM, Wei Z (2012) The sulfur-isotopic comppositions of benzothiophenes and dibenzothiophenes as a proxy for thermochemical sulfate reduction. Geochim Cosmochim Acta 84:152–164
Amrani A, Said-Ahmad W, Shaked Y, Kiene RP (2013) Sulfur isotope homogeneity of oceanic DMSP and DMS. PNAS 110:18413–18418
An Y, Huang JX, Griffin WL, Liu C, Huang F (2017) Isotopic composition of Mg and Fe in garnet peridotites from the Kaapvaal and Siberian cratons. Geochim Cosmochim Acta 200:167–185
Anbar AD, Rouxel O (2007) Metal stable isotopes in paleoceanography. Ann Rev Earth Planet Sci 35:717–746
Ancour AM, Sheppard SMF, Guyomar O, Wattelet J (1999) Use of $^{13}C$ to trace origin and cycling of inorganic carbon in the Rhone river system. Chem Geol 159:87–105
Anderson AT, Clayton RN, Mayeda TK (1971) Oxygen isotope thermometry of mafic igneous rocks. J Geol 79:715–729
Angert A, Cappa CD, DePaolo DJ (2004) Kinetic O-17 effects in the hydrologic cycle: indirect evidence and implications. Geochim Cosmochim Acta 68:3487–3495
Angert A, Said-Ahmad W, Davidson C, Amrani A (2019) Sulfur isotopes ratio of atmospheric carbonyl sulfide constrains its sources. Sci Rep 9:741. https://doi.org/10.1038/s41598-018-37131-3
Anoshkina Y, Costas-Rodriguez M, Speeckaert M, Van Biesen W, Delanghe J, Vanhaecker D (2017) Iron isotopic composition of blood serum in anemia of chronic kidney disease. Metallomics 24:517–524
Antler G, Turchyn AV, Rennie V, Herut B, Sivan O (2013) Coupled sulphur and oxygen isotope insight to bacterial sulphate reduction in the natural environment. Geochim Cosmochim Acta 118:98–117
Antonelli MA, Kim ST, Peters M, Labidi J, Cartigny P, Walker RJ, Lyons JR, Hoek J, Farquhar J (2014) Early inner solar system origin for anomaleous sulfur isotopes in differentiated protoplanets. PNAS 111:17749–17754
Aranda R, Stern LA, Dietz ME, McCormick MC, Barrow JA, Mothershead RF (2011) Forensic utility of isotope ratio analysis of the explosive urea nitrate and its precursors. Forensic Sci Int 206:143–149
Archer C, Vance D (2006) Coupled Fe and S isotope evidence for Archean microbial Fe(III) and sulphate reduction. Geology 34:153–156
Arnold M, Sheppard SMF (1981) East Pacific rise at 21°N: isotopic composition and origin of the hydrothermal sulfur. Earth Planet Sci Lett 56:148–156
Arthur MA, Dean WE, Claypool CE (1985) Anomalous $^{13}C$ enrichment in modern marine organic carbon. Nature 315:216–218

Asael D, Matthews A, Oszczepalski S, Bar-Matthews M, Halicz L (2009) Fluid speciation controls of low temperature copper isotope fractionation applied to the Kupferschiefer and Timna ore deposits. Chem Geol 262:147–158

Augustin L, many others (2004) Eight glacial cycles from an Antarctic ice core. Nature 429:623–628

Ayliffe LK, Chivas AR (1990) Oxyen isotope composition of the bone phosphate of Australian kangaroos: potential as a palaeoenvironmental recorder. Geochim Cosmochim Acta 54:2603–2609

Ayliffe LK, Lister AM, Chivas AR (1992) The preservation of glacial-interglacial climatic signatures in the oxygen isotopes of elephant skeletal phosphate. Palaeo Palaeo Palaeo 99:179–191

Ayliffe LK, Chivas AR, Leakey MG (1994) The retention of primary oxygen isotope compositions of fossil elephant skeletal phosphate. Geochim Cosmochim Acta 58:5291–5298

Bacastow RB, Keeling CD, Lueker TJ, Wahlen M, Mook WG (1996) The $\delta^{13}C$ Suess effect in the world surface oceans and its implications for oceanic uptake of $CO_2$: analysis of observations at Bermuda. Global Biochem Cycles 10:335–346

Bajnai D, Fiebig J, TomasovychA, Garcia SM, Rollion-Bard C, Raddatz J, Löffler N, Primo-Ramos C, Brand U (2018) Assessing kinetic fractionation in brachiopod calcite using clumped isotopes Sci Rep 8:533

Bajnai D, 13 others (2020) Dual clumped isotope thermometry resolves kinetic biases in carbonate formation temperatures. Nat Commun 11:4005

Baker AJ, Fallick AE (1989) Heavy carbon in two-billion-year-old marbles from Lofoten-Vesteralen, Norway: implications for the Precambrian carbon cycle. Geochim Cosmochim Acta 53:1111–1115

Baker A, 13 others (2019) Global analysis reveals climatic controls on the oxygen isotope composition of cave drip water. Nat Comm 10:2984

Baker J, Matthews A (1995) The stable isotope evolution of a metamorphic complex, Naxos, Greece. Contr Mineral Petrol 120:391–403

Baker JA, Macpherson CG, Menzies MA, Thirlwall MF, Al-Kadasi M, Mattey DP (2000) Resolving crustal and mantle contributions to continental flood volcanism, Yemen: constraints from mineral oxygen isotope data. J Petrol 41:1805–1820

Balci N, Shanks WC, Mayer B, Mandernack KW (2007) OOxygen and sulphur isotope systematics of sulfate produced by bacterial and abiotic oxidation of pyrite. Geochim Cosmochim Acta 15:3769–3811

Balter V, Zazzo A, Moloney AP, Moynier F, Schmidt O, Monahan FJ, Albarede F (2010) Bodily variability of zinc natural isotope abundances in sheep. Rapid Communications Mass Spectr 24:605–612

Balter V, Lamboux A, Zazzo A, Telouk P, Leverrier Y, Marcel J, Moloney AP, Monahan FJ, Schmidt O, Albarede F (2013) Contrasting Cu, Fe, and Zn isotope patterns in organs and body fluids of mice and sheep, with emphasis on cellular fractionation. Metallomics 5:1470–1482

Balter V, 14 others (2015) Natural variations of copper and sulfur stable isotopes in blood of hepatocellular carcinoma patients. PNAS 112:982–985

Banner JL, Wasserburg GJ, Dobson PF, Carpenter AB, Moore CH (1989) Isotopic and trace element constraints on the orgin and evolution of saline groundwaters from central Missouri. Geochim Cosmochim Acta 53:383–398

Bao H (2015) Sulfate: a time capsule for Earth's $O_2$, $O_3$ and $H_2O$. Chem Geol 395:108–118

Bao H, Gu B (2004) Natural perchlorate has a unique oxygen isotope signature. Environ Sci Tech 38:5073–5077

Bao H, Koch PL (1999) Oxygen isotope fractionation in ferric oxide-water systems: low temperature synthesis. Geochim Cosmochim Acta 63:599–613

Bao H, Thiemens MH, Farquahar J, Campbell DA, Lee CC, Heine K, Loope DB (2000) Anomalous $^{17}O$ compositions in massive sulphate deposits on the Earth. Nature 406:176–178

Bao H, Thiemens MH, Heine K (2001) Oxygen-17 excesses of the Central Namib gypcretes: spatial distribution. Earth Planet Sci Lett 192:125–135

Bao H, Cao X, Hayles JA (2016) Triple oxygen isotopes: fundamental relationships and applications. Ann Rev Earth Planet Sci 44:463–492

Barkan E, Luz B (2005) High precision measurements of $^{17}O/^{16}O$ and $^{18}O/^{16}O$ ratios in $H_2O$. Rapid Commun Mass Spectr 19:3737–3742

Barkan E, Luz B (2007) Diffusivity fractionations of H216O/H217OH216O/H217O and H216O/H218O in air and their implications for isotope hydrology. Rapid Commun Mass Spectrom 21:2999–3005

Barkan E, Luz B (2011) The relationship among the three stable isotopes of oxygen in air, seawater and marine photosynthesis. Rapid Commun Mass Spectrom 25:2367–2369

Barnes I, Irwin WP, White DE (1978) Global distribution of carbon dioxide discharges and major zones of seismicity. US Geol Survey, Water-Resources Investigation 78–39, Open File Report

Barnes JJ, Franchi IA, Anand M, Tartese R, Starkey NA, Koike M, Sano Y, Russell SS (2013) Accurate and precise measurements of the D/H ratio and hydroxyl content in lunar apaites using NanoSIMS. Chem Geol 337–338:48–55

Barnes JJ, Franchi IA, McCubbin FM, Anand M (2019) Multiple reservoirs of volatiles in the Moon revealed by the isotope composition of chlorine in lunar basalts. Geochim Cosmochim Acta 266:144–162

Baroni M, Thiemens MH, Delmas RJ, Savarino J (2007) Mass-independent sulfur isotopic composition in stratospheric volcanic eruptions. Science 315:84–87

Bartelink EJ, Chesson LA (2019) Recent applications of isotope analysis to forensic anthropology. Forensic Sci Res 4:29–44

Batenburg AM, Walter S et al (2011) Temporal and spatial variability of the stable isotope composition of atmospheric molecular hydrogen. Atm Chem Phys Discuss 11:10087–10120

Baumgartner LP, Rumble D (1988) Transport of stable isotopes. I. Development of a kinetic continuum theory for stable isotope transport. Contr Mineral Petrol 98:417–430

Baumgartner LP, Valley JW (2001) Stable isotope transport and contact metamorphic fluid flow. In: Stable Isotope Geochemistry. Rev Mineral Geochem 43:415–467

Bauska TK, Baggenstos D, Brook EJ, Mix AC, Marcott SA, Petrenko VV, Schaefer H, Severinghaus J, Lee JE (2016) Carbon isotopes characterize rapid changes in atmospheric carbon dioxide during the last deglaciation. PNAS 113:3465–3470

Beaty DW, Taylor HP (1982) Some petrologic and oxygen isotopic relationships in the Amulet Mine, Noranda, Quebec, and their bearing on the origin of Archaean massive sulfide deposits. Econ Geol 77:95–108

Bechtel A, Hoernes S (1990) Oxygen isotope fractionation between oxygen of different sites in illite minerals: a potential geothermometer. Contrib Mineral Petrol 104:463–470

Bechtel A, Sun Y, Püttmann W, Hoernes S, Hoefs J (2001) Isotopic evidence for multi-stage base metal enrichment in the Kupferschiefer from the Sangershausen Basin, Germany. Chem Geol 176:31–49

Becker RH, Epstein S (1982) Carbon, hydrogen and nitrogen isotopes in solvent-extractable organic matter from carbonaceous chondrites. Geochim Cosmochim Acta 46:97–103

Bell DR, Ihinger PD (2000) The isotopic composition of hydrogen in nominally anhydrous mantle minerals. Geochim Cosmochim Acta 64:2109–2118

Bemis BE, Spero HJ, Bijma J, Lea DW (1998) Reevaluation of the oxygen isotopic composition of planktonic foraminifera: experimental results and revised paleotemperature equations. Paleoceanography 13:150–160

Bender ML, Keigwin LD (1979) Speculations about upper Miocene changes in abyssal Pacific dissolved bicarbonate $\delta^{13}C$. Earth Planet Sci Lett 45:383–393

Bender M, Sowers T, Labeyrie L (1994) The Dole effect and its variations during the last 130000 years as measured in the Vostok ice core. Global Biogeochem Cycles 8:363–376

Bennett SA, Rouxel O, Schmidt K, Garbe-Schönberg D, Statham PJ, German CR (2009) Iron isotope fractionation in a buyant hydrothermal plume, 5°S Mid-Atlantic Ridge. Geochim Cosmochim Acta 73:5619–5634

Bergmann KD, Al Balushi SA, Mackey TJ, Grotzinger JP, Eiler J (2018) A 600-million-year carbonate clumped-isotope record from the Sultanate of Oman. J Sediment Res 88:960–979

Berndt ME, Seal RR, Shanks WC, Seyfried WE (1996) Hydrogen isotope systematics of phase separation in submarine hydrothermal systems: experimental calibration and theoretical models. Geochim Cosmochim Acta 60:1595–1604

Berner RA (1990) Atmospheric carbon dioxide levels over Phanerozoic time. Science 249:1382–1386

Berner U, Faber E, Scheeder G, Panten D (1995) Primary cracking of algal and landplant kerogens: kinetic models of isotope variations in methane, ethane and propane. Chem Geol 126:233–245

Beucher CP, Brzezinski MA, Jones JL (2008) Sources and biological fractionation of silicon isotopes in the Eastern Equatorial Pacific. Geochim Cosmochim Acata 72:3063–3073

Bickle MJ, Baker J (1990) Migration of reaction and isotopic fronts in infiltration zones: assessments of fluid flux in metamorphic terrains. Earth Planet Sci Lett 98:1–13

Bidigare RR et al (1997) Consistent fractionation of $^{13}C$ in nature and in the laboratory: growth-rate effects in some haptophyte algae. Global Biogeochem Cycles 11:279–292

Bilenker LD, Simon AC, Reich M, Lundstrom CC, Gajos N, Bindeman I, Barra F, Munizaga R (2016) Fe–O stable isotope pairs elucidate a high-temperature origin of Chilean iron oxide-apatite deposits. Geochim Cosmochim Acta 177:94–104

Bindeman IN (2008) Oxygen isotopes in mantle and crustal magmas as revealed by single crystal analysis. Rev Mineral Geochem 69:445–478

Bindeman IR (2021) Triple oxygen isotopes in evolving continental crust, granites and clastic sediments. Rev Mineral Geochem 86:241–290

Bindeman IN, Eiler JN et al (2005) Oxygen isotope evidence for slab melting in modern and ancient subduction zones. Earth Planet Sci Lett 235:480–496

Bindeman IN, Ponomareva VV, Bailey JC, Valley JW (2004) Volcanic arc of Kamchatka: a province with high-$\delta^{18}O$ magma sources and large scale $^{18}O/^{16}O$ depletion of the upper crust. Geochim Cosmochim Acta 68:841–865

Bindeman IN, Eiler JM, Wing BA, Farquhar J (2007) Rare sulfur and triple oxygen isotope geochemistry of volcanogenic sulfate aerosols. Geochim Cosmochim Acta 71:2326–2343

Bindeman IN, Gurenko A, Sigmarsson O, Chaussidon M (2008) Oxygen isotope heterogeneity and disequilibria of olivine crystals in large volume Holocene basalts from Iceland: evidence for magmatic digestion and erosion of Pleistocene hyaloclastites. Geochim Cosmochim Acta 72:4397–4420

Bindeman IN, Serebryakov NS (2011) Geology, petrology and O and H isotope geochemistry of remarkably $^{18}O$ depleted Paleoproterozoic rocks of the Belomorian belt, Karelia, Russia, attributed to global glaciation 2.4 Ga. Earth Planet Sci Lett 306:163–174

Bindeman IN, Bekker A, Zakharov DO (2016) Oxygen isotope perspective on crustal evolution on early earth: a record of precambrian shales with emphasis on Paleoproterozoic glaciations and Great Oxidation event. Earth Planet Sci Lett 437:101–113

Bindeman IN, Zakharov DO, Palandri J, Greber ND, Dauphas N, Retallack GJ, Hofmann A, Lackey JS, Bekker A (2018) Rapid emergence of subaerial landmasses and onset of a modern hydrologic cycle 2.5 billion years ago. Nature 557:545–548

Bindeman IN, Bayon G, Palandri J (2019) Triple oxygen isotope investigation of fine-grained sediments from major world's rivers: insight into weathering processes and global fluxes into the hydrosphere. Earth Planet Sci Lett 528:115951

Bird MI, Ascoughz PL (2012) Isotopes in pyrogenic carbon: a review. Org Geochem 42:1529–1539

Bird MI, Chivas AR (1989) Stable-isotope geochronology of the Australian regolith. Geochim Cosmochim Acta 53:3239–3256

Bird MI, Longstaffe FJ, Fyfe WS, Bildgen P (1992) Oxygen isotope systematics in a multiphase weathering system in Haiti. Geochim Cosmochim Acta 56:2831–2838
Bhattacharyya A, Cambell KM, Kelly SD, Roebbert Y, Weyer S, Bernier-Latmani R, Borch T (2017) Biogenic non-crystalline U(IV) revealed as major component in uranium ore deposits. Nat Commun 8:15538
Black JR, Epstein E, Rains WD, Yin Q-Z, Casey WD (2008) Magnesium isotope fractionation during plant growth. Environ Sci Technol 42:7831–7836
Blair N, Leu A, Munoz E, Olsen J, Kwong E, Desmarais D (1985) Carbon isotopic fractionation in heterotrophic microbial metabolism. Appl Environ Microbiol 50:996–1001
Blake RE, O'Neil JR, Garcia GA (1997) Oxygen isotope systematics of biologically mediated reactions of phosphate: I Microbial degradation of organophosphorus compounds. Geochim Cosmochim Acta 61:441–4422
Blake RE, O'Neil JR, Surkov A (2005) Biogeochemical cycling of phosphorus: insights from oxygen isotope effects of phosphoenzymes. Am J Sci 305:596–620
Blättler CL, Miller NR, Higgins JA (2015) Mg and Ca isotope signatures of authigenic dolomite in siliceous deep-sea sediments. Earth Planet Sci Lett 419:32–42
Blattner P, Dietrich V, Gansser A (1983) Contrasting $^{18}O$ enrichment and origins of High Himalayan and Transhimalayan intrusives. Earth Planet Sci Lett 65:276–286
Blisnink PM, Stern LA (2005) Stable isotope altimetry: a critical review. Am J Sci 305:1033–1074
Blum TB, Kitajima K, Nakashima D, Strickland A, Spicuzza MJ, Valley JW (2016) Oxygen isotope evolution of the Lake Owyhee volcanic field, Oregon, and implications for the low-$\delta^{18}O$ magmatism of the Snake River Plain—Yellowstone hotspot and other low- $\delta^{18}O$ large igneous provinces. Contr Mineral Petrol 171:92
Boctor NZ, Alexander CM, Wang J, Hauri E (2003) The sources of water in Martian meteorites: clues from hydrogen isotopes. Geochim Cosmochim Acta 67:3971–3989
Boehme SE, Blair NE, Chanton JP, Martens CS (1996) A mass balance of $^{13}C$ and $^{12}C$ in an organic-rich methane-producing marine sediment. Geochim Cosmochim Acta 60:3835–3848
Bogard DD, Johnson P (1983) Martian gases in an Antarctic meteorite. Science 221:651–654
Böhlke JK, Sturchio NC, Gu B, Horita J, Brown GM, Jackson WA Jr B, Hatzinger PB (2005) Perchlorate isotope forensics. Anal Chem 77:7838–7842
Bolliger C, Schroth MH, Bernasconi SM, Kleikemper J, Zeyer J (2001) Sulfur isotope fractionation during microbial reduction by toluene-degrading bacteria. Geochim Cosmochim Acta 65:3289–3299
Böttcher ME, Brumsack HJ, Lange GJ (1998) Sulfate reduction and related stable isotope ($^{34}S$, $^{18}O$) variations in interstitial waters from the eastern Mediterranean. Proc Ocean Drill Program, Sci Res 160:365–373
Böttcher ME, Thamdrup B, Vennemann TW (2001) Oxygen and sulfur isotope fractionation during anaerobic bacterial disproportionation of elemental sulfur. Geochim Cosmochim Acta 65:1601–1609
Bottinga Y, Craig H (1969) Oxygen isotope fractionation between $CO_2$ and water and the isotopic composition of marine atmospheric $CO_2$. Earth Planet Sci Lett 5:285–295
Bottomley DJ, Katz A, Chan LH, Starinsky A, Douglas M, Clark ID, Raven KG (1999) The origin and evolution of Canadian Shield brines: evaporation or freezing of seawater? New lithium isotope and geochemical evidence from the Slave craton. Chem Geol 155:295–320
Bouchez J, von Blanckenburg F, Schüssler J (2014) Modeling novel stable isotope ratios in the weathering zone. Am J Sci 313:267–308
Bourdon B, Roskosz M, Hin RC (2018) Isotope tracers of core formation. Earth Sci Rev 181: 61–81
Bowen GJ (2010) Isoscapes: spatial pattern in isotopic biogeochemistry. Ann Rev Earth Planet Sci 38:161–187
Bowen GJ, Good SP (2015) Incorporating water isoscapes in hydrological and water resource investigations. Wires Water 2:107–119. https://doi.org/10.1002/wat2.1069

Bowers TS, Taylor HP (1985) An integrated chemical and isotope model of the origin of midocean ridge hot spring systems. J Geophys Res 90:12583–12606
Bowman JR, O'Neil JR, Essene EJ (1985) Contact skarn formation at Elkhorn, Montana. II. Origin and evolution of C-O-H skarn fluids. Am J Sci 285:621–660
Boyce JW, Treiman AH, Guan Y, Ma C, Eiler JM, Gross J, Greenwood JP, Stolper EM (2015) The chlorine isotope fingerprint of the lunar magma ocean. Sci Adv 1(8):e1500380
Boyd SR, Pillinger CT (1994) A preliminary study of $^{15}N/^{14}N$ in octahedral growth from diamonds. Chem Geol 116:43–59
Boyd SR, Pillinger CT, Milledge HJ, Mendelssohn MJ, Seal M (1992) C and N isotopic composition and the infrared absorption spectra of coated diamonds: evidence for the regional uniformity of $CO_2$–$H_2O$ rich fluids in lithospheric mantle. Earth Planet Sci Lett 109:633–644
Bradley AS, Summons RE (2010) Multiple origins of methane at the Lost City hydrothermal field. Earth Planet Sci Lett 297:34–41
Brandriss ME, O'Neil JR, Edlund MB, Stoermer EF (1998) Oxygen isotope fractionation between diatomaceous silica and water. Geochim Cosmochim Acta 62:1119–1125
Brenninkmeijer CAM (1993) Measurement of the abundance of $^{14}CO$ in the atmosphere and the $^{13}C/^{12}C$ and $^{18}O/^{16}O$ ratio of atmospheric CO with applications in New Zealand and Australia. J Geophys Res 98:10595–10614
Brenninkmeijer CAM, Lowe DC, Manning MR, Sparks RJ, van Velthoven PFJ (1995) The $^{13}C$, $^{14}C$ and $^{18}O$ isotopic composition of CO, $CH_4$ and $CO_2$ in the higher southern latitudes and lower stratosphere. J Geophys Res 100:26163–26172
Brenninkmeijer CAM, Janssen C, Kaiser J, Röckmann T, Rhee TS, Assonov SS (2003) Isotope effects in the chemistry of atmospheric trace compounds. Chem Rev 103:5125–5161
Brzezinski MA, Jones JL (2015) Coupling of the distribution of silicon isotopes to the meridional overturning circulation of the North Atlantic Ocean. Deep-Sea Res II 116:79–88
Bridgestock LJ, Williams H et al (2014) Unlocking the zinc isotope systematics of iron meteorites. Earth Planet Sci Lett 400:153–164
Broecker WS (1974) Chemical oceanography. Harcourt Brace Jovanovich, New York
Brüchert V, Pratt LM (1996) Contemporaneous early diagenetic formation of organic and inorganic sulfur in estuarine sediments from the St Andrew Bay, Florida, USA. Geochim Cosmochim Acta 60:2325–2332
Brumsack HJ, Zuleger E, Gohn E, Murray RW (1992) Stable and radiogenic isotopes in pore waters from Leg 1217. Japan Sea. Proc Ocean Drill Program 127(128):635–649
Brunner B, Bernasconi SM, Kleikemper J, Schroth MH (2005) A model of oxygen and sulfur isotope fractionation in sulfate during bacterial sulfate reduction. Geochim Cosmochim Acta 69:4773–4785
Bryant JD, Koch PL, Froelich PN, Showers WJ, Genna BJ (1996) Oxygen isotope partitioning between phosphate and carbonate in mammalian apatite. Geochim Cosmochim Acta 60:5145–5148
Buhl D, Neuser RD, Richter DK, Riedel D, Roberts B, Strauss H, Veizer J (1991) Nature and nurture: environmental isotope story of the river Rhine. Naturwissenschaften 78:337–346
Bullen TD (2014) Metal stable isotopes in weathering and hydrology. In: Treatise on geochemistry, vol 7, 2nd edn. pp 329–359
Burdett JW, Arthur MA, Richardson A (1989) A neogene seawater sulfate isotope age curve from calcareous pelagic microfossils. Earth Planet Sci Lett 94:189–198
Burke A, Present TM, 10 others (2018) Sulfur isotopes in rivers: insights into global weathering budgets, pyrite oxidation, and the modern sulfur cycle. Earth Planet Sci Lett 496:168–177
Burke A, Moore KA, Sigl M, Nita DC, McConnell JR, Adkins JF (2019) Stratospheric eruptions from tropical and extra-tropical volcanoes constrained using high-resolution sulfur isotopes in ice cores. Earth Planet Sci Lett 521:113–119
Burruss RC, Laughrey CD (2010) Carbon and hydrogen isotope reversal in deep basin gas: evidence for limits to the stability of hydrocarbons. Org Geochem 41:1285–1296

Busigny V, Cartigny P, Philippot P (2011) Nitrogen isotopes in ophiolitic metagabbros: a re-evaluation of modern nitrogen fluxes in subduction zones and implication for the early Earth atmosphere. Geochim Cosmochim Acta 75:7502–7521

Butler IB, Archer C, Vance D, Oldroyd A, Rickard D (2005) Fe isotope fractionation on FeS formation in ambient aqueous solution. Earth Planet Sci Lett 236:430–442

Cabral RA, Jackson MG, Rose-Koga EF, Koga KT, Whitehouse MJ, Antonelli MA, Farquhar J, Day JM, Hauri EH (2013) Anomaleous sulphur isotopes in plume lavas reveal deep mantle storage of Archaean crust. Nature 496:490–493

Cai C, Worden RH, Bottrell SH, Wang LS, Yang CS (2003) Thermochemical sulfate reduction and the generation of hydrogen sulphide and thiols (mercaptans) in Triassic carbonate reservoirs from the Sichuan Basin, China. Chem Geol 202:39–57

Cai C, Zhang C, Worden RH, Xiao Q, Wang T, Gvirtzman Z, Li H, Said-Ahmad W, Lianqi J (2016) Sulfur isotopic compositions of individual organosulfur compounds and their genetic links in the Lower Plaleozoic petroleum pools of the Tarim Basin, NW China. Geochim Cosmochim Acta 182:88–108

Caldelas C, Weiss DJ (2017) Zinc homoeostasis and isotopic fractionation in plants: a review. Plant Soil 411:17–46

Calmels D, Gaillerdet J, Brenot A, France-Lanord C (2007) Sustained sulfide oxidation by physical erosion processes in the Mackenzie River basin: climatic perspectives. Geology 35:1003–1006

Came RE, Eiler JM, Veizer J, Azmy K, Brand U, Weidmann CR (2007) Coupling of surface temperatures and atmospheric $CO_2$ concentrations during the Paleozoic era. Nature 449:198–201

Cameron EM (1982) Sulphate and sulphate reduction in early Precambrian oceans. Nature 296:145–148

Cameron EM, Hall GEM, Veizer J, Krouse HR (1995) Isotopic and elemental hydrogeochemistry of a major river system: Fraser River, British Columbia, Canada. Chem Geol 122:149–169

Canfield DE, Teske A (1996) Late Proterozoic rise in atmospheric oxygen concentration inferred from phylogenetic and sulphur-isotope studies. Nature 382:127–132

Canfield DE, Thamdrup B (1994) The production of $^{34}S$ depleted sulfide during bacterial disproportion to elemental sulfur. Science 266:1973–1975

Cano E, Sharp ZD, Shearer CK (2020) Distinct oxygen isotope compositions of the Earth and Moon. Nature Geosciences 13:270–274

Cao X, Bao H (2021) Small triple oxygen isotope variations in sulfate: mechanisms and applications. Rev Mineral Geochem 86:463–488

Cartigny P (2005) Stable isotopes and the origin of diamond. Elements 1:79–84

Cartigny P (2010) Mantle-related carbonados? Geochemical insights from diamonds from the Dachine komatiite (French Guiana). Earth Planet Sci Lett 296:329–339

Cartigny P, Marty B (2013) Nitrogen isotopes and mantle geodynamics: the emergence of life and the atmosphere-crust-mantle connection. Elements 9:359–366

Cartigny P, Boyd SR, Harris JW, Javoy M (1997) Nitrogen isotopes in peridotitic diamonds from Fuxian, China: the mantle signature. Terra Nova 9:175–179

Cartigny P, Harris JW, Javoy M (1998) Subduction related diamonds? The evidence for a mantle-derived origin from coupled delta $^{13}C$-delta $^{15}N$ determinations. Chem Geol 147:147–159

Cartigny P, Palot M, Thomassot E, Harris JW (2014) Diamond formation: a stable isotope perspective. Ann Rev Earth Planet Sci 42:699–732

Cartwright I, Valley JW (1991) Steep oxygen isotope gradients at marble-metagranite contacts in the NW Adirondacks Mountains, NY. Earth Planet Sci Lett 107:148–163

Casciotti KL (2016) Nitrogen and oxygen isotopic studies of the marine nitrogen cycle. Ann Rev Mar Sci 8:379–407

Cerling TE (1991) Carbon dioxide in the atmosphere: evidence from Cenozoic and Mesozoic paleosols. Am J Sci 291:377–400

Cerling TE, Sharp ZD (1996) Stable carbon and oxygen isotope analyses of fossil tooth enamel using laser ablation. Palaeo Palaeo Palaeo 126:173–186

Cerling TE, Brown FH, Bowman JR (1985) Low-temperature alteration of volcanic glass: hydration, Na, K, $^{18}O$ and Ar mobility. Chem Geol 52:281–293

Cerling TE, Wang Y, Quade J (1993) Expansion of C4 ecosystems as an indicator of global ecological change in the late Miocene. Nature 361:344–345

Cerling TE, Harris JM, MacFadden BJ, Leakey MG, Quade J, Eisenmann V, Ehleringer JR (1997) Global vegetation change through the Miocene/Pliocene boundary. Nature 389:153–158

Cerling T, Barnette JE, Bowen GJ, Chesson LA, Ehleringer JR, Remien CH, Shea P, Tipple BJ, West JB (2016) Forensic stable isotope biogeochemistry. Ann Rev Earth Planet Sci 44:175–206

Cesar J, Nightingale M, Becker V, Mayer B (2020) Stable carbon isotope systematics of methane, ethane and propane from low-permeability hydrocarbon reservoirs. Chem Geol 558:119907

Cikomola JC, 9 others (2017) Whole blood Fe isotope signature in a sub-Saharan African population. Metallomics 9:1142–1149

Chakrabarti R, Knoll AH, Jacobsen SB, Fischer WW (2012) Si isotope variability in Proterozoic cherts. Geochim Cosmochim Acta 91:187–201

Chakraborty S, Muskatel BH, Jackson TL, Ahmed M, Levine RD, Thiemens MH (2014) Massive isotopic effect in vacuum of $N_2$ and implications for meteorite data. PNAS 111:14704–14709

Chamberlain CP, Poage MA (2000) Reconstructing the paleotopography of mountain belts from the isotopic composition of authigenic minerals. Geology 28:115–118

Channon MB, Gordon GW, Morgan JL, Skulan JL, Smith SM, Anbar AD (2015) Using natural stable calcium isotopes of human blood to detect and monitor changes in bone mineral balance. Bone 77:69–74

Chapligin B, Leng MJ et al (2011) Inter-laboratory comparison of oxygen isotope compositions from biogenic silica. Geochim Cosmochim Acta 75:7242–7256

Chapman T, Cui Y, Schubert B (2019) Stable carbon isotopes of fossil plant lipids support moderately high $pCO_2$ in the early Paleogene. ACS Earth Space Chem 3:1966–1973

Chaussidon M, Marty B (1995) Primitive boron isotope composition of the mantle. Science 269:383–386

Chaussidon M, Albarede F, Sheppard SMF (1987) Sulphur isotope heterogeneity in the mantle from ion microprobe measurements of sulphide inclusions in diamonds. Nature 330:242–244

Chaussidon M, Albarede F, Sheppard SMF (1989) Sulphur isotope variations in the mantle from ion microprobe analysis of microsulphide inclusions. Earth Planet Sci Lett 92:144–156

Chazot G, Lowry D, Menzies M, Mattey D (1997) Oxygen isotope compositions of hydrous and anhydrous mantle peridotites. Geochim Cosmochim Acta 61:161–169

Chen H, Savage PS, Teng FZ, Helz RT, Moynier F (2013) Zinc isotopic fractionation during magmatic differentiation and the isotopic composition of bulk Earth. Earth Planet Sci Lett 369–370:34–42

Chesson LA, Tipple BJ, Howa JD, Bowen GJ, Barnette JE, Cerling TE, Ehleringer JR (2014) Stable isotopes in forensics applications. Treatise Geochem (sec Ed) 14:285–317

Chesson LA, 17 others (2018) Applying the principles of isotope analysis in plant and animal ecology to forensic sciences in the Americas. Oecologia 187:1077–1094

Chiba H, Sakai H (1985) Oxygen isotope exchange rate between dissolved sulphate and water at hydrothermal temperatures. Geochim Cosmochim Acta 49:993–1000

Chivas AR, Andrew AS, Sinha AK, O'Neil JR (1982) Geochemistry of Pliocene-Pleistocene oceanic arc plutonic complex, Guadalcanal. Nature 300:139–143

Ciais P, Tans PP, Trolier M, White JWC, Francey RJ (1995) A large northern hemisphere terrestrial $CO_2$ sink indicated by the $^{13}C/^{12}C$ ratio of atmospheric $CO_2$. Science 269:1098–1102

Cifuentes LA, Fogel ML, Pennock JR, Sharp JR (1989) Biogeochemical factors that influence the stable nitrogen isotope ratio of dissolved ammonium in the Delaware Estuary. Geochim Cosmochim Acta 53:2713–2721

Claypool GE, Holser WT, Kaplan IR, Sakai H, Zak I (1980) The age curves of sulfur and oxygen isotopes in marine sulfate and their mutual interpretation. Chem Geol 28:199–260
Clayton RN (2002) Self-shielding in the solar nebula. Nature 451:860–861
Clayton RN (2004) Oxygen isotopes in meteorites. Treatise on geochemistry, vol 1. Elsevier, Amsterdam, pp 129–142
Clayton RN, Mayeda TK (1996) Oxygen isotope studies of achondrites. Geochim Cosmochim Acta 60:1999–2017
Clayton RN, Mayeda TK (1999) Oxygen isotope studies of carbonaceous chondrites. Geochim Cosmochim Acta 63:2089–2104
Clayton RN, Mayeda TK, Rubin AE (1984) Oxygen isotopic compositions of enstatite chondrites and aubrites. J Geophys Res Solid Earth 89:C245–C249
Clayton DD, Nittler LR (2004) Astrophysics with presolar stardust. Ann Rev Astron Astrophys 42:39–78
Clayton RN, Steiner A (1975) Oxygen isotope studies of the geothermal system at Warakei, New Zealand. Geochim Cosmochim Acta 39:1179–1186
Clayton RN, Friedman I, Graf DL, Mayeda TK, Meents WF, Shimp NF (1966) The origin of saline formation waters. 1. Isotopic composition. J Geophys Res 71:3869–3882
Clayton RN, Muffler LJP, White (1968) Oxygen isotope study of calcite and silicates of the River Branch No. I well, Salton Sea Geothermal Field, California. Am J Sci 266:968–979
Clayton RN, Grossman L, Mayeda TK (1973) A component of primitive nuclear composition in carbonaceous meteorites. Science 182:485–488
Clayton RN, Hurd JM, Mayeda TK (1973b) Oxygen isotopic compositions of Apollo 15, 16 and 17 samples and their bearing on lunar origin and petrogenesis. In: Proceedings of 4th lunar science conference, geochimica cosmochimica acta supplement, vol 2, pp 1535–1542
Cliff SS, Thiemens MH (1997) The $^{18}O/^{16}O$ and $^{17}O/^{16}O$ ratios in atmospheric nitrous oxide: a mass independent anomaly. Science 278:1774–1776
Cliff SS, Brenninkmeijer CAM, Thiemens MH (1999) First measurement of the $^{18}O/^{16}O$ and $^{17}O/^{16}O$ ratios in stratospheric nitrous oxide: a mass-independent anomaly. J Geophys Res 104:16171–16175
Cline JD, Kaplan IR (1975) Isotopic fractionation of dissolved nitrate during denitrification in the eastern tropical North Pacific Ocean. Mar Chem 3:271–299
Clog M, Aubaud C, Cartigny P, Dosso L (2013) The hydrogen isotopic composition and water content of southern Pacific MORB: a reassessment of the D/H ratio of the depleted mantle reservoir. Earth Planet Sci Lett 381:156–165
Clor LE, Fischer TP, Hilton DR, Sharp ZD, Hartono U (2005) Volatile and N isotope chemistry of the Molucca Sea collision zone: tracing source components along the Sangihe arc, Indonesia. Geochem Geophys Geosys 6:Q03J14. https://doi.org/10.1029/2004GC00825
Cobert F, Schmitt AD, Bourgeade P, Labolle F, Badot PM, Chabaux F, Stille P (2011) Experimental identification of Ca isotopic fractionations in higher plants. Geochim Cosmochim Acta 75:5467–5482
Cohen AS, Coe Al, Kemp DB (2007) The Late-Paleocene-Early Eocene and Toarcian (Early Jurassic) carbon isotope excursions: a comparison of their time scales, associated environmental changes, causes and consequences. J Geol Soc 164:1093–1108
Cole JE, Fairbanks RG (1990) The southern oscillation recorded in the $\delta^{18}O$ of corals from Tarawa atoll. Paleoceanography 5:669–683
Cole JE, Fairbanks RG, Shen GT (1993) The spectrum of recent variability in the southern oscillation: results from a Tarawa atoll. Science 260:1790–1793
Colon DP, Bindeman IN, Wotzlaw JF, Christiansen FH, Stern RA (2018) Origins and evolution of rhyolitic magmas in the central Snake River Plain: insights fromcoupled high-precision geochronology, oxygen isotope, and hafnium isotope analyses of zircon. Contr Mineral Petrol 173:11
Colman AS, Blake RE, Karl DM, Fogel ML, Turekian KK (2005) Marine phosphate oxygen isotopes and organic matter remineralization in the oceans. PNAS 102:13023–13028

Connolly CA, Walter LM, Baadsgaard H, Longstaffe F (1990) Origin and evolution of formation fluids, Alberta Basin, western Canada sedimentary basin: II. Isotope systematics and fluid mixing. Appl Geochem 5:397–414

Conway TM, John SG (2014) Quantification of dissolved iron sources in the North Atlantic Ocean. Nature 511:212–215

Conway TM, John SG (2015) The cycling of iron, zinc and cadmium in the North East Pacific ocean—insights from stable isotopes. Geochim Cosmochim Acta 164:262–283

Cook N, Hoefs J (1997) Sulphur isotope characteristics of metamorphosed Cu-(Zn) volcanogenic massive sulphide deposits in the Norwegian Caledonides. Chem Geol 135:307–324

Cooper KM, Eiler JM, Asimov PD, Langmuir CH (2004) Oxygen isotope evidence for the origin of enriched mantle beneath the mid-Atlantic ridge. Earth Planet Sci Lett 220:297–316

Coplen TB (2007) Calibration of the calcite-water oxygen-isotope geothermometer at Devils Hole, Nevada, a natural laboratory. Geochim Cosmochim Acta 71:3948–3957

Coplen TB, Hanshaw BB (1973) Ultrafiltration by a compacted clay membrane. I. Oxygen and hydrogen isotopic fractionation. Geochim Cosmochim Acta 37:2295–2310

Corsetti FA (2017) Carbon isotope stratigraphy of the Neoproterozoic—Cambrian transition: an introduction. Cambridge University Press

Corsetti FA, Baud A, Marenco PJ, Richoz S (2005) Summary of early Triassic carbon isotope records. CR Palevol 4:473–486

Costa-Rodriguez M, Anoshkina Y, Lauwens S, Van Vlierberghe H, Delanghe J, Vanhaecke F (2015) Isotopic analysis of Cu in blood serum by multi-collector ICP-mass-spectrometry: a new approach for the diagnosis and prognosis of liver cirrhosis. Metallomics 7:491–498

Craddock PR, Dauphas N (2010) Iron isotopic compositions of geological reference materials and chondrites. Geostand Geoanal Res 35:101–123

Craddock PR, Warren JM, Dauphas N (2013) Abyssal peridotites reveal the near-chondritic Fe isotope composition of the earth. Earth Planet Sci Lett 365:63–76

Craig H (1953) The geochemistry of the stable carbon isotopes. Geochim Cosmochim Acta 3: 53–92

Craig H (1961) Isotopic variations in meteoric waters. Science 133:1702–1703

Craig H, Gordon L (1965) Deuterium and oxygen-18 variations in the ocean and the marine atmosphere. In: Symposium on marine geochemistry. Graduate School of Oceanography, vol 3. University of Rhode Island, OCC Publications, p 277

Craig H, Boato G, White DE (1956) Isotopic geochemistry of thermal waters. In: Proceedings of 2nd conference on nuclear process geological settings, p 29

Craig H, Chou CC, Welhan JA, Stevens CM, Engelkemeier A (1988) The isotopic composition of methane in polar ice cores. Science 242:1535–1539

Criss RE, Taylor HP (1986) Meteoric-hydrothermal systems. Stable isotopes in high temperature geological processes. Rev Mineral 16:373–424

Criss RE, Champion DE, McIntyre DH (1985) Oxygen isotope, aeromagnetic and gravity anomalies associated with hydrothermally altered zones in the Yankee Fork Mining District, Custer County, Idaho. Econ Geol 80:1277–1296

Criss RE, Fleck RJ, Taylor HP (1991) Tertiary meteoric hydrothermal systems and their relation to ore deposition, Northwestern United States and Southern British Columbia. J Geophys Res 96:13335–13356

Crowson RA, Showers WJ, Wright EK, Hoering TC (1991) Preparation of phosphate samples for oxygen isotope analysis. Anal Chem 63:2397–2400

Cui Y, Schubert BA, Jahren AH (2020) A 23 m.y. record of low atmospheric $CO_2$. Geology 48:888–892

Cummins RC, Finnegan S, Fike DA, Eiler JM, Fischer WW (2014) Carbonate clumped isotope constraints on Silurian ocean temperature and seawater $\delta^{18}O$. Geochim Cosmochim Acta 140:241–258

Curry WB, Duplessy JC, Labeyrie LD, Shackleton NJ (1988) Quaternary deep-water circulation changes in the distribution of $\delta^{13}C$ of deep water $\Sigma CO_2$ between the last glaciation and the Holocene. Paleoceanography 3:317–342

Cypionka H, Smock A, Böttcher MA (1998) A combined pathway of sulfur compound disproportionation in Desulfovibrio desulfuricans. FEMS Microbiol Lett 166:181–186

Daeron M, Guo W et al (2011) $^{13}C^{18}O$ clumping in speleothems: observations from natural caves and precipitation experiments. Geochim Cosmochim Acta 75:3303–3317

Daniels WC, Russell JM, Giblin AE, Welker JM, Klein ES, Huang Y (2017) Hydrogen isotope fractionation in leaf waxes in the Alaskan Arctic tundra. Geochim Cosmochim Acta 213:216–236

Dansgaard W (1964) Stable isotope in precipitation. Tellus 16:436–468

Dansgaard W et al (1993) Evidence for general instability of past climate from a 250 kyr ice-core record. Nature 364:218–220

Dauphas N, Marty B (1999) Heavy nitrogen in carbonatites of the Kola peninsula: a possible signature of the deep mantle. Science 286:2488–2490

Dauphas N, Teng NZ, Arndt NT (2010) Magnesium and iron isotopes in 2.7 Ga Alexo komatiites: mantle signatures, no evidence for Soret diffusion, identification of diffusive transport in zoned olivine. Geochim Cosmochim Acta 74:3274–3291

Dauphas N, John SG, Rouxel O (2017) Iron Isotope Systematics. Rev Mineral Geochem 82:415–510

Davidson C, Amrani A, Angert A (2021) Troposheric carbonyl sulfide mass balance based on direct measurements of sulfur isotopes. PNAS 118(6):e2020060118

Day JM, Moynier F (2014) Evaporative fractionation of volatile stable isotopes and their bearing on the origin of the Moon. Phil Trans Roy Soc A 372:20130259

De Hoog JCM, Taylor BE, Van Bergen MJ (2009) Hydrogen-isotope systematics in degassing basaltic magma and application to Indonesian arc basalts. Chem Geol 266:256–266

De Hoog JCM, Savov IP (2018) Boron isotopes as a tracer of subduction zone processes. In: Marschall H, Foster G (eds) Boron isotopes. Springer, pp 217–247

De La Rocha CL, De Paolo DJ (2000) Isotopic evidence for variations in the marine calcium cycle over the Cenozoic. Science 289:1176–1178

De Moor JM, Fischer TP et al (2013) Sulfur degassing at Erta Ale (Ethiopia) and Masaya (Nicaragua) volcanoes: implications for degassing processes and oxygen fugacities of basaltic systems. Geochem Geophys Geosys 14(10). https://doi.org/10.1002/ggge.20255

De Moor JM, Fischer TP, Sharp ZD, Hauri EH, Hilton DR, Atudorei V (2010) Sulfur isotope fractionation during the May 2003 eruption of Anatahan volcano, Mariana Islands: implications for sulfur sources and plume processes. Geochim Cosmochim Acta 74:5382–5397

De Souza GF, Reynolds BC, Johnson GC, Bullister JL, Bourdon B (2012) Southern Ocean control of silicon stable isotope distribution in the deep Atlantic Ocean. Global Biogeochem Cycl 26: GB2035. https://doi.org/10.1029/2011GB004141

De Souza GF, Slater RD, Hain MP, Brzezinski MA (2015) Distal and proximal controls on the silicon stable isotope signature of North Atlantic Deep Water. Earth Planet Sci Lett 432:342–353

Debret B, Bouilhol P, Pons ML, Williams H (2018) Carbonate transfer during the onset of slab devolatilization: new insights from Fe and Zn stable isotopes. J Petrol 59:1145–1166

Defouilloy C, Cartigny P, Assayag N, Moynier F, Barrat JA (2016) High-precision sulfur isotope composition of enstatite meteorites and implications of the formation and evolution of their parent bodies. Geochim Cosmochim Acta 172:393–409

Degens ET, Epstein S (1962) Relationship between $^{18}O/^{16}O$ ratios in coexisting carbonates, cherts and diatomites. Bull Am Assoc Pet Geol 46:534–535

Deines P (1980) The isotopic composition of reduced organic carbon. In: Fritz P, Fontes JC (eds) Handbook of environmental geochemistry, vol 1. Elsevier, New York, Amsterdam, pp 239–406

Deines P (1989) Stable isotope variations in carbonatites. In: Bell K (ed) Carbonatites, genesis and evolution. Unwin Hyman, London, 619 p
Deines P, Gold DP (1973) The isotopic composition of carbonatite and kimberlite carbonates and their bearing on the isotopic composition of deep-seated carbon. Geochim Cosmochim Acta 37:1709–1733
Deines P, Haggerty SE (2000) Small-scale oxygen isotope variations and petrochemistry of ultradeep (>300 km) and transition zone xenoliths. Geochim Cosmochim Acta 64:117–131
Deines P, Gurney JJ, Harris JW (1984) Associated chemical and carbon isotopic composition variations in diamonds from Finsch and Premier Kimberlite, South Africa. Geochim Cosmochim Acta 48:325–342
Delavault H, Chauvel C, Thomassot E, Devey CW, Dazas B (2016) Sulfur and lead isotopic evidence of relic Archean sediments in the Pitcairn mantle plume. PNAS 113:12952–12956
Delaygue G, Jouzel J, Dutay JC (2000) Oxygen-18—salinity relationship simulated by an oceanic general simulation model. Earth Planet Sci Lett 178:113–123
Deloule E, Robert F (1995) Interstellar water in meteorites? Geochim Cosmochim Acta 59:4695–4706
Deloule E, Robert F, Doukhan JC (1998) Interstellar hydroxyl in meteoritic chondrules: implications for the origin of water in the inner solar system. Geochim Cosmochim Acta 62:3367–3378
DeNiro MJ, Epstein S (1978) Influence of diet on the distribution of carbon isotopes in animals. Geochim Cosmochim Acta 42:495–506
DeNiro MJ, Epstein S (1979) Relationship between the oxygen isotope ratios of terrestrial plant cellulose, carbon dioxide and water. Science 204:51–53
DeNiro MJ, Epstein S (1981) Isotopic composition of cellulose from aquatic organisms. Geochim Cosmochim Acta 45:1885–1894
Dennis KJ, Schrag DP (2010) Clumped isotope thermometry of carbonatites as an indicator of diagenetic alteration. Geochim Cosmochim Acta 74:4110–4122
Dennis KJ, Cochran JK, Landman NH, Schrag DP (2013) The climate of the Late Cretaceous: new insights from the application of the carbonate clumped isotope thermometer to Western Interior Seaway macrofossil. Earth Planet Sci Lett 362:51–65
Dennis PF, Rowe PJ, Atkinson TC (2001) The recovery and isotopic measurement of water from fluid inclusions in speleothems. Geochim Cosmochim Acta 65:871–884
Derry LA, Kaufmann AJ, Jacobsen SB (1992) Sedimentary cycling and environmental change in the Late Proterozoic: evidence from stable and radiogenic isotopes. Geochim Cosmochim Acta 56:1317–1329
Des Marais DJ (2001) Isotopic evolution of the biogeochemical carbon cycle during the Precambrian. In: Valley J, Cole D (eds) Stable isotope geochemistry. Rev Mineralogy 43:555–578
Des Marais DJ, Moore JG (1984) Carbon and its isotopes in mid-oceanic basaltic glasses. Earth Planet Sci Lett 69:43–57
Deutsch S, Ambach W, Eisner H (1966) Oxygen isotope study of snow and firn of an Alpine glacier. Earth Planet Sci Lett 1:197–201
Dickens GR (2003) Rethinking the global carbon cycle with a large dynamic and micrmediated gas hydrate capacitor. Earth Planet Sci Lett 213:169–182
Dickson JAD, Coleman ML (1980) Changes in carbon and oxygen isotope composition during limestone diagenesis. Sedimentology 27:107–118
Dickson JAD, Smalley PC, Raheim A, Stijfhoorn DE (1990) Intracrystalline carbon and oxygen isotope variations in calcite revealed by laser micro-sampling. Geology 18:809–811
Dietzel M, Tang J, Leis A, Köhler SJ (2009) Oxygen isotopic fractionation during inorganic calcite precipitation—effects of temperature, precipitation rate and pH. Chem Geol 268:107–115
Dipple GM, Ferry JM (1992) Fluid flow and stable isotope alteration in rocks at elevated temperatures with applications to metamorphism. Geochim Cosmochim Acta 56:3539–3550

Dixon JE, Bindeman IN, 10 others (2017) Light stable isotopic compositions of enriched mantle sources: resolving the dehydration paradox. Geochem Geophys Geosys 18:3801–3839
Dobson PF, O'Neil JR (1987) Stable isotope composition and water contents of boninite series volcanic rocks from Chichi-jima, Bonin Islands, Japan. Earth Planet Sci Lett 82:75–86
Dobson PF, Epstein S, Stolper EM (1989) Hydrogen isotope fractionation between coexisting vapor and silicate glasses and melts at low pressure. Geochim Cosmochim Acta 53:2723–2730
Dodd JP, Sharp ZD (2010) A laser fluorination method for oxygen isotope analysis of biogenic silica and a new oxygen isotope calibration of modern diatoms in freshwater environments. Geochim Cosmochim Acta 74:1381–1390
Dodd JP, Sharp ZD, Fawcett PJ, Brearley AJ, McCubbin FM (2013) Rapid post-mortem maturation of diatom silica oxygen isotope values. Geochem Geophys Geosys 13(9). https://doi.org/10.1029/2011GC004019
Dodson MH (1973) Closure temperature in cooling geochronological and petrological systems. Contr Mineral Petrol 40:259–274
Dole M, Lange GA, Rudd DP, Zaukelies DA (1954) Isotopic composition of atmospheric oxygen and nitrogen. Geochim Cosmochim Acta 6:65–78
Donahue TM, Hoffman JH, Hodges RD, Watson AJ (1982) Venus was wet: a measurement of the ratio of deuterium to hydrogen. Science 216:630–633
Dorendorf F, Wiechert U, Wörner G (2000) Hydrated sub-arc mantle: a source for the Kluchevskoy volcano, Kamchatka, Russia. Earth Planet Sci Lett 175:69–86
Douglas PM, Stolper DA, 17 others (2017) Methane clumped isotopes: progress and potential for a new isotopic tracer. Org Geochem 113:262–282
Douthitt CB (1982) The geochemistry of the stable isotopes of silicon. Geochim Cosmochim Acta 46:1449–1458
Drake MJ, Righter K (2002) Determining the composition of the earth. Nature 416:39–44
Driesner T (1997) The effect of pressure on deuterium-hydrogen fractionation in high-temperature water. Science 277:791–794
Driesner T, Seward TM (2000) Experimental and simulation study of salt effects and pressure/density effects on oxygen and hydrogen stable isotope liquid-vapor fractionation for 4–5 molal aqueous NaCl and KCl solutions to 400 °C. Geochim Cosmochim Acta 64:1773–1784
Dunbar RB, Wellington GM, Colgan MW, Glynn PW (1994) Eastern sea surface temperature since 1600 A.D.: the $\delta^{18}O$ record of climate variability in Galapagos corals. Paleoceanography 9:291–315
Duplessy JC, Shackleton NJ, Fairbanks RG, Labeyrie L, Oppo D, Kallel N (1988) Deepwater source variations during the last climatic cycle and their impact on the global circulation. Paleoceanography 3:343–360
Durka W, Schulze ED, Gebauer G, Voerkelius S (1994) Effects of forest decline on uptake and leaching of deposited nitrate determined from $^{15}N$ and $^{18}O$ measurements. Nature 372:765–767
Dutton A, Wilkinson BH, Welker JM, Bowen GJ, Lohmann KC (2005) Spatial distribution and seasonal variation in $^{18}O/^{16}O$ of modern precipitation and river water across the conterminous USA. Hydrol Process 19:4121–4146
Ehhalt D, Rohrer F (2009) The tropospheric cycle of $H_2$: a critical review. Tellus 61:500–535
Eiler JM (2007) The study of naturally-occuring multiply-substituted isotopologues. Earth Planet Sci Lett 262:309–327
Eiler J (2011) Paleoclimate reconstruction using carbonate clumped isotope thermometry. Quat Sci Rev 30:3575–3588
Eiler JM (2013) The isotopic anatomies of molecules and minerals. Ann Rev Earth Planet Sci 41:411–441
Eiler JM, Kitchen N (2004) Hydrogen isotope evidence for the origin and evolution of the carbonaceous chondrites. Geochim Cosmochim Acta 68:1395–1411
Eiler JM, Baumgartner LP, Valley JW (1992) Intercrystalline stable isotope diffusion: a fast grain boundary model. Contr Mineral Petrol 112:543–557

Eiler JM, Valley JW, Baumgartner LP (1993) A new look at stable isotope thermometry. Geochim Cosmochim Acta 57:2571–2583
Eiler JM, Farley KA, Valley JW, Hofmann A, Stolper EM (1996) Oxygen isotope constraints on the sources of Hawaiian volcanism. Earth Planet Sci Lett 144:453–468
Eiler JM, Crawford A, Elliott T, Farley KA, Valley JW, Stolper EM (2000) Oxygen isotope geochemistry of oceanic-arc lavas. J Petrol 41:229–256
Eiler JM, Stolper EM, McCanta M (2011) Intra- and inter-crystalline oxygen isotope variations in minerals from basalts and peridotites. J Petrol 52:1393–1413
Eiler JM et al (2014) Frontiers of stable isotope geoscience. Chem Geol 372:119–143
Eisenhauer A, 10 others (2019) Calcium isotope ratios in blood and urine: a new biomarker for the diagnosis of osteoporosis. Bone Rep 10:100200
Ek M, Hunt AC, Schönbächler M (2017) A new method for high-precision palladium isotope analyses of iron meteorites and other metal samples. JAAS 32:647–656
Elardo SM, Shahar A (2017) Non-chondritic iron isotope ratios in planetary mantles as a result of core formation. Nat Geosci 10:317–321
Eldridge CS, Compston W, Williams IS, Harris JW, Bristow JW (1991) Isotopic evidence for the involvement of recycled sediments in diamond formation. Nature 353:649–653
Elkins LJ, Fischer TP, Hilton DR, Sharp ZD, McKnight S, Walker J (2006) Tracing nitrogen in volcanic and geothermal volatiles from the Nicaraguan volcanic front. Geochim Cosmochim Acta 70:5215–5235
Elliot M, Labeyrie L, Duplessy JC (2002) Changes in North Atlantic deep-water formation associated with the Dansgaard-Oeschger temperature oscillations (60–10 ka). Quat Sci Rev 21:1153–1165
Elliott T, Jeffcoate AB, Bouman C (2004) The terrestrial Li isotope cycle: light-weight constraints on mantle convection. Earth Planet Sci Lett 220:231–245
Emiliani C (1955) Pleistocene temperatures. J Geol 63:538–578
Emiliani C (1978) The cause of the ice ages. Earth Planet Sci Lett 37:349–354
Engel MH, Macko SA, Silfer JA (1990) Carbon isotope composition of individual amino acids in the Murchison meteorite. Nature 348:47–49
Epica community members (2004) Eight glacial cycles from an Antarctic ice core. Nature 429:623–628
Epstein S, Yapp CJ, Hall JH (1976) The determination of the D/H ratio of non-exchangeable hydrogen in cellulose extracted from aquatic and land plants. Earth Planet Sci Lett 30:241–251
Epstein S, Thompson P, Yapp CJ (1977) Oxygen and hydrogen isotopic ratios in plant cellulose. Science 198:1209–1215
Epstein S, Krishnamurthy RV, Cronin JR, Pizzarello S, Yuen GU (1987) Unusual stable isotope ratios in amino acid and carboxylic acid extracts from the Murchison meteorite. Nature 326:477–479
Erez J, Luz B (1983) Experimental paleotemperature equation for planktonic foraminifera. Geochim Cosmochim Acta 47:1025–1031
Eslinger EV, Savin SM (1973) Oxygen isotope geothermometry of the burial metamorphic rocks of the Precambrian Belt Supergroup, Glacier National Park, Montana. Bull Geol Soc Am 84:2549–2560
Estep MF, Hoering TC (1980) Biogeochemistry of the stable hydrogen isotopes. Geochim Cosmochim Acta 44:1197–1206
Etiope G, Sherwood-Lollar B (2013) Abiotic methane on earth. Rev Geophys 51:276–299
Etiope G, Schoell M (2014) Abiotic gas: a typical, but not rare. Elements 10:291–296
Evans BW, Hattori K, Baronnet A (2013) Serpentinite: what, why where? Elements 9:99–106
Evans KA (2012) The redox budget of subduction zones. Earth Sci Rev 113:11–32
Exley RA, Mattey DP, Boyd SR, Pillinger CT (1987) Nitrogen isotope geochemistry of basaltic glasses: implications for mantle degassing and structure. Earth Planet Sci Lett 81:163–174
Fairbanks RG (1989) A 17000 year glacio-eustatic sea level record: influence of glacial melting rates on the Younger Dryas event and deep ocean circulation. Nature 342:637–642

Fantle MS (2010) Evaluating the Ca isotope proxy. Am J Sci 310:194–210
Fantle MS, De Paolo DJ (2005) Variations in the marine Ca cycle over the past 20 million years. Earth Planet Sci Lett 237:102–117
Fantle MS, De Paolo DJ (2007) Ca isotopes in carbonate sediment and pore fluid from ODP Site 807A: the $Ca^{2+}{}_{aq}$-calcite equilibrium fractionation factor and calcite recrystallization rates in Pleistoce sediments. Geochim Cosmochim Acta 71:2524–2546
Fantle MS, Higgins J (2014) The effects of diagenesis and dolomitization on Ca and Mg isotopes in marine platform carbonates: implications for the geochemical cycles of Ca and Mg. Geochim Cosmochim Acta 142:458–481
Fantle MS, Barnes BD, Lau KV (2020) The role of diagenesis in shaping the geochemistry of the marine carbonate record. Ann Rev Earth Planet Sci 48:549–583
Farkas J, Buhl D, Blenkinsop J, Veizer J (2007) Evolution of the oceanic calcium cycle during the late Mesozoic: evidence from $\delta^{44/40}$ Ca of marine skeletal carbonates. Earth Planet Sci Lett 253:96–111
Farquhar J, Thiemens MH (2000) The oxygen cycle of the Martian atmosphere-regolith system: $\Delta^{17}O$ of secondary phases in Nakhla and Lafayette. J Geophys Res 105:11991–11998
Farquhar GD et al (1993) Vegetation effects on the isotope composition of oxygen in atmospheric $CO_2$. Nature 363:439–443
Farquhar J, Chacko T, Ellis DJ (1996) Preservation of oxygen isotopic compositions in granulites from Northwestern Canada and Enderby Land, Antarctica: implications for high-temperature isotopic thermometry. Contr Mineral Petrol 125:213–224
Farquhar J, Thiemens MH, Jackson T (1998) Atmosphere-surface interactions on Mars: $\Delta^{17}O$ measurements of carbonate from ALH 84001. Science 280:1580–1582
Farquhar J, Bao H, Thiemens M (2000) Atmospheric influence of Earth's earliest sulfur cycle. Science 289:756–759
Farquhar J, Wing B, McKeegan KD, Harris JW (2002) Insight into crust-mantle coupling from anomalous $\Delta^{33}S$ of sulfide inclusions in diamonds. Geochim Cosmochim Acta Spec Suppl 66: A225
Farquhar J, Johnston DT, Wing BA, Habicht KS, Canfield DE, Airieau S, Thiemens MH (2003) Multiple sulphur isotope interpretations for biosynthetic pathways: implications for biological signatures in the sulphur isotope record. Geobiology 1:27–36
Farquhar J, Kim ST, Masterson A (2007) Implications from sulfur isotopes of the Nakhla meteorite for the origin of sulfate on Mars. Earth Planet Sci Lett 264:1–8
Feng X, Epstein S (1995) Carbon isotopes of trees from arid environments and implications for reconstructing atmospheric $CO_2$ concentrations. Geochim Cosmochim Acta 59:2599–2608
Ferry JM (1992) Regional metamorphism of the Waits River Formation: delineation of a new type of giant hydrothermal system. J Petrol 33:45–94
Ferry JM, Dipple GM (1992) Models for coupled fluid flow, mineral reaction and isotopic alteration during contact metamorphism: the Notch Peak aureole, Utah. Am Mineral 77:577–591
Ferry JM, Passey BH, Vasconcelos C, Eiler JM (2011) Formation of dolomite at 40–80 °C in the Latemar carbonate buildup, Dolomites, Italy from clumped isotope thermometry. Geology 39:571–574
Ferry JM, Kitajima K, Strickland A, Valley JW (2014) Ion microprobe survey of the grain-scale oxygen isotope geochemistry of minerals in metamorphic rocks. Geochim Cosmochim Acta 144:403–433
Fiebig J, Chiodini G, Caliro S, Rizzo A, Spangenberg J, Hunziker JC (2004) Chemical and isotopic equilibrium between $CO_2$ and $CH_4$ in fumarolic gas discharges: generation of $CH_4$ in arc magmatic-hydrothermal systems. Geochim Cosmochim Acta 68:2321–2334
Fiebig J, Bajnai D, Löffler N, Methner K, Krsnik E, Mulch A, Hofmann S (2019a) Combined high-precision $\Delta_{48}$ and $\Delta_{47}$ analysis of carbonates. Chem Geol 522:186–191

Fiebig J, Stefansson A, Ricci A, Tassi F, Viveiros F, Silva C, Lopez TM, Schreiber C, Hofmann S (2019b) Mountain BW (2019) Abiogenesis not required to explain the origin of volcanic-hydrothermal hydrocarbons. Geochem Persp Lett 11:23–27

Field CW, Gustafson LB (1976) Sulfur isotopes in the porphyry copper deposit at El Salvador, Chile. Econ Geol 71:1533–1548

Finnegan S, Bergmann K, Eiler JM, 6 others (2011) The magnitude and duration of Late Ordovician-Early Silurian glaciation. Science 331:903–906

Fiorentini E, Hoernes S, Hoffbauer R, Vitanage PW (1990) Nature and scale of fluid-rock exchange in granulite-grade rocks of Sri Lanka: a stable isotope study. In: Vielzeuf D, Vidal Ph (eds) Granulites and crustal evolution. Kluwer, Dordrecht, pp 311–338

Fischer TP, Giggenbach WF, Sano Y, Williams SN (1998) Fluxes and sources of volatiles discharged from Kudryavy, a subduction zone volcano, Kurile Islands. Earth Planet Sci Lett 160:81–96

Fischer TP, Hilton DR, Zimmer MM, Shaw AM, Sharp ZD, Walker JA (2002) Subduction and recycling of nitrogen along the Central American margin. Science 297:1154–1157

Fischer-Gödde M, Kleine T (2017) Ruthenium isotopic evidence for an inner solar system origin of the late veneer. Nature 541:525

Fittoussi C, Bourdon B, Kleine T, Oberli F, Reynolds BC (2009) Si isotope systematics of meteorites and terrestrial peridotites: implications for Mg/Si fractionation in the solar nebula and for Si in the Earth's core. Earth Planet Sci Lett 287:77–85

Fogel ML, Cifuentes LA (1993) Isotope fractionation during primary production. In: Macko SA (ed) Engel MH. Organic Geochemistry, Plenum Press, pp 73–98

Fogel ML, Griffin PC, Newsome SD (2016) Hydrogen isotopes in individual amino acids reflect differentiated pools of hydrogen from food and water in *Escherichia coli*. PNAS 113:E4648–E4653

Francey RJ, Tans PP (1987) Latitudinal variation in oxygen-18 of atmospheric $CO_2$. Nature 327:495–497

Franchi IA, Wright IP, Sexton AS, Pillinger T (1999) The oxygen isotopic composition of Earth and Mars. Meteorit Planet Sci 34:657–661

Franz HB et al (2014) Isotopic links between atmospheric chemistry and the deep sulphur cycle on Mars. Nature 508:364–368

Frape SK, Fritz P (1987) Geochemical trends from groundwaters from the Canadian Shield. In: Fritz P, Frape SK (eds) Saline water and gases in crystalline rocks, vol 33. Geological Association of Canada, Special Paper, pp 19–38

Frape SK, Fritz P, McNutt RH (1984) Water-rock interaction and chemistry of groundwaters from the Canadian Shield. Geochim Cosmochim Acta 48:1617–1627

Freeman KH, Hayes JM (1992) Fractionation of carbon isotopes by phytoplankton and estimates of ancient $CO_2$ levels. Global Biogeochem Cycles 6:185–198

Freeman KH, Hayes JM, Trendel JM, Albrecht P (1990) Evidence from carbon isotope measurements for diverse origins of sedimentary hydrocarbons. Nature 343:254–256

Frei R, Gaucher C, Poulton SW, Canfield DE (2009) Fluctuations in Precambrian atmospheric oxygenation recorded by chromium isotopes. Nature 461:250–253

Freyer HD (1979) On the $^{13}C$-record in tree rings. I. $^{13}C$ variations in northern hemisphere trees during the last 150 years. Tellus 31:124–137

Freyer HD, Belacy N (1983) $^{13}C/^{12}C$ records in northern hemispheric trees during the past 500 years—anthropogenic impact and climatic superpositions. J Geophys Res 88:6844–6852

Freymuth H, Andersen MB, Elliott T (2019) Uranium isotope fractionation during slab dehydration beneath the Izu arc. Earth Planet Sci Lett 522:244–254

Fricke HC, O'Neil JR (1999) The correlation between $^{18}O/^{16}O$ ratios of meteoric water and surface temperature: its use in investigating terrestrial climate change over geologic time. Earth Planet Sci Lett 170:181–196

Fricke HC, Wickham SM, O'Neil JR (1992) Oxygen and hydrogen isotope evidence for meteoric water infiltration during mylonitization and uplift in the Ruby Mountains—East Humboldt Range core complex, Nevada. Contr Mineral Petrol 111:203–221

Fricke HC, Clyde WC, O'Neil JR (1998a) Intra-tooth variations in $\delta^{18}O$ ($PO_4$) of mammalian tooth enamel as a record of seasonal variations in continental climate variables. Geochim Cosmochim Acta 62:1839–1850

Fricke HC, Clyde WC, O'Neil JR, Gingerich PD (1998b) Evidence for rapid climate change in North America during the latest Paleocene thermal maximum: oxygen isotope compositions of biogenic phosphate from the Bighorn Basin (Wyoming). Earth Planet Sci Lett 160:193–208

Fripiat F, Cavagna AJ, Delairs F, de Brauwere A, Andre L, Cardinal D (2012) Processes controlling the Si isotopic composition in the Southern Ocean and application for paleoceanography. Biogeosciences 9:2443–2457

Frost CD, von Blanckenburg F, Schoenberg R, Frost BR, Swapp SM (2007) Preservation of Fe isotope heterogeneities during diagenesis and metamorphism of banded iron formation. Contr Mineral Petrol 153:211–235

Fry B (1988) Food web structure on Georges Bank from stable C, N and S isotopic compositions. Limnol Oceanogr 3:1182–1190

Fu B, Kita NT, Wilde SA, Liu X, Cliff J, Greig A (2012) Origin of the Tongbai-Dabie-Sulu Neoproterozoic low-$d^{18}O$ igneous province, east-central China. Contr Mineral Petrol 165:641–662

Fu Q, Sherwood Lollar B, Horita J, Lacrampe-Couloume G, Seyfried WE (2007) Abiotic formation of hydrocarbons under hydrothermal conditions: constraints from chemical and isotope data. Geochim Cosmochim Acta 71:1982–1998

Gagan MK, Ayliffe LK, Beck JW, Cole JE, Druffel ER, Schrag DP (2000) New views of tropical paleoclimates from corals. Quat Sci Rev 19:45–64

Galewsky J, Steen-Larsen HC, Field RD, Worden J, Risi C, Schneider M (2016) Stable isotopes in atmospheric water vapour and applications to the hydrologic cycle. Rev Geophys 54:809–865

Galili N, Shemesh A, 16 others (2019) The geologic history of seawater oxygen isotopes from marine iron oxides. Science 365:469–473

Galimov EM (1985) The relation between formation conditions and variations in isotope compositions of diamonds. Geochem Int 22(1):118–141

Galimov EM (1988) Sources and mechanisms of formation of gaseous hydrocarbons in sedimentary rocks. Chem Geol 71:77–95

Galimov EM (1991) Isotopic fractionation related to kimberlite magmatism and diamond formation. Geochim Cosmochim Acta 55:1697–1708

Galimov EM (2006) Isotope organic geochemistry. Org Geochem 37:1200–1262

Gao X, Thiemens MH (1993a) Isotopic composition and concentration of sulfur in carbonaceous chondrites. Geochim Cosmochim Acta 57:3159–3169

Gao X, Thiemens MH (1993b) Variations of the isotopic composition of sulfur in enstatite and ordinary chondrites. Geochim Cosmochim Acta 57:3171–3176

Gao Y, Hoefs J, Przybilla R, Snow JE (2006) A complete oxygen isotope profile through the lower oceanic crust, ODP hole 735B. Chem Geol 233:217–234

Gargano A, Sharp Z, Shearer C, Simon JI, Halliday A (2020) The Cl isotope composition and halogen contents of Apollo-return samples. PNAS 117:23418–23425

Garlick GD, Epstein S (1967) Oxygen isotope ratios in coexisting minerals of regionally metamorphosed rocks. Geochim Cosmochim Acta 31:181

Gaschnig RM, Reinhard CT, Planavsky NJ, Wang X, Asael D, Chauvel C (2017) The Molybdenum isotope system as a tracer of slab input in subduction zones: an example from Martinique, Lesser Antilles Arc. Geochem Geophys Geosys 18:4674–4689

Gat JR (1971) Comments on the stable isotope method in regional groundwater investigation. Water Resour Res 7:980

Gat JR (1984) The stable isotope composition of Dead Sea waters. Earth Planet Sci Lett 71:361–376

Gat JR, Issar A (1974) Desert isotope hydrology: water sources of the Sinai desert. Geochim Cosmochim Acta 38:1117–11131

Gautier E, Savarino J, Hoek J, Erbland J, Caillon N, Hattori S, Yoshida N, Albalat E, Albarede F, Farquhar J (2019) 2600-years of stratospheric volcanism through sulfate isotopes. Nature Commun 10:466

Gazquez F, Evans NP, Hodell DA (2017) Precise and accurate isotope fractionation factors ($\alpha^{17}O$, $\alpha^{18}O$ and $\alpha D$) for water and $CaSO_4 \cdot 2H_2O$ (gypsum). Geochim Cosmochim Acta 198:259–270

Gazquez F, Morellon M, Bauska T, Herwatz D, Surma J, Moreno A, Staubwasser M, Valero-Garces B, Delgado-Huertas A, Hodell DA (2018) Triple oxygen and hydrogen isotopes of gypsum hydration water for quantitative paleo-humidity reconstruction. Earth Planet Sci Lett 481:177–188

Geng L, Alexander B, Cole-Dai J, Steig EJ, Savarino J, Sofen EE, Schauer AJ (2014) Nitrogen isotopes in ice core nitrate linked to anthropogenic atmospheric acidity change. PNAS 111:5808–5812

Georg RB, Reynolds BC, Frank M, Halliday AN (2006) Mechanisms controlling the silicon isotopic compositions of river water. Earth Planet Sci Lett 249:290–306

Georg RB, Halliday AN, Schauble EA, Reynolds BC (2007) Silicon in the Earth's core. Nature 447:1102–1106

Gerdes ML, Baumgartner LP, Person M, Rumble D (1995) One- and two-dimensional models of fluid flow and stable isotope exchange at an outcrop in the Adamello contact aureole, Southern Alps, Italy. Am Mineral 80:1004–1019

Gerlach TM, Taylor BE (1990) Carbon isotope constraints on degassing of carbon dioxide from Kilauea volcano. Geochim Cosmochim Acta 54:2051–2058

Gerlach TM, Thomas DM (1986) Carbon and sulphur isotopic composition of Kilauea parental magma. Nature 319:480–483

Gerst S, Quay P (2001) Deuterium component of the global molecular hydrogen cycle. J Geophys Res 106:5021–5031

Geske A, Goldstein RH, Mavromatis V, Richter DK, Buhl D, Kluge T, John CM, Immenhauser A (2015a) The magnesium isotope ($\delta^{26}Mg$) signature of dolomites. Geochim Cosmochim Acta 149:131–151

Geske A, Lokier S, Dietzel M, Richter DK, Buhl D, Immenhauser A (2015b) Magnesium isotope composition of sabkha porewater and related sub-recent stoichiometric dolomites, Abu Dabi (UAE). Chem Geol 393–394:112–124

Ghosh P et al (2006a) $^{13}C$–$^{18}O$ bonds in carbonate minerals: a new kind of paleothermometer. Geochim Cosmochim Acta 70:1439–1456

Ghosh P, Garzione CN, Eiler JM (2006b) Rapid uplift of the Altiplano revealed through C-13-O-18 bonds in paleosol carbonates. Science 311:511–515

Giggenbach WF (1992) Isotopic shifts in waters from geothermal and volcanic systems along convergent plate boundaries and their origin. Earth Planet Sci Lett 113:495–510

Gilbert A, Yamada K, Suda K, Ueno Y, Yoshida N (2016) Measurement of position-specific $^{13}C$ isotopic composition of propane at the nanomole level. Geochim Cosmochim Acta 177:205–216

Gilbert A, Sherwood Lollar B, Musat F, Giunta T, Chen S, Kajimoto Y, Yamada K, Boreham CJ, Yoshida N, Ueno Y (2019) Intramolecular isotopic evidence for bacterial oxidation of propane in subsurface natural gas reservoirs. PNAS 116:6653–6658

Giletti BJ (1986) Diffusion effect on oxygen isotope temperatures of slowly cooled igneous and metamorphic rocks. Earth Planet Sci Lett 77:218–228

Gilg HA (2000) D/H evidence for the timing of kaolinization in Northeast Bavaria, Germany. Chem Geol 170:5–18

Girard JP, Savin S (1996) Intercrystalline fractionation of oxygen isotopes between hydroxyl and non-hydroxyl sites in kaolinite measured by thermal dehydroxylation and partial fluorination. Geochim Cosmochim Acta 60:469–487

Giunta T, Young E, 11 others (2019) Methane sources and sinks in continental sedimentary systems: new insights from paired clumped isotopologues $^{13}CH_3D$ and $^{12}CH_2D_2$. Geochim Cosmochim Acta 245:327–351

Giunta T, Labidi J, Kohl IE, Ruffine L, Donval JP, Geli L, Cagatay MN, Lu H, Young ED (2021) Evidence for methane isotopic bond re-ordering in gas reservoirs sourcing cold seeps from the Sea of Marmara. Earth Planet Sci Lett 553:116619

Given RK, Lohmann KC (1985) Derivation of the original isotopic composition of Permian marine cements. J Sediment Petrol 55:430–439

Goericke R, Fry B (1994) Variations of marine plankton $\delta^{13}C$ with latitude, temperature and dissolved $CO_2$ in the world ocean. Global Geochem Cycles 8:85–90

Goldberg SL, Present TM, Finnergan S, Bergmann KD (2021) A high-resolution record of early Paleozoic climate. PNAS 118 6:e2013083118

Goldhaber MB, Kaplan IR (1974) The sedimentary sulfur cycle. In: Goldberg EB (ed) The sea, vol IV. Wiley, New York

Goldhammer T, Brunner B, Bernasconi SM, Ferdelman TG, Zabel M (2011) Phosphate oxygen isotopes: insights into sedimentary phosphorus cycling from the Benguela upwelling system. Geochim Cosmochim Acta 75:3741–3756

Goldstein TP, Aizenshat Z (1994) Thermochemical sulfate reduction—a review. J Therm Anal 42:241–290

Gonfiantini R (1986) Environmental isotopes in lake studies. In: Fritz P, Fontes J (eds) Handbook of environmental isotope geochemistry, vol 2, p 112

Gordon GW, Monge J, Channon MB, Wu Q, Skulan JL, Anbar AD, Fonseca R (2014) Predicting multiple myeloma disease activity by analyzing natural calcium isotope composition. Leukemia 28:2112–2115

Goto KT, 11 others (2014) Uranium isotope systematics of ferromanganese crusts in the Pacific Ocean: implications for the marine $^{238}U/^{235}U$ isotope system. Geochim Cosmochim Acta 146:43–58

Grachev AM, Severinghaus JP (2003) Laboratory determination of thermal diffusion constants for $^{29}N/^{28}N_2$ in air at temperatures from −60 to 0 °C for reconstruction of magnitudes of abrupt climate changes using the ice core fossil-air paleothermometer. Geochim Cosmochim Acta 67:345–360

Graham S, Pearson N, Jackson S, Griffin W, O'Reilly SY (2004) Tracing Cu and Fe from source to porphyry: in situ determination of Cu and Fe isotope ratios in sulfides from the Grasberg Cu-Au deposit. Chem Geol 207:147–169

Grasse P, Ehlert C, Frank M (2013) The influence of water mass mixing on the dissolved Si isotope composition in the eastern Equatorial Pacific. Earth Planet Sci Lett 380:60–71

Green GR, Ohmoto D, Date J, Takahashi T (1983) Whole-rock oxygen isotope distribution in the Fukazawa-Kosaka Area, Hokuroko District, Japan and its potential application to mineral exploration. Econ Geol Monogr 5:395–411

Greenop R, Hain MP, Sosdian SM, Oliver KI, Goodwin P, Chalk TB, Lear CH, Wilson PA, Foster GL (2017) A record of Neogene seawater $\delta^{11}B$ reconstructed from paired $\delta^{11}B$ analyses on benthic and planktic foraminifera. Clim past 13:149–170

Greenwood JP, Riciputi LR, McSween HY (1997) Sulfide isotopic compositions in shergottites and ALH 84001, and possible implications for life on Mars. Geochim Cosmochim Acta 61:4449–4453

Greenwood JP, Itoh S, Sakamoto N, Vicenzi EP, Yurimoto H (2008) Hydrogen isotope evidence for loss of water from Mars through time. Geophys Res Lett 35:5203

Greenwood RC, Franchi IA, Jambon A, Barrat JA, Burbine TH (2006) Oxygen isotope variation in stony-iron meteorites. Science 313:1763–1765

Greenwood JP, Itoh S, Sakamoto N, Warren P, Taylor L, Yurimoto H (2011) Hydrogen isotope ratios in lunar rocks indicate delivery of cometary water to the Moon. Nat Geosci 4:79–82

Greenwood PF, Mohammed L, Grice K, McCulloch M, Schwark L (2018a) The application of compound-specific sulfur isotopes to the oil-source rock correlation of Kurdistan petroleum. Org Geochem 117:22–30

Greenwood RC, Barrat JA, Miller MF, Anand M, Dauphas N, Franchi IA, Sillard P, Starkey NA (2018b) Oxygen isotope evidence for accretion of Earth's water before a high-energy Moon-forming giant impact. Sci Adv 4:eaao5928

Gregory MJ, Mathur R (2017) Understanding copper isotope behavior in the high-temperature magmatic hydrothermal porphyry environment. Geochem Geophys Geosys 18:4000–4015

Gregory RT, Taylor HP (1981) An oxygen isotope profile in a section of Cretaceous oceanic crust, Samail Ophiolite, Oman: evidence for $\delta^{18}O$ buffering of the oceans by deep (>5 km) seawater-hydrothermal circulation at Mid-Ocean Ridges. J Geophys Res 86:2737–2755

Gregory RT, Taylor HP (1986) Possible non-equilibrium oxygen isotope effects in mantle nodules, an alternative to the Kyser-O, Neil-Carmichael $^{18}O/^{16}O$ geothermometer. Contr Mineral Petrol 93:114–119

Griffith EM, Paytan A, Kozdon R, Eisenhauer A, Ravelo AC (2008a) Influences on the fractionation of calcium isotopes in planktonic foraminifera. Earth Planet Sci Lett 268:124–136

Griffith EM, Schauble EA, Bullen TD, Paytan A (2008b) Characterization of calcium isotopes in natural and synthetic barite. Geochim Cosmochim Acta 72:5641–5658

Griffith EM, Payton A, Caldeira K, Bullen TD, Thomas E (2008c) A dynamic marine calcium cycle during the past 28 million years. Science 322:1671–1674

Grimes CB, Ushikubo T, John BE, Valley JW (2011) Uniformly mantle-like $\delta^{18}O$ in zircons from oceanic plagiogranite and gabbros. Contr Mineral Petrol 161:13–33

Grootes PM, Stuiver M, White JWC, Johnsen S, Jouzel J (1993) Comparison of oxygen isotope records from the GISP-2 and GRIP Greenland ice cores. Nature 366:552–554

Grossman EL (1984) Carbon isotopic fractionation in live benthic foraminifera—comparison with inorganic precipitate studies. Geochim Cosmochim Acta 48:1505–1512

Grottoli AG, Eakin CM (2007) A review of modern coral $\delta^{18}O$ and $\Delta^{14}C$ proxy records. Earth Sci Rev 81:67–91

Gruber N (1998) Anthropogenic $CO_2$ in the Atlantic Ocean. Global Biogeochem Cycles 12:165–191

Gruber N et al (1999) Spatiotemporal patterns of carbon-13 in the global surface oceans and the oceanic Suess effect. Global Biogeochem Cycles 13:307–335

Gueguen B, Rouxel O, Rouget ML, Bollinger C, Ponzevera E, Germain Y, Fouquet Y (2016) Comparative geochemistry of four ferromanganese crusts from the Pacific Ocean and significance for the use of Ni isotopes as paleoceanographic tracers. Geochim Cosmochim Acta 189:214–235

Guelke M, von Blanckenburg F (2007) Fractionation of stable iron isotopes in higher plants. Environ Sci Tech 41:1896–1901

Guilbaud R, Butler IB, Ellam RM (2011) Abiotic pyrite formation produces a large Fe isotope fractionation. Science 332:1548–1551

Guilpart E, Vimeux F, Evan S, Brioude J, Metzger JM, Barthe C, Risi C, Cattani O (2017) The isotopic composition of near-surface water vapor at the Maido observatory (Reunion Island, southwestern Indian Ocean) documents the controls of the humidity of the subtropical troposphere. J Geophys Res Atmospheres 122:9628–9650

Guo W, Eiler JM (2007) Temperatures of aqueous alteration and evidence for methane generation on the parent bodies of the CM chondrites. Geochim Cosmochim Acta 71:5565–5575

Guo W (2020) Kinetic clumped isotope fractionation in the DIC-$H_2O$-$CO_2$ system: patterns, controls and implications. Geochim Cosmochim Acta 268:230–257

Gupta P, Noone D, Galewsky J, Sweeney C, Vaughn BH (2009) Demonstration of high-precision continuous measurements of water vapor isotopologues in laboratory and remote field deployment using wavelength-scanned cavity ring-down spectroscopy (WS-CRDS) technology. Rapid Commun Mass Spectrom 23:2534–2542

Guy RD, Fogel ML, Berry JA (1993) Photosynthetic fractionation of the stable isotopes of oxygen and carbon. Plant Phys 101:37–47
Haack U, Hoefs J, Gohn E (1982) Constraints on the origin of Damaran granites by Rb/Sr and $\delta^{18}O$ data. Contrib Mineral Petrol 79:279–289
Hackley KC, Anderson TF (1986) Sulfur isotopic variations in low-sulfur coals from the Rocky Mountain region. Geochim Cosmochim Acta 50:703–1713
Hahm D, Hilton DR, Castillo PR, Hawkins JW, Hanan BB, Hauri EH (2012) An overview of the volatile systematics of the Lau Basin—resolving the effects of source variation, magmatic degassing and crustal contamination. Geochim Cosmochim Acta 85:88–113
Haimson M, Knauth LP (1983) Stepwise fluorination-a useful approach for the isotopic analysis of hydrous minerals. Geochim Cosmochim Acta 47:1589–1595
Halama R, Bebout GE, John T, Scambelluri M (2011) Nitrogen recycling in subducted mantle rocks and implications for the global nitrogen cycle. Int J Earth Sci 103:2081–2099
Halbout J, Robert F, Javoy M (1990) Hydrogen and oxygen isotope compositions in kerogen from the Orgueil meteorite: clues to a solar origin. Geochim Cosmochim Acta 54:1453–1462
Halder J, Terzer S, Wassenaar LI, Araguas-Araguas LJ, Aggarwal PK (2015) The Global Network of Isotopes in Rivers (GNIR): integration of water isotopes in watershed observation and riverine research. Hydrol Earth Syst Sci 19:3419–3431
Hallis LJ, Anand M, Greenwood RC, Miller MF, Franchi IA, Russell SS (2010) The oxygen isotope composition, petrology and geochemistry of mare basalts: evidence for large-scale compositional variation in the lunar mantle. Geochim Cosmochim Acta 74:6885–6899
Halverson GP, Poitrasson F, Hoffman PE, Nedelec A, Montel JM, Kirby J (2011) Fe isotope and trace element geochemistry of the Neoproterozoic syn-glacial Rapitan iron formation. Earth Planet Sci Lett 309:100–112
Hamza MS, Epstein S (1980) Oxygen isotope fractionation between oxygen of different sites in hydroxyl-bearing silicate minerals. Geochim Cosmochim Acta 44:173–182
Han X, Guo Q, Strauss H, Liu C, Hu J, Guo Z, Wei R, Peters M, Tian L, Kong J (2017) Multiple sulfur isotope constraints on sources and formation processes of sulfate in Beijing $PM_{2.5}$ aerosol. Environ Sci Tech 51:7794–7803
Hare VJ, Loftus E, Jeffrey A, Ramsey CB (2018) Atmospheric $CO_2$ effect on stable carbon isotope composition of terrestrial fossil archives. Nature Commun 9:252
Harmon RS, Hoefs J (1995) Oxygen isotope heterogeneity of the mantle deduced from global $^{18}O$ systematics of basalts from different geotectonic settings. Contr Mineral Petrol 120:95–114
Harmon RS, Hoefs J, Wedepohl KH (1987) Stable isotope (O, H, S) relationships in Tertiary basalts and their mantle xenoliths from the Northern Hessian Depression, W.Germany. Contr Mineral Petrol 95:350–369
Harte B, Otter M (1992) Carbon isotope measurements on diamonds. Chem Geol 101:177–183
Hartmann M, Nielsen H (1969) $\delta^{34}S$-Werte in rezenten Meeressedimenten und ihre Deutung am Beispiel einiger Sedimentprofile aus der westlichen Ostsee. Geol Rundsch 58:621–655
Hastings MG, Sigman DM, Steig EJ (2005) Glacial/interglacial changes in the isotopes of nitrate from the Greenland Ice Sheet Project 2 (GISP2) ice core. Global Biogeochem Cycl 19:GB4024
Hauri EH, Wang J, Pearson DG, Bulanova GP (2002) Microanalysis of $\delta^{13}C$, $\delta^{15}N$ and N abundances in diamonds by secondary ion mass spectrometry. Chem Geol 185:149–163
Hauri EH, Weinreich T, Saal AE, Rutherford MC, Van Orman JA (2011) High pre-eruptive water contents preserved in melrt inclusions. Science 333:213–215
Hauri EH, Saal AE, Nakajima M, Anand M, Rutherford MJ, Van Orman JA, Le Voyer M (2017) Origin and evolution of water in the Moon's interior. Ann Rev Earth Planet Sci 45:89–111
Hawkesworth CJ, Kemp AIS (2006) Using hafnium and oxygen isotopes in zircons to unravel the record of crustal evolution. Chem Geol 226:144–162
Hayes JM (2001) Fractionation of carbon and hydrogen isotopes in biosynthetic processes. In: Valley JW, Cole DR (eds) Stable isotope geochemistry. Reviews in mineralogy and geochemistry, vol 43. pp 225–277

Hayes JM, Waldbauer JR (2006) The carbon cycle and associated redox processes through time. Phil Trans R Soc B 361:931–950
Hayes JM, Kaplan IR, Wedeking KW (1983) Precambrian organic chemistry, preservation of the record. In: Schopf JW (ed) Earth's earliest biosphere: its origin and evolution, Chap. 5. Princeton University Press, pp 93–132
Hayes JM, Popp BN, Takigiku R, Johnson MW (1989) An isotopic study of biogeochemical relationships between carbonates and organic carbon in the Greenhorn Formation. Geochim Cosmochim Acta 53:2961–2972
Hayes JM, Strauss H, Kaufman AJ (1999) The abundance of $^{13}C$ in marine organic matter and isotopic fractionation in the global biogeochemical cycle of carbon during the past 800 Ma. Chem Geol 161:103–125
Hays PD, Grossman EL (1991) Oxygen isotopes in meteoric calcite cements as indicators of continental paleoclimate. Geology 19:441–444
Hays JD, Imbrie J, Shackleton NJ (1976) Variations in the earth's orbit: pacemaker of the ice ages. Science 194:943–954
He Z, Zhang Y, Deng X, Hu H, Li Y, Yu H, Archer C, Li J, Huang F (2020) The behavior of Fe and S isotopes in porphyry copper systems: constraints from the Tongshankou Cu-Mo deposit, eastern China. Geochim Cosmochim Acta 270:61–83
Heaton THE (1986) Isotopic studies of nitrogen pollution in the hydrosphere and atmosphere: a review. Chem Geol 59:87–102
Heimann A, Beard BL, Johnson CM (2008) The role of volatile exsolution and sub-solidus fluid/rock interactions in producing high $^{56}Fe/^{54}Fe$ ratios in siliceous igneous rocks. Geochim Cosmochim Acta 72:4379–4396
Helffrich G, Shahar A, Hirose K (2018) Isotopic signature of core derived $SiO_2$. Am Mineral 103:1161–1164
Helman Y, Barkan E, Eisenstadt D, Luz B, Kaplan A (2005) Fractionation of the three stable oxygen isotopes by oxygen producing and consuming reactions in photosynthetic organisms. Plant Phys 2005:2292–2298
Hemingway JD, Olsen H, Turchyn AV, Tipper ET, Bickle MJ, Johnston DT (2020) Triple oxygen isotope insight into terrestrial pyrite oxidation. PNAS 117:7650–7657
Hendy CH, Wilson AT (1968) Paleoclimatic data from speleothems. Nature 219:48–51
Henkes GA, Passey BH, Grossman EL, Shenton BJ, Yancey TE, Perez-Huerta A (2018) Temperature evolution and the oxygen isotope composition of Phanerozoic oceans from clumped isotope thermometry. Earth Planet Sci Lett 490:40–50
Heraty LJ, Fuller ME, Huang L, Abrajano T, Sturchio NC (1999) Isotopic fractionation of carbon and chlorine by microbial degradation of dichlormethane. Org Geochem 30:793–799
Herwartz D (2021) Triple oxygen isotope variations in Earth's crust. Rev Mineral Geochem 86
Herwartz D, Pack A, Friedrichs B, Bischoff A (2014) Identification of the giant impactor Theia in lunar rocks. Science 344:1146–1150
Herwartz D, Pack A, Krylov D, Xiao Y, Muehlenbachs K, Sengupta S, Di Rocco T (2015) Revealing the climate of snowball Earth from $\Delta^{17}O$ systematics of hydrothermal rocks. PNAS 112:5337–5341
Herwatz D, Surma J, Voigt C, Assonov S, Staubwasser M (2017) Triple oxygen isotope systematics of structurally bonded water in gypsum. Geochim Cosmochim Acta 209:254–266
Heuser A, Eisenhauer A (2009) A pilot study on the use of natural calcium isotope ($^{44}Ca/^{40}Ca$) fractionation in urine as a proxy for the human body calcium balance. Bone. https://doi.org/10.1016/j.bone.2009.11037
Higgins JA, Schrag DP (2012) Records of neogene seawater chemistry and diagenesis in deep-sea carbonate sediments and pore fluids. Earth Planet Sci Lett 357–358:386–396
Hin RC, Schmidt MW, Bourdon B (2012) Experimental evidence for the absence of iron isotope fractionation between metal and silicate liquids at 1GPA and 1250–1300 °C and its cosmochemical consequences. Geochim Cosmochim Acta 93:164–181

Hin RC, Burkhardt C, Schmidt MW, Bourdon B (2013) Experimental evidence for Mo isotope fractionation between metal and silicate liquids. Earth Planet Sci Lett 379:38–48
Hin RC, Fitoussi C, Schmidt MW, Bourdon B (2014) Experimental determination of the Si isotope fractionation factor between liquid metal and liquid silicate. Earth Planet Sci Lett 387:55–66
Hinrichs KU, Hayes JM, Sylva SP, Brewer PG, DeLong EF (1999) Methane-consuming archaebacteria in marine sediments. Nature 398:802–805
Hitchon B, Friedman I (1969) Geochemistry and origin of formation waters in the western Canada sedimentary basin. 1. Stable isotopes of hydrogen and oxygen. Geochim Cosmochim Acta 33:1321–1349
Hitchon B, Krouse HR (1972) Hydrogeochemistry of the surface waters of the Mackenzie River drainage basin, Canada. III. Stable isotopes of oxygen, carbon and sulfur. Geochim Cosmochim Acta 36:1337–1357
Hoag KJ, Still CJ, Fung IY, Boering KA (2005) Triple oxygen isotope composition of tropospheric carbon dioxide as a tracer of terrestrial gross carbon fluxes. Geophys Res Lett 32: L02802
Hoefs J (1970) Kohlenstoff-und Sauerstoff-Isotopenuntersuchungen an Karbonatkonkretionen und umgebendem Gestein. Contrib Mineral Petrol 27:66–79
Hoefs J (1992) The stable isotope composition of sedimentary iron oxides with special reference to Banded Iron Formations. In: Isotopic signatures and sedimentary records. Lecture notes in earth science, vol 43. Springer, pp 199–213
Hoefs J, Emmermann R (1983) The oxygen isotope composition of Hercynian granites and pre-Hercynian gneisses from the Schwarzwald, SW Germany. Contrib Mineral Petrol 83:320–329
Hoefs J, Sywall M (1997) Lithium isotope composition of quaternary and tertiary biogene carbonates and a global lithium isotope balance. Geochim Cosmochim Acta 61:2679–2690
Hoering T (1975) The biochemistry of the stable hydrogen isotopes. Carnegie Inst Washington Yearb 74:598
Hoernes S, Van Reenen DC (1992) The oxygen isotopic composition of granulites and retrogressed granulites from the Limpopo Belt as a monitor of fluid-rock interaction. Precambrian Res 55:353–364
Hoffman JH, Hodges RR, McElroy MB, Donahue TM, Kolpin M (1979) Composition and structure of the Venus atmosphere: results from Pioneer Venus. Science 205:49–52
Hoffman PE, Kaufman AJ, Halverson GP, Schrag DP (1998) Neoproterozoic snowball earth. Science 281:1342–1346
Hofmann ME, Horvath B, Schneider L, Peters W (2017) Atmospheric measurements of $\Delta^{17}O$ in $CO_2$ in Göttingen, Germany reveal a seasonal cycle driven by biospheric uptake. Geochim Cosmochim Acta 199:143–163
Hofstetter TB, Scharzenbach RP, Bernasconi SM (2008) Assessing transformation processes of organic compounds using stable isotope fractionation. Environ Sci Technol 42:7737–7743
Holloway JR, Blank JG (1994) Application of experimental results to C-O-H species in natural melts. In: Carroll MR, Holloway JR (eds) Volatiles in magmas. Reviews Mineral 30:187–230
Holmden C (2009) Ca isotope study of Ordovician dolomite, limestone, and anhydrite in the Williston basin: implications for subsurface dolomitization and local Ca cycling. Chem Geol 268:180–188
Holser WT (1977) Catastrophic chemical events in the history of the ocean. Nature 267:403–408
Holser WT, Kaplan IR (1966) Isotope geochemistry of sedimentary sulfates. Chem Geol 1:93–135
Homoky WB, Severmann S, Mills RA, Statham PJ, Fones GR (2009) Pore-fluid Fe isotopes reflect the extent of benthic Fe redox recycling: evidence from continental shelf and deep-sea sediments. Geology 37:751–754
Homoky WB, John SG, Conway TM, Mills RA (2013) Distinct iron isotope signatures and supply from marine sediment solution. Nat Commun 4. https://doi.org/10.1038/ncomms3143
Hoppe P, Zinner E (2000) Presolar dust grains from meteorites and their stellar sources. J Geophys Res Space Phys 105:10371–10385

Horita J (1989) Stable isotope fractionation factors of water in hydrated salt minerals. Earth Planet Sci Lett 95:173–179

Horita J (2014) Oxygen and carbon isotope fractionation in the system dolomite-water-$CO_2$ to elevated temperatures. Geochim Cosmochim Acta 129:111–124

Horita J, Berndt ME (1999) Abiogenic methane formation and isotope fractionation under hydrothermal conditions. Science 285:1055–1057

Horita J, Polyakov VB (2015) Carbon-bearing iron phases and the carbon isotope composition of the deep earth. PNAS 112:31–36

Horita J, Cole DR, Wesolowski DJ (1995) The activity-composition relationship of oxygen and hydrogen isotopes in aqueous salt solutions: III. Vapor-liquid water equilibration of NaCl solutions to 350 °C. Geochim Cosmochim Acta 59:1139–1151

Horita J, Driesner T, Cole DR (1999) Pressure effect on hydrogen isotope fractionation between brucite and water at elevated temperatures. Science 286:1545–1547

Horita J, Cole DR, Polyakov VB, Driesner T (2002a) Experimental and theoretical study of pressure effects on hydrous isotope fractionation in the system brucite-water at elevated temperatures. Geochim Cosmochim Acta 66:3769–3788

Horita J, Zimmermann H, Holland HD (2002b) Chemical evolution of seawater during the Phanerozoic: implications from the record of marine evaporates. Geochim Cosmochim Acta 66:3733–3756

Horn I, von Blanckenburg F, Schoenberg R, Steinhöfel G, Markl G (2006) In situ iron isotope ratio determination using UV-femtosecond laser ablation with application to hydrothermal ore formation processes. Geochim Cosmochim Acta 70:3677–3688

Horner TJ, Williams HM, Hein JR, Saito MA, Burton KW, Halliday AN, Nielsen SG (2015) Persistence of deeply sourced iron in the Pacific Ocean. PNAS 112:1292–1297

Horvath B, Hofmann M, Pack A (2012) On the triple oxygen isotope composition of carbon dioxide from some combustion processes. Geochim Cosmochim Acta 95:160–168

Hotz K, Krayenbuehl PA, Walczyk T (2012) Mobilization of storage iron is reflected in the iron isotopic compositionof blood in humans. J Biol Inorg Chem 17:301–309

Hotz K, Walczyk T (2013) Natural iron isotopic composition of blood is an indicator of dietary iron absorption efficiency in humans. J Biol Inorg Chem 18:1–7

Hu S, Lin Y, Zhang J, Hao J, Feng L, Xu L, Yang W, Yang J (2014) NanoSIMS analysis of apatite and melt inclusions in the GRV 020090 Martian meteorite: hydrogen isotope evidence for recent past underground hydrothermal activity on Mars. Geochim Cosmochim Acta 140:321–333

Hu Y, Teng FZ, Zhang HF, Xiao Y, Su BX (2016) Metasomatism-induced magnesium isotopic heterogeneity: evidence from pyroxenites. Geochim Cosmochim Acta 185:88–111

Huang B, Xiao X, Hu Z, Yi P (2005a) Geochemistry and episodic accumulation of natural gases from the Ledong gas field in the Yinggehai basin, offshore South China. Org Geochem 36:1689–1702

Huang Y, Wang Y, Alexandre M, Lee T, Rose-Petruck C, Fuller M, Pizzarello S (2005b) Molecular and compound-specific isotopic characterization of monocarboxylic acids in carbonaceous chondrites. Geochim Cosmochim Acta 69:1073–1084

Huang L, Strurchio NC, Abrajano T, Heraty LJ, Holt BD (1999) Carbon and chlorine isotope fractionation of chlorinated aliphatic hydrocarbons by evaporation. Org Geochem 30:777–785

Huang KJ, Shen B, Lang XG, Tang WB, Peng Y, Ke S, Kaufman AJ, Ma HR, Li FB (2015) Magnesium isotope compositions of the Mesoproterozoic dolostones: implications for Mg isotope systematics of marine carbonates. Geochim Cosmochim Acta 164:333–351

Huang S, Jacobsen SB (2017) Calcium isotopic compositions of chondrites. Geochim Cosmochim Acta 201:364–376

Hudak MR, Bindeman I (2018) Conditions of pinnacle formation and glass hydration in cooling ignimbrite sheets from H and O isotope systematics at Crater Lake and the Valley of Ten Thousand Smokes. Earth Planet Sci Lett 500:56–66

Hudson JD (1977) Stable isotopes and limestone lithification. J Geol Soc London 133:637–660

Hui H, 10 others (2017) A heterogeneous lunar interior for hydrogen isotopes as revealed by the lunar highland samples. Earth Planet Sci Lett 473:14–23

Hulston JR (1977) Isotope work applied to geothermal systems at the Institute of Nuclear Sciences, New Zealand. Geothermics 5:89–96

Hulston JR, Thode HG (1965) Variations in the $^{33}S$, $^{34}S$ and $^{36}S$ contents of meteorites and their relations to chemical and nuclear effects. J Geophys Res 70:3475–3484

Huntington KW, Wernicke BP, Eiler JM (2010) Influence of climate change and uplift on Colorado Plateau paleotemperatures from carbonate clumped isotope thermometry. Tectonics 29:TC3005. https://doi.org/10.1029/2009TC002449

Huntington KW, Budd DA, Wernicke BP, Eiler JM (2011) Use of clumped-isotope thermometry to constrain the crystallization temperature of diagenetic calcite. J Sediment Res 81:656–669

Hüri E, Marty B (2015) Nitrogen isotope variations in the solar system. Nat Geosci 8:515–522

Hutchison W, Babiel RJ, 8 others (2019) Sulphur isotopes of alkaline magmas unlock long-term records of crustal recycling on Earth. Nature Commun 10:4208

Hutchinson W, Finch AA, Boyce AJ (2020) The sulfur isotope evolution of magmatic-hydrothermal fluids: insights into ore-forming processes. Geochim Cosmochim Acta 288:176–198

Iacumin P, Bocherens H, Marriotti A, Longinelli A (1996) Oxygen isotope analysis of coexisting carbonate and phosphate in biogenic apatite; a way to monitor diagenetic alteration of bone phosphate? Earth Planet Sci Lett 142:1–6

Ibanez-Mejia M, Tissot FL (2019) Extreme Zr stable isotope fractionation during magmatic fractional crystallization. Sci Adv 5:eaax8648

Ikehata K, Notsu K, Hirata T (2011) Copper isotope characteristics of copper-rich minerals from Besshi-type volcanogenic massive sulfide deposits, Japan, determined using a Femtosecond La-MC-ICP-MS. Econ Geol 106:307–316

Ikehata K, Hirata T (2012) Copper isotope characteristics of copper-rich minerals from the Horoman peridotite complex, Hokkaido, northern Japan. Econ Geol 107:1489–1497

Ionov DA, Hoefs J, Wedepohl KH, Wiechert U (1992) Contents and isotopic composition of sulfur in ultramafic xenoliths from Central Asia. Earth Planet Sci Lett 111:269–286

Ionov D, Cooper K, Brooker R (2007) Li isotope fractionation in peridotites and mafic melts. Geochim Cosmochim Acta 71:202–218

Ireland TR, Avila J, Greenwood RC, Hicks LJ, Bridges JC (2020) Oxygen isotopes and sampling of the Solar system. Space Sci Rev 216. https://doi.org/10.1007/s11214-020-0645-3

Irwin H, Curtis C, Coleman M (1977) Isotopic evidence for the source of diagenetic carbonate during burial of organic-rich sediments. Nature 269:209–213

Ishibashi J, Sano Y, Wakita H, Gamo T, Tsutsumi M, Sakai H (1995) Helium and carbon geochemistry of hydrothermal fluids from the Mid-Okinawa trough back arc basin, southwest of Japan. Chem Geol 123:1–15

Jaffrés JB, Shields GA, Wallmann K (2007) The oxygen isotope evolution of seawater: a critical review of a long-standing controversy and an improved geological water cycle model for the past 3.4 billion years. Earth Sci Rev 83:83–122

James DE (1981) The combined use of oxygen and radiogenic isotopes as indicators of crustal contamination. Ann Rev Earth Planet Sci 9:311–344

James AT (1983) Correlation of natural gas by use of carbon isotopic distribution between hydrocarbon components. Am Assoc Petrol Geol Bull 67:1167–1191

James AT (1990) Correlation of reservoired gases using the carbon isotopic compositions of wet gas components. Am Assoc Petrol Geol Bull 74:1441–1458

Jaouen K, Pons ML, Balter V (2014) Iron, copper and zinc isotopic fractionation up mammal trophic chains. Earth Planet Sci Lett 374:164–172

Jaouen K, Beasley M, Schoeninger M, Hublin JJ, Richards MP (2016) Zinc isotope ratios of bones and teeth as new dietary indicators: results from a modern food web (Koobi For a, Kenya). Sci Reports 6:26281

Jasper JP, Hayes JM (1990) A carbon isotope record of $CO_2$ levels during the late Quaternary. Nature 347:462–464
Jasper JP, Hayes JM, Mix AC, Prahl FG (1994) Photosynthetic fractionation of C-13 and concentrations of dissolved $CO_2$ in the central equatorial Pacific. Paleoceanography 9:781–798
Javoy M, Pineau F, Delorme H (1986) Carbon and nitrogen isotopes in the mantle. Chem Geology 57:41–62
Javoy M, 10 others (2010) The chemical composition of the Earth: enstatite chondrite models. Earth Planet Sci Lett 293:259–268
Jeffcoate AB, Elliott T, Kasemann SA, Ionov D, Cooper K, Brooker R (2007) Li isotope fractionation in peridotites and mafic melts. Geochim Cosmochim Acta 71:202–218
Jeffrey AW, Pflaum RC, Brooks JM, Sackett WM (1983) Vertical trends in particulate organic carbon $^{13}C/^{12}C$ ratios in the upper water column. Deep Sea Res 30:971–983
Jenden PD, Kaplan IR, Poreda RJ, Craig H (1988) Origin of nitrogen-rich natural gases in the California Great Valley: evidence from helium, carbon and nitrogen isotope ratios. Geochim Cosmochim Acta 52:851–861
Jenden PD, Drazan DJ, Kapan IR (1993) Mixing of thermogenic natural gases in northern Appalachian Basin. Am Assoc Petrol Geol Bull 77:980–998
Jendrzejewski N, Eggenkamp HGM, Coleman ML (2001) Characterisation of chlorinated hydrocarbons from chlorine and carbon isotopic compositions: scope of application to environmental problems. Appl Geochem 16:1021–1031
Jenkyns HC (2010) Geochemistry of oceanic anoxic events. Geochem Geophys Geosys 11: Q03004
Jiang J, Clayton RN, Newton RC (1988) Fluids in granulite facies metamorphism: a comparative oxygen isotope study on the South India and Adirondack high grade terrains. J Geol 96:517–533
Jiang G, Wang X, Shi X, Xiao S, Zhang S, Dong J (2012) The origin of decoupled carbonate and carbon isotope signatures in the early Cambrian (ca 542–520 Ma) Yangtze platform. Earth Planet Sci Lett 317–318:96–110
Joachimski MM, Pancost RD, Freeman KH (2002) Carbon isotope geochemistry of the Frasnian-Famennian transition. Palaeo Palaeo Palaeo 181:91–104
Joachimski MM, van Geldern R, Breisig S, Buggisch W, Day J (2004) Oxygen isotope evolution of biogenic calcite and apatite during the Middle and Late Devonian. Int J Earth Sci 93:542–553
Joachimski MM, Simon L, van Geldern R, Lecuyer C (2005) Boron isotope geochemistry of Paleozoic brachiopod calcite: implications for secular change in the boron iosotope geochemistry of seawater over the Phanerozoic. Geochim Cosmochim Acta 69:4035–4044
Joachimski MM, Breisig S, Buggisch W, Talent JA, Mawson R, Gereke M, Morrow JR, Day J, Weddige K (2009) Devonian climate and reef evolution: insights from oxygen isotopes in apatite. Earth Planet Sci Lett 284:599–609
Joachimski MM, Lai X, Shen S, Jiang H, Luo G, Chen J, Sun Y (2012) Climate warming in the latest Permian and the Permian-Triassic mass extinction. Geology 40:195–198
Johnsen SJ, Dansgaard W, White JW (1989) The origin of Arctic precipitation under present and glacial conditions. Tellus 41B:452–468
Johnsen SJ, Clausen HB, Dansgaard W, Gundestrup N, Hammer CU, Tauber H (1995) The Eem stable isotope record along the GRIP ice core and ist interpretation. Quat Res 43:117–124
Johnson DG, Jucks KW, Traub WA, Chance KV (2001) Isotopic composition of stratospheric water vapour: measurements and photochemistry. J Geophys Res 106D:12211–12217
Johnson CM, Skulan JL, Beard BL, Sun H, Neals andon KH, Braterman PS (2002) Isotopic fraction between Fe(III) and Fe(II) in aqueous solutions. Earth Planet Sci Lett 195:141–153
Johnson CM, Beard BL, Beukes NJ, Klein C, O'Leary JM (2003) Ancient geochemical cycling in the Earth as inferred from Fe-isotope studies of banded iron formations from the Transvaal craton. Contr Mineral Petrol 114:523–547

Johnson CM, Beard BL, Roden EE (2008) The iron isotope fingerprints of redox and biogeochemical cycling in modern and ancient Earth. Ann Rev Earth Planet Sci 36:457–493
Johnston DT (2011) Multiple sulfur isotopes and the evolution of Earth's surface sulfur cycle. Earth Sci Rev 106:161–183
Jørgensen BB, Böttcher MA, Lüschen H, Neretin LN, Volkov II (2004) Anaerobic methane oxidation and a deep $H_2S$ sink generate isotopically heavy sulfides in Black Sea sediments. Geochim Cosmochim Acta 68:2095–2118
Jouzel J, Merlivat L, Roth E (1975) Isotopic study of hail. J Geophys Res 80:5015–5030
Jouzel J, Merlivat L, Lorius C (1982) Deuterium excess in an East Antarctic ice core suggests higher relative humidity at the oceanic surface during the last glacial maximum. Nature 299:688–691
Jouzel J, Lorius C, Petit JR, Barkov NI, Kotlyakov VM, Petrow VM (1987) Vostok ice core: a continuous isotopic temperature record over the last climatic cycle (160000 years). Nature 329:403–408
Juranek LW, Quay PD (2010) Basin-wide photosynthetic production rates in the subtropical and tropical Pacific Ocean determined from dissolved oxygen isotope ratio measurements. Global Biogeochem Cycles 24:GB2006. https://doi.org/10.1029/2009GB003492
Kampschulte A, Strauss H (2004) The sulfur isotope evolution of Phanerozoic seawater based on the analyses of sructurally substituted sulfate in carbonates. Chem Geol 204:255–280
Kaplan IR, Hulston JR (1966) The isotopic abundance and content of sulfur in meteorites. Geochim Cosmochim Acta 30:479–496
Kaplan IR, Rittenberg SC (1964) Microbiological fractionation of sulphur isotopes. J Gen Microbiol 34:195–212
Kasting JF, Howard MT, Wallmann K, Veizer J, Shields G, Jaffrés J (2006) Paleoclimates, ocean depth and the oxygen isotopic composition of the ocean. Earth Planet Sci Lett 252:82–93
Kato C, Moynier F, Valdes MC, Dhaliwal JK, Day JM (2015) Extensive volatile loss during formation and differentiation of the Moon. Nature Commun 6:7617. https://doi.org/10.1038/ncomms8617
Kaufman AJ, Knoll GM (1995) Neoproterozoic variations in the C-isotopic composition of seawater: stratigraphic and biogeochemical implications. Precambrian Res 73:27–49
Kaye J (1987) Mechanisms and observations for isotope fractionation of molecular species in planetary atmospheres. Rev Geophysics 25:1609–1658
Keeling CD (1958) The concentration and isotopic abundance of atmospheric carbon dioxide in rural areas. Geochim Cosmochim Acta 13:322–334
Keeling CD (1961) The concentration and isotopic abundances of carbon dioxide in rural and marine air. Geochim Cosmochim Acta 24:277–298
Keeling CD, Mook WG, Tans P (1979) Recent trends in the $^{13}C/^{12}C$ ratio of atmospheric carbon dioxide. Nature 277:121–123
Keeling CD, Carter AF, Mook WG (1984) Seasonal, latitudinal and secular variations in the abundance and isotopic ratio of atmospheric carbon dioxide. II. Results from oceanographic cruises in the tropical Pacific Ocean. J Geophys Res 89:4615–4628
Keeling CD, Bacastow RB, Carter AF, Piper SC, Whorf TR, Heimann M, Mook WG, Roeloffzen H (1989) A three-dimensional model of atmospheric $CO_2$ transport based on observed winds. 1. Analysis of observational data. Geophys Monogr 55:165–236
Keeling CD, Whorf TP, Wahlen M, van der Plicht J (1995) Interannual extremes in the rate of rise of atmospheric carbon dioxide since 1980. Nature 375:666–670
Keeling RF, Graven HD, Welp LR, Resplandy L, Bi J, Piper SC, Sun Y, Bollenbacher A, Meijer HA (2017) Atmospheric evidence for a global secular increase in carbon isotopic discrimination of land photosynthesis. PNAS 114:10361–10366
Kelly J, Fu B, Kita N, Valley J (2007) Optically continuous silcrete quartz cements in the St. Peter sandstone. Geochim Cosmochim Acta 71:3812–3832

Kelly WC, Rye RO, Livnat A (1986) Saline minewaters of the Keweenaw Peninsula, Northern Michigan: their nature, origin and relation to similar deep waters in Precambrian crystalline rocks of the Canadian Shield. Am J Sci 286:281–308

Kelson JR, Huntington KW, Schauer AJ, Saenger C, Lechler AR (2017) Toward a universal carbonate clumped isotope calibration: diverse synthesis and preparatory methods suggest a single temperature relationship. Geochim Cosmochim Acta 197:104–131

Kelts K, McKenzie JA (1982) Diagenetic dolomite formation in quaternary anoxic diatomaceous muds of DSDP Leg 64, Gulf of California. Initial Rep DSDP 64:553–569

Kempton PD, Harmon RS (1992) Oxygen isotope evidence for large-scale hybridization of the lower crust during magmatic underplating. Geochim Cosmochim Acta 56:971–986

Kendall B, Dahl TW, Anbar AD (2017) The stable isotope geochemistry of molybdenum. Rev Mineral Geochem 82:683–732

Kennicutt MC, Barker C, Brooks JM, De Freitaas DA, Zhu GH (1987) Selected organic matter indicators in the Orinoco, Nile and Changjiang deltas. Org Geochem 11:41–51

Kerrich R, Rehrig W (1987) Fluid motion associated with Tertiary mylonitization and detachment faulting: $^{18}O/^{16}O$ evidence from the Picacho metamorphic core complex, Arizona. Geology 15:58–62

Kerrich R, Latour TE, Willmore L (1984) Fluid participation in deep fault zones: evidence from geological, geochemical and to $^{18}O/^{16}O$ relations. J Geophys Res 89:4331–4343

Kerridge JF (1983) Isotopic composition of carbonaceous-chondrite kerogen: evidence for an interstellar origin of organic matter in meteorites. Earth Planet Sci Lett 64:186–200

Kerridge JF, Haymon RM, Kastner M (1983) Sulfur isotope systematics at the 21°N site, East Pacific Rise. Earth Planet Sci Lett 66:91–100

Kerridge JF, Chang S, Shipp R (1987) Isotopic characterization of kerogen-like material in the Murchison carbonaceous chondrite. Geochim Cosmochim Acta 51:2527–2540

Kharaka YK, Berry FAF, Friedman I (1974) Isotopic composition of oil-field brines from Kettleman North Dome, California and their geologic implications. Geochim Cosmochim Acta 37:1899–1908

Kharaka YK, Cole DR, Hovorka SD, Gunter WD, Knauss KG, Freifeld BM (2006) Gas-water-rock interactions in Frio formation following $CO_2$ injection: implications to the storage of greenhouse gases in sedimentary basins. Geology 34:577–580

Kiczka M, Wiederhold JG, Kraemer SM, Bourdon B, Kretzschmar R (2010) Iron isotope fractionation during Fe uptake and translocation in Alpine plants. Environ Sci Technol 44:6144–6150

Kim KR, Craig H (1990) Two isotope characterization of $N_2O$ in the Pacific ocean and constraints on its origin in deep water. Nature 347:58–61

Kim KR, Craig H (1993) Nitrogen-15 and oxygen-18 characteristics of nitrous oxide. Science 262:1855–1858

King PL, McLennan SM (2009) Sulfur on Mars. Elements 6:107–112

Kirkley MB, Gurney JJ, Otter ML, Hill SJ, Daniels LR (1991) The application of C-isotope measurements to the identification of the sources of C in diamonds. Appl Geochemistry 6:477–494

Kloppmann W, Girard JP, Négrel P (2002) Exotic stable isotope composition of saline waters and brines from the crystalline basement. Chem Geol 184:49–70

Knauth LP (1988) Origin and mixing history of brines, Palo Duro Basin, Texas, USA. Appl Geochem 3:455–474

Knauth LP, Lowe DR (1978) Oxygen isotope geochemistry of cherts from the Onverwacht group (3.4 billion years), Transvaal, South Africa, with implications for secular variations in the isotopic composition of chert. Earth Planet Sci Lett 41:209–222

Knauth LP, Beeunas MA (1986) Isotope geochemistry of fluid inclusions in Permian halite with implications for the isotopic history of ocean water and the origin of saline formation waters. Geochim Cosmochim Acta 50:419–433

Knoll AH, Hayes JM, Kaufman AJ, Swett K, Lambert IB (1986) Secular variation in carbon isotope ratios from Upper Proterozoic successions of Svalbard and East Greenland. Nature 321:832–838

Kohl IE, Bao H (2011) Triple oxygen isotope determination of molecular oxygen incorporation in sulfate produced during abiotic pyrite oxidation (pH 2–11). Geochim Cosmochim Acta 75:1785–1798

Kohn MJ (1996) Predicting animal $\delta^{18}O$: accounting for diet and physiological adaptation. Geochim Cosmochim Acta 60:4811–4829

Kohn MJ (1999) Why most "dry" rocks should cool "wet." Am Mineral 84:570–580

Kohn MJ (2010) Carbon isotope fractionation of terrestrial C3 plants as indicator of (paleo) ecology and (paleo)climate. PNAS 107:19691–19695

Kohn MJ (2016) Carbon isotope discrimination in C3 plants is independent of natural variations in $pCO_2$. Geochem Perspectives Lett 2:35–43

Kohn MJ, Valley JW (1994) Oxygen isotope constraints on metamorphic fluid flow, Townshend Dam, Vermont, USA. Geochim Cosmochim Acta 58:5551–5566

Kohn MJ, Cerling TE (2002) Stable isotope compositions of biological apatite. Rev Mineral Geochem 48:455–488

Kohn MJ, Valley JW, Elsenheimer D, Spicuzza M (1993) Oxygen isotope zoning in garnet and staurolite: evidence for closed system mineral growth during regional metamorphism. Am Mineral 78:988–1001

Kolodny Y, Kerridge JF, Kaplan IR (1980) Deuterium in carbonaceous chondrites. Earth Planet Sci Lett 46:149–153

Kolodny Y, Luz B, Navon O (1983) Oxygen isotope variations in phosphate of biogenic apatites, I. Fish bone apatite—rechecking the rules of the game. Earth Planet Sci Lett 64:393–404

Kolodny Y, Luz B, Sander M, Clemens WA (1996) Dinosaur bones: fossils or pseudomorphs? The pitfalls of physiology reconstruction from apatitic fossils. Palaeo Palaeo Palaeo 126:161–171

Kool DM, Wrage N, Oenema O, Harris D, Van Groenigen JW (2009) The $^{18}O$ signature of biogenic nitrous oxide is determined by O exchange with water. Rapid Commun Mass Spectrom 23:104–108

Koren G, 14 others (2019) Global 3-D simulations of the triple oxygen isotope signature $\Delta^{17}O$ in atmospheric $CO_2$. J Geophys Res Atmospheres 124:8808–8836

Krishnamurthy RV, Epstein S, Cronin JR, Pizzarello S, Yuen GU (1992) Isotopic and molecular analyses of hydrocarbons and monocarboxylic acids of the Murchison meteorite. Geochim Cosmochim Acta 56:4045–4058

Kroopnick P (1985) The distribution of $^{13}C$ of $\Sigma CO2$ in the world oceans. Deep Sea Res 32:57–84

Kroopnick P, Craig H (1972) Atmospheric oxygen: isotopic composition and solubility fractionation. Science 175:54–55

Kroopnick P, Weiss RF, Craig H (1972) Total $CO_2$, $^{13}C$ and dissolved oxygen-$^{18}O$ at Geosecs II in the North Atlantic. Earth Planet Sci Lett 16:103–110

Krouse HR (1977) Sulfur isotope studies and their role in petroleum exploration. J Geochem Explor 7:189–211

Krouse HR, Case JW (1983) Sulphur isotope abundances in the environment and their relation to long term sour gas flaring, near Valleyview, Alberta. Final report, Research Management Division, University Alberta RMD Rep 83/18

Kueter N, Lilley MD, Schmidt MW, Bernasconi SM (2019) Experimental carbonatite/graphite carbon isotope fractionation and carbonate/graphite geothermometry. Geochim Cosmochim Acta 253:290–306

Kump LR (1989) Alternative modeling approaches to the geochemical cycles of carbon, sulfur and strontium isotopes. Am J Sci 289:390–410

Kump LR (2005) Ironing out biosphere oxidation. Science 307:1058–1059

Kump LR, Arthur MA (1999) Interpreting carbon-isotope excursions: carbonates and organic matter. Chem Geol 161:181–198

Kvenvolden KA (1995) A review of the geochemistry of methane in natural gas hydrate. Org Geochem 23:997–1008

Kyser TK, O'Neil JR (1984) Hydrogen isotope systematics of submarine basalts. Geochim Cosmochim Acta 48:2123–2134

Kyser TK, O'Neil JR, Carmichael ISE (1981) Oxygen isotope thermometry of basic lavas and mantle nodules. Contrib Mineral Petrol 77:11–23

Kyser TK, O'Neil JR, Carmichael ISE (1982) Genetic relations among basic lavas and mantle nodules. Contrib Mineral Petrol 81:88–102

Labeyrie LD, Juillet A (1982) Oxygen isotope exchangeability of diatom valve silica; interpretation and consequences for paleoclimatic studies. Geochim Cosmochim Acta 46:967–975

Labeyrie LD, Duplessy JC, Blanc PL (1987) Deep water formation and temperature variations over the last 125000 years. Nature 327:477–482

Labidi J, Cartigny P, Birck JL, Assayag N, Bourrand JJ (2012) Determination of multiple sulphur isotopes in glasses: a reappraisal of the MORB $\delta^{34}$S. Chem Geol 334:189–198

Labidi J, Cartigny P, Moreira M (2013) Non-chondritic sulphur isotope composition of the terrestrial mantle. Nature 501:208–211

Labidi J, Cartigny P, Hamelin C, Moreira M, Dosso L (2014) Sulfur isotope budget ($^{32}$S, $^{33}$S, $^{34}$S, $^{36}$S) in Pacific-Antarctic ridge basalts: a record of mantle source heterogeneity and hydrothermal sulfide assimilation. Geochim Cosmochim Acta 133:47–67

Labidi J, Shahar A, Le Losq C, Hillgren VJ, Mysen BO, Farquhar J (2016) Experimentally determined sulfur isotope fractionation between metal and silicate and implications for planetary differentiation. Geochim Cosmochim Acta 175:181–194

Labidi J, Young ED, Giunta T, Kohl IE, Seewald J, Tang HT, Lilley MD, Früh-Green GL (2020) Methane thermometry in deep-sea hydrothermal systems: evidence for re-ordering of doubly-substituted isotopologues during fluid cooling. Geochim Cosmochim Acta 288:248–261

Lachniet MS (2009) Climatic and environmental controls on speleothem oxygen-isotope values. Quat Sci Rev 28:412–432

Land LS (1980) The isotopic and trace element geochemistry of dolomite: the state of the art. In: Concepts and models of dolomitization. Soc Econ Paleontol Min Spec Publ 28:87–110

Landais A, Barkan E, Luz B (2008) Record of $\delta^{18}$O and $^{17}$O excess in ice from Vostok, Antarctica during the last 150000 years. Geophys Res Lett 35:L02709

Lane GA, Dole M (1956) Fractionation of oxygen isotopes during respiration. Science 123:574–576

Larner F, 8 others (2015) Zinc isotopic compositions of breast cancer tissue. Metallomics 7:112–117

Laskar AH, Mahata S, Liang MC (2016) Identification of anthropogenic $CO_2$ using triple oxygen and clumped isotopes. Environ Sci Tech 50:11806–11814

Lau KV, Maher K, 11others (2017) The influence of seawater carbonate chemistry, mineralogy, and diagenesis on calcium isotope variations in Lower-Middle Triassic carbonate rocks. Chem Geol 471:13–37

Lawrence JR (1989) The stable isotope geochemistry of deep-sea pore water. In: Handbook of environmental isotope geochemistry, vol 3. Elsevier Publ Co, pp 317–356

Lawrence JR, Gieskes JM (1981) Constraints on water transport and alteration in the oceanic crust from the isotopic composition of the pore water. J Geophys Res 86:7924–7934

Lawrence JR, Taviani M (1988) Extreme hydrogen, oxygen and carbon isotope anomalies in the pore waters and carbonates of the sediments and basalts from the Norwegian Sea: methane and hydrogen from the mantle? Geochim Cosmochim Acta 52:2077–2083

Lawrence JR, Taylor HP (1971) Deuterium and oxygen-18 correlation: clay minerals and hydroxides in quaternary soils compared to meteoric waters. Geochim Cosmochim Acta 35:993–1003

Lawrence JR, White JWC (1991) The elusive climate signal in the isotopic composition of precipitation. In: Stable isotope geochemistry: a tribute to Samuel Epstein, vol 3. Special Publication, The Geochemical Society, pp 169–185

Laws EA, Popp BN, Bidigare RR, Kennicutt MC, Macko SA (1995) Dependence of phytoplankton carbon isotopic composition on growth rate and $CO_{2aq}$: theoretical considerations and experimental results. Geochim Cosmochim Acta 59:1131–1138

Lazar C, Young ED, Manning CE (2012) Experimental determination of equilibrium nickel isotope fractionation between metal and silicate from 500 °C to 950 °C. Geochim Cosmochim Acta 86:276–395

Leclerc AJ, Labeyrie LC (1987) Temperature dependence of oxygen isotopic fractionation between diatom silica and water. Earth Planet Sci Lett 84:69–74

Lécuyer C, Grandjean P, Reynard B, Albarede F, Telouk P (2002) $^{11}B/^{10}B$ analysis of geological materials by ICP-MS Plasma 54: application to boron fractionation between brachiopod calcite and seawater. Chem Geol 186:45–55

Leder JL, Swart PK, Szmant AM, Dodge RE (1996) The origin of variations in the isotopic record of scleractinian corals: I. Oxygen. Geochim Cosmochim Acta 60:2857–2870

Lemarchand D, Gaillardet J, Lewin E, Allegre CJ (2000) The influence of rivers on marine boron isotopes and implications for reconstructing past ocean pH. Nature 408:951–954

Lemarchand D, Gaillardet J, Lewin E, Allègre CJ (2002) Boron isotope systematics in large rivers: implications for the marine boron budget and paleo-pH reconstruction over the Cenozoic. Chem Geol 190:123–140

Leng MJ, Marshall JD (2004) Palaeoclimate interpretation of stable isotope data from lake sediment archives. Quaternary Sci Rev 23:811–831

Lepot K, Williford KH, Ushikubo T, Sugitani K, Mimura K, Spicuzza MJ, Valley JW (2013) Texture-specific isotopic compositions in 3.4 Gyr old organic matter support selective preservation in cell-like structures. Geochim Cosmochim Acta 112:66–86

Leshin LA, Epstein S, Stolper EM (1996) Hydrogen isotope geochemistry of SNC meteorites. Geochim Cosmochim Acta 60:2635–2650

Leshin LA, McKeegan KD, Carpenter PK, Harvey RP (1998) Oxygen isotopic constraints on the genesis of carbonates from Martian meteorite ALH 84001. Geochim Cosmochim Acta 62:3–13

Leuenberger M, Siegenthaler U, Langway CC (1992) Carbon isotope composition of atmospheric $CO_2$ during the last ice age from an Antarctic ice core. Nature 357:488–490

Lewan MD (1983) Effects of thermal maturation on stable carbon isotopes as determined by hydrous pyrolysis of Woodford shale. Geochim Cosmochim Acta 47:1471–1480

Li JL, Schwarzenbach EM, John T, Ague JJ, Huang F, Gao J, Klemd R, Whitehouse MJ, Wang XS (2020) Uncovering and quantifying the subduction zone sulfur cycle from the slab perspective. Nat Commun 11:514

Li JX, Qin KZ, Li GM, Evans NJ, Huang F, Zhao JX (2018) Iron isotope fractionation during magmatic-hydrothermal evolution: a case from the Duolong porphyry Cu–Au deposit, Tibet. Geochim Cosmochim Acta 238:1–15

Li L, Cartigny P, Ader M (2009) Kinetic nitrogen isotope fraction associated with thermal decomposition of $NH_3$: experimental results and potential applications to trace the origin of $N_2$ in natural gas and hydrothermal systems. Geochim Cosmochim Acta 73:6282–6297

Li W, Jackson SE, Pearson NJ, Graham S (2010) Copper isotope zonation in the Northparkes porphyry Cu–Au deposit, SE Australia. Geochim Cosmochim Acta 74:4078–4096

Li WY, Teng FZ, Ke S, Rudnick R, Gao S, Wu FY, Chappell B (2010) Heterogeneous magnesium isotopic composition of the upper continental crust. Geochim Cosmochim Acta 74:6867–6884

Li YF, Marty B, Shcheka S, Zimmermann L, Keppler H (2016) Nitrogen isotope fractionation during terrestrial core-mantle separation. Geochem Persp Lett 2:138–147

Liang MC, Mahata S, Laskar AH, Thiemens MH, Newman S (2017) Oxygen isotope anomaly in tropospheric $CO_2$ and implications for $CO_2$ residence time in the atmosphere and gross primary productivity. Scientific Rep 7:13180

Liotta M, Rizzo A, Paonita A, Caracausi A, Martelli M (2012) Sulfur isotopic compositions of fumarolic and plume gases at Mount Etna (Italy) and inferences on their magmatic source. Geochem Geophys Geosys 13(5). https://doi.org/10.1029/2012GC0042118

Lister GS, Kelts K, Chen KZ, Yu JQ, Niessen F (1991) Lake Qinghai, China: closed-basin lake levels and the oxygen isotope record for ostracoda since the latest Pleistocene. Palaeo Palaeo Palaeo 84:141–162

Little SH, Vance D, Walker-Brown C, Landing WM (2014) The oceanic mass balance of copper and zinc isotopes, investigated by analysis by their inputs and outputs to ferromanganese oxide sediments. Geochim Cosmochim Acta 125:673–693

Lisiecki LE, Raymo ME (2005) A Pliocene-Pleistocene stack of 57 globally distributed benthic $\delta^{18}O$ records. Paleoceanography 20:PA1003. https://doi.org/10.1029/2004PA001071

Liu SA, Teng FZ, Yang W, Wu FY (2011) High temperature inter-mineral magnesium isotope fractionation in mantle xenoliths from the North China craton. Earth Planet Sci Lett 308:131–140

Liu Y, Spicuzza MJ, Craddock PR, Day JM, Valley JW, Dauphas N, Taylor LA (2010) Oxygen and iron isotope constraints on near-surface fractionation effects and the composition of lunar mare basalt source regions. Geochim Cosmochim Acta 74:6249–6262

Lloyd MR (1967) Oxygen-18 composition of oceanic sulfate. Science 156:1228–1231

Lloyd MR (1968) Oxygen isotope behavior in the sulfate-water system. J Geophys Res 73:6099–6110

Loewen MW, Graham DW, Bindeman IN, Lupton JE, Garcia MO (2019) Hydrogen isotopes in high 3He/4He submarine basalts: primordial vs. recycled water and the veil of mantle enrichment. Earth Planet Sci Lett 508:62–73

Longinelli A (1966) Ratios of oxygen-18: oxygen-16 in phosphate and carbonate from living and fossil marine organisms. Nature 211:923–926

Longinelli A (1984) Oxygen isotopes in mammal bone phosphate: a new tool for paleohydrological and paleoclimatological research? Geochim Cosmochim Acta 48:385–390

Longinelli A, Craig H (1967) Oxygen-18 variations in sulfate ions in sea-water and saline lakes. Science 156:56–59

Longinelli A, Edmond JM (1983) Isotope geochemistry of the Amazon basin. A Reconnaissance. J Geophys Res 88:3703–3717

Longinelli A, Nuti S (1973) Revised phosphate-water isotopic temperature scale. Earth Planet Sci Lett 19:373–376

Longstaffe FJ (1989) Stable isotopes as tracers in clastic diagenesis. In: Hutcheon IE (ed) Short course in burial diagenesis. Mineralogical Association of Canada short course series, vol 15, pp 201–277

Longstaffe FJ, Schwarcz HP (1977) $^{18}O/^{16}O$ of Archean clastic metasedimentary rocks: a petrogenetic indicator for Archean gneisses? Geochim Cosmochim Acta 41:1303–1312

Lorius C, Jouzel J, Ritz C, Merlivat L, Barkov NI, Korotkevich YS, Kotlyakov VM (1985) A 150000 year climatic record from Antarctic ice. Nature 316:591–596

Lücke A, Moschen R, Schleser G (2005) High-temperature carbon reduction of silica: a novel approach for oxygen isotope analysis of biogenic opal. Geochim Cosmochim Acta 69:1423–1433

Luz B, Barkan E (2000) Assessment of oceanic productivity with the triple-isotope composition of dissolved oxygen. Science 288:2028–2031

Luz B, Barkan E (2005) The isotopic ratios $^{17}O/^{16}O$ and $^{18}O/^{16}O$ in molecular oxygen and their significance in biogeochemistry. Geochim Cosmochim Acta 69:1099–1110

Luz B, Barkan E (2010) Variations of $^{17}O/^{16}O$ and $^{18}O/^{16}O$ in meteoric waters. Geochim Cosmochim Acta 74:6276–6286

Luz B, Kolodny Y (1985) Oxygen isotope variations in phosphate of biogenic apatites, IV: mammal teeth and bones. Earth Planet Sci Lett 75:29–36

Luz B, Kolodny Y, Horowitz M (1984) Fractionation of oxygen isotopes between mammalian bone-phosphate and environmental drinking water. Geochim Cosmochim Acta 48:1689–1693

Luz B, Cormie AB, Schwarcz HP (1990) Oxygen isotope variations in phosphate of deer bones. Geochim Cosmochim Acta 54:1723–1728
Luz B, Barkan E, Bender ML, Thiemens MH, Boering KA (1999) Triple-isotope composition of atmospheric oxygen as a tracer of biosphere productivity. Nature 400:547–550
Lyons TW, Reinhard CT, Planavsky NJ (2014) The rise of oxygen in Earth's early ocean and atmosphere. Nature 506:307–315
Mader M, Schmidt C, van Geldern R, Barth JA (2017) Dissolved oxygen in water and its stable isotope effects: a review. Chem Geol 473:10–21
Magna T, Wiechert U, Halliday AN (2006) New constraints on the lithium isotope composition of the moon and terrestrial planets. Earth Planet Sci Lett 243:336–353
Magyar PM, Orphan VJ, Eiler JM (2016) Measurement of rare isotopologues of nitrous oxide by high-resolution multi-collector mass spectrometry. Rapid Comm Mass Spectr 30:1923–1940
Mahaffy PR, Webster CR et al (2013) Abundance and isotopic composition of gases in the Martian atmosphere from the Curiosity Rover. Science 341:263–266
Mahaffy PR, Conrad PG, and the MSL Team (2015) Volatile and isotopic imprints of ancient Mars. Elements 11:51–56
Maher K, Larson P (2007) Variation in copper isotope ratios and controls on fractionation in hypogene skarn mineralization at Coroccohuayco and Tintaya, Peru. Econ Geol 102:225–237
Mandeville CW, Webster JD, Tappen C, Taylor BE, Timbal A, Sasaki A, Hauri E, Bacon CR (2009) Stable isotope and petrologic evidence for open-system degassing during the climactic and pre-climactic eruptions of Mt. Mazama, Crater Lake, Oregon. Geochim Cosmochim Acta 73:2978–3012
Mane P, Hervig R, Wadhwa M, Garvie LA, Balta JB, McSween HY (2016) Hydrogen isotopic composition of the Martian mantle inferred from the newest Martian meteorite fall, Tissint. Meteor Planet Sci 51:2073–2091
Mansor M, Fantle MS (2019) A novel framework for interpreting pyrite-based Fe isotope records of the past. Geochim Cosmochim Acta 253:39–62
Marin J, Chaussidon M, Robert F (2010) Microscale oxygen isotope variations in 1.9 Ga Gunflint cherts: assessments of diagenetic effects and implications for oceanic paleotemperature reconstructions. Geochim Cosmochim Acta 74:116–130
Marin-Carbonne J, Chaussidon M, Boiron MC, Robert F (2011) A combined in situ oxygen, silicon and fluid inclusion study of a chert sample from Onverwacht Group (3.35 Ga, South Africa): new constraints on fluid circulation. Chem Geol 286:59–71
Marin-Carbonne J, Chaussidon M, Robert F (2012) Micrometer-scale chemical and isotopic criteria (O and Si) on the origin and history of Precambrian cherts: implications for paleo-temperature reconstructions. Geochim Cosmochim Acta 92:129–147
Marin-Carbonne J, Robert F, Chaussidon M (2014a) The silicon and oxygen isotope compositions of Precambrian cherts: a record of oceanic paleo-temperatures? Precam Res 247:223–234
Marin-Carbonne J et al (2014b) Coupled Fe and S isotope variatiions in pyrite nodules from Archaen shale. Earth Planet Sci Lett 392:67–79
Marin-Carbonne J, 13 others (2020) In situ Fe and S isotope analyses in pyrite from the 3.2 Ga Mendon Formation (Barberton Greenstone Belt, South Aftica): evidence for early microbial iron reduction. Geobiology 18:306–325
Markl G, Lahaye Y, Schwinn G (2006a) Copper isotopes as monitors of redox processes in hydrothermal mineralization. Geochim Cosmochim Acta 70:4215–4228
Markl G, von Blanckenburg F, Wagner T (2006b) Iron isotope fractionation during hydrothermal ore deposition and alteration. Geochim Cosmochim Acta 70:3011–3030
Marowsky G (1969) Schwefel-, Kohlenstoff-und Sauerstoffisotopenuntersuchungen am Kupfer-schiefer als Beitrag zur genetischen Deutung. Contrib Mineral Petrol 22:290–334
Marschall HR, Tang M (2020) High temperature processes: is it time for lithium isotopes? Elements 16:247–252

Marschall HR, Wanless D, Shimizu N, Pogge von Strandmann P, Elliot T, Monteleone B (2017) The boron and lithium isotopic composition of mid-ocean ridge basalts and the mantle. Geochim Cosmochim Acta 207:102–138
Martin E (2018) Volcanic plume impact on the atmosphere and climate: O- and S-isotope insight into sulfate aerosol formation. Geosciences 8:198
Martin AP, Condon DJ, Prave AR, Lepland A (2013) A review of temporal constraints for the Palaeoproterozoic large, positive carbonate carbon isotope excursion (the Lomagundi-Jatuli event). Earth Sci Rev 127:242–261
Martin E, Bindeman I (2009) Mass-independent isotopic signatures of volcanic sulfate from three supereruption ash deposits in Lake Tecopa, California. Earth Planet Sci Lett 282:102–114
Martin JE, Vance D, Balter V (2015) Magnesium stable isotope ecology using mammal tooth enamel. PNAS 112:430–435
Martinson DG, Pisias NG, Hays JD, Imbrie J, Moore TC, Shackleton NJ (1987) Age dating and the orbital theory of the ice ages: development of a high resolution 0 to 300000 year chronostratigraphy. Quat Res 27:1–29
Marty B (2012) The origins and concentrations of water, carbon, nitrogen and noble gases on Earth. Earth Planet Sci Lett 313–314:56–66
Marty B, Humbert F (1997) Nitrogen and argon isotopes in oceanic basalts. Earth Planet Sci Lett 152:101–112
Marty B, Zimmermann L (1999) Volatiles (He, C, N, Ar) in mid-ocean ridge basalts: assesment of shallow-level fractionation and characterization of source composition. Geochim Cosmochim Acta 63:3619–3633
Marty B, Chaussidon M, Wiens RC, Jurewicz Burnett DS (2011) A $^{15}$N-poor isotopic composition for the solar system as shown by Genesis solar wind samples. Science 332:1533–1536
Mason TFD et al (2005) Zn and Cu isotopic variability in the Alexandrinka volcanic-hosted massive sulphide (VHMS) ore deposit, Urals, Russia. Chem Geol 221:170–187
Mason E, Edmonds M, Turchyn AV (2017) Remobilization of crustal carbon may dominate arc emissions. Science 357:290–294
Masson-Delmotte V, Jouzel J et al (2005) GRIP deuterium excess reveals rapid and orbital-scale changes in Greenland moisture origin. Science 309:118–121
Mastalerz M, Schimmelmann A (2002) Isotopically exchangeable organic hydrogen in coal relates to thermal maturity and maceral composition. Org Geochem 33:921–931
Matheney RK, Knauth LP (1989) Oxygen isotope fractionation between marine biogenic silica and seawater. Geochim Cosmochim Acta 53:3207–3214
Mathur R, Dendas M, Titley S, Phillips A (2010) Patterns in the copper isotope composition of minerals in porphyry copper deposits in southwestern United States. Econ Geol 105:1457–1467
Mathur R, Wang D (2019) Transition metal isotopes applied to exploration geochemistry: insights from Fe, Cu, and Zn. In: Ore deposits: origin, exploration and exploitation. Geophysical Monograph, vol 242, pp 163–183
Matsubaya O, Sakai H (1973) Oxygen and hydrogen isotopic study on the water of crystallization of gypsum from the Kuroko-type mineralization. Geochem J 7:153–165
Matsuhisa Y (1979) Oxygen isotopic compositions of volcanic rocks from the east Japan island arcs and their bearing on petrogenesis. J Volcanic Geotherm Res 5:271–296
Matsumoto R (1992) Causes of the oxygen isotopic depletion of interstitial waters from sites 798 and 799, Japan Sea, Leg 128. Proc Ocean Drill Program, Sci Results 127(128):697–703
Matsuo S, Friedman I, Smith GI (1972) Studies of quaternary saline lakes. I. Hydrogen isotope fractionation in saline minerals. Geochim Cosmochim Acta 36:427–435
Mattey DP, Carr RH, Wright IP, Pillinger CT (1984) Carbon isotopes in submarine basalts. Earth Planet Sci Lett 70:196–206
Mattey DP, Lowry D, MacPherson C (1994) Oxygen isotope composition of mantle peridotites. Earth Planet Sci Lett 128:231–241

Mauersberger K (1981) Measurement of heavy ozone in the stratosphere. Geophys Res Lett 8:935–937

Mauersberger K (1987) Ozone isotope measurements in the stratosphere. Geophys Res Lett 14:80–83

Mazza SE, Stracke A, Gill JB, Kimura JI, Kleine T (2020) Tracing dehydration and melting of the subducted slab with tungsten isotopes in arc lavas. Earth Planet Sci Lett 530:115942

McCaig AM, Wickham SM, Taylor HP (1990) Deep fluid circulation in Alpine shear zones, Pyrenees, France: field and oxygen isotope studies. Contr Mineral Petrol 106:41–60

McClelland JW, Montoya JP (2002) Trophic relationships and the nitrogen isotope composition of amino acids in plankton. Ecology 83:2173–2180

McCollom TM, Seewald JS (2006) Carbon isotope composition of organic compounds produced by abiotic synthesis under hydrothermal conditions. Earth Planet Sci Lett 243:74–84

McConnaughey T (1989a) $^{13}C$ and $^{18}O$ disequilibrium in biological carbonates. II. In vitro simulation of kinetic isotope effects. Geochim Cosmochim Acta 53:163–171

McConnaughey T (1989b) $^{13}C$ and $^{18}O$ disequilibrium in biological carbonates. I. Patterns. Geochim Cosmochim Acta 53:151–162

McCorkle DC, Emerson SR (1988) The relationship between pore water isotopic composition and bottom water oxygen concentration. Geochim Cosmochim Acta 52:1169–1178

McCorkle DC, Emerson SR, Quay P (1985) Carbon isotopes in marine porewaters. Earth Planet Sci Lett 74:13–26

McCrea JM (1950) On the isotopic chemistry of carbonates and a paleotemperature scale. J Chem Phys 18:849–857

McDermott F (2004) Palaeo-climate reconstruction from stable isotope variations in speleothems: a review. Quaternary Sci Rev 23:901–918

McGarry S, Bar-Matthews M, Matthews A, Vaks A, Schilman B, Ayalon A (2004) Constraints on hydrological and paleotemperature variations in the eastern Mediterranean region in the last 140 ka given by the δD values of speleothem fluid inclusions. Quat Sci Rev 23:919–934

McGregor ID, Manton SR (1986) Roberts Victor eclogites: ancient oceanic crust. J Geophys Res 91:14063–14079

McInerney FA, Wing SL (2011) The Paleocene-Eocene thermal maximum: a perturbation of carbon cycle, climate and biosphere with implications for the future. Ann Rev Earth Planet Sci 39:489–516

McKay DS et al (1996) Search for past life on Mars: possible relic biogenic activity in martian meteorite ALH 84001. Science 273:924–930

McKeegan KD, Kallio AP, Heber VS et al (2011) The oxygen isotopic composition of the Sun inferred from captured solar wind. Science 332:1528–1532

McKenzie J (1984) Holocene dolomitization of calcium carbonate sediments from the coastal sabkhas of Abu Dhabi, U.A.E.: a stable isotope study. J Geol 89:185–198

McKibben MA, Riciputi LR (1998) Sulfur isotopes by ion microprobe. In: Applications of microanalytical techniques to understanding mineralizing processes. Rev Econ Geol 7:121–140

McLaughlin K, Chavez F, Pennington JT, Paytan A (2006) A time series investigation of the oxygen isotope composition of dissolved inorganic phosphate in Monterey Bay, California. Limnol Oceanogr 51:2370–2379

McSween HY, Taylor LA, Stolper EM (1979) Allan Hills 77005: a new meteorite type found in Antarctica. Science 204:1201–1203

Meier-Augustein W (2010) Stable isotope forensics. Wiley-Blackwell, Chichester

Mengel K, Hoefs J (1990) Li–$\delta^{18}O$–$SiO_2$ systematics in volcanic rocks and mafic lower crustal xenoliths. Earth Planet Sci Lett 101:42–53

Merlivat L, Jouzel J (1979) Global climatic interpretation of the deuterium-oxygen 18 relationship for precipitation. J Geophys Res 84:5029–5033

Meshoulam A, Ellis GS, Ahmad WS, Deev A, Sessions AL, Tang Y, Adkins JF, Jinzhong L, Gilhooly WP, Aizenshat Z, Amrani A (2016) Study of thermochemical sulfate reduction

mechanism using compound specific sulfur isotope analysis. Geochim Cosmochim Acta 188:73–92
Michalski G, Bhattacharya SK, Mase DF (2011) Oxygen isotope dynamics of atmospheric nitrate and its precursor molecules. In: Baskaran M (ed) Handbook of environmental isotope geochemistry. Springer, pp 613–635
Mikaloff-Fletcher SE et al (2006) Inverse estimates of anthropogenic $CO_2$ uptake, transport and storage by the ocean. Global Biogeochem Cycles 20:GB2002. https://doi.org/10.1029/2005GB002532
Mikhail S, Guillermier C, Franchi IA, Beard AD, Crispin K, Verchovsky AB, Jones AP, Milledge HJ (2014) Emperical evidence for the fractionation of carbon isotopes between diamond and iron carbide from the Earth's mantle. Geochem Geophys Geosys 15:855–866
Milkov AV (2005) Molecular and stable isotope compositions of natural gas hydrates: a revised global dataset and basic interpretations in the context of geological settings. Org Geochem 36:681–702
Miller HB, Farley KA, Vasconcelos PM, Mostert A, Eiler J (2020) Intracrystalline site preference of oxygen isotopes in goethite: a single-mineral paleothermometer. Earth Planet Sci Lett 539:116237
Ming T, Anders E, Hoppe P, Zinner E (1989) Meteoritic silicon carbide and its stellar sources, implications for galactic chemical evolution. Nature 339:351–354
Minigawa M, Wada E (1984) Stepwise enrichments of $^{15}N$ along food chains: further evidence and the relation between $\delta^{15}N$ and animal age. Geochim Cosmochim Acta 48:1135–1140
Misra S, Froelich PN (2012) Lithium isotope history of Cenozoic seawater: changes in silicate weathering and reverse weathering. Science 335:818–823
Mix HT, Chamberlain CP (2014) Stable isotope records of hydrologic change and paleotemperature from smectite in Cenozoic western North America. Geochim Cosmochim Acta 141:532–546
Moine BN, Bolfan-Casanova N, Radu IB, Ionov DA, Costin G, Korsakov AV, Golovin AV, Oleinikov OB, Deloule E, Cottin JY (2020) Molecular hydrogen in minerals as a clue to interpret δD variations in the mantle. Nat Commun 11:3604
Moldovanyi EP, Lohmann KC (1984) Isotopic and petrographic record of phreatic diagenesis: lower cretaceous Sligo and Cupido formations. J Sediment Petrol 54:972–985
Mondillo N, Wilkinson JJ, Boni M, Weiss DJ, Mathur R (2018) A global assessment of Zn isotope fractionation in secondary Zn minerals from sulfide and non-sulfide ore deposits and model for fractionation control. Chem Geol 500:182–193
Monster J, Anders E, Thode HG (1965) $^{34}S/^{32}S$ ratios for the different forms of sulphur in the Orgueil meteorite and their mode of formation. Geochim Cosmochim Acta 29:773–779
Monster J, Appel PW, Thode HG, Schidlowski M, Carmichael CW, Bridgwater D (1979) Sulphur isotope studies in early Archean sediments from Isua, West Greenland: implications for the antiquity of bacterial sulfate reduction. Geochim Cosmochim Acta 43:405–413
Montoya JP, Horrigan SG, McCarthy JJ (1991) Rapid, storm-induced changes in the natural abundance of $^{15}N$ in a planktonic ecosystem, Chesapeake Bay, USA. Geochim Cosmochim Acta 55:3627–3638
Mook WG, Koopman M, Carter AF, Keeling CD (1983) Seasonal, latitudinal and secular variations in the abundance and isotopic ratios of atmospheric carbon dioxide. I. Results from land stations. J Geophys Res 88:10915–10933
Morgan JL, Skulan JL, Gordon GW, Romaniello SJ, Smith SM, Anbar AD (2012) Rapidly assessing changes in bone mineral balance using natural stable calcium isotopes. PNAS 109:9989–9994
Morin S, Savarino J, Frey MF, Yan N, Bekki S, Bottenheim JW, Martins JM (2008) Tracing the origin and fate of $NO_x$ in the arctic atmosphere using stable isotopes in nitrate. Science 322:730–732
Moschen R, Lücke A, Parplies U, Radtke B, Schleser GH (2006) Transfer and early diagenesis of biogenic silica oxygen isotope signals during settling and sedimentation of diatoms in a

temperate freshwater lake (Lake Holzmaar, Germany). Geochim Cosmochim Acta 70:4367–4379

Mossmann JR, Aplin AC, Curtis CD, Coleman ML (1991) Geochemistry of inorganic and organic sulfur in organic-rich sediments from the Peru Margin. Geochim Cosmochim Acta 55:3581–3595

Moynier F, Blichert-Toft J, Telouk P, Luck JM, Albarede F (2007) Comparative stable isotope geochemistry of Ni, Cu, Zn and Fe in chondrites and iron meteorites. Geochim Cosmochim Acta 71:4365–4379

Moynier F, Pichat S, Pons ML, Fike D, Balter V, Albarède F (2008) Isotope fractionation and transport mechanisms of Zn in plants. Chem Geol 267:125–130

Moynier F, Agranier A, HezelDC BA (2010) Sr stable isotope composition of Earth, the Moon, Mars, Vesta and meteorites. Earth Planet Sci Lett 300:359–366

Moynier F, Paniello R, Beck P, Gounelle M, Podosek F, Albarede F, Zanda B (2011) Nature of volatile depletion and genetic relationships in enstatite chondrites and aubrites inferred from Zn isotopes. Geochim Cosmochim Acta 75:297–307

Moynier F, Fujii T, Shaw AS, Le Borgne M (2013) Heterogeneous distribution of natural zinc isotopes in mice. Metallomics 5:693–699

Moynier F, Foriel J, Shaw AS, Le Borgne M (2017) Distribution of Zn isotopes during Alzheimer's disease. Geochem Persp Lett 3:142–150

Moynier F, Bourgne L, Lahoud E, Mahan B, Mouton-Ligier F, Hugon J, Paquet C (2020) Copper and zinc isotopic excursions in the human brain affected by Alzheimer's disease. Alzheimer's and Dementia 12:e12111

Muehlenbachs K (1998) The oxygen isotopic composition of the oceans, sediments and the seafloor. Chem Geol 145:263–273

Muehlenbachs K, Byerly G (1982) $^{18}O$ enrichment of silicic magmas caused by crystal fractionation at the Galapagos Spreading Center. Contr Mineral Petrol 79:76–79

Muehlenbachs K, Clayton RN (1972) Oxygen isotope studies of fresh and weathered submarine basalts. Can J Earth Sci 9:471–479

Muehlenbachs K, Clayton RN (1976) Oxygen isotope composition of the oceanic crust and its bearing on seawater. J Geophys Res 81:4365–4369

Nabelek PI (1991) Stable isotope monitors. In: Contact metamorphism. Rev Mineral 26:395–435

Nabelek PI, Labotka TC, O'Neil JR, Papike JJ (1984) Contrasting fluid/rock interaction between the Notch Peak granitic intrusion and argillites and limestones in western Utah: evidence from stable isotopes and phase assemblages. Contr Mineral Petrol 86:25–43

Neretin LN, Böttcher ME, Jørgensen BB, Volkov II, Lüschen H, Hilgenfeldt K (2004) Pyritization processes and greigite formation in the advancing sulfidization front in the Upper Pleistone sediments of the Black Sea. Geochim Cosmochim Acta 68:2081–2094

Ni P, Chabot NL, Ryan CJ, Shahar A (2020) Heavy iron isotope composition of iron meteorites explained by core crystallization. Nat Geosci 13:611–615

Niedermeyer EM, Forrest M, Beckmann B, Sessions AL, Mulch A, Scheefuß E (2016) The stable hydrogen isotopic composition of sedimentary plant waxes as quantitative proxy for rainfall in the West African Sahel. Geochim Cosmochim Acta 184:55–70

Nielsen H, Ricke W (1964) S-Isotopenverhaltnisse von Evaporiten aus Deutschland. Ein Beitrag zur Kenntnis von $\delta^{34}S$ im Meerwasser Sulfat. Geochim Cosmochim Acta 28:577–591

Niles PB, Leshin LA, Guan Y (2005) Microscale carbon isotope variability in ALH84001 carbonates and a discussion of possible formation environments. Geochim Cosmochim Acta 69:2931–2944

Nisbet EG, 22 others (2019) Very strong atmospheric methane growth in the 4 years 2014–2017: implications for the Paris agreement. Global Biogeoch Cycles 33:318–342

Nishio Y, Sasaki S, Gamo T, Hiyagon H, Sano Y (1998) Carbon and helium isotope systematics of North Fiji basin basalt glasses: carbon geochemical cycle in the subduction zone. Earth Planet Sci Lett 154:127–138

Norris RD, Röhl U (1999) Carbon cycling and chronology of climate warming during the Paleocene/Eocene transition. Nature 401:775–778

Norton D, Taylor HP (1979) Quantitative simulation of the hydrothermal systems of crystallizing magmas on the basis of transport theory and oxygen isotope data: an analysis of the Skaergaard intrusion. J Petrol 20:421–486

Nriagu JO, Coker RD, Barrie LA (1991) Origin of sulphur in Canadian Arctic haze from isotope measurements. Nature 349:142–145

Ochoa Gonzalez R, Strekopytov S, Amato F, Querol X, Reche C, Weiss D (2016) New insights from zinc and copper isotopic compositions into the sources of atmospheric particulate matter from two major European cities. Environ Sci Tech 50:9816–9824

Ockert C, Gussone N, Kaufhold S, Teichert BM (2013) Isotope fractionation during Ca exchange on clay minerals in a marine environment. Geochim Cosmochim Acta 112:374–388

O'Neil JR, Roe LJ, Reinhard E, Blake RE (1994) A rapid and precise method of oxygen isotope analysis of biogenic phosphate. Israel J Earth Sci 43:203–212

Ohkouchi N, Chikaraishi Y, 16 others (2017) Advances in the application of amino acid nitrogen isotopic analysis in ecological and biogeochemical studies. Org Geochem 113:150–174

Ohmoto H (1972) Systematics of sulfur and carbon isotopes in hydrothermal ore deposits. Econ Geol 67:551–578

Ohmoto H (1986) Stable isotope geochemistry of ore deposits. Rev Mineral 16:491–559

Ohmoto H, Goldhaber MB (1997) Sulfur and carbon isotopes. In: Barnes HL (ed) Geochemistry of hydrothermal ore deposits, 3rd edn. Wiley Interscience, New York, pp 435–486

Ohmoto H, Rye RO (1979) Isotopes of sulfur and carbon. In: Geochemistry of hydrothermal ore deposits, 2nd edn. Holt Rinehart and Winston, New York

Ohmoto H, Kakegawa T, Lowe DR (1993) 3.4 billion year old biogenic pyrites from Barberton, South Africa: sulfur isotope evidence. Science 262:555

Ongley JS, Basu AR, Kyser TK (1987) Oxygen isotopes in coexisting garnets, clinopyroxenes and phlogopites of Roberts Victor eclogites: implications for petrogenesis and mantle metasomatism. Earth Planet Sci Lett 83:80–84

Ono S, Shanks WC, Rouxel OJ, Rumble D (2007) S-33 constraints on the seawater sulphate contribution in modern seafloor hydrothermal vent sulfides. Geochim Cosmochim Acta 71:1170–1182

Ono S, Wang DT, Gruen DS, Sherwood-Lollar B, Zahniser MS, McManus BJ, Nelson DD (2014) Measurement of a doubly substituted methane isotopologue, $^{13}CH_3D$, by tunable infrared laser direct absorption spectroscopy. Anal Chem 86:6487–6494

Onuma N, Clayton RN, Mayeda TK (1970) Oxygen isotope fractionation between minerals and an estimate of the temperature of formation. Science 167:536–538

Ott U (1993) Interstellar grains in meteorites. Nature 364:25–33

Owen T, Maillard JP, DeBergh C, Lutz BL (1988) Deuterium on Mars: the abundance of HDO and the value of D/H. Science 240:1767–1770

Pack A (2021) Isotopic traces of atmospheric $O_2$ in rocks, minerals and melts. Rev Mineral Geochem 86:217–240

Pack A, Herwartz D (2014) The triple oxygen isotope composition of the Earth mantle and $\Delta^{17}O$ variations in terrestrial rocks. Earth Planet Sci Lett 390:138–145

Pack A, Gehler A, Süssenberger A (2013) Exploring the usability of isotopically anomaleous oxygen in bones and teeth as palaeo-$CO_2$-barometer. Geochim Cosmochim Acta 102:306–317

Pagani M, Arthur MA, Freeman KH (1999a) Miocene evolution of atmospheric carbon dioxide. Paleoceanography 14:273–292

Pagani M, Freeman KH, Arthur MA (1999b) Late Miocene atmospheric $CO_2$ concentrations and the expansion of $C_4$ grasses. Science 285:876–879

Page B, Bullen T, Mitchell M (2008) Influences of calcium availability and tree species on Ca isotope fractionation in soil and vegetation. Biogeochemistry 88:1–13

Palmer MR, Pearson PN, Conbb SJ (1998) Reconstructing past ocean pH-depth profiles. Science 282:1468–1471

Paniello RC, Day JM, Moynier F (2012) Zinc isotopic evidence for the origin of the Moon. Nature 490:376–379

Park S, Perez T, Boering KA, Trumbore SE, Gil J, Marquina S, Tyler SC (2011) Can $N_2O$ stable isotopes and isotopomers be useful tools to characterize sources and microbial pathways of $N_2O$ production and consumption in tropical soils? Global Biogeochem Cycles 25. https://doi.org/10.1029/2009GB003615

Pawellek F, Veizer J (1994) Carbon cycle in the upper Danube and its tributaries: $\delta^{13}C_{DIC}$ constraints. Israel J Earth Sci 43:187–194

Payne JL, Kump LR (2007) Evidence for recurrent early triassic massive volcanism from quantitative interpretation of carbon isotope fluctuations. Earth Planet Sci Lett 256:264–277

Paytan A, Kastner M, Campbell D, Thiemens MH (1998) Sulfur isotope composition of Cenozoic seawater sulfate. Science 282:1459–1462

Paytan A, Luz B, Kolodny Y, Neori A (2002) Biologically mediated oxygen isotope exchange between water and phosphorus. Global Biogeochem Cycles 16–13:1–70

Paytan A, Kastner M, Campbell D, Thiemens M (2004) Seawater sulfur isotope fluctuations in the Cretaceous. Science 304:1663–1665

Pearson PN, Palmer MR (2000) Atmospheric carbon dioxide concentrations over the past 60 million years. Nature 406:695–699

Pearson PN, Foster GI, Wade BS (2009) Atmospheric carbon dioxide through the Eocene-Oligocene climate transition. Nature 461:1110–1113

Peckmann J, Thiel V (2005) Carbon cycling at ancient methane-seeps. Chem Geol 205:443–467

Pedentchouk N, Freeman KH, Harris NB (2006) Different response of δD-values of n-alkanes, isoprenoids and kerogen during thermal maturation. Geochim Cosmochim Acta 70:2063–2072

Perez T, Garcia-Montiel D, Trumbore SE, Tyler SC, de Camargo P, Moreira M, Piccolo M, Cerri C (2006) Determination of $N_2O$ isotopic composition ($^{15}N$, $^{18}O$, and $^{15}N$ intramolecular distribution) and $^{15}N$ enrichment factors of $N_2O$ formation via nitrification and denitrification from incubated Amazon forest soils. Ecol Appl 16:2153–2167

Perry EA, Gieskes JM, Lawrence JR (1976) Mg, Ca and $^{18}O/^{16}O$ exchange in the sediment-pore water system, Hole 149, DSDP. Geochim Cosmochim Acta 40:413–423

Peters MT, Wickham SM (1995) On the causes of $^{18}O$ depletion and $^{18}O/^{16}O$ homogenization during regional metamorphism, the east Humboldt Range core complex, Nevada. Contr Mineral Petrol 119:68–82

Peters KE, Rohrbach BG, Kaplan IR (1981) Carbon and hydrogen stable isotope variations in kerogen during laboratory-simulated thermal maturation. Am Assoc Petrol Geol Bull 65:501–508

Petersen SV, 29 others (2019) Effects of improved $^{17}O$ correction on interlaboratory agreement in clumped isotope calibrations, estimates of mineral-specific offsets, and temperature dependence of acid digestion fractionation. Geochem Geophys Geosys 20:3495–3519

Peterson BJ, Fry B (1987) Stable isotopes in ecosystem studies. Ann Rev Ecol Syst 18:293–320

Phillips FM, Bentley HW (1987) Isotopic fractionation during ion filtration: I. Theory. Geochim Cosmochim Acta 51:683–695

Philp RP (2007) The emergence of stable isotopes in environmental and forensic geochemistry studies: a review. Eviron Chem Lett 5:57–66

Piani L, Marrocchi Y, Rigaudier T, Vacher LG, Thomassin D, Marty B (2020) Earth's water may have been inherited from material similar to enstatite chondrite meteorites. Science 369:1110–1116

Piasecki A, Sessions A, Lawson M, Ferreira AA, Neto EVS, Eiler JM (2016) Analysis of the site-specific carbon isotope composition of propane by gas source isotope ratio mass spectrometry. Geochim Cosmochim Acta 188:58–72

Piasecki A, Sessions A, Lawson M, Ferreira AA, Neto EVS, Ellis GS, Lewan MD, Eiler JM (2018) Position-specific $^{13}C$ distributions within propane from experiments and natural gas samples. Geochim Cosmochim Acta 220:110–124

Piasecki A, Bernasconi SM, Grauel AL, Hannisdal B, Ho SL, Leutert TM, Meinicke N, Tisserand A, Meckler N (2019) Application of clumped isotope thermometry to benthic foraminifera. Geochem Geophys Geosys 20(4)
Pineau F, Javoy M (1983) Carbon isotopes and concentrations in mid-ocean ridge basalts. Earth Planet Sci Lett 62:239–257
Pineau F, Javoy M, Bottinga Y (1976) $^{13}C/^{12}C$ ratios of rocks and inclusions in popping rocks of the Mid-Atlantic Ridge and their bearing on the problem of isotopic composition of deep-seated carbon. Earth Planet Sci Lett 29:413–421
Plank T, Manning CE (2019) Subducting carbon. Nature 574:343–352
Poage MA, Chamberlain CP (2001) Empirical relationships between elevation and the stable isotope composition of precipitation and surface waters: considerations for studies of paleoelevation change. Am J Sci 301:1–15
Poitrasson F, Freydier R (2005) Heavy iron isotope composition of granites determined by high resolution MC-ICP-MS. Chem Geol 222:132–147
Poitrasson F, Levasseur S, Teutsch N (2005) Significance of iron isotope mineral fractionation in pallasites and iron meteorites for the core-mantle differentiation of terrestrial planets. Earth Planet Sci Lett 234:151–164
Poitrasson F, Roskosz M, Corgne A (2009) No iron isotope fractionation between molten alloys and silicate melt to 2000 °C and 7.7 GPa: experimental evidence and implications for planery differentiation and accretion. Earth Planet Sci Lett 278:376–385
Poitrasson F, Delpech G, Gregoire M (2013) On the iron isotope heterogeneity of lithospheric mantle xenoliths: implications for mantle metasomatism, the origin of basalts and the iron isotope composition of the Earth. Contr Mineral Petrol 165:1243–1258
Pons ML, Debret B, Bouilhol P, Delacour A, Williams H (2016) Zinc isotope evidence for sulfate-rich fluid transfer across subduction zones. Nat Commun 7:13794
Pope E, Bird DK, Rosing MT (2012) Isotope composition and volume of Earth's early oceans. PNAS 109:4371–4376
Popp BN, Takigiku R, Hayes JM, Louda JW, Baker EW (1989) The post Paleozoic chronology and mechanism of $^{13}C$ depletion in primary organic matter. Am J Sci 289:436–454
Popp BN, Laws EA, Bidigare RR, Dore JE, Hanson KL, Wakeham SG (1998) Effect of phytoplankton cell geometry on carbon isotope fractionation. Geochim Cosmochim Acta 62:69–77
Popp BN, 10 others (2002) Nitrogen and oxygen isotopomeric constraints on the origins and sea-to-air flux of $N_2O$ in the oligotrophic subtropical North Pacific gyre. Global Biogeochem Cycles 16. https://doi.org/10.1029/2001GB001806
Poreda R (1985) Helium-3 and deuterium in back arc basalts: Lau Basin and the Mariana trough. Earth Planet Sci Lett 73:244–254
Poreda R, Schilling JG, Craig H (1986) Helium and hydrogen isotopes in ocean-ridge basalts north and south of Iceland. Earth Planet Sci Lett 78:1–17
Price FT, Shieh YN (1979) The distribution and isotopic composition of sulfur in coals from the Illinois Basin. Econ Geol 74:1445–1461
Pringle EA, Moynier F (2017) Rubidium isotopic composition of the Earth, meteorites and the Moon: evidence for the origin of volatile loss during planetary accretion. Earth Planet Sci Lett 473:62–70
Prokoph A, Shields GA, Veizer J (2008) Compilation and time-series analysis of a marine carbonate $\delta^{18}O$, $\delta^{13}C$, $^{87}Sr/^{86}Sr$ and $\delta^{34}S$ database through Earth history. Earth Sci Rev 87:113–133
Prokopiou M, Sapart CJ, Rosen J, Sperlich P, Blunier T, Brook E, van de Wal RS, Röckmann T (2018) Changes in the isotopic signature of atmospheric nitrous oxide and ist gobal average source during the last three millenia. J Geophys Res: Atmospheres 123:10757–10773
Puceat E, Joachimski MM et al (2010) Revised phosphate-water fractionation equation reassessing paleotemperatures derived from biogenic apatite. Earth Planet Sci Lett 298:135–142

Quade J et al (1992) A 16-Ma record of paleodiet using carbon and oxygen isotopes in fossil teeth from Pakistan. Chem Geol 94:183–192
Quade J, Cerling TE (1995) Expansion of C4 grasses in the late Miocene of northern Pakistan: evidence from stable isotopes in paleosols. Palaeo Palaeo Palaeo 115:91–116
Quade J, Breecker DO, Daeron M, Eiler J (2011) The paleoaltimetry of Tibet: an isotopic perspective. Am J Sci 311:77–115
Quast A, Hoefs J, Paul J (2006) Pedogenic carbonates as a proxy for palaeo-$CO_2$ in the Paleozoic atmosphere. Palaeo Palaeo Palaeo 242:110–125
Quay PD, Tilbrook B, Wong CS (1992) Oceanic uptake of fossil fuel $CO_2$: carbon-13 evidence. Science 256:74–79
Quay PD, Emerson S, Wilbur DO, Stump S (1993) The $\delta^{18}O$ of dissolved $O_2$ in the surface waters of the subarctic Pacific: a tracer of biological productivity. J Geophys Res 98:8447–8458
Quay PD, Wilbur DO, Richey JE, Devol AH, Benner R, Forsberg BR (1995) The $^{18}O/^{16}O$ of dissolved oxygen in rivers and lakes in the Amazon Basin: determining the ratio of respiration to photosynthesis in freshwaters. Limnol Oceanogr 40:718–729
Quay PD, Stutsman J, Wibur D, Snover A, Dlugokencky E, Brown T (1999) The isotopic composition of atmospheric methane. Global Geochem Cycles 13:445–461
Raab M, Spiro B (1991) Sulfur isotopic variations during seawater evaporation with fractional crystallization. Chem Geol 86:323–333
Rabinovich AL, Grinenko VA (1979) Sulfate sulfur isotope ratios for USSR river water. Geochemistry 16(2):68–79
Radke J, Bechtel A, Gaupp R, Püttmann W, Schwark L, Sachse D, Gleixner D (2005) Correlation between hydrogen isotope ratios of lipid biomarkers and sediment maturity. Geochim Cosmochim Acta 69:5517–5530
Rahn T, Wahlen M (1997) Stable isotope enrichment in stratospheric nitrous oxide. Science 278:1776–1778
Rahn T, Eiler JM, Boering KA, Wennberg PO, McCarthy MC, Tyler S, Schauffler S, Donnelly S, Atlas E (2003) Extreme deuterium enrichment in stratospheric hydrogen and the global atmospheric budget of $H_2$. Nature 424:918–921
Rai VK, Thiemens MH (2007) Mass independently fractionated sulphur components in chondrites. Geochim Cosmochim Acta 71:1341–1354
Rai VK, Jackson TL, Thiemens MH (2005) Photochemical mass-independent sulphur isotopes in achondritic meteorites. Science 309:1062–1065
Raiswell R, Berner RA (1985) Pyrite formation in euxinic and semi-euxinic sediments. Am J Sci 285:710–724
Raitzsch M, Hönisch B (2014) Cenozoic boron isotope variations in benthic foraminifera. Geology 41:591–594
Rau GH, Sweeney RE, Kaplan IR (1982) Plankton $^{13}C/^{12}C$ ratio changes with latitude: differences between northern and southern oceans. Deep Sea Res 29:1035–1039
Rau GH, Takahashi T, DesMarais DJ (1989) Latitudinal variations in plankton $^{13}C$: implications for $CO_2$ and productivity in past ocean. Nature 341:516–518
Rau GH, Takahashi T, DesMarais DJ, Repeta DJ, Martin JH (1992) The relationship between $\delta^{13}C$ of organic matter and $\Sigma CO_2(aq)$ in ocean surface water: data from a JGOFS site in the northeast Atlantic Ocean and a model. Geochim Cosmochim Acta 56:1413–1419
Raven MR, Adkins JF, Werne JP, Lyons TW, Sessions AL (2015) Sulfur isotopic composition of individual organic compounds from Cariaco Basin sediments. Org Geochem 80:53–59
Redding CE, Schoell M, Monin JC, Durand B (1980) Hydrogen and carbon isotopic composition of coals and kerogen. In: Douglas AG, Maxwell JR (eds) Phys Chem Earth 12:711–723
Rees CE, Jenkins WJ, Monster J (1978) The sulphur isotopic composition of ocean water sulphate. Geochim Cosmochim Acta 42:377–381
Regelous M, Elliott T, Goath CD (2008) Nickel isotope heterogeneity in the early solar system. Earth Planet Sci Lett 272:330–338

Rice DD, Claypool GE (1981) Generation, accumulation and resource potential of biogenic gas. Am Assoc Petrol Geol Bull 65:5–25
Rice A, Dayalu A, Quay P, Gammon R (2011) Isotopic fractionation during soil uptake of atmospheric hydrogen. Biogeosciences 8:763–769
Richet P, Bottinga Y, Javoy M (1977) A review of H, C, N, O, S, and Cl stable isotope fractionation among gaseous molecules. Ann Rev Earth Planet Sci 5:65–110
Riciputi LR, Cole DR, Machel HG (1996) Sulfide formation in reservoir carbonates of the Devonian Nishu Formation, Alberta, Canada: an ion microprobe study. Geochim Cosmochim Acta 60:325–336
Righter K (2019) Volatile element depletion of the Moon—the roles of precursors, post impact disc dynamics and core formation. Sci Adv 5:1, eaau7658
Rindsberger MS, Jaffe S, Rahamin S, Gat JR (1990) Patterns of the isotopic composition of precipitation in time and space; Data from the Israeli storm water collection program. Tellus 42B:263–271
Ripley EM, Li C (2003) Sulfur isotope exchange and metal enrichment in the formation of magmatic Cu-Ni-(PGE)-deposits. Econ Geol 98:635–641
Ripperger S, Rehkämper M, Porcelli D, Halliday AN (2007) Cadmium isotope fractionation in seawater—a signature of biological activity. Earth Planet Sci Lett 261:670–684
Robert F (2001) The origin of water on Earth. Science 293:1056–1058
Robert F, Epstein S (1982) The concentration and isotopic composition of hydrogen, carbon and nitrogen carbonaceous meteorites. Geochim Cosmochim Acta 46(8):1–95
Robert F, Merlivat L, Javoy M (1978) Water and deuterium content in ordinary chondrites. Meteoritics 12:349–354
Robert F, Gautier D, Dubrulle B (2000) The solar system D/H ratio: observations and theories. Space Sci Rev 92:201–224
Röckmann T et al (1998) Mass independent oxygen isotope fractionation in atmospheric CO as a result of the reaction CO + OH. Science 281:544–546
Röckmann T, Jöckel P, Gros V, Bräunlich M, Possnert G, Brenninkmeijer CAM (2002) Using $^{14}C$, $^{13}C$, $^{18}O$ and $^{17}O$ isotopic variations to provide insights into the high northern latitude surface CO inventory. Atmos Chem Phys 2:147–159
Röckmann T, Kaiser J, Brenninkmeijer CAM, Brand WA (2003) Gas chromatography/isotope ratio mass spectrometry method for high-precision position-dependent $^{15}N$ and $^{18}O$ measurements of atmospheric nitrous oxide. Rapid Commun Mass Spectrom 17:1897–1908
Röhl U, Norris RD, Bralower TJ, Wefer G (2000) New chronology for the late Paleocene thermal maximum and its environmental implications. Geology 28:927–930
Rolison JM, Stirling CH, Middag R, Gault-Ringold M, George E, Rijkenberg MJ (2018) Iron isotope fractionation during pyrite formation in a sulfidic Precambrian ocean analogue. Earth Planet Sci Lett 488:1–13
Romanek CS et al (1994) Record of fluid-rock interaction on Mars from the meteorite ALH 84001. Nature 372:655–657
Romer R, Meixner A, Förstner HJ (2014) Lithium and boron in late-orogenic granites—isotopic fingerprints for the source of crustal melts? Geochim Cosmochim Acta 131:98–114
Rooney MA, Claypool GE, Chung HM (1995) Modeling thermogenic gas generation using carbon isotope ratios of natural gas hydrocarbons. Chem Geol 126:219–232
Rouxel O, Fouquet Y, Ludden JN (2004a) Copper isotope systematics of the Lucky Strike, Rainbow and Logatschev seafloor hydrothermal fields on the Mi-Atlantic Ridge. Econ Geol 99:585–600
Rouxel O, Fouquet Y, Ludden JN (2004b) Subsurface processes at the Lucky Strike hydrothermal field, Mid-Atlantic Ridge: evidence from sulfur, selenium and iron isotopes. Geochim Cosmochim Acta 68:2295–2311
Rouxel O, Bekker A, Edwards KJ (2005) Iron isotope constraints on the Archean and Proterozoic ocean redox state. Science 307:1088–1091

Rouxel O, Ono S, Alt J, Rumble D, Ludden J (2008a) Sulfur isotope evidence for microbial sulfate reduction in altered oceanic basalts at ODP Site 801. Earth Planet Sci Lett 268:110–123
Rouxel O, Shanks WC, Bach W, Edwards KJ (2008b) Integrated Fe- and S-isotope study of seafloor hydrothermal vents at East Pacific Rise 9–10°N. Chem Geol 252:214–227
Rozanski K, Sonntag C (1982) Vertical distribution of deuterium in atmospheric water vapour. Tellus 34:135–141
Rozanski K, Araguas-Araguas L, Gonfiantini R (1993) Isotopic patterns in modern global precipitation. In: Climate change in continental isotopic records. Geophys Monograph 78:1–36
Rumble D, Yui TF (1998) The Qinglongshan oxygen and hydrogen isotope anomaly near Donghai in Jiangsu Province, China. Geochim Cosmochim Acta 62:3307–3321
Rumble D, Young ED, Shahar A, Guo W (2011) Stable isotope cosmochemistry and the evolution of planetary systems. Elements 7:23–28
Russell AK, Kitajima K, Strickland A, Medaris LG, Schulze DJ, Valley JW (2013) Eclogite-facies fluid infiltration: constraints from $\delta^{18}O$ zoning in garnet. Contr Mineral Petrol 165:103–116
Rye RO (2005) A review of stable isotope geochemistry of sulfate minerals in selected igneous environments and related hydrothermal systems. Chem Geol 215:5–36
Rye RO, Schuiling RD, Rye DM, Jansen JBH (1976) Carbon, hydrogen and oxygen isotope studies of the regional metamorphic complex at Naxos, Greece. Geochim Cosmochim Acta 40:1031–1049
Rye RO, Bethke PM, Wasserman MD (1992) The stable isotope geochemistry of acid sulfate. Econ Geol 87:227–262
Saal AE, Hauri EH, Van Orman JA, Rutherford MJ (2013) Hydrogen isotopes in lunar volcanic glasses and melt inclusions reveal a carbonaceous chondrite heritage. Science 340:1317–1320
Saccocia PJ, Seewald JS, Shanks WC (2009) Oxygen and hydrogen isotope fractionation in serpentine-water and talc-water systems from 250 to 450°C, 50 MPa. Geochim Cosmochim Acta 73:6789–6804
Sachse D, Billault I et al (2012) Molecular paleohydrology: interpreting the hydrogen-isotopic composition of lipid biomarkers from photosynthesizing organisms. Ann Rev Earth Planet Sci 40:221–249
Sackett WM (1988) Carbon and hydrogen isotope effects during the thermocatalytic production of hydrocarbons in laboratory simulation experiments. Geochim Cosmochim Acta 42:571–580
Sackett WM, Thompson RR (1963) Isotopic organic carbon composition of recent continental derived clastic sediments of Eastern Gulf Coast, Gulf of Mexico. Bull Am Assoc Petrol Geol 47:525
Sackett WM, Eadie BJ, Exner ME (1973) Stable isotope composition of organic carbon in recent Antarctic sediments. Adv Org Geochem 1973:661
Safarian AR, Nielsen SG, Marschall HR, McCubbin FM, Monteleone BD (2014) Early accretion of water in the inner solar system from a carbonaceous-like source. Science 346:623–626
Saino T, Hattori A (1980) $^{15}N$ natural abundance in oceanic suspended particulate organic matter. Nature 283:752–754
Saino T, Hattori A (1987) Geophysical variation of the water column distribution of suspended particulate organic nitrogen and its $^{15}N$ natural abundance in the Pacific and its marginal seas. Deep Sea Res 34:807–827
Sakai H (1968) Isotopic properties of sulfur compounds in hydrothermal processes. Geochem J 2:29–49
Sakai H, Casadevall TJ, Moore JG (1982) Chemistry and isotope ratios of sulfur in basalts and volcanic gases at Kilauea volcano, Hawaii. Geochim Cosmochim Acta 46:729–738
Sakai H, DesMarais DJ, Ueda A, Moore JG (1984) Concentrations and isotope ratios of carbon, nitrogen and sulfur in ocean-floor basalts. Geochim Cosmochim Acta 48:2433–2441
Saltzman MR, Thomas E (2012) Carbon isotope stratigraphy. Elsevier, The Geologic Time Scale
Sano Y, Marty B (1995) Origin of carbon in fumarolic gas from island arcs. Chem Geol 119:265–274

Sarntheim M et al (2001) Fundamental modes and abrupt changes in North Atlantic circulation and climate over the last 60 ky—concepts, reconstruction and numerical modeling. In: Schäfer P, Ritzau W, Schlüter M, Thiede J (eds) The northern North Atlantic. Springer, Heidelberg, pp 365–410

Sass E, Kolodny Y (1972) Stable isotopes, chemistry and petrology of carbonate concretions (Mishash formation, Israel). Chem Geol 10:261–286

Satish-Kumar M, So H, YoshinoT KM, Hiroi Y (2011) Experimental determination of carbon isotope fractionation between iron carbide melt and carbon: $^{12}C$-enriched carbon in the Earth's core? Earth Planet Sci Lett 310:340–348

Sauzeat L, 8 others (2018) Isotopic evidence for disrupted copper metabolism in Amyotrophic Lateral Sclerosis. iScience 6264–6271

Savage PS, Georg RB, Williams HM, Burton KW, Halliday AN (2011) Silicon isotope fractionation during magmatic differentiation. Geochim Cosmochim Acta 75:6124–6139

Savage PS, Georg RB, Williams HM, Turner S, Halliday AN, Chappell BW (2012) The silicon isotope composition of granites. Geochim Cosmochim Acta 92:184–202

Savin SM, Epstein S (1970a) The oxygen and hydrogen isotope geochemistry of clay minerals. Geochim Cosmochim Acta 34:25–42

Savin SM, Epstein S (1970b) The oxygen and hydrogen isotope geochemistry of ocean sediments and shales. Geochim Cosmochim Acta 34:43–63

Savin SM, Lee M (1988) Isotopic studies of phyllosilicates. Rev Mineral 19:189–223

Schauble EA (2004) Applying stable isotope fractionation theory to new systems. Rev Mineral Geochem 55:65–111

Schidlowski M (2001) Carbon isotopes as biochemical recorders of life over 3.8 Ga of Earth history. Evolution of a concept. Precam Res 106:117–134

Schiegl WE, Vogel JV (1970) Deuterium content of organic matter. Earth Planet Sci Lett 7:307–313

Schilling K, Halliday AN, Lamb A, Crnogorac-Jurevic T, Larner F (2020) Urinary zinc stable isotope signature as indicator for cancer types with disrupted zinc metabolism. Goldschmidt (abstracts)

Schimmelmann A, Lewan MD, Wintsch RP (1999) D/H ratios of kerogen, bitumen, oil and water in hydrous pyrolysis of source rocks containing kerogen types I, II, IIS and III. Geochim Cosmochim Acta 63:3751–3766

Schimmelmann A, Sessions AL, Mastalerz M (2006) Hydrogen isotopic (D/H) composition of organic matter during diagenesis and thermal maturation. Ann Rev Earth Planet Sci 34:501–533

Schleicher NJ, Dong S, Packman H, Little SH, Ochoa Gonzalez R, Najorka J, Sun Y, Weiss DJ (2020) A global assessment of copper, zinc, and lead isotopes in mineral dust sources and areosols. Front Earth Sci 8:167

Schmidt M, Botz R, Rickert D, Bohrmann G, Hall SR, Mann S (2001) Oxygen isotopes of marine diatoms and relations to opal-A maturation. Geochim Cosmochim Acta 65:201–211

Schmidt TC, Zwank L, Elsner M, Berg M, Meckenstock RU, Haderlein SB (2004) Compound-specific stable isotope analysis of organic contaminants in natural environments: a critical review of the state of the art, prospects and future challenges. Anal Bioanal Chem 378:283–300

Schmitt J, Schneider R et al (2012) Carbon isotope constraints on the deglacial $CO_2$ rise from ice cores. Science 336:711–714

Schmitt AD, Stille P, Vennemann T (2003) Variations of the $^{44}Ca/^{40}Ca$ ratio in seawater during the past 24 million years: evidence from $\delta^{44}Ca$ and $\delta^{18}O$ values of Miocene phosphates. Geochim Cosmochim Acta 67:2607–2614

Schoell M (1980) The hydrogen and carbon isotopic composition of methane from natural gases of various origins. Geochim Cosmochim Acta 44:649–661

Schoell M (1983) Genetic characterization of natural gases. Bull Am Assoc Petrol Geol 67:2225–2238

Schoell M (1984) Recent advances in petroleum isotope geochemistry. Org Geochem 6:645–663
Schoell M (1988) Multiple origins of methane in the Earth. Chem Geol 71:1–10
Schoell M, McCaffrey MA, Fago FJ, Moldovan JM (1992) Carbon isotope compositions of 28,30-bisnorhopanes and other biological markers in a Monterey crude oil. Geochim Cosmochim Acta 56:1391–1399
Schoell M, Schouten S, Sinninghe Damste JS, de Leeuw JW, Summons RE (1994) A molecular organic carbon isotope record of Miocene climatic changes. Science 263:1122–1125
Schoenberg R, von Blanckenburg F (2006) Modes of planetary-scale Fe isotope fractionation. Earth Planet Sci Lett 252:342–359
Schoenemann SW, Steig EJ, Ding Q, Markle BR, Schauer AJ (2014) Triple water-isotopologue record from WAIS divide Antarctica: controls on glacial-interglacial changes in $^{17}O$ excess of precipitation. J Geophys Res Atmos 119:8741–8763
Schoeninger MJ, DeNiro MJ (1984) Nitrogen and carbon isotopic composition of bone collagen from marine and terrestrial animals. Geochim Cosmochim Acta 48:625–639
Schrag DP (1999) Effects of diagenesis on the isotopic record of late Paleogene tropical sea surface temperature. Chem Geol 161:2265–2278
Schrag DP, Hampt G, Murry DW (1996) Pore fluid constraints on the temperature and oxygen isotopic composition of the Glacial Ocean. Science 272:1930–1932
Schubert BA, Jahren AH (2012) The effect of atmospheric $CO_2$ concentration on carbon isotope fractionation in C3 land plants. Geochim Cosmochim Acta 96:29–43
Schwalb A, Burns SJ, Kelts K (1999) Holocene environments from stable isotope stratigraphy of ostracods and authigenic carbonate in Chilean Altiplano lakes. Palaeo Palaeo Palaeo 148:153–168
Schwarcz HP, Melbye J, Katzenberg MA, Knyf M (1985) Stable isotopes in human skeletons of southern Ontario: reconstruction of palaeodiet. J Archaeol Sci 12:187–206
Schuessler JA, Schoenberg R, Sigmarsson O (2009) Iron and lithium isotope systematics of the Hekla volcano, Iceland—evidence for Fe isotope fractionation during magma differentiation. Chem Geol 258:78–91
Seccombe PK, Spry PG, Both Ra, Jones MT, Schiller JC (1985) Base metal mineralization in the Kaumantoo Group, South Australia: a regional sulfur isotope study. Econ Geol 80:1824–1841
Seitz HM, Brey GP, Lahaye Y, Durali S, Weyer S (2004) Lithium isotope signatures of peridotite xenoliths and isotope fractionation at high temperature between Olivine and Pyroxene. Chem Geol 212:163–177
Sengupta S, Pack A (2018) Triple oxygen isotope mass balance for the Earth's ocean with applications to Archean cherts. Chem Geol 495:18–26
Sengupta S, Peters ST, Reitner J, Duda JP, Pack A (2020) Triple oxygen isotopes of cherts through time. Chem Geol 554:119789
Sessions AL, Sylva SP, Summons RE, Hayes JM (2004) Isotopic exchange of carbon-bound hydrogen over geologic time scales. Geochim Cosmochim Acta 68:1545–1559
Severinghaus JP, Brook EJ (1999) Abrupt climate change at the end of the last glacial period inferred from trapped air in polar ice. Science 286:930–934
Severinghaus JP, Bender ML, Keeling RF, Broecker WS (1996) Fractionation of soil gases by diffusion of water vapor, gravitational settling and thermal diffusion. Geochim Cosmochim Acta 60:1005–1018
Severinghaus JP, Sowers T, Brook EJ, Alley RB, Bender ML (1998) Timing of abrupt climate change at the end of the Younger Dryas interval from thermally fractionated gases in polar ice. Nature 391:141–146
Severinghaus JP, Beaudette R, Headly MA, Taylor K, Brook EJ (2009) Oxygen-18 of $O_2$ records the impact of abrupt climate change on terrestrial biosphere. Science 324:1431–1434
Severmann S, Johnson CM, Beard BL, German CR, Edmonds HN, Chiba H, Green DRH (2004) The effect of plume processes on the Fe isotope composition of hydrothermally derived Fe in the deep ocean as inferred from the Rainbow vent site, Mid-Atlantic Ridge, 36°14′N. Earth Planet Sci Lett 225:63–76

Severmann S, Johnson CM, Beard BL, McManus J (2006) The effect of early diagenesis on the Fe isotope composition of porewaters and authigenic minerals in continental margin sediments. Geochim Cosmochim Acta 70:2006–2022

Severmann S, McManus J, Berelson WM, Hammond DE (2010) The continental shelf benthic iron flux and its isotope composition. Geochim Cosmochim Acta 74:3984–4004

Sessions AL (2016) Factors controlling the deuterium contents of sedimentary hydrocarbons. Org Geochem 96:43–64

Shackleton NJ, Kennett JP (1975) Paleotemperature history of the Cenozoic and initiation of Antarctic glaciation: oxygen and carbon isotope analyses in DSDP sites 277, 279 and 281. Initial Rep DSDP 29:743–755

Shackleton NJ, Hall MA, Line J, Cang S (1983) Carbon isotope data in core V19–30 confirm reduced carbon dioxide concentration in the ice age atmosphere. Nature 306:319–322

Shahar A, Young ED (2007) Astrophysics of CAI formation as revealed by silicon isotope LA-MC-ICPMS of an igneous CAI. Earth Planet Sci Lett 257:497–510

Shahar A, Ziegler K, Young ED, Ricollaeu A, Schauble E, Fei Y (2009) Experimentally determined Si isotope fractionation between silicate and Fe metal and implications for the Earth's core formation. Earth Planet Sci Lett 288:228–234

Shahar A, Hillgren VJ, Young ED, Fei Y, Macris CA, Deng L (2011) High-temperature Si isotope fractionation between iron metal and silicate. Geochim Cosmochim Acta 75:7688–7697

Shahar A, Schauble EA, Caracas R, Gleason AE, Reagan MM, Xiao Y, Shu J, Mao W (2016) Pressure-dependent isotopic composition of iron alloys. Science 352:580–582

Shaheen R, Albaunza MM, Jackson TL, McCabe J, Savarino J, Thiemens MH (2014) Large sulfur-isotope anomalies in nonvolcanic sulfate aerosol and its implications for the Archean atmosphere. PNAS 111:11979–11983

Shanks WC (2001) Stable isotopes in seafloor hydrothermal systems: vent fluids, hydrothermal deposits, hydrothermal alteration, and microbial processes. Rev Mineral Geochem 43:469–525

Sharp ZD (1995) Oxygen isotope geochemistry of the $Al_2SiO_5$ polymorphs. Am J Sci 295:1058–1076

Sharp ZD, Shearer CK, McKeegan KD, Barnes JD, Wang YQ (2010) The chlorine isotope composition of the Moon and implications for an anhydrous mantle. Science 329:1050–1053

Sharp ZD, Wostbrock JA, Pack A (2018) Mass-dependent triple oxygen isotope variations in terrestrial materials. Geochem Perspec Lett 7:27–31

Shaw AM, Hilton DR, Fischer TP, Walker JA, Alvarado GE (2003) Contrasting He-C relationshipsin Nicaragua and Costa Rica: insights into C cycling through subduction zones. Earth Planet Sci Lett 214:499–513

Shaw AM, Hauri EH, Fischer TP, Hilton DR, Kelley KA (2008) Hydrogen isotopes in Mariana arc melt inclusions: implications for subduction dehydration and the deep-earth water cycle. Earth Planet Sci Lett 275:138–145

Shaw AM, Hauri EH, Behn MD, Hilton DR, Macpherson CG, Sinton JM (2012) Long-term preservation of slab signatures in the mantle inferred from hydrogen isotopes. Nat Geosci 5:224–228

Shawar L, Halevy I, Said-Ahmad W, Feinstein S, Boyko V, Kamyshny A, Amrani A (2018) Dynamics of pyrite formation and organic matter sulfurization in organic-rich carbonate sediments. Geochim Cosmochim Acta 241:219–239

Shelton KL, Rye DM (1982) Sulfur isotopic compositions of ores from Mines Gaspe, Quebec: an example of sulfate-sulfide isotopic disequilibria in ore forming fluids with applications to other porphyry type deposits. Econ Geol 77:1688–1709

Shemesh A, Kolodny Y, Luz B (1983) Oxygen isotope variations in phosphate of biogenic apatites, II. Phosphorite Rocks. Earth Planet Sci Lett 64:405–441

Shen Y, Buick R (2004) The antiquity of microbial sulfate reduction. Earth Sci Rev 64:243–272

Sheppard SMF (1986) Characterization and isotopic variations in natural waters. In: Stable isotopes in high temperature geological processes. Rev Mineral 16:165–183

Sheppard SMF, Epstein S (1970) D/H and $O^{18}/O^{16}$ ratios of minerals of possible mantle or lower crustal origin. Earth Planet Sci Lett 9:232–239

Sheppard SMF, Gilg HA (1996) Stable isotope geochemistry of clay minerals. Clay Mineral 31:1–24

Sheppard SMF, Harris C (1985) Hydrogen and oxygen isotope geochemistry of Ascension Island lavas and granites: variation with crystal fractionation and interaction with sea water. Contrib Mineral Petrol 91:74–81

Sheppard SMF, Schwarcz HP (1970) Fractionation of carbon and oxygen isotopes and magnesium between coexisting metamorphic calcite and dolomite. Contr Mineral Petrol 26:161–198

Sheppard SMF, Nielsen RL, Taylor HP (1971) Hydrogen and oxygen isotope ratios in minerals from Porphyry Copper deposits. Econ Geol 66:515–542

Sherwood Lollar B, Frape SK, Weise SM, Fritz P, Macko SA, Welhan JA (1993) Abiogenic methanogenesis in crystalline rocks. Geochim Cosmochim Acta 57:5087–5097

Sherwood Lollar B, Westgate TD, Ward JA, Slater GF, Lacrampe-Couloume G (2002) Abiogenic formation of alkanes in the Earth's crust as a minor source for global hydrocarbons reservoirs. Nature 416:522–524

Sherwood Lollar B et al (2006) Unravelling abiogenic and biogenic sources of methane in the Earth's deep subsurface. Chem Geol 226:328–339

Sherwood OA, Schwietzke S, Arling VA, Etiope G (2017) Global inventory of gas geochemistry data from fossil fuel, microbial and burning sources, version 2017. Earth Syst Sci Data 9:639–656

Shieh YN, Schwarcz HP (1974) Oxygen isotope studies of granite and migmatite, Grenville province of Ontario, Canada. Geochim Cosmochim Acta 38:21–45

Shields G, Veizer J (2002) Precambrian marine carbonate isotope database: version 1.1. Geochem Geophys Geosyst 300. https://doi.org/10.1029/2001GC000266

Shmulovich KI, Landwehr D, Simon K, Heinrich W (1999) Stable isotope fractionation between liquid and vapour in water-salt systems up to 600 °C. Chem Geol 157:343–354

Siebert C, Nägler TF, von Blanckenburg F, Kramers JD (2003) Molybdenum isotope records as a potential new proxy for paleoceanography. Earth Planet Sci Lett 211:159–171

Sidkar J, Rai VK (2020) Si-Mg isotopes in enstatite chondrites and accretion of reduced planetary bodies. Sci Reports 10:1273

Simon K (2001) Does δD from fluid inclusions in quartz reflect the original hydrothermal fluid? Chem Geol 177:483–495

Simon JI, DePaolo DJ (2010) Stable calcium isotopic composition of meteorites and rocky planets. Earth Planet Sci Lett 289:457–466

Simon L, Lecuyer C, Marechal C, Coltice N (2006) Modelling the geochemical cycle of boron: implications for the long-term $d^{11}B$ evolution of seawater and oceanic crust. Chem Geol 225:61–76

Sio CK, Dauphas N, Teng FZ, Chaussidon M, Helz RT, Roskosz M (2013) Discerning crystal growth from diffusion profiles in zoned olivine by in-situ Mg–Fe isotopic analysis. Geochim Cosmochim Acta 123:302–321

Skauli H, Boyce AJ, Fallick AE (1992) A sulphur isotope study of the Bleikvassli Zn–Pb–Cu deposit, Nordland, northern Norway. Mineral Deposita 27:284–292

Skirrow R, Coleman ML (1982) Origin of sulfur and geothermometry of hydrothermal sulfides from the Galapagos Rift, 86°W. Nature 249:142–144

Skulan JL, DePaolo (1999) Calcium isotope fractionation between soft and mineralized tissues as a monitor of calcium use in vertebrates. PNAS 96:13709–13713

Skulan JL, Bullen TD, Anbar AD, Puzas JE, Shackelford L, LeBlanc A, Smith SM (2007) Natural calcium isotope composition of urine as a marker of bone mineral balance. Clin Chem 53:1155–1158

Smit KV, Shirey SB, Hauri EH, Stern RA (2019) Sulfur isotopes in diamonds reveal differences in continent construction. Science 364:383–385

Smith JW, Batts BD (1974) The distribution and isotopic composition of sulfur in coal. Geochim Cosmochim Acta 38:121–123
Smith JW, Gould KW, Rigby D (1982) The stable isotope geochemistry of Australian coals. Org Geochem 3:111–131
Snyder G, Poreda R, Hunt A, Fehn U (2001) Regional variations in volatile composition: isotopic evidence for carbonate recycling in the Central American volcanic arc. Geochem Geophys Geosystems 2:2001GC000163
Sofer Z (1984) Stable carbon isotope compositions of crude oils: application to source depositional environments and petroleum alteration. Am Assoc Petrol Geol Bull 68:31–49
Sofer Z, Gat JR (1972) Activities and concentrations of oxygen-18 in concentrated aqueous salt solutions: analytical and geophysical implications. Earth Planet Sci Lett 15:232–238
Sonnerup RE, Quay PD, McNichol AP, BullisterJL Westby TA, Anderson HL (1999) Reconstructing the oceanic $^{13}C$ Suess effect. Global Biogeochem Cycles 13:857–872
Soudry D, Glenn CR, Nathan Y, Segal I, VonderHaar D (2006) Evolution of Tethyan phosphogenesis along the northern edges of the Arabian–African shield during the Cretaceous-Eocene as deduced from temporal variations of Ca and Nd isotopes and rates of P accumulation. Earth Sci Rev 78:27–57
Sowers T (2001) The $N_2O$ record spanning the penultimate deglaciation from the Vostok ice core. J Geophys Res 106:31903–31914
Sowers T (2010) Atmospheric methane isotope records covering the Holocene period. Quaternary Sci Rev 29:213–221
Sowers T, Bender M, Raynaud D, Korotkevich YS, Orchardo J (1991) The $\delta^{18}O$ of atmospheric $O_2$ from air inclusions in the Vostok ice core: timing of $CO_2$ and ice volume changes during the Penultimate deglaciation. Paleoceanography 6:679–696
Sowers T, Bender M, Raynaud D, Korotkevich YS (1992) $\delta^{15}N$ of $N_2$ in air trapped in polar ice: a tracer of gas transport in the firn and a possible constraint on ice age-gas age differences. J Geophys Res 97:15683–15697
Sowers T et al (1993) A 135000-year Vostock-SPECMAP common temporal framework. Paleoceanography 8:737–766
Spandler C, Pirard C (2013) Element recycling from subducting slabs to arc crust: a review. Lithos 170:208–223
Spencer CJ, Partin CA, Kirkland CL, Raub TD, Liebmann J, Stern RA (2019) Paleoproterozoic increase in $\delta^{18}O$ driven by rapid emergence of continental crust. Geochim Cosmochim Acta 257:16–25
Spero HJ, Bijma J, Lea DW, Bemis BE (1997) Effect of seawater carbonate concentration on foraminiferal carbon and oxygen isotopes. Nature 390:497–500
Spicuzza M, Day J, Taylor L, Valley JW (2007) Oxygen isotope constraints on the origin and differentiation of the Moon. Earth Planet Sci Lett 253:254–265
Spooner PT, Guo W, Robinson LF, Thiagarajan N, Hendry KR, Rosenheim BE, Leng MJ (2016) Clumped isotope composition of cold-water corals: a role for vital effects? Geochim Cosmochim Acta 179:123–141
Stachel T, Harris JW, Muehlenbachs K (2009) Sources of carbon in inclusion bearing diamonds. Lithos 112:625–637
Stahl W (1977) Carbon and nitrogen isotopes in hydrocarbon research and exploration. Chem Geol 20:121–149
Steele RC, Elliott T, Coath CD, Regelous M (2011) Confirmation of mass-independent Ni isotopic variability in iron meteorites. Geochim Cosmochim Acta 75:7906–7925
Steele RC, Coath CD, Regelous M, Russell S, Elliott T (2012) Neutron-poor nickel isotope anomalies in meteorites. Astrophys J 758:59–80
Stefurak EJ, Woodward WF, Lowe DR (2015) Texture-specific Si isotope variations in Barberton Greenstone Belt cherts record low temperature fractionations in early Archean seawater. Geochim Cosmochim Acta 150:26–52

Steinhoefel G, Horn I, von Blanckenburg F (2009) Micro-scale tracing of Fe and Si isotope signatues in banded iron formation using femtosecond laser ablation. Geochim Cosmochim Acta 73:5343–5360

Steinhoefel G, von Blanckenburg F, Horn I, Konhauser KO, Beukes NJ, Gutzmer J (2010) Deciphering formation processes of banded iron formations from the Transvaal and Hamersley successions by combined Si and Fe isotope analysis using UV femtosecond laser ablation. Geochim Cosmochim Acta 74:2677–2696

Stephant A, Anand M, Zhao X, Chan Q H, Bonifacie M, Franchi IA (2019) The chlorine isotopic composition of the Moon: insights from melt inclusions. Earth Planet Sci Lett 523:115715

Stern LA, Chamberlain CP, Reynolds RC, Johnson GD (1997) Oxygen isotope evidence of climate change from pedogenic clay minerals in the Himalayan molasse. Geochim Cosmochim Acta 61:731–744

Sternberg LS, Anderson WT, Morrison K (2002) Separating soil and leaf water $^{18}O$ isotope signals in plant stem cellulose. Geochim Cosmochim Acta 67:2561–2566

Steuber T, Buhl D (2006) Calcium-isotope fractionation in selected modern and ancient marine carbonates. Geochim Cosmochim Acta 70:5507–5521

Stevens CM (1988) Atmospheric methane. Chem Geol 71:11–21

Stevens CM, Krout L, Walling D, Venters A, Engelkemeier A, Ross LE (1972) The isotopic composition of atmospheric carbon monoxide. Earth Planet Sci Lett 16:1457–2165

Stewart MK (1974) Hydrogen and oxygen isotope fractionation during crystallization of mirabilite and ice. Geochim Cosmochim Acta 38:167–172

Stolper DA, Sessions AL, Ferreira AA, Santos Neto EV, Schimmelmann A, Shusta SS, Valentine DL, Eiler JM (2014) Combined $^{13}C$-D and D-D clumping in methane: methods and preliminary results. Geochim Cosmochim Acta 126:169–191

Strauß H (1997) The isotopic composition of sedimentary sulfur through time. Palaeo Palaeo Palaeo 132:97–118

Strauß H (1999) Geological evolution from isotope proxy signals—sulfur. Chem Geol 161:89–101

Strauß H, Peters-Kottig W (2003) The Phanerozoic carbon cycle revisited: the carbon isotope composition of terrestrial organic matter. Geochem Geophys Geosys 4:1083. https://doi.org/10.1029/2003GC000555

Stueber AM, Walter LM (1991) Origin and chemical evolution of formation waters from Silurian–Devonian strata in the Illinois basin. Geochim Cosmochim Acta 55:309–325

Sturchio NC, Böhlke JK, Beloso AD, Streger SH, Heraty LJ, Hatzinger PB (2007) Oxygen and chlorine isotopic fractionation during perchlorate biodegradation: laboratory results and implications for forensics and natural attenuation studies. Environ Sci Techn 41:2796–2802

Sturchio NC, Beloso A, Heraty LJ, Wheatcraft S, Schumer R (2014) Isotopic tracing of perchlorate sources in groundwater from Pomona, California. Appl Geochem 43:80–87

Styrt MM, Brackmann AJ, Holland HD, Clark BC, Pisutha-Arnold U, Eldridge CS, Ohmoto H (1981) The mineralogy and the isotopic composition of sulfur in hydrothermal sulfide/sulfate deposits on the East Pacific Rise, 21°N latitude. Earth Planet Sci Lett 53:382–390

Sugawara S, Nakazawa T, Shirakawa Y, Kawamura K, Aoki S, Machida T, Honda H (1998) Vertical profile of the carbon isotope ratio of stratospheric methane over Japan. Geophys Res Lett 24:2989–2992

Sullivan-Ojeda A, Philipps E, Sherwood-Lollar B (2020) Multi-element (C, H, Cl, Br) stable isotope fractionation as a tool to investigate transformation processes for halogenated hydrocarbons. Environ Sci: Proc Impacts 22:567–582

Summons RE, Jahnke LL, Roksandic Z (1994) Carbon isotopic fractionation in lipids from methanotrophic bacteria: relevance for interpretation of the geochemical record of biomarkers. Geochim Cosmochim Acta 58:2853–2863

Surma J, Assonov S, Staubwasser M (2021) Triple oxygen isotope systematics in the hydrologic cycle. Rev Mineral Geochem 86:401–428

Swart PK (2015) The geochemistry of carbonate diagenesis: the past, present and future. Sedimentology 62:1233–1304

Sweeney RE, Kaplan IR (1980) Natural abundance of $^{15}N$ as a source indicator for near-shore marine sedimentary and dissolved nitrogen. Mar Chem 9:81–94

Sweeney RE, Liu KK, Kaplan IR (1978) Oceanic nitrogen isotopes and their use in determining the source of sedimentary nitrogen. In: Robinson BW (ed) DSIR Bull 220:9–26

Talbot MR (1990) A review of the palaeohydrological interpretation of carbon and oxygen isotopic ratios in primary lacustrine carbonates. Chem Geol 80:261–279

Tanaka Y, Hirata T (2018) Stable isotope composition of metal elements in biological samples as tracers for element metabolism. Analytical Sci 34:645–655

Tang Y, Perry JK, Jenden PD, Schoell M (2000) Mathematical modeling of stable carbon isotope ratios in natural gases. Geochim Cosmochim Acta 64:2673–2687

Tang Y, Huang Y, Ellis GS, Wang Y, Kralert PG, Gillaizeau B, Ma Q, Hwang R (2005) A kinetic model for thermally induced hydrogen and carbon isotope fractionation of individual n-alkanes in crude oil. Geochim Cosmochim Acta 69:4505–4520

Tang YJ, Zhang HF, Nakamura E, Moriguti T, Kobayashi K, Ying JF (2007) Lithium isotope systematics of peridotite xenoliths from Hannuoba, North China craton: implications for melt-rock interaction in the considerably thinned lithospheric mantle. Geochim Cosmochim Acta 71:4327–4341

Taran YA, Kliger GA, Sevastianov VS (2007) Carbon isotope effect in the open system Fischer Trosch synthesis. Geochim Cosmochim Acta 71:4474–4487

Taylor HP (1968) The oxygen isotope geochemistry of igneous rocks. Contr Mineral Petrol 19:1–71

Taylor HP (1974) The application of oxygen and hydrogen isotope studies to problems of hydrothermal alteration and ore deposition. Econ Geol 69:843–883

Taylor BE, O'Neil JR (1977) Stable isotope studies of metasomatic Ca–Fe–Al–Si skarns and associated metamorphic and igneous rocks, Osgood Mountains, Nevada. Contr Mineral Petrol 63:1–49

Taylor HP (1977) Water/rock interactions and the origin of $H_2O$ in granite batholiths. J Geol Soc 133:509

Taylor HP (1978) Oxygen and hydrogen isotope studies of plutonic granitic rocks. Earth Planet Sci Lett 38:177–210

Taylor HP (1980) The effects of assimilation of country rocks by magmas on $^{18}O/^{16}O$ and $^{87}Sr/^{86}Sr$ systematics in igneous rocks. Earth Planet Sci Lett 47:243–254

Taylor BE, Eichelberger JC, Westrich HR (1983) Hydrogen isotopic evidence of rhyolitic magma degassing during shallow intrusion and eruption. Nature 306:541–545

Taylor BE (1986a) Magmatic volatiles: isotopic variation of C, H and S. Rev Mineral 16:185–225

Taylor HP (1986b) Igneous rocks: II. Isotopic case studies of circumpacific magmatism. In: Stable isotopes in high temperature geological processes. Rev Mineralogy 16:273–317

Taylor BE (1987a) Stable isotope geochemistry of ore-forming fluids. In: Stable isotope geochemistry of low-temperature fluids, vol 13. Short Course Mineralogical Association, Canada, pp 337–445

Taylor HP (1987b) Comparison of hydrothermal systems in layered gabbros and granites, and the origin of low-$\delta^{18}O$ magmas. In: Magmatic processes: physicochemical principles, vol 1. The Geochemical Society Special Publication, pp 337–357

Taylor BE, Bucher-Nurminen K (1986) Oxygen and carbon isotope and cation geochemistry of metasomatic carbonates and fluids—Bergell aureole, Northern Italy. Geochim Cosmochim Acta 50:1267–1279

Taylor BE, Wheeler MC (1994) Sulfur- and oxygen isotope geochemistry of acid mine drainage in the Western United States. In: Environmental geochemistry of sulphide oxidation. American Chemical Society Symposium Series, vol 550. American Chemical Society, Washington, DC, pp 481–514

Taylor HP (1988) Oxygen, hydrogen and strontium isotope constraints on the origin of granites. Trans Royal Soc Edinburgh: Earth Sci 79:317–338

Taylor HP (1997) Oxygen and hydrogen isotope relationships in hydrothermal mineral deposits. In: Barnes HL (ed) Geochemistry of hydrothermal ore deposits, 3rd edn. Wiley-Interscience p, New York, pp 229–302

Taylor HP, Forester RW (1979) An oxygen and hydrogen isotope study of the Skaergaard intrusion and its country rocks: a description of a 55 M.Y. Old fossil hydrothermal system. J Petrol 20:355–419

Taylor HP, Sheppard SMF (1986) Igneous rocks: I. Processes of isotopic fractionation and isotope systematics. In: Stable isotopes in high temperature geological processes. Rev Mineralogy 16:227–271

Taylor HP, Turi B, Cundari A (1984) $^{18}O/^{16}O$ and chemical relationships in K-rich volcanic rocks from Australia, East Africa, Antarctica and San Venanzo Cupaello, Italy. Earth Planet Sci Lett 69:263–276

Teece MA, Fogel ML (2007) Stable carbon isotope biogeochemistry of monosaccharides in aquatic organisms and terrestrial plants. Org Geochem 38:458–473

Teichert BM, Gussone N, Torres ME (2009) Controls on calcium isotope fractionation in sedimentary porewaters. Earth Planet Sci Lett 279:373–382

Telmer KH, Veizer J (1999) Carbon fluxes, $pCO_2$ and substrate weathering in a large northern river basin, Canada: carbon isotope perspective. Chem Geol 159:61–86

Telouk P, 8 others (2015) Copper isotope effect in serum of cancer patients; a pilot study. Metallomics 7:299–308

Teng FZ, Dauphas N, Helz R (2008) Iron isotope fractionation during magmatic differentiation in Kilauea Iki lava lake. Science 320:1620–1622

Teng FZ, Dauphas N, Helz RT, Gao S, Huang S (2011) Diffusion-driven magnesium and iron isotope fractionation in Hawaiian olivine. Earth Planet Sci Lett 308:317–324

Thiagarajan N, Adkins J, Eiler J (2011) Carbonate clumped isotope thermometry of deep-sea corals and implications for vital effects. Geochim Cosmochim Acta 75:4416–4425

Thiagarajan N, Xie H, Ponton C, Kitchen N, Peterson B, Lawson M, Fornolo M, Xiao Y, Eiler J (2020) Isotopic evidence for quasi-equilibrium chemistry in thermally mature natural gases. PNAS 117:3989–3995

Thiel V, Peckmann J, Seifert R, Wehrung P, Reitner J, Michaelis W (1999) Highly isotopically depleted isoprenoids: molecular markers for ancient methane venting. Geochim Cosmochim Acta 63:3959–3966

Thiemens MH (1988) Heterogeneity in the nebula: evidence from stable isotopes. In: Kerridge JF, Matthews MS (eds) Meteorites and the early solar system. University of Arizona Press, pp 899–923

Thiemens MH (1999) Mass-independent isotope effects in planetary atmospheres and the early solar system. Science 283:341–345

Thiemens MH (2006) History and applications of mass-independent isotope effects. Ann Rev Earth Planet Sci 34:217–262

Thiemens MH, Heidenreich JE (1983) The mass independent fractionation of oxygen—a novel isotope effect and its cosmochemical implications. Science 219:1073–1075

Thiemens MH, Jackson T, Zipf EC, Erdman PW, van Egmond C (1995) Carbon dioxide and oxygen isotope anomalies in the mesophere and stratosphere. Science 270:969–972

Thode HG, Monster J (1964) The sulfur isotope abundances in evaporites and in ancient oceans. In: Vinogradov AP (ed) Proceedings of geochemistry conference commemorating the centenary of V I Vernadskii's birth, vol 2, 630 p

Thomassot E, Cartigny P, Harris JW, Lorand JP, Rollion-Bard C, Chaussidon M (2009) Metasomatic diamond growth: a multi-isotope study ($^{13}C$, $^{15}N$, $^{33}S$, $^{34}S$) of sulphide inclusions and their host diamonds from Jwaneng (Botswana). Earth Planet Sci Lett 282:79–90

Thompson P, Schwarcz HP, Ford DE (1974) Continental Pleistocene climatic variationa from speleothem age and isotopic data. Science 184:893–895

Thompson LG, Mosley-Thompson E, Henderson KA (2000) Ice-core palaeoclimate records in tropical South America since the last glacial maximum. J Quat Sci 15:377–394

Thompson LG et al (2006) Abrupt tropical climate change: past and present. PNAS 103:10536–10543
Tian Z, Jolliff BL, Korotev RL, Fegley B, Lodders K, Day JM, Chen H, Wang K (2020) Potassium isotope composition of the Moon. Geochim Cosmochim Acta 280:268–280
Tiedemann R, Sarntheim M, Shackleton NJ (1994) Astronomic timescale for the Pliocene Atlantic $\delta^{18}O$ and dust flux records of Ocean Drilling Program site 659. Paleoceanography 9:619–638
Tilley B, Muehlenbachs K (2013) Isotope reversals and universal stages and trends of gas maturation in sealed self-contained petroleum systems. Chem Geol 339:194–204
Todd CS, Evans BW (1993) Limited fluid-rock interaction at marble-gneiss contacts during Cretaceous granulite-facies metamorphism, Seward Peninsula, Alaska. Contr Mineral Petrol 114:27–41
Tostevin R, Turchyn AV, Farquhar J, Johnston DT, Eldridge DL, Bishop JK, McIlvin M (2014) Multiple sulfur isotope constraints on the modern sulfur cycle. Earth Planet Sci Lett 396:14–21
Trail D, Boehnke P, Savage PS, Liu MC, Miller ML, Bindeman I (2018) Origin and significance of Si and O isotope heterogeneities in Phanerozoic, Archean and Hadean zircon. PNAS 115:10287–10292
Trinquier A, Elliott T, Uffbeck D, Coath C, Krot AN, Bizarro M (2009) Origin of nucleosynthetic isotope heterogeneity in the solar protoplanetary disk. Science 324:374–376
Tripati AK, Eagle RA, Thiagarajan N, Gagnon AC, Bauch H, Halloran PR, Eiler JM (2010) $^{13}C$–$^{18}O$ isotope signatures and “clumped isotope” thermometry in foraminifera and coccoliths. Geochim Cosmochim Acta 74:5697–5717
Troll VR, 10 others (2019) Global Fe–O isotope correlation reveals magmatic origin of Kiruna-type apatite-iron oxide ores. Nat Commun 10:1712
Trudinger CM, Enting IG, Francey RJ, Etheridge DM, Rayner PJ (1999) Long-term variability in the global carbon cycle inferred from a high-precision $CO_2$ and $\delta^{13}C$ ice-core record. Tellus 51B:233–248
Truesdell AH, Hulston JR (1980) Isotopic evidence on environments of geothermal systems. In: Fritz P, Fontes J (eds) Handbook of environmental isotope geochemistry, vol I. Elsevier, New York Amsterdam, pp 179–226
Trust BA, Fry B (1992) Stable sulphur isotopes in plants: a review. Plant Cell Environ 15:1105–1110
Tucker ME, Wright PV (1990) Carbonate sedimentology. Blackwell Scientific Publishing, London, pp 365–400
Tudge AP (1960) A method of analysis of oxygen isotopes in orthophosphate—its use in the measurement of paleotemperatures. Geochim Cosmochim Acta 18:81–93
Turchyn AV, Schrag DP (2004) Oxygen isotope constraints on the sulfur cycle over the past 10 million years. Science 303:2004–2007
Turchyn AV, Schrag DP (2006) Cenozoic evolution of the sulphur cycle: insight from oxygen isotopes in marine sulphate. Earth Planet Sci Lett 241:763–779
Turner AJ, Frankenberg C, Kort EA (2019) Interpreting contemporary trends in atmospheric methane. PNAS 116:2805–2813
Uemura R, Abe O, Motoyama H (2010) Determining the $^{17}O/^{16}O$ ratio of water using a water—$CO_2$ equilibration method: application to glacial-interglacial changes in $^{17}O$ excess from the Dome Fuji ice core Antarctica. Geochim Cosmochim Acta 74:4919–4936
Urey HC (1947) The thermodynamic properties of isotopic substances. J Chem Soc 1947:562
Usui T, Alexander CM, Wang J, Simon JI, Jones JH (2012) Origin of water and mantle-crust interactions on Mars inferred from hydrogen isotopes and volatile element abundances of olivine-hosted melt inclusions of primitive shergottites. Earth Planet Sci Lett 357–358:119–129
Usui T, Alexander CM, Wang J, Simon JI, Jones JH (2015) Meteoritic evidence for a previously unrecognized hydrogen reservoir on Mars. Earth Planet Sci Lett 410:140–151
Valdes MC, Moreira M, Foriel J, Moynier F (2014) The nature of Earth's building blocks as revealed by calcium isotopes. Earth Planet Sci Lett 394:135–145
Valley JW (1986) Stable isotope geochemistry of metamorphic rocks. Rev Mineral 16:445–489

Valley JW (2001) Stable isotope thermometry at high temperatures. Rev Mineral Geochem 43:365–413

Valley JW (2003) Oxygen isotopes in zircon. Rev Mineral Geochem 53:343–385

Valley JW, Bohlen SR, Essene EJ, Lamb W (1990) Metamorphism in the Adirondacks. II. J Petrol 31:555–596

Valley JW, Eiler JM, Graham CM, Gibson EK, Romanek CS, Stolper EM (1997) Low temperature carbonate concretions in the martian meteorite ALH 84001: evidence from stable isotopes and mineralogy. Science 275:1633–1637

Valley JW et al (2005) 4.4 billion years of crustal maturation: oxygen isotope ratios in magmatic zircon. Contr Mineral Petrol 150:561–580

Van der Merve NJ, Lee-Thorp J, Thackery F, Hall-Martin A (1990) Source area determination of elephant ivory by isotopic analysis. Nature 346:744–746

Vasconcelos C, Mackenzie JA, Warthmann R, Bernasconi S (2005) Calibration of the $\delta^{18}O$ paleothermometer for dolomite precipitated in microbial cultures and natural environments. Geology 33:317–320

Vazquez R, Vennemann TW, Kesler SE, Russell N (1998) Carbon and oxygen isotope halos in the host limestone, El Mochito Zn, Pb (Ag) skarn massive sulfide/oxide deposit, Honduras. Econ Geol 93:15–31

Veizer J, Hoefs J (1976) The nature of $^{18}O/^{16}O$ and $^{13}C/^{12}C$ secular trends in sedimentary carbonate rocks. Geochim Cosmochim Acta 40:1387–1395

Veizer J et al (1997) Oxygen isotope evolution of Phanerozoic seawater. Palaeo Palaeo Palaeo 132:159–172

Veizer J et al (1999) $^{87}Sr/^{86}Sr$, $\delta^{13}C$ and $\delta^{18}O$ evolution of Phanerozoic seawater. Chem Geol 161:37–57

VennemannTW Smith HS (1992) Stable isotope profile across the orthoamphibole isograd in the Southern Marginal Zone of the Limpopo Belt, S Africa. Precambrian Res 55:365–397

Vennemann TW, Kesler SE, O'Neil JR (1992) Stable isotope composition of quartz pebbles and their fluid inclusions as tracers of sediment provenance: implications for gold- and uranium-bearing quartz pebble conglomerates. Geology 20:837–840

Vennemann TW, Kesler SE, Frederickson GC, Minter WEL, Heine RR (1996) Oxygen isotope sedimentology of gold and uranium-bearing Witwatersrand and Huronian Supergroup quartz pebble conglomerates. Econ Geol 91:322–342

Vennemann TW, Fricke HC, Blake RE, O'Neil JR, Colman A (2002) Oxygen isotope analysis of phosphates: a comparison of techniques for analysis of $Ag_3PO_4$. Chem Geol 185:321–336

Ventura GT, Gall L, Siebert C, Prytulak J, Szatmari P, Hürlimann M, Halliday AN (2015) The stable isotope composition of vanadium, nickel and molybdenum in crude oils. Appl Geochem 59:104–117

Vho A, Lanari P, Rubatto D, Hermann J (2020) Tracing fluid transfers in subduction zones: an integrated thermodynamic and $\delta^{18}O$ fractionation modeling approach. Solid Earth 11:307–328

Viers J et al (2007) Evidence of Zn isotope fractionation in a soil-plant system of a pristine tropical watershed (Nsimi, Cameroon). Chem Geol 239:124–137

Villanueva GL, Mumma MJ, Novak RE, Käufl HU, Hartogh P, Encrenaz T, Tokunaga A, Khayat A, Smith MD (2015) Strong water anomalies in the martian atmosphere: probing current and ancient reservoirs. Science 348:218–221

Voegelin AR, Nägler TF, Beukes NJ, Lacassie JP (2010) Molybdenum isotopes in late Archean carbonate rocks: implications for early Earth oxygenation. Precambr Res 182:70–82

Von Blanckenburg F, von Wiren N, Guelke M, Weiss D (2009) Fractionation of metal stable isotopes by higher plants. Elements 5:375–380

Von Grafenstein U, Erlenkeuser H, Trimborn P (1999) Oxygen and carbon isotopes in fresh-water ostracod valves: assessing vital offsets and autoecological effects of interest for paleoclimate studies. Palaeo Palaeo Palaeo 148:133–152

Wacker U, Fiebig J, Tödter J, Schöne BR, Bahr A, Friedrich O, Tütken T, Gischler E, Joachimski MM (2014) Emperical calibration of the clumped isotope paleothermometer using calcites of various origins. Geochim Cosmochim Acta 141:127–144

Wada E, Hattori A (1976) Natural abundance of $^{15}N$ in particulate organic matter in North Pacific Ocean. Geochim Cosmochim Acta 40:249–251

Wagner S, Brandes J, Spencer RG, Ma K, Rosengard SZ, Moura JM, Stubbins A (2019) Isotopic composition of oceanic dissolved black carbon reveals non-riverine source. Nature Comm 10:5064

Waldeck AR, Covie BR, Bertran E, Wing BA, Halevy I, Johnston DT (2019) Deciphering the atmospheric signal in marine sulfate oxygen isotope composition. Earth Planet Sci Lett 522:12–19

Wallmann K (2001) The geological water cycle and the evolution of marine $\delta^{18}O$ values. Geochim Cosmochim Acta 65:2469–2485

Walter JB, Cruz-Uribe AM, Marschall HR (2019) Isotopic compositions of sulfides in exhumed high-pressure terranes: implications for sulfur cycling in subduction zones. Geochem Geophys Geosys 20:3347–3374

Walter S, 10 others (2016) Isotopic evidence for biogenic molecular hydrogen production in the Atlantic ocean. Biogeosciences 13:323–340

Walczyk T, von Blanckenburg F (2002) Natural iron isotope variation in human blood. Science 295:2065–2066

Wang K, Savage PS, Moynier F (2014) The iron isotope composition of enstatite chondrites: implications for their origin and the metal/sulfide fractionation factor. Geochim Cosmochim Acta 142:149–165

Wang K, Jacobsen (2016) Potassium isotopic evidence for a high-energy giant impact origin of the Moon. Nature 538:487–489

Wang Y, Sessions AL, Nielsen RJ, Goddard WA (2009) Equilibrium $^{2}H/^{1}H$ fractionations in organic molecules. II: Linear alkanes, alkenes, ketones, carboxylic acids, esters, alcohols and ethers. Geochim Cosmochim Acta 73:7076–7086

Wang Y, Zhu XK, Mao JW, Li ZH, Cheng YB (2011a) Iron isotope fractionation during skarn-type metallogeny: a case study of Xinqiao Cu–S–Fe–Au deposit in the middle-lower Yangtze valley. Ore Geol Rev 43:194–202

Wang Z, Chapellaz J, Park K, Mak JE (2011b) Large variations in southern biomass burning during the last 650 years. Science 330:1663–1666

Wang XL, Planavsky NJ, Reinhard CT, Hein JR, Johnson TM (2016) A Cenozoic seawater redox record derived from $^{238}U/^{235}U$ in ferromanganese crusts. Am J Sci 316:64–83

Wanner C, Sonnenthal EL, Liu XM (2014) Seawater $\delta^{7}Li$: a direct proxy for global $CO_2$ consumption by continental silicate weathering? Chem Geol 381:154–167

Warr O, Young ED, Giunta T, Kohl IE, Ash JL, Sherwood-Lollar B (2021) High-resolution, long-term isotopic and isotopologue variation identifies the sources and sinks of methane in a deep subsurface carbon cycle. Geochim Cosmochim Acta 294:315–334

Warren CG (1972) Sulfur isotopes as a clue to the genetic geochemistry of a roll-type uranium deposit. Econ Geol 67:759–767

Washington KE, West AJ, Kaldeon-Asael B, Katchinoff AR, Stevenson EI, Planavsky NJ (2020) Lithium isotope composition of modern and fossilized Cenozoic brachiopods. Geology 48:1058–1061

Waterhouse JS, Cheng S, Juchelka D, Loader NJ, McCarroll D, Switsur R, Gautam L (2013) Position-specific measurement of oxygen isotope ratios in cellulose: isotope exchange during heterotrophic cellulose synthesis. Geochim Cosmochim Acta 112:178–192

Watson LL, Hutcheon ID, Epstein S, Stolper EM (1994) Water on Mars: clues from deuterium/hydrogen and water contents of hydrous phases in SNC meteorites. Science 265:86–90

Wawryk CM, Foden JD (2015) Fe-isotope fractionation in magmatic-hydrothermal deposits: a case study from the Renison Sn-W deposit, Tasmania. Geochim Cosmochim Acta 150:285–298

Weber JN, Raup DM (1966a) Fractionation of the stable isotopes of carbon and oxygen in marine calcareous organisms-the Echinoidea. I. Variation of $^{13}C$ and $^{18}O$ content within individuals. Geochim Cosmochim Acta 30:681–703

Weber JN, Raup DM (1966b) Fractionation of the stable isotopes of carbon and oxygen in marine calcareous organisms-the Echinoidea. II. Environmental and genetic factors. Geochim Cosmochim Acta 30:705–736

Webster CR, Mahaffy PR et al (2013) Isotope ratios of H, C, and O in $CO_2$ and $H_2O$ of the Martian atmosphere. Science 341:260–263

Wefer G, Berger WH (1991) Isotope paleontology: growth and composition of extant calcareous species. Mar Geol 100:207–248

Wei Z, 26 others (2019) A global database of water vapor isotopes measured with high temporal resolution infrared laser spectroscopy. Sci Data 6:180302

Welch SA, Beard BL, Johnson CM, Braterman PS (2003) Kinetic and equilibrium Fe isotope fractionation between aqueous Fe(II) and Fe(III). Geochim Cosmochim Acta 67:4231–4250

Welhan JA (1987) Stable isotope hydrology. In: Short course in stable isotope geochemistry of low-temperature fluids, vol 13. Mineral Association, Canada, pp 129–161

Welhan JA (1988) Origins of methane in hydrothermal systems. Chem Geol 71:183–198

Well R, Flessa H (2009) Isotopogue enrichment factors of $N_2O$ reduction in soils. Rapid Commum Mass Spectrom 23:2996–3002

Wenzel B, Lecuyer C, Joachimski MM (2000) Comparing oxygen isotope records of Silurian calcite and phosphate—$\delta^{18}O$ composition of brachiopods and conodonts. Geochim Cosmochim Acta 69:1859–1872

Westerhausen L, Poynter J, Eglinton G, Erlenkeuser H, Sarntheim M (1993) Marine and terrigenous origin of organic matter in modern sediments of the equatorial East Atlantic: the $\delta^{13}C$ and molecular record. Deep Sea Res 40:1087–1121

Weyer S, Ionov D (2007) Partial melting and melt percolation in the mantle: the message from Fe isotopes. Earth Planet Sci Lett 259:119–133

Weyer S, Anbar AD, Brey GP, Münker C, Mezger K (2005) Iron isotope fractionation during planetary differentiation. Earth Planet Sci Lett 240:251–264

White JWC (1989) Stable hydrogen isotope ratios in plants: a review of current theory and some potential applications. Stable isotopes in ecological research, Ecological Studies, vol 68. Springer, New York, pp 142–162

White JWC, Lawrence JR, Broecker WS (1994) Modeling and interpreting D/H ratios in tree rings: a test case of white pine in the northeastern United States. Geochim Cosmochim Acta 58:851–862

Whiticar MJ (1999) Carbon and hydrogen isotope systematics of bacterial formation and oxidation of methane. Chem Geol 161:291–314

Whiticar MJ, Faber E, Schoell M (1986) Biogenic methane formation in marine and freshwater environments: $CO_2$ reduction vs. acetate fermentation-Isotopic evidence. Geochim Cosmochim Acta 50:693–709

Whittacker SG, Kyser TK (1990) Effects of sources and diagenesis on the isotopic and chemical composition of carbon and sulfur in Cretaceous shales. Geochim Cosmochim Acta 54:2799–2810

Wickham SM, Taylor HR (1985) Stable isotope evidence for large-scale seawater infiltration in a regional metamorphic terrane; the Trois Seigneurs Massif, Pyrenees, France. Contrib Mineral Petrol 91:122–137

Wickman FE (1952) Variation in the relative abundance of carbon isotopes in plants. Geochim Cosmochim Acta 2:243–254

Widory D, Minet JJ, Barbe-Leborgne M (2009) Sourcing explosives: a multi-isotope approach. Sci Justice 49:62–72

Wiechert U, Halliday AN (2007) Non-chondritic magnesium and the origin of the inner terrestrial planets. Earth Planet Sci Lett 256:360–371

Wiechert U, Halliday AN, Lee D-C, Snyder GA, Taylor LA, Rumble D (2001) Oxygen isotopes and the moon forming giant impact. Science 294:345–348

Willbold M, Elliott T (2017) Molybdenum isotope variations in magmatic rocks. Chem Geol 449:253–268

Wille M, Kramers JD, Nägler TF, Beukes NJ, Schröder S, Meiser T, Lacassie JP, Voegelin AR (2007) Evidence for a gradual rise of oxygen between 2.6 and 2.5 Ga from Mo isotopes and Re-PGE signatures in shales. Geochim Cosmochim Acta 71:2417–2435

Wille M, Nebel O, Van Kranendonk MJ, Schoenberg R, Kleinhanns IC, Ellwood MJ (2013) Mo-Cr evidence for a reducing Archean atmosphere in 3.46–2.76 Ga black shales from the Pilbara, western Australia. Chem Geol 340:68–76

Williams HM, Archer C (2011) Copper stable isotopes as tracers of metal-sulphide segregation and fractional crystallization processeson iron meteorite parent bodies. Geochim Cosmochim Acta 75:3166–3178

Williams HM, Markowski A, Quitte G, Halliday AN, Teutsch N, Levasseur S (2006) Fe isotope fractionations in iron meteorites: new insight into metal-sulphide segregation and planetary accretion. Earth Planet Sci Lett 250:486–500

Williams HM, Wood BJ, Wade J, Frost DJ, Tuff J (2012) Isotopic evidence for internal oxidation of the Earth's mantle during accretion. Earth Planet Sci Lett 321–322:54–63

Williams HM, Bizimis M (2014) Iron isotope tracing of mantle heterogeneity within the source regions of oceanic basalts. Earth Planet Sci Lett 404:396–407

Williford KH, Ushikubo T, Lepot K, Kitajima K, Hallman C, Spicuzza MJ, Kozdon R, Eigenbrode JL, Summons RE, Valley JW (2016) Carbon and sulfur isotopic signatures of ancient life and environment at the microbial scale: Neoarchean shales and carbonates. Geobiology 14:105–128

Wing BA, Farquhar J (2015) Sulfur isotope homogeneity of lunar mare basalts. Geochim Cosmochim Acta 170:266–280

Wong WW, Sackett WM (1978) Fractionation of stable carbon isotopes by marine phytoplankton. Geochim Cosmochim Acta 42:1809–1815

Wortmann UG, Chernyavsky B, Bernasconi SM, Brunner B, Böttcher ME, Swart PK (2007) Oxygen isotope biogeochemistry of pore water sulfate in the deep biosphere: dominance of isotope exchange reactions with ambient water during microbial sulfate reduction (ODP site 1130). Geochim Cosmochim Acta 71:4221–4232

Wostbrock JA, Sharp ZD (2021) Triple oxygen isotopes in silica-water and carbonater-water systems. Rev Mineral Geochem 68:367–400

Wright I, Grady MM, Pillinger CT (1990) The evolution of atmospheric $CO_2$ on Mars: the perspective from carbon isotope measurements. J Geophys Res 95:14789–14794

Wu L, Beard BL, Roden EE, Johnson CM (2011) Stable iron isotope fractionation between aqueous Fe(II) and hydrous ferric oxide. Environ Sci Technol 45:1845–1852

Xia J, Ito E, Engstrom DE (1997a) Geochemistry of ostracode calcite: Part I. An experimental determination of oxygen isotope fractionation. Geochim Cosmochim Acta 61:377–382

Xia J, Engstrom DE, Ito E (1997b) Geochemistry of ostracode calcite: Part 2. The effects of water chemistry and seasonal temperature variation on *Candona rawsoni*. Geochim Cosmochim Acta 61:383–391

Xia X, Chen J, Braun R, Tang Y (2013) Isotopic reversals with respect to maturity trends due to mixing of primary and secondary products in source rocks. Chem Geol 339:205–212

Xiao Y, Hoefs J, van den Kerkhof AM, Simon K, Fiebig J, Zheng YF (2002) Fluid evolution during HP and UHP metamorphism in Dabie Shan, China: constraints from mineral chemistry, fluid inclusions and stable isotopes. J Petrol 43:1505–1527

Xiao Y, Zhang Z, Hoefs J, van den Kerkhof A (2006) Ultrahigh pressure rocks from the Chinese Continental Scientific Drilling Project: II Oxygen isotope and fluid inclusion distributions through vertical sections. Contr Mineral Petrol 152:443–458

Xie H, Ponton C, Formolo MJ, Lawson M, Peterson BK, Lloyd MK, Sessions AL, Eiler JM (2018) Position-specific hydrogen isotope equilibrium in propane. Geochim Cosmochim Acta 238:193–207

Yang J, Epstein S (1984) Relic interstellar grains in Murchison meteorite. Nature 311:544–547

Yang C, Telmer K, Veizer J (1996) Chemical dynamics of the "St Lawrence" riverine system: $\delta D_{H2O}$, $\delta^{18}O_{H2O}$, $\delta^{13}C_{DIC}$, $\delta^{34}S_{SO4}$ and dissolved $^{87}Sr/^{86}Sr$. Geochim Cosmochim Acta 60:851–866

Yapp CJ (1983) Stable hydrogen isotopes in iron oxides—isotope effects associated with the dehydration of a natural goethite. Geochim Cosmochim Acta 47:1277–1287

Yapp CJ (1987) Oxygen and hydrogen isotope variations among goethites (α-FeOOH) and the determination of paleotemperatures. Geochim Cosmochim Acta 51:355–364

Yapp CJ (2007) Oxygen isotopes in synthetic goethite and a model for the apparent pH dependence of goethite-water $^{18}O/^{16}O$ fractionation. Geochim Cosmochim Acta 71:1115–1129

Yapp CJ, Epstein S (1982) Reexamination of cellulose carbon-bound hydrogen δD measurements and some factors affecting plant-water D/H relationships. Geochim Cosmochim Acta 46:955–965

Yeung LY, Young ED, Schauble EA (2012) Measurement of $^{18}O^{18}O$ and $^{17}O^{18}O$ in the atmosphere and the role of isotope exchange reactions. J Geophys Res 117:D18306. https://doi.org/10.1029/2012JD017992

Yeung LY, Ash JL, Young ED (2014) Rapid photochemical equilibration of isotope bond ordering in $O_2$. J Geophys Res Atmos 119:10552–10566

Yeung LY, Ash JL, Young ED (2015) Biological signatures in clumped isotopes of $O_2$. Science 348:431–434

Yokochi R, Marty B, Chazot G, Burnard P (2009) Nitrogen in perigotite xenoliths: lithophile behaviour and magmatic isotope fractionation. Geochim Cosmochim Acta 73:4843–4861

Yoshida N, Toyoda S (2000) Constraining the atmospheric $N_2O$ budget from intramolecular site preference in $N_2O$ isotopomers. Nature 405:330–334

Yoshida N, Hattori A, Saino T, Matsuo S, Wada E (1984) $^{15}N/^{14}N$ ratio of dissolved $N_2O$ in the eastern tropical Pacific Ocean. Nature 307:442–444

Young ED (1993) On the $^{18}O/^{16}O$ record of reaction progress in open and closed metamorphic systems. Earth Planet Sci Lett 117:147–167

Young ED, Rumble D (1993) The origin of correlated variations in in-situ $^{18}O/^{16}O$ and elemental concentrations in metamorphic garnet from southeastern Vermont, USA. Geochim Cosmochim Acta 57:2585–2597

Young ED, Ash RD, England P, Rumble D (1999) Fluid flow in chondritic parent bodies: deciphering the compositions of planetesimals. Science 286:1331–1335

Young ED, Tonui E, Manning CE, Schauble E, Macris CA (2009a) Spinel-olivine magnesium isotope thermometry in the mantle and implications for the Mg isotopic composition of Earth. Earth Planet Sci Lett 288:524–533

Young ED, Yeung LY, Kohl IE (2014) On the $\Delta^{17}O$ budget of atmospheric $O_2$. Geochim Cosmochim Acta 135:102–125

Young ED, Kohl IE, Warren PH, Rubie DC, Jacobson SA, Morbidelli A (2016) Oxygen isotope evidence for vigorous mixing during the Moon-forming giant impact. Science 351:493–496

Young ED, Kohl IE, 22 others (2017) The relative abundances of resolved $^{12}CH_2D_2$ and $^{13}CH_3D$ and mechanisms controlling isotopic bond ordering in abiotic and biotic methane gases. Geochim Cosmochim Acta 203:235–264

Young ED, Manning CE, Schauble EA, Shahar A, Macris CA, Lazar C, Jordan M (2015) High-temperature equilibrium isotope fractionation of non-traditional isotopes: experiments, theory and applications. Chem Geol 395:176–195

Young MB, McLaughlin K, Kendall C, Stringfellow W, Rollow M, Elsbury K, Donald E, Payton A (2009b) Characterizing the oxygen isotopic composition of phosphate sources to aquatic ecosystems. Environ Sci Techn 43:5190–5196

Yurimoto A, Krot A, Choi BG, Aléon J, Kunihiro T, Brearly AJ (2008) Oxygen isotopes in chondritic components. Rev Mineral Geochem 68:141–186

Yurtsever Y (1975) Worldwide survey of stable isotopes in precipitation. Rep Sect Isotope Hydrol IAEA, November 1975, 40 pp

Zaarur S, Affek HP, Stein M (2016) Last glacial-Holocene temperatures and hydrology of the Sea of Galilee and Hula Valley from clumped isotopes in *Melanopsis* shells. Geochim Cosmochim Acta 179:142–155

Zaback DA, Pratt LM (1992) Isotopic composition and speciation of sulfur in the Miocene Monterey Formation: reevaluation of sulfur reactions during early diagenesis in marine environments. Geochim Cosmochim Acta 56:763–774

Zachos J, Pagani M, Sloan L, Thomas E, Billups K (2001) Trends, rhythms and aberrations in global climate 65 Ma to present. Science 292:686–693

Zakharov DO, Bindeman IN, Slabunov AV, Ovtcharov M, Coble M, Serebryakov NS, Schaltegger U (2017) Dating the Paleoproterozoic snowball earth glaciations using contemporaneous subglacial hydrothermal systems. Geology 45:667–670

Zakharov DO, Bindeman IN, Tanaka R, Fridleifsson G, Reed MH, Hampton RL (2019) Triple oxygen isotope systematics as a tracer of fluids in the crust: a study from modern geothermal systems of Iceland. Chem Geol 530:119312

Zambardi T, Poitrasson F, Corgne A, Meheut M, Quitte G, Anand M (2013) Silicon isotope variations in the inner solar system: implications for planetary formation, differentiation and composition. Geochim Cosmochim Acta 121:67–83

Zeebe RE (1999) An explanation of the effect of seawater carbonate concentration on foraminiferal oxygen isotopes. Geochim Cosmochim Acta 63:2001–2007

Zeebe RE (2010) A new value for the stable oxygen isotope fractionation between dissolved sulfate ion and water. Geochim Cosmochim Acta 74:818–828

Zhao Y, Vance D, Abouchami W, de Baar HJ (2014) Biogeochemical cycling of zinc and its isotopes in the Southern Ocean. Geochim Cosmochim Acta 125:653–672

Zhao X, Zhang H, Zhu X, Tang S, Tang Y (2010) Iron isotope variations in spinel peridotite xenoliths from North China craton: implications for mantle metasomatism. Contr Mineral Petrol 160:1–14

Zhao XM, Cao HH, Mi X, Evans NJ, Qi YH, Huang F, Zhang HF (2017) Combined iron and magnesium isotope geochemistry of pyroxenite xenoliths from Hannuaba, North China Craton: implications for mantle metasomatism. Contr Mineral Petrol 172:40

Zhang HF et al (2000) Recent fluid processes in the Kapvaal craton, South Africa: coupled oxygen isotope and trace element disequilibrium in polymict peridotites. Earth Planet Sci Lett 176:57–72

Zhang T, Krooss BM (2001) Experimental investigation on the carbon isotope fractionation of methane during gas migration by diffusion through sedimentary rocks at elevated temperature and pressure. Geochim Cosmochim Acta 65:2723–2742

Zhang R, Schwarcz HP, Ford DC, Schroeder FS, Beddows PA (2008a) An absolute paleotemperature record from 10 to 6 ka inferred from fluid inclusion D/H ratios of a stalagmite from Vancouver Island, British Columbia, Canada. Geochim Cosmochim Acta 72:1014–1026

Zhang T, Amrani A, Ellis GS, Ma Q, Tang Y (2008b) Experimental investigation on thermochemical sulfate reduction by $H_2S$ initiation. Geochim Cosmochim Acta 72:3518–3530

Zheng YF, Böttcher ME (2016) Oxygen isotope fractionation in double carbonates. Isotopes Environ Health Stud 52:29–46

Zheng YF, Hoefs J (1993) Carbon and oxygen isotopic vovariations in hydrothermal calcites. Theoretical modeling on mixing processes and application to Pb–Zn deposits in the Harz Mountains, Germany. Mineral Deposita 28:79–89

Zheng YF, Fu B, Li Y, Xiao Y, Li S (1998) Oxygen and hydrogen isotope geochemistry of ultra-high pressure eclogites from the Dabie mountains and the Sulu terrane. Earth Planet Sci Lett 155:113–129

Zheng YC, Liu SA, Wu CD, Griffin WL, Li ZQ, Xu B, Yang ZM, Hou ZQ, O'Reilly SY (2019) Cu isotopes reveal initial Cu enrichment in sources of giant porphyry deposits in a collisional setting. Geology 47:135–138

Zhu Y, Shi B, Fang C (2000a) The isotopic compositions of molecular nitrogen: implications on their origins in natural gas accumulations. Chem Geol 164:321–330

Zhu XK, O'Nions RK, Guo Y, Belshaw NS, Rickard D (2000b) Determination of natural Cu-isotope variations by plasma-source mass spectrometry: implications for use as geochemical tracers. Chem Geol 163:139–149

Zhu XK, O'Nions K, Guo Y, Reynolds BC (2000c) Secular variations of iron isotopes in North Atlantic Deep Water. Science 287:2000–2002

Ziegler K, Young ED, Schauble E, Wasson JT (2010) Metal-silicate silicon isotope fractionationin enstatite meteorites and constraints on Earth's core formation. Earth Planet Sci Lett 295:487–496

Ziegler S, Merker S, Streit B, Boner M, Jacob DE (2016) Towards understanding isotope variability in elephant ivory to establish isotopic profiling and source-area determination. Biol Cons 197:154–163

Zierenberg RA, Shanks WC, Bischoff JL (1984) Massive sulfide deposit at 21°N, East Pacific Rise: chemical composition, stable isotopes, and phase equilibria. Bull Geol Soc Am 95:922–929

Zimmer MM, Fischer TP, Hilton DR, Alvaredo GE, Sharp ZD, Walker JA (2004) Nitrogen systematics and gas fluxes of subduction zones: insights from Costa Rica arc volatiles. Geochem Geophys Geosys 5:Q05J11. https://doi.org/10.1029/2003GC000651

Zinner E (1998) Stellar nucleosynthesis and the isotopic composition of presolar grains from primitive meteorites. Ann Rev Earth Planet Sci 26:147–188

Zwicker J, Birgel D, Bach W, Richoz S, Smrzka D, Grasemann B, Gier S, Schleper C, Rittmann SK, Kosun E, Peckmann J (2018) Evidence for archaeal methanogenesis within veins at the onshore serpentinite hosted Chimaera seeps, Turkey. Chem Geol 483:567–580

# Index

**A**
Ab initio methods, 93
Abiogenic methane, 397
Absorption, 23, 33, 34
Acetate fermentation, 396
Achondrites, 134, 273
Acid mine drainage, 155
Acid-Volatile Sulfides (AVS), 420
Adsorption, 98
Aerosols, 147, 378
Albite, 120
Aliphatic hydrocarbons, 391
Alkaline earth elements, 105
α-position, 366
Alunite, 132
Alzheimer's disease, 437
Amino acids, 65
Ammonium, 68
Amphibole, 78
Anaerobic sulfate reduction, 87
Anammox, 66
Andesites, 118
Anemia, 438
Anorthite, 123
Anthropogenic contamination, 99
Apatite, 18, 110
Aragonite, 60
Arc lavas, 96
Aromatic hydrocarbons, 392
Asteroids, 112, 267
Atmospheric dust, 341
Atmospheric hydrogen, 377
Atmospheric water vapour, 363
A-type granite, 120

**B**
Baddeleyite, 168
Ba-isotopes, 119
Banded iron formations, 126
Barite, 92
Basalts, 107
Batch volatilization, 307
Benthic foraminifera, 409
β-position, 366
Biogenic deposits, 326
Biogenic gas, 396
Biogenic silica, 125, 405
Biomarkers, 424
Biosignatures, 146
Biosynthesis, 56
Biotite, 68, 120
Black carbon, 394
Black shales, 172
Boiling fluid, 318
Bond lengths, 103
Bond strength, 77
Bones, 111
Boron isotopes, 99
Brachiopods, 418
Br isotopes, 131
Bronze, 178
Bulk silicate earth, 94
Burial diagenesis, 17
Butane, 394

**C**
Ca isotopes, 111
Calcite, 60
Calcium-aluminium inclusions, 274
Cancer, 437
Canyon Diablo iron meteorite, 83
Carbohydrates, 62, 384
Carbonaceous chondrites, 112, 175, 268
Carbonados, 285
Carbonate concretions, 413
Carbonate-graphite thermometer, 315
Carbonate precipitation, 114
Carbonates, 97


J. Hoefs, *Stable Isotope Geochemistry*, Springer Textbooks in Earth Sciences, Geography and Environment, https://doi.org/10.1007/978-3-030-77692-3

Carbonatites, 17, 18, 113
Carbon isotope events, 358
Carbon isotopes, 58
Carbon isotope stratigraphy, 413
Carbon monoxide, 375
Carbonyl sulfide (COS), 378
Carboxylic acids, 272
Carnivores, 159
Cassiterite, 178
Cave carbonates, 110
Cavity ring-down spectroscopy, 34
Cd isotope, 174
Ce isotopes, 180
Cellulose, 384
$C_4$plants, 61
Chalcopyrite, 152
Channelized fluid, 308
Cherts, 126
Chlorinated hydrocarbons, 127, 388
Chlorine isotopes, 127
Chlorite, 109
Chlorophyll, 110
Chondrites, 107
Chondrules, 144, 274
Chromites, 139, 187
Clastic sediments, 108, 402
Clay minerals, 57, 98, 403
Clinopyroxen, 68
Closed system, 88
Clumped isotopes, 16, 17
Clumped isotope temperatures, 269
Clumping in methane, 398
$CO_2$-reduction, 396
Coals, 100, 162
Collagen, 388
Compound-specific isotope analyses, 382
Conodonts, 432
Contact metamorphism, 309
Continental crust, 97
Continuous flow—isotope ratio monitoring mass spectrometers, 30
Coordination number, 93
Corals, 114
Core formation, 123
Cr isotopes, 138
Crude oils, 137
Crystal fractionation, 113
Crystal-melt fractionation, 107
$C_3$plants, 61
Cu isotopes, 152

**D**

Dansgaard-Oeschger events, 427
Deep-water upwelling, 148
Degassing, 157
Dehydration reaction, 103
Delta value (δ), 8
Denitrification, 65
Deuterium, 50
Deuterium excess, 55, 332
Devolatilisation, 68
Diagenesis, 115
Diamonds, 68
Diatoms, 125
Differentiation, 107
Diffusion, 19, 20, 96
Diffusion coefficient, 19, 20
Dissimilatory Fe(III) reduction, 146
Dissimilatory sulfate reduction, 86
Dissolved Inorganic Carbon (DIC), 346
Dole effect, 366
Dolomite, 109
Dolomitization, 17

**E**

Early life, 152, 399
Eclogite, 97, 282
Enamel, 159, 417
ε-value, 8
Enstatite chondrites, 112, 275
Equilibrium fractionations, 24, 26
Ethane, 394
Eu isotopes, 181
Euxinic conditions, 173
Evaporation/precipitation, 53
Evaporation-condensation, 9, 156
Evaporites, 83
Even Hg isotopes, 189
Exploration, 154
Explosives, 435

**F**

Facies indicator, 389
Fayalite, 144
Fe isotopes, 142
Fe-Mn crusts, 149, 420
Fe–Mn oxides, 94
Femtosecond laser, 122
Fe oxy/hydroxides, 146
Fertilizers, 99, 435
First principles calculations, 93
Fischer-Tropsch, 272
Fluid inclusions, 316
Fluid-rock interactions, 80
Food webs, 110
Foraminifera, 17, 102
Forensic isotope geochemistry, 435
Formation waters, 131, 320

Forsterite, 107
Fractional crystallization, 107, 118
Fractionation factor (α), 7
Fractionations during evaporation, 337
Freshwater carbonates, 416

**G**
Gabbro, 300
Ga isotopes, 160
Garnet, 78
Gas hydrates, 395
GC-C-IRMS, 58
Ge isotopes, 162
Geospeedometer, 95
Geothermometers, 24
GEOTRACES, 340
Goethite, 94
Graham's law, 19
Granulites, 313
Graphite, 60
Gravitational fractionation, 362
Great Oxidation Event (GOE), 141, 368
Groundwater, 125, 335
Gypsum, 345

**H**
Hailstones, 330
Halite, 129
Harzburgites, 165
Heavy Rare Earth Elements isotopes, 181
HED meteorites, 277
Hematite, 93
Hemochromatosis, 438
Herbivores, 159
Hg isotopes, 188
Homoestatic regulation, 438
Hopanes, 393
Hot springs, 301
Hydration sphere, 57
Hydrogen isotopes, 50
Hydrothermal fluids, 98
Hydrothermal systems, 97

**I**
Ice cores, 91, 334
Ilmenite, 135
Infrared spectroscopy, 34
Interglacial, 335
Intraplate basalts, 135
Iridium isotopes, 187
Iron isotopes, 142
Iron meteorites, 144, 274
Iron oxide, 418
Isoprenoids, 391
Isoscapes, 435
Isotope effects, 4
Isotope geothermometers, 23
Isotope Ratio Mass Spectrometer (IRMS), 49
I-type granites, 104

**J**
Juvenile water, 283

**K**
Kerogen, 271
Kimberlites, 282
Kinetic effects, 10
K isotopes, 133
Komatiites, 113
Kyanite, 315

**L**
Land plants, 381
Laser absorption spectroscopy, 52
Lherzolite, 165
Limestone, 412
Lipids, 62, 384
Lithium isotopes, 94
Lower crustal rocks, 313

**M**
Mackinawite, 422
Magmatic ore deposits, 325
Magnesite, 107, 109
Magnesium isotopes, 105
Magnetite, 93, 137
Mammals, 418
Marbles, 18, 307
Marine plants, 380
Mars, 112
Martian meteorites, 278
Mass balance, 49
Mass dependent fractionations, 12
Mass extinctions, 191, 414
Mass independent fractionations, 13
MC-ICP-MS technique, 84
Medical studies, 436
Melt inclusions, 276
Metal isotopes, 93
Metamorphic dehydration, 133
Metamorphic rocks, 307
Metamorphic water, 320
Metasomatism, 107, 113
Meteoric water line, 55, 332
Meteoric waters, 318
Methane, 18, 58
Methanogenesis, 391
Methyl bromide, 131

Methylmercury, 189
Mica, 68
Micronutrient, 147, 155
Mineral dissolution, 97
Mn oxyhydroxide, 156
Mo isotopes, 169
Molybdenite, 170
Moon, 117
MORB, 97
MORB glasses, 68
Multicollector-ICP-Mass Spectrometry, 38
Multicollector ICP-MS, 95
Multiply substituted isotopologues, 15
Muscovite, 120

**N**
N-alkanes, 391
Nd isotopes, 181
Ni isotopes, 149
Nitrification, 65
Nitrogen cycle, 64
Nitrogen fixation, 68
Nitrogen isotopes, 64
Nitrous oxide ($N_2O$), 365
Nominally anhydrous minerals, 283
$NO_x$ production, 365
Nuclear Magnetic Resonance (NMR) Spectroscopy, 34
Nuclear volume effects, 14

**O**
Ocean acidification, 102
Oceanic Anoxic Events (OAE), 414
Oceanic crust, 97
Ocean water, 100
Odd Hg isotopes, 189
Oddo-Harkins" rule, 2
OIB, 96
Oil migration, 392
Oils, 391
Olivine, 68
Open system, 88
Ophiolite, 107
Ore deposits, 153
Organic contaminants, 388
Organic sulfur, 394
Organic sulfur compounds, 421
Orthopyroxene, 113
Osmium isotopes, 187
Osteoporosis, 116, 437
Ostracodes, 425
Oxygen fugacity, 295
Oxygen isotopes, 71
Oxygen isotope thermometers, 313
Oxygen minimum zones, 176
Ozone, 50

**P**
Palaeoclimatology, 17, 423
Paleocene-Eocene-Thermal-Maximum (PETM), 432
Paleoclimate, 405
Paleoprecipitation, 334
Paleo-redox proxy, 141
Paleotemperature scale, 407
Palladium isotopes, 185
Partial-exchange technique, 26
Partial melting, 113
Particulate Organic Matter (POM), 348
Partition functions, 6, 25
PDB standard, 58
Peat, 160
Perchlorate, 130, 379
Peridotite, 68
Pervasive fluid, 308
Phlogopites, 68, 283
Phosphate, 73, 354
Phosphoric acid, 58
Phosphorites, 198
Photolysis, 273
Photosynthesis, 50
PH-value, 101
Phytan, 393
Phytoliths, 126
Phytoplankton, 62
Plagioclase, 113
Planktonic foraminifera, 409
Platinum isotopes, 186
Plutonic rocks, 290
Pore waters, 109
Porphyry copper deposits, 154
Position specific isotope composition, 383
Potassium isotopes, 132
Precipitation/dissolution processes, 147
Precipitation rates, 118
Pressure effects, 21
Pristane, 393
Propane, 62, 394
Protein, 387
Provenance studies, 402

PT-paths, 312
Pyrite, 85
Pyrolysis, 58
Pyroxenites, 157
Pyrrhotite, 420

**Q**
Quadruple sulfur isotopes, 90
Quartz, 122

**R**
Radiolaria, 125
Rainwaters, 99
Rayleigh fractionation, 115
Rb isotopes, 134
Redox reactions, 93
Regional Metamorphism, 310
Re isotopes, 182
Relative humidities, 55, 333
Residence times, 98, 341
Respiration, 351
Rhyolites, 118
Rivers, 98, 337
Rubisco, 61
Ru isotopes, 187
Rutile, 315

**S**
Saline deep waters, 344
Sandstones, 124
Sb isotopes, 179
Scavenging, 341
Seawater, 102
Secondary Ion Mass Spectrometry (SIMS), 20, 39
Sedimentary rocks, 400
Selenate reduction, 165
Selenium isotopes, 163
Serpentinite, 103
Serpentinization process, 130
$^{17}$O excess, 333
Shales, 124, 400
Shear zones, 311
Si isotopes, 122
Silcrete, 124
Silicon carbide, 271
Sillimanite, 315
Silver isotopes, 173
Site-specific isotope fractionations, 18
Smectite, 109
Smithsonite, 329
SNC-meteorites, 277
Sn isotopes, 177
Snowball Earth, 311
Soils, 119
Solar system, 267
Solar wind, 269
Soret diffusion, 296
Sorption, 23
Source amino acids, 67
Speleothems, 17, 110, 425
Spinel, 107
Sponges, 126
Stable isotope forensics, 388
Standard Mean Ocean Bromine (SMOB), 131
Standard Mean Ocean Chloride (SMOC), 128
Standards, 35
Stannite, 178
Stratosphere, 362
S-type granites, 104
Subduction metamorphism, 312
Subduction zone metamorphism, 97
Subduction zones, 103, 133
Sulfur isotopes, 83
Surface ocean water, 151
Symmetry rule, 1

**T**
θ-value, 79
Tellurium isotopes, 167
Terrestrial fractionation line, 79
Terrestrial Fractionation Line, 268
Thermal diffusion, 19, 20
Thermal maturation, 385
Thermochemical sulfate reduction, 89
Thermogenic gas, 396
Thermometers in geothermal systems, 306
Three-isotope method, 26
Ti isotopes, 135
Tin isotopes, 178
Titanite, 315
Tl isotopes, 193
Tourmaline, 104
Tree rings, 423
Troilite, 144, 273
Trophic amino acids, 67
Trophic level, 67, 116
Troposphere, 92, 362
Tungsten isotopes, 183
Two-direction approach, 26

**U**
Ultrafiltration, 57
Ultra-High Pressure (UHP)-rocks, 311
Ultramafic rocks, 123
Ultramafic xenoliths, 281
Uranium isotopes, 196
UV self shielding, 270

**V**
Vapor pressure, 53
V-CDT standard, 83
Vent fluids, 121
Venus, 280
Vesta, 144
Vibration frequencies, 93
V isotopes, 136
"Vital" effects, 60
Volatile depletion, 276
Volcanic ash, 92
Volcanic gases, 130, 301
Volume diffusion, 19, 20, 24
V-PDB-standard, 58
VSMOW standard, 74

**W**
Wall-rock alteration, 320
Water-mass mixing, 121
Water/rock ratio, 302
Weathering profiles, 401
Willemite, 329
W isotopes, 184

**Z**
Zero point energy, 5
Zircon, 98
Zn isotopes, 156
Zr isotopes, 168